Cognition Through Understanding

Cognition Through Understanding

Self-Knowledge, Interlocution, Reasoning, Reflection

Philosophical Essays, Volume 3

TYLER BURGE

OXFORD
UNIVERSITY PRESS

OXFORD
UNIVERSITY PRESS

Great Clarendon Street, Oxford, OX2 6DP,
United Kingdom

Oxford University Press is a department of the University of Oxford. It furthers the University's objective of excellence in research, scholarship, and education by publishing worldwide. Oxford is a registered trade mark of Oxford University Press in the UK and in certain other countries

First Edition published in 2013

Published in the United States of America by Oxford University Press
198 Madison Avenue, New York, NY 10016, United States of America

British Library Cataloguing in Publication Data
Data available

ISBN 978–0–19–967202–8

Contents

Preface

I do not remember when I first came to think of rationalism as an important line of thought in philosophy. From my philosophical beginnings, I did find the rationalists more interesting than the empiricists. Descartes and Kant, in particular, struck me as striving in attractive ways for philosophical understanding. By contrast, Locke and Hume seemed more oriented toward debunking overreaches in philosophy. I realized that both approaches have their place. But I thought that the first approach held out the better prospect of depth and discovery in philosophy. The idea that one could gain knowledge and insight purely through reasoning or understanding seemed one that should not be given up without very powerful reason to do so. Descartes and Kant did overreach. They also discovered more. They did so, partly, by finding limited ways in which reasoning and understanding could be trusted to yield insight and cognition.

In my early education in philosophy, I was faced with a near taboo against any sort of rationalism. The taboo was almost as strong as the one against questioning materialism. (See the conclusion of 'Modest Dualism'.) Although logical positivism had fallen, a dominant idea at the time was that philosophers should transform themselves either into commentators on empirical science or, taking Wittgenstein as model, into therapists whose aim was to show philosophical questions to be confused or pointless. I recognized the importance of empirical science as a topic and resource for philosophy. I was never convinced by the reductionistic impulses that drove the dominant idea. Fortunately, the taboo no longer holds sway.

I learned from writings consecrated by the taboo, however. In particular, I learned from Quine and Putnam's sophisticated empiricism (or in the case of Putnam, near empiricism—he held out, as he thought bravely, for the apriority of the principle of non-contradiction). Their work taught me how empirical warrant can be unconsciously embedded in what we take for granted, in what we may seem to know by intuition or reflection alone. Their work also showed the dependence of many definitions and meaning explications on empirically supported background beliefs. I think that I began to take in how limited and hedged a fruitful rationalism would have to be, at least beyond its most obvious applications to pure logic and pure mathematics.

What kept me from being convinced by the prevailing empiricism, apart from a natural aversion to ideology, was its patently inadequate account of logic and mathematics, its denial or failure to reflect on cognition in ordinary value judgments, and its failure to give a convincing account of some aspects of philosophy itself.

Quine had decisively defeated the positivists' attempts to show logic and mathematics to be empty of substantive cognition. Yet Quine's and others' efforts to show that logic and mathematics are justified purely through their role in natural science seemed to me to conflate pure logic and mathematics with applied logic and mathematics. Or else, they failed, without good reason, to take epistemic practice in these disciplines as the primary ground for accounts of the disciplines' epistemology. I later came to take Frege's sarcastic remark about Mill in *Foundations of Arithmetic* as a pithy summation of the mistake: 'It might well be supposed that numerical formulae would be synthetic or analytic, a posteriori or a priori, according as the general laws on which their proofs depend are so. John Stuart Mill, however, is of the opposite opinion.'

Regarding ordinary value judgments, I thought it clear that such judgments as that slavery is wrong or torturing children for the fun of it is wrong are cognitively sound and warranted not by sense experience, but by understanding what right and wrong are. Attempts to fit such judgments into empiricist molds seemed, and seem, hopeless.

Regarding philosophy itself, my thinking is harder to chart in retrospect. The terrain is much more complex. I mention two landmarks. I came to be impressed by Frege's example in philosophy, through teaching his work in my early years at UCLA. Some of his insights into mathematics, logic, and language—for example, his anti-Millian point, his account of knowledge of axioms, his distinction between negation and denial, his distinction between sense and reference—seemed to owe nothing to empirical sources for their warrant. I also remember, perhaps two years past graduate school, asking Quine—during a long, late-night car trip in New England—whether he thought that his principle of charity is empirically grounded. After a full minute of silence, he replied with no elaboration, 'Perhaps it's not'. I do not know how he took the exchange. I do know that he sometimes did not bother to defend orally views that he nevertheless firmly held. But to me, at that formative time, it seemed that the key witness had failed to support the defense.

These essays reflect attempts to understand certain types of cognition at least some of whose aspects seem to be warranted apriori. The attempts center mostly on issues that are orthogonal to the more standard epistemological issues regarding mathematics, logic, and ethics. I discuss self-knowledge, interlocution, certain types of inference in reasoning, and reflection. I think that discussing these different, but still central, issues might enrich our sense of the roles of reasoning and understanding in cognition.

As I note in the Introduction, my interest in these areas of cognition is fundamentally motivated by the idea that each area helps bring out underlying psychological capacities that are, very likely, distinctive to human beings. A major theme in the essays is to try to understand cognitive capacities that constitutively make selves—and persons—the beings that they are. I think that epistemology should be pursued, more often than it actually is, as a close partner to philosophy of mind and to psychology. One cannot understand epistemic norms except as norms for using *well* certain psychological capacities that yield cognition, knowledge, science, or rational belief. The norms are best understood in conjunction with understanding psychological capacities to meet them—the very capacities for which the norms are standards for good use.

Through my work on anti-individualism and especially on perception, I have come to appreciate the animal underpinnings of the central rational capacities—understanding and reason. This appreciation has led to qualifications of some of my earlier rationalist claims—for example, the claim that one can know something from interlocution, or from a computer readout, purely apriori. (See 'Postscript: "Content Preservation"'.) I, too, overreached. Moreover, I have what I regard as a new appreciation of how inference itself can support a conclusion without doing so through reason or understanding. (See 'Epistemic Warrant: Humans and Computers'.)

I remain impressed by the warranting power of understanding and reasoning, and by their substantial warranting roles in the types of cognition discussed in these essays. I remain convinced that these capacities are among those that make human beings special—not only distinctive among terrestrial animals, but worthy of being specially valued and respected. The capacities to understand and to reason—especially to reason in a reflective, self-critical way—underlie science. I believe that these capacities are constitutive aspects of morality (see 'Self and Constitutive Norms'), which Kant rightly took in *Critique of Practical Reason* to be an object of wonder and awe. Science and morality are two of the things in which human beings can rightly take most pride.

Many vivid memories attend the development of these essays. The basic ideas of the first essay on self-knowledge, 'Individualism and Self-Knowledge', were presented in 1984, while I was giving, in German, summer lectures and seminars in Munich. I met Andreas Kemmerling, Willi Vossenkuhl, Maria Eugenie Rivas Monroy, and (at a conference in Kirchberg, Austria) Brian O'Shaughnessy, long-term friends; and I benefited from connection with Dieter Henrich and the German tradition of interest in self-consciousness. I have marvelous memories of wandering Munich's enormous park, Englische Garten, and of an epic hail storm that shattered all the windows in my apartment, and left its marks on cars in the city for many years afterwards. 'Our Entitlement to Self-Knowledge' and 'Content Preservation' were developed before and during my time at All Souls, when I gave the Locke Lectures at Oxford, in 1993. I was fortunate in interacting fruitfully with Georg Kreisal, P. F. Strawson, and Martin Davies; and I found a friend and congenial fellow rationalist in Chris Peacocke. The beauty and warm glow of Oxford during that stay remain with me. I associate 'Reason and the First Person' with a conference in St. Andrews, Scotland, in 1995, where I first gave the main part of the paper. I recall the environment as bleakly, menacingly, and darkly beautiful. A conference on my work on self-knowledge in Granada in

1996 furthered my understanding of the main topic and drew me into more serious reflection on reflection. The exchange with Benejam in this volume is too small an indication of the stimulation I received throughout that conference. I was wonderfully hosted and befriended by Juan Acero, Maria Frapolli, and Manuel Carpenteiro. I had the immense good fortune of experiencing the other-worldly beauty of the Alhambra during both day and night—experiences I made sure to live again since then—and of seeing the mosque in Cordoba, the strangest important building that I have ever visited. I gave a version of 'Memory and Persons', I believe, in Paris in 2001, where I enjoyed the eccentricity of David Wiggins and became friends with Lucy O'Brien. The rose garden in the Bois de Boulogne, the Val de Grace cathedral, the luminous, underrated, Renaissance church St. Etienne du Mont, and a solo drive to Burgundy remain etched in memory. I associate giving the Dewey Lectures in New York in 2007 with the friendly hosting of Akeel Bilgrami and Carol Rovane. I also associate the event with observations of street life during arduous, but fascinating, trudges up and down Amsterdam Avenue in light, early winter snow. During those painful walks, I wondered whether, in view of the intensity of what was to become my nearly four-year bout with plantar fasciitis, I would ever be able to walk again normally. I connect revision of the lectures for publication with the joy of having finally returned to normal.

My thanks for their perseverance, love, support, and companionship go, as always, to my wife Dorli and my sons, Johannes and Daniel.

Tyler Burge

Acknowledgments

'Reasoning about Reasoning', *Philosophia* 8 (1979), 651–656

'Individualism and Self-Knowledge', *The Journal of Philosophy* 85 (1988), 649–653

'Content Preservation', *The Philosophical Review* 102 (1993), 457–488

'Our Entitlement to Self-Knowledge', *Proceedings of the Aristotelian Society* 96 (1996), 91–116

'Interlocution, Perception, and Memory', *Philosophical Studies* 86 (1997), 21–47

'Computer Proof, Apriori Knowledge, and Other Minds', *Philosophical Perspectives* 12 (1998), 1–37

'Memory and Self-Knowledge', in Peter Ludlow and Norah Martin (eds.), *Externalism and Self-Knowledge* (Stanford: CSLI Publications, 1998), 351–370

'Reason and the First-Person', in Crispin Wright, Barry C. Smith, and Cynthia Macdonald (eds.), *Knowing Our Own Minds: Essays on Self-Knowledge* (New York and Oxford: Oxford University Press, 1998), 243–270

'A Century of Deflation and a Moment about Self-Knowledge', Presidential Address at Pacific APA, April, *Proceedings of the APA* 73 (1999), 25–46

'Comprehension and Interpretation', in L. Hahn (ed.), *The Philosophy of Donald Davidson* (Chicago: Open Court Publishers, 1999), 229–550

'Concepts, Conceptions, Reflective Understanding: Reply to Peacocke', in Martin Hahn and Bjorn Ramberg (eds.), *Reflections and Replies: Essays on the Philosophy of Tyler Burge* (Cambridge, Mass.: MIT Press, 2003), 383–396

'Memory and Persons', *The Philosophical Review* 112:3 (2003), 289–337

'Mental Agency and Authoritative Self-Knowledge: Reply to Kobes', in Martin Hahn and Bjorn Ramberg (eds.), *Reflections and Replies: Essays on the Philosophy of Tyler Burge* (Cambridge, Mass.: MIT Press, 2003), 417–434

'Reply to Benejam: "Thought Experiments and Semantic Competence"', in Maria J. Frapolli and Esther Romero (eds.) *Meaning, Basic Self-Knowledge, and Mind: Essays on Tyler Burge* (Stanford, Ca.: CSLI Publications, 2003)

'Modest Dualism', in Robert C. Koons and George Bealer (eds.), *The Waning of Materialism* (New York: Oxford University Press, 2010), 233–250

'The Dewey Lectures 2007: Self and Self-Understanding': Lecture I: 'Some Origins of Self'; Lecture II: 'Self and Constitutive Norms'; Lecture III: 'Self-Understanding', *The Journal of Philosophy* 108 (2011), 287–383

'Living Wages of *Sinn*', *The Journal of Philosophy* 109 (2012), 40–84

1 *Introduction*

In the essays collected here, I reflect on some types of cognition that rest primarily or entirely on understanding of thought or language.[1] Most of the understanding-based routes to cognition that I discuss are supported by non-empirical epistemic warrants. I focus on types of understanding, and types of non-empirical epistemic warrants, associated with self-knowledge, interlocution, reasoning, and reflection.

Three considerations motivate my interest in epistemic warrants associated with these types of cognition. First, the warrants that I discuss differ in interesting ways from epistemic warrants commonly scrutinized in philosophy—warrants for perceptual beliefs and for beliefs inferred from perceptual beliefs, and warrants for logical and mathematical beliefs. Second, I believe that the four types of cognition are, either in themselves or in significant respects, distinctive of persons, and (on earth) of human beings. Third, the warrants shed light on cognitive powers that underlie each type of cognition. The relevant cognitive powers involve a fascinating variety of kinds of understanding.

A *warrant* for a belief or other type of cognition is a right to rely on the cognition as cognition. A warrant derives from meeting standards for operating well in the service of fulfilling representational functions of the relevant type of cognition. The standards allow for limitation in available information and limitation in the individual's representational powers. Thus, even though being true is the fundamental representational function of belief, a belief can be warranted even if it is false—assuming that the belief is formed through a cognitive power that is reliably conducive to yielding true beliefs, given available information. So being warranted is fulfilling a less fundamental, but still important representational function—roughly the function of representing reliably and as well as is psychologically possible, given available information. Understanding a warrant

[1] I intend '*cognition*' as a somewhat open-ended term for levels of representation above the level of perception that are informative about the world. Above the level, in regards to sophistication and abstractness. Propositional attitudes are paradigm cases of cognition. I count lower levels of representation—such as allocentric, amodal spatial maps in memory—as cognition as well. Here I focus on propositional capacities. Since I take practical representation to have a cognitive dimension, I believe that there is both theoretical cognition about the physical, psychological, or mathematical world, and practical cognition about what is practically functional or good. Here I focus on "theoretical" cognition, commonsensical or scientific—that is, cognition whose primary representational function is to yield veridical representation of the world.

thus helps in understanding basic facts about the cognitive power and the power's representational functions. Understanding warrants can also help in understanding the natures of individuals with those powers. In the cases that I discuss, the main powers are powers of understanding. So the unifying project of these essays is to understand certain powers of understanding associated with the four types of cognition.

The first, second, and fourth types of cognition that I discuss (self-knowledge, interlocution, and reflection) are among the important ways in which human beings are psychologically distinct from other terrestrial animals.[2]

The third type (propositional reasoning) is probably shared with other animals. However, I think that it yields non-empirical, or apriori, cognition only in humans. I believe that some non-human terrestrial animals can reason. They depend for some of their reasoning on deductive inferential transitions that are warranted non-empirically. But I think it likely that all the starting points and endpoints of their reasoning are *empirical*. Their cognition is too closely connected to perception, and to action that serves their sensed needs, to rely on resources that are warranted purely by intellection.

Colloquially, a type of cognition is *non-empirical*, or *apriori*, if one's warrant for the cognition—one's right to rely on it as cognition—does not derive from sense experience. Apriori types of cognition rest on reason or understanding, that is, on intellection.

I am interested in non-empirical, or apriori, warrants for cognition because I believe that they are interesting in themselves and because I believe that they yield special insight into powers of cognition that are distinctive of human beings, or more generally persons. Most of the non-empirical warrants that I discuss with respect to all four types of cognition are, very likely, special to human beings. I think that these types of cognition are so central and fundamental that they help determine the psychological natures of individuals that have them. They certainly help determine the natures of selves and persons. Having non-empirical or apriori warrants depends on using powers of intellection that mark the distinctiveness of persons.

Leibniz thought that apriori cognition of necessary truths is the distinctive mark of human cognition.[3] It may well be one such mark. But there are others. Most of the cognition that I discuss here is not cognition of necessary truths. Much of it is nonetheless apriori. The essays seek to broaden awareness of the

[2] There may be an near ancestor of self-knowledge in higher animals—knowledge that they are in pain, for example. I think that such animals do not have a full-blown first-person I-concept, a concept of self. So the knowledge is not, strictly, self-knowledge. But they have some lower-level ego-centric analog of the first-person I-concept. I believe that conceptualization of such sensations as pain may well be a very primitive matter. I doubt that any such reflexive, knowledgeable, self-attribution is non-empirical. At any rate, the types of non-empirical self-knowledge that I center upon seem very probably not to occur in non-human terrestrial animals.

[3] See, for example, G. W. Leibniz, *New Essays on Human Understanding*, ed. and trans. P. Remnant and J. Bennett (New York: Cambridge University Press, 1981), I, i, 25.

range of cognition that rests on intellection, beyond the preserve of logic, mathematics, and other disciplines that bear on necessity.

I turn from colloquial specification of apriority to a more nearly precise specification. A type of cognition is apriori if and only if some instances of that type have non-empirical warrants. An apriori (non-empirical) warrant is one whose warranting force does not derive from perception or from sensing. Typically, the warranting force of an apriori warrant derives from intellection—reason or understanding. A psychological state, or a psychological transition, is apriori if the individual in that state, or making that transition, has and relies on an apriori warrant for it.[4]

Warrants derive from meeting norms, or standards, for operating well in the service of realizing a function or purpose. *Epistemic warrants* derive from a psychology's meeting norms or standards that govern good routes for realizing the representational function of belief formation—the function of forming true beliefs—or for realizing the representational function—preservation of truth and warrant—of certain transitions (inferences) that serve true belief. The relevant standard is for operating *representationally well* cognitively, well enough to have the right to hold the belief, or make the transition, given relevant information and relevant cognitive resources.[5] An epistemic warrant is a right to hold a belief, or to make an inference that serves true belief. Meeting the relevant norm, or standard, yields a right to hold belief or to make an inference. To *have* an epistemic warrant is to meet an epistemic norm or standard in a way that suffices for its being right to hold a belief or make an inference.

There are two types of warrant—justification and entitlement. A *justification* is a warrant that consists partly in the operation or possession of a reason. An individual is justified if and only if the reason is operative or relied upon in the individual's psychology.[6] An *entitlement* is a warrant whose force does not

[4] I understand apriori *truths* to be truths that can be believed with apriori warrant. It may be that any apriori truth can also be believed with empirical warrant. But not all truths believed with empirical warrant can be believed with an apriori warrant. I think that warrant for a cognitive state instance, or (even more fundamentally) for a cognitive state type relative to a way of having or forming that cognitive state type—together with warrants for inferential transitions—are the basic sorts subjects of apriority. Apriori warrants for ways of having psychological states or ways of making inferential transitions are basic. Apriori truth is to be explicated in terms of apriori warrant. I believe that traditional rationalists, who often wrote of apriori truths, conceived such truths as apriori because those rationalists took such truths to be associated with canonical, warranted ways of establishing them—through a proof, for example. For further discussion of these matters, see my 'Frege on Apriority' in *Truth, Thought, Reason: Essays on Frege* (Oxford: Clarendon Press, 2005), 357–364.

[5] A cognition's operating well representationally is to be distinguished from its operating well practically. Its operating well representationally is serving the function of forming, or preserving, true belief. Its operating well practically is serving the function of forming states that benefit the individual's survival for mating, or that serve some other practical—perhaps moral—function or purpose. See my *Origins of Objectivity* (Oxford: Oxford University Press, 2010), chapter 8.

[6] The simplest type of availability is presence and accessibility within the prospective believer's own psychology. But individuals can have justifications even though they lack the justifying reason as the content of a fully formed state in their psychologies–for example, if they could through prompting or reflection come to form such states relatively easily. I discuss such matters in 'Some Origins of Self'

consist, even partly, in the individual's using or having a reason. Entitlements can be rationalized—associated with a justifying argument that explains the entitlement. But entitlements do not require the individual to have such reasons or explanations. An individual has an entitlement if and only if the entitlement attaches to the psychological processes that yield the relevant state or transition. Entitlements need not be understood, understandable, or even thinkable, by the warranted individual. They need not reside in any reasoning that is operative or relied upon in the warranted individual. They derive purely from the individual's psychological system's operating well with regard to realizing or fulfilling its representational functions.

Each type comes in epistemic and practical forms. I shall focus on epistemic forms. *Epistemic* entitlements and justifications derive from operating well in the service of realizing the representational function of forming true beliefs, or carrying out inferences that function to support the truth of beliefs.

Many epistemic warrants suffice for *knowledge*, if they are undefeated by stronger warrants and if there are no other countervailing circumstances.

Entitlements as well as justifications can be apriori. An entitlement is apriori if the correct account of the source of the entitlement adverts entirely to powers of understanding or reason, even though the entitlement does not depend on the warranted individual's having the understanding or reason. For example, the account may say that an individual is entitled to rely on an inference in reasoning, because the inference rule that in fact governs the inference is apriori truth- and warrant-preserving, even though the individual could not think the inference rule, and thus is unable to use it as a reason. Or an individual may be apriori entitled to rely on what he or she is told as being true because the form of the telling is an apriori sign of rationality in the source, even though the individual recipient does not and could not have this account as a reason for relying on what he or she is told.

I emphasize two traditional points about apriority. I do so because they are still often missed.

First, to be apriori is not *eo ipso* to be infallible, indubitable, or unrevisable. An apriori warrant is one that does not derive its force from sense experience. It does not follow that the warrant *infallibly* guarantees that a warranted belief is true. One can have an apriori warrant—short of a proof—to believe a complex mathematical proposition; but the belief may be mistaken.

For similar reasons, one can have an apriori warrant, but *doubt* the proposition that one is warranted in holding. So having the warrant for believing a proposition

and 'Self and Constitutive Norms', reprinted later in Part I. There may be cases in which fully formed psychological states operate as reasons in a psychology, but are not accessible, even in principle, to the reasoner's consciousness. That is, there may be modular reasons. In any such cases, the reasoner is, of course, not responsible for the modular reasons or reasoning. I think of such cases of warrant as nearly at the borderline as between justification and entitlement. I see them, however, as justifications within the individual's psychology even if they are not justifications available to the individual. If one wanted, one could say that the *individual* is entitled to the justification that occurs within the inaccessible parts of the individual's psychology.

does not entail that the proposition is indubitable. One may mistakenly not rely on one's warrant, either because one is psychologically timid, or because one incompletely understands the proposition that one is warranted in believing, or because one has mistaken views about the warrant itself. Or one may find countervailing considerations that defeat the considerations that yielded a warranted belief. Then one might *rationally* doubt the proposition that was once apriori warranted.

Because some apriori warrants do not guarantee the truth of propositions that they warrant belief in, they are vulnerable to possible counter-considerations. That is why some beliefs that have apriori warrant are rationally *revisable*. I think that apriori warranted beliefs can be revised even on *empirical* grounds. Cases of this sort arise in certain types of reflection.

The main point to bear in mind here is that a claim that a type of cognition, or an epistemic warrant, is apriori is a claim purely about the nature of the standard that is met in having a right to hold a belief. Although claims of apriority in philosophy have often been associated with infallibility, indubitability, or unrevisability—indeed, with close-mindedness—apriority *per se* has no such associations.

I do believe that a *few* apriori warrants are infallible and cannot be rationally revised. They cannot be defeated by any other warranted consideration. I think that our warrants to believe instances of pure *cogito* thoughts and some warrants to believe certain simple logical and mathematical truths, or to rely on certain simple forms of inference, have these features. Most of the apriori warrants that I discuss are not even close to infallibly guaranteeing truth, or to being rationally unrevisable.

I doubt that *any* proposition is strictly indubitable. Psychologies are very malleable. In any case, apriority concerns the nature of one's right to believe, not one's power to resist belief.

There may be propositions that are not *rationally* dubitable. But these are not co-extensive with apriori knowable propositions. One can be apriori warranted—for example, in believing a complex, inductively supported mathematical conjecture—and rationally doubt whether one has gotten things right.

The second traditional point about apriority is that one does not show a warrant not to be apriori by showing that the warranted belief or thought-transition depends on sense experience. A warrant is apriori if the force of its warrant is independent of sense experience. That sort of independence is compatible with various other ways in which a warranted belief or thought-transition may depend on sense experience, besides depending on sense experience for the force or nature of the relevant warrant. For example, to acquire a given belief or to understand the believed proposition, one might have to read a book, study a diagram, or hear an explanation. The relevant sort of independence of sense experience concerns the force or nature of the warrant, not whether having the warrant, or the belief, depends causally on having sense experience. The warrant itself must not advert to sense experience, even if having the warrant psychologically depends on having had sense experience.

An apriori warranted belief can also be empirically warranted. So even showing that a belief has a warrant whose force derives from sense experience does not show that the belief is not apriori warranted. Apriority ultimately concerns the nature of a specific warrant, not the nature of a truth, or a belief. We have apriori warrant to believe simple arithmetical truths, through understanding them. There may be a secondary empirical warrant for such beliefs—from counting, or from the role of arithmetic in empirical science—as well.

The focus in most of these essays is epistemological. My main motivations lie, however, in a desire to understand certain psychologically relevant differences between persons, including human beings, and other animals. As I noted in the Preface, I think that the apriori warranted human cognition that I discuss here is not shared with other animals. Fulfilling epistemic norms—and having epistemic, including apriori, warrants—depends on good use of relevant cognitive powers. By reflecting on good, warranted use to the underlying powers, one can learn something about the natures and representational functions of the powers. Thus the essays point toward understanding distinctively human psychological capacities. More generally, they point toward understanding psychological capacities distinctive of persons.

Of course, there are many distinctively human, cognitive—or cognitively dependent—capacities beyond those that I discuss here. I barely touch on human knowledge of language, mathematics, and history. I have almost nothing to say about human capacities to create art, to form societies governed by legal norms, or to maintain personal relationships that involve understanding others' points of view. Most of these capacities employ one or another distinctively human type of cognition. I hope to say more about some of these matters eventually. I do not, however, propose, now or later, to survey all ways in which human beings are psychologically or socially distinctive.

Humans share perception with a vast range of other animals. I think it likely that humans also share perceptual belief and simple propositional inferences involving such beliefs. In other work, I explore these shared types of representation and cognition.[7]

Here, I center on four types of cognition that are non-empirical, or have substantial non-empirical aspects. Each of these types also has uses and applications that are empirical. We know a lot about ourselves, for example, by making inductions from self-observation. Much of our non-inferential introspective self-knowledge is empirically grounded.

Similarly, the vast bulk of what we know from being told things by others utilizes empirical warrants for trusting others' expertise, or utilizes empirical background knowledge to separate questionable from trustworthy elements in others' reporting.

[7] For background on shared types of cognition, see my *Origins of Objectivity*; 'Perceptual Entitlement', *Philosophy and Phenomenological Research* 67 (2003), 503–548; 'Origins of Perception', *Disputatio* 4:29 (2011), 1–38; 'Steps Toward Origins of Propositional Thought', *Disputatio* 4:29 (2011), 39–67.

I now believe that *all* interlocution is warranted empirically. I believe that our warranted comprehension of what others say or write inevitably has an empirical element. (I discuss this change of view below and in 'Postscript: Content Preservation'.) I continue to think, however, that there are substantial and interesting non-empirical components to warrants for relying on what others tell one.

Inductive reasoning commonly has empirical warrant, even apart from any empirical premises used in the reasoning. That is, many—indeed most—inductive inferential transitions are warranted empirically. Clearly, however, some reasoning rests on inferential transitions that are apriori warranted—for example, deductive reasoning or some inductive, inference-to-the-best-explanation reasoning in mathematics.

Much—probably most—successful reflection, including philosophical reflection, has broad empirical warrant. Still, some reflection depends on cognitive states with apriori warrant.

So, in considering non-empirical cognition in these four areas, one must isolate lines of warrant that are non-empirical from other lines that are empirical.

Some philosophers are empiricists. They believe that there is no significant knowledge or belief that is warranted non-empirically. Empiricism dominated philosophy in the Anglo-American tradition from the 1920s into the latter part of the twentieth century. But empiricism never gave a plausible account of logical, mathematical, or moral knowledge—each constituting a large area of human cognitive life. This fact came to be recognized increasingly in the last two decades of the twentieth century.

Empiricism is no longer a dominant doctrine. I do not focus on criticizing it. In these essays, I *assume* that empiricism is mistaken. Its inability to account for logical, mathematical, and moral knowledge, and certain *cogito* types of self-knowledge, seems to me clear. I explore types of non-empirical cognition that go beyond the most obvious types. Since some of the understanding that I discuss *is* empirical (in particular, comprehension of language utterances by others), it is understanding, not apriority, that provides the unifying theme in the four areas indicated in the volume's title. Still, the main focus of these essays is on types of understanding that are warranted non-empirically.

I

Part I of this volume centers on self-knowledge.[8] As noted, a lot of our self-knowledge is empirical. Descartes' *cogito* (I am now thinking) exemplifies, however, a type of self-knowledge that is non-empirical. Descartes' famous

[8] For a source of writing on self-knowledge that I do not reprint in these essays, see M. J. Frapolli and E. Romero (eds.), *Meaning, Basic Self-Knowledge, and Mind: Essays on Tyler Burge*, (Stanford: CSLI Publications, 2003). The volume contains my replies to eleven essays on my work, many of which center on issues regarding self-knowledge. I do reprint one reply that concerns reflection. See Part IV in this volume.

example suggests a fairly wide range of further types of non-empirical self-knowledge. I believe that these types help mark human beings—especially their moral powers and powers for critical reasoning—as distinctive. The essays in Part I gradually develop a unified account of non-empirical types of self-knowledge.

'Individualism and Self-Knowledge' (1988) was written with a single objective in mind. It was to show that anti-individualism and non-empirical types of self-knowledge are compatible. Anti-individualism is the view that the natures of many psychological or mental states are partly determined by relations between individuals and specific attributes in the environment.[9] Non-empirical types of self-knowledge hinge on warrants that support knowledge, but that do not rely on sense experience for such support. For the most part, we can know what the environment is only through empirical warrants—warrants that depend for their force ultimately on warranted perceptual belief and on warranted inferences from such belief. So prima facie, it can seem puzzling that we can know non-empirically what our mental states are—which involves knowing their representational natures, their representational contents—when those natures depend for being what they are on relations to specific attributes in the environment that can be known about only empirically.

The basic idea for resolving the puzzle is that self-knowledge, no less than ordinary knowledge of the environment, depends on utilizing a standpoint whose representational resources have already been made into what they are by the environment. In knowing the representational contents of our mental states we commonly use those very contents in specifying themselves and in specifying the mental states that they help type-individuate. We know how to use and specify those contents by understanding them, not by investigating their constitutive conditions. In other words, at the point where we use reflection for the relevant sort of self-knowledge, the environment has inevitably already had its effect in making the psychological reflection (and the psychological states which are reflected upon) what they are.

To understand those contents in a way sufficient to know what they are, we need not distinguish them from alternatives that might have been our contents if we had come to have our psychological states and contents in a different environment. So although we can know only empirically some or all of the features of our environment that determine the natures (including representational contents) of our psychological states, we can know what our psychological states (and their contents) are non-empirically. The essay brings out the fact that we have no standpoint to know ourselves beyond the standpoint that is marked by being embedded in the environment that we find ourselves.

[9] For fuller discussion, see the Introduction to my *Foundations of Mind Philosophical Essays, Volume II* (New York: Oxford University Press, 2007), and several essays in that volume, and *Origins of Objectivity*, chapter 3.

In writing the essay, I categorized certain types of non-empirical self-knowledge, radiating out from Descartes' pure *cogito*: I am hereby engaging in thinking. I stated that I thought that these pure cases contain seeds for developing an understanding of other non-empirical types of self-knowledge. Many readers—understandably interested in the wider topic of self-knowledge, non-empirical and otherwise—leapt to the conclusion that I was attempting to give a general account of self-knowledge, or at least of non-empirical types of self-knowledge. This view led to many observations of at least prima facie difficulties in taking what I said in the essay to provide such an account. Many of these observations are correct. In fact, I articulate several of these difficulties in the essay. Presented as criticisms, they were irrelevant. For my objective in the essay was not to give a general account of non-empirical types of self-knowledge. It was to show why the non-empiricality of certain types of self-knowledge is compatible with anti-individualism.

It was often pointed out, in response to the article, that to know that one believes that cats are animals, one must not only be warranted in one's specification of the representational content of the belief (the content—cats are animals). One must also be warranted in specifying one's mental state as a belief, as distinguished from a hope or a wish. I said nothing about the warrant for this latter specification.

To take another example, my account centers on self-knowledge of states with representational content, particularly propositional attitudes. As I noted myself, it says little about knowledge of one's sensations, such as knowledge that one is in pain. I believe that such self-knowledge is empirical. Still, it shares many striking features with non-empirical self-knowledge—its directness, relative certainty, and so on.

I did not discuss such matters, except to note them. I did not try to fit them into a general account of self-knowledge. My aim was to show the compatibility of anti-individualism and non-empirical types of self-knowledge. The puzzle about such compatibility derives from the way the *representational contents* of one's first-order mental states depend on the environment. Determining the mode (belief, hope, wish) of one's psychological states raises its own questions. But it does not raise questions about compatibility with anti-individualism. The apparent threat to compatibility derives from the way in which representational content of mental states depends on the environment.

Similarly, I did not try to fit self-knowledge of sensations into a general account, because knowing what one's sensations are does not hinge on specification of representational content, and does not (at least in some key respects) depend for its nature on relations to an environment beyond the individual. I took accounting for the *general* phenomenon of authoritative and (as a sub-set) non-empirical types of self-knowledge to be a further project.

I did take myself to be setting a direction for developing a more general account. I thought—and still think—that by reflecting on Descartes' *cogito* case, and mining it for its deeper implications, one can gain insight into other types of self-knowledge that do not share some of its particular features. *Cogito* is commonly taken to be a curiosity. It is infallible and self-verifying in ways that no other types of self-knowledge are. However, it either embodies or points to characteristics of the broader range of other non-empirical types of self-knowledge. I return to this point at the end of this section.

'Our Entitlement to Self-Knowledge' (1996) discusses another feature of non-empirically warranted self-knowledge. This feature is its symbiotic relation to critical reasoning. I believe that Descartes' reflexive *cogito* and other instances of non-empirical self-knowledge are necessarily associated with a structure of critical reasoning. Critical reasoning is the process of evaluating reasons as reasons, and implementing those evaluations in thought and action. The structure of critical reasoning consists in apriori connections—such as inferential transitions—among different types of states constitutively involved in critical reasoning. I believe that the structure type-individuates the type of thinking that constitutes critical reasoning, and the structure provides a type of unity that marks an individual as a critical reasoner.

The essay sets out a structure of transitions within critical reasoning. When thought follows procedures associated with this structure, the thought is non-empirically warranted. I claim that the structure is an integral part of, constitutive to, critical reasoning. Following procedures associated with the structure requires making transitions based on *understanding* reasons and their applicability to propositional attitudes. The relevant warrants derive partly from the minimal-competence understanding that is involved in engaging in critical reasoning.

Discussed in the abstract, the structure can sound distant and formidable. In fact, it is familiar. It is simply what is common to instances of reasoning like the following: I believe that *p*; but wait, I have obtained reason to believe that *r* and that *s*. I believe that *r* and that *s*. Taken together, *r* and *s* count as strong reasons against believing *p*. I know of no sufficient reasons that counteract or outweigh these reasons against believing *p*. So I have sufficient reason to believe not-*p*. So not-*p*.

What is distinctive about the structure is that it cites reasons *as* reasons for or against one's own attitudes, which one also reasons about; and following it involves implementing one's reasoning about reasons, for or against one's attitudes, into the attitudes one reasons about. One implements the second-order reasoning about first-order attitudes in the first-order attitudes (the last step). I believe that we know by apriori reflection that our reasoning sometimes has this structure. And I think that we know apriori that reasoning that follows this structure is (critically) reasonable, in this sense: If one is reasonable (and warranted) in accepting the individual steps in reasoning that has this structure, then one one is reasonable in one's conclusion.

The argument of the essay hinges on distinguishing the sort of *self-knowledge* that is a key element in critical reasoning from self-knowledge that is grounded in empirical self-observation. A key distinguishing feature is that the exercise of capacities underlying empirically grounded self-knowledge allows for *brute error*. Brute error is the type of error that is compatible with being warranted in one's belief and that does not derive from any sort of psychological malfunction. Intuitively, brute error in self-attributions derives not from internal shortcomings, from normal inductive errors, or from misperceptions of behavior that derive from abnormal observation conditions. The type of self-attribution that underwrites the self-knowledge that is integral to critical reasoning does not allow *brute* error: all errors derive from some sort of misuse of the competence that enables self-knowledge to figure in the distinctive structure of critical reasoning. Such self-attribution is not empirically grounded; it does not rest on self-observation, much less inference from self-observation.

The idea is that if a type of self-attribution yields the relevant self-knowledge when the self-attribution is successful, and if that type is constitutively necessary for the rational review of one's attitudes in critical reasoning, then that type must be immune to brute error. Otherwise, there would be a dissociation between the rational review of one's attitudes and the implementation of the reasons that are marshaled in the review within the set of attitudes that are under review (the lower-level, first-order attitudes). The last step in the structure that I laid out earlier would not follow apriori from the preceding steps (assuming them warranted/reasonable). This point is worked out in some detail in the article. The argument emphasizes rational, non-empirical connections within the structure of critical reasoning. Given that it is apriori reasonable to believe in critical reasoning, with the structure of reason-yielding transitions that the argument lays out, the argument yields an apriori reason to believe that some self-knowledge is warranted non-empirically.

The argument of the essay was meant to support not only the view that to figure in critical reasoning in the way that some self-knowledge does, such self-knowledge must be warranted non-empirically. It was also meant to support the view that part of the non-empirical warrant for the relevant type of self-knowledge derives its force from the role of such self-knowledge within the rational structure of critical reasoning.

Christopher Peacocke accepted the idea that some self-knowledge has non-empirical warrant, and must have such warrant to figure in critical reasoning. He maintained, correctly, that this point is prima facie independent of the further point that the non-empirical warrant for the relevant self-knowledge derives its force from something about critical reasoning. Peacocke was not persuaded by the second point. He held that the relevant self-knowledge derives its non-empirical warrant from sources that are completely independent from the role that that self-knowledge has in critical reasoning.[10]

I thought that the competence understanding involved in the relevant self-knowledge gains some of its warrant from its functioning within the structure of critical reasoning. Indeed, I thought that the understanding draws on fundamental transitional capacities that are integral to being a critical reasoner. This point invited further development. Later essays engaged in such development,

[10] Christopher Peacocke, 'Our Entitlement to Self-Knowledge: Entitlement, Self-Knowledge, and Conceptual Re-Deployment', *Proceedings of the Aristotelian Society* 96 (1996), 117–158.

culminating in the third of the Dewey Lectures. I return to the issue at the end of this section.

'Our Entitlement to Self-Knowledge' connects closely with the Dewey Lectures and with essays in Part III of this collection. It is a starting point for 'Reason and the First-Person'.

'Memory and Self-Knowledge' criticizes errors in Paul Boghossian's attempt to show that anti-individualism and non-empirical self-knowledge are incompatible. Much of the discussion turns on a distinction, which I had made earlier in 'Content Preservation' (see Part II of these essays), between *substantive memory* and *purely preservative memory*. Purely preservative memory is a type of memory that functions to preserve past thoughts for current use, without introducing new contents or attitudes (for example, as premises), with their own warrants and subject matter, into current cognition. Substantive memory is memory that does introduce new premises, with their own warrants and subject matter, into current cognition.

I discuss several errors in Boghossian's argument. The key error is the unargued and question-begging assumption that lacking an ability to identificationally discriminate one past content from another, one cannot know one's yesterday's thought. But knowing one's yesterday's thought does not, in general, depend on identificational discrimination of one past content from another. It is not an identificational representational competence. It commonly depends only on competently preserving the past thought in memory. If purely preservative memory operates properly, it preserves knowledge automatically.

In addition to the unsound abstract argument that anti-individualism and non-empirical self-knowledge are incompatible, Boghossian presents a parallel argument that anti-individualism is incompatible with our assumption of an ability 'to tell apriori whether any particular inference of ours satisfies one of [the forms of valid inference]'. Boghossian's view rests on use of the slow-switching case that I introduced in 'Individualism and Self-Knowledge'.

Let us suppose, in accord with my case, that an individual might have "twin" concepts that the individual is not in a position to discern from one another, because the individual does not realize that he or she switched environments. The idea is that such an individual might carry out an inference that begins with one of the twin concepts in the first premise, introduces the other twin concept in the second premise, and applies an inference rule that requires a single concept. Such reasoning would yield an invalid argument, whose invalidity depends on the equivocation. Boghossian concludes that the possibility of such a case undermines our ability to tell apriori whether any of our arguments is valid.

This case raises more interesting issues than the first argument. It should immediately strike one, however, that the conclusion is overwrought. The mere possibility of such a highly unusual case hardly undermines our normal ability to determine non-empirically that our inferences are valid. At most it might show that such ability is in principle fallible, and that a default apriori warrant could in principle be overturned by empirical considerations (considerations that showed that the circumstances of the argument were abnormal). It seems to me that relatively little of the work in the cottage industry that has developed out of discussion of these issues shows a perspective on the *outre* character of the cases that are discussed. The value of the cases lies in sharpening our conception of the powers of reason, not in adjudicating the truth of large doctrines such as anti-individualism, the apriority of reasoning, or the existence of non-empirical self-knowledge.

In 'Memory and Self-Knowledge', I note that there is always a route to avoiding equivocation. One can resolutely rely on purely preservative memory to connect later uses of a concept anaphorically to the first use in the argument. I certainly do not think that in every imagined case, an individual thus relies on purely preservative memory. I take the existence of purely preservative memory, and our ability to resolutely rely on it, to show that the slow switching cases cannot prevent a carefully directed reasoning process from yielding valid reasoning. There is no across-the-board undermining of the control of reason in processes of reasoning.

I think, however, that the relevant type of equivocation *can* occur. Individuals do not always rely on purely preservative memory. That point is obvious from the fact that in *ordinary* cases, reasoners sometimes fall into fallacies of equivocation. The switching cases elicit the fact that there can be a subtle interplay between the external circumstances that motivate use of a premise in an argument and the employment of purely preservative memory.

Again, this sort of interplay occurs in ordinary cases, particularly cases involving occurrence-based, context-dependent representational devices. For example, one can think that ball is red, looking at a red ball; then one can think, after blinking but again looking at a ball, that ball is round; then one concludes, that ball is red and round. Suppose that while one blinked, a different ball is substituted. The new ball is, in the context, indiscernible from the first ball. Suppose, however, that the new ball is not red, though it looks just as red as the original ball. Then the truth value of the conclusion and the validity of the argument hinge on what one's demonstrative thoughts in the second premise and conclusion refer to.

The issue illustrates tension between purely preservative memory—which would anaphorically retain reference between the first and second premises—and environmental pull, from perception, on reference—which would divert demonstrative reference in the second premise from the referent in the first premise to a different referent, the new ball that is perceived. If one resolutely attaches the second occurrence of the form that ball (the occurrence in the second premise) to the occurrence in the first premise, relying on the first occurrence to establish the referent in the second, then the argument is valid. In such a case, anaphora overrides perception in establishing the referent of that ball in the second premise. If, as is more common and natural, one relies on perception to establish the referent of each occurrence of the form that ball, then the argument is, as it stands, invalid; and the conclusion is false.

In such a case, one would be presuming that the referent of that ball in the second premise would be anaphorically connected to the occurrence of that ball in the first premise. Otherwise, one would not have taken oneself to be producing a deductive argument. Environmental pull, deriving from the context-dependent referential force of perception, would render one's presumption mistaken. One would then have engaged in an unnoticed switch in referents that would render one's argument invalid—despite one's logical competence and apparent reasonability in carrying out the argument.

On reflection, one should realize that *whenever* one relies on a sequence of demonstrative- or indexical referential devices in establishing a referent in a context, one is in principle liable to undiscerned reference shifts. This is a recurrent, normal source of possible invalidity in deductive argumentation. We commonly handle such cases by presuming on our normal, highly reliable capacities for contextual object tracking. Such capacities in effect presume identities of reference linking different occurrent applications of demonstrative or indexical representational contents in thought. Such identities are normally not consciously added as separate premises. Even when the identities fail, presuming upon them yields something like warranted, valid, but unsound reasoning. Cashing out this term "something like" is non-trivial. The reasoner is normally warranted, and reasons competently, even when unrecognized referent switches occur.

The switching cases introduced in 'Individualism and Self-Knowledge' are more radical than the switching case just discussed. Assuming that enough time passes after a switch, and assuming other specific conditions are met (such as that memories of the old environment are not lost), the changes in reference that are involved in switches of whole environments—not just contextual referents in a given environment—yield a different sort of change in reference. The change is not simply a change in the referents of demonstrative- or indexical referential devices.[11] The change is in ability-general representational contents, concepts.

In the most interesting switching cases, the individual has two concepts that are not introspectively discernible to the individual. In the relevant argumentation, the reasoner does not resolutely rely on purely preservative memory. One premise calls up one of the concepts; another premise calls up another. And deductive argument fails because of the switch.

The idea that an individual might have two concepts that the individual cannot discern by simple introspection is sometimes presented as if it were a peculiar, or even non-credible situation. However, concepts mark conceptual modes of presentation that mark psychological abilities. Many differences in psychological abilities are not open to immediate introspective discernment. There are many aspects of our psychological states that we cannot cognize easily and through

[11] See 'Other Bodies' in Andrew Woodfield (ed.), *Thought and Object: Essays on Intentionality* (Oxford: Clarendon Press, 1982), 97–120, reprinted in *Foundations of Mind* (Oxford: Clarendon Press, 2007), for extended discussion of this point.

simple reflection. The idea, often presented as common sense, that the nature of our thoughts is transparent to us is, I think, simply naive.

The cases in which concepts are switched are more unusual, and hypothetical, than the cases in which the referents of demonstrative or indexical representations are switched. Nevertheless, the same sort of tension between purely preservative memory and environmental pulls on reference occur. In the concept switching cases, purely preservative memory has a stronger pull than in the contextual-referent switching cases. But one can, I think, certainly imagine cases in which concept switches occur in the course of a deductive argument, undermining the argument.

In 'Memory and Self-Knowledge' I note that there is a significant difference between the "equivocations" that are imagined as a result of slow-switching and the equivocations that exemplify ordinary errors in inference. I stated that "equivocations" in the slow-switching cases are always cases in which the reasoner 'tacitly and mistakenly presupposes that the [twin] concepts apply to the same objects'. This tacit presupposition, or presumption, is shown by three facts. First, the reasoner does not realize that there are different concepts; the concepts apply to different entities that are indiscernible to the reasoner at the time of the reasoning. Second, the reasoner would accept the results of substituting either concept for the other in a generalization, or descriptive predication. Third, if the reasoner were in circumstances in which use of one of the "twin" concepts were be elicited by a question, and the reasoner were presented with an object that satisfies the other concept, the reasoner would (mistakenly) apply the concept to that object. In short, except for the important fact that the reasoner's psychology may never unite the two concepts in an equivalence claim, the reasoner's psychology shows every sign of being disposed to accept 'For all objects x, Fx if and only if F_tx', where the term that 'F_t' stands in for expresses the "twin" concept of the concept expressed by the term that 'F' stands in for.

Exactly how to characterize this "presupposition" or "presumption" seems to me an interesting psychological question. I do not try to give a precise answer to the question. What seems to me clear, however, is that the presumption is warranted, given the thinker's information. The thinker would be entitled to take instances of the twin concept (say, Tw-water) to be instances of the home concept (instances of water). The existence of the warranted "presumption" allows mistakes (the analogs of equivocation) in argumentation that derive from slow-switches to be assimilated more to empirical errors than to unreasonable inferences. The form of the inference would involve an ordinary equivocation, *if* one omitted to include the presumption of identity, or sameness in tracking. The presumption makes the inference more like a valid enthymeme with a mistaken tacit "premise" than like an ordinary fallacy of equivocation.

As in the case of demonstrative tracking, I do not claim that there *is* a mistaken tacit premise. The presupposition or presumption must be characterized more subtly. Still, the errors appear to me to be warranted. And it appears that the mistakes are not failures in a specifically logical competence. They are more like warranted, unfortunate failures in tracking. They are warranted, empirically based failures that affect the application of one's logical competence.

Their possibility brings out in-principle limits to our non-empirical capacities to understand our own minds infallibly. They do not undermine the intuitive point that in normal cases we can determine non-empirically the validity of our arguments—even when we are *not* resolutely relying on purely preservative memory. In normal cases, we correctly understand the sequence of steps in our deductive arguments as involving univocality, or sameness, of concepts through the steps. Unless we have specific grounds to doubt that the cases are normal, we are apriori warranted in relying on our intellectual competence.

Determining the exact normative status of the equivalence presupposition or presumption is an interesting project.[12] The normative status of the reasoning in slow-switching, "equivocating" cases is quite general, not local to particular imaginable cases. The role of the equivalence presupposition, or presumption, shows that, apart from resolute reliance on purely preservative memory, we have no infallible, transparent competence to avoid such errors. Reasoners that incur the imagined "equivocations" are not thereby irrational. Intuitively, their inferences remain apriori warranted. Apriority consists in warranted, not infallible, reliance on understanding or reason. In normal cases, and even in most switching cases, competent reasoners are warranted in relying on their understanding in producing formally valid arguments.

'A Century of Deflation and a Moment about Self-Knowledge' was the last Presidential address to the Pacific Division of the American Philosophical Association in the twentieth century. In the first half of the address, I used the occasion to spoof deflationary tendencies in philosophy during the century.

Philosophy can no longer arrogate to itself a central role in yielding knowledge. The sciences have long separated themselves as autonomous disciplines and eclipsed it in driving cognitive progress. I think that philosophy still makes progress, and does sometimes yield knowledge, on its own. It makes progress in contributing to scientific advances—especially in frontier sciences—by clarifying concepts and isolating presuppositions in scientific knowledge. It yields knowledge of its own in areas of cognition that are not systematized by a science. Epistemology is one such area. For example, understanding at least some types of self-knowledge is something philosophy is perhaps uniquely fitted for. Moreover, philosophy can contribute clarity and depth of understanding—including understanding unclarities and difficulties—at levels of abstraction rarely confronted in the sciences. So it has a unique and valuable role, even though, because of the sheer complexity of what is known, its place in intellectual culture cannot be as central as it once was.

[12] My students Mikkel Gerken and Luca Struble have produced interesting and different work on this issue.

The analytical part of the essay takes up the issue raised by Peacocke that I discuss above in connection with 'Our Entitlement to Self-Knowledge'. I explain why I think that Peacocke's account of the special "authoritative" nature of non-empirical self-knowledge is inadequate. I claim that some of the special authority of non-empirical self-knowledge lies in another feature implicit in Descartes' *cogito*—mastery of a mature first-person concept. I maintain that 'the ability to move freely, rationally, and immediately, from first-level attitudes to second-level self-attributions of them, and back'—a crucial aspect of the structure of critical reason set out in 'Our Entitlement'—is part of the understanding involved in mastering the first-person concept. Such understanding is part of the source of warrant for non-empirical self-attribution.

There are two aspects to this source. One is the capacity to understand and adjudicate threats to one's prima facie warranted self-attributions. A person is an authoritative judge partly because persons "know the law"—at least some of the rational norms by which hard cases are adjudicated. The other aspect is that mastery of the first-person concept partly consists in applications of it that involve acknowledging intellectual and moral responsibility for certain instances of one's psychological states. This aspect is the genus of which the performative aspect in Descartes' *cogito* is a species. The element of being responsible for a thought within a system of critical reason is non-empirical and authoritative, other things equal.

Of course, neither of these aspects of a source for authority in non-empirical self-knowledge can make a self-attribution warranted unless the understanding being exercised is reliable in yielding veridical cognition. However, exercise of the relevant type of understanding within the structure of critical reason owes nothing for its warrant to sense experience.

I think that although the discussion in this essay is right as far as it goes, it does not sufficiently articulate wherein non-empirically warranted self-knowledge is immune to brute error, or wherein that sort of self-knowledge gets its warrant partly through its association with critical reason. There are elements of rational structures that inform self-knowledge of representational states that deserve more explication. I make glancing reference to these elements in the third-to-last paragraph of the essay. I do not elaborate them until the last of the Dewey Lectures, 'Self-Understanding'.

In 'Mental Agency in Authoritative Self-Knowledge: Reply to Kobes', I respond to Bernard Kobes's fine development of a connection between the performative element in *cogito*-like instances of self-knowledge (instances that I call 'basic self-knowledge') and knowledge of what one will do in intentional action.[13] I emphasize a point that I develop much further in the Dewey Lectures. The point is that whereas non-empirical self-knowledge is not subject to brute error, knowledge of what one will intentionally do *is* subject to brute error. I also

[13] See Bernard Kobes, 'Mental Content and Hot Self-Knowledge' in M. Hahn and B. Ramberg (eds.), *Reflections and Replies: Essays on the Philosophy of Tyler Burge* (Cambridge, Mass.; MIT Press, 2003), 201–228.

emphasize that not all non-empirical self-knowledge is knowledge of active or action-motivating states. Some non-empirical self-knowledge is of passive psychological states, like memories or standing perceptual beliefs. Here is a respect in which the performative aspects of Descartes' *cogito* cannot model all instances of non-empirical self-knowledge.

Both Kobes and I (section II) highlight the role of preservational mechanisms in the non-empirical warrants for types of self-knowledge that go beyond basic self-knowledge. I do not, however, take all instances of non-empirical self-knowledge to derive, by way of such preservational mechanisms, from instances of basic self-knowledge. I think that some non-empirical self-knowledge and the psychological states that are self-known are irreducibly non-active. Some is irreducibly knowledge of non-performative mental states. In this view, I follow Spinoza, not Descartes.

One other large theme in this essay is that warrants for non-empirical self-knowledge are immediate—non-inferential. This point is, of course, another one that Descartes highlighted regarding *cogito*.

Even though it is immediate, the self-knowledge that is illustrated in the pure, self-verifying instances of *cogito* is warranted by justification. (A pure instance is illustrated by an actual thinking of a thought that has the form I am hereby thinking [entertaining] the thought that snow is not white.) That is, the proposition, as thought on the occasion, is self-evident in the sense that it constitutes a reason for itself. Indeed, the self-verifying character of pure instances of *cogito* makes such thoughts infallible. If someone thinks a thought of that form, the thought is guaranteed to be true. There are also impure *cogito*-like instances of self-knowledge. For example, a performative judgment of the form I hereby judge that writing requires concentration is impure inasmuch as it is not self-verifying by virtue of its form: instances can be false. I could judge that I am hereby judging that writing requires concentration, but not in fact do so. I could be distracted, so that I did not in fact engage in the judgment that I judge that I am making. Such errors are pathological, but possible. Although the form of the judgment is not strictly self-verifying, I am inclined to count the occurrent thoughts as reasons for themselves in cases where they are warranted and the individual's understanding encompasses not only the content but an appropriate, performative (as opposed to purely descriptive) use of the content. Thus I think that self-knowledge of this sort is warranted through a justification—as pure cases of *cogito* are.

By contrast, non-*cogito*-like, non-empirical instances of self-knowledge are warranted as entitlements. For example, a judgment I believe that I am older than my sister is not in any sense a reason for itself. There is nothing here analogous to the performative make-itself-true aspect of the *cogito*-like cases that suggests that the nature or content of the belief supports its own truth. In my reply to Kobes, I emphasize that most instances of self-knowledge are of this non-make-itself-true sort. They are non-inferential, but they are not reasons for themselves. The warrants for these cases of self-knowledge do not lie in justifications. They reside in the reliable, minimal, competence-understanding that fits within a rational structure, such as the structure of critical reasoning. The preservational mechanisms mentioned earlier are the bedrock of this structure. I do not elaborate this point, however, until 'Self and Self-Understanding'.

I said at the beginning of this Introduction that I use warrants for non-empirical cognition as a way of investigating cognitive powers distinctive of persons. Human beings are the only persons we know of. My interest in the distinctiveness of persons is, of course, not driven by interest in the biological distinctiveness of human beings. I want to understand powers, governed by basic cognitive and practical norms, that mark human beings as psychologically special, indeed, I think, especially valuable. The cognitive norms are associated with critical reason. Critical reason is the power to engage in meta-reasoning that marshals reasons and evidence, and that checks and reviews itself in the face of actual or possible criticism. It is the sort of reason that underlies the empirical and mathematical sciences. It drives philosophy itself. Practical norms for critical reasoning cluster around the notions of freedom and responsibility, and are epitomized by moral norms. Critical reason plays an essential role in the powers that underlie these practical norms.

Individuals that have the psychological powers to be subject to these norms are traditionally called 'persons'. Human beings may not be the only persons. Perhaps there are other critically rational beings capable of critical reason and morality in the universe. I am interested in the psychological powers that make persons persons.

The Dewey Lectures, given at Columbia University in December 2007 and substantially revised for the 2011 publication, explore a key element among such psychological powers—the power of self-understanding exercised in non-empirical self-knowledge. Given that self-knowledge is constitutively necessary for being subject to norms of critical reason and morality, I ask what psychological powers, and what psychological and normative structures, underlie this constitutive necessity. The answers provide an outline of a theory of the psychological and normative bases for non-empirical self-knowledge.

'Some Origins of Self', the first of the lectures, begins by discussing notions of person and self. I take *persons* to be something like what Strawson took them to be—individuals with both physical properties and certain distinctive psychological capacities. Very young children and demented adults lack the relevant psychological capacities—centrally, capacities for critical reason and for moral responsibility. They are persons by virtue of being the kind of individual that *naturally* has the relevant psychological powers in mature, undamaged states. I take *selves* to be individuals, or phases of individuals, that actually have the psychological capacities that make persons persons. Kant characterized persons in a way that comes close to capturing what I mean by 'self'. He followed Locke in centering on a capacity for self-consciousness with a diachronic reach—a capacity to be conscious of oneself as oneself, and as one is at different times.

Most of this first lecture consists in reflections on phenomena that occur in the prehistory of two features of selves—self-consciousness and a certain sort of memory of one's own psychological past. This discussion should be construed as a detailed development of transitions from simple versions of ego-centric, or *de se*, indexes, which I discuss in 'Memory and Persons' and '*De Se* Preservation and Personal Identity: Reply to Shoemaker' (both in Part III of this volume), toward a full first-person concept.

The three milestones that I discuss in the prehistory of the relevant sort of self-consciousness are passing the mirror test, imitation, and joint attention. In each of these cases, I discuss crossmodal (visual/proprioceptive) and intramodal (visual) couplings of ego-centric indexes—a pre-conceptual representation that is a precursor of the first-person concept I̲. I try to show how these psychological couplings provide pre-conceptual precursors that are psychological bases for the eventual capacity to coordinate first- and third-person conceptual points of view on oneself. I believe that this latter sort of conceptual capacity is part of the minimal basis for having a mature self-consciousness capable of objectification. Such a capacity is expressible in uses of the mature self-concept that enters into critical reason and moral responsibility.

In the latter part of the lecture, I discuss various types of memory that are known to occur in human psychology. This taxonomic work leads up to what I call '*meta-psychological, autobiographical episodic memory of one's past from the inside*'. The moniker is, of course, a mouthful. But being careful about the exact nature of the memory pays dividends. For it is a type of memory that is central to the most important psychological kinds. I believe that this is the type of memory that Kant had in mind in his characterization of persons (selves). I think that the same type of memory was what Locke had in mind in his famous attempt to explain in terms of memory the identity of persons over time. Locke was trying not only to produce an account of the constitutive diachronic essence of persons. He was also trying to explain a necessary condition on moral responsibility.

Locke fails to provide an acceptable account for either purpose. I point out relatively obvious (and frequently cited) difficulties with his accounts. In particular, I think that in his account of moral responsibility, he takes memory to have a role at too specific a level. One can be morally responsible for acts that one cannot at all remember engaging in.

Darwin highlights memory in his remarks on wherein human beings are distinctive. I point out relatively obvious difficulties with his account as well. I believe, however, that Kant, Locke, and Darwin are all onto the fact that a certain type of memory—I think, *meta-psychological, autobiographical episodic memory of one's past from the inside*—does figure centrally in our conception of selves, and our conceptions of critical reason and moral responsibility. In the subsequent lecture, I try to show what this role is. This first lecture concludes by forecasting that this and other types of memory figure centrally in a kind of self-understanding that is partly constitutive of selves.

In 'Self and Constitutive Norms' I expand the methodological strategy outlined in 'Our Entitlement to Self-Knowledge'. I assume that a minimal type of self-understanding is constitutive to being a self. By reflecting on the applicability of two types of norms—those of critical reason and morality—I try to make progress in understanding the particular sort of self-understanding that is central. I think that certainly the first of these two norms, and arguably the second, are constitutive to being a self. By reflecting on the nature of the applicability and applications of these norms, I hope to gain insight into the nature of the relevant self-understanding, and ultimately into the nature of selves and persons.

In the course of exploring the applicability and application of these norms, I develop a notion of an *apperceptive core point of view*. Apperceptive core points of view are the aspects of self-understanding that set standards for successful realization of the relevant norms. In the moral case, they also set a baseline for mitigating negative evaluation. Apperceptive core points of view are starting points for inferences that are to be evaluated or that figure in evaluations.

After motivating and developing these points, I return to *diachronic* powers of self-consciousness. I derive the conclusion that meta-psychological autobiographical episodic memory from the inside is constitutive to being a self. I believe that this derivation constitutes substantial progress on Kant and Locke's remarks about the constitutive role of diachronic self-consciousness in being a self.

I carry out such a derivation with respect to three diachronic constitutive features of selves—elements in ordinary inference, elements in practical decision-making, and dialectical elements of both critical reasoning and moral thinking. For example, in the case of ordinary inference, I carry out the derivation from two assumptions about selves. First, I assume that selves, critical reasoners, and moral beings must be capable of propositional inference. Second, I assume that selves, critical reasoners, and moral beings must be able to think, meta-psychologically and consciously, about their own propositional inferences, as their own.

In 'Self-Understanding' I give a unified account of how immunity to brute error in self-knowledge is constitutively necessary for the applicability of norms of critical reason and morality. A large subspecies of the relevant type of self-knowledge—the subspecies that does not involve knowledge of non-representational sensory states—is non-empirical.

I emphasize that except in *cogito* cases, the self-attributions whose warrant guarantees truth *cannot* be determined, or shown to be thus warranted, by appeal to the contents and modes of the attributed states. I argue, however, that constitutively for selves, there are some cases of immunity to brute error. These are the cases that ground applicability of moral norms and norms of critical reason.

The self-knowledge that is integral to critical reason and morality is not deep. It contrasts with wisdom, which we admire and is hard to come by. If the self-knowledge integral to critical reason and morality were deep, it could not be a condition on being critical reasoner. There are many unwise critical reasoners that have not used their critical reason very well. If it were hard to come by, we would admire it more. For all that, the relevant self-knowledge is warrant-factive, or immune to brute error. And it is philosophically important because it is constitutive to being a critical reasoner, and to being a moral agent.

These points are intuitively plausible. I argue for them by appealing to the base-line conditions for evaluations of critical reasonability and morality. If an individual's self-understanding of a psychological element can be warranted but mistaken, the individual's self-understanding is too detached from that element to count as understanding it from the inside. In such cases, the individual is more fundamentally accountable for the meta-representational attitude than for the self-understood psychological element. The relevant norms require that the individual be able to understand what he or she does or undergoes *in* the doing or undergoing. Such understanding lacks the contingent relation between subject matter and point of view that makes for liability to brute error.

The key underlying idea is that immunity to brute error resides in use of the capacities whose structures and natures are constitutive to a type of point of view. The relevant capacities are mostly preservational capacities—such as inference or purely preservative memory or meta-level redeployment of concepts in specifying themselves—that hold a point of view together. I think of these preservational capacities as fundamental structural components of a point of view. They are like the skeletal transit lines of a transportational system, except that the system here is a representational psychology. The main types of these skeletal lines make a point of view, and an individual with the point of view, the kind of point of view, and individual, that it is. More complex psychologies, and individuals, are marked by special types of preservational capacities. For example, the preservational capacities in self-knowledge, communication, and reflection mark human beings as cognitively special—and different in kind from other types of individual.

Thus, the discussion of preservative capacities in this lecture, 'Self-Understanding', should be understood in the context of the discussion of purely preservative memory that I first introduced in 'Content Preservation' (Part II). Preservation in communication and purely preservative memory in inference are instances of a deeper preservational phenomenon. My discussion of the role of preservative capacities in this lecture is my fullest one. It best elicits the constitutive role of such capacities in determining the natures of basic types of individuals with representational psychologies.

In order to understand immunity to brute error in self-knowledge, I consider immunity to brute error in three other cases: (a) warrant for first-order deductive inference; (b) warrant for beliefs in simple, self-evident truths on the basis of understanding them; and (c) warrant for non-inferential beliefs that one attitude (or content together with mode-type) is a reason to believe another. I use these comparison cases to provide a context for reflecting on the warrants involved in the relevant type of self-understanding. I see the most basic warrants as attaching to certain preservational powers that are basic to any representational mind. These warrants combine with relevant meta-representational competencies to mark selves as distinctive.

Here I discuss again the way in which the role of the relevant type of self-knowledge in critical reasoning bears on the warrant for the self-knowledge. This is the issue raised by Peacocke with regard to 'Our Entitlement to Self-Knowledge'.

One is warranted in non-empirical self-knowledge through understanding relevant self-attributions. To be epistemically warranted, the understanding of these self-attributions must be reliable in yielding veridical cognition. There are three key aspects to this understanding of self-attributions that connect to critical reason. Each of these aspects is shared with critical reason and is psychologically more basic than—hence independent of—the self-knowledge itself.

First, all instances of non-empirical self-knowledge—including both pure *cogito* instances and self-attributions of standing beliefs—depend on inter-level preservational structures. These are structures that transmit warrant between object-level and meta-representational levels of cognition. In all instances of non-empirical self-knowledge, these structures are shared with critical reasoning. Fulfilling epistemic norms associated with use of these preservational structures is shared between the relevant self-knowledge and critical reasoning. These structures are not more fundamental to self-knowledge than to critical reason. For they are equally basic to inter-level reasoning and inter-level self-attribution. So warrants for relying on these structures in self-understanding are the same warrants for relying on these structures in critical reasoning. The understanding involved in self-knowledge and the understanding involved in critical reasoning have a common source—which is not proprietary to self-knowledge—in the minimal understanding involved in making transitions along these preservational routes.

To summarize the point: There is a set of inter-level structures, which include as subspecies both purely preservative memory and use of object-level contents in meta-level specifications of them, that are constitutive to a mature first-person point of view. These structures and the understanding necessary to make use of them in thought are constitutive to both non-empirical self-knowledge and critical reasoning. The relevant understanding is not more basic to the warrant for self-knowledge than it is to the warrant for critical reasoning. So one's warrant for non-empirical self-knowledge gains some of its authority from an aspect of understanding that is constitutive to critical reasoning and self-attribution, and not more basic to one than the other.

Second, the aspect of the self-understanding that consists in understanding the first-person concept constitutively involves recognizing and implementing responsibility for supporting one's first-level propositional attitudes with reasons, where one can.[14] Part of implementing responsibility for rationally supporting one's first-level propositional attitudes is mastering the rationally immediate inter-level application of reasons.[15] This recognition of the role of the first-person concept in marking responsibility for implementing reason support for one's own attitudes is part of what is distinctive to the understanding involved in critical reason. It is also part of the understanding of the first-person concept. Hence it is part of the understanding that warrants self-attributions constitutive of self-knowledge. Here the force of the warrant for self-attribution comes from critical reason, not a more basic common source.

To summarize the point: using the first-person concept in a knowledgeable way constitutively involves the understanding of reasons employed in critical reasoning. Whatever warrant such understanding provides partly depends on warrants associated with the understanding of reasons.

Third, the understanding employed in the self-attribution in non-empirical self-knowledge constitutively includes a capacity to specify the self-attributed contents in a way that relies on thinking and understanding those contents *in* their specification. This specificational capacity depends on a first-level understanding of relevant reasons for or against commitments to the relevant content. Understanding a content involves understanding reasons that support or undermine attitudes with that content. This understanding takes a meta-representational form in critical reason. One not only understands *how to* marshal reasons for or against a content. One also understands such reasons as reasons. Insofar as non-empirical self-knowledge figures in critical reason, the self-understanding that generates warrant for the self-knowledge is necessarily enhanced by this understanding of the way any given self-attributed content is embedded in a network of possible supporting- or undermining reasons. This point—made most fully at the beginning of section VII of 'Self-Understanding', the third Dewey Lecture—elaborates the remark in 'A Century of Deflation and a Moment about Self-Knowledge' that a person is an authoritative judge regarding some of the attitudes he or she has because persons "know the law". They know the rational norms and rational connections that help individuate the attitudes that they are self-attributing. They are more reliable at specifying and self-attributing contents insofar as they can locate the contents in a network of reasons. For a position in such a network is an important aspect of the very identity and individuation of the content. So the warrant for the self-understanding that underlies non-empirical self-knowledge gains some of its force from powers of understanding that are constitutive to and distinctive of critical reason.

To summarize the point: The warrant for non-empirical self-knowledge in critical reasoners gets some of its force from the way a critical reasoner must use

[14] I emphasize these points mainly in 'A Century of Deflation and a Moment of Self-knowledge' and in 'Reason and the First-Person'.

[15] I am referring to the inter-level structure discussed toward the end of 'Our Entitlement to Self-Knowledge'.

understanding of reason relations, as such, in identifying and canonically specifying the representational contents of his or her attitudes.

In the latter sections of 'Self-Understanding', I sketch a relatively full account of non-empirical self-knowledge—the sort of account merely intimated in 'Individualism and Self-Knowledge'. Specifically, I extend the account of authoritative self-knowledge beyond knowledge of one's occurrent propositional attitudes to knowledge of one's standing attitudes (such as beliefs), to meta-representational episodic memory, to anticipations of actions, and to certain non-propositional psychological states, including one's sensations.[16] In all these cases, I take the non-empirical warrant for the relevant self-knowledge to consist in the understanding involved in thinking the known contents.

I hope that the fuller account vindicates my emphasis in 'Individualism and Self-Knowledge' on instances of *cogito* as keys to understanding authoritative self-knowledge more generally. Although *cogito* cases are very special cases, they contain, in germ, many of the key features that are central to and constitutive to all types of authoritative self-knowledge. *Cogito* instances are special and peculiar in their self-evidence, self-justification, and self-verification. But they point beyond themselves in other respects: their non-inferential immediacy; their first-person character; their use of canonical specification of contents that requires understanding of the referred-to contents as well as the specification; their use of a betokening understanding of the attitude mode; their reliance on inter-level representational relations that are routes for preserving warrant; their immunity to brute error; their being warranted through understanding.

The essay ends with a contrast between *de re* understanding—immediate, non-inferential understanding of particulars—and generalized understanding of laws or essences. Self-understanding is a prominent case of *de re* understanding. I think that being a self, and being a person, rest as much on understanding *particulars* in ways that are immune to brute error as they do on uses of reason and generalized understanding that are immune to brute error. Although the latter loom larger in the history of philosophy, the former—as Descartes realized—are equally important for understanding cognition that is distinctive of persons and selves.

II

Part II centers on cognition that derives from interlocution—linguistic communication with others. Communication lies at the heart of most of our knowledge that goes beyond perceptual belief. It is central to what makes possible a shared culture and a progressive science in a complex world. Oddly, it has not been

[16] Knowledge of one's sensations is empirical. However, this form of self-knowledge shares important features with the non-empirical cases—especially the immediacy, the direct role of the referred-to state in understanding the self-attribution, and the immunity to brute error.

very central in the history of epistemology. It has been treated as a specialized topic—more a curiosity than the hugely central matter that it actually is.

My discussion of interlocution is part of systematic reflection on the higher powers of mind—powers of intellection. Relations between self-knowledge and reflection, on one hand, and interlocution, on the other, connect closely to relations between first-person and third-person specifications of psychological matters.

The reader should not allow the compartmentalization of the essays in this volume into the four parts to obscure these relations. I have noted earlier the important connections between 'Content Preservation' and some of the essays on self-knowledge in Part I. 'Content Preservation' is also closely related to 'Reason and the First Person' in Part III. To understand much of what goes on in the volume, one might do well to read 'Content Preservation' first.

'Content Preservation' introduced two notions that I think important for epistemology and philosophy of mind. One is purely preservative memory. To repeat my earlier explication: *purely preservative memory* is a type of memory that functions to preserve past attitudes, with their cotents, for current use, without introducing new contents or attitudes (for example, as premises), with their own warrants and subject matter, into current cognition.[17] This notion figures not only as a model for understanding communication in 'Content Preservation'. It also figures in the account of self-knowledge in the Dewey Lecture 'Self-Understanding' and the account of the unity of persons in 'Memory and Persons' and '*De Se* Preservation and Personal Identity'.

The other notion that I introduce, and that I will focus on first, is entitlement. To repeat my earlier explication: An *entitlement* is a warrant whose force does not consist, even partly, in a reason. An *epistemic* entitlement derives from an individual's meeting an epistemic standard for using a good route to truth, where the standard is not that of having a reason. A *practical* entitlement derives from the individual's meeting a practical standard for operating well in the service of realizing some purpose or practical function, where the standard is not that of having a reason.

Entitlement is one side of the distinction, cited earlier, between two types of warrant. The other side, justification, is traditional. The notion of entitlement, isolated as applying to a distinct type of warrant, is relatively new. Attempts to understand warrants purely in terms of individuals' reasons inevitably hyper-intellectualize warrant.

Hyper-intellectualization emerges vividly in accounts of the some of the most basic types of knowledge. For example, the idea that young children (say, under the age of three) or higher non-human animals either do not know anything or must have reasons for their perceptual beliefs to have knowledge through them is not credible. A reason for a perceptual belief would have to be a propositional content that makes reference to perception as such. There is evidence that very

[17] In 'Content Preservation', I explicate the notion purely preservative memory in terms of preservation of thoughts and their contents. Here I broaden the explication to include non-propositional states and their contents.

young children and higher animals do not make reference to their perceptual states as such. They certainly do not refer to them, even unconsciously, in propositional attitudes that operatively sustain or support their perceptual beliefs. Yet excluding young children and higher animals from the circle of knowers seems narrow and provincial.

Even if it turns out that children can refer to perceptions as such at very young ages, it seems clear that being warranted in having perceptual beliefs does not depend on having a reasoned justification that cites the belief's deriving from perception as a ground for the belief. It is enough that the belief does in fact derive from perception in the right way. An individual need not know that it does.

Some of the over-concentration on justification, and the neglect of entitlement, derives from a tendency in the history of philosophy to conflate ordinary knowledge and warranted belief with *scientia*—a type of knowledge grounded in a deliberative, self-conscious methodology, as epitomized by science. Much of the history of philosophy is, understandably, centered on *scientia*, not on ordinary, relatively unsophisticated knowledge. Still, the systematic neglect of the less sophisticated types of knowledge distorts epistemology. Indeed, it ultimately distorts the epistemic ground for empirical *scientia*.

Similarly, our reliance on inference—in induction or deduction—and on general cognitive capacities such as perception and empirical belief formation is fundamentally warranted not through meta-justifications, but through entitlements. Eliminating philosophy's long-term tendency to hyper-intellectualize warrant seems to me to be a long-term project that requires the cooperation of many minds to be successful.[18]

Some subtle issues in some of my previous explications of the notion of entitlement need to be noted here. I write, '. . . entitlements are epistemic rights or warrants that need not be understood by or even accessible to the subject. . . . Justifications . . . involve reasons that people have and have access to'.[19] Although the formulation is not explicit, other writing in the context makes it clear that to have a justification an individual need not have access to a reason considered *as* a reason. Animals and very young children do not have the concept reason and cannot think of anything as a reason. They *can* have reasons—justifications—inasmuch as they engage in propositional inference. I think that higher animals and young children do have reasons for some of their beliefs and activities. What I doubt is that they have reasons for their perceptual beliefs, and more generally that being warranted amounts to having a reason. The reasons that they have use perceptual beliefs as premises, as support for other beliefs. They

[18] I do not mean to suggest that the introduction of the notion of entitlement is the only current in philosophy that opposes the endemic hyper-intellectualization in epistemology. Quine's attempt to naturalize epistemology and various reliabilist programs point in the same direction. I think that many of these other programs underplay the role of norms that psychological states and processes must meet in order to contribute to knowledge. I think that the notion of entitlement is important for understanding the normative standards that underlie knowledge.

[19] 'Content Preservation', *The Philosophical Review* 102 (1993), 230.

lack reasons for their starting points—the perceptual beliefs themselves. They are nonetheless epistemically entitled to those beliefs.

The subtlety that I believe that I was insufficiently sensitive to when I introduced the notion of entitlement concerns the notions of having a reason and having access to a warrant. The initial explication of the notion of entitlement in 'Content Preservation', just quoted, centers on access. If access is understood in one common way, modular states are not accessible to an individual, even though the states are in the individual's psychology. The individual cannot bring them up, even on reflection, even under psychoanalysis, just by thinking hard. On such an understanding, if there is modular reasoning, the reasons involved would count as entitlements rather than justifications. I think that it is legitimate to understand *an* entitlement–justification distinction in terms of access—accessibility to being made conscious through hard thinking, without learning new information.

I came to think that a functional-structural conception is more basic than a conception in terms of accessibility to consciousness. As I now use the terms, *justification* consists in *warrants by reasons*; an individual has a justification only if a relevant reason is *present in an individual's psychology*; *entitlements* are *warrants that do not consist, even partly, in reasons*; an individual can have an entitlement even if no rationalizing explanation of the entitlement is present in the individual's psychology. As always, I take reasons to be propositional. If there are modular reasons, they are justifications in an individual's psychology, although not justifications accessible to the individual's consciousness. The justification is accessible only in the weak sense that it occurs in the individual's psychology.

The terminological issue is not fundamentally important. What is basic is being clear about the different ideas involved: the form and function of the warrant in the individual's psychology, and the accessibility to consciousness of the warrant.

The other notion introduced in 'Content Preservation' is purely preservative memory. When memory, long- or short-term, functions as purely preservative, it functions to retain the mode and representational content of a past psychological state, without introducing that content (with its mode) as contributing a new premise with its own warrant into an inference, or as constituting a new topic in the cognitive context.

Recall that lines of preservation form the skeleton of a representational psychology. Types of preservation are a large factor in determining the *nature* of any reprsentational psychology. 'Content Preservation' is the first place where I introduce the most basic type of preservation—purely preservative memory. Although I emphasize the role of purely preservative memory in propositional inference, it takes different forms. In one form or another, it is one of the functions of memory in any representational psychology. Thus, for example, preservation of perceptual contents in memory is a type of purely preservative memory.

The notion of purely preservative memory is important for understanding inference. In any inference, past steps must be redeployed in the inference. When they are reinvoked, they must have been preserved by memory from the time when they were first invoked. If the reinvocation required a new warrant, one could never rely on the previously invoked step. One would need to introduce a new step with its own warrant. Such a procedure would yield an infinite regress. To understand the epistemology of inference, one must conceive the role of memory as that of an enabling condition, not a source of warrant in the inference. One can be warranted (by entitlement) in relying on memory, just as one can be warranted in relying on a capacity to infer according to an inference rule. The memory does not, however, figure in warranting any step in the inference, even steps that involve reinvoking past steps and thus that rely on memory.

This point is relatively obvious when applied to inference. I think that the interest of the point lies in isolating this particular preservative function for memory—a function that is central to understanding the psychology and epistemology of a wide range of cases that go beyond obvious cases of inference. Preservation over time of the mode and content of psychological states is a (I am tempted exaggerate and say 'the') fundamental unifier in a representational psychology. As noted in my discussion of essays in Part I, my account of self-knowledge makes central reference to this type of preservation.

The primary objective of 'Content Preservation' is to develop a conception of interlocution (commonly called 'testimony') that emphasizes the unreasoned primitivity of passage of information through communication. I explore the idea that a very primitive entitlement for accepting the word of others is non-empirical. The entitlement is often overridden or overdetermined by considerations that center on the particular context, content, or author of a piece of communication. Still, default acceptance of the word of another, other things equal, is an epistemological starting point. I believe that this point fits our actual practice, when that practice is examined in a careful way.

Thus although the idea that we have a non-empirical entitlement to accept what we comprehend as being told can seem to be an unusual idea, it seems to me to apply quite naturally to our actual practice—if one is careful to distinguish general aspects of the practice from aspects that are particular to occurrent communications. These latter aspects can affect and ultimately determine the nature of our warrant (or lack of warrant) for accepting what others say. Such aspects may qualify the default entitlement—either by raising warranted doubts, or by supplying further empirical support. The warrants that derive from particular features of a given communication are usually empirical. I believe, however, that we are entitled to the actual bias that we have—as young children and as adults in unloaded communicative situations—toward acceptance-as-true of what we understand others as presenting-as-true, other things equal. I believe that the default entitlement to this bias is apriori—independent from sense experience for its warranting force.

The main point of the essay is to explain the entitlement to the bias. The idea, stated very summarily, is as follows. The intelligibility of a propositional assertion is apriori related to the assertion's having an origin in a being with reason. For propositional capacities are constitutively associated with propositional uses, which include uses as reasons. Such reasons either support or sustain belief. A function of reason is to support and sustain truth impersonally. So reasons are reliable supporters and sustainers of truth—including sincere, communicative, reliable supporters and sustainers—other things equal. To have a capacity with a function, an individual must be competent, other things equal, to realize the capacity. Any failure to realize the capacity can be assumed to be an aberration, which could not have figured in the normal conditions for obtaining the competence. So a reasonable being is, other things equal, a reliable source for truth. Unless one has reason to think that the conditions of a particular communication act are not normal conditions, one is entitled to rely upon their being normal. So one can rely on what a being with reason says unless one knows something that countermands such reliance.

I further argued that although understanding a communication event as an intelligible propositional assertion always requires the use of perception, in some cases (cases of understanding that do not involve perception-backed *de re* cognition—cases of what I call 'intellectual understanding') perception figures as an enabling condition rather than as a contributor to the warrant for the understanding. For example, to be intellectual understanding, the understanding does not involve seeing some particular that the interlocutor is pointing to. So I limited my claim to utterances of truths such as cats are smaller than elephants and 2 is a prime number that do not involve perceptually-backed *de re* applications.

Here I used the analogy to purely preservative memory as it functions in inference. I maintained that perception can function as a trigger for understanding, as perception of diagrams in mathematics often does. I held that perception can function to bring into operation an understanding that is warranted intellectually.[20] There are differences between the mathematical case and the case of utilizing understanding of what is said. I discuss these in both 'Content Preservation' and 'Interlocution, Perception, and Memory'.

The idea of an understanding that is causally dependent on sensory-causal relation to what another says, but that is not warranted through sense perception or through the nature of the sensory registrations, is obviously a delicate one. The understanding must be reliable to be warranted. So the causal chain relating the individual's understanding to the psychological states of the interlocutor must be reliable, and the chain must go through sensory channels. I wanted to emphasize

[20] In 'Comprehension and Interpretation' in L. Hahn (ed.), *The Philosophy of Donald Davidson* (Chicago: Open Court Publishers, 1999), I criticize the view that understanding in linguistic interchanges is fundamentally a matter of interpretation—which consists in an inference to the best explanation from perception of linguistic events that provide the basis for the inference. In such an inference the perceived linguistic events are initially construed as non-contentful, or at least as lacking specific content: the content must be inferred by the individual from those events together with background assumptions about the most plausible meaning. It may be that some analog of such inferences occurs in comprehension. The process that leads from linguistic perception to comprehension is certainly a complex one, with many transitions. I believe, however, that it is important not to assimilate this process to person-level inference. It is sub-personal, and it may well be, in major respects, non-propositional.

that apart from the sheer reliability of the sensory channels in connecting to other individuals' thoughts, nothing about the nature of the sensory registrations or the perceptual representation, is essential for the understanding's being warranted. I claimed that although the understanding must make use of a causal-sensory channel, the causal relation is an enabling condition, not a contributor to the warrant.

I now believe that these claims in the "further argument" were mistaken. I believe that human beings do not have a non-empirical warrant for their comprehension of what others say. I stand by the claim that we have an apriori entitlement to rely on the Acceptance Principle. I believe that we do have an apriori default (overridable) entitlement to believe what we understand others to be telling us. Contrary to what I maintained in 'Content Preservation' and elsewhere, I think that the comprehension that is needed to bring pieces of communication from others under the Acceptance Principle is inevitably warranted empirically. The force of one's warrant for one's comprehension depends on perceiving others' linguistic output competently and reliably. In 'Postscript: Content Preservation', I diagnose and criticize my mistake. I try to explain what led me to it, why some replies to it seemed unconvincing, and why it is mistaken.

The mistake was a significant one. It occurs in several of the essays in this section. It affects the argument for an apriori route to knowledge of other minds, sketched in 'Reason and the First Person' (Part III of this volume) and intimated in 'Computer Proof, Apriori Knowledge, and Other Minds'. However, it seems to me that much of what I wrote on interlocution retains value and is unaffected. Despite its defects, the defence of the mistaken claim does, I think, bring out the relatively small role that perception plays in the whole process. The role of perception in comprehension is crucial. In humans, it is ineliminable. But language perception is deeply informed by intellection in its contribution to comprehension. And I continue to think that our default warrant for accepting as true what we comprehend as being told to us is thoroughly non-empirical. Our overall default warrant for belief based on interlocution is empirical, but *barely empirical.*

The epistemic status of belief and knowledge that are derived from interlocution is less important than two other matters emphasized in the essay. The first is the primitivity of the entitlement to accept what others say. The second is the importance in communication of relying on others as rational sources for belief. I believe that 'Content Preservation' opened a valuable way of thinking about a rich set of issues.

'Postscript: Content Preservation' serves as postscript not only to 'Content Preservation' but to the other essays on interlocution that are included here. It centers on three issues. One is the role of knowledge in an antecedent chain of communication in making knowledge by a recipient possible. A second is the Acceptance Principle and its role in articulating a default prima facie entitlement to believe an interlocutor. A third is the epistemic status of the initial comprehension of another's utterance. On the second issue, I defend the position of 'Content

Preservation' against some recent criticisms. On the first and third issues, I criticize my earlier views. I regard the first issue as a relatively minor one, at least in the context of the discussion in 'Content Preservation'. The third issue is much more central. It concerns the mistake that I mentioned two and three paragraphs back.

First, I discuss a counterexample by Peter Graham to a claim in 'Content Preservation' that for a recipient to gain knowledge from interlocution, someone in the antecedent chain must have the knowledge. With some qualifications, I accept the counterexample. I point out ways in which it is not a counterexample to a broader (non-equivalent) principle governing the relation between the recipient's knowledge and knowledge in the antecedent chain that I also state in 'Content Preservation'.

The main value of this discussion, apart from acknowledging and correcting a mistake of mine, lies in my development of the nature of the recipient's entitlement in the counterexample case. This account differs from Graham's. It emphasizes the role of sociological natural kinds in grounding the recipient's entitlement. The role of reliance on sociological institutions supplements and overlays the role of reliance on the interlocutor's rationality in grounding the recipient's entitlements. The former type of entitlement grounds knowledge in the case under discussion, whereas the latter does not. I think that the case illustrates the variety of entitlements that an individual can have on a particular occasion. It also illustrates ways in which epistemic norms are grounded in psychological and sociological kinds.

The second topic of the Postscript is the keystone position of the Acceptance Principle in the structure of warrant for relying on interlocution. I discuss some criticism of taking the principle as basic to the epistemology of interlocution. The criticism underestimates the abstractness of the structure of the relevant default entitlement, and indeed mistakes the very nature of default entitlement. Although the Acceptance Principle is rarely the last word in an adult's warrant for relying on the truth of a piece of communication, it is the first word. And its voice both dominates the structure of additional pros and cons and commonly remains in the mix that constitutes a final warrant. It retains a presence even when other types of entitlements, grounded less in the rational nature of one's interlocutor than in the empirically determinable social role of the interlocutor, overlay it.

The third topic of the Postscript is the epistemic status of the initial intake of a report—the exercise of the recipient's capacity for comprehension. As indicated above, I criticize my view—presented not only in 'Content Preservation' but also in several other essays—that one's epistemic entitlement to rely on one's comprehension of what another says as a correct comprehension can be non-empirical. I discuss the four main ideas that had seemed to me to support this claim. I explain in considerable detail what seems to me to be good about these four ideas, but also why they fail collectively to support the counter-intuitive claim. Fortunately, the error can be isolated in such a way that it affects little else that I have claimed about the epistemology of interlocution. In particular, the structure of the default entitlements to rely on interlocutors' word, and the function and nature of the Acceptance Principle, remain in place.

'Interlocution, Perception, and Memory' answers objections by David Christensen and Hilary Kornblith to views presented in 'Content Preservation'. Although I still do not accept the grounds that they give for their objections, I accept the conclusion of the objection to my view that belief based on interlocution—in particular the comprehension on which such belief is based—can be apriori warranted.

In retrospect, I value the part of the essay that centers on a distinction between ordinary instruments in science and human beings as interlocutors. I believe that the fact that information is passed through a source (a human interlocutor) that can be expected to use reason makes that source an epistemically different "instrument" than an artifact that has not been programmed to go through rational propositional procedures. I allow that in principle artifacts might have the rational standing of human interlocutors. Indeed, computers that carry out proofs in mathematics have something like that standing. However, ordinary scientific instruments, like thermometers or microscopes or measuring instruments, even those that have linguistic labels on their indicators, are more like amplifications of our perceptual capacities than interlocutors whose propositional output can be expected to be backed by rational procedures. The difference corresponds to a difference between natural meaning—tracking natural events in the world and their correlation with other states and events—and representational meaning, the product of a source with propositional attitudes. There is an apriori connection between having propositional attitudes and truth. The connection between the reading on an ordinary instrument's dial and a natural event in the world is the ordinary causal connection that perception relies upon. However, as the next essay to be discussed emphasizes, some artifacts do function as (artifactual) interlocutors. They are constructed to be resources for reason that provide representational content. That content, and the procedures that systematize it, constitute reasons.

In 'Computer Proof, Apriori Knowledge, and Other Minds', I discuss epistemic issues that arise from the fact that mathematicians rely on computer proofs that are too long and complex for any human to check. For the sake of the argument, I make four assumptions: that pure mathematics can be known apriori; that computers are not autonomous thinkers (thus they cannot be regarded as ordinary interlocutors); that some mathematical propositions can be known apriori, even without knowing a proof; and (from 'Content Preservation') that although commonly empirical, knowledge obtained from interlocution can be apriori. As noted, I now reject this fourth assumption. It can nonetheless play the limitative, for-the-sake-of-argument role that it plays in the essay.

These assumptions serve to isolate the main issue of the essay: Does the fact that a mathematical proof is carried out by a computer, and is too long for a mathematician to check, in itself make it impossible for a competent mathematician to know the theorem of the proof apriori? The primary point of the essay is to argue for a negative answer. I continue to think that this answer is correct, even though I no longer believe that, strictly speaking, it *is* possible, by relying essentially on a computer, to know a theorem apriori. I think that one's warrant inevitably depends for at least a small part of its force on warrant for one's perceptual beliefs regarding what the computer's outputs are. I believe, however, that the details of the discussion retain epistemic interest.

I consider several ideas that may seem to show that reliance on a computer to carry out a humanly uncheckable proof *in itself* forces knowledge of the proved theorem to be empirical. I discuss the following ideas: (a) that reliance on the computer's reliability as a physical machine can be warranted only empirically; (b) that the source's being a computer rather than a thinker renders empirical any warrant for accepting its outputs; (c) that the difficulty and unsurveyability of the proof, and the need to check whether the computer has the power to carry out such a proof, render any warrant for accepting the theorem empirical; (d) that the need to reidentify the computer as a single source of mathematical outputs requires empirically warranted check.

I argue that none of these ideas is correct. The goal of the argument is to elicit non-empirically warranted mathematical powers that can certify the computer as a source sufficiently worthy of trust to take it to yield a successful proof. I discuss analogies between a great mathematician's non-empirical capacity to appreciate his or her own mathematical power and such a mathematician's capacity to appreciate mathematical power from another source, here a computer. The value of the essay lies, I think, in the details of the discussion—focusing on neglected ways in which our non-empirical powers of reasoning can be applied.

'Comprehension and Interpretation' criticizes uses of Quine and Davidson's theories of translation and interpretation as accounts of our normal understanding of others' utterances. Although Quine and Davidson present their pictures as idealized rational reconstructions—not as accounts of actual linguistic understanding—they focus on sophisticated inferences from evidence, especially in difficult cases. Quine centers on radical translation of a foreigner. Davidson centers on on-the-fly interpretation of non-standard utterances, such as malapropisms and irony. I think that the basic cases for theorizing about comprehension are simpler. They are cases in which we reliably and correctly take in what another says without thinking.

The essay develops the point in 'Content Preservation' that our basic non-meta-representational understanding—our comprehension—of what others say or write is not a matter of person-level propositional inference from evidence. (See note 20.) And it is not warranted through reason or reasoning. It is normally warranted through an entitlement that resides in the reliability of a competence to comprehend another's utterance immediately—without going through propositional inference based on beliefs about evidence. I confine this point to standing linguistic understanding—the sort of understanding that does not depend on understanding context-dependent devices, such as devices of *de re* demonstrative reference.

I compare, in some detail, the immediacy of such comprehension to the immediacy of perceptual belief. I argue that both are normally warranted through entitlement, not justification. An individual can comprehend another's utterance without even having the concepts needed to justify the comprehension. For example, a child can comprehend its parents' utterances without conceiving them as utterances and without having concepts of meaning or truth.

The child's entitlement to rely on its comprehension rests on the reliability of the child's non-inferential comprehension competence. Similar points apply to default warrants governing standing linguistic comprehension capacities in adults. The warrants are analogous to the basic entitlement that we have to rely on perceptual belief: this latter entitlement resides in the reliability of our perceptual systems in producing veridical perceptual states and in the reliability of our psychological transitions from veridical perceptual states to veridical belief states.

The essay expresses differences with Davidson on the role of the social in psychology. I hold, and still hold, that it is metaphysically possible for an individual to have beliefs without language. I think that very young children and non-human animals are probably actual examples.

I think that it is metaphysically, though not psychologically, possible for an individual to have language without social relations. And I think that one (metaphysically) could have the concept of belief without social relations. On the other hand, I think that dependence on others is metaphysically necessary for having certain types of beliefs, if one lacks certain types of background information. For example, one cannot believe that arthritis can occur in one's thigh, unless one depends partly on others for one's concept arthritis.

The essay also contains two considerations that were meant to further support the view that our comprehension of others can be, and under very restricted conditions is, apriori. One consideration emphasizes the way that our apriori comprehension of our own thoughts is intertwined with comprehension of others' utterances. The other consideration appeals to the possibility of injecting, or mainlining, one individual's thought contents into another individual's psychology, using brain-to-brain causal processes that do not go through perception or the senses.

As indicated, I no longer accept the conclusion that these considerations were supposed to support. The first consideration states something correct about actual enabling conditions for some first-person understanding of some of one's own thoughts. It does not show that comprehending others' utterances can ever be apriori. I discuss the second consideration in some detail in 'Postscript: Content Preservation'.

The value of the essay seems to me to lie in its evocation of the immediate, non-inferential nature of much linguistic comprehension,[21] and of the similarities

[21] I re-emphasize that being non-inferential does not mean not involving processes. *Inference*, as I use the term, is processing that connects propositional contents. Similarly, immediacy does not

between comprehension of others' utterances and perceptual belief about ordinary non-linguistic entities. In particular, the primary warrant for both is an entitlement. I believe that the similarities are closer than I believed at the time I wrote the essay: both entitlements are inevitably empirical. Still, the essay elicits how much of our warrant for relying on computers in mathematics is non-empirical. It seems to me that the mistake is instructive, but the main claims of the essay stand.

'A Warrant for Belief in Other Minds' was written in the late 1990s. The last early version of it is dated 1999. Originally, it was titled 'A Non-Empirical Warrant for Belief in Other Minds'. When I first wrote it, I intended to develop the argument that can be found, in truncated form, in 'Computer Proof, Apriori Knowledge, and Other Minds' (section V) and in the last section of 'Reason and the First Person'. That argument supported and employed positions that I no longer accept. As noted above and in 'Postscript: Content Preservation', I no longer believe that our warrant for relying on comprehension of others' utterances is ever non-empirical. In 2010–2011, I revised the essay to its present form, and revised its title. An alternative revised title would be 'A Barely Empirical Warrant for Belief in Other Minds'.

I believe that the argument brings out a distinctive way of knowing other minds—through comprehending speech—that has been under-appreciated in the history of philosophy. I discuss traditional arguments from analogy and arguments from inference-to-the-best-explanation. I argue that although they are part of the account of how we know other minds, most of them focus on too high a cognitive level to account for the primitiveness of one of the warrants we have for believing in other minds. I think that the depth of our reliance on propositional form as a sign of mind needs far more attention. This essay takes a step in that direction.

III

The essays in Part III center on relations between reason, or reasoning, and being a reasoning individual. The individual may be a self—a critical reasoner, capable of self-evaluation—or just a non-reflective, first-level reasoner. Being an individual requires having some unifying condition. The unifying conditions that I investigate are various types of propositional forms and propositional interconnections. The work in Part III is closely connected with the work in Parts I and II on preservational structures in psychologies of all levels—pre-rational, rational, and critically rational. (The connections are closest with 'Content Preservation', 'Our Entitlement to Self-Knowledge', and the last of the Dewey

entail lack of processing. A representational state is *immediate* if it is not the product of an inference.

Lectures, 'Self-Understanding'.) These structures are sources of unity that are constitutive to being egos, thinkers, and selves. The preservational structures discussed in Part III are psychologically more complex than the structures that underlie the ego-centric indexes that designate whole psychological individuals. The types of individuals discussed in this part are at least reasoners, and in some cases selves or persons.

'Reason and the First Person' argues that there is a constitutive connection between use of the mature first-person concept and understanding what a reason is. I begin by considering Lichtenberg's doubt that Descartes' *cogito* needs to use, or can support justification of a use for, the first-person concept. I argue that a use for that concept is guaranteed and legitimated by understanding what a reason is.

I distinguish two aspects of understanding what a reason is. One is understanding the role of the concept reason in evaluating or appraising. The other is understanding the role of the concept reason in moving one to apply the evaluations or appraisals in reasoning—to affect attitudes in accord with reasons. Understanding the concept reason requires understanding this implementational role, as well as the evaluative role. Understanding the implementational role requires an ability to mark conceptually, in particular instances, the attitudes or acts for which rational evaluation of an attitude or act enjoins shaping it in accord with the rational evaluation. The first-person concept marks those attitudes that are subject to rationally immediate implementation.

I explore a sense in which theoretical reason has a practical aspect through its connection to rational agency. A critical reasoner would be deficient if the reasoner lacked the apriori understanding marked by the first-person concept. Use of that concept is underwritten by reason.

As noted, the work in this essay connects to work in'Our Entitlement to Self-Knowledge' and the second and third Dewey Lectures. It spells out connections between one's right to use the first-person concept in self-knowledge and the use of critical reason. And it spells out the constitutive role of responsible commitment to norms of critical reason and morality that comes with being a self or person. The work in the essay should also be regarded in the light of the extensive discussion in the first Dewey Lecture and in 'Memory and Persons' of ego-representations that are ontogenetic predecessors of the first-person concept. (There is some discussion of these predecessors of the mature first-person concept in the nineteenth and twentieth paragraphs of 'Reason and the First Person'.) The first-person concept is part of the capacity to understand and acknowledge the normative standards—practical and theoretical—that are already present in pre-critical reasoning.

Although written later, 'Memory and Persons' and its postscript, '*De Se* Preservation and Personal Identity: Reply to Shoemaker' are natural predecessors for 'Reason and the First Person'. In 'Memory and Persons', I discuss a traditional issue regarding personal identity. Most of the discussion centers on the fundamental role of ego-centric or *de se* representation in some psychological competencies that are fundamental for any reasoner, including non-critical reasoners that are not persons. Thus the essay discusses psychological predecessors—pre-personal ego-centric representations—for the mature first-person concept. These predecessors are embedded in unifying competencies and structures that mark specific psychological kinds of individuals at various levels of representational competence.

The essay begins by distinguishing experiential memory, substantive content memory, and purely preservative memory. Experiential memory always has *de se* (or ego-centric) content, marking various grades of psychological complexity. The discussion here is an antecedent of the more complex discussion of couplings of ego-centric (*de se*) markers and of autobiographical episodic memory from the inside, which figures in the first two Dewey Lectures. The discussion of purely preservative memory has its antecedent in 'Content Preservation'.

I argue for a fundamental constitutive role for preservative memory with either *de se* content or *de se* presuppositions in three psychological activities: use of perception in action or to serve other needs; carrying out intentions; and engaging in inference. I show how each of these activities is constitutively inter-related with *de se* markers and preservation of such markers in memory. The first two involve a sensitivity to and privileging of the individual's own needs, aims, and point of view. This sensitivity and this privileging are marked by *de se* representational content of states with those competencies. The third activity, inference, presupposes such *de se* sensitivity and privileging in a more complex way.

I maintain that these three activities are constitutively necessary for being a person, indeed, for being any individual with propositional attitudes. The three activities constitutively depend on memory, with either *de se* content or *de se* presuppositions, or both. No individual with propositional attitudes could lack the sensitivity and privileging involved in these *de se* capacities. Having the capacity is part of what it is to be an individual with propositional attitudes (and more specifically, to be a person). Having these *de se* competencies requires applying them veridically. Veridical applications in memory require sameness of individual over time. Memory, with its *de se* presuppositions of transtemporal agent identity, is a condition on the possibility of an individual's having a representational mind. Memory, with its *de se* presuppositions of transtemporal identity is a condition on the possibility of an individual with propositional capacities. Hence it is a condition of the possibility of being a person.

The foregoing sets the stage for a discussion of the traditional dispute over the Lockean idea that the concept of personal identity can be analyzed in terms of the concept of memory. The dispute is over Locke's apparent approach to explaining personal identity over time in terms of memory. Butler objected that such an explanation is circular, for the concept of memory presupposes that the rememberer and the individual with the past psychological states that memory retains are the same individual. Shoemaker and Parfit proposed a way of circumventing Butler's objection. They appealed to a notion *quasi-memory* that allows connection between an individual and the past psychological states of another individual. They tried to explain personal identity in terms of sufficient psychological connectedness illustrated by quasi-memory, together with a condition that the connections do not branch in certain ways. This latter condition is introduced to handle imaginable cases of fission in which two individuals are equally psychologically connected to the states of the same past individual.

In 'Memory and Persons' I claim that the approaches of Shoemaker and Parfit merely avoid definitional circularity, and that a deeper sort of circularity—explanatory circularity—remains. The point is that the very sorts of psychological competencies whose connections over time are constitutive to individuals, including persons, are themselves essentially marked by *de se* preservational capacities. One cannot get underneath these capacities by appealing to neutral capacities like quasi-memory. For to be the sort of individual that could have quasi-memory, one must have the *de se* capacities as core competencies. Thus having quasi-memory constitutively requires having *de se* capacities. These points are worked out in some detail for each of the three basic psychological activities—use of perception, intentional action, and inference.

The main value of the essay seems to me to reside not so much in refuting a rather special and peculiar reductionist strategy, albeit a variant on an old and famous one. It resides in delineating apriori knowable, constitutively necessary connections among psychological capacities that are constitutive to representational mind, on one hand, and *de se* representation, on the other. *De se* representation marks sensitivity by the individual to his own needs, aims, and point of view. *De se* representation is a phylogenetic and constitutive ancestor of the first-person concept. It marks something central to what it is to be an individual with a psychology—a capacity in such an individual to privilege, be sensitive to, and mark psychological states and their perspectives as individual's own, by relating them to the individual's own needs, aims, and point of view. Psychological states are not specifiable apart from contents that indicate in a *de se* manner the psychological states as the individual's own. All representational psychological states must be connected to representational psychological states that function to privilege the individual as a distinct being.

Shoemaker responded to the argument in 'Memory and Persons'. Unfortunately, the response misconstrues the argument. The response serves to illustrate the power of my argument, rather than serving to reply to it, much less refute it. In '*De Se* Preservation and Personal Identity: Reply to Shoemaker', I show that a scenario that Shoemaker designs to support his approach actually confirms the line of 'Memory and Persons'. His reply fails to come to grips with my claim that having quasi-memories requires and must be explained in terms of *de se* memory. A detailed discussion of his scenario shows that the individuals in it have ordinary *de se* capacities, and that the additional quasi-memory capacities operate in a psychology that has *de se* capacities at its core. These core capacities are constitutively necessary for quasi-memory, whereas the contrary is clearly not so. Herein lies the explanatory circularity in Shoemaker and Parfit's accounts. There is a circle in the explanation of the constitutive natures of the relevant capacities.

I highlight two basic functions for *de se* capacities. These functions must be present in any representational individual psychology—hence in the psychology of any person. One function is to be sensitive to the individual's own needs, goals, and point of view and to privilege them. The second function is to serve those needs, goals, and point of view through representational powers. Part of what it is to have an individual mind is to be sensitive to one's own needs, projects, and perspectives, and to do things for oneself. Powers to carry out the latter function require, as 'Memory and Persons' shows, having *de se* preservational powers. These powers are unifiers of individual minds over time. They mark and presuppose the identity of the individual.

Since quasi-memory is unspecific in its function as to whether it operates intra-individually or connects one individual with another, it cannot realize or explain the unity of persons, or other individuals with representational minds. No other element in the Shoemaker–Parfit position can do so. A special feature of representational psychologies is that the core psychological states make reference to the individual, or some aspect of the individual, in a way that marks the centrality of that individual's core needs and projects for that individual. Psychologies contain *de se* specifications of the individuals who have them, because those specifications mark fundamental functions and powers constitutive of being an individual with a representational psychology. The constitutive nature of such specifications makes it impossible to account for personal identity in terms of psychological states that do not contain them.

This account provides a de-intellectualized and elaborated counterpart of Kant's unity of apperception. Kant insightfully maintained that psychologies contain powers that function to provide a unifying condition for the psychology. These powers, in turn, provide a unifying condition for the individual with that psychology. Kant also took reflexive representation to be an essential aspect of the unification that interested him most (the unification that marks critical reasoners). Thus he recognized that the unifying psychological processes, such as memory and inference, cannot be specified independently of specification, within the unified psychology, of the individual that authors the psychology. His expressions of these insights focused, however, too exclusively on a very high level of psychological sophistication. Because he was interested in *scientia*—the high-level cognition exhibited in science and moral deliberation by persons and selves—he couched his point in terms of self-consciousness. He really meant *self* here—a reasoner capable of deliberation, reflection, critical reasoning.

The basic point, however, does not depend on a high-level capacity for self-consciousness. *De se* states and *de se* preservational powers are present in any representational mind—even animals that have only perception and no propositional attitudes, perhaps even animals that lack consciousness altogether. These powers take on more sophisticated forms in more complex minds. They re-occur in propositional inference—as I explain in some detail in 'Memory and Persons'. They become apperceptive in the self-consciousness and self-knowledge of persons and selves, as I discuss in the Dewey Lectures. But the basic *de se* unifying functions go all the way down. Tracing their realizations at different levels of representation is a large part of understanding both the essence and the varieties of representational mind.

'Modest Dualism' deals with the classical mind–body problem. This essay's ontological focus distinguishes it from most of the other essays in this volume. The ontological discussion centers, however, on formal structures that mark capacities to reason—capacities that ground the psychological kinds and epistemic norms that are central topics in the collection.

I begin by emphasizing the importance of doing metaphysics in such a way that it hews closely to what we know, especially in the sciences. The essay criticizes a cultural and professional tendency to assume more reason to accept materialist ontologies than we actually have. I review my argument from 'Individualism and the Mental' against a particular version of materialism—the token identity theory.[22] Then I discuss a less committal version of materialism—compositional materialism.

A paradigm that has served the natural sciences well is to take more complex entities that are studied in the special sciences to be composed out of physical entities that are studied in physics. I emphasize that the applicability of this paradigm for physics, chemistry, and biology always had more basis in the actual kinds studied by those sciences than it does for psychology or the social sciences. I maintain that although the paradigm should continue to be explored in trying to understand the relation between psychology and neuroscience, it has no solid, comprehensive evidence to support it.

The main contribution of the essay is a pair of closely related arguments for the conclusion that the natures, the basic kinds, of states studied in cognitive psychology—kinds whose natures involve representational contents with propositional structures—are not kinds whose instances are material composites of physical entities.

One argument centers on causation. According to the natural sciences, reason relations and propositional structures are not structural features of material composites. The causation by material parts of material composites, operating in their physical relations to one another, must suffice to compose causation by the material composites. Given what we currently know, the causal powers and causal structure of material parts do not seem alone to compose the causal powers and causal structure of causal events that involve rational, propositional structures of propositional states and events. Material causation is not itself rationally structured. Rational structures are causally relevant. So rational causes do not appear to be a mere composite of causes by physical parts.

The other argument centers on constitutive structure, and goes as follows. The physical structure of material composites consists in physical bonds (physical relations) among the parts. According to modern natural science, there is no place

[22] 'Individualism and the Mental', *Midwest Studies in Philosophy* 4 (1979), 73–121, reprinted in *Foundations of Mind.*

in physical structure of material composites for rational, propositional bonds or relations. The structure of propositional psychological states and events constitutively includes representational content with propositional structure—the sort of content and structure that can constitute reasons. So propositional states and events are not material composites.

Both key ideas in the arguments are deeply ingrained in the relevant sciences. The idea that the physical world, as studied by the natural sciences, does not have the structure of a text—propositional structure—is deeply entrenched since the advent of modern science. The relation between physical properties and relations, on one hand, and the physical particulars that have them, on the other, is not the relation of predication. Propositional negation and quantification are not to be found in the physical world by the physical sciences. The idea that many psychological states and events, at least among the higher animals, are fundamentally propositional attitudes capable of engaging in reason-supporting relations to other attitudes is fundamental in both common sense and cognitive psychology. The depth of these prima facie opposing commitments in the natural and psychological sciences makes it, at least, prima facie implausible to regard propositional attitude states and events as material composites out of physical states and events.

Rejecting materialist paradigms is not rejecting the dependence of the psychological on the physical. I see no ground for thinking that propositional states and events could ever float free of a physical basis. The point of the argument is to bring out that the relations between the physical and the psychological are likely to be more subtle and complex than simple materialist or physicalist positions propose. I believe that philosophical and scientific investigation of these relations should proceed in an exploratory way. I think, however, that it is well for such investigations to bear in mind that in propositional structure and in reason, we are dealing with something that has a quite distinctive form and function, and that is subject to quite distinctive practical and cognitive norms.

'Epistemic Warrant: Humans and Computers' begins by distinguishing again between justification—warrant by reason—and entitlement—warrant without reason. The essay then discusses warrants for transitions in inferences. I argue that many deductive and most inductive inferential transitions are warranted by entitlement, not by justification (reason). Thus, when premises form reasons for conclusions, they often do so through a mixture of justification and entitlement. The premises are reasons for the conclusions. Hence, they are justifying elements. The individual is, however, often warranted in the inferential transitions not by reason, but by entitlement. Since the full warrant for the inferred conclusion involves both premises and inferential transitions, the full warrant is a mix of justification and entitlement. Since justification occurs if a reason plays any role in the warrant, it is correct to call the conclusion justified.

In such cases, the individual's premises count as reasons, justifications, for the individual's conclusion if and only if the combination of the premises and the inferential rules that are actually relied upon yields a rationalizing explanation of the belief-worthiness of the conclusion. (The conclusion can count as justified whether or not the transitions are cognized in the psychology in a way that makes the warrant for relying on them a justification or an entitlement.) Thus even many deductive and nearly all inductive conclusion-supporting inferences are warranted through a mixture of justification and entitlement.

I then discuss the likelihood that there are cases in which the conditions on premises of a warranted inference counting as justifications of the conclusion is not met. I center on empirically warranted inferences. I assume that the inference tracks likely connections in nature. So it supports the truth of the conclusion. But the inferential transition does not follow a rule that provides any insight at all into the nature or existence of the connections. The transition is the product of brute evolutionary adaptation. So the combination of the premises and whatever rule governs the inferential transition does *not* yield a rationalizing explanation of the belief-worthiness of the conclusion. Still, the inference is a good route to truth. The goodness of the route is stamped into the inferrer, perhaps, by evolution because it does connect environmental properties in a patterned way. Then, I think, the premises and inferential transition entitle one to the conclusion. The premises do not provide a reason or justification for the conclusion. I believe that this account provides insight into the status of blind, but competent, truth-tracking, associative inference.

I turn to reliance on computers in mathematics. I recap some of what I wrote in 'Computer Proof, Apriori Knowledge, and Other Minds' about human warrant for relying on computers to prove mathematical theorems. I argue that in normal cases of computer proofs, the computers function as proxies for reasoners. I also compare reliance on computers for carrying out a proof with the reliance by an intuitive mathematician, like Ramanujan, on unconscious powers to produce theorems, without carrying out a proof. Whereas warranted reliance on computers inevitably involves some inductive reasoning about the computers' reliability and power, Ramanujan's warrant is likely to have been an unconscious entitlement, analogous to entitlements to perceptual beliefs, except that Ramanujan's entitlement probably derived from non-perceptual sources—sources of "intuitive" understanding.

Ramanujan's practice suggests, what is in fact the case, that there are various ways in which successful mathematics goes beyond deductive proof. I center on some epistemic differences between standard deductive computer proofs and probabilistic computer proofs. I argue that the key epistemic difference between the two types of proofs is not that the latter is inevitably less certain or less strongly warranted than the former, but that the latter gives less insight into necessary connections between premises and conclusion. The type of reason that deductive proofs provide for believing their conclusions provides a deeper understanding of the necessity of the truth of the conclusion, and of the necessary relations between premises and conclusion. In all cases in which a mathematician relies essentially on a computer's mathematical operations, deductive or inductive, the mathematician's ultimate warrant for believing the conclusion of the computer's work is inductive and empirical. Still, in understanding the deductive transition types that the computer goes through, the mathematician gains a deeper understanding of the necessary ways that premises support conclusion—and a deeper understanding of the necessity of the conclusion.

I sketch three types of reliance on computers in empirical work: Bayesian modeling of empirical phenomena, genetic algorithms in modeling empirical phenomena, and genetic algorithms in discovering purported empirical laws. I argue that only the first type of computer transitions model human reason-giving inferences. Genetic algorithms are not proxies for reasoning. The randomized mutations and recombinations are not analogs of reasoning—inferential transitions that provide reasons, justifications, for conclusions.

These differences do not bear on the power of our warrants for believing the computer's results. Inductive warrants can be equally strong in the different cases. My reflections bear on the relations between the different types of transitions in the computer's processing, on one hand, and corresponding steps in human inferential transitions, on the other. It is possible that genetic algorithms provide approximate models for certain types of human inferential transitions. In such cases, previous steps may enhance credibility of later steps—yielding entitlements to rely upon them—without providing explanatory rationalization of the later steps—that is, without being reasons for them.

The third type of reliance on computers in empirical work, use of computers to discover empirical generalizations or laws, may model a type of inference that sometimes occurs in science. Such inferences to good explanations might support their conclusions in the sense that they make the conclusions reliably more credible. But they fail to provide even a partial rationalizing explanation of the enhanced credibility of the conclusions. Such inferences yield entitlements to their conclusions, all the way down, not justifications. It seems to me certain that inductive (or abductive) inferences in science are not all of this sort. On the other hand, I think that genetic algorithms strongly suggest that some inductive reasoning in science may be more like blind Humean association than like the deliberative evidence-explaining inferences that provide the ideal of reasoned scientific explanation.

Understanding non-deductive inferential connections in science is incomplete understanding. It may well be that there is an irreducible reliance on environmentally, evolutionarily stamped-in associative patterns in our inferential transitions, even in science. We would be entitled to rely on such transitions well before they took a form that provides a rationalizing explanation of conclusions that they warrant. The paper uncovers, I think, several levels of incomplete understanding of inductive inferential transitions that suggest a role for blind entitlement, even in science. Although such transitions cannot yield ideal scientific understanding, they seem to be a necessary tool in the development of understanding that produces scientific reasons.

The essay does not explicitly discuss ways in which these types of inference bear on psychological types of individuality. I think it clear, however, that, along with memory, inferential patterns are the prime unifiers of a higher-level psychology. By considering different types of warrant for inferential transitions, one clarifies different types of psychological competence. Deliberative, critical, reasoned inference in practical reasoning, common sense reasoning, science, and mathematics helps mark the most sophisticated psychological personality. This high-level nature is inevitably dependent on lower-level, more "animal" types of inferential transition that mark less sophisticated levels of psychological individuality. I think that this essay contributes to thinking about relations among these levels.

IV

The essays in Part IV are more exploratory, more "initial", even than most of the other essays in the volume. They barely scratch the surface of a large and fascinating topic—reflection. They touch on reflection as it occurs in philosophy, mathematics, logic, and sometimes in the empirical sciences. They should be read in conjunction with my discussion of the epistemology of thought experiments in philosophy, particularly the thought experiments that led to modern arguments for anti-individualism.[23] The essays reprinted here highlight apriori aspects of reflection. In other work, I highlight empirical aspects.

It is crucially important, in understanding reflection, that one not assume that all warranted reflection is apriori. Very frequently, reflection employs uncontroversial empirical truths—especially general truths that would not, in common parlance, be counted 'theoretical'. I believe that some "armchair" reflection is one natural, and not *always* inappropriate or fruitless, occupation for philosophers. Some armchair reflection is like the reflection in elementary mathematics that relies on apriori understanding. Much of it involves a mix of apriori and empirical elements. Some of it is fundamentally empirical. Much of my work on reflection has sought to develop insights in the rationalist tradition, while emphasizing that reflection is a much more varied phenomenon, both psychologically and epistemically, than traditional rationalist models presented it as being.

It is also crucially important that one not assimilate even apriori aspects of reflection to conceptual analysis. I believe that very little analysis, strictly speaking, occurs in fruitful philosophical or mathematical reflection. I think that the key to making progress on this complex topic is to think about specific cases of successful reflection, remaining open to their differences from traditional paradigms of analysis (the analysis of the concepts bachelor and knowledge), and from one another.

[23] See, for example, *Foundations of Mind*, *op. cit.*, 18, 27–28, 162–179, 291–306; 'Some Reflections on Scepticism: Reply to Stroud' in *Reflections and Replies Essays on the Philosophy of Tyler Burge*, M. Hahn and B. Ramberg eds. (Cambridge, Mass.: MIT Press, 2003).

'Reasoning about Reasoning' is a critical discussion of David Lewis's account of iteration in reasoning about reasons in his book *Convention*. The topic comprises both iterative reasoning about one's own reasons and iterative reasoning about another's reasons. I accept this slight refinement on Lewis's gloss on governing *having a reason*: if a person has a reason to believe something and does not already believe it, the person would incur some deficiency in rationality if the person were to disbelieve it; and the person should be able to come to believe it with only a little reflection—or at least with reflection, however much or little, that does not make use of information that the individual does not already have.

The essay makes the point that one cannot iterate reasoning about having a reason infinitely, if each iteration is inductive or has strength less than 100%. Because there is a real difference between having a reason about a subject matter at a given level and having a reason to believe one has that reason (which is a reason at a different representational level), having reason at the lower level does not guarantee that one has reason at the higher level. This point is especially applicable to transitions from first-order reasons (reasons that are not about reasons) to second-order reasons (reasons about having first-order reasons). If one takes seriously the condition on deficiency in rationality, in the gloss on 'having a reason', the point seems to apply in any iteration from one higher-order level to the next: having a reason at one level does not guarantee that (at the next higher level) one has a reason to believe—or is rationally deficient in not believing—that one has that reason.

In fact, it seems certain that at a sufficiently high level of iteration, an individual *i* incurs no deficiency in rationality if *i* has a reason to believe p, but cannot 'with only a little reflection' come to believe that *i* has reason to believe that *i* has reason to believe . . . that *i* has reason to believe p. Combining beliefs whose warranting strength is less than 100% in a piece of reasoning diminishes the warranting strength of the conclusion.

Since all judgments about another person's reasons are inductive and have strength less than 100%, the case of interpersonal reasoning about reasons adds another possible source of friction that drags against infinite hierarchies of having reasons about having reasons.

Looking back, I think that the value of the essay is two-fold. It cautions against over-idealized theories about having reasons—driven more by logic than by a realistic epistemology. And it insists on distinguishing between levels of reasoning. Some reasoners can reason, and thus *have* reasons, but lack a concept of reason. In insisting on being realistic about what it is to have a reason and in emphasizing level differences, the essay constitutes a criticism of hyper-intellectualization in epistemology.[24]

[24] See also 'On Knowledge and Convention', *The Philosophical Review* 84 (1975), 249–255; reprinted in *Foundations of Mind*, *op. cit.*, for another critical discussion of Lewis that opposes a different, but related, sort of hyper-intellectualization in epistemology.

I include 'Thought Experiments and Semantic Competence: Reply to Benejam' not because it makes a distinctive contribution, but because it contains in very compressed form many of my criticisms of the traditional conception of reflection. It might serve as an introductory orientation to the other essays on reflection.

'Concepts, Conceptions, Reflective Understanding: Reply to Peacocke' is part of a long, on-going constructive interchange with Christopher Peacocke on the nature and scope of apriori knowledge, an interchange that began in 1993 when he graciously helped host my Locke Lectures at Oxford. This essay highlights several areas of agreement, and explicitly shares an appreciation of the contributions of the great traditional rationalists—Descartes, Leibniz, Kant, and Frege. The essay also brings out a significant area of disagreement. Peacocke develops the classical conception of reflection, most fully present in Leibniz and Kant. According to this conception, reflection consists in bringing to clear consciousness an understanding that is already present in the reflector's psychology. In fact, I think that the classical conception was somewhat more liberal than the conception that Peacocke elaborates. It allowed that reflection puts together an explication for the first time from conceptual components that are present, perhaps unconsciously, in the individual's psychology.

I think that even this liberal version of the classical conception fails to do justice to some very important cases of reflection. A thinker may have an incomplete understanding of a concept and may not have the concepts to provide an adequate explication of it. Leibniz's infinitesimal-based understanding of his concepts of limit and differentiation is, I think, an example.

A central difficulty with Peacocke's account is that it overrates the role of unconscious *but fully formed* conceptions in guiding reflection to definition-like explications of concepts (or terms) that are initially not clearly and distinctly understood. I believe that although such guidance can happen, it is uncommon. I think that most reflection is guided by memory of examples—often socially inculcated—senses of similarity among examples, inductive capacities, and inferential patterns that are either too low-level and specific or too generic to provide satisfactory explications or definitions. I think that explications or definitions are commonly discovered—and enter into the psychology for the first time—through the process of reflection. Reflection is rarely merely a dialectical bringing-to-consciousness of an unconscious, "implicit", but fully formed explicational, criterial conception.[25]

The role of dependence on, and perhaps deference to, others for a fuller conception plays a significant role in everyday life. Sometimes one simply has to learn more from others in order to obtain an adequate, reflective understanding of one's own concepts. Sometimes even the experts, or the whole community

[25] See 'Postscript: Individualism and the Mental', 'Intellectual Norms and Foundations of Mind', 'Concepts, Definitions, and Meanings'–all reprinted in *Foundations of Mind*, *op. cit.*, and 'Living Wages of *Sinn*' in this volume.

combined, lack sufficient understanding to provide adequate explication of one's concepts. Mere reflection even by experts will not yield sufficient understanding of what guides, justifies, and explicates one's concepts. I make this point with respect to Newton and Leibniz's respective understandings of key concepts in the calculus.[26] I take this point further in the last section of 'Living Wages of *Sinn*'. I make a similar point elsewhere with respect to reflection on empirical concepts. Dalton's understanding of his concept atom and erstwhile communal explications that situated tomato under the super-ordinate concept vegetable are examples.

The essay concludes with a detailed discussion of the relation between non-inferential understanding of simple, self-evident logical truths and the inferential mastery that is necessary for having such understanding. Our justification for believing simple logical truths does not derive from inferring them from premises that are more basic, or better justified, than the simple logical truths. The justification seems non-inferential, and resides in understanding the truths themselves. On the other hand, as Frege emphasizes, understanding the truths requires a background of inferential capacities. One could not understand the truths unless one could connect them to some other truths via inference. One can both infer the simple truths as conclusions and use them to infer to other truths. Such inferential capabilities are essential to understanding the truths in such a way as to be non-inferentially justified in believing them. I explore this combination of ideas through reflection on Frege's account of his axioms.[27]

'Reflection' was written for a conference in Fribourg, Switzerland, in summer 2004. I gave the paper at the conference as a public lecture, but never submitted it for publication. It begins with a review of criticisms of classical rationalist views of reflection—points made in the reply to Peacocke. I pay some attention to differences among the classical rationalists, especially ways in which Kant anticipates a more modern conception of reflection. I then take some small exploratory steps toward identifying positive aspects of reflection.

I consider three examples of what I consider to be successful reflection—one in meta-logic and two in philosophy. The example from meta-logic is the clarification of the notions of logical validity and logical consequence in the work of Skolem, Gödel, and Tarski.[28] The examples from philosophy are the attempts to clarify the error-presupposes-veridicality principle and the thought experiments that led to modern anti-individualism.

I try to isolate prominent features of these examples that yield clues to sorts of things reflection is especially well suited to understanding and clues to the ways

[26] See also the discussion of these matters in 'Postscript: Individualism and the Mental' in *Foundations of Mind*, *op. cit.*, 162–179.

[27] See also 'Frege on Knowing the Foundations', *Mind* 107 (1998), 305–347, reprinted in *Truth, Thought, Reason*, *op. cit.*

[28] A more detailed discussion of this case is set out in 'Logic and Analyticity' *Grazer Philosophische Studien*, 66 (2003), 199–249. I intend to include this paper in a later collection of papers.

in which reflection operates. All the cases center on a representational function of a cognitive activity, and on ways of fulfilling such functions. Two of the cases utilize insights into the form of a representational activity and into relations among representational abilities. All cases rely on judgments about necessary implications regarding the relevant subject matter. And all involve insight into explanatory priority. I believe that it is no coincidence that these features show up in the *meta-representational* enterprise of reflection. I think that reflection is better suited (though not uniquely well-suited) to be a distinctive and successful form of cognition when it focuses on representational powers than when it focuses on other matters.

The essay concludes by reviewing one of the primary oversights of the classical rationalist conception of reflection—its view that at least implicit understanding of the norms of reasoning is available to reflection for all reasoners. The classical idea was that the basic principles of reasoning are present, at least subliminally, in the psychologies of all reasoners. I believe that this view is clearly mistaken. I claim that this mistake applies even to critical reasoners. I reflect on what minimal resources must be present in any critical reasoner.

'Living Wages of *Sinn*' would have been included in the volume of my essays on Frege, *Truth, Thought, Reason: Essays on Frege*, if it had been written before that volume came out. I count the present volume as only second-best for inclusion of the essay, because the essay's explicit task is to discuss ways in which Frege's notion of *Sinn* figures in current philosophy of mind, philosophy of language, and epistemology. However, the main contribution of the essay, in my judgment, is its discussion of reflective understanding of representational contents. So the piece has some claim to be included here.

The essay begins by distinguishing Frege's notion of sense from modern notions of linguistic meaning. Senses are the ability-general aspects of thought contents that are associated, in certain specific ways, with linguistic expressions or linguistic uses.[29] Modern notions of linguistic meanings are contents of understanding linguistic expressions, contents that are associated with rules for using the expressions in idiolects or communal languages. The two notions, sense and linguistic meaning, are complementary theoretical tools for understanding language. They differ in their theoretical foci.

[29] Ability-general representational contents are those that individuate representational capacities (abilities) that are individuated independently of any particular, specific, occurrent exercise of it. Such representational abilities are freely repeatable in the sense that any exercise of the relevant ability need not be tied, anaphorically or in memory, to any other occurrent exercise of the ability in order for the ability to be the ability that it is. Conceptual and perceptual attributives are always ability-general. One could have the ability—either acquire or have it innately—in any number of occurrently different situations. For example, one could learn the concept piano in an isolated town in Russia or in New York City, in the nineteenth, twentieth, or twenty-first centuries. Ability-general representational contents contrast with occurrence-based *applications*—for example particular *occurrent uses* of a demonstrative or indexical word or representational content.

I defend this distinction, which I first articulated in 'Sinning Against Frege' (1979),[30] against criticisms by Saul Kripke. Here the issues primarily concern Frege's texts. Textual considerations help bring out how different Frege's conception is from modern conceptions.

I then criticize some of Frege's judgments about the senses of indexicals and proper names, using some of Kripke's important work on these matters. I maintain, however, that especially in understanding uses of demonstratives and in understanding the phenomenon of incomplete understanding, Frege's notion of sense is important to modern theorizing about language and thought.

The essay then focuses on Frege's seemingly disparate views on the individuation of senses. I argue that the main line of his thinking involves a very nuanced and sophisticated approach to understanding the form and content of psychological states, particularly propositional attitudes. The approach centers on recognizing the truth-conditional content and logical form involved in judgments and propositional inferences. I discuss various respects in which he notes that determining the content and structure of psychological states is a fallible, highly theoretical enterprise.

First, the sense associated with a linguistic utterance—for example, with demonstrative uses backed by perceptual beliefs—can vary significantly with context. There is no recipe for determining a contextually associated sense. Asking the individual what he or she was thinking is just a start to determining the nature of the thought. Second, the logical form of a sense can be determined only by considering many patterns of inferences that make use of it, as Frege's revolutionary method for determining truth-conditional logical form illustrated. Mere introspection does not suffice for understanding the truth-conditional structure of one's thought contents. Third, Frege's distinction between sense and coloring is valuable in separating truth-conditional aspects of a thought from psychologically relevant, collateral effects that are always associated with language usage. I argue that Frege was right not to allow coloring as a canonical, constitutive parameter in the structure of every thought. Commonly, coloring of one thought is the content of another thought. Applying the distinction in reflection is a theoretically delicate matter. Fourth, Frege's rationalist insights into *incomplete understanding* provide a further limit on reflective understanding of one's own thoughts, or those of others.

I have been asked, in a sceptical way, what difference is there between incomplete conceptual understanding and any other factually mistaken belief. I think that any defective conceptual understanding is factually defective. But not all factually mistaken beliefs constitute incomplete understanding, of a concept. I think it clear that some true beliefs that are expressible with a concept are not constitutive to understanding the concept. For example, mistakenly believing that

[30] 'Sinning Against Frege', *The Philosophical Review* 88 (1979), 398–432; reprinted in *Truth, Thought, Reason*.

there are thirteen rather than twelve pianos in a given region of space at a given time is not a defect in understanding the concept piano.

I believe that only radical philosophical ideology would maintain that incomplete understanding of a concept is *just any* sort of factual ignorance of a proposition containing it. One such ideology is an extreme extension of Quinean holism, according to which there is no distinction at all between change of belief and change of concept (as distinguished from no sharp distinction). Another such ideology is Leibniz's strange view that all facts (true propositions) expressible with a concept are embedded in the nature of the concept.

It is not part of my position that some, or any, concepts *are* embedded in a concept. My view on incomplete understanding is compatible with holding that all concepts are atomic. What I do believe is that connecting *some* (factual) propositions with a concept is constitutively relevant to understanding the concept, on an ordinary view of understanding. (There may be a family of such propositions, any group of which suffices.) Other propositions involving a concept are not constitutive to any level of understanding the concept.

At least with a non-defective concept—one that is not constitutively associated with a mistaken theory or a mistaken set of beliefs—the relevant propositions are true; they provide unifying generalizations about instances of correct applications of the concept; and they tend to justify and explain the unities—the fact that the instances are instances of the relevant concept.

I am not committed to thinking that, for finite minds, there is always such a thing as a complete mastery of the concept. Incomplete mastery could always be relative to a better, more deeply explanatory explication. Or the notion of complete mastery can be too vague to be fruitful. I am also not committed to believing that there are *sharp* boundaries marking any of the distinctions among levels of understanding. Moreover, I think that in most cases, including both empirical and mathematical cases, no explication, no matter how good, *fixes* the *representatum* or extension of the explicated concepts. Nearly always, the *representatum* is fixed by attributional applications to cases, together with objective similarity relations among cases.

I want to highlight two values to taking incomplete understanding seriously as an important idea in philosophy. One is that doing so underlines the important fact that having a constitutive explication (a general criterial understanding) is almost never necessary for *having* and using a concept. One can have a concept without fully understanding it. Having concepts is grounded in applications to cases, and in not fully articulated or generalized senses of patterns of similarities among cases. That is why, in philosophical understanding, examples tend to drive progress in reflection. The second value of taking incomplete understanding seriously is that it combats the deeply entrenched but deeply mistaken idea that we have a transparent grip on—understanding of—the nature of our own concepts. The deepest progress in factual knowledge is often concomitantly a progress in conceptual understanding.

In 'Living Wages of *Sinn*', I distinguish different types of incomplete understanding and discuss limitations on reflection. I have discussed such limitations in other work. I do so more fully and systematically here. By reference to the post-Newtonian elaboration and refinement of Newton's concepts in the calculus, I provide a relatively detailed account of issues that complicate our understanding of reflective conceptual explications. I highlight interplay between reflection and new mathematical work. And I further develop criticism of the traditional conception of reflection. According to this conception, reflection brings to consciousness conceptual explications that are unconscious but already fully formed, or at least fully available in the unconscious psychology of the reflective thinker, in the sense that all elements for forming the explication are present in the psychology.

This collection invites the reader to make connections and recognize systematic structures that were not fully in mind when the essays were written. I have sketched some of these connections and structures in this introduction. I think that in the best philosophical work, a philosophical unconscious guides the author's thinking in incompletely understood ways, ways that point fruitfully beyond what is realized, consciously or even unconsciously, at the time the work is produced. This phenomenon is a natural consequence of the way in which thought is molded by a subject matter that is larger and more complex than can be fully grasped at any given time, or indeed, sometimes, ever. I hope that, in reading these essays together, some readers will have the pleasure of thinking constructively, not just critically, beyond where any given essay manages to reach. I hope myself to continue to be one such reader.

PART I

SELF-KNOWLEDGE

2 *Individualism and Self-Knowledge*

The problem I want to discuss derives from the juxtaposition of a restricted Cartesian conception of knowledge of one's own thoughts and a non-individualistic conception of the individuation of thoughts. Both conceptions are complex and controversial. But I shall not explain them in detail, much less defend them. I shall explicate them just enough to make the shape of the problem vivid. Then I shall say something about solving the problem.

Descartes held that we know some of our propositional mental events in a direct, authoritative, and not merely empirical manner. I believe that this view is correct. Of course, much of our self-knowledge is similar to the knowledge of others' mental events. It depends on observation of our own behavior and reliance on others' perceptions of us. And there is much that we do not know, or even misconstrue, about our own minds. Descartes tended to underrate these points. He tended to overrate the power of authoritative self-knowledge and its potential for yielding metaphysical conclusions. Characterizing the phenomenon that interested Descartes is a substantial task. I shall not take on this task here. I think, however, that Descartes was right to be impressed with the directness and certainty of some of our self-knowledge. This is the point I shall rely on.

Descartes' paradigm for this sort of knowledge was the cogito. The paradigm includes not only this famous thought, but fuller versions of it—not merely 'I am now thinking', but 'I think (with this very thought) that writing requires concentration' and 'I judge (or doubt) that water is more common than mercury'. This paradigm goes further toward illuminating knowledge of our propositional attitudes than has generally been thought. But I note it here only to emphasize that Descartes' views about the specialness of some self-knowledge are not merely abstract philosophical doctrine. It is certainly plausible that these sorts of judgments or thoughts constitute knowledge, that they are not products of ordinary empirical investigation, and that they are peculiarly direct and authoritative. Indeed, these sorts of judgments are self-verifying in an obvious way: making these judgments itself makes them true. For mnemonic purposes, I shall call such judgments *basic self-knowledge*.

Presented in an APA symposium on Individuation and Self-Knowledge, December 30, 1988. Donald Davidson commented; see *The Journal of Philosophy* 85:11 (1988), 664–5. Substantially this paper was the Nelson Lecture, University of Michigan, February 1986. I benefited from the occasion.

Let us turn from knowledge of one's thoughts to individuation of one's thoughts. My view on this matter is that many thoughts are individuated non-individualistically: individuating many of a person or animal's mental kinds—certainly including thoughts about physical objects and properties—is necessarily dependent on relations that the person bears to the physical, or in some cases social, environment. This view is founded on a series of thought experiments, which I shall assume are familiar.[1] Their common strategy is to hold constant the history of the person's bodily motion, surface stimulations, and internal chemistry. Then, by varying the environment with which the person interacts while still holding constant the molecular effects on the person's body, one can show that some of the person's thoughts vary. The details of the thought experiments make it clear that the variation of thoughts is indicative of underlying principles for individuating mental kinds. The upshot is that which thoughts one has—indeed, which thoughts one can have—is dependent on relations one bears to one's environment.

Our problem is that of understanding how we can know some of our mental events in a direct, non-empirical manner, when those events depend for their identities on our relations to the environment. A person need not investigate the environment to know what his thoughts are. A person does have to investigate the environment to know what the environment is like. Does this not indicate that the mental events are what they are independently of the environment?

By laying aside certain contrary elements in Descartes' views, one can reconstruct a tempting inference to an affirmative answer from his conception of self-knowledge.

In reflecting on the demon thought experiment, one might think that, since we can know our thoughts authoritatively, while doubting whether there is any physical world at all, the natures of our thoughts—our thought kinds—must be independent of any relation to a physical world. A parallel inference is presupposed in Descartes' discussion of the real distinction between mind and body. In *Meditations* vi, he argues that the mind can exist independently of any physical entity. He does so by claiming that he has a "clear and distinct idea" of himself as only a thinking and unextended thing, and a "clear and distinct idea" of body as only an extended and unthinking thing. He claims that it follows that the mind that makes him what he is can exist independently of any physical body. The argument also occurs in *Principles* I, LX:

> ... because each one of us is conscious [through clear and distinct ideas] that he thinks, and that in thinking he can shut off from himself all other substance, either thinking or

[1] Compare 'Individualism and the Mental', *Midwest Studies in Philosophy* 4 (1979), 73–121; 'Other Bodies', in Andrew Woodfield (ed.), *Thought and Object: Essays on Intentionality* (Oxford: Clarendon Press, 1982), 97–120; 'Individualism and Psychology', *The Philosophical Review* 95:1 (1986), 3–45; 'Cartesian Error and the Objectivity of Perception', in P. Pettitt and J. McDowell (eds.), *Subject, Thought, and Context* (New York: Oxford University Press, 1986), 117–136; 'Intellectual Norms and Foundations of Mind', *The Journal of Philosophy* 83:12 (1986), 697–720.

extended, we may conclude that each of us . . . is really distinct from every other thinking substance and from every corporeal substance.[2]

Descartes also believed that he had "clear and distinct ideas" of his thoughts. One might argue by analogy that, since one can "shut off" these thoughts from all corporeal substance, they are independent for their natures from physical bodies in the environment, and presumably from other thinkers. This line of argument implies that knowledge of one's own thoughts guarantees the truth of individualism.[3]

The root mistake here has been familiar since Arnauld's reply. It is that there is no reason to think that Descartes' intuitions or self-knowledge give him sufficient clarity about the nature of mental events to justify him in claiming that their natures are independent of relations to physical objects. Usually, this point has been made against Descartes' claim to have shown that mental events are independent of a person's body. But it applies equally to the view that mental kinds are independent of the physical environment. One can know what one's mental events are and yet not know relevant general facts about the conditions for individuating those events. It is simply not true that the cogito gives us knowledge of the individuation conditions of our thoughts which enables us to "shut off" their individuation conditions from the physical environment. Our thought experiments, which have directly to do with conditions for individuation, refute the independence claim.[4]

It is one thing to point out gaps in inferences from self-knowledge to individualism. It is another to rid oneself of the feeling that there is a puzzle here. Why is our having non-empirical knowledge of our thoughts not impugned by the fact that such thoughts are individuated through relations to an environment that we know only empirically?

Let us assume that our thoughts about the environment are what they are because of the nature of entities to which those thoughts are causally linked. According to our thought experiments, a person with the same individualistic physical history could have different thoughts if the environment were appropriately different. One senses that such a person could not, by introspection, tell the difference between the actual situation (having one set of thoughts) and the counterfactual situation (having another).

This intuition must be articulated carefully. What do we mean by 'introspection'? In each situation, the person knows what his thoughts are; and in each

[2] *The Philosophical Works of Descartes*, trans. E. S. Haldane and G. R. T. Ross, 2 vols. (New York: Dover, 1955), vol. I, 243–244.

[3] Compare *Philosophical Works of Descartes*, trans. Haldane and Ross, vol. I, 190.

[4] I have discussed this and other features of the inference in 'Intellectual Norms and Foundations of Mind'. See also 'Perceptual Individualism and Authoritative Self-Knowledge', in R. Grimm and D. Merrill (eds.), *Contents of Thought* (Tucson: Arizona University Press, 1988), 86–98. I now think that Descartes' views have more anti-individualistic elements than I realized in writing those articles. I hope to discuss these matters elsewhere.

situation the thoughts are different. If 'introspection' were explicated in terms of self-knowledge, there would be an introspectible difference.

Certainly, if one were stealthily shifted back and forth between actual situations that modeled the counterfactual situations, one would not notice some feature in the world or in one's consciousness which would tell one whether one was in the "home" or the "foreign" situation. But this remark does not capture the idea that the two lives would feel the same. The thoughts would not switch as one is switched from one actual situation to another twin actual situation. The thoughts would switch only if one remained long enough in the other situation to establish environmental relations necessary for new thoughts. So quick switching would not be a case in which thoughts switched but the introspection remained the same.

But slow switching could be such a case. Suppose that one underwent a series of switches between actual earth and actual twin earth so that one remained in each situation long enough to acquire concepts and perceptions appropriate to that situation. Suppose occasions where one is definitely thinking one thought, and other occasions where one is definitely thinking its twin.[5] Suppose also that the switches are carried out so that one is not aware that a switch is occurring. The continuity of one's life is not obviously disrupted. So, for example, one goes to sleep one night at home and wakes up in twin home in twin bed—and so on. (Your standard California fantasy.) Now suppose that, after decades of such switches, one is told about them and asked to identify when the switches take place. The idea is that one could not, by making comparisons, pick out the twin periods from the "home" periods.

I grant these ideas. The person would have no signs of the differences in his thoughts, no difference in the way things "feel". The root idea is that at least some aspects of one's mental life are fixed by the chemical composition of one's body. One might call these aspects *pure phenomenological feels*. If one were uncomfortable with this notion, one could explicate or replace it in terms of an abstraction from the person's inability to discriminate between different mental events under the stated switching situations.

The upshot of all this is that the person would have different thoughts under the switches, but the person would not be able to compare the situations and note when and where the differences occurred. This point easily, though I think mistakenly, suggests the further point that such a person could not know what thoughts he had unless he undertook an empirical investigation of the environment which would bring out the environmental differences. But this is absurd. It is absurd to think that, to know which thoughts we think, we must investigate the empirical environment

[5] Of course, there can arise difficult questions about whether one is still employing thoughts from the departed situation or taking over the thoughts appropriate to the new situation. I think that general principles govern such transitions, but such principles need not sharply settle all borderline cases. Insofar as one finds problems associated with actual switches distracting, one could carry out the objection I am articulating in terms of counterfactual situations.

in such a way as to distinguish our actual environment from various twin environments.

In basic self-knowledge, a person does individuate his thoughts in the sense that he knows the thought tokens as the thought tokens, and types, that they are. We know which thoughts we think. When I currently and consciously think that water is a liquid, I typically know that I think that water is a liquid. So much is clear.

How can one individuate one's thoughts when one has not, by empirical methods, discriminated the empirical conditions that determine those thoughts from empirical conditions that would determine other thoughts?

It is uncontroversial that the conditions for thinking a certain thought must be presupposed in the thinking. Among the conditions that determine the contents of first-order empirical thoughts are some that can be known only by empirical means. To think of something as water, for example, one must be in some causal relation to water—or at least in some causal relation to other particular substances that enable one to theorize accurately about water. In the normal cases, one sees and touches water. Such relations illustrate the sort of conditions that make possible thinking of something as water. To know that such conditions obtain, one must rely on empirical methods. To know that water exists, or that what one is touching is water, one cannot circumvent empirical procedures. But to *think* that water is a liquid, one need not *know* the complex conditions that must obtain if one is to think that thought. Such conditions need only be presupposed.

Now let us turn to knowledge of one's thoughts. Knowing what one is thinking when one has thoughts about physical entities presupposes some of the same conditions that determine the contents of the empirical thoughts one knows one is thinking. This is a result of the second-order character of the thoughts. A knowledgeable judgment that one is thinking that water is a liquid must be grounded in an ability to think that water is a liquid.

When one knows that one is thinking that *p*, one is not taking one's thought (or thinking) that *p* merely as an object. One is thinking that *p* in the very event of thinking knowledgeably that one is thinking it. It is thought and thought about in the same mental act. So any conditions that are necessary to thinking that *p* will be equally necessary to the relevant knowledge that one is thinking that *p*. Here again, to think the thought, one need not know the enabling conditions. It is enough that they actually be satisfied.

Both empirical thoughts and thinking that one is thinking such thoughts presuppose conditions that determine their contents. In both cases, some of these conditions can be known to be satisfied only by empirical means. Why do these points not entail that one cannot know that one is thinking that such and such unless one makes an empirical investigation that shows that the conditions for thinking such and such are satisfied? The answer is complex, but it can be seen as a series of variations on the point that one must start somewhere.

It is helpful in understanding self-knowledge to consider parallel issues regarding perceptual knowledge. It is a fundamental mistake to think that perceptual knowledge of physical entities requires, as a precondition, knowledge of the conditions that make such knowledge possible. Our epistemic right to our perceptual judgments does not rest on some prior justified belief that certain enabling conditions are satisfied. In saying that a person knows, by looking, that there is food there, we are not required to assume that the person knows the causal conditions that make his perception possible. We certainly do not, in general, require that the person has first checked that the light coming from the food is not bent through mirrors, or that there is no counterfeit food in the vicinity. We also do not require that the person be able to recognize the difference between food and every imaginable counterfeit that could have been substituted.

In fact, it is part of our common conception of the objectivity of perception that there is no general guarantee that the perceiver's beliefs, dispositions, and perceptions could in every context suffice to discriminate the perceived object from every possible counterfeit. The possibility of unforeseeable misperceptions and illusions is fundamental to objectivity. So the very nature of objective perception insures that the perceiver need not have a perfect, prior mastery over the conditions for his perceptual success.

This point is obvious as applied to common practice. But it is the business of philosophy and the pleasure of scepticism to question common practice. My discussion of knowledge and individualism has proceeded on the unargued assumption that scepticism is mistaken. Granted this assumption, the point that perceptual knowledge does not require knowledge of its enabling conditions is obvious.

I shall not overburden this essay with an attempt to disarm scepticism. But it is worth noting that nearly all currently defended responses to scepticism, other than transcendental ones, agree in denying that perceptual knowledge must be justified by separately insuring that the enabling conditions hold and the sceptic's defeating conditions do not hold.[6] And since transcendental responses provide at

[6] This remark applies to reliabilist theories, Moorean theories that insist on the directness of perception, Quinean theories that attempt to show that the sceptic's doubt is covertly a bad empirical doubt, and Carnapian theories that attempt to show that the sceptic's question is somehow irrelevant to actual empirical claims. The words 'first' and 'separately' are crucial in my formulations. As against some reliabilist views that try to block scepticism by denying closure principles, I think that we can know that no demon is fooling us. But we know this by inferring it from our perceptual knowledge.

Several philosophers have thought that anti-individualism, combined with the view that we are authoritative about what thoughts we think, provides a "transcendental" response to scepticism. Compare Hilary Putnam, Hilary, *Reason, Truth, and History* (New York: Cambridge University Press, 1981). Putnam's argument is criticized by Anthony L. Brueckner, 'Brains in a Vat', *The Journal of Philosophy* 83:3 (1986), 148–167. I agree with Brueckner that Putnam's arguments do not do much to undermine scepticism. But Brueckner seems to hold that, if anti-individualism and the authority of self-knowledge are accepted, one would have an anti-sceptical argument. He suggests that the assumption of anti-individualism undercuts the assumption of authoritative self-knowledge. I do not accept this suggestion. I believe, however, that there is no easy argument against scepticism from anti-individualism and authoritative self-knowledge. This is a complicated matter best reserved for other occasions.

most general guarantees against scepticism, the only tenable responses, which I know of, that attempt to justify particular perceptual knowledge claims in the face of scepticism take this route. I think that it is the right route.

I have maintained that perceptual knowledge of physical objects does not presuppose that one has first checked to insure that the background enabling conditions are fulfilled. The same point applies to knowledge of one's own mental events, particularly knowledge of the sort that interested Descartes. Such knowledge consists in a reflexive judgment which involves thinking a first-order thought that the judgment itself is about. The reflexive judgment simply inherits the content of the first-order thought.

Consider the thought, 'I hereby judge that water is a liquid'. What one needs in order to think this thought knowledgeably is to be able to think the first-order, empirical thought (that water is a liquid) and to ascribe it to oneself, simultaneously. Knowing one's thoughts no more requires separate investigation of the conditions that make the judgment possible than knowing what one perceives.

One knows one's thought to be what it is simply by thinking it while exercising second-order, self-ascriptive powers. One has no "criterion", or test, or procedure for identifying the thought, and one need not exercise comparisons between it and other thoughts in order to know it as the thought one is thinking. Getting the "right" one is simply a matter of thinking the thought in the relevant reflexive way. The fact that we cannot use phenomenological signs or empirical investigation to discriminate our thoughts from other thoughts that we might have been thinking if we had been in a different environment in no way undermines our ability to know what our thoughts are. We "individuate" our thoughts, or discriminate them from others, by thinking those and not the others, self-ascriptively. Crudely put, our knowledge of our own thoughts is immediate, not discursive. Our epistemic right rests on this immediacy, as does our epistemic right to perceptual beliefs. For its justification, basic self-knowledge in no way needs supplementation from discursive investigations or comparisons.[7]

So far I have stressed analogies between basic self-knowledge and perceptual belief. But there are fundamental differences. A requirement that, to know what thoughts we are thinking, we must be able first to discriminate our thoughts from twin thoughts is, in my view, even less plausible than the analogous position with regard to perceptual knowledge.

Why? In developing an answer to this question, I want to dwell on some fundamental ways in which perceptual knowledge of physical entities differs from the sort of self-knowledge that we have been featuring. We commonly regard perceptual knowledge as *objective*. For our purposes, there are two relevant notions of objectivity. One has to do with the relation between our perceptions and the physical entities that are their objects. We commonly think

[7] I shall not develop the issue of one's epistemic right to one's authoritative self-ascriptions here. It is an extremely complex issue, which deserves separate attention.

that there is no necessary relation between any one person's abilities, actions, thoughts, and perceptions up to and including the time of a particular perception, on one hand, and the natures of those entities which that person perceptually interacts with at that time, on the other. On any given occasion, our perceptions could have been misperceptions. The individual physical item that one perceptually interacts with at any given time is fundamentally independent from any one person's perceptions—and conceptions. The nature of the physical entity could have been different even while one's perceptual states, and other mental states, remained the same.

This fact underlies a normative point about perception. We are subject to certain sorts of possible errors about empirical objects—misperceptions and hallucinations that are "brute". Brute errors do not result from any sort of carelessness, malfunction, or irrationality on our part. A person can be perceptually wrong without there being anything wrong with him. Brute errors depend on the independence of physical objects' natures from how we conceive or perceive them, and on the contingency of our causal relations to them. The possibility of such errors follows from the fact that no matter what one's cognitive state is like (so, no matter how rational or well-functioning one is) one's perceptual states could in individual instances fail to be veridical—if physical circumstances were sufficiently unfortunate.

There is a second sense in which perceptual knowledge is objective. This sense bears on the relation between one person's perceptions of an object and other person's perceptions of the same object. The idea is that perceptual knowledge, like all other empirical knowledge, is impersonal. Any observer could have been equally well placed to make an observation. Others could have made an observation with the same type of presentation of the scene, if they had been in the same position at the relevant time. And this possible observation could have had the same justificatory status as the original observation. Even though empirical commitments must be made by persons, nothing relevant to the justification of any empirical commitment regarding the physical world has anything essentially to do with any particular person's making the commitment.

The paradigmatic cases of self-knowledge differ from perceptual knowledge in both of these respects. To take the first: in the case of cogito-like judgments, the object, or subject matter, of one's thoughts is not contingently related to the thoughts one thinks about it. The thoughts are self-referential and self-verifying. An error based on a gap between one's thoughts and the subject matter is simply not possible in these cases. When I judge: I am thinking that writing requires concentration, the cognitive content that I am making a judgment about is self-referentially fixed by the judgment itself; and the judgment is self-verifying. There is a range of cases of self-knowledge which extend out from this paradigm. I think that, in all cases of authoritative knowledge, brute mistakes are impossible. All errors in matters where people have special authority about themselves are errors which indicate something wrong with the thinker. Dealing with the whole range requires subtlety. But the point as applied to what I take to be the basic cases is straightforward. No errors at all are possible in strict cogito judgments; they are self-verifying.[8]

The paradigmatic cases of self-knowledge also differ from perceptual knowledge in that they are essentially personal. The special epistemic status of these cases depends on the judgments' being made simultaneously from and about one's first-person point of view. The point of view and time of the judgment must be the same as that of the thought being judged to occur. When I judge: I am thinking that writing requires concentration, the time of the judgment and that of the thought being judged about are the same; and the identity of the first-person pronouns signals an identity of point of view between the judge and the thought being judged about. In all cases of authoritative self-knowledge, even in those cases which are not "basic" in our sense, it is clear that their first-person character is fundamental to their epistemic status.

These differences between perceptual knowledge and authoritative self-knowledge ground my claim that it is even less plausible than it is in the case of perceptual knowledge to think that basic self-knowledge requires, as a precondition, knowledge of the conditions that make such knowledge possible.

Let us think about the difference as regards objectivity in the relation to an object. In the case of perceptual knowledge, one's perception can be mistaken because some counterfeit has been substituted. It is this possibility which tempts one into the (mistaken) view that, to have perceptual knowledge, one must first know something that rules out the possibility of a counterfeit. But in the cases of the cogito-like self-verifying judgments there is no possibility of counterfeits. No abnormal background condition could substitute some other object in such a way as to create a gap between what we think and what we think about. Basic self-knowledge is self-referential in a way that insures that the object of reference just is the thought being thought. If background conditions are different enough so that there is another object of reference in one's self-referential thinking, they are also different enough so that there is another thought. The person would remain in the same reflexive position with respect to this thought, and would again know, in the authoritative way, what he is thinking.

For example, imagine a case of slow switching between actual home and actual twin-home situations. In the former situation, the person may think "I am thinking that water is a liquid". In the latter situation, the person may think "I am thinking that twater is a liquid". In both cases, the person is right and as fully justified as ever. The fact that the person does not know that a switch has

[8] Mistakes about the *res* in *de re* judgments are not counterexamples to the claim that basic cogito-like judgments are self-verifying (hence infallible). Suppose I judge: I am thinking that my aunt is charming; and suppose that the person that I am judging to be charming is not my aunt (I have some particular person in mind). It is true that I am making a mistake about the identity of the person thought about; I have no particular authority about that, or even about her existence. But I am not making a mistake about what I am thinking about that person; there is no mistake about the intentional act and intentional content of the act. Authority concerns those aspects of the thought which have intentional (about-ness) properties. For me, those are the only aspects of the content of a thought.

occurred is irrelevant to the truth and justified character of these judgments. Of course, the person may learn about the switches and ask "Was I thinking yesterday about water or twater?"—and not know the answer. Here knowing the answer may sometimes indeed depend on knowing empirical background conditions. But such sophisticated questions about memory require a more complex story. If a person, aware of the fact that switching has occurred, were to ask "Am I now thinking about water or twater?", the answer is obviously "both". Both concepts are used. Given that the thought is fixed and that the person is thinking it self-consciously, no new knowledge about the thought could undermine the self-ascription—or therefore its justification or authority.

In basic self-knowledge, one simultaneously thinks through a first-order thought (that water is a liquid) and thinks about it as one's own. The content of the first-order (contained) thought is fixed by non-individualistic background conditions. And by its reflexive, self-referential character, the content of the second-order judgment is logically locked (self-referentially) onto the first-order content which it both contains and takes as its subject matter. Since counterfeit contents logically cannot undermine such self-knowledge, there should be no temptation to think that, in order to have such knowledge, one needs to master its enabling conditions.

The view I constructed on Descartes runs contrary. On that view, since basic self-knowledge is more certain than perceptual knowledge, it is more imperative that one be master of all its enabling conditions. One temptation toward this sort of reasoning may derive from construing self-knowledge as a perfected perceptual knowledge. If one thinks of one's relation to the subject matter of basic self-knowledge on an analogy to one's relation to objects of empirical investigation, then the view that one's thoughts (the subject matter) are dependent for their natures on relations to the environment will make it appear that one's knowledge of one's thoughts cannot be any more direct or certain than one's knowledge of the environment. If one begins by thinking of one's thoughts as objects like physical objects, except that one cannot misperceive or have illusions about them, then to explicate authoritative self-knowledge, one makes one of two moves. Either one adds further capacities for ruling out the possible sources of misperception or illusion in empirical perception, or one postulates objects of knowledge whose very nature is such that they cannot be misconstrued or misconceived. In the first instance, one grants oneself an omniscient faculty for discerning background conditions whose independence from us, in the case of perceptual knowledge, is the source of error. In the second instance, one imagines objects of thought (propositions that can be thought only if they are completely understood, or ideas whose *esse* is their *percipi*) whose natures are such that one cannot make any mistakes about them—objects of thought which one can "see" from all sides at once. In either case, one takes oneself to have ultimate insight into the natures of one's thoughts.

This line of reasoning is deeply misconceived. One need only make it explicit to sense its implausibility. The source of our strong epistemic right, our justification, in our basic self-knowledge is not that we know a lot about each thought we know we have. It is not that we can explicate its nature and its enabling conditions. It is that we are in the position of thinking those thoughts in the second-order, self-verifying way. Justification lies not in the having of supplemental background knowledge, but in the character and function of the self-evaluating judgments.

Let us turn to the point that self-knowledge is personal. The view that anti-individualism is incompatible with authoritative self-knowledge is easily engendered by forgetting the essentially first-person character of self-knowledge. We switch back and forth between thinking our thoughts and thinking about ourselves from the point of view of another person who knows more about our environment than we do. This is a key to Descartes' sceptical thought experiments. And it would not be surprising if he tended to think about self-knowledge in such a way as to give it a sort of omniscience from the third-person point of view—in order to protect the first-person point of view from the fallibilities to which impersonal or third-person judgments (especially empirical judgments) are prone. Since we are not omniscient about empirical matters, it is natural to reduce the scope of the relevant third-person perspective so that the character of one's thoughts is independent of an environment about which we cannot be omniscient. Individualism ensues.

To illustrate the train of thought in a more concrete way: we think that we are thinking that water is a liquid. But then, switching to a third-person perspective, we imagine a situation in which the world is not as we currently think it is—a situation, say, in which there is no water for us to interact with. We take up a perspective on ourselves from the outside. Having done this, we are easily but illegitimately seduced into the worry that our original first-person judgment is poorly justified unless it can somehow encompass the third-person perspective, or unless the third-person perspective on empirical matters is irrelevant to the character of the first-person judgment. In this fallen state, we are left with little else but a distorted conception of self-knowledge and a return to individualism.[9]

[9] My knowledge that I am thinking that mercury is an element depends on an ability to think—not explicate—the thought that mercury is an element. Compare my knowledge that my words 'mercury is an element' are true if and only if mercury is an element. This knowledge depends on understanding the words 'mercury is an element' well enough to say with them, or think with them, that mercury is an element. It is this ability which distinguishes this knowledge from mere knowledge that the disquotation principle as applied to 'mercury is an element' is true (mere knowledge that the sentence '"mercury is an element" is true if and only if mercury is an element' is true). I know that my word 'mercury' applies to mercury (if to anything), not by being able to provide an explication that distinguishes mercury from every conceivable twin mercury, but by being a competent user of the word, whose meaning and reference are grounded in this environment rather than in some environment where the meaning of the word-form would be different. The fact that one may not be able to explicate the difference between mercury and every possible twin mercury should not lead one to assimilate one's use of 'mercury' to knowledge of purely formal relationships (e.g., knowledge that all instances of the disquotation principle are true).

One other comparison: I know that I am here (compare: on earth) rather than somewhere else (compare: twin earth). My knowledge amounts to more than knowing I am wherever I am. I have

As one thinks a thought reflexively, it is an object of reference and knowledge, but simultaneously a constituent of one's point of view. The essential role that the first-person singular plays in the epistemic status of authoritative self-knowledge differentiates this knowledge not only from empirical knowledge, but also from most apriori knowledge, the justification of which does not depend on the first-person point of view in the same way.

The tendency to blur distinctions between apriori knowledge (or equally, knowledge involved in explication of one's concepts) and authoritative self-knowledge is, I think, an instance of Descartes' central mistake: exaggerating the implications of authoritative self-knowledge for impersonal knowledge of necessary truths. One clearly does not have first-person authority about whether one of one's thoughts is to be explicated or individuated in such and such a way. Nor is there any apparent reason to assume that, in general, one must be able to explicate one's thoughts correctly in order to know that one is thinking them.

Thus, I can know that I have arthritis, and know I think I have arthritis, even though I do not have a proper criterion for what arthritis is. It is a truism that to think one's thoughts, and thus to think cogito-like thoughts, one must understand what one is thinking well enough to think it. But it does not follow that such understanding carries with it an ability to explicate correctly one's thoughts or concepts via other thoughts and concepts; nor does it carry an immunity to failures of explication. So one can know what one's thoughts are even while one understands one's thoughts only partially, in the sense that one gives incomplete or mistaken explications of one's thoughts or concepts. One should not assimilate 'knowing what one's thoughts are' in the sense of basic self-knowledge to 'knowing what one's thoughts are' in the sense of being able to explicate them correctly—being able to delineate their constitutive relations to other thoughts.[10]

For its justification, basic self-knowledge requires only that one think one's thoughts in the self-referential, self-ascriptive manner. It neither requires nor by

normal ability to perceive and think about my surroundings. I have this knowledge because I perceive my surroundings and not other conceivable surroundings, and I have it even though other places that I could not distinguish by perception or description from here are conceivable. For a variety of reasons, one should not assimilate terms like 'water' to indexicals like 'here'. Compare 'Other Bodies'. But these analogies may be helpful here.

[10] Davidson's views about self-knowledge have some crucial points in common with mine. But he may be making this mistake when he writes that, if one concedes the possibility of partial understanding as I do, one must concede that anti-individualism undermines the authority of self-knowledge. Compare Davidson, 'Knowing One's Own Mind', *Proceedings and Addresses of the American Philosophical Association* 60:3 (1987), 441–458 (p. 448). Compare also Davidson, 'First Person Authority', *Dialectica* 38 (1984), 101–112. It is unclear to me why Davidson says this. I have discussed the distinction between the sort of understanding necessary to think and the sort of understanding necessary to explicate one's thoughts in 'Individualism and the Mental'; 'Intellectual Norms and Foundations of Mind'; 'Wherein is Language Social?' in A. George (ed.), *Reflections on Chomsky* (Oxford: Blackwell, 1989), 175–192; and 'Frege on Sense and Linguistic Meaning', in David Bell and Neil Cooper (eds.), *The Analytic Tradition: Meaning, Thought, and Knowledge* (Oxford: Blackwell, 1991), 30–60.

itself yields a general account of the mental kinds that it specifies. Conceptual explication—knowledge of how one's thought kinds relate to other thought kinds—typically requires more objectification: reasoning from empirical observation or reflection on general principles. It requires a conceptual mastery of the conditions underlying one's thoughts and a conceptual mastery of the rules one is following. These masteries are clearly beyond anything required to think thoughts in the second-order, self-ascriptive way. Explicative knowledge is neither self-verifying nor so closely tied to particular mental events or particular persons' points of view.[11]

Despite, or better because of, its directness and certainty, basic self-knowledge is limited in its metaphysical implications. It is nonetheless epistemically self-reliant. By itself it yields little of metaphysical interest; but its epistemic credentials do not rest on knowledge of general principles, or on investigation of the world.

[11] As I indicated earlier, basic self-knowledge is at most an illuminating paradigm for understanding a significant range of phenomena that count as self-knowledge. Thus, the whole discussion has been carried out under a major simplifying assumption. A full discussion of authoritative self-knowledge must explicate our special authority, or epistemic right, even in numerous cases where our judgments are not self-verifying or immune to error. I think, however, that reflection on the way that errors can occur in such cases gives not the slightest encouragement to the view that anti-individualism (as regards either the physical or social environments) is a threat to the authority of our knowledge of the contents of our thoughts.

3 *Our Entitlement to Self-Knowledge*

I want to understand our epistemic warrant for a certain range of judgments about our own thoughts and attitudes. I am guided by two hypotheses. One is that there are certain sorts of self-knowledge that are epistemically special. The other is that the epistemic right or warrant we have to these sorts of self-knowledge is, in a sense, environmentally neutral. I want to understand this specialness and this environmental neutrality.

The hypothesis of epistemic specialness will be argued for in this paper. The hypothesis of environmental neutrality is relevant to a project that deals with scepticism and the nature and functions of reason. I will not develop this latter hypothesis in depth here, but I will comment on it for the sake of orientation.

Most of our empirical thoughts and our thoughts about our empirical thoughts depend for their individuation conditions on relations that we bear to a particular environment. But, on my guiding hypothesis, our epistemic warrant for our judgments about our thoughts does not depend on particular relations to a particular environment. It is common to any environment and derives from the nature of the thinker as a critical reasoner. This point is relevant to showing that certain claims to self-knowledge which are among the premises in a certain anti-sceptical argument do not beg the question by depending on presumptions about the environment that the sceptic calls into question. In this paper I will not discuss scepticism. But I begin with the sort of *cogito*-like judgments that figured in traditional anti-sceptical arguments. I believe these judgments relevant not only to scepticism, but to the epistemic specialness of some self-knowledge. Although some striking features of *cogito*-like judgments are not shared by all members of the wider range of judgments about one's thoughts whose epistemic status interests me, *cogito*-like judgments provide a useful paradigm for reflection.

So I begin with some remarks about a judgment that:

(1) I am thinking that there are physical entities.

This judgment is an instance of *cogito*-like thoughts, an elaboration of Descartes' *I am thinking*. Let us construe 'thinking' in (1) minimally—as *engaging in thought* or *having a thought*, regardless of whether it is merely entertaining a thought, making a judgment, or whatever. In this sense, one 'thinks' all propositional components of any thought one thinks (including negated ones, antecedents of conditionals, and so on). (1) is the content of my judgment. I accept it as true. To be true, (1) requires only that I am engaging in some thought whose content is that there are physical entities.

We do not rest this judgment upon any observation or perception such as was traditionally called 'inner sense'. The judgment is direct, based on nothing else. Making the judgment requires sufficient understanding to think (1). But once one makes the judgment, or indeed just engages in the thought, one makes it true. The thought is *contextually self-verifying*. One cannot err if one does not think it, and if one does think it one cannot err. In this sense, such thinkings are *infallible*.

I do not claim that judgments like (1) are indubitable. The scope for human perversity is very wide. One could be so far gone as to think to oneself: 'I do not know whether I am now thinking or not; maybe I am dead or unconscious; my mantra may have finally made me blissfully free of thought'. Such mistaken doubt would evince cognitive pathology, but I think it possible. It is an error, however, that most people would avoid without swerving.

Key features of (1) are shared by judgments of

(2) I judge, herewith, that there are physical entities.

When *judge* in (2) is used to execute not merely describe a judgment, judgments of (2) are contextually self-verifying. (2) is not made true by the mere thinking of it, nor does it have quite the same quasi-logical self-evident status that (1) does. These are subtleties that I will have to discuss on another occasion.

(1) and (2) are not mere philosophical curiosities. I think that they represent the form of many ordinary self-aware judgments (at least when (1) is taken to have the 'herewith' reflexivity of (2)). When one makes a judgment and is conceptually aware of one's so doing, whether or not one spells out this conceptual awareness, one's judgments have a reflexive form like that of (2). Such conceptual self-awareness goes beyond simply consciously thinking a thought, but it is not an unusual phenomenon among people with normal second-order abilities. Thus I believe that *cogito*-like judgments constitute a significant segment of our everyday mental activity.

To remark that (1) and (2) are *contextually self-verifying* is to remark on their truth conditions, not on our justification or epistemic warrant in thinking them. It does seem that understanding (1) suffices for knowing that it is true. And the relevant understanding requires no great perspicacity. But noting that it is self-evidently self-verifying (supposing that this needed no more comment—which of course it would) would not capture fully what is involved in its epistemic status. For I think that *cogito*-like judgments share an interesting epistemic status with a number of types of self-knowledge that are not contextually self-verifying or infallible, and that lack the quasi-logical status of (1). I have in mind a wider class of judgments about states, not just reflexive occurrences—judgments about what one believes, wants, intends.

When we make judgments about many of our mental states and events, our judgments commonly constitute knowledge. I know very well that I believe that there are physical entities—if I judge that I do. Such judgments do not merely evince an inner state in the way that a yelp evinces a pain; nor are they avowals or conventional practices without cognitive value. What is the epistemic status of such judgments? What epistemic warrant do we have to make them?

I take the notion of epistemic warrant to be broader than the ordinary notion of justification. An individual's epistemic warrant may consist in a justification that the individual has for a belief or other epistemic act or state. But it may also be an *entitlement* that consists in a status of operating in an appropriate way in accord with norms of reason, even when these norms cannot be articulated by the individual who has that status. We have an entitlement to certain perceptual beliefs or to certain logical inferences even though we may lack reasons or justifications for them. The entitlement could in principle presumably—though often only with extreme philosophical difficulty—be articulated by someone. But this articulation need not be part of the repertoire of the individual that has the entitlement.

Our epistemic warrant to much of our self-knowledge is of this sort. Most of us have no justifying argument or evidence backing the relevant judgments. The judgments are immediate, non-inferential. Although *cogito*-like judgments may count as self-evident or self-verifying, most judgments that interest me do not. Wherein are we being reasonable—in the sense of operating under norms sanctioned by reason—in making judgments about our own minds?

As I have intimated, the remarks about self-verification suggest an initial analogy between *cogito*-like judgments and knowledge of simple logical truths. The truth of judgments of (1) and (2) *is*, in a broad sense, present in the form and logic of the thought. There is something of the same self-evident and obvious features here as there are in simple logical truths. The main differences are that *cogito*-like judgments are dependent on being thought for being true, and are in their specially direct way self-verifying.

Another analogy to knowledge of simple logical truths is this: The key to the epistemic status of *cogito*-like judgments seems to reside in ordinary understanding, not in some mechanism connecting the knower with a sensed object.

This point will be one of the key elements in my account of the environmental neutrality and specialness of self-knowledge. Perceptual experiences particular to a given environment inevitably figure in the acquisition of understanding of almost any given content. But one's epistemic warrant for believing the content may not incorporate the perceptual experiences or beliefs that go into understanding it. This is the traditional view of knowledge of logical or mathematical truths. One may need perceptual experience to come to understand simple logical or arithmetical notions and truths. (This is surely the case with such logical truths as 'nothing is both a dog and not a dog'.) But on the traditional view such experience is not a constituent of one's justification or entitlement in believing simple logical or arithmetical truths.

I am not arguing for the traditional view—just recalling it. The element in it relevant to our purposes is the following. The account of epistemic justification or entitlement may presuppose understanding, which may be dependent on particular perceptual relations to a given environment. But the account need not include perceptual beliefs or experiences as *constituents* in the individual's justification or entitlement. The account can allow attribution of concepts to the individual which could be acquired only in a limited range of possible environments, while itself taking a form that is applicable to any critical reasoner, regardless of the particular environmentally dependent contents of his or her thought.

I want to illustrate the relevance of this idea to our discussion by reconsidering the scenario of one's being switched between different environments unawares—a scenario I discussed in a paper some years back.[1] Let us assume for the sake of argument that my thinking that there are physical entities (hence my thinking that I am thinking that there are physical entities) is the thought that it is because of relevant causal relations I bear to actual physical objects in my environment. Let us also assume that an individual with a chemically identical body could have been brought up in a situation in which such relations were lacking—and in which the concept of physical object could not be acquired—but in which different, counterpart thoughts occurred. (I doubt that *physical object* is a concept universal to all possible critical reasoners; but if one did not doubt, another concept could be chosen.) Finally, let us grant that if at any time one were switched unawares from one's actual situation into such a counterpart situation, one would have no resources that would tip one off to the difference.

Unless memory and learning connections to the original environment were broken, it is hard to describe a switch of actual situations that would produce a new twin set of the concepts, with no residue from the past experiences. So in the case I am imagining one's thoughts do not switch to twin thoughts. Because of a switch one's thoughts might, however, change content, broadening their extensions without one's being aware of their doing so.

I take it that this observation is sufficient to prompt the following question. Given that we are insensitive to such alleged possible changes in content, how can we know what we are thinking?

I will not try to deal with this question in all its ramifications here. But as I noted some years back, some of the negative force of the question can be shown to be illusory by this consideration: There is no way for one to make a mistake about the content of one's present-tensed thought in the relevant cases.

Suppose that I think that I am engaging in a thought that there are physical objects. In thinking this, I have to engage in the very thought I am referring to and ascribing to myself. The reference to the content—expressed in the that-clause—cannot be carried out unless I actually engage in the thought. The intentional content mentioned in the that-clause is not merely an object of reference or

[1] 'Individualism and Self-Knowledge', *The Journal of Philosophy* 85 (1988), 649–663. Reprinted in Quassim Cassam (ed.), *Self-Knowledge* (Oxford: Oxford University Press, 1994).

cognition; it is part of the cognition itself. It is thought and thought about in the same act. If background conditions are different enough so that I am thinking different thoughts, they will be different enough so that the objects of reference and self-ascription will also be different. So no matter how my thoughts are affected, no matter how I am switched around, I will be correct in self-ascriptions of content that are correctly expressed in *cogito*-that-clause form.

It would be a mistake to reply that because one's correct reference does not give one any grasp of what one is referring to, this reference is empty. For to self-ascribe thoughts in the way expressed by that-clauses, one has to understand the thoughts one is referring to well enough to think them. One need not have any more explicatory understanding of one's thoughts than is necessary to think them. One need not master anti-individualism, much less have an empirical mastery of the conditions that have established the identity of the thoughts one thinks. Such mastery is emphatically not guaranteed by mastery of *cogito*-self-ascriptions. But one *is* guaranteed that one ascribes something of which one has the ordinary understanding involved in using concepts and thinking thoughts.

This understanding presupposes the causal-perceptual relations to a particular environment that help determine what content is available for being understood. What one *can* think is partly dependent on relations to one's environment. And one's second-order self-ascriptions inherit both the content and the background environmental content-determining conditions from one's first-order understanding.

I have granted that one need not be sensitive to actual or counterfactual changes in what one understands under transportations into environments where the content of one's understanding changes or would be different. One need not be capable of detecting such changes. But in any situation in which a person can think the relevant *cogito*-like judgments, the person would think them with understanding—and to all appearances, knowledgeably. At any rate, there is no obvious reason why knowledge in such judgments would be prevented by such changes, much less such possible changes.[2] In any such twin situation, the person

[2] Are there switching situations in which one would have reasonable ground for doubting what contents one is thinking, so that a *cogito*-type judgment would not constitute knowledge? This is very complex, but I will make a few remarks here. The self-ascription in the that-clause way cannot involve a mistake about the intentional content. So the possibility of a switch does not threaten a mistake. I think therefore that such possibilities pose no relevant alternative threat to one's entitlement to one's judgment about the that-clause content of one's thoughts. I believe that the relevant minimal understanding suffices for knowledge in *cogito*-like judgments. Even in non-*cogito*-like judgments, switches in content cannot, for the same reason, undermine knowledgeability of the content of self-ascriptions. Compare my 'Individualism and Self-Knowledge', p. 659. A fuller story has to be told about the propositional-attitude concepts in non-*cogito*-like judgments. I think the possibility of switching, or of errors of incomplete understanding, do not by themselves undermine knowledge; but I will have to discuss these matters further elsewhere.

Some worries about switching situations have focused on memory. I think that they tend to confuse preservative memory with memory of objects or with comparisons within memory, and to overrate the extent to which the content retrieved in memory is sensitive to immediate environmental context. For a discussion of preservative memory, see my 'Content Preservation', *The Philosophical Review* 102 (October 1993), 457–488.

would understand the self-ascribed contents and would self-ascribe them with a justice and reliability that is equal to that in any ordinary situation. Some entitlement attaching to understanding seems to be what the self-knowledge depends upon, not on some knowledge of what the understanding consists in, or whether it differs from understanding that is past or possible.

The person's epistemic entitlement to the self-ascriptions presupposes understanding. Understanding is, as I have noted, dependent on and local to causal-perceptual relations to a given environment. But the entitlement that underlies knowledgeable *cogito*-like thoughts and other self-ascriptions does not seem local and seems to survive such switches. It seems to be carried somehow by the fact that we correctly self-ascribe any content at all with understanding. Where does the entitlement derive from? And what makes it capable of surviving such environmental switches?

I think that the relevant entitlement derives not from the reliability of some causal-perceptual relation between cognition and its object. It has two other sources. One is the role of the relevant judgments in critical reasoning. The other is a constitutive relation between the judgments and their subject matter—or between the judgments about one's thoughts and the judgments' being true. Understanding and making such judgments is constitutively associated both with being reasonable and with getting them right.

Briefly drawn, my line of thought will be this. To be capable of critical reasoning, and to be subject to certain rational norms necessarily associated with such reasoning, some mental acts and states must be *knowledgeably* reviewable.[3] The specific character of this knowledgeable reviewability requires that it be associated with an epistemic entitlement that is distinctive. The entitlement must be stronger than that involved in perceptual judgments. There must be a non-contingent, rational relation, of a sort to be explained, between relevant first-person judgments and their subject matter or truth.

All of us, even sceptics among us, recognize a practice of critical reasoning.[4] Critical reasoning is reasoning that involves an ability to recognize and effectively employ reasonable criticism or support for reasons and reasoning. It is reasoning guided by an appreciation, use, and assessment of reasons and reasoning as such. As a critical reasoner, one not only reasons. One recognizes

[3] I think that the following necessity also holds: To think the relevant first-person present tense thoughts about one's thoughts and attitudes, one must be capable of critical reasoning. Indeed, I think that to have a fully formed first-person concept or fully formed concepts of propositional attitudes, one must be capable of critical reasoning. To master concepts of propositional attitudes in a suitably rich sense, one must be capable of appreciating the force and relevance of reasons to attitudes as such, which amounts to being able to reason critically about reasons and reasoning. And to master a fully formed first-person concept, one must have concepts of propositional attitudes.

[4] In actual practice, critical reasoning approximates what I call *reflective* reasoning. Reflective reasoning makes use of all the main concepts necessary to a full understanding of essential or fundamental elements in reasoning. Critical reasoning is simply reasoning that is sufficiently articulate to appreciate reasons as reasons and to employ articulated criticism of reasons and reasoning (as reasons and reasoning).

reasons as reasons. One evaluates, checks, weighs, criticizes, supplements one's reasons and reasoning. Clearly, this requires a second-order ability to think about thought contents or propositions, and rational relations among them.

When one carries out a proof, one checks the steps of the reasoning, making sure that the inferences are valid. Any activity of proof requires some conception of validity, which requires an ability to think of the propositions in a proof as constituting reasons for what follows from them. Indeed, it is arguable that use of *therefore* in reasoning—deductive or otherwise—constitutes an exercise of this meta-cognitive ability. When one engages in practical deliberation, one articulates and weighs considerations on each side, goes over possible sources of bias, thinks through consequences. Essential to carrying out critical reasoning is using one's knowledge of what constitutes good reasons to guide one's actual first-order reasoning.

A non-critical reasoner reasons blind, without appreciating reasons as reasons. Animals and small children reason in this way. But reasoning under rational control of the reasoner is critical reasoning. Not all reasoning by critical reasoners is critical. Much of *our* reasoning is blind, poorly accessible, and unaware. We change attitudes in rational ways without having much sense of what we are doing. Often we are poor at saying what our reasoning is. Still, the ability to take rational control of one's reasoning is crucial in many enterprises—in giving a proof, in thinking through a plan, in constructing a theory, in engaging in debate. For reasoning to be critical, it must sometimes involve actual awareness and review of reasons; and such a reviewing standpoint must normally be available.[5]

Critical reasoning involves an ability not merely to assess truth, falsity, evidential support, entailment, and non-entailment among *propositions* or *thought contents*. It also involves an ability to assess the truth and reasonability of reasoning—hence *attitudes*. This is not to say that critical reasoning must *focus* on attitudes, as opposed to their subject matter. Normally we reason not about ourselves but about the world or about practical goods. But to be fully a critical reasoner, one must be able to—and sometimes actually—identify, distinguish, evaluate propositions as asserted, denied, hypothesized or merely considered.[6] Such abilities and activities are central to argumentation. Similarly, in critical practical reasoning, one must be able to—and sometimes actually—evaluate propositions conceptualized as expressing pro-attitudes, to distinguish them explicitly from those that express beliefs, and to evaluate relations of reason

[5] I think Kant neglected distinctions between reasoning, critical reasoning, and reflective reasoning. But he clearly saw that it is the possibility of applications of 'I think' to our thoughts—not our being self-aware in this way all the time—that is basic to full reflective rationality. Of course, the form of 'I think' does not by itself make the relevant contribution to reflective rationality. One could dream *cogito*-thoughts. It is the ability to be conceptually aware of oneself as thinking with a certain control and agency that is crucial.

[6] In effect, Frege's use of the assertion sign is an acknowledgment of a minimal use of these abilities. Without an ability to recognize that a proposition should be and is judged to be true, one cannot reason critically. Having a concept of judgment and using it in reasoning meets my requirement.

among such propositions as so conceptualized. Such evaluation constitutes minimal evaluations of propositional attitudes.

To be a critical reasoner, one must also be able to, and sometimes actually, use one's knowledge of reasons to make, criticize, change, confirm *commitments* regarding propositions—to engage explicitly in reason-induced changes of mind. Critical reasoning here involves an ability to distinguish subjectivities from more objectively supportable commitments and to explicitly alter the former in favor of the latter. Its point is reasonably to confirm and correct attitudes and reasoning (not merely assess propositional connections), by reference to rational standards.

Critical reasoning must be exercised on itself. Any critical reasoning, even about abstract propositional relations or about the reasoning of others, involves commitments by the reasoner. And genuinely critical reasoning requires an application of rational standards to *those* commitments. A being that assessed good and bad reasoning in others or in the abstract, but had no inclination to apply such standards to the commitments involved in those very assessments, would not be a critical reasoner. To reason critically—to consider reasons bearing on the truth of some matter, to suspend belief or desire, to weigh values under a conception of the good—one must treat one's own commitments as matters to be considered and evaluated. Critical evaluation of one's own commitments is central to forming them and to rationally changing one's mind or standing fast.

So critical reasoning requires thinking about one's thoughts. But it further requires that that thinking be normally knowledgeable. To appreciate one's reasons as reasons—to check, weigh, criticize, confirm one's reasons—one must know what one's reasons, thoughts, and reasoning are. One need not always be knowledgeable, or even right. But being knowledgeable must be the normal situation when one reflects on one's reasons in the course of carrying out reasonable inquiry or deliberation. The interest here is less in the requirement of normal knowledgeability—which is shared with other cognitive activities, such as perception. The interest lies in the ground of the requirement. Why must we be normally knowledgeable about our thoughts when we reflect upon them?

I will answer this question in three stages. First, I want to show that to evaluate reasons critically, one must have an *epistemic entitlement* to one's judgments about one's thoughts, reasons, and reasoning. Second, I want to support the stronger thesis that critical reasoning requires that one *know* one's thoughts, reasons, and reasoning. Third, I will try to show that this knowledge must take a distinctive, non-observational form.

So I begin with the matter of entitlement. The basic idea is simple. Put crudely: since one's beliefs or judgments about one's thoughts, reasons, and reasoning are an integral part of the overall procedures of critical reasoning, one must have an epistemic right to those beliefs or judgments. To be reasonable in the whole enterprise, one must be reasonable in that essential aspect of it.

Less crudely, consider the process of reasoning which involves the confirming and weighing of one's reasons. One must make judgments about one's attitudes and inferences. If one's judgments about one's attitudes or inferences were not reasonable—if one had no epistemic entitlement to them—one's reflection on one's attitudes and their interrelations could add no rational element to the reasonability of the whole process. But reflection does add a rational element to the reasonability of reasoning. It gives one some rational control over one's reasoning.

To put the point somewhat more fully: if one lacked entitlement to judgments about one's attitudes, there could be no norms of reason governing how one ought check, weigh, overturn, confirm reasons or reasoning. For if one lacked entitlement to judgments about one's attitudes, one could not be subject to rational norms governing how one ought to alter those attitudes given that one had reflected on them. If reflection provided no reason-endorsed judgments about the attitudes, the rational connection between the attitudes reflected upon and the reflection would be broken. So reasons could not apply to how the attitudes should be changed, suspended, or confirmed *on the basis of* reasoning depending on such reflection. But critical reasoning just is reasoning in which norms of reason apply to how attitudes should be affected partly on the basis of reasoning that derives from judgments about one's attitudes. So one must have an epistemic entitlement to one's judgments about one's attitudes.

I turn now to the stronger thesis. One might imagine some gap between epistemic entitlement and knowledge. Might one have an epistemic entitlement but be systematically mistaken? Or might failure of some third Gettier-type condition (beyond truth and epistemic entitlement) undermine knowledge?

It is possible in given cases for reflection to be disconnected in these ways from the attitudes purportedly reflected upon. But both possibilities if generalized are incompatible with our having the sort of entitlement to the reflection just argued for. That entitlement rested on the assumption that reflection added a rational element to the reasonability of the whole process of critical reasoning—a process whereby object-level attitudes are guided by reflection on their reasonability. If reflective judgments were not normally true, reflection could not add to the rational coherence or add a rational component to the reasonability of the whole process. It could not rationally control and guide the attitudes being reflected upon (even though one could imagine situations in which such disconnected reflection would be mechanically or instrumentally beneficial in forming true or rational beliefs). So reflection would not add in the relevant way to the reasonability of the process, and therefore would not have the source of entitlement just argued for.

The same point applies to the possible failure of some Gettier-type condition. Again, if reflection were connected to the truth of our judgments about our thoughts in an accidental or non-knowledge-yielding way, the reason-guiding and rational-coherence-making functions of rational review would be broken. Since part of our entitlement to reflective judgments about our attitudes derives from their functions in critical reasoning, the entitlement itself would be undermined.

Not only the relevant entitlement to reflective judgments that derives from their functions within critical reasoning, but critical reasoning itself is constitutively dependent on the truth- and Gettier conditions being met. If a being had an epistemic entitlement to its judgments about its attitudes but were systematically mistaken about them—never got them right—it would not be a critical reasoner. Or if our entitlement were always connected to the truth of our judgments about our thoughts in an accidental or non-knowledge-yielding way, critical reasoning would not be possible. For critical reason requires rational integration of one's higher-order evaluations with one's first-order, object-oriented reasoning. The former must be reason-guided and reason-guiding. And they must cement the rational coherence between the two levels. If the two came radically apart, or were only accidentally connected, critical reasoning would not occur.

So if we failed normally to know our thoughts and attitudes, in ordinary reasoning about reasons, either through systematic falsity of our judgments or through systematic mismatch between our entitlement and truth, critical reasoning would not occur among us. Indeed, the entitlement to reflective judgments that derives from those judgments' place in critical reasoning would lapse. But critical reasoning does occur among us; and we are entitled to reflective judgments by virtue of their contribution to the reasonability of critical reasoning. So as critical reasoners we must know our thoughts and attitudes.

Symptomatic of the connection I have noted between the rationality of reflection in critical reasoning and the truth of reflective judgments is the fact there are severe limits on brute errors in judgments about one's present ordinary, accessible propositional attitudes. A brute error is an error that indicates no rational failure and no malfunction in the mistaken individual.[7] Brute perceptual errors commonly result from misleading natural conditions or look-alike substitutes. One can be perceptually wrong without there being anything wrong with one. Such brute perceptual errors are unremarkable. But errors about what one's thoughts and attitudes are normally seem to involve some malfunction or rational deficiency. There are exceptions—the cases of unconscious, modular attitudes that are not accessible to reflection. One could easily make brute errors about these. There are judgments about one's emotions, character, or deep motives, that seem hard to get right. I leave open whether these might sometimes involve brute errors. But it seems that we make mistakes about many attitudes that are accessible to reflection primarily when we are subject to some failure of rationality or defect in our cognitive powers.[8]

[7] I introduced the notion of brute error in 'Individualism and Self-Knowledge', 657. I intend rational failures to include any failure of entitlement or justification, not just ones that are epistemically culpable. I intend malfunctions to cover not only mechanical or biological failures in, say, the individual's perceptual apparatus, but also failures of normal understanding—as for example when an individual believes arthritis can occur in the thigh. The idea is that a brute error would have occurred even if the individual's epistemic warrants were in order and the individual's perception and ordinary understanding were functioning optimally.

[8] Our epistemic entitlement to judgments about our present attitudes is a general right and is compatible with our making various mistakes about our attitudes even in the course of critical reasoning. (Of course, then we are, in a sense, not critically reasoning *with* the attitudes we are mistaken about.) We make mistakes of haste, bias, and self-deception. Some attitudes are hard to get

I stated that I would argue that the specific role of knowledge of our thoughts in critical reasoning requires that it be associated with a distinctive sort of epistemic entitlement that necessitates a non-contingent, rational relation between the relevant first-person present-tense judgments and their subject matter or truth. Why need self-knowledge be in any way special? Why is it not enough that it be pretty reliable observation? Some knowledge of our own mental states and events is empirical in the sense that it is based either on imaging, remembering, or reasoning about *sensed* inner-goings-on, or on observing our own behavior and hearing about it from others. Simplicity tempts some to hold that all self-knowledge is like that.

Let me elaborate this temptation. It is commonly held that beliefs about others' attitudes must be based on inferences from or criteria for observation.[9] On the model at issue, beliefs about one's own attitudes differ only in that one need not always infer those beliefs, because one is the closest witness. There is no authority in self-knowledge, other than the authority of inner observational presence, practice, and familiarity.

This simple observational model does not account plausibly for *cogito*-like thoughts. Such thoughts are logically special in their self-verification and epistemically special in their clear dependence for entitlement on intellection and understanding, not on any sort of observation. But *cogito* thoughts do not constitute the full range of thoughts that enter essentially into critical reasoning.

The simple observational model is encumbered with the obscurity of the notion of inner observation as applied to thoughts and attitudes. Unlike sensations or images, thoughts and attitudes lack distinctive presentations or phenomenologies. The model is phenomenologically implausible for many immediate judgments about one's own beliefs or current thoughts. But I want to show that there is a deeper problem if the model is taken to cover all cases.[10]

at, except with discipline, and even maturation or therapy. In some cases, other people are better at knowing our attitudes than we are. So one might demand further specification of our entitlement. One might ask under what conditions it is overturned or insufficient to give us knowledge. And one might inquire in more depth into the conditions under which errors arise. These issues are complex. I think that when our judgments about a certain class of our thoughts and attitudes are in a certain sense immediate (which entails that they are neither inferred nor otherwise biased by other attitudes), and when our minds are not subject to malfunction, we do not make errors. But there is no recipe for insuring that our judgments are immediate or that they are about the relevant class. There is no internal recipe for avoiding error.

[9] I do not accept this view, but I need not question it here. Compare 'Content Preservation'. Certainly one's beliefs about others' thoughts are often based this way.

[10] Hume is, I think, a proponent of the simple observational model. A more recent proponent is D. M. Armstrong, *A Materialist Theory of the Mind* (London: Routledge & Kegan Paul, 1968), 323–38. The rationalist tradition, in its emphasis on the role of self-knowledge in rationality, and the role of understanding (not sensory observation) in self-knowledge, is the source of my view. Kant develops this tradition in a particularly deep way, although his epistemology left him with what was, in my opinion, an implausibly restrictive account of cognition of one's own thoughts, one indeed that overrates the role of inner sense. A more recent non-observational account that emphasizes the role of self-knowledge in reasoning may be found in Sidney Shoemaker, 'On Knowing One's Own Mind', *Philosophical Perspectives* 2 (1988), 183–209.

Before presenting the argument, I will say what I take to be fundamental to the simple observational model. The model need not claim any phenomenological presentation in self-knowledge, though waiving such a claim weakens the analogy to observation. The fundamental claim is that one's epistemic warrant for self-knowledge always rests partly on the existence of a pattern of veridical, but brute, contingent, non-rational relations—which are plausibly always causal relations—between the subject matter (the attitudes under review) and the judgments about the attitudes. This claim is compatible with holding that from the point of view of epistemology, observational judgments are often immediate and non-inferential, requiring no background causal hypothesis on the part of the individual about their source.

My view about perceptual entitlement is more specific than the fundamental claim of the simple observational model. I believe that our entitlement rests partly on our being perceivers, which entails that we—or our species-perceptual systems—are or have been in brute, contingent, non-rational but veridical relations to objects of perception, and the kinds that our perceptual judgments specify.[11] It is necessarily constitutive of the content of our observational or perceptual beliefs about physical objects, and of the very nature of our perceptual systems, that we be veridically attuned to the environment through causal relations to it—either in our learning histories or indirectly in the evolution of our perceptual systems.[12] Entitlement to observational physical object beliefs rests

[11] A tempting oversimplification is to claim that these constitutive veridical causal relations are always *reliably* veridical. Such a claim is tempting because in so many cases our perceptions are reliable. Perhaps many types of perception must be. But the claim is oversimplified because some perceptual intentional types in some perceptual systems are established through the systems' reliable avoidance of false negatives rather than through their reliable achievement of true positives. It is more critical to a hare's perceptual system that it not fail to register a predator when one is there than that it be reliable in its registration of predators. So the system could commonly indicate the presence of predators falsely—and be broadly unreliable in its perceptions—as long as it was reliable in correctly indicating present predators. It remains, however, constitutive of the systems' perceiving predators as predators that some veridical perceptions played a role in the evolution-fashioned function or in the actual use of the system. Of course, reliability is more important for perceptions of safety than for perceptions of danger.

These qualifications on reliability, of course, complicate any account of the relation between perceptual-content constitution and perceptual entitlement. For presumably epistemic entitlements are prima facie comprised by constitutively unreliable perceptual deliverances. I believe that a perceptual system in any agent, however, is constitutively associated with reliable perceptions in a range of cases. But these are issues for another occasion.

[12] I am inclined to think that it is a *conceptual* necessity that there be *causal* relations in perception. But for purposes of my argument, the fundamental feature is that the entitlement to observational beliefs necessarily rests on some pattern of brute, contingent, non-rational relations between observed and observer, regardless of whether the contingent relations are causal. It is common to my view and the opposed observational view of self-knowledge that in many of the cases under dispute, there is a causal mechanism that relates attitudes to judgments about them. What is in dispute is the nature of the epistemic entitlement that one has to such judgments, not the existence of a psychological mechanism. On the simple observational model, our entitlement to self-knowledge always rests partly on the brute, contingent, non-rational causal relations. On my view, in some important cases, it does not: Christopher Peacocke has pointed out to me that in some, though I think not all, cases of special self-knowledge, the entitlement may specify some causal relation between subject matter and

partly on this necessity. But in particular instances of perception, the relations on which one's entitlement to perceptual judgments rests are brute, contingent, non-rational.[13] It is this claim that forms the paradigm for the simple observational model.

The existence of veridical beliefs involving *de re* causally based relations to the environment is necessary to and constitutive of something's being a perceptual belief system. But the individual relations are brute, contingent, non-rational. The brute contingency of these relations in individual instances of veridical perceptual judgment is fundamental to observation. Different conditions could have caused a perceptual judgment that was internally indistinguishable (indeed I think the same judgment-type) but non-veridical, without loss of entitlement, even as the system functioned optimally well.[14] Thus perception is always

judgment. But not all causal relations are brute, contingent, non-rational ones. (For example those involved in a person's deductive inference are not.) Where a causal relation is not merely a background enabling condition, but an element in the relevant entitlement to self-knowledge, it will on my view never be a brute, contingent, non-rational one. It will be associated in the entitlement with norms for transfer of reasons.

[13] The simple observational model is inspired by a comparison of self-knowledge to observational judgments about physical objects. I operate with a commonsensical conception of such judgments. There are non-commonsensical conceptions that take the basic relation that underlies our epistemic right to be one between the observer and some sort of mental item, a sense datum or an appearance. I ignore such theories not because I regard them as mistaken (although I *do* regard them as mistaken). I ignore them because they model observation of physical objects on knowledge of one's mental events. The model I am attacking proposes to illumine self-knowledge through an independent model. I do think that knowledge of our pains and other sensations—as contrasted with knowledge of our propositional states and events—is empirical in the sense that it depends for its entitlement on sensory experience or sensory beliefs. Judgments that constitute such knowledge just are sensory beliefs. Although I believe that brute error is possible in certain judgments of this sort, such cases are marginal. Understanding even these empirical judgments will, I think, owe more to the kinds of considerations I am elaborating than to reflection on ordinary perceptions of physical objects. But I regard knowledge of one's sensations as requiring separate treatment from knowledge of one's thoughts and attitudes.

It is worth noting that a view that we must 'inferentially' base judgments about physical objects on observations of sense data would also normally be committed to holding that one's entitlement to those judgments rests on brute, contingent, non-rational relations to the physical objects that always allow for brute error. The same point applies to inference-to-the-best-explanation views of our warrants for perceptual beliefs about physical objects.

[14] This gloss on the brute *contingency* of the relations, apart from the parenthetical remark, is less committal than my own view of the contingency involved in observational relations. I think the same perceptual object could, with different external auxiliary conditions, have caused a different non-veridical judgment. And I think that a different perceptual object, or perhaps none at all, could have combined with different external auxiliary conditions to cause a perception or perceptual judgment of the same type, though perhaps one with a different token demonstrative element, making it non-veridical. (I do not rely on these views in my argument here.) These different conditions, in individual cases, need not affect the individual's entitlement to the perceptual judgment; nor need they affect the well-functioning of the individual's perceptual-cognitive apparatus. The sense in which the relevant relations are *non-rational* is complex. Perhaps it can suffice here to note that since in the case of ordinary perception the perceptual objects are physical kinds or physical individuals, there can in *that* case be no question of a rational relation between them—which have no intentional content at all—and perceptions or perceptual judgments.

subject to brute error. The object or conditions of perception could lead us into misperception without there being any failure of entitlement and without there being any malfunction of our cognitive or perceptual systems.

The objectivity of perception depends on the possibility of epistemically entitled misperception. Perceptual justification and criticism necessarily presuppose a distinction between a person's cognitive perspective and the objective, physical subject matter. They further presuppose this unremarkable possibility of contingent mismatches in individual cases that in no way impugn the individual's epistemic entitlements or perceptual-cognitive functioning. Rational and epistemic evaluation fixes on the individual's perceptual judgments and perspective, not on their physical subject matter. For this is only brute contingently related, in individual cases, to epistemic entitlement.

A consequence of interpreting all self-knowledge on the simple observational model is that in any given case brute errors—errors that do not reflect on the rationality or sound functioning of the reviewing judgment—are possible. I intimated earlier that brute errors do not seem to threaten some instances of judgments about attitudes. I propose to show why this must be so.

Not all one's knowledge of one's propositional attitudes can fit the simple observational model. For general application of the model is incompatible with the function of knowledge of one's own attitudes in critical reasoning. The main idea is that such application would entail a dissociation between cognitive review and the thoughts reviewed that is incompatible with norms of epistemic reasonability that are basic to all critical inquiry, including empirical, mathematical, philosophical, and practical inquiry.

Rational evaluation of attitudes commonly applies to and within a perspective or point of view. The argument will make reference to this fact. Different people have different points of view. My judgment that your beliefs are irrational may be reasonable from my point of view. But it does not follow that there is reason from your perspective to change your beliefs. I may have made some brute error about what your beliefs are, or your perspective may have different associated reasons or background information from mine.

There can be different perspectives or points of view within a given person. What is reasonable for a person at a given time may be different from what is reasonable from the perspective of the person's memory back on that time. What is reasonable on reflection may differ from what is reasonable in modular cognitive processes, or in an instant practical reasoning, or in subconscious reasoning. My argument hinges on how reasons transfer across points of view.

Suppose that all one's knowledge of one's propositional mental events and states fit the simple observational model. Then one's entitlement to instances of such knowledge would always rest on purely contingent relations between any given judgment about one's mental states and the subject matter of the judgment. What is more, brute error would be possible in any given case. Normative evaluations of reasonability and epistemic entitlement in critical reasoning—in checking and evaluating one's reasoning—would apply within the perspective of the judgments, but not immediately within the perspective of the subject matter of the judgments, except insofar as it contingently conformed to those judgments, and except insofar as it happened to be embedded in a perspective relevantly similar to the perspective from which the judgments were made. For the subject matter might, in any given case, fail to conform to the judgments through no failure of justification or entitlement in the judge, and through no malfunction of the relevant faculties.

But this picture is nonsense if it is applied to all judgments about one's own propositional attitudes. For it is constitutive of critical reasoning that if the reasons or assumptions being reviewed are justifiably found wanting by the reviewer, it *rationally follows immediately* that there is prima facie reason for changing or supplementing them, where this reason applies within the point of view of the reviewed material (not just within the reviewing perspective). If the relation between the reviewing point of view and the reasons or assumptions being reviewed always fit the simple observational model, there would never be an immediate rationally necessary connection between justified rational evaluation within the review, on one hand, and its being prima facie reasonable within the reviewed perspective to shape attitudes in accord with that evaluation, on the other. For the relation between the perspective of the review and that of the reviewed attitudes would always be purely contingent, even under canonical descriptions of them, for purposes of rational evaluation. (The attitudes reviewed would be to the reviews as physical objects are to our observational judgments. They would be purely 'objects' of one's inquiry, not part of the perspective of the inquiry.) It would be reasonable for the person from the point of view of the review that a change in the reviewed material be made. But this reason would not necessarily transfer to within the point of view of the attitudes under review, even though that is a point of view of the same person. Its transferring would depend on brute, contingent, non-rational relations between the two points of view.

In critical reasoning, however, the connection is rationally immediate and necessary. Justifiably finding one's reasons invalid or one's thoughts unjustified, is normally *in itself* a paradigmatic reason, from the point of view of the thoughts being reviewed (as well as from the perspective of the review), to alter them.

If in the course of critical reasoning I reasonably conclude that my belief that a given person is guilty rests entirely on unreasonable premises or bad reasoning, then it normally follows immediately both for the perspective of the review and for the perspective of the reviewed belief that it is reasonable to give up my belief about guilt or look for new grounds for it. In such second-order reasoning, I am not normally reasonable in altering my first-order views about guilt or innocence only with the proviso that they are embedded in reasons that contingently match those associated with my reviewing perspective. I do not normally have the sort of excusing condition that allows for rational error that hinges on the contingent relation that the subject matter bears to my judgments about it. Rather my checking my belief and finding it wanting normally *itself* provides immediate prima facie reason to change it from within the perspective of the review. This is because the first- and second-order perspectives are the *same* point of view.

The reviewing of reasons that is integral to critical reasoning includes the review and the reviewed attitudes in a single point of view. The simple observational model treats the review and the system being reviewed as dissociated in a way incompatible with the norms of critical reasoning. It makes the reviewed system an object of investigation, but not part of the investigation's point of view. So the model fails to account for the norms of critical reasoning.

A closely related point centers on epistemic responsibility. We are epistemically responsible only because we are capable of reviewing our reasons and reasoning. And we are paradigmatically responsible for our reasons when we check and review them in the course of critical reasoning. But the simple observational model implies that in carrying out reviews of one's reasoning, one is epistemically responsible not primarily for the thoughts being reviewed but primarily for the review. The model implies that we are in reviewing our reasons only derivatively responsible for objects of review, as one might be responsible for the actions of one's child or dog—but fully and primarily responsible only where one's knowledge and control contingently matched what one is justified in believing about them.

But one is not epistemically responsible for the thoughts one reflects upon in critical reasoning in the way one is responsible for something one owns or parents. One's responsibility in reflecting on one's thoughts is immediately for the whole point of view. The simple observational model fails to account for the fact that critical reasoning is carried out within a single multi-level point of view.

Of course, we are sometimes disunified. Sometimes to our own good and efficiency, sometimes to our misfortune, we fail to know our motives or reasons, or know them only through observation and empirical reasoning. Sometimes from the point of view of our self-conscious reviewing selves, we are indeed epistemically responsible only derivatively for attitudes that we know only empirically. But in these cases, we are not reasoning critically with those aspects of ourselves that we know only in these ways.

Theoretical knowledge of one's modular attitudes is one sort of purely observationally based 'self-knowledge'. When attitudes *cannot* be known in a direct non-observational way, one commonly enters some qualification on the sense in which the attitudes are *one's own*. In these cases failure to know the attitudes non-observationally is no sign of dissociation. But when one knows only observationally unconscious attitudes which are in principle accessible to non-observational self-knowledge, there is some dissociation of self, constituted by a divide between the point of view of one's critical reasoning and the attitude known only observationally.

Knowledge through therapy of one's unconscious, before full integration of that knowledge, provides one sort of example. One may know the attitudes on the basis of observationally based therapy, but the unconscious attitudes may provide a point of view of their own into which the meta-evaluations of them may not transfer. Those evaluations may not speak to unconscious considerations that are integral to the unconscious pathology; or the unconscious point of view may not have 'taken in' matters that are integral to the rationality of the meta-, therapeutic point of view.

Psychoanalytic cases are not the only sort that illustrate the relevant dissociation. One may know from experience or theory that one will act a certain way, and yet rational meta-considerations may not penetrate to the system of attitudes that motivate the action. Some self-admitted compulsions provide examples. One knows one has or will have the relevant intention, but knows the intention only as object; it is then not the product of critical deliberation. Sometimes rational considerations from the meta-point of view may not have the same rational force and relevance within the point of view that includes the observationally known attitudes.

There are cases of knowledge of one's beliefs like this as well. One may know from self-observation that 'underneath' one believes something because one needs to believe it, while feeling sincere rational urges to assert the contrary. The system of underlying practical beliefs that motivate the needed belief may form a point of view that does not recognize as sufficient the rational meta-reasons that one can offer oneself for giving up the need-based belief. The person's meta-perspective may correctly condemn the need-based belief as epistemically irrational. But the belief may be dissociated from the point of view of his observational knowledge of that belief. It may be rational 'overall' for the person to give up the belief, but the practical rationality of the limited, need-driven perspective may exclude or outweigh the considerations that count against the belief.

The relevant psychological dissociation is, I think, sometimes partly to be explained in terms of the fact that a second 'point of view', or system of attitudes with its own internal coherence, has gotten set up within the person, in such a way that reasons from the point of view of the person's critical rationality do not automatically transfer to within the second point of view, rather as reasons from my perspective do not always apply as reasons from yours. Being known from the perspective of the critical reasoner *only* as an object, on the basis of observation and theory, is sufficient for an attitude to be dissociated in this way. Attitudes that are part of such a dissociated point of view may provide us with reasons, even operative ones, for doing things. But insofar as we know them only observationally, they are not part of our critical reasoning.

Where we know our thoughts or attitudes only by observation, the question of *means* of control—of effective application of reasons to them—arises, at least from the perspective of our observational knowledge. Where we know our attitudes only as empirical objects, not only are our rational evaluations of those attitudes relativized to contingencies associated with the knowledge. But our ability to apply our reasons (those associated with the point of view of the observational knowledge) must acknowledge the contingency of our rational control over those attitudes. We must face a question of how, by what means, to make those reasons effective in view of the contingent relation between the point of view of the self-knowledge and rational evaluation, on one hand, and the observationally known attitudes, on the other. Again, this is not critical reasoning. In critical reasoning, such questions of means and control do not arise, since one's relation to the known attitudes is rationally immediate: they are part of the perspective of the review itself.

Sometimes observation-based self-knowledge enables one to assimilate an attitude into one's critical point of view, and to take direct critical control and responsibility for the attitude. I may learn through observing my behavior or through reasoning in therapy that I believe that a friend is untrustworthy. I may 'internalize' this belief so that it is no longer merely an object of observational knowledge. This process is sometimes immediate, sometimes hard, requiring deep personal change.

Although much reasoning and rational attitude formation occurs outside the purview of critical self-knowledge, or indeed any self-knowledge, critical reasoning remains central to our identities as persons. So no reasonable account of self-knowledge can ignore the role and entitlements critical reasoning gives to self-knowledge.

The argument I have given against the simple observational model indicates that the relations between knowledge and subject matter on which one's entitlement rests cannot always be causally brute, contingent, non-rational ones. In some particular instances of self-knowledge, the connection must be a rational one. For conclusions about the reasonability of one's thoughts based on self-review *directly* yield reasons within the point of view of those reviewed thoughts to alter or confirm them. The relation between self-knowledge and subject matter is that they must normally and rationally *be part of the same theoretical and practical point of view*—elements of a single theory or plan.

Connections between reviews and thoughts under review that are fully open to reason and that allow immediate transmission of reasons are necessary to the rational coherence of a point of view. A merely observation-based relation between attitudes insures that they are parts of different points of view. Indeed, it is constitutive of a point of view that failure to follow or understand its connections by the holder of the point of view is a failure of rationality. Thus a point of view is not closed under deductive consequence, and contains rational connections other than deductive ones.

So entitlement to knowledge of one's own thoughts and attitudes is not purely a matter of what one does. It has to do with who one is. One's status as a person and critical reasoner entails epistemic entitlement to some judgments about one's propositional attitudes. It entails some non-observational knowledge of them.

Cogito-like thoughts illustrate one non-contingent rational relation between knowledge and subject matter. In those cases, the reviewed thought is simply a logical part of the review. But *cogito*-like thoughts are in many ways special cases. If we are to understand critical reasoning, the entitlement that I have discussed must apply more broadly. It must include judgments about beliefs, intentions, wants, as well as occurrent thoughts.

So far, I have put little weight on the first-person present tense form of the relevant pieces of self-knowledge. Clearly, for the review and the reviewed thoughts to be part of a practice of critical reasoning, the reviewed thoughts must be capable of becoming part of the reasoner's present array of attitudes. And the special features of *cogito* cases do depend on present tense. But much of what I have said about the dependence of an entitlement on its role in critical reasoning, and about the non-observational character of this entitlement, applies to preservative memory—that type of memory that preserves propositions and our commitments to them in reasoning.[15] I think that this sort of memory provides us with some non-observational knowledge of our past mental states and events, and is epistemically underwritten by its role in critical reasoning.

The first-person point of view is clearly basic to self-knowledge in critical reasoning. The self-knowledge that I featured differs from observational knowledge of physical objects in that the first-person point of view is deeply relevant to the epistemic status of the knowledge. In observations of physical objects, anyone could have made substantially the same observation with equal right, if the same angle of perception had been available at the same time. But self-ascriptions constitute an epistemic angle in themselves.

What does this metaphor come to? If the reviewing knowledge is to be integral to critical reasoning, if it is to provide immediate rational ground for change in the reviewed material, the review must take up the same perspective or point of view as the act under review—the reasoner's own object-level point of view. The first-person point of view bears a distinctive relation to the relevance of rational norms to rational activity. For a review of a propositional mental event or state to yield an immediate rational ground to defend or alter the attitude, the point of view of the review and that of the attitude reviewed must be the same and must be first-personal.

In evaluating reasoning critically, one must make commitments to attitudes partly on the basis of critical evaluations of them. If one is to fully articulate the rational basis for the application of rational norms within critical reasoning, the commitments to both reviews and reviewed attitudes must be conceptually acknowledged as one's own. For acknowledging them as one's own *is* taking them as attitudes that one could rationally and directly change or confirm. Acknowledging first-order attitudes as one's own is necessary to articulating the direct rational relevance of one's critical reasons to first-order reasoning (or more generally, reasonable activity). I intend to say more about this matter on another occasion.

I have sketched the environmental neutrality of our entitlement to self-knowledge. The entitlement remains constant under possible unnoticeable variations in

[15] Compare 'Content Preservation'.

environmental circumstances or cognitive content. For it does not depend on the empirical content of the judgments. It does not depend on checking whether our judgments meet certain conditions. It depends on the judgments' being instances of a kind essential to critical reasoning. Critical reasoning presupposes that people are entitled to such judgments. Since we are critical reasoners, we are so entitled.

Epistemic entitlement derives from jurisdiction—from the place of the judgments in reasoning. In *cogito*-like thoughts, this place is coded in the content of the judgments themselves. In other relevant sorts of self-knowledge, which are fallible, the entitlement, indeed one's knowledge, depends only on one's not misusing the judgments and on one's remaining a sane critical reasoner.[16]

[16] Versions of this essay were given as the third of six Locke Lectures at Oxford in 1993, and the first of two Whitehead Lectures at Harvard in 1994, as well as on several other occasions. The key idea and first draft of the essay dates from 1985. I am grateful for helpful comments on drafts or talks based on this essay to Robert Adams, Kent Bach, Tony Brueckner, Phil Clark, David Kaplan, Christopher Peacocke, Marleen Rozemond, Hilary Putnam, Nathan Salmon, Houston Smit, Barry Stroud, Patrick Suppes, and Corliss Swain.

4 *Memory and Self-Knowledge*

In 'Individualism and Self-Knowledge' I argued that immediate, authoritative, non-empirically warranted self-knowledge is compatible with anti-individualism about the individuation of propositional attitudes. Paul Boghossian uses my slow-switching cases to argue that such self-knowledge and anti-individualism are incompatible. I will try to show why this argument does not succeed.[1]

I postulated as possible a case like this: An individual grows up in an environment like ours with a normal set of experiences of a particular sort of object or stuff, say aluminum. The individual does not know anything about the microstructural features of metals. But he has seen aluminum and made use of it. Perhaps he has heard things about it from others. The individual grows to maturity with this learning history. The individual has a normal lay concept of aluminum. It applies to aluminum and nothing else. I presumed that the individual believes that look-alike metals can be different metals. So his conception of aluminum allows that something could be superficially indistinguishable from aluminum and not be aluminum.

I then imagined that the individual is switched unawares to another planet (either forever, or gradually back and forth staying at each planet a substantial amount of time before switching). Given what he knows and can discern perceptually, he cannot distinguish the second environment from his home environment. The second environment contains twaluminum in all the places aluminum occupies in the original environment. Aluminum is lacking altogether. Twaluminum is a different metal that is indistinguishable for the individual from aluminum. I claimed that the individual's original concept does not apply to twaluminum. But I claimed that if the individual had enough experiences with the new metal comparable to those of the old, or if the individual interacted sufficiently with a community that had a term that applied to twaluminum, the individual would normally eventually acquire a concept more appropriate to the new environment than the concept of aluminum.

In 'Individualism and Self-Knowledge' I assumed that the newly acquired concept would be a "twin" concept. So if one had an aluminum concept in the first environment, one would acquire a twaluminum concept in the second. The

[1] Tyler Burge, 'Individualism and Self-Knowledge', *The Journal of Philosophy* 85 (1988), 649–663; Paul Boghossian, 'Content and Self-Knowledge', *Philosophical Topics* 17 (1989), 5–26.

twaluminum concept does not apply to aluminum. In these cases, the individual's word-form "aluminum" expresses on different occasions two concepts whose extensions are disjoint. Let use call such cases "*Disjoint Type* cases".

I will continue to suppose that Disjoint Type cases are possible. But they are not the only sorts of cases that can arise. The new set of concepts may be broadenings of the old. Thus the concept in the second environment may apply both to aluminum and twaluminum. One pressure in this direction is that these metals are indistinguishable to the individual, and instances of both (let us assume) enter into normal, paradigmatic perceptual applications of the individual's word-form "aluminum". If the individual's concept broadened in this way, the individual would acquire a non-natural-kind concept rather like our actual concept of jade.[2] Call these "*Amalgam Type* cases".

Whether in slow switching we have an Amalgam Type case depends partly, probably mainly, on how the individual is committed to the standards of the communities in the two environments. The relevant individual lacks knowledge of the metals' substructures. In the absence of commitments to communal standards and to communal understanding, and given extensive, prolonged new experiences with twaluminum, the relevant individual will lack the cognitive resources for the referent of his word-form "aluminum" to be fixed as a single natural kind. In the absence of other constraints, only normal experience with metals that are in fact instances of a single natural kind would so fix the referent. In the face of normal, constant perceptual application of the term to different natural kinds (where there is no countervailing pressure having, for example, to

[2] This point does not in the least constitute an admission that in the first environment, apart from actual switching, the individual must express this broadened concept in using the word form "aluminum". The concept is fixed partly by actual causal interactions. I take it that the individual's belief that metals may differ while appearing the same helps prevent his concept from applying from the beginning to just anything that looks to him like aluminum. And in the absence of constraining factors, there is no basis for its applying just to aluminum and twaluminum—as opposed to aluminum and all possible look-alike metals in addition to twaluminum. Before the switches, twaluminum played no more role in the individual's concept acquisition than any other possible metal that might look like aluminum.

These matters are complex, however. Some have held that the individual's future constrains his past, so that if he later comes into contact with twaluminum in such a way as to yield a broadened concept, he always has the broader concept—one that applies to both aluminum and twaluminum. There are interesting questions here about how to determine the limits of normal circumstances. But I do not accept this view as a general analysis. I think there is a constraint in the account of concept determination that for most normal human beings requires a concept to be one of a natural kind if one's experience in normal circumstances is overwhelmingly that of instances of a natural kind. And there is a constraint that allows concepts to change extension with sufficient changes in what is normal within the individual's experience. Allowing the whole future to count in determining an individual's concept would not plausibly accommodate conceptual change through change of normal circumstances. It would also underestimate the role of causation in determining an individual's present cognitive abilities, and would pointlessly complicate accounts of communal sharing of concepts. But I need not go into these matters here. For if the amalgam concept is always the individual's concept, there would be no conceptual change. And this would ease rather than threaten the compatibilist position that I am outlining.

do with differential values or purposes to which the metals might be put), the extension will commonly broaden.[3]

The individual's specific intentions may also play a role. The individual might resist such broadening by intentionally and specifically limiting the original concept to the original paradigms in the first environment. Such resistance might yield a Disjoint Type case or even no change at all. But this sort of resistance is certainly not the norm. In the absence of regular application to a single kind, or special intentions by the individual, it appears that something like relations of communal dependence would be needed to prevent the expression "aluminum" from coming, through contact with the second environment, to express a concept that would apply equally to aluminum and twaluminum. How the individual relates to the two communities bears on whether we have a Disjoint Type or an Amalgam Type case. The relevant parameters are complex and contextual, and I will not try to discuss them here.

In any event, I will suppose that after sufficient switching, it is possible for the individual to have a single concept expressed by the word-form "aluminum" that applies to aluminum and twaluminum. These are Amalgam Type cases.

In both types of case the individual has undergone a conceptual change that is unknown to the individual. In Disjoint Type cases, the individual has taken on two concepts for disjointly different sorts of things without knowing it. In Amalgam Type cases, a concept that comes to be expressed by the individual's word-form includes the extension of the original concept before the switching, but is broader. And again the individual is unaware of the change. I leave open whether in Amalgam Type cases, the individual retains the original concept in his repertoire. In both Disjoint Type and Amalgam Type cases the individual's lack of awareness of the change derives from his inability to distinguish aluminum from twaluminum. The individual cannot explain or articulate a distinction between the concepts in the first and second environments. He is unaware that there are two concepts. So there may be times when the individual is unable to determine whether he was thinking, at some earlier time, about aluminum alone or twaluminum or some amalgam.

Nevertheless, I believe that the individual will commonly have immediate, non-empirically warranted self-knowledge of the form *I think (believe, judge) that p*, where *that p* includes a relevant concept (*aluminum, twaluminum*, or the amalgam concept). Before the switches, the individual might know a *cogito*-like thought in thinking it: *I am hereby thinking that aluminum is a light metal.* And the individual can have self-knowledge of this sort—though the content will be

[3] Regular, frequent switching at relatively short intervals—as distinguished from a single switch, or very infrequent switches with long intervals at each place—seem to make Amalgam Type cases more intuitive for some. In effect, such switching might make the conjunction of the two planets seem more plausibly a single "normal" environment. But I think the main issue has to do not with speed or frequency but with whether additional factors, such as communal factors, serve to distinguish the environments in cognitively relevant ways, once the new planet becomes a factor. All of these matters deserve deeper reflection. Needless to say, there will be many don't cares and borderline cases.

different—after the switches as well. In a Disjoint Type case, for example, the individual can think and authoritatively know after a switch: *I am hereby thinking that twaluminum is a light metal.*

There are many complicated issues associated with this view. I will not go into many of these. And I will not repeat the considerations that support the view, except for one reminder. It is not disputed that the individual can on given occasions think *cogito-* or other self-attributional thoughts with a definite concept, say, the aluminum concept or the twaluminum concept. In relevant self-attributions, the individual simultaneously uses and self-attributes concepts (in the reflexive, that-clause way). In these cases the individual cannot get *the content* wrong.[4] For the attributed intentional content is fixed by what the individual thinks. It is not something that he identifies independently.

Much of the literature on this subject deals with problems that arise from the assumption that we need to *identify* the content of our thoughts in such a way as to be able to rule out relevant alternatives to what the content might be. Boghossian, unlike many of those who write on this subject, seems to recognize that this assumption is not acceptable on my view. One's relation to one's content, when one is non-empirically self-attributing in the reflexive, that-clause way is not analogous to a perceptual, identificational relation to which alternatives would be relevant. In present tense self-attributions of the relevant kind, alternatives are irrelevant. Boghossian's strategy is to consider cases of memory and argue that these cases reflect badly on my views about the present tense cases.

Boghossian writes:

> [Burge's claims] amount to saying that, although [the subject] S will not know tomorrow what he is thinking right now, he does know right now what he is thinking right now. For any given moment in the present, say t1, S is in a position to think a self-verifying judgment about what he is thinking at t1. By Burge's criteria, therefore, he counts as having direct and authoritative knowledge at t1 of what he is thinking at that time. But it is quite clear that tomorrow he won't know what he thought at t1. No self-verifying judgment concerning his thought at t1 will be available to him then. Nor, it is perfectly clear, can he know by any other non-inferential means. To know what he thought at t1 he must discover what environment he was in at that time and how long he had been there. But there is a

[4] Compare Tyler Burge, 'Our Entitlement to Self-Knowledge', *Proceedings of the Aristotelian Society* 96 (1996), 91–116. Boghossian rightly points out what I myself had indicated—that self-verifying judgments are just a small sub-class of the self-knowledge to which we have special authoritative, non-empirical entitlements. I think, however, that they provide a paradigm that is suggestive of many of the key features of the larger class. I have discussed aspects of these matters in the above cited article. All I do here is to deal with his objections to my taking anti-individualism about content to be compatible with seeing self-verifying judgments to be cases of non-observational self-knowledge. But the main line of his argument would apply to all cases of ordinary non-empirical self-knowledge, and my reply does also. So neither my examples of *cogito* thoughts nor Boghossian's remarks about self-verification are central to the main issues about memory that I will be discussing. My points above about the individual's inability to get the content wrong in non-empirical present-tense self-attributions (in the that-clause way) apply not only to self-verifying *cogito* thoughts, but to all non-empirical that-clause type self-attributions. I will argue that the same point carries over to certain types of memory of past self-attributions.

mystery here. For the following would appear to be a platitude about memory and knowledge: if S knows that p at t1, and if at (some later time) t2, S remembers everything S knew at t1, then S knows that p at t2. Now, let us ask: *why* does S not know today whether yesterday's thought was a *water* thought or *twater* thought? The platitude insists that there are only two possible explanations: either S has forgotten or he *never* knew. But surely memory failure is not to the point. In discussing the epistemology of relationally individuated content, we ought to be able to exclude memory failure by stipulation. It is not as if thoughts with widely individuated contents might be easily known but difficult to remember. The only explanation, I venture to suggest, for why S will not know tomorrow what he is said to know today is not that he has forgotten but that he never knew. Burge's self-verifying judgments do not constitute genuine knowledge.[5]

Let the thought *that p* be what individual S believes before the environmental switches occur. For example, S may believe *I am thinking that aluminum is a light metal*. Boghossian's argument is as follows:

(1) If S does not forget anything, then whatever S knows at time t1, S knows at time t2.
(2) In the cases at hand S does not forget anything.
(3) S does not know that p at time t2.
(4) So S does not know that p at time t1.

Let us consider Disjoint Type cases first. So assume that at t2, the individual has a new set of concepts disjoint from the ones that he had before the switching began. He need not have lost the old set, however. I do not concede that the individual 'will not know tomorrow what he is thinking right now' (i.e. at t1), at least in the sense of "knowing what" that is relevant to my view. Moving to the other environment and acquiring new concepts will not normally obliterate old concepts or memories that derive from the first environment. If one always lost all past concepts by acquiring new ones after a switch, one would never be able to remember or report accurately what one had said or thought. The old abilities will normally still be there; and there are situations, such as invocation of memory, or reasoning based on memory, or return to the first environment with acts of deference to its communal norms, that can bring these abilities into play.

Boghossian defends (3) by saying 'it is quite clear' that it is true. But it is not clear. In fact, if S has forgotten nothing, I see no reason to think that S will not know (in the relevant sense) at time t2 what he knew at t1. (As I will soon indicate, I think that the phrase "know what he thought" covers two different sorts of "knowing what".) S can at t2 remember his thinking at t1, and his memory can link the content of the earlier thought to that of the memory-induced one, by fixing the memory induced content as that of the remembered one. Merely being

[5] Boghossian, 'Content and Self-Knowledge', 22–23.

in the second environment, with concepts appropriate to that environment, does not prevent him from retaining and thinking thoughts appropriate to the first. Nor does it automatically prevent his retaining knowledge that he had before.

I will concentrate on cases where knowledge is activated through memory. I have maintained that the individual may not know whether yesterday he had an aluminum or twaluminum thought. He does not have discriminative knowledge of this form. But memory need not work by discrimination; it can work through preservation. The memory need not set out to identify or pick out an aluminum rather than a twaluminum thought, trying to find one by working through the obstacles set by the switches. Preservative memory normally retains the content and attitude commitments of earlier thinkings, through causal connections to the past thinkings. That is one of its functions – maintaining and preserving a point of view over time. It need not take a past thought as an object of investigation, in need of discrimination from other thoughts. Memory need not use the form "Yesterday I was thinking a _____ type of thought", where the memory attempts to *identify* the thought content as an object. Again, if it did, the individual might perhaps err by using a thought appropriate to the second environment in making an attribution to a thought event in the first environment.

The memory need not be *about* a past event or content at all. It can simply link the past thought to the present, by preserving it. Such cases involve a particular type and function of memory—preservative memory—which preserves propositional contents and attitudes toward them, rather than *referring* to objects, attitudes, contents, images, or events. The memory content is fixed by the content of the thinking that it recalls. Similarly, the "referent" of the past tense in the memory is fixed not by an independent identification of the past event, but through the memory connection to the event itself. The individual reasons from the past thought, takes it up again, without the memory's taking it or anything else as an object (as, by contrast, the memory does in substantive memory).[6]

There is a broad but qualified analogy between preservative memory and certain aspects of pronomial back-reference. The analogy must be used with caution. I do not model preservative memory on pronomial back-reference. I believe that preservative memory is more basic (both ontogenetically and in explanations of epistemology and rationality) than anaphora in language. In fact, it seems to me that a linguistic theory of anaphora has to be able to account for anaphora supported by preservative memory. Still, the analogy may be helpful in the respects in which it holds.

[6] For discussion of the distinction between the different types of memory, see Tyler Burge, 'Content Preservation', *The Philosophical Review* 102 (1993), 457–488. A psychological analog of my distinction can be found in E. Tulving, 'Episodic and Semantic Memory', in E. Tulving and W. Donaldson (eds.), *Organization of Memory* (New York: Academic Press, 1972), 382–402; and *Elements of Episodic Memory* (Oxford: Oxford University Press, 1983).

In using pronouns, the speaker need not be able to identify the referent of a pronoun, or even its antecedent, in order to secure the antecedent. The speaker might get the antecedent or its referent wrong if he were asked to identify it independently of the pronoun. For to secure an antecedent for the pronoun, it is enough for him to rely on chains inherent in the discourse. The causal chains in preservative memory do a similar job in connecting later thoughts, including later self-attributions, to earlier ones. The same faculty is fundamental to ordinary reasoning, which preserves previous steps of reasoning to make the coherence of reasoning possible. Given appropriate reliance upon preservative memory, and given the existence of causal memory chains back to the states which carried intentional content, preservative memory takes up the "antecedent" content automatically, without having to identify it. As with anaphora, the thinker need not be able to identify the antecedent, much less its referent. He may rely on the mechanisms of memory to do the job.

In the case of anaphora the interpretation or referent of the pronoun is not always the same as that of the antecedent. Anaphora is a syntactic device, whose semantic interpretations may vary, depending on the type of anaphora and the linguistic context of the pronoun. To this degree, anaphora and preservative memory differ. Preservative memory is not primarily a syntactic matter. It is a preservation of content, fundamental to the coherence of rational activity. In the memory case, the content and referent of the remembered material is not distinct from that of the antecedent thought content, which in ordinary that-clause-type self-attributions is both thought and referred to. The point of preservative memory is to fix the content in present mental acts or states as the same as the content of those past ones that are connected by causal-memory chains to the present ones.[7] If the individual relies primarily upon preservative memory, and if the causal-memory chains are intact, the individual's self-attribution is a reactivation of the content of the past one, held in place by a causal memory chain linking present to past attributions.

Memory could preserve the content of a past thought in either of two ways. The individual could remember the past thinking as an event; and only the content of the thinking could be remembered in the preservative way. His

[7] With important qualifications, there is an analogy to pronouns of laziness, pronouns which can be expanded into rewrites of the antecedent. Just as pronouns of laziness are in effect exact reproductions of their syntactic antecedents, preservative memory produces an exact reproduction of its content "antecedent". The difference between syntax and semantics is, however, crucial. Perhaps in thinking about this analogy, it would help to imagine that the antecedent of the pronoun of laziness is a term that is both used and self-referentially mentioned. So the prounoun as rewritten must do the same. *Ordinary* pronouns of laziness can, of course, carry different referents from their antecedents, because of contextual or scope elements of the discourse in which the pronoun is embedded. The function of preservative memory is to preserve content, which is by no means the function of pronouns of laziness. Pronouns are syntactic devices with a variety of semantic interpretations, whereas preservative memory is a feature of thought. In fact, preservative memory is typically a reactivation of earlier material not a pronomial shorthand for it. Purely preservative memory can be expressed in language, of course. But it is not a linguistic device.

memory would then tie current conceptual use to the concepts of that past thinking. Here substantive event memory and preservative memory would work together—the former identifying an event, the latter preserving the content and attitude-modality of the event. Or, second, memory could just preserve and bring up the attitudinal-commitment and the content of the earlier thought, making use of a file that tracks tenses and indexicals within it, without referring to the past thinking event. Here the memory is not referentially *of* the earlier event. It simply carries forward the content and force of the earlier thought for later use. In neither case need the individual *identify* the *content*. Memory functions to allow him to employ it again. It preserves the content regardless of what the individual thinks about it or knows about the world.

When activated, the remembered thought content will no longer be associated with a cognitive state that is indexed with present tense. And any new thought occurrences produced by the memory will not themselves be self-verifying. Thus instead of *I am hereby thinking that aluminum is a light metal*, one would be remembering what can be approximately expressed as *I was thereby thinking that aluminum is a light metal*. But—this is important—the "was" and "thereby" have the special preservative character involved in preservative memory. They relate to elements in the original thought preservatively rather than referentially. There is no independent reference, from the time-perspective of the present memory, to the time of the past thinking. Nor is there independent reference to the act originally expressed in "hereby". The preserved thought content will be preservatively linked to the tense and self-verification of the original thought event.[8]

Despite these differences between the memory and the remembered thought, I have no objections to thinking of the thought content of the memory as being the same content as that of the original self-attribution. At any rate, the memory can remember the self-verification and reflexivity of the original thought. The linkages are made by the cognitive system to the original applications of concepts. They do not depend for their operation—or, I will argue, for their being epistemically warranted—on any new cognitive relations to the environment. The causal-preservative linkages simply allow for redeployment of old concepts.[9]

[8] There is some analogy between these anaphorical uses of tense and Castenada's "I*". Compare H. N. Castaneda, 'Indicators and Quasi-Indicators', *American Philosophical Quarterly* 4 (1967), 85–100, and Castaneda, 'On the Logic of Attributions of Self-Knowledge to Others', *The Journal of Philosophy* 65 (1968), 439–456. Thus although the preserved thought content no longer picks out the present in its tense, it keeps track of the present-tense character as well as the time of the original thought. In this respect "was" is misleading. Similarly for "thereby" and "hereby". The self-verification of the original thought is preserved, even though the remembering may not itself be self-verifying.

[9] Boghossian appears to place weight on the supposed later unavailability of the self-verification of a self-verifying thought. As noted (note 8), I think the past self-verifying thought is available through preservative memory, even though the activated memory is not itself self-verifying. But self-verification is not a necessary feature of our non-empirical entitlement to self-knowledge. I used *cogito* cases primarily for expositional purposes in the original article; I think that they carry clues to understanding the wider array of non-empirically known self-attributions. In any case, the respects in which self-verification is not re-enacted in preservative memory are not respects in which the

Boghossian writes:

> Now, let us ask: *why* does S not know today whether yesterday's thought was an [*aluminum*] thought or a [*twaluminum*] thought? The platitude insists that there are only two possible explanations: either S has forgotten or he never knew . . .

Boghossian is asking the wrong question here. So doing yields a misleading application of the "platitude". In preserving knowledge, S (or S's memory) need not be in the third-person position of solving the problem of whether yesterday's knowledge had one content rather than another. That would be to take the past thought as an object of identification. The knowledge to be preserved did not have the form that Boghossian's question implies, something like: *I am thinking that aluminum is a light metal; and the thought just thought is an aluminum thought as distinct from a twaluminum thought.* S did not think a thought that looked on a content from the outside and opened itself to questions of comparative discrimination.

The right question is whether knowledge first expressed by *I am thinking that aluminum is a light metal* can be preserved. I have always maintained that thinking back to yesterday, S might be unable to discriminate the aluminum thought from a twaluminum thought. My view has been that S need not make such discriminations, except insofar as they are made by what he actually thinks—and by what is preserved in memory from those thoughts. If the individual tries to *identify* in memory a past thought as an object of investigation, of course, he may misattribute thoughts appropriate to the second environment to uses in the first.

The form of the question is important. The question to be asked on behalf of the individual's memory is not whether the original thought contained a concept of aluminum (or twaluminum). Such a question takes past thoughts as objects of investigation. The question is whether the original thought or knowledge can be preserved. To connect with my position, Boghossian should have asked why S does not know today that he thought[10] that aluminum is a light metal. I think

environment bears on the individual's concepts. Thus the issue of self-verification is irrelevant to the nature of the individual's concept *aluminum*. (Compare note 4.) That concept is redeployed in preservative memory.

In switching cases, especially if there are multiple switches back and forth, the individual's memory may have difficulty separating the causal-anaphoric files from one another. Many past individual thinkings may be lost to memory for all practical purposes. But insofar as one can remember an individual thinking event through preservative memory, one is in a position to retain the knowledge. Moreover, the beliefs (and other standing attitudes) associated with the thinkings are there to be accessed, and will cause fewer practical problems than remembering individual thinkings in the overgrown past. Here again I am assuming that the individual has not lost the concepts appropriate to the original environment. If the individual has lost those concepts, or if there are problems accessing past thoughts, we cannot assume that the individual has forgotten nothing.

[10] The past tense here refers back, anaphorically, to the time of yesterday's thought. But "yesterday" is not specified as such in the content of the preservative memory. If it were, the individual would go beyond preservation to identification, and would be vulnerable to an error of identification.

that S *does* know this insofar as the knowledge derives from preservative memory rather than from a third-person perspective on his past thoughts. If he tried to access the knowledge in the ways Boghossian's question suggests, he might fail to know by failing to identify the right thought. He might lack any special entitlement to his conclusions. When the individual thinks *I thought that aluminum is a light metal*, where a relevant thought is preserved from the original environment, preservative memory, working properly, automatically links the present thought with the remembered thought, with its aluminum concept, in the original environment.[11] It does so by relying on causal-preservative relations to the past thought event.

The differences in the form of the question about the individual's knowledge of his past correspond to two ways of understanding the question of whether the individual "knows what he was thinking yesterday". If one has identification in mind, perhaps it is natural to infer that in the slow switching cases, the individual does not know what he was thinking yesterday, since he cannot discriminate between two seemingly relevant possibilities. One is inclined to think that at least in many relevant switching cases, the existence of alternative contents accessible to the individual will make the alternative contents relevant to whether the individual knows, in the sense of "knows how to pick out", what—that is, which—thought he was thinking.

But identification is not at issue in preservation of self-knowledge, as distinct from third-person identificatory thinking back on it. If the "what" indicates preservation, then where the individual does not forget past thoughts, he will continue to know what he thought. He will continue to know the same content that he knowledgeably thought before.

Is the individual epistemically entitled to the products of preservative memory? One is, I think, entitled to rely upon such products, as long as it is in fact working properly, except perhaps in certain cases where one has reason to think one's memory is slipping and not maintaining the causal memory chains. To be entitled to rely on such memory, one need not supplement it with discriminatory identifications. One need not be able to defend it against potentially confusing challenges that would require one to distinguish the belief one actually calls up from beliefs that are similar to it. In fact, I think that as long as the causal memory links are in place, preservative memory is authoritative in something like the way much immediate present-tense self-knowledge is.[12]

[11] It is easy to confuse the situation being discussed with situations in which the individual knows about the switches but does not know when they occurred. This is easy because we who are thinking about the case know about the switches. Where the individual knows about the switches, there are special opportunities for confusion for that individual. He may confusedly despair of relying on memory, for example. But I think that where he does rely upon preservative memory, he will remain epistemically warranted. Since the argument I am considering does not rest on examples in which the individual knows about the switches, I have not gone into them.

[12] I think that since we are not dealing with identification, issues about relevant alternatives do not arise in anything like same way that they arise with perception or other forms of identification. As long as preservative memory is working properly, there is no possibility of error. There are no issues about "look alikes", since the memory is not "looking". Several issues here need detailed discussion, but since the argument we are considering does not raise them, I will not go into them on this occasion.

Perhaps Boghossian thinks that one is not entitled to rely on memory because of one's inability to distinguish original thoughts from the new twin thoughts, or inability to distinguish a thought actually remembered from a twin thought. If so, he needs to argue for this view. For assuming it in effect begs the question. I began with the claim that one need not be able to distinguish aluminum thoughts from twaluminum thoughts to have knowledge in certain self-ascriptions, such as the *cogito*-like judgment of *I am hereby thinking that aluminum is a light metal*. This same view applies to memory, at least purely preservative memory—which works by simply preserving thoughts already thought and making them available for reactivation and reasoning.

Preservative memory is necessary to any reasoning that takes place over time, hence any reasoning. We are as fundamentally entitled to rely upon it as we are entitled to rely upon reasoning. In fact, if we were not entitled to rely upon preservative memory, we would not be entitled to rely upon reasoning.

In the case of preservative memory, as in the case of direct self-knowledge, entitlement depends only on ordinary understanding and on the normal working of one's cognitive faculties. In neither case is discursive defense needed to safeguard the entitlement.

Consider the role of preservative memory in deductive reasoning. Such memory is needed just to carry the argument along over time. Such memory does not constitute or enhance the justificational force of the individual's justification for believing a deduced theorem. But the individual must be entitled to rely on preservative memory to be entitled to rely upon the deductive reasoning. One is entitled to rely on one's memory in such reasoning if it is working properly—if the thoughts are preserved in the course of the reasoning. Even if there are alternative thoughts that one cannot distinguish from those one is in fact thinking, one is entitled to the reasoning one is actually carrying out, as long as the reasoning is understood and deductively sound. And one is entitled to rely on the preservative memory on which the reasoning depends—as long as it preserves the thoughts thought earlier in the argument. It does so not by identifying the earlier thoughts or discriminating them from similar thoughts, but simply by preserving them for later employment.

Similarly, the individual may not be very good at distinguishing the content of past events, taken as objects for identification. But purely preservative memory of the past contents remains; and that is sufficient for knowledge through memory. Where self-attributions rely upon preservative memory that is working properly, the content will be knowledgeably preserved. Entitlement to memory-dependent thought or reasoning depends on the proper working of preservative memory, not on the individual's ability to specify what the memory is doing or retrieving, or on the individual's checking to verify that the memory is in good working order.

Considerations about preservative memory help undermine another of Boghossian's criticisms of anti-individualism. This criticism concerns reasoning, not self-knowledge. Boghossian considers some switching cases in which a person has two concepts expressed by the same word-form.[13] He claims that externalism (which we shall not distinguish from anti-individualism) opens an unattractive possibility of undetectable equivocations in reasoning. He draws the stronger conclusion that "externalism undermines our ability to tell apriori whether any particular inference of ours satisfies one of [the forms of valid inference]".[14]

Boghossian seems to assume that apriori warrants must be "internally detectable". He seems to assume further that it follows from a certain argument's being vulnerable to possible undetectable equivocation that we can never tell apriori, even in normal cases, that they are justified. I think that both assumptions are

[13] Paul Boghossian, 'Externalism and Inference', *Philosophical Issues* 2 (1992), 11–28; Boghossian, 'Reply to Schiffer', *Philosophical Issues* 2 (1992), 39–42. Boghossian claims ('Reply', p. 39) that it has been widely overlooked that according to anti-individualism (or "externalism") a thinker might have Earthly and Twin Earthly contents cohabit in his system without there being any internal indication to him that this is so. He even claims that "no one seems to have noticed" the point ('Externalism and Inference', p. 17). In fact this was the central construal of the switching case I explain in 'Individualism and Self-Knowledge (p. 652; 58 in this volume). Much of the interest and challenge of the issues about self-knowledge derive from the assumption of cohabitation. In setting up the thought experiment, I wrote about acquiring new concepts in the new situation, and said nothing about losing old ones. Displacement was never part of the switching cases, at least in my understanding of them. Cohabitation was always the assumed case. I did not and do not consider the displacement model (as a general model for switching cases) a plausible account. I did not discuss and criticize the displacement model in 'Individualism and Self Knowledge', but largely because I thought it clearly implausible. I did consider it, and I included elements in my account that were meant to suggest its implausibility. Displacement raises obvious problems about memory in many standard cases. And it has no basis, as far as I know, in our ordinary understanding of how concept acquisition works in ordinary "moving" cases, especially among relatively mature language users with good memories and recurrent uses for the old concepts. Merely being switched stealthily from one place to another and gaining new concepts will not in general cause one to lose conceptual abilities. Part of the idea behind my postulation of switching was to indicate that the individual would retain a prospective use for direct applications of the old concepts (in addition to being able, I think, to remember objects, events, and thoughts from the departed environment). Concepts mark abilities; *just* moving around and acquiring new concepts will not in general obliterate such abilities, especially given that one still has uses for the old concepts and a perfectly good memory.

I wrote 'Now suppose that, after decades of such switches, one is told about them and asked to identify when the switches take place. The idea is that one could not, by making comparisons, pick out the twin periods from the "home" periods.' ('Individualism and Self-Knowledge', p. 653; 58 in this volume). One person has thought that only if the concepts are displacing each other will the question *what concepts a person has* track when the switches took place. My supposition was not as fully characterized as it should have been. But this interpretation misses my reasoning. What I think tracks when the switches take place is not what concepts the individual has at given times, but what uses the person makes of the concepts. I assumed that one would mostly use earth concepts on earth and mostly use twin earth concepts on twin earth—once one had acquired each set. So if one knew when the various thinkings took place and if one had some internal sign (which I think we in fact lack) that would enable one to *compare* and distinguish the earth and twin-earth thinkings (concept-uses) in retrospect, one could tell (more or less) when the switches took place. Although there may be some uses on twin-earth of earth concepts, they would not normally be dominant. If one could distinguish in memory ("by comparison") between uses on twin-earth that were governed by memory of earthean uses, on one hand, and twin-earth uses that were more tied to the immediate circumstances (and similarly for uses on earth), one could sort out the differences of location *through* memories of those uses even more exactly. Of course, I think that we cannot do these things. The relevant memory works through preservation, not comparison.

[14] Boghossian, 'Externalism and Inference', 22n.

mistaken.[15] But my discussion here challenges even the view that according to anti-individualism we must be subject to possible undetectable equivocations producing invalid reasoning, because of the possibility of switching cases.

An example of Boghossian's argument adapted to our case goes as follows. Assume anti-individualism to be true. Then Alice might remember an event of picking up, and feeling the light weight of, some aluminum on earth, before she was switched; and in remembering the event, she might think correctly that she picked up some aluminum at that time. Then remembering a sample of twaluminum on twin earth that she saw yesterday, she might think that yesterday there was some twaluminum beside her. (Both of these are uses of substantive, object-oriented memory.) She might reason from these premises, fallaciously, to the conclusion that she once picked up the same sort of thing that was beside her yesterday. The word form "aluminum" undetectably expresses for Alice two different concepts. The concepts used in the reasoning are supposed to be different because whereas the earth concept is evoked by the memory of the long past event of picking up aluminum, the twin earth concept is supposed to be evoked by the memory of what is in fact twaluminum on twin earth. The inference appears to Alice to be valid; but because of the switch in concepts, it is invalid.

The weakness of examples of this sort is that they overlook the centrality of preservative memory in reasoning, and fail to note the particular character of the relation between the different concepts within the individual's cognitive system. As Boghossian insists, there is little ground to think that Alice made a mistake in reasoning. But contrary to his claims, there is no necessity on anti-individualistic grounds to attribute one to her.[16]

Alice's argument is carried out in thought. I think it natural to agree with Boghossian's account of the first premise. In determining what Alice is thinking in the second premise, one must remember that both the substantive memory of yesterday's experience, and the preservative memory of the use of the concept in the first premise, are operating in Alice's thinking. What it is to carry out valid arguments in thought is to connect premises, holding them together in a way that

[15] For a discussion of apriority that entails the falsity of both assumptions, see Burge, 'Content Preservation'.

[16] Although I do not agree with all of his remarks, Stephen Schiffer seems to me to make essentially the right points in his reply (Schiffer, 'Boghossian on Externalism and Inference', *Philosophical Issues* 2 (1992), 29–38). I simply bring out a generalization of Schiffer's points, and invoke the notion of preservative memory. Boghossian claims regarding the second premise of an analog of the example about Alice that it is "entirely independent" of the first premise. But in actual reasoning we typically tie key terms in premises together through preservative memory, as Schiffer in effect points out. Schiffer also shows that various rephrasings of arguments, using relative clauses instead of separate premises, would elicit the non-independence of relevant components of the two premises, given the reasoner's intentions. One might make the same sort of point with respect to other examples by appeal to pronouns. These tyings together are particularly strong in these cases, in view of the fact that the reasoner would, if the opportunity arose, identify the same objects using either concept. But even if an individual does not tie premises together in this way, Boghossian's argument fails, as I shall show.

supports the conclusion. And preservative memory—even in short arguments that we idealize as occurring in a specious present—is essential to this enterprise. Insofar as we think intuitively that Alice is not making a mistake in reasoning, it is natural, and in most cases I think correct, to take her to be holding constant, through preservative memory within the argument, the concept used in the first premise in her thinking the second premise. The role of the concept aluminum in the reasoning is primary in her thinking, and preservative memory takes the occurrence of the concept in the first premise as a basis for its reuse in the second premise.

Anti-individualism does not say that every thought's content is fixed by the type of object that occasions the thought. Although free-standing memories normally evoke the concepts utilized in or appropriate to the remembered context, the exigencies of reasoning will often take precedence. One commonly utilizes concepts used earlier in an argument to identify objects in memories invoked in later steps.

Given this usage by Alice, the second premise is false: She is mistaken in applying her concept aluminum, preserved from the first premise and originally evoked by her experience on earth, to her experience yesterday on twin earth. She is using the concept obtained from the first premise to *identify* the metal she remembers seeing, as expressed in the second premise. The mistake is a mistake of memory identification, not one of reasoning. Variations on this point apply to all of Boghossian's examples.[17]

Perhaps there are cases where the reasoner does not tie the parts of an argument together in this way. That is, the reasoner's intended reasoning does not close the question of whether the concepts expressed by the same word-sound in an argument that is otherwise syntactically valid are the same. Insofar as the reasoner's intentions in reasoning are not dominant in requiring "anaphorically" that the same concept be used through the reasoning, and insofar as we think that there is a gap between the premises that the reasoner has not made explicit, it would seem obvious that the reasoner tacitly and mistakenly presupposes that the concepts apply to the same objects. This presupposition is not present in cases of equivocation that occasion invalid reasoning. In such cases the individual overlooks the difference between concepts expressed by the same term, but has no tendency to treat the concepts as interchangeable in general beliefs, and no tendency to apply them to the same objects. So to fully capture the reasoner's

[17] As I have noted if someone who has undergone switches relies upon memory to *identify* a past object or event—including a past thought—he is subject to error. But supposing that Alice thought yesterday that twaluminum is beside her, she is in a position, relying on *preservative* memory, to remember what she thought then. And if this memory were to generate reasoning, then it would normally be held constant through an argument, and it might generate misidentifications of other past thoughts in later premises. Then although the reasoner knows what those past thoughts were, and could call them up if she relied purely on preservative memory, the reasoner might make mistakes about those past thoughts through reliance on identificatory substantive memory. These cases bring out again that accessing past knowledge in switching cases involves more parameters than in normal cases.

cognitive state in a case where the reasoner does presuppose (mistakenly) that the concepts apply to the same objects, one would have to supply for the reasoner the mistaken presupposition that twaluminum is aluminum. Again, there is no mistake in reasoning, only a mistake in presupposition.

I am doubtful that there are any clear cases of invalid equivocation deriving from switching cases. But if there are, they are marginal. [PS, 2011: I have given up this doubt.] And one can avoid any such cases by firmly and intentionally relying upon preservative memory in maintaining the same concepts throughout the course of one's arguments. In view of the fact that switching cases leave one inclined to apply different concepts to the same objects, this intention will accord with both *de re* and general memories, beliefs, and desires involving the different concepts. Applications of concepts governed by connection to other premises may then express misidentifications, or other mistaken beliefs, that the switched individual is already inclined to.

Let us return to self-knowledge. So far I have focused criticism on step (3) of Boghossian's argument: The assumption that in slow switching situations, assuming anti-individualism, we cannot know at the later time, without empirical investigation, what was known at the earlier time. All of my discussion of self-knowledge has simply granted step (2)—that nothing is forgotten. I have granted that the individual has, without realizing it, both the original concept and a new concept after slow-switching. And I have assumed that this situation does not (in general) result in the individual's actually forgetting anything. So the original beliefs are not forgotten. At worst, they can fail to be accessed in certain situations.

But some cases may run contrary to the assumption that forgetting is irrelevant. It may be that in certain Disjoint Type cases, in which an individual does not reactivate the concepts from the original environment (say, if he forgets specific events from the earlier environment), the individual eventually loses the original concept. He loses the ability to think specifically about aluminum. I see no reason to think that in such cases the individual's knowledge of his thoughts before he lost the concept is threatened. The argument Boghossian gives will simply be short-circuited at a different point, step (2).

Let us finally turn to Amalgam Type cases. It will be recalled that these are cases where, in addition to having the original concept, the individual acquires a jade-like concept that applies equally to aluminum and twaluminum. So assume that S has a broadened concept which includes the extension of the original one. If S has not lost the original concept, then the response I gave for Disjoint Type cases reapplies. I see no reason to think that S will ordinarily lose the original concept, even in Amalgam Type cases. Normally S will be able to access through memory the old conceptual ability to think about aluminum, about past-aluminum experiences, or about past aluminum thoughts. But if S does lose the original concept, then a different reply is appropriate. If losing a concept is a form of forgetting, then premise (2) of the argument is mistaken. Clearly if one loses a concept when it is replaced by a new one, and for that reason one has no access to beliefs one once had, one may lose knowledge one once had.[18] Whether (1), (2), or (3) is at fault, I see no reason here to doubt that one's authoritative, non-empirically warranted self-knowledge is compatible with the truth of anti-individualism.[19]

[18] There is reason to think that step (1) is false in any case. There appear to be cases in which one knows something at one time but loses the knowledge at a later time, not because one forgets anything, but because one's original warrant is (misleadingly but reasonably) outweighed by prima facie defeating alternatives that emerge at a later time. Suppose that one has true warranted belief amounting to knowledge at a given time. For example, suppose one sees a lectern and thereby knows that the lectern is in the room. Suppose at some later time one acquires good reason to believe that one's apparent experience of the lectern was or may well have been the product of an illusion-causing hologram. Then even though one had in fact seen the lectern, one could at the later time perhaps lose the knowledge that the lectern was in the room, despite the fact that one forgets nothing. Such cases might seem to be relevant to some of the switching cases at hand. One might think that after the switching, one's original warrant is undermined by relevant alternatives, even though one's original self-knowledge was in place. So one could be prevented from having knowledge of one's past (originally known) thoughts, even though one forgets nothing.

I will not discuss here whether there are any such cases in which step (1) fails. I am not inclined to press this point because I believe that in neither the present-tense self-knowledge case nor the case where preservative memory preserves self-knowledge, is the relevant-alternatives epistemology applicable. And whenever preservative memory is relied upon, the original self-knowledge can be retained regardless of what switching has occurred. I am inclined to think that in certain basic epistemic functions, knowledge is immune to threat from alternatives as long as the basic processes are working properly (and perhaps additionally, those processes' proper working is not called specifically and reasonably into question), and as long as those processes are not subverted by irrational interferences from within the cognitive system itself. Authoritative self-knowledge and preservative memory are among these functions. I have not defended this view in appropriate generality. But it has informed my picture of self-knowledge from the beginning, and has so far not been explicitly attacked head-on. In fact, as I have noted, Boghossian's criticism in effect acknowledges the view and tries to defeat my position without questioning it.

[19] Thanks to Glenn Branch and Peter Ludlow for comments on a draft, and to Youichi Matsusaka whose seminar paper on this subject helped clarify a key point.

5 *A Century of Deflation and a Moment about Self-Knowledge*[1]

The person who notified me of my selection for this office tried to entice me into making this speech by saying that mine would be the last address of the century, indeed millennium, to this body. Maybe he thought I went in for apocalypses. Or maybe it was a set-up for impeachment. Or maybe he thought that I'd think I'd get the last word. I'm confident that President-Elect Professor Herman will not be speechless next century. Some of the rest of you may have something to say as well. Despite my sense that the occasion lacks quite the drama I was promised, I have succumbed to some temptation to hazard something synoptic.

Our subject got a good start in this century. There were some lasting ideas about language, and ferment over mathematical logic. But early on, philosophy fell into a negative mood that it has had a hard time shaking off. Self-nihilating movements have come in waves. Positivism, Wittgensteineanism, Ordinary Language Philosophy, Naturalism (as in Naturalized Epistemology), and many other isms have tried to show that philosophical problems are the result of some simple oversight. Quine's attack on positivism seemed to clear the air, and did in fact free the subject to develop in a fruitful way. But a self-destructive streak persists.

This streak resonates with a tendency, shared with movements outside of philosophy, to deflate older conceptions of the specialness of human beings—a specialness that philosophy had been wont to explore and extol. It is as if philosophy has recently measured itself and humanity, and found both gravely wanting.

I brought with me a transcript I recently found of an interview between a prospective graduate student who is seeking advice about a career in philosophy from a philosopher—Professor Carwittup. Although the professor seems to have done nearly all the talking, one can make out a *virtual* dialog:

"Professor, I am thinking about going into philosophy as a way of finding meaning in my life."

[1] Given as Presidential Address to the Pacific Division of the American Philosophical Association in Berkeley, April 1999. I would like to thank Ned Block, Dorli Burge, Paul Coppock, Deborah Brown, Chris Peacocke, and Seana Shiffrin for discussions that improved various aspects of this paper.

"Oh come off it! There isn't any meaning. There is only choice among radical translation manuals. Translating the heathen is making up a system of claims we already agree with. For missionaries and philosophers, disagreement and insight are complementary illusions."

"Sir professor, I want at least to know what *I* am talking about."

"Reference/schmeference. Your word 'mama' refers to a momentary stage of a physical object on one interpretation, and to the number seventeen on another equally good interpretation. You might as well stop being sentimental about your mother."

"Don't philosophers pursue the truth? Isn't truth referring to something and characterizing it as it is?"

"A benighted view: Calling something true is merely paying it a solidarity compliment. It is the compliment of removing quotation marks. A job in editing would better equip you to understand truth."

"But science and philosophy make progress . . . "

"What an old-fashioned idea. So-called progress is just a series of power-driven paradigm shifts, or aimless redirections of the conversation. Perhaps you should consider becoming a talk-show host.

"You think at least mathematics is a stable science? Rationalist superstition! Mathematics just makes it up like the rest of us. It tells useful stories about fictional objects. Philosophy of math is literary criticism. An English department would train you for that."

"Maybe I should go into ethics. I would like to understand my reasons for counting actions right or wrong."

"Give it up! You would just be emoting in an evolutionarily inevitable way. There are schools of acting that would serve your purposes just as well."

"But Professor, I can't think of myself as an actor or some biologically programmed gush of emotion. Maybe I should work on the free will problem."

"Even with close study of the brain, science seems not to have found the will. We now think the will may not exist. But if this result does not hold up, the free will problem and its solutions are unverifiable, hence meaningless. And do not presume that you understand what I just said."

"Philosophy of mind seems intriguing . . . "

"The mind? Are you kidding? You must mean an automaton, or a quivering blob of grey matter, overlaid with a functionally specifiable grid of causal relationships. You can stop worrying about going out of your mind at least. These are not the sorts of things you can go out of."

"But there are conferences and magazine articles on consciousness. They seem kind of exciting . . . "

"No, consciousness is for engineers and sociologists to study. Consciousness is just a self-monitoring feedback mechanism, common to guided missiles and Asian nations. Philosophy does show us a new range of things we can be touchy-feely with."

"You mean missiles and nations can feel things and I can get cozy with them? My friends in California have been saying this all along. I want to get in touch with my own feelings too. Aren't there things philosophers call 'qualitative feels'?"

"How quaint of you to ask. In us, pain is just a way of representing neural damage. In the Chinese nation, it is a report that the communist party has been questioned. Pain has no other properties than the news it provides. The pain is bad news, but it's the damage that is bad. Don't blame the messenger!

"And pleasure is no better than pain. Orgasm is the internal sentence 'something pleasurable is going on down there.' True or false? It doesn't matter. Sex, too, is a text.

"There *may* be an opportunity for social activism in this theory of orgasm. The theory is a recipe for impotence. By espousing it you might help with the problem of overpopulation."

"Well if I am going to be a social activist, I don't want to be lost in the crowd. I want to be able to distinguish myself from others, so I can avoid an identity crisis."

"The theory of personal identity casts doubt on identity crises. Persons are just fissions or fusions in a continuum of psychological states, impersonally specified and beamable to any receptive material in the universe. A politician could use the fissions. The prospects for voter registration are endless. Maybe politics is your field."

"But Professor, on such a theory it seems hard to know who is alive and who is dead . . ."

"Philosophy frees us from caring seriously about our deaths. Since we are just functional patterns common to the Asian nations and quivering blobs, we need not care much whether we live or die. What matters in personal identity is whether our patterns of quiver survive."

"Sir, Herr Professor, how does one know such strange things?"

"Knowledge and justification have been naturalized: they are just reliable connections to the way things are. A thermostat has it all. Knowledge is going with the flow. Perhaps you should consider a job in heating and air-conditioning."

"Isn't there more to it than that? My friends are constantly upsetting me with their sceptical worries."

"Scepticism is an obtuse failure to recognize standards of discussion appropriate to the context. If the sceptic would just learn some manners, he'd stop misleading the youth. Defusing scepticism is devising a system of etiquette, or keeping score in a game. Have you thought about being a water polo referee?"

"Professor, I wouldn't be happy blowing whistles; I know myself that well . . ."

"Self-knowledge? Psychological experiments have now shown that we never get anything right about ourselves. But this is a good thing. Ignorance has been selected for. You can best cultivate collective ignorance by advising any number of Congressmen."

"But those guys are so irrational . . ."

"We all are. We are gloriously irrational animals. Psychological experiments have also shown that our reasoning is entirely made up of fallacies. Like ignorance, irrationality has been selected for. We'd best nourish our irrationality, on pain of getting mowed down by beings crazier than we are. There are some self-help groups in Southern California that might help you become more irrational.

"So you still want to go into philosophy? You need not submit a writing sample. We base admission decisions on a personality test to determine whether you are sufficiently sympathetic with masochism. Philosophy is in the business of showing that it has no business being in business."

The interview seems to be concluded there.

A few months ago, I was speaking with an accomplished philosopher who had been reading philosophers writing at the end of the nineteenth century. He marveled at the things they would say. They would blandly make the most ridiculous claims, he noted. He championed our enlightened age of precision and rigor, and conjectured that no one nowadays makes such absurd claims. I agree with his observations about the past. But I think that this satisfaction with the present doctrinal state of the profession shows a certain historical blindness to the near. Our age has produced as many absurdities as any other. But a large bulk of ours are overkill in the service of self-immolation, quite a different matter from the wild fictions commonly told by revisionary metaphysicians. The excesses at the ends of the last three centuries were excesses of hubris in philosophy. Leibniz's, Fichte's, and F. H. Bradley's absurdities are the products of self-inflation. It is not just the individual philosopher's self-inflation; that is perennial. Philosophy, and with it humankind, were also seen in the warm glow of self-satisfaction.

Some of this inflation of humankind, which in philosophy commonly takes on some form of idealism, survives in our century in scientized guises. I have in mind various revivals of idealism, which present themselves as antirealism, instrumentalism, and various forms of *so-called* "realism" which aren't realism at all (for example, critical realism or realism with a sweet face). But the idealisms are not the excesses that are distinctive in our century. What is distinctive is the array of simplistic deflations of philosophy and humankind.

The social causes of these parallel deflations are not hard to find. The evolution of science has shown that philosophy can no longer see itself as the ultimate synthesis of everything. The critique of the special status of human beings has a broader array of cultural causes. The rise of psychoanalysis and neuroscience—and to go more basic, the advent of methods of mass destruction and a series of regimes bent on genocide—have made it hard to see human beings as special in any way worth extolling. It is true that they are capable of special evil—private as well as public. Unlike the animals, humans can be shallow, calculating monsters of serial betrayal and destructiveness. But these are not recent discoveries.

Philosophy is not the queen of the sciences. But it need not act as a docile slave. Philosophy *can* do science at a specially abstract level, and it can function as midwife. But it does too many things to fit in those boxes. Philosophy opens lines of inquiry, and occasionally provides knowledge, that no other subject is on to. Some of this knowledge is not, and never will be, systematized in any science. Some of it is an instance of part of what is special and valuable about human beings. A central task of philosophy is to articulate this specialness in ways that avoid the inflations of the past while firmly retaining a perspective on the wonder of humankind and its values, intellectual capacities, and institutions.

I have presumed on your patience and on the license associated with my happily temporary office. I have presented a scatter of unargued attitudes—not good philosophical practice. I have also come close to joining the ranks of naysayers about philosophy, even in naysaying its naysaying. Before leaving the meta-philosophical mode, I want to say something good about our subject in this century. I believe that we have some things to be proud of.

One is the commitment to clarity and argument. For this we can thank the positivists, Quine, Church, Strawson, Grice, Davidson, Putnam, and Kripke, as well as Frege, Moore, and Russell. Clarity and commitment to argument do not produce creativity, but they aid it. Moreover, they enhance communication. They have nourished the growth of a vital philosophical community. This is an element of science that philosophy has succeeded in borrowing. Let it not go the way of scholasticism, herd journalism, or advertising.

There are substantive achievements as well as methodological ones. A better understanding of science and a concern to integrate philosophy with it is one. Scientific knowledge and modeling have checked metaphysical excess better than in previous centuries. A better understanding of language is another substantive achievement. Philosophy of science and philosophy of language have raised new philosophical issues, transforming and enriching old ones.

Another achievement is an offshoot of the denigration of traditional views of the specialness of humankind. The anti-descriptivist picture of reference, the anti-individualist view of mind, and the recognition of externalist aspects of epistemic warrant, have helped us see how our natures are determined by norms that reach beyond what we as individuals control. We can better understand the ways that rational beings depend on a universe that is not made up of structures of reason at all. These developments should help us understand our specialness, without falling into the idealisms and hyperintellectualizations that distorted early modern philosophy.

One route to understanding what is intellectually special about human beings is through types of knowledge and warrant distinctive of us. By understanding the relevant types of knowledge and warrant, we better understand the intellectual capacities that help ground the warrant. In earlier work I have explored apriori knowledge, knowledge through interlocution, and self-knowledge. Here I want to discuss some issues regarding self-knowledge.

It is empirically clear that higher non-human animals have propositional attitudes. They therefore have concepts. I believe that all higher animals have a concept of pain. A primary use of this concept is to recognize when they are in pain. We share such self-knowledge with higher animals. Yet we seem to have types of self-knowledge that are more distinctively ours. One distinctive aspect of our self-knowledge concerns our first-person concept. The ape that knows it is in pain attributes the pain ego-centrically. Some sort of mental file is associated with the ego-centric element in the attribution, and is maintained over time. Higher animals commonly believe that they are in pain when they are.

But I think that they lack a *first-person* concept. This is a way in which some aspects of our self-attributions of pain, not to speak of more sophisticated self-attributions, are distinctive. I shall say more about this concept later.

A second way in which our self-knowledge is distinctive lies, I think, in our *understanding* of propositional attitudes. Higher animals have propositional attitudes. But although it is currently an empirically open question whether higher animals make rudimentary attributions of simple attitudes to other animals, I doubt whether any subhuman animals have concepts of propositional attitudes that include the full range of fundamental inferential connections—particularly connections to normative notions like *reason*—that ours do. Although animals have reasons, they do not understand them as such. I shall develop this point later.

As do several others, I hold that we have a non-perceptual, non-empirical self-knowledge that has a kind of authority. I have developed this view in the context of my anti-individualism about the mind. I have argued that many of our propositional attitudes are constitutively dependent on non-conceptual relations we bear to our physical and social environments.[2] The non-empirical, authoritative warrant for such self-attributions is partly independent of the content of the first-level propositional attitudes. The warrant is independent of environmental variations that could only be known empirically. I shall not defend this position here. I simply take it for granted.

I have also argued that the nature of the warrant has something to do with the fact that the content of the relevant second-level self-attributions cannot come apart from the content of the first-level propositional attitudes. Intuitively the knowledgeable self-attribution *I believe that water is a liquid* includes the content of the first-level belief that *water is a liquid*.[3] I have further argued that the nature of the warrant has something to do with the role that self-attributions play in critical reason.[4] Critical reason is reason that is capable of being (and sometimes is actually) guided by a conception and evaluation of reasons as such. Critical reason is a paradigmatically distinctive activity of human beings. Animals think and have reasons, but they do not think in the self-guiding way that involves the evaluation of reasons as such and the implementation of reason in accord with such evaluations. Thus critical reason necessarily has a second-order aspect. It can be theoretical. But it is also what is present in practical deliberation and in having a conception of a life or life plan. It is part of what is essential to practical and intellectual responsibility.

The argument that the warrant for our self-attributions of certain propositional attitudes has to do with their role in critical reason turns on the existence of a certain immediacy in the rational applicability to first-level reasoning of reasons acknowledged in second-level self-evaluations. For example, if I am prima facie warranted in judging that my belief that A is guilty is a poorly grounded belief, then it follows immediately that I am prima facie warranted in changing the belief (the belief at the first level). It is rationally immediate that the warrant for a

[2] Tyler Burge, 'Individualism and the Mental', *Midwest Studies in Philosophy* 4 (1979), 73–121; 'Other Bodies', in Andrew Woodfield (ed.), *Thought and Object: Essays on Intentionality* (Oxford: Clarendon Press, 1982), 97–120; 'Individualism and Psychology', *The Philosophical Review* 95:1 (1986), 3–45; 'Cartesian Error and the Objectivity of Perception', in P. Pettitt and J. McDowell (eds.), *Subject, Thought, and Context* (New York: Oxford University Press, 1986), 117–136; 'Intellectual Norms and Foundations of Mind', *The Journal of Philosophy* 83:12 (1986), 697–720; 'Perceptual Individualism and Authoritative Self-Knowledge', in R. Grimm and D. Merrill (eds.), *Contents of Thought* (Tucson: Arizona University Press, 1988), 86–98; 'Individualism and Self-Knowledge', *The Journal of Philosophy* 85 (1988), 649–663; 'Our Entitlement to Self-Knowledge', *Proceedings of the Aristotelian Society* 96 (1996), 91–116.

[3] 'Individualism and Self-Knowledge'.

[4] 'Our Entitlement to Self-Knowledge'.

second-level evaluation judgment applies, at least prima facie, to the belief being evaluated. This immediacy is not present in evaluations of others' attitudes, or even in certain evaluations of oneself "from the outside". This is so only because there is an immediate rational connection, partly analogous to that of a logical connection, between the second-level judgment about the belief and the belief itself. I theorized that the immediate connection is one within a single point of view. I maintained that this multi-level point of view's being a system of critical reason helps explain the authoritative character of the warrant for self-attributions within it.[5]

I now want to return to what was for me an illuminating exchange with Christopher Peacocke here in the Bay Area four years ago. Peacocke sketched a different conception of the warrant in self-attributions. He agreed that critical reason essentially involves self-attributions that carry a non-empirical warrant. But he held that the source of the warrant lies not in the role of self-attribution in critical reason, but in a prior source. Critical reason is to be explained, on this view, in terms of an independently warranted capacity for self-attributions, together with certain other features.

Authoritative instances of self-knowledge comprise a range of cases.[6] There are *cogito* cases like *I am now thinking that presidential speeches are a peculiar phenomenon*, or *I hereby judge that examples need elaboration.* These self-attributions are either logically or performatively self-verifying. Their warrant clearly depends on understanding, not on any sort of perception. Neither the warrant nor the truth of the judgment depends on any causal relation between judgment and subject matter. The subject matter of the judgment and the judgment about it occur in the same act, and in some cases are one and the same.

The point that self-knowledge is not perceptual and has some special warranted status is obvious in these cases. But other types of self-knowledge—some knowledge of standing propositional attitudes, for example—are also authoritative. Presumably these instances of self-knowledge depend on some causal relation between judgment and subject matter.[7] For example, my judgment that

[5] 'Our Entitlement to Self-Knowledge'.

[6] This is something that I have noted from my first writings on the subject. Compare 'Individualism and Self-Knowledge', pp. 649, 658, 663 n11; 55, 62, 67 n11 in this volume. There is another range of cases involving knowledge of one's non-propositional states—sensations, or perceptual representations, for example—that does not fit directly into the picture I have just outlined. I believe that this sort of knowledge also gets its authority from the role of such states in a point of view. But the relation to critical reason is more complex. I hope to develop these aspects of self-knowledge on another occasion.

[7] Peacocke argues strongly and plausibly for the role of causation in some of the entitlements. Cf. Christopher Peacocke, 'Our Entitlement to Self-Knowledge: Entitlement, Self-Knowledge and Conceptual Redeployment', *Proceedings of the Aristotelian Society* 96 (1996), 117–158, especially sections I and II. His arguments may give the impression that I deny such a role. But I only deny the role of causation in *cogito*-like cases. I maintained that in none of the authoritative cases (including non-basic, non-*cogito* cases) does the entitlement depend on a causal-*perceptual* mechanism. The presence of "perceptual" is crucial. I cited the case of non-*cogito* attributions of standing states and indicated that they were authoritative as well. I grant that there is a causal element in some of these self attributions, and in correct formulations of the entitlement, at least at some levels of

I believe that California is south of Oregon is about a standing state, which together with some triggering occurrence presumably helps cause the self-attribution.

Peacocke cites two other types of authoritative self-knowledge. In one, I have a conscious memory impression that Napoleon lost at Waterloo. I endorse this impression as indicating that Napoleon did lose at Waterloo. The memory impression and the endorsement cause my self-attributing judgment: *I believe that Napoleon lost at Waterloo*.

In the other type, the process leading to the self-attribution is one of making up one's mind.[8] When I am asked, "Do you intend to stay up late after the speech?" or "Do you believe that you can afford that purchase?" I can answer the question by putting into operation whatever decision-making or belief-forming procedure I have for deciding the answers to the first-level questions—whether to stay up late and whether I can afford the purchase.[9]

Peacocke offers accounts of the entitlements underlying the three types of non-*cogito*-like self-attribution. He does not discuss the *cogito*-like cases. He

abstraction. Perception, I argued, is a poor paradigm for *any* authoritative self-knowledge because it is subject to brute error and is in a certain sense impersonal, whereas authoritative self-knowledge lacks both of these features. Compare 'Individualism and Self-Knowledge', pp. 657–659; 61–64 in this volume, 'Our Entitlement to Self-Knowledge', pp. 91–94, 106 n 12; 68–70, 79 n 12 in this volume.

There are yet other cases which are authoritative but not *cogito*-like, in the sense that they are not performatively or logically self-verifying. For example, in *I am wondering whether I should lengthen this footnote*, I may be tracking an occurrent wondering that is conceived as independent of the second-level judgment. Then the judgment is not self-verifying, and is not *cogito*-like. The second-level judgment need not constitute the wondering; it does not make itself true. One could certainly mistake the modality—thinking one is wondering, when one has already decided. But judgments of this sort can still be authoritative, in something like the way judgments about standing states are. (Whether they are authoritative depends on more than their form.) I am indebted to Ned Block for remarks that led to this point.

[8] This type of case was originally emphasized by Gareth Evans in *The Varieties of Reference*, ed. J. McDowell (Oxford: Oxford University Press, 1982), chapter 7. Such cases have been usefully discussed by Richard Moran, 'Making Up Your Mind: Self-Interpretation and Self-Constitution', *Ratio* 1:2 (1988), 135–151.

[9] This type of case may or may not differ from the *cogito*-like cases. In many instances, one will reason about the first-level question and conclude with a performative-like self-attribution: *I do (in this very act) form the intention to stay up* or *I (now) think that I can afford it*. In fact, as so far described, there is not here a single type of case. The making-up-one's-mind example is commonly represented as following a conscious procedure. Peacocke describes it this way: "When you are asked 'Do you intend to go to next year's Joint Session?' you may be considering that question for the first time. You can answer the question by putting into operation whatever procedure you have for deciding whether to go to next year's Joint Session, and answering 'I do intend to go' if and only if you do then decide to go." Compare Peacocke, 'Our Entitlement to Self-Knowledge', pp. 121–122. But to answer the explicit question whether one intends to stay up late by following the procedure "answer that one does so intend if and only if one decides to stay up late," one must know whether one decides to stay up—which is exactly the matter that is supposed to be illuminated. To insure that making-up-one's-mind examples constitute a new type of case, it must be stipulated that one follows the procedure unconsciously and one is not making the self-attribution in a *cogito*-like or performative way. That is, the second-level judgment is not in any way self-verifying, or constitutive of the first-level attitude, but is fallible and conceived as independent; yet it does not regard the first-level judgment from an outsider's point of view.

holds that the entitlements mention the causal character of the belief-forming mechanisms. The entitlements are said to consist in the reliability of the mechanisms, and the lack of a certain sort of intervening layer of representation between judgment and subject matter. Let me explain his account in more detail.

With regard to self-attributions of standing attitudes without intervening conscious memory states, the attitude is reliably the "rational cause" of the self-attribution. Peacocke adds that if formation of such self-attributions were not caused by the relevant state, and did not accord with self-attributions based on conscious attitudes (like the conscious memory state), the thinker would abandon that way of forming self-attributions.[10]

In self-attributions "rationally caused" by a conscious memory and the endorsement of the memory, the entitlement consists in the self-attribution's having those causes and in the fact that those causes are sufficient for the truth of the self-attribution.

In the case of the mechanism in which the self-attribution is formed through making up one's mind, the entitlement is supposed to consist in the self-attribution's being "rationally caused" by the same states that cause the thinker to have the first-level state (for example, the intention to stay up late).

A key *explanandum* that I proposed for any account of authoritative self-knowledge is that authoritative self-attributions do not appear to be subject to *brute error*.[11] A brute error is one that occurs when the subject's cognitive systems are functioning well and the subject is performing in a fully warranted way. Brute errors occur in perception, but not, I claimed, in authoritative self-attributions. I think that understanding this feature of self-knowledge helps elicit the special authoritative, non-perceptual character of some of our self-knowledge.

Where the content of first-level attitudes is simply reused in second-level attributions, there is no possibility of error regarding content.[12] But in non-*cogito* cases (and indeed, in a sense, in performative-type-*cogito* cases), error is possible about whether one holds the attributed attitude toward the specified content.

Peacocke offers an explanation of the impossibility of brute error in the case of the conscious memory that Napoleon lost at Waterloo. He points out that the memory and its endorsement are sufficient for, indeed constitutive of, one's believing that Napoleon lost at Waterloo. Thus they are sufficient for the truth of the self-attribution they cause. He holds that this constitutes a disanalogy with perceptual knowledge. A perceptual experience is never sufficient for the correctness of a perceptual belief, even if the perceiver is operating well and with optimal warrant. He holds that the lack of a layer of experience between the memory representation and the self-attribution is the key to explaining the

[10] I think that this ability to check against other methods of self-attribution is quite inessential to the authority of our primary warrant for self-knowledge. As I shall note below, conscious representations of the self-attributed attitudes are usually not available.

[11] 'Individualism and Self-Knowledge', pp. 657ff; 61ff. in this volume.

[12] Compare 'Individualism and Self-Knowledge' and Peacocke, 'Our Entitlement to Self-Knowledge'.

impossibility of brute error of self-attributions "rationally caused" by endorsements of the memories. He writes, "For self-ascriptions made in this way, brute error is impossible. It is impossible precisely because, in these psychological self-ascriptions, there is nothing that plays the role that experience plays in genuine observational knowledge of physical objects."[13]

Now I think that all of this is close to being right, as far as it goes. But I do not find the explanation of our immunity to brute error, much less the explanation of the authority of our warrant for self-knowledge, satisfying. The intervening conscious memory case is fixed as one in which the self-attribution is veridical: It is appropriately caused by the state that it attributes. Perceptions that are appropriately caused by the objects or states that they represent are guaranteed (trivially) to be veridical also. Peacocke and I agree that there remains a disanalogy between the self-attribution case and the perceptual case.

What distinguishes this case of self-attribution from cases in which one makes a self-attribution that does not properly connect to the memory trace? We cannot usefully say that in that case there is necessarily malfunction or irrationality. For that is what needs to be explained. It is not specific enough to say, as Peacocke does, that "nothing plays the role that experience plays in genuine observational knowledge of physical objects". That role is perceptual and is subject to brute error. Granted, but the difference is what needs to be explained. The memory and endorsement are, unlike the objects of perception, causes that provide warrant. We may call them "rational causes". But again the nature of the rationality needs explanation.

The core of Peacocke's explanation consists of two points: There is a reliable mechanism that is associated with an attributed conscious state. And there is no intervening layer of representation between the conscious state and the self-attributive judgment it causes.

I think that these points do not suffice to explain what is normatively special or authoritative about the warrant. The role of consciousness of the attributed state in the explanation is perhaps significant in the case of the conscious intervening memory state.[14] I am sceptical, however, about whether appeal to consciousness is *sufficient* to explain what is normatively special in the case.

[13] Here is a fuller rendering of his account: 'If one's conscious memory representation were somehow made available to one only through a further layer of experience of the memory, for which the question of the veridicality of the experience in respect of the memory arises, then there would apparently be a possibility of brute error. But . . . the conscious memory representation is already, as a conscious state, something capable of giving reasons for forming beliefs. No further layer of experience of the memory exists, and none is necessary. In the presence of a willingness to take its deliverances at face value, a conscious memory representation can make reasonable self-ascriptions of attitudes. . . . For self-ascriptions made in this way, brute error is impossible. It is impossible precisely because, in these psychological self-ascriptions, there is nothing that plays the role that experience plays in genuine observational knowledge of physical objects.' Peacocke, 'Our Entitlement to Self-Knowledge', pp. 125–126.

[14] In conversation Peacocke has informed me that the feature of consciousness is fundamental to his view. I am quite sure that the issues involved are more complex than the remarks that follow may suggest.

In the first place, I doubt that it is a sufficiently normative notion to carry the burden of explaining the special normative character of the relevant authoritative warrant. To be explanatory of authoritative access to one's mental states, the relevant notion of consciousness cannot be explained in terms of access, on pain of circularity. So it must be something like phenomenal consciousness. But phenomenal consciousness together with reliability—apart from some further explanatory supplement—do not seem to me to carry the normative elements needed for explaining the relevant authority.

In the second place, there are conscious states that we self-attribute with no intervening representational state but with no authority. We are conscious of certain of our moods, but not authoritative about their nature. We may even be reliable at getting them right, but not authoritative in our judgments. For example, I may be rather reliable at telling whether my unease constitutes anxiety rather than anger or depression, and I may be conscious of the relevant self-attributed state. It does not follow that my self-attributions of my mood are authoritative. I believe that whether they are authoritative depends on further issues regarding the nature of the relation between the mood state and perspective of the self-attribution. This same sort of point applies to construals of the nature of many conscious propositional attitudes. In many ordinary senses, we can be conscious of a state and reliable at characterizing its real nature, but not authoritative about our characterizations.

Moreover, consciousness does not help with most cases of authoritative self-attribution. It does not seem a *necessary* feature of explanations of the special authoritative warrant. Normally when one authoritatively self-attributes a standing belief, as when I judge that I believe that California is south of Alaska, the attributed belief is not conscious in any other sense than that one brings it to consciousness in self-attributing it. It is the special rational status of this self-attribution that needs explaining.

Many such beliefs, like many perceptual beliefs that can be authoritatively self-attributed, are not acquired through storing any previous conscious act of judgment. Many of the propositional attitudes (especially many perceptual beliefs and beliefs unconsciously derived from perceptual beliefs) that we acquire and are authoritative about enter our system through no conscious endorsement.[15]

Consciousness plays a central role in understanding mind and even rational deliberation. But I doubt that it is ubiquitous or fine-grained enough to play a central role in explaining all instances of authoritative self-attribution. It does not seem to be sufficient to explain the authority of even those cases in which the self-attributing individual is conscious of the attributed state.

[15] There is a psychological literature that supports this common sense point: Daniel T. Gilbert, 'How Mental Systems Believe', *American Psychologist* 46 (1991), 107–119; 'The Assent of Man: Mental Representation and the Control of Belief', in D. M. Wegner and J. Pennebaker (eds.) *Handbook of Mental Control*, (Englewood Cliffs, NJ: Prentice Hall, 1993), 57–87; Daniel T. Gilbert, Douglas S. Krull, and Patrick S. Malone, 'Unbelieving the Unbelievable: Some Problems in the Rejection of False Information', *Journal of Personality and Social Psychology* 59 (1990), 601–613.

Recall that in the case in which a standing state is directly self-attributed without intervening conscious state, Peacocke adds the condition that the reliable mechanism is such that it would be given up if it came in conflict with other conscious bases for attributing the underlying standing beliefs. This suggests that the judgments about unconscious cases are warranted only because the individual is disposed to make them cohere with cases in which the self-attributed state is already conscious. I think that such a suggestion would be mistaken. The condition that Peacocke adds is not essential to the explanation of the entitlement: The self-attribution of standing beliefs, like my belief that California is south of Alaska, would be just as authoritative if I had *no* conscious, separated memories, or other conscious bases for attributing such beliefs, to check the attributions against, and *no* inclination to let the attributions of unconscious states be affected by attributions based on conscious ones.

Perhaps the *self-attribution* must be conscious to be authoritative. Perhaps for a self-attribution to be authoritative, it must be possible to make the attributed state conscious, in the sense that it must be susceptible to being consciously self-attributed.[16] But I do not see that the self-attribution need rely on any further conscious states, actual or potential, to make it warranted or authoritative. A disposition by the individual to require coherence with such conscious bases is not an essential factor in the authoritative warrant (though absence of actual incoherence with such conscious bases, may be essential). The case of intervening conscious memory traces is, I think, relatively unusual, not the central type.

What I think is important to the warrant is not intervening conscious memory traces—actual or potential. It is the role that the attributed states play in the rational system in which the attribution is carried out. Also important is the individual's ability to withdraw commitment to memories on the basis of understood rational norms. I shall elaborate these points shortly.

There remains Peacocke's point that the mechanism is reliable, and that there is no intervening layer of representation between the first-level state and the self-attribution it causes. Normal reliability of a causal route to self-attribution is surely necessary but not sufficient to the authoritative warrant.[17]

So what of the lack of an intervening layer of representation? This lack does not, I think, suffice to explain the authoritativeness of the warrant or the immunity to brute error either.

It must be remembered that there *is* a layer of representation, present in the attribution but not present in the attributed state. The conceptual analog of the that-clause in authoritative self-attributions redeploys a content already present in the attributed state. But the representation of the type of propositional attitude

[16] Cf. Immanuel Kant, *Critique of Pure Reason*, B131.

[17] One might allow that a person remains warranted and authoritative in self-attributions, at least temporarily, if some diseased memory mechanism that normally funds the self-attributions begins to slip into unreliability. In normal functioning individuals, however, the causal route must be reliable in normal circumstances if it is to be warranted—if it is to carry the relevant entitlement.

that is attributed (e.g., that it is a belief) and the attribution of the attitude *in the first-person way* are additional layers of representation. The former, the representation of the type of propositional attitude, can go awry. Moreover, we authoritatively redeploy the content of the first-order attributed state only when we are correctly and authoritatively related to the state. It is certainly possible to get the contents of some of our attitudes wrong—to misrepresent them, in non-*cogito*-like self-attributions. So except in pure *cogito* cases, redeployment is not necessitated by the form of the attributions, but is an aspect of the authority. So it too needs to be explained.

I believe that brute error, hence lack of authority, is compatible with high reliability and lack of intervening layer of representation. Suppose that we were wired through evolution so that in attributing standing states, we were able about 98 percent of the time—entirely unreflectively and without going through any procedure—to correctly self-attribute otherwise unconscious states. Suppose that these attributed states are counterparts to states that in our actual psychologies are modular.[18] Suppose that we were unable to control or integrate these early-perceptual states any better than we can in the actual circumstances. Suppose that we were also wired to get our mothers' propositional attitude states right about 96 percent of the time. Suppose that these are natural, evolved cognitive capacities. I stipulate that in neither case is there an intervening layer of representation. (Some extrasensory causal vibrations put us on to our mothers' states.) Suppose that mistakes occur because a "wrong" standing state under certain conditions triggers the causal mechanism in a way that is indiscernible from the inside. Such errors do not necessitate any malfunction of our capacity, certainly not in discerning one's mother's states. The reliable mechanism is only *so* reliable because of interfering conditions external to its operation. No failure of warrant need occur when we make mistakes. I believe that brute error would clearly occur in attributions of one's mother's states, and might well occur in attributions of one's own.[19]

The source of our authoritative warrant in actual self-attributions cannot be fully explained in terms of reliability together with an absence of representational layer. These mechanistic and reliabilist resources are *part* of the account of our warrant for *some* authoritative attributions. But they do not come to grips with the essential matter. One must say more about the epistemic norms governing the transitions between the relevant intentional states.[20] To prepare development of

[18] I do not assume that the reliability has to be this high, or that it has to have any particular number attached to it. As I have noted, I am open to allowing special circumstances when individuals are unreliable but still warranted in their attributions.

[19] It does not matter whether one is warranted in relying on the mechanisms, although I am inclined to think that one could be. Peacocke agrees about the cases just described and would take his stand on the consciousness provision that I discussed previously.

[20] I think that the presence of additional representative layers also does not entail vulnerability to brute error. To requote a passage from Peacocke: 'If one's conscious memory representation were somehow made available to one only through a further layer of experience of the memory, for which the question of the veridicality of the experience in respect of the memory arises, then there would

this point, I turn to Peacocke's primary criticism of my position. He agrees that anyone who has critical reason must have an authoritative warrant for self-attributions. But he holds that the role in a system of critical reason is not a source of such a warrant.

Peacocke's own view is that critical reason is the result of combining two more primitive capacities, which are warranted independently of their role in critical reason. One capacity is thought which is guided by evidence, but which involves no assessment of beliefs or reasons for belief as such.[21] The other capacity is the ability to self-attribute beliefs. The warrant for the latter is supposed to be independently explained by the mechanisms that we have already discussed. I have maintained that these mechanisms are insufficient to explain the authoritative nature of the relevant warrants. They are insufficient because they do not touch the nature of the rational norms governing mature, authoritative self-attribution.

I believe that Peacocke is pressing an important question about the source of our warrant for authoritative self-knowledge. There are levels of self-attribution, associated with different levels of cognitive ability. Underlying all levels is the reflexive ego-centric sensitivity present in the senses of bodily awareness, touch, and spatially oriented vision, that is the primordial base of self-consciousness.

apparently be a possibility of brute error.' Cf. Peacocke, 'Our Entitlement to Self-Knowledge', pp. 125–126. The qualification 'for which the question of the veridicality of the experience in respect of the memory arises' is bearing all the weight. For intervening memory representations do play an actual psychological role in our reasoning in long proofs. Such supplemental representational layers do not entail vulnerability to brute error in preservative memory. Everything depends on their normative role in reasoning. If they play the role of perceptual experience of the memory—taking the memory as a perceptual object—then of course brute error is possible. Perhaps Peacocke is alluding to this point in the phrase 'for which the question of the veridicality of the *experience* in respect of memory arises'. The reference to experience and the point that it is an open question whether the experience is veridical allude to a difference in normative or rational role. But the difference with experience is the starting point of an explanation, not the end point. And what is meant by a question of veridicality arising needs to be developed.

Peacocke comes closer to what I think is needed in an almost parenthetical paragraph in which he notes the possibility of holding that necessarily if one has the concept of belief and a proper mastery of the first person, then if the question arises, one will judge that one has the relevant first-order belief when one has the standing belief or when one endorses a conscious memory. 'Our Entitlement to Self-Knowledge', pp. 126–127. The issue here concerns "the relevant beliefs" in the standing cases. In some cases we are authoritative; in others we are not. I think that the differences cannot be explained purely in terms of relative reliability of correct self-attributions of conscious states. They need to be explained in terms of the special authority that exempts us from brute error. But I think that Peacocke is on to something important in connecting the attributions to conceptual understanding of belief and the first person.

[21] 'Our Entitlement to Self-Knowledge', pp. 127–131. Peacocke calls this "second-tier thought" because it involves thought about relations of support, evidence, or consequence between contents. I think that this sort of thought is an abstraction from thought about norms governing propositional attitudes. Children do not have the concept of intentional content (or of truth, evidence, and consequence) independently of the concept of propositional attitudes. Peacocke's cases of someone engaging in belief revision at a stage prior to the acquisition of propositional attitude concepts are, in my view, improperly described, and should be redescribed without assuming that the person conceives of evidence as evidence. Such persons are sensitive to evidence in belief revision without conceiving it as evidence for propositions or for beliefs.

This sensitivity is present even in lower animals. It is developmentally prior to propositional thought. It is certainly prior to the conceptual abilities appropriately associated with the self-consciousness of persons. As I indicated earlier, higher animals have beliefs about their pains. This is a propositional self-attribution, involving some sort of ego-centric self-representation by the animal. I am sure that the epistemology of this sort of self-attribution is special. Perhaps it is free from brute error.[22] Presumably the causal mechanisms involved are analogous to mechanisms involved in us. Such mechanisms may provide the primitive basis for the reliability of our own self-attributions. But the animal's self-attribution of pain lacks the authoritative warrant that I am pursuing.

Perhaps at a yet higher stage—but before understanding the concept of warrant for attitudes—higher animals and young children have rudimentary propositional attitude concepts that play a role in the anticipation of others' behavior. Perhaps they even have an analog of a representation of false belief. But I believe that they lack an ability that is central to a mature mastery of our concept of belief. Such mastery involves a rudimentary conception of good and bad warrant for belief—some ability to think *I believe it because*, where there follows a specification of not *merely* a cause, but a specification of a reason backed by some conception of norms of reason.

The chimp cannot come to doubt its belief that it is in pain by thinking that a reason to doubt is that the sensation was caused by the sudden surprising application of ice, rather than by a cut. The chimp might perhaps retrospectively change its former belief (so that it does not affect memory). Such a change would be rational, but it would not be on the basis of an acknowledgment of reasons as such.

An understanding that extended only as far as the functional role of beliefs in being caused by other matters and in causing behavior, even in oneself, would not, I think, be a mature mastery of our concept of belief. One needs a rudimentary understanding of norms governing beliefs and inferences, a conceptualization of warrant—whether the warrant is entitlement or reason.[23]

In my view, the authority of the warrant for mature self-attributions of belief by persons lies partly in the role such attributions play in a system of conceptualized reasoned revision of propositional attitudes. It lies in the adjudication of error according to *understood* norms for confronting error. I believe that knowing what a reason is, and understanding normative grounds for belief revision (not merely thinking and changing attitudes in accord with rational evidential procedures) are so basic to understanding belief that these abilities are criterial for having a full or mature concept of belief (or other propositional attitude). So mature self-attribution of belief requires understanding the concept of belief.

[22] Immunity to brute error is, in any case, necessary but not sufficient for the relevant authoritative warrant for self-attribution.

[23] One cannot understand warrants (entitlements or reasons) without understanding reasons. My term "justification", which I often contrast with "entitlement", is to be understood in terms of reasons.

Such understanding requires not only reliable unreasoned self-attribution but having the capabilities integral to critical reason. The authority of our warrant for mature self-attributions of belief is associated with this conceptual mastery, and more basically with the ability to utilize self-attributions as baselines for the application of critical reason. Authoritative mature self-attributions cannot precede critical reason.

This point about relations between the concept of belief, critical reason, and the authority of our warrant in self-attribution is associated with a further point about abilities associated with the first-person concept. The further point seems to me more strongly evident. Having the first-person concept that expresses the self-consciousness of persons and is associated with having the relevant authoritative warrant itself requires having a concept of a warrant or reason for propositional attitudes.

Young children, and old people in certain stages of dementia, are persons even though they lack critical reason. But they are persons partly by virtue of having a history of or potential for critical rational abilities. The authoritative self-attributions whose warrant we are discussing are the self-attributions of mature persons. I assume that at some stage beyond what higher animals can do, but prior to having the concept of warrant for a belief, a child can, by understanding language, understand a prototype of the full first-person concept. But the self-consciousness of persons goes beyond marking an ego-centric file associated with bodily sensitivity, or even mental agency. The self-attribution of mature persons requires that one have the intellectual abilities fundamental to being a person.

These abilities are so basic to marking an important kind that it seems to me that they should be taken as criterial for having the first-*person* concept that is fully expressive of personhood. Having such a concept and having the capacities that are central to being a person go together. Both require having a concept of a system of evaluable beliefs—of a point of view—held together with reasons and subject to revision under the force of reasons. They also require having some conception of rational plans that span non-trivial time periods.

I have argued elsewhere that what is central and unique to our first-person concept is that it marks an ability and disposition to acknowledge the immediate relevance of reasons to implementation of the reasons in thought or action—and a disposition to actually implement reasons where appropriate.[24] I believe further that having the full first-*person* concept involves having a rudimentary understanding of its function. It requires not merely understanding the rule that the concept applies to the author of the thought containing it. It requires a rudimentary ability to reason critically—applying reasons and recognizing through first-person acknowledgment of beliefs and plans that reasons governing one's beliefs and plans are immediately relevant to changing or maintaining them. This ability

[24] Tyler Burge, 'Reason and the First Person', in Crispin Wright, Barry C. Smith and Cynthia Macdonald (eds.) *Knowing Our Own Minds* (Oxford: Clarendon Press, 1998), 243–270.

requires having the concepts of propositional attitude and reason, and an ability to acknowledge and implement reasons when they are relevant to guiding thought. Such abilities carry warrants associated with critical reason. So utilizing the first-person concept that marks the abilities and self-consciousness of mature persons cannot precede the ability to reason critically. So self-attributions using this concept cannot precede the ability to reason critically.

I think that this fact is connected with our recognition that certain first-person self-attributions are authoritative—not merely immediate and highly reliable. More basically, the fact that the intellectual abilities definitive of critical reason are necessarily associated with mature self-attributions by persons underlies the authoritative nature of the self-attributions. In my view, attributions involving subpersonal ego-centric sensitivities or concepts, though warranted—perhaps even epistemically special, perhaps even free of brute error[25]—lack the authoritative warrant that marks mature persons' self-attributions.

Self-attributions by mature persons have a normative authority that is associated with both understanding rational procedures for dealing with error and with understanding one's own responsibility to rational norms in the implementation of reasons. The authority is in large measure epistemic, but it is associated with the inviolacy and respect owed to persons. Acknowledging in the first-*person* way that attitudes are one's own is fitting them into a structure of reflective, critical reason that helps constitute personhood.

An account of epistemic warrant should explain why a warranted attitude is in normal circumstances a reliable guide to truth. This is why at certain levels of explanation and in some cases, it is correct, as Peacocke emphasizes, to mention causal mechanisms that mediate first-level mental states and some authoritative self-attributions. But an account of the authority of self-attributions must explain what is common to both causally-based and non-causal, *cogito*-like cases.

Transitions between first-level propositional attitudes and self-attributions provide basic links in a critically rational structure that help constitute our self-concepts, and indeed our personhood. The ability to move freely, rationally, and immediately, from first-level attitudes to second-level self-attributions of them, and back, is an element in self-conscious critical reason.

The authoritative status of warrants in self-attribution presupposes, for their being warrants at all, a capacity to make reliable self-attributions. But the warrants' authority depends on the role of self-attribution in critical reason. This dependence is explained by the fact that our mature first-person concept—and, I think, our mature propositional attitude concepts—are constitutively associated with critical reason.

[25] As noted earlier, I believe that freedom from brute error, though a necessary feature of the authoritative warrant of mature self attributions, is not sufficient. Of course, certain types of reasoning that are not pieces of self-knowledge at all are immune to brute error—deductive reasoning for example.

More basically, the warrant's authority depends on critical reason because that status derives from two aspects of critical reason.[26] It derives, first, from our using self-attributions as a baseline for applying understood rational procedures to deal with error. One's status as a judge of one's mental states is authoritative partly because in addition to being reliable, *one knows the law*. One understands procedures for adjudicating difficult cases, even including self-attributions.

The authority derives, secondly, from the role of our acknowledging responsibility to rational norms in constituting our special status as persons. The employment of the mature first-person concept in its paradigmatic function of acknowledging intellectual and practical responsibility for thoughts and reasons generates the authoritative status of the warrants that we have been discussing. The self-attributions are authoritative partly because they help constitute our concept of ourselves as *persons*, as rational, deliberative beings. This concept in turn is fixed by its function in acknowledging intellectual and practical responsibility for thoughts and reasons.

A reliable mechanism of self-attribution in a being that lacked any conception of reasons for beliefs could be warranted. The warrant might be interestingly different from the warrant for perceptual beliefs. But it would lack the authority of mature self-attributions by persons because it would not be an epistemic constituent in personhood and critical reason. To play that role, the attributions must be associated with the full first-person concept, or with the abilities I associated with such a concept. Being part of a system that underwrites the transfer of reasons immediately between first- and second-levels within a critically rational point of view is part of what is epistemically distinctive about authoritative self-attributions. Acknowledgment of one's attitudes is authoritative because the acknowledgment is essential to a system that constitutes one's personhood.

This account of the authority of mature self-attributions (which I do not claim to be exhaustive) helps explain why self-knowledge has never been a serious target for scepticism about either the truth of or the warrant for self-attributions of mental states and events. The explanation does not derive purely from the presence of a reliable mechanism without an intermediate perceptual layer of representation. For the reliability of the fallible non-*cogito* self-attributions would itself be a potential target for scepticism. I believe that scepticism has tended to avoid targeting self-knowledge for three reasons: One is the sheer obviousness of the relevant judgments. A second is the fact that warrant for them is implicit in mastery of the first-person concept and propositional-attitude concepts, used in the relevant self-attributions. But I think the central reason is our implicit ability to connect the warrant for the self-attributions to critical

[26] I say "more basically" because I think that it is less important to the argument exactly what the boundaries of the mature first-person concept and mature propositional attitude concepts are. What is important is that these concepts are related in their mature use to the relevant aspects of critical reason. These aspects are what fund the authoritative character of the relevant warrants.

reason. Sceptics must acknowledge critical reason to urge scepticism as a reasonable position. A theory that appeals only to reliability and consciousness, and to the lack of intervening perceptual representations, does not answer sceptical worries about whether the causes of our self-attributions are truly rational, or warrant-giving, causes. It does not call on all the resources that we have in explaining our confidence that scepticism about self-knowledge is rationally self-defeating.

What I have said here is too cursory to do justice to these complex issues. I wanted just to illustrate the kind of problem about the specialness of human beings that I think philosophy can confront and make progress on. You will be glad to know that I shall postpone further efforts in this direction, and that you will get to dinner before the century expires.

6 *Mental Agency in Authoritative Self-Knowledge: Reply to Kobes*

Bernard W. Kobes's creative and closely reasoned paper ('Mental Content and Hot Self-Knowledge') connects the performative element in basic self-knowledge to knowledge of what one will do in intentional action. I think this an illuminating comparison. In both cases, the knowledge derives from intellectual control over what one is doing, and in both cases the warrant derives partly from one's status as a rational agent. Some knowledge of what one will do is partly constitutive of being an intentional or rational agent. Some performative knowledge of one's propositional attitudes is partly constitutive of being a critically rational agent.[1]

I

There are differences between the cases. When one engages in intentional physical action, knowledge of what one will intentionally do depends partly, in a relatively immediate way, on matters outside one's control. One can know that one will pick up a fork, when one intends to do so. But the knowledge depends on the veridicality of one's perception of the fork and on one's using the perception to guide one's hands in the normal way to grasp the fork. Knowledge of one's thoughts in relevant performative cases is not hostage to the brute contingencies on which perception and physical action rely.

One can also know that one will raise one's arm. Here the perception of an object is not necessary in quite the way it is in the case of picking up a fork. One's perceptual relation to one's arm may be through proprioception. Proprioception is subject to fewer brute contingencies than visual perception. Still, I think that the dependence on a sense renders one liable to brute errors. In knowledge of what one will intentionally do through physical action, one is generally subject to

[1] See my 'Our Entitlement to Self-Knowledge', *Proceedings of the Aristotelian Society* 96 (1996), 91–116. The notion of brute error that is used in what follows was introduced in my 'Individualism and Self-Knowledge', *The Journal of Philosophy* 85 (1988), 649–663. Some of the points made regarding the relation between authoritative self-knowledge and brute error are also made in the same article.

such errors. One's intentions can be thwarted in ways that undermine true belief, hence knowledge, of what one will do. Some of these ways do not undermine one's epistemic warrant, nor do they involve malfunction of one's cognitive capacities. That is to say, brute error seems always to be possible. By contrast, authoritative self-knowledge of one's own mental states and events, including self-knowledge that involves a performative element, is not subject to brute error.

Self-knowledge that involves a performative element includes a broader range of cases than those that are *logically* self-verifying. The thought *I am hereby entertaining the thought that writing requires concentration* is logically self-verifying. I call such cases *pure cogito cases.* The logic and meaning of the intentional thought content requires that if the content is thought, it is true. But there are cases that normally involve self-verification and a performative element, even though the intentional content does not strictly entail that a thinking of the thought be self-verifying, or even true. These are *impure cogito cases.* For example, one can conceive of a case in which one commits oneself to the whole content of the thought *I am hereby thinking (in the sense of committing myself to the view) that writing requires concentration* without committing oneself to the truth of the thought that writing requires concentration, thus without making the whole content (*I am thinking . . .*) true. One might believe that one is committing oneself to the component thought, but one simply absent-mindedly thinks it through. But such a case would be highly abnormal, even pathological. The intentional content is such that its normal use requires a performative, reflexive, self-verifying thought. It is an (impure) *cogito* case only where the thought is performatively self-verifying.

Or on writing out a check for a charity organization, one might think *I intend* (i.e., *hereby form the intention) to give to Oxfam*, where making the judgment about one's intention *is* the formation of the intention. The same act constitutes both. Thoughts of this form are fallible. Performative cases of this sort also count as impure *cogito* cases. One can not only think through this content and leave it false. One can even judge-true the content (*I intend* [i.e., *hereby form the intention*] *to give to Oxfam*) yet fail to form the intention, thus leaving the judgment false. But such a judgment would again be pathological. The normal use is to form the intention performatively in the formation of the judgment. So the act of making the judgment makes the judgment true.

In all of these pure and impure *cogito* cases, intellectual control is coordinate with the act of thinking a thought—whether forming a belief or making a judgment—that both commits one to the truth of the intentional content and carries out the thought that the content is about, thereby making the thought true.

I believe that most performative cases of authoritative self-knowledge involve a reflexive element. Although I think that there is something to Kobes's description of performative cases as 'telic', I am not sure that there is full agreement on these matters. Kobes's explanation of this notion is somewhat metaphorical. But it seems to me that in most paradigmatic cases of judgments with performative elements, including basic self-knowledge, the relation is reflexive or reciprocal between judgment and subject matter, rather than unidirectional. For example, when I judge, in the performative way, *I hereby form the intention to give to Oxfam*, as I make the decision and begin to write the check, the judgment is normally reflexive. It normally constitutes, rather than being caused by, the formation of the intention. In performative cases, a mental act of self-attribution makes itself true.[2]

Of course, the judgment about the intention *can* be caused by an antecedent intention. The intention can form just beforehand; or it can be a standing state. (Or perhaps the causation can be simultaneous, but between distinct events.) I believe that self-attributions of such intentions and beliefs can be authoritative. But in most cases they are not, in my view, strictly performative. Since cases of authoritative self-knowledge in which the attributed propositional attitudes are occurrent thoughts or intentions that cause the self-attribution are similar to cases of authoritative self-knowledge of standing propositional attitudes, I shall discuss the two types later under the rubric of authoritative self-knowledge of standing states.

Kobes holds (p. 204) that we have self-knowledge of what we are thinking even when we are thinking ordinary conscious thoughts about the world, and that in many such cases we are not thinking "explicitly" reflexive thoughts. I agree that self-conscious self-knowledge is present in many ordinary first-level thoughts about the world. But I think that there is a reflexive element in more such self-knowledge than most people realize. I do not know what he means by 'explicitly'. We do not often verbalize such thoughts with 'hereby' or 'in this very thought'. But I think that when we have authoritative propositional self-knowledge of what we are presently thinking when we are thinking about the world, particularly when there is a performative element in the self-knowledge, there is normally—or at any rate, very frequently—a reflexive second-order element in the logical form of the first-order thoughts.[3] Although these performative or reflexive cases form a larger class than one might first think, I believe that they do not constitute the whole of what I see as authoritative self-knowledge. Kobes places heavy emphasis on the activity of belief-formation. I agree that agency is at the heart of our understanding of first-person authority. But I doubt that it can bear weight in just the way that Kobes or Descartes require it to.

Here Spinoza seems to me to provide a salutary qualification on the Cartesian view to which Kobes's emphasis is congenial. Descartes maintained that belief is always the product of an act—in fact a "willed" act—to assent to a proposition. Descartes seems to me right in maintaining that some instances of belief-formation are instances of a type of agency, and some of these cases are subject to considerations of intellectual responsibility. But I think that this model should not be fully

[2] I think that such performatives as *I promise to give it to you* are also self-verifying. I reject analyses that claim that they lack a truth-value.

[3] See elaboration of this point in my discussion of examples (1) and (2) in 'Our Entitlement to Self-Knowledge'.

generalized. Opposing Descartes, Spinoza maintained that belief is the default position, not an activity on a proposition that one noncommittally understands. He maintained, as a thesis of philosophical psychology, that belief is concomitant with understanding. Spinoza held that doubt and suspension of belief are acts. One can undo the initial non-active default position. But formation of belief is, on his view, not an act. It is automatic if not checked.[4]

Spinoza's view, in its fully general even if vague form, seems exaggerated. But he seems to me right in maintaining that the formation of a great number of beliefs, particularly perceptual beliefs, is not strictly an activity by the whole mental agent. Some of our authoritative self-knowledge resides in our self-attribution of such beliefs. These considerations help motivate my view that the performative model cannot fully explain first-person authority.[5]

Let us lay aside modular and otherwise inaccessible beliefs. There are still beliefs and other propositional attitudes whose formations do not constitute acts of ours. They seem rather to form in us. They may be part of a functional organization that is essential to being an agent. But not all such beliefs that are accessible to self-conscious self-attribution are themselves the products of acts of commitment. At least, I can think of no natural sense of agency that applies to their formation. Most perceptual beliefs are of this sort. Many beliefs that rest on interlocution, especially in childhood, are as well. Beliefs that derive from perceptual beliefs by way of hard-wired inductive mechanisms—as opposed to active, person-level inferences—are also not in any obvious sense products of agency.

Kobes sometimes speaks of non-modular beliefs as "up to us." There may be something in this. They make up a point of view that we as doxastic agents can claim as ours. Once we become critical reasoners, we are epistemically responsible for them. But it does not follow that they are products of agency. Rather, we may be responsible for revising them if counter-considerations arise, and we may have a parallel responsibility for their maintenance. It would seem to me mistaken to hold that all non-modular propositional attitudes—or even all attitudes that we are potentially authoritative about—are formed through intellectual agency. Thus I think that our authority about occurrent and standing attitudes is not captured fully by appealing to the model of "hot" or "smoldering" telic self-knowledge.

Nor do all cases of authoritative self-knowledge exhibit an ability to bestow the content on the propositional attitudes that form the topic of the knowledge. Often those attitudes have a nature and existence that is prior to and independent

[4] Descartes, *Meditation* IV; Spinoza, *Ethics* II, 49. For psychological evidence bearing on the matter, see Daniel T. Gilbert, 'How Mental Systems Believe', *American Psychologist* 46 (1991), 107–119; and Daniel M. Wegner and James W. Pennebaker (eds.), *Handbook of Mental Control* (Englewood Cliffs, NJ: Prentice Hall, 1993).

[5] I believe that Kobes's brief remarks about our being the author of our thoughts (p. 208) tend to overrate our authorship. Many of our ordinary beliefs are not self-conscious or the products of agency, but we can be authoritative in our reflective self-attributions of them.

of authoritative judgments about them. Authoritative knowledge of our standing or occurrent perceptual beliefs, and of many of our other standing propositional attitudes, needs, I think, a broader account.

Of course, in authoritatively attributing propositional attitudes to ourselves, we normally commit ourselves to those attitudes. Claiming them as one's own may seem in effect to endorse them, at least when they are not viewed as foreign "objects" inside one's self. And in those "foreign object" cases, the self-attribution is not authoritative. This may suggest that agency is pervasive in the propositional attitudes thus self-attributed. I think that this is an important point.

But the truth of some authoritative self-attributions does not depend on these endorsements. In making some authoritative self-attributions, one is making a judgment that is not intended to be made true by one's endorsing the attitude on the spot. It is intended to capture a stable attitude that was present antecedent to the self-attribution, and whatever re-endorsement that might involve. As example, consider: *I believe that my sister is younger than I am.* Such self-attributions are fallible. The ways that they are corrected—for example, by reference to past statements or behavior—show that endorsements of the first-level propositional attitude that are implicit in the self-attributions themselves are not in general taken to be sufficient to guarantee the truth of the self-attributions.

Thus a full account of the specialness and authority of some of our self-knowledge needs to go beyond both Kobes's telic model and beyond my paradigm of self-verification and performative acts in *cogito* cases, pure or impure. When I introduced my self-verifying model as a paradigm, I called the sort of self-knowledge that it encompasses "basic self-knowledge". But I was fully aware, as Kobes recognizes, that there is a range of other cases of authoritative self-knowledge that does not exhibit self-verification. There is first-person knowledge of sensations, of occurrent perceptual beliefs, of nearly all standing states that predate the formation of judgments or even knowledge about them. There are certain cases of memory. There is knowledge of some of one's feelings or emotions. And so on.[6]

[6] Although all of these extended cases involve the possibility of fallible self-attributions, I do not believe that they are any less authoritative than the performative or self-verifying cases, in my sense of 'authoritative'. There is at least the appearance of disagreement with Kobes on this point. See p. 215. I might also say that I find the talk of substantiality and insubstantiality, which derives from a very odd and misleading technical use of this term in Boghossian's article, unfruitful in many ways. See Paul Boghossian, 'Content and Self-Knowledge', *Philosophical Topics* 17 (1989), 5–26, section III. I regard authoritative self-knowledge, even in the self-verifying cases, as *substantial* in all normal senses of the word. Kobes's explanations of the term are different from Boghossian's, and I have no criticism of what he says on this score—except insofar as use of the term, with yet more different special meanings, remains a source of possible confusion.

II

My view in 'Individualism and Self-Knowledge' was that there are features of self-knowledge—other than self-verification—that are dramatically and paradigmatically realized in what I called the basic cases and that provide a key to understanding the whole range. I have yet to carry out this strategy fully. But I have taken some further steps.[7] And I still believe that it is a viable and promising enterprise.

I want to stress that 'Individualism and Self-Knowledge' was not intended as a full account of authoritative self-knowledge. Not only does it indicate that the paradigmatic "basic" cases do not constitute the full range of authoritative self-knowledge. It is focused on the semantical or logical role of attribution of intentional content in self-knowledge and on the easiest cases of a self-verifying attitude relation. It does not discuss the notion of warrant, in any depth, *even for the basic cases*.

The point of the article was to raise, and provide an initial response to, an apparent problem about the relation between authoritative self-knowledge and anti-individualism as an account of the nature of propositional attitudes. Although I made a number of comments about differences between authoritative self-knowledge and perceptual knowledge, I concentrated on pointing in a direction for an account of authoritative self-knowledge. Criticism of the paper for being "thin" as an account of authoritative self-knowledge is off the mark. The points, which Kobes cites others as making, about differences between the basic cases and other cases of authoritative self-knowledge were points that I anticipated and in several cases explicitly noted myself. I thought that those were matters to be taken up later. I still expect to deliver on the promissory note. I hope that this reply will advance matters.

Kobes's discussion of extensions from the paradigmatic cases is sympathetic and discerning. I think that he is right that some cases of authoritative memory and some cases of authoritative knowledge of standing states can be understood in terms of mechanisms preserving traces of earlier states known or knowable in the paradigmatic ways.[8] I think that these mechanisms should be seen as part of the rational apparatus that is constitutive of being a critical reasoner. As I have noted, however, I do not think that all of these cases of preservation can be assimilated to the act-paradigm that Kobes develops.

[7] Burge, 'Individualism and Self-Knowledge'. For the further steps, see my 'Our Entitlement to Self-Knowledge'; 'Memory and Self-Knowledge', in Peter Ludlow and Norah Martin (eds.), *Externalism and Self-Knowledge* (Stanford: CSLI Publications, 1998), 351–370; 'Reason and the First Person', in Crispin Wright, Barry C. Smith, and Cynthia Macdonald (eds.), *Knowing Our Own Minds: Essays on Self-Knowledge* (New York: Oxford University Press, 1998), 243–270; and 'A Century of Deflation and a Moment about Self-Knowledge', Presidential Address at Pacific APA, April, *Proceedings of the APA* 73 (1999), 25–46.

[8] See my 'Content Preservation', *The Philosophical Review* 102 (1993), 457–488; 'Computer Proof, Apriori Knowledge, and Other Minds', *Philosophical Perspectives* 12 (1998), 1–37; and 'A Century of Deflation and a Moment about Self-Knowledge'.

III

I turn now to the two objections Kobes discusses. His treatments of the Loar and slow-switching cases are well reasoned, and broadly plausible to me. My views differ in some significant ways, however. There are also some matters I would like to try to clarify.

Kobes's discussion of slow-switching cases makes some points in common with my discussion of such cases in my 'Memory and Self-Knowledge'.[9] In that recent paper I focused on unaware slow switching rather than knowledgeable slow switching. I emphasized the non-reflective role of preservative memory in knowledge of one's past thoughts.[10] Even in knowledgeable slow switching, I see no reason to think that the individual cannot remember the past thoughts, even if he cannot distinguish them from cohabiting twin thoughts. The individual is not forced to ask the question that Kobes and Boghossian focus on. He need not ask whether he was thinking about water or twater. In fact, where he asks this question in this way, I believe that he loses his authority in his application of memory (see 'Memory and Self-Knowledge').

The individual can instead rely on preservative memory to take up the content and attitude-type of the thought that he in fact thought at the earlier time. Relying on memory to individuate rather than to preserve would be a mistake. Such reliance in the switching cases could indeed undermine knowledge of the past. It would treat the remembered event as an object, rather than anaphorically—as part of a single point of view. In such a case, the memory would not be authoritative.

If one uses information about having been switched between earth and Twin Earth to try, through memory, to discriminate earth- from Twin Earth-type thoughts, one loses one's authority over one's past thoughts. Errors deriving from such uses of memory are not naturally counted cases of forgetting. But if they are not so counted and are seen rather as unfortunate use of newly acquired information, then, as Kobes and I both point out, Boghossian's "platitude" that if one forgets nothing one cannot lose knowledge is shown to be false. Indeed, the "platitude" is false for a wide range of cases, many of which have nothing to do with switching scenarios. New information, misleading information, can drive out old knowledge. So no pressure is generated on the anti-individualist by these means.

I am uneasy about Kobes's appeal to equivocal thoughts as the entire basis for his account of reasoning in the relevant switching cases. He is, I think, quite right to reject the idea that anti-individualism commits us to an unacceptable susceptibility to equivocation in deductive reasoning, even by the most rational reasoners.

[9] Neither of us had read the other's account.

[10] Kobes's "smoldering self-knowledge" is I think a special case of preservative memory—the case where the antecedent involves a performative. I also see the role of agency in preservative memory somewhat differently. But I believe that we are on to the same phenomenon.

It may be that in some instances one thinks equivocal thoughts. Or as I would prefer to see it, one may think multiple thoughts, on given occasions, without distinguishing them. It may be that Kobes's idea is part of a necessary solution. But all of the reasoning cases that I know of can be accounted for by noting that even someone who has switched and has "twin" concepts in his repertoire is usually not using both concepts when he uses either of them. Rather, features about the cognitive context and cognitive point of a thought determine which of the concepts is employed. I refer the interested reader to a fuller account of the matter in 'Memory and Self-Knowledge'.[11]

Thus I do not accept Kobes's assumption (p. 217) that because the switched individual is on Twin Earth and has the twin concept, he 'cannot escape including an exercise of the Twin Earth concept' in any given thought in which he remembers something about the twin object back on Earth. This assumption suggests some magical effect that merely being on Twin Earth has on one's thinking Twin Earth thoughts (once one has acquired them). I believe that, at least in the cases that have been discussed in the literature so far, one does not have to appeal to equivocal or multiple thoughts to block unacceptable results about reasoning in slow-switching cases.

An analogue of the reasoning cases occurs for demonstratives, without appeal to anything as complex as Twin Earth. A person can be looking at a tomato and think *that is healthy so that is healthy*. In place of a healthy tomato, a perceptually indiscernible, rotten one could be substituted so quickly between the antecedent and the consequent of the thought that the reasoner would not notice. Then a rational agent *could* become committed to an invalid proposition through no irrationality—if he used the demonstrative twice and independently to indicate the object before him. He might even assume mistakenly that the thought is a logical truth.

A correct account of the logical form of the invalid, false thought will not treat 'that' in its two occurrences as being syntactically the same. The token applications count, from the point of view of a syntax relevant to logical form, as formally different. If the thinker treats both occurrences as deictic, as independently applied demonstratives, then he is liable to error with regard to what could *seem* to be a very safe logical truth. But the thinker would not be committed to the syntactic or logical form of a logical truth. If the thinker does not treat both occurrences as deictic, then there need be no difficulty. If the thinker is to avoid any susceptibility to difficulty, the second occurrence of 'that' must be tied anaphorically to the first. Obviously, one could avoid switching the tomatoes, and still get the contrast between the logical truth and the non-logical truth. The difference does not lie in the objects or *res*. The difference lies in the uses or

[11] The fundamental point that I make in this article is anticipated by Stephen Schiffer, 'Boghossian on Externalism and Inference', *Philosophical Issues* 2 (1992), 29–38, in his comment on the paper by Boghossian that Kobes discusses. Of course, reference to the contextual determination of which thought is thought is only part of the full account.

applications in thought that are counterparts of linguistic uses of the demonstrative 'that'. These are under the potential cognitive control of the thinker.

Although the twin concepts that I have discussed in slow-switching cases are not demonstrative or indexical, the twin concepts contribute differently to logical form or logical syntax. One can engage in equivocation without realizing it, in a way analogous to the way one can engage in a demonstrative shift of reference without realizing it.

To avoid susceptibility to equivocation, one must tie one's concepts together "anaphorically" in one's reasoning. But this is the normal way that we implicitly understand steps in a piece of deductive reasoning as fitting together anyway. We implicitly understand such steps in that way even apart from anti-individualist considerations. When I think "Every man is mortal; Socrates is a man; so Socrates is mortal", I allow no equivocation on 'man', 'mortal', or 'Socrates' by implicitly relying on a sameness of conceptual and indexical use.

If I understand him correctly, Kobes makes substantially this point (pp. 211–212). But I think that the links between steps in reasoning are formed anaphorically. Sometimes there is a telic line that is "forged", but it seems to me that commonly the anaphora is correctly seen as simply preserving a content that was already unequivocally in place at the earlier step in the reasoning. Such preservation seems more "thetic" than "telic", but I am quite happy to dispense with this terminology.

I turn now to Kobes's discussion of Loar's doubt. There are elements in the very posing of the doubt that seem to me to be odd and in some respects significantly off the mark. The problem is supposed to be to explain how someone could assure himself of the apriority of an inference from the existence presuppositions that Socrates and hemlock exist, to the reflexive judgment *I am now thinking that Socrates drank some hemlock*.

In the first place, the relevance of the existence assumptions seems to me quite unclear. Kobes (on Loar's behalf) writes, 'Now let us suppose that *S* knows that such reflexive thoughts [as "I am now thinking that Socrates drank some hemlock"] are always true, given the existence presuppositions [that Socrates and hemlock exist]' (p. 203). The authority of authoritative self-knowledge does not extend to the *res* in *de re* judgments. The relevant self-attributions in authoritative self-knowledge are to be seen as about the intentional content and the attitude-type of the attributed attitude. There are intentional elements in thought referring to Socrates and hemlock. These are trivially not identical with Socrates or hemlock. They have aboutness or representational properties and functions. Socrates and hemlock do not. The intentional or representational content does not even *include* the referents of the conceptual elements. It includes only the "senses" or modes of presentation, or conceptual and applicational elements of the thoughts.

Thus it is epistemically possible that one think mistakenly that Socrates drank some hemlock even if Socrates and hemlock were to turn out not to exist. If we were to find out that Socrates did not exist, I would still have thought a thought properly expressed in such terms. And I would have authoritative self-knowledge of my thought (of the thought that I would have thought) even if Socrates and hemlock were to turn out not to exist. In that case, my thinking the relevant thought would not, of course, depend on my bearing causal relations to Socrates himself or to hemlock. It would depend on another causal complex connecting me to other relevant uses of the name 'Socrates' or the noun 'hemlock'.[12] Given that Socrates and hemlock do exist, of course, my thinking the intentional content of the thought—not merely my thinking things *de re* about Socrates—depends on my actually bearing those causal relations to those objects. But from an epistemic as opposed to individuative point of view—which is the point of view at issue—the existence or non-existence of Socrates and hemlock is irrelevant to the self-knowledge.

In the second place, and more centrally, it seems to me that the challenge to produce an apriori *inference* is misconceived. Kobes takes up this challenge to show how a thinker *S* "could demonstrate to himself a priori that he is thinking that Socrates drank some hemlock". But I think that the challenge itself needs some scrutiny.

Loar's idea appears to be that, in light of externalist claims that thinking the reflexive thought depends on bearing causal relations to some external objects, one should worry that one could know one's thoughts only empirically, because one can know these relations to the environment only empirically. So one needs to be able to reproduce some apriori inference from the existence presuppositions to the reflexive thought.

That this way of understanding the problem that I set up in 'Individualism and Self-Knowledge' is offtrack is suggested by the following fact. No inference to a conclusion constituted by one of the reflexive, self-verifying *cogito*-cases could possibly be needed to justify the relevant thought. Consider the thought *I am in this very thought entertaining the thought that Socrates drank some hemlock.* Because of the performative, self-verifying character of the thought, no inference to it could possibly be needed to provide support for it. It is clearly a starting point, a basic thought, whose justification lies in its own performance and content, not in the content of other thoughts from which it might be inferred. So no inference to it is needed for it to have epistermic support. This point seems to me to be completely independent of the truth or falsity of anti-individualism. So the demand that one be able to support it through some inference is misconceived.

Of course, not all instances of authoritative self-knowledge are reflexive, performative, or self-verifying. In fact, the real issue is not one of justifying the truth of the whole self-attribution. It concerns the warrant for the attribution of the *intentional content* of the attitude that one attributes.

[12] Such cases are discussed in my 'Russell's Problem and Intentional Identity', in James B. Tomberlin (ed.), *Agent, Language, and the Structure of the World* (Indianapolis: Hackett, 1983), 79–110.

This remark brings me to a third respect in which the problem is misconceived as posed. In 'Individualism and Self-Knowledge', I asked why the fact that we have only empirical access to causal relations that fix the nature of our thoughts does not entail that we cannot know that we are thinking such and such unless we engage in empirical investigation that shows the conditions for thinking such and such are satisfied. I said that the answer "can be seen as a series of variations on the point that one must start somewhere" ('Individualism and Self-Knowledge', p. 654).

I think that Loar and Kobes overlook the starting point that I took to be basic. They suggest that the key to my "reconciliation strategy" is a schematic generalization: Kobes writes:

> What *S* knows is that all thoughts of a certain form—including "I am now thinking that Socrates drank some hemlock"—are true. But it would seem that in order to use that information to resolve his empiricist doubt, *S* would already have to know that he is thinking a thought of that reflexive form! And that knowledge may, for all we have said, depend on empirical knowledge of external causal or historical relations. (p. 204)

He later calls the schematic generalization my reconciliation strategy "unadorned".

There are two things wrong with this account of my view. One is that the *fundamental* relevant generalization is *not* that thoughts of a certain form are true. Only pure *cogito* cases—not even all performative or reflexive cases—of authoritative self-knowledge are true in virtue of their form, together with the fact that they are actually thought. For example, the thought *I hereby intend to give to Oxfam* is not true in virtue of its form, as *I am hereby entertaining the thought that writing requires concentration* is. Only the latter is a *pure cogito* case. I discussed both sorts of cases in my article. But the key generalization fixes on the intentional content attributed in the relevant thoughts, not their truth. Thus, the main issue concerns the contents *to give to Oxfam* and *that writing requires concentration*. The point, as I stated it, is that performative or reflexive cases are such that the intentional content that they attribute is thought and thought about at the same time. So the content of the attributed bottom-level attitude and the content attributed in the self-attributional thought are locked together. Thoughts of that reflexive form cannot mistake the intentional content of the attributed thought, though some of them—the performative cases that are not pure *cogito* cases—are fallible. Thus one could be mistaken in holding that one is *intending* in the relevant case. But one could not be mistaken because of some mismatch in content between self-attribution and an attributed intention. I think that one can know, by simply understanding one's thought, when one's thoughts have a reflexive form and require the relevant locking.

The apparent threat in the switching cases is that the content of one's self-attribution and the content of the attributed propositional attitude will come apart. The schematic generalization shows that this apparent possibility is illusory, at least in a large number of cases of authoritative knowledge. It is illusory in all *cogito* cases, whether pure or impure—all cases of reflexive performatives. Successful reflexive performatives, whether pure *cogito* cases or other sorts of reflexive performatives, are self-verifying. Some thoughts with the form of reflexive performatives—impure *cogito* cases—can be false. Yet in all such cases, the content attributed in the self-attribution cannot fail to be the content of the attitude that is attributed. One may mistake the attitude. But one cannot get the attitude right and mistake its content—at least not in the performative, reflexive cases. More will need to be said about authoritative instances of self-knowledge that are not performative or reflexive, as for example, self-attributions of standing states. I shall return to these cases in a short while.

The second thing wrong with the account of my reconciliation strategy is more fundamental. The exposition Kobes gives of my account understates what my "reconciliation story" says. My account does not merely appeal to a schematic generalization. It emphasizes that people must understand the content well enough to think and attribute it. In particular cases of thinking the relevant performative or reflexive thoughts, the content is thought and thought about at the same time. The same content is deployed at the lower level and at the higher level of self-attribution at the same time.

I wrote that my answer to the problem I raised would be a series of variations on the theme that one must start somewhere. The basic starting point that I alluded to is one's understanding of the intentional content of one's own thought. The starting point is not a generalization about the form of thoughts. It is the minimal understanding necessary to think the self-attribution, and to raise sceptical scenarios with respect to it, in the first place.[13]

Thus thinkers who self-attribute in the relevant way must be taken to understand their contents well enough to think them. In thinking them in the reflexive, performative cases that I centered on, they must think and think about their contents in the self-attribution itself in such a way that the intentional content at the different levels cannot come apart. It is the same, understood content at both levels.

The problem as Loar poses it ignores the reflexive nature of the relevant thinking. By treating the thinker as if he must wonder what content he is thinking (in the conditional, 'If I am now thinking that I am now thinking that Socrates drank some hemlock . . . ', p. 204), Loar treats the thinker's understanding of his own thought as if it were understanding from the third-person point of view. The thinker is not in a position to wonder about the content of his thought, in the relevant way, given that he makes the self-attribution. He must minimally understand the content in order to think the self-attribution in the first place.

[13] In the last paragraph of his essay, Kobes seems to me to state very well an essential aspect of the starting point—that one starts with a mental act. My account emphasizes this as well, but also emphasizes the type of understanding commonly involved in performative cases, the reflexive understanding that is present in most performative-type self-attributions. The intentional content associated with the that-clause is thought and thought about at the same time.

The problem as posed fails to acknowledge that the thinker must be able to understand, grasp, the particular intentional content in thinking it. In thinking reflexively, the thinker thinks the content and self-attributes it in the way that is expressed linguistically in that-clauses.

Kobes is right (p. 222) to make the important point that my account of reconciliation is from the point of view of a theorist constructing an account of the thinker's epistemic *entitlement* to the self-attribution. Or rather I am constructing an account of one element in the entitlement. An entitlement is an epistemic warrant that the individual has but need not have the concepts or abilities to explain or understand, even on reflection. So in my view there is no need for the individual to be able to give an apriori, or any other kind of, account of why he is non-empirically warranted in his self-attributions. The individual need know nothing of anti-individualism or of any reconciliation strategy to have the relevant warrant.

In thinking that *I am now (in this very act of thought) thinking that Socrates drank hemlock*, I understand the content of my thought well enough to satisfy the condition on understanding that is relevant to making it a reasonable question whether the thought that I am thinking constitutes knowledge. Given that the understanding is in place, and given that I actually think the relevant thought, I cannot be mistaken in my self-attribution of the content. No empirical knowledge is needed to establish an understanding of the intentional content of the thought that one is thinking.

But suppose, as Loar and Kobes do, that the individual knows about anti-individualism and the reconciliation strategy. What are we to say about the challenge to produce an apriori inference to the self-attribution as conclusion? What I think we should say depends on the particular type of authoritative self-knowledge.

In pure *cogito* cases, no justificatory inference is possible, and none is needed. The judgment *I am hereby entertaining a thought that Socrates drank some hemlock*, when made, is self-evidently self-verifying. It is obvious that empirical issues and issues about switching are irrelevant. For from the point of view of the person making the judgment, the thought is understood by him and understood to be self-verifying, hence obviously true. It is at least as much a piece of non-empirical knowledge as is a self-evident truth of logic.

In cases of reflexive performatives that are not pure *cogito* cases—like *I hereby judge that Socrates drank some hemlock* or *I hereby intend to give to Oxfam*—the self-verification is not formally guaranteed merely by thinking the thought. But their truth is guaranteed, in normal circumstances, by the clearly understood performance of the act. As I indicated earlier, however, the mistake that is supposed to be threatened by the slow switching cases is not just any sort of mistake. It is supposed to be a mistake that derives specifically from some dislocation of content between that of the self-attribution and that of the attributed attitude. But again, no justificatory inference here is necessary, or, as far as I can see, possible. If one understands the content of one's mental act of self-attribution, one understands that there is no room for a transition between the content of the self-attribution and that of the attributed attitude. For the content of the attributed attitude is fixed as that of the self-attribution. One can think the first thought and fail to make a judgment, and thus think something false. One can make the judgment articulated in the second thought and fail to have the relevant intention, and thus judge falsely. But one cannot indicate through the thought a judgment or intention that lacks the content that is attributed—the intentional content: *that Socrates drank some hemlock*. One can understand these points on reflection. That is what the reconciliation point emphasizes. It simply elicits something present in one's ordinary understanding. Again, empirical considerations and issues about switching are obviously irrelevant, once one reflects on one's own thoughts. So again no apriori inference is needed, or as far as I can see, possible.

It seems to me that Kobes may himself fail to appreciate the implications of reflexivity when he writes:

> For all *S* has is the thought that he, *S*, thinks that *p*, and this higher-order thought, even if it is a belief, is not yet presented as something that *S* can think about.... From the thought *I think that p*, *S* is not in a position straightforwardly to infer that he thinks any sort of reflexive or higher-order thought, and that is what he needs. (pp. 222–223)

This seems wrong, or at least misleading. At any rate, it is wrong for the cases that I explicitly discussed, which are all reflexive, dual-level ('hereby', 'in this very thought') cases—performative analogues of the *cogito*. The relevant thought is *I think with this very thought that p*. Insofar as the performance is reflexive, as the cases I discussed explicitly were, at least the content is both thought and thought about in the same act. In the pure *cogito* cases (*I am hereby entertaining the thought that*...), the attitude relation *and* the content of the attitude attributed are both in the position of being both thought and thought about. This is true even in the non-pure reflexive case *I hereby judge that*.... In many of the other reflexive cases—for example, the intending case—at least one of the attitude relations lacks this dual-level role. Thus, the judgment about the intention is not being thought about as a judgment. But in all such cases, the intentional content that is attributed to a propositional attitude has the dual-level role. Only the intentional content is really at issue in the switching scenarios.

As far as I can see, for reflexive cases no hierarchy is relevant to the justification. One needs no engagement with a hierarchy to justify to oneself that one has not made a mistake that results from a disengagement of content between the self-attribution and the attributed attitude. And no inference is appropriate. It is self-evident from the understanding present in one's making the judgment itself that the problem cannot arise.

In these cases, the very posing of the problem results from ignoring the implications of the sort of reflexive understanding necessary to thinking the relevant intentional contents. The puzzle misleads one into treating those contents as objects of identification or potential investigation, or as otherwise separable from the content of the self-attribution. But in fact they are already understood and "individualized" in the only way necessary for the relevant self-knowledge.

The thoughts at both the self-attribution level and the attributed level are dependent for their content on environmental relations. But given that one is thinking the thoughts—and understands them *in* thinking them, as opposed to understanding them through empirical investigation or in some explicatory way—no switching and nothing about the environmental relations could possibly lead one into error. This realization is not a product of an inference from premises. It is the result of reflection on the nature of one's understanding.

I have left open whether some self-knowledge that can reasonably be counted "performative" might not be reflexive. Kobes's solution goes through an ingenious discussion of how a hierarchy is generated through the telic mechanism. Perhaps sometimes a hierarchy *is* generated in some non-reflexive performative cases. I would like to understand this better. If there are such cases, I think his solution is likely to be, in its main outline, correct. I do not reject Kobes's account as wholly inapplicable to the problem. There *might* be cases that can be handled in Kobes's way at any level up an infinite hierarchy. But I do not think that the account covers the most common cases, or the cases that I centered my discussion on. In cases where one thinks the intentional content reflexively as the content of the attribution and of the attributed state or act, no hierarchy arises in dealing with the switching cases.

The explicit articulation of 'hereby' or 'in this very thought' is not necessary for reflexivity. Most self-attributions of occurrent attitudes are reflexive. Moreover, many cases of self-aware conscious thoughts that *p*—that is, self-aware occurrent propositional attitudes that do not explicitly formulate a self-attribution of the propositional attitude—are nevertheless reflexive self-attributions of the relevant attitude toward the content that *p*. The self-awareness often involves an unarticulated reflexive self-attribution. In judging consciously and explicitly that *p*, at whatever level, one commonly implicitly believes that one thereby judges that *p*, as a component of the judgment that *p*. Whether one also believes that one believes that one judges that *p* is a matter of the subtlety of one's self-awareness; it is not required by the self-awareness of the bottom-level judgment. Language often suggests a hierarchical separation that is not present in the actual thinking.

IV

Everything that I have said so far about the problem Kobes poses centers on reflexive cases, or at least performative cases. But I have emphasized that some authoritative self-knowledge is not performative or reflexive at all. I have in mind knowledge of one's standing states and of certain of one's past standing states—for example, perceptual beliefs—through preservative memory. In these cases, there is nothing in merely understanding the self-attribution itself that prevents a disengagement between the content of the self-attribution and the content of the attributed state. Is an apriori inference needed in these cases?

No, an entitlement to self-knowledge holds in these cases as well. So no inferential justification by the subject is *needed.*

But suppose the subject were apprised of anti-individualism and of the reconciliation strategy. Is a non-empirical justificatory inference possible?

It seems to me that an inference is not in place even in these cases. One needs to explain to oneself a relation of non-inferential transition between lower-level attitude and self-attribution that is not subject to environmental vicissitudes or in need of empirical support. One starts by postulating some belief with a certain content—say, that Roberto drank some hemlock. This belief may not be the product of any mental *act.* It might be acquired perceptually, or through interlocution, in a non-active way. It may or may not be warranted. Suppose that this belief has been residing in one. Something brings this belief to consciousness; or something causes one to remember the belief. One comes to employ the first-person concept and one's concept of belief in judging: *I believe (or remember) that Roberto drank some hemlock.*

Why, in light of the way that one's concepts of Roberto, drinking, and hemlock depend on relations to an external environment, is one entitled non-empirically to one's self-attribution? There can, in these cases, be dislocations between the content of one's initial standing state and the content of one's self-attribution. But insofar as the relation between standing state and self-attribution is not dependent on investigation of or other reliance on the environment, beyond the causal input that made the standing state with its particular content possible in the first place, no dislocation would be affected by switches. The self-attribution would simply inherit the content of the environmentally determined standing state.

I have elsewhere explained rational non-inferential relations that fit this description. In the case of self-attribution of a present standing belief, the relation is necessary for rational deliberation. Individuals come to be reliable in making self-attributions through such relations. In the case of authoritative self-attributions of past standing states, the relation is a combination of purely preservative memory and the bringing to consciousness of the standing state that I just described. Purely preservative memory is the more basic rational relation of the two. It is necessary for engaging in any kind of reasoning in time, not just self-conscious critical reasoning. It preserves the content between different attitudinal states over time.

Both relations are or are supported by causal relations. Both can be broken in ways that would lead to error. But the breaks would be internal to the thinker's cognitive system. They would not involve brute errors. They would not in any way depend, for their accuracy in preserving the content of the attributed stated, on relations to or input from the environment. They are rational relations internal to a cognitive point of view and practice. Only the initial standing state, at least in the case of the belief involving Roberto, would depend for its nature, content, veridicality on relations to the environment. That content would then be inherited and operated on in the rational relations on which self-attributions are founded. Similar points could be made in moving from first-level, standing self-attributions to mental acts that self-attribute, at the second level, the standing first-level self-attributions. And so on.

I have elsewhere characterized both such relations as essentially transitions *within* a point of view.[14] In this respect, they are like inferences. The transitions do not require that the initial subject matter for the self-attribution be mental acts. No reference is made to performatives in explaining them. There is no element of self-verification. The self-attributions are acts, to be sure. But they conform to an antecedently established subject matter. They are not established reflexively in the self-attribution itself. They are fallible. They may be thrown offtrack by bias, top-down reasoning, internal malfunction. But in cases of *authoritative* self-attributions they are not subject to brute error. Their relation to the content of the standing states is not hostage to vicissitudes of the environment. Their warrants derive not from relations to the environment that could be known only empirically. Their warrants derive from the reliability of the causal connections and from the roles of the relevant relations, and the associated self-attributions, in various aspects of rational systems. The warrants are thus not perceptually based. They are non-empirical.

[14] See my 'Our Entitlement to Self-Knowledge' and 'A Century of Deflation and a Moment about Self-Knowledge'.

7 *Self and Self-Understanding**

Lecture I: Some Origins of Self

I will reflect on constitutive features of selves—especially a certain sort of self-understanding. This self-understanding is the main topic of these lectures.

I

'Self' is a technical term, refined from ordinary usage. Ordinary usage is, however, very close to what I want. A definition from the *Oxford English Dictionary* runs, 'Self: a person's essential being that distinguishes the person from others, especially considered as the object of introspection or reflexive action; a person's particular nature.' Kant characterized a *person* as 'what is conscious of its numerical identity, of its self, in different times'.[1] This

* This essay is a revision, with some expansion, of the Dewey Lectures, given at Columbia University, December 2007. I am grateful to Christopher Peacocke for valuable criticisms in spring 2011 of the last section of Lecture I and all of Lecture III; and to Denis Bühler for saving me from an error and prompting an argument in Lecture II.

[1] Immanuel Kant, *Critique of Pure Reason*, A361; see also *Metaphysics Mrongovius* 29: 911 in Karl Ameriks and Steve Naragon (eds.), *Lectures on Metaphysics* (New York: Cambridge University Press, 1997), p. 276. Kant's formulation in the *Critique* is not ideally specific and might qualify as ambiguous. It is clear from context that Kant means the consciousness to include consciousness at a given time of the self as it is at different times. It is also clear that Kant intends the consciousness to be noninferential, and in my terms *de re*. Kant shows less interest in the *diachronic* implications of his formulation than one might hope. He relies on these implications in the Third Paralogism to argue that cognition of self over time is necessarily empirical and cannot meet requirements of rationalist theories of self. In my 'Memory and Persons', *The Philosophical Review* 112: 3 (July 2003), 289–337, note 50, I sketch why I reject Kant's argument that self-cognition of self over time is necessarily empirical. In notes from his lectures, some characterizations of *person* or *personality* omit reference to transtemporal self-consciousness altogether, citing only a being conscious of its identity in different states. See *Metaphysics L_1* 28: 276–277; *Metaphysics Dohna* 28: 680. Both passages are in Ameriks and Naragon (eds.), *Lectures on Metaphysics*, 87–88, 381. See also *Critique of Practical Reason* 5: 87, in Mary J. Gregor (ed), *Practical Philosophy* (New York: Cambridge University Press, 1996), 210. Kant thought that self-consciousness necessarily involves inner sense. There is ample evidence that he assumed that inner sense necessarily has past and future as well as present applications. So these failures to mention transtemporal capacities do not reflect a different position from the one in *Critique of Pure Reason*. But they do show that he does little to develop this aspect of the notion. In *The Metaphysics of Morals* 6: 223—in Gregor (ed.), *Practical Philosophy*, 378—Kant

conception of *person* is very close to the notion of a *self* that interests me. It is a purely psychological notion that specifies a self-consciousness with diachronic reach. Regardless of the ontology of selves—whether or not they are purely psychological beings—the *concept* self is a psychological concept.

Kant's concept contrasts with Strawson's concept person. Strawson's concept is close to the common-sense notion of person—roughly, the sort of bodily being—paradigmatically human being—that normally matures into a self-conscious critical reasoner with moral capabilities. Strawson's *concept* is *not* purely a psychological concept. It *entails* that persons have physical as well as psychological characteristics. These two notions—the OED/Kant notion and Strawson's notion—typify a distinction that I draw between selves and persons.

Selves are *loci*, indeed agents, of psychological activities, and agents that can engage in, and are subject to, certain valuations—the distinctive valuations associated with persons. Selves evaluate themselves and other selves for rationality, critical rationality, morality, social cooperation, character, creativity, grace. Being an agent and topic of such valuations is part of what makes them selves. Since selves constitute the particular psychological natures of persons, selves' being agents and topics of such valuations is part of what makes persons *persons*.

In engaging in these types of valuation, selves become the subject matter of self-knowledge and self-understanding. So the valuational and cognitive powers of selves are intertwined. Some of the types of evaluation that I listed are constitutive to being a self. Selves and persons are constitutively capable of evaluating, and engaging in, critical reason. Arguably, they are constitutively capable of being moral. A certain reflexive cognition is constitutive to these evaluations. Critical reason and morality constitutively depend on self-understanding. My topic is the psychology and epistemology of the kind of self-understanding that is required for critical reason and morality, and that is itself constitutive to being a self, and a person.

First, some introductory points. I am interested in *constitutive* matters—in the *nature* of selves—those features of selves that make them selves.[2] I characterized selves as certain *loci* of *psychological* realities. The fact that self is a purely psychological notion does not entail, or even much encourage, the view that selves lack physical natures. The psychological nature of the *concept* cannot legislate these ontological matters. I will, however, be focusing on the *psychological* nature of selves. I bracket any further, non-psychological aspects of their nature.

distinguishes moral personality—a free being with reason—from psychological personality, explained in terms of consciousness of self in different states. (See also *Metaphysics* L_1 28: 276–277) The two notions are connected, however, inasmuch as a free being with reason must be capable of remembering and anticipating acts if acts are to be imputed to the individual's moral personality.

[2] I discuss natures in my *Foundations of Mind: Philosophical Essays, Volume II* (New York: Oxford University Press, 2007), 1–3 and *passim*; and in *Origins of Objectivity* (Oxford: Oxford University Press, 2010), 57–67 and *passim*. For working through *passims*, see the indexes.

It is natural and right to connect psychological discussion with discussion of the brain. I will have almost nothing to say about the brain. I think that a psychological framework is necessary to finding anything in the brain that is of psychological interest. Ultimately, psychology and neuroscience are collaborative enterprises. Each must provide checks and balances on the other. But a psychological framework will provide us with more than enough material to work with here. The issues about understanding and value that I will be discussing *must* be connected to psychological inquiry. We cannot understand these issues in non-psychological terms. Perhaps this situation will change. But I doubt it.

The tradition of discussing selves is complex. There are several closely related concepts that figure in the tradition—soul, spirit, mind, subject, conscious subject, self-conscious subject, rational animal, rational being, critically rational being, ego, person. Several of these will appear in my discussion. First, I will return to contrasting my notion of self with that of person.

Strawson's elaboration of the notion of person has strongly influenced philosophy in the last half-century. I think that this influence is deserved. But it has led to misconceptions about relations between psychological and physical attributions. I will not have time for detailed discussion of Strawson's views. I will just describe them and state some attitudes toward them.

Strawson developed his concept of persons to combat dualist conceptions of mind.[3] According to his concept, a person essentially has both psychological and corporeal attributes. Strawson's key claim is that use of the concept person, with its essential application to a corporeal being, is necessary for understanding psychological concepts. In other words, application of psychological concepts would be incoherent and unintelligible if they were not applied to something, a person, that has corporeal as well as psychological attributes.

I believe that none of Strawson's arguments, or any by his followers, justifies this claim.[4] All depend on views about knowledge or individuation of content that are neither plausible nor well supported. I do think that Strawson's notion person is useful. What I reject is a set of arguments for holding that all psychological notions are conceptually dependent on a prior notion of a corporeal being, a person. I will take psychological notions on their own terms. I emphasize that these are conceptual, constitutive points. I will not be defending an ontological position.

It is only because mature persons naturally have certain psychological capacities that we evaluate them for critical rationality, take them to have a special

[3] P. F. Strawson, *Individuals* (1959) (London: Routledge, 2002), chapter 3.

[4] Strawson argues for his views in *Individuals*. See also his *The Bounds of Sense* (1966) (London: Routlege, 2006), 163ff. For followers, see Gareth Evans, *The Varieties of Reference* (New York: Oxford University Press, 1982), 208ff., 237ff.; John McDowell, *Mind and World* (Cambridge, Mass.: Harvard University Press, 1994), 100ff.; John Campbell, *Past, Space, and Self* (Cambridge, Mass.: MIT, 1995), 92ff.; Quassim Cassam, *Self and World* (New York: Oxford University Press, 1997), passim; and most of the essays in José L. Bermúdez, Anthony Marcel, and Naomi Eilan (eds.), *The Body and the Self* (Cambridge, Mass.: MIT, 1995).

moral status, and so on. Not all persons have these states and capacities at all times. Six-month-olds are persons. But if a being naturally and constitutively did not have or develop psychological states and capacities beyond those *exhibited* by six-month-old human beings, they would not be persons.[5]

Perhaps persons *develop* into having or being selves. They would do so by developing the competencies that meet certain conditions for falling under norms constitutive of being selves, such as norms for engaging in critical reasoning or for being morally responsible. Then persons would not have or be selves during all times when they, the persons, exist. Demented human individuals that lack psychological capacities beyond chimps are persons, but only because their natural capacities have been distorted by age, damage, or disease. We understand our concept person partly by reference to certain natural psychological competencies. The reference is to naturally having the competencies in a mature state, not to having them throughout a person's existence. I think of selves as beings that actually have the competencies that make persons the valuable beings that they are. Thus selves are types that set standards for being a person. The standards allow for prolepsis and retrospection.

It is compatible with the main part of my discussion to hold that mature selves are just phases in the lives of persons. There are intuitions that pull in that direction. On the other hand, I find it natural to reify—to say that someone has gone out of existence when a person still exists, but has irrevocably lost those psychological capacities that made the person a person. I am inclined to count selves as individuals. I think that natural instincts and even explanatory and normative considerations suggest being liberal in allowing existence to different types of being, even where they overlap in seemingly messy ways. Here, these issues will not matter. I focus on constitution, nature, essence. I shall speak of selves, for the sake of exposition, as individuals.

Like most psychological notions, the notion of self bears complex constitutive relations to a wider physical reality. As far as I know, selves may necessarily depend on a physical body. I think, however, that there is no evident *conceptual* or even apriori necessity that selves have physical properties, although all selves that we know of depend in some way on their bodies.

Selves are necessarily conscious, at least some of the time. A type of consciousness presupposed by all other types is phenomenal consciousness.[6] Phenomenal consciousness consists in there being some way that it is like to be in a mental state. Phenomenal consciousness is an occurrent condition. It is the psychological bedrock of selfhood. An individual's stream of consciousness is

[5] I write of *naturally* having competencies, or of *natural* competencies, to allow that a self or person could be damaged or diseased so that performance that would realize the competencies is blocked. If a self permanently loses such competencies, it no longer exists or the person no longer is a self.

[6] The notion of phenomenological consciousness is vividly illustrated in Thomas Nagel, 'What Is It Like to Be a Bat?' *The Philosophical Review* 83:4 (October 1974), 435–450.

interrupted, in sleep for example. But phenomenally conscious psychological life is where the acts, cognitions, and evaluations that most matter to us occur.[7]

Underlying conscious life is, of course, a vast system of unconscious psychological states and activities. In empirical explanation, unconscious elements are primary. But unconscious psychological states and occurrences are constitutively relevant to selves only insofar as they bear certain relations to conscious states and occurrences.

I will concentrate here on the *representational* nature of selves. Selves are subjects with representational competencies, states, events, and acts. They are subjects with perceptual systems, capacities for inference, beliefs, intentions, perceptions, occurrent thoughts, decisions.

Now I introduce a technical term. The *point of view* of an individual is the system of representational states and occurrences that are, in a certain sense, correctly imputable to the individual.[8] Non-human animals as well as persons and selves have points of view. Imputability is a partly intuitive, partly theoretical notion. I start with intuitive points. *Non-psychological* entities are not part of a point of view. Irretrievably unconscious modular *psychological* states and processes are also not part of a point of view. They are not imputable to the individual, though they occur in the individual's psychological subsystems. Potentially or actually conscious perceptions, memories, beliefs, intentions, emotions, decisions, imaginings are part of a point of view. So are attitudes in the Freudian unconscious.

I believe that the key to the nature of selves lies in reflecting on powers distinctive to their points of view. These powers include competence for a certain type of reflexive self-attribution of elements in the point of view. Such self-attributions give the type of point of view that is distinctive of selves a *multi-tiered* structure. I shall explore this structure. It is the structure of self-consciousness and representation of self.

[7] Compare William James, *The Principles of Psychology*, vol. 1 (Cambridge, Mass.: Harvard University Press, 1981 [1890]), chapter 10, 284–285, which identifies the self with '. . . either the entire stream of our personal consciousness, or the present "segment" or "section" of that stream, according as whether we take a broader or a narrower view.' He calls the narrower, more present segment the 'innermost centre within the circle, or sanctuary within the citadel, constituted by the subjective life as a whole. Compared with this element of the stream, the other parts, even of the subjective life, seem transient external possessions, of which each in turn can be disowned, whilst that which disowns them remains. . . . It is the home of interest. . . . It is the source of effort and attention, and the place from which appear to emanate the fiats of the will.'

[8] By extension, I count the representational contents and non-representational conscious features of the relevant psychological states the individual's point of view. I emphasize that '*point of view*' and '*perspective*', which I use interchangeably, are *not* to be construed as indicating anything essentially perceptual or empirical. One can have a purely mathematical point of view or perspective. Points of view and perspectives are ways of representing (thinking about, perceiving, remembering) a subject matter. I confine these terms to "ways," and states marked by those ways, that are imputable to an individual, rather than subsystems of individuals.

II

I begin by discussing some background conditions and some representational antecedents of the representation of self that is distinctive of selves and persons.

Every living system can differentiate itself from other things. It expels foreign matter as waste, ingests other bodies and not its own, protects itself from external threats, and so on. These capacities are not in themselves either representational or psychological. They form the functional background for self-interest and representation of self.

Perception is the lower border of sensory representation.[9] Not all sensory systems are perceptual. Hence not all are representational. Sensings of heat or light by amoebae or worms and the sensors in muscle tissue that signal stress are not representational in my sense. The explanatory paradigms that drive the theory of perception differ fundamentally from those that drive explanations of the sensitivities in these sensory systems. Explanation of perception makes essential reference to veridicality conditions. Explanation centers on how the system represents the surrounding environment accurately, to the extent that it does. Of course, even the simplest sensory systems exhibit functional success and failure. A sensory system is functionally successful when its response to stimulus is associated in the environment with the macro-qualities that it evolved to respond to. But in non-perceptual sensory systems, success is only functional good fortune. There is no explanatory value in invoking veridicality conditions.

By contrast, perceptual systems are explained in terms of the formation of states that are about the distal environment and that are veridical or not. The mark of a perceptual system is *perceptual constancy*. Perceptual constancies are abilities systematically to represent given distal features, by sensory means, *as* those same features, even though the proximal stimulations caused by those features vary radically. For example, most perceptual visual systems can represent something as having a given size as it moves closer or farther away—thus as it causes radically different proximal stimulations. Perceptual constancies occur in the visual, auditory, and touch systems of a wide variety of animals, including the visual systems of certain insects.

As noted, all living organisms show self-interest, broadly understood. I am interested in antecedents of *representation* of self. The most primitive representational antecedent is constitutive to all perceptual and actional representational content. This antecedent is an *ego-centric index*.[10] In their most primitive forms,

[9] I defend this claim and the associated conception of representation in *Origins of Objectivity*, especially chapters 8 and 9. A state is representational in my sense if and only if specification of the state in terms of veridicality conditions could ground serious, correct explanation.

[10] Another term for such representation is '*de se* index'. For discussion of such indexes, see 'Memory and Persons', and *Origins of Objectivity*, 199–201, 287–88, and passim. For other discussions, see Marc Jeannerod, *The Cognitive Neuroscience of Action* (Cambridge, Mass.: Blackwell, 1997); Jeannerod, 'To Act or Not to Act: Perspectives on the Representation of Actions', *Quarterly Journal of Experimental Psychology* 52A:1 (1999), 1–29; J. D. Crawford,

such indexes are not conceptual. Applications of ego-centric indexes anchor all perception and all primitive actional states. In primitive forms, they are not used or applied *by the individual*. They are, however, a primitive type of ego-related representation. These primitive forms become incorporated into propositional perceptual beliefs and propositional intentions. Then they are conceptual.

In perception, ego-centric indexes refer to the individual—or, more commonly, some position on the individual or some time at which the perception occurs—as an *origin* with respect to which representation of other entities is mapped. Ego-centric indexes single out an individual, position, or time in a context-dependent way. For example, an index might mark a spatial origin in relation to which other entities are spatially related in a perception. A nest-perception might represent the nest as to the left at such and such a distance from the origin of the perception. The origin is perhaps a position between the eyes of an insect. Or a perceptual memory of a food source might represent it as having been visited at some time interval from the time represented by an ego-centric index as the present time of the memory representation.

Like the self-concept, ego-centric indexes are immune to reference failure. They automatically mark an origin of reference that occurs whenever a state with the relevant index is instantiated. Unlike applications of the first-person concept, occurrent applications of ego-centric indexes are normally not acts. Applications depend purely on their role in a larger system of representation. Unlike applications of the first-person concept, not all ego-centric indexes *refer* to an individual or ego. Many refer only to a position or a time.

Ego-centric indexes have two constitutive functions. One is the function just discussed—to index an origin for a framework of representation. The other is to type-individuate an aspect of psychological states that treats the origin as having immediate ego-relevance. All representational psychologies function to privilege the individual's needs, goals, and perspective, and function to provide the representational basis for serving the individual's needs, goals, and perspective—doing things for the individual from the individual's own motivations or other powers. The applications that realize the first of the two constitutive functions of ego-centric indexes—applications to spatial or temporal positions—are always themselves privileged in an individual's psychology. They function as privileged in that they function in the psychology to have psychologically immediate ego-relevance. They mark spatial origins of perception and spatial positions from which the individual's actions are initiated. They mark positions that the individual must protect in serving the individual's own needs, goals, and perspective. They mark times, present times, for initiating action and calibrating memories and anticipations. They mark times that are privileged with respect to serving the individual's needs, goals, and perspective.

W. P. Medendorp, and J. J. Marotta, 'Spatial Transformations for Eye–Hand Coordination', *Journal of Neurophysiology* 92 (2004), 10–19.

Like all psychological representational content, the ego-relevant content of spatial and temporal ego-centric indexes marks representational functions that are in turn grounded in representational competencies. The ego-relevant content marks competencies to treat indexically anchored positions as privileged in realizing functions to distinguish and serve the individual's own needs, goals, and perspective. All representational psychologies have these functions and competencies. All mark temporal or spatial positions indexically in an ego-relevant way. The competencies that underlie these primitive ego-related representations are enriched in individuals that have complex psychologies, until they are part of the suite of competencies that are constitutive of persons and selves. Ego-centric indexes are the ur-antecedents of concepts of self.

Of course, an insect cannot spell out this ego-relevance. It has no attributives that apply to psychological states or content. Ego-centric indexes, however, are part of the representational content of—and help individuate—psychological states that systematically relate a framework origin to the individual's needs or perspective, non-inferentially and as immediately as the system allows. This functional relation to the individual's needs, goals, or perspective is what makes an index—and the psychological state that it helps individuate—ego-centric.

The relation to the individual's needs, goals, or perspective usually runs through capacities mediating perception and action. For example, the psychological system might be such that if an object is perceived to be approaching the origin, the animal will take protective measures. Or certain feeding actions might be initiated if a memory marks a food source as having been visited at a certain temporal interval from the present.

Ego-centric indexes and their two functions are constitutive to every representational psychology. Every representational psychology contains some states that include them. Powers to realize these functions help mark an individual with a representational mind as an individual.

Fulfilling these functions involves a representational capacity to distinguish and privilege the individual's own needs, goals, and perspective. By distinguishing and privileging the individual's needs, goals, and perspective, ego-centric indexes provide a representational basis for serving them. The unity of an individual's representational psychology consists partly in having representational powers that mark off and unify that psychology by fulfilling the individual's needs and goals, and otherwise serving its representational perspective.

As noted, ego-centric indexes are ancestors of the first-person concept, expressed by the word 'I', that represents persons and selves. The indexes' infallible reference, their anchoring a representational framework, and their direct ego-relevance all presage aspects of the first-person concept. Having ego-centric indexes does not require a capacity to *represent* anything *as* having psychological states. In this respect, they differ from a full concept of self.

III

In the remainder of this lecture, I discuss four capacities that mark milestones in the developmental prehistory of distinctively *self*-representation. The first three are more complex exercises of ego-centric indexes. The exercises are psychological couplings of ego-centric indexes between and within sensory modalities. These couplings are, I think, early precursors of a certain capacity for objectification that is characteristic of selves and persons. I have in mind the capacity to think of oneself from first- and third-person points of view. In their most primitive forms, these couplings occur among sensory capacities. They are non-conceptual. Non-human animals exhibit at least the first capacity, and possibly all three. They are thus nodes in the "animal" roots of selves. The fourth capacity is a type of memory that yields another kind of objectification constitutive to selves and persons.

Let us consider the first of these four capacities. In the last three decades, ethology and developmental psychology have been astir over beginnings of self-awareness. Some of this discussion is onto important junctures in the prehistory of personhood and selfhood. The literature has, however, persistently blurred important psychological distinctions. Much of it vastly overstates the directness of connection between the phenomena being theorized about and self-awareness in any sense that involves representation of a representer's own psychology.

I start with what is called '*the mirror test*'. A mark is applied to a bodily surface without an individual's being aware of the application. The mark is placed so that the individual can see it only in a mirror. An individual passes the mirror test if after a short time, it acts as if its own behavior is the source of the behavior shown in the mirror.[11] For example, the individual might touch the mark. Chimpanzees, elephants, and dolphins pass the test. Only a few gorillas in captivity have passed it. It was long thought that although monkeys can learn to use mirrors to spot other objects, they cannot pass the test. Recently, there is some evidence that rhesus macaques pass it. Human children first pass it between 16 and 24 months.

Passing the test has been taken as evidence for self-awareness.[12] The relevant notions of *self* and *awareness* need clarification. There is no clear sign that in passing the test an animal represents itself as having psychological states. The

[11] Gordon G. Gallup, Jr., 'Chimpanzees: Self-Recognition', *Science* 167: 3914 (Jan. 2, 1970), 341–343; Gallup, 'Self Recognition in Primates: A Comparative Approach to the Bidirectional Properties of Consciousness', *American Psychologist* 32:5 (May 1977), 329–338; Sue T. Parker, Robert W. Mitchell, and Maria L. Boccia (eds.), *Self-Awareness in Animals and Humans: Developmental Perspectives* (New York: Cambridge University Press, 1994); Diana Reiss and Lori Marino, 'Mirror Self-Recognition in the Bottlenose Dolphin: A Case of Cognitive Convergence', *Proceedings of the National Academy of Sciences* 98:10 (May 8, 2001), 5937–5942.

[12] Gallup, 'Self-Recognition: Research Strategies and Experimental Design', in Parker, Mitchell, and Boccia (eds.). *Self-Awareness in Hun and Animals*. Gallup, the originator of the test, gives the following experimental condition on exhibition of self-awareness: 'An organism is self-aware to the extent that it can be shown capable of becoming the object of his own attention.'

experiments do not indicate possession of what I count a self-concept or first-person concept. Still, there are steps in those directions.

Passing the mirror test rests on bidirectional couplings in the individual's psychology between visual perceptions and representations in kinesthetic proprioception. The individual can match what it sees with what it senses in its own movements. The individual's psychology translates between visual representations and kinesthetic/proprioceptive representations as of the same movements—tokens and types. The translation is systematic. The mapping seems to use a continuing image of the whole body, built through proprioception in the individual's learning history. Mirrors are not essential for such couplings, but passing the mirror test evinces them.

The systematic character of the couplings and the fact that they operate on a whole-body image differentiate the behavior exhibited in passing the mirror test from the behavior of feeling an irritant on a body part, finding it visually, and brushing it off. This latter type of behavior also involves couplings between different sense modalities. It does not, however, suggest visual identification of something, codified in the body schema, as the whole individual.[13] Visually locating and dealing with an irritant does not depend on systematic whole-body coordination between vision and proprioception. Visually identifying as such what has been mapped proprioceptively *as of a whole individual* is a kind of reflexive "self" identification. Here the "self" is the whole individual. The individual is identified in terms of bodily characteristics. The identification is, as far as the experiments show, at the level of intermodal connections between perceptual capacities, not necessarily effected through propositional thought.

The connection between vision and *kinesthetic* representation makes possible not only the individual's representation of its whole body, but makes possible a kind of representation of the body *as the individual's own*—that is, something like an *attribution* of *ego-indexing* to the individual's own bodily actions. Let me explain this point step by step.

First, ordinary visual representational content is ego-centrically indexed. But such content *attributes* to *seen* particulars only physical or functional characteristics (shape, position, color, danger, food, and so on). Primate vision represents body parts and bodily movements as such. It probably also differentiates purposive from non-purposive movements.[14]

[13] For scientific discussion of the notion of body schema, see J. Paillard, 'Body Schema and Body Image: A Double Dissociation in Deafferented Patients', in Gantcho N. Gantchev, Shigemi Mori, and Jean Massion (eds.), *Motor Control: Today and Tomorrow* (Sofia, Bulgaria: Academic Publishing House, 1999), 197–214. For philosophical discussion, see Brian O'Shaughnessy, 'Proprioception and the Body Image', in Bermúdez, Marcel, and Eilan (eds.), *The Body and the Self*, 175–204.

[14] B. Hare et al., 'Chimpanzees Know What Conspecifics Do and Do Not See', *Animal Behaviour* 59:4 (April 2000), 771–786; Laurie R. Santos, Jonathan I. Flombaum, and Webb Phillips, 'The Evolution of Human Mindreading: How Nonhuman Primates Can Inform Social Cognitive Neuroscience', in Steven M. Platek, Julian Paul Keenan, and Todd K. Shackelford (eds.). *Evolutionary Cognitive Neuroscience* (Cambridge, Mass.: MIT, 2007), 433–456. A lot of theory in

Second, the bodily movement seen in the mirror causes not only visual representation but a kinesthetic sense of the movement. The representational content of kinesthetic perception, like that of visual perception, is ego-centrically indexed—marking the origin of the perception's perspective. But there is an important difference between proprioception and vision. In proprioception, ego-centric indexes also tag the *perceived* entities. Proprioception marks the perceived bodily movement as having the same relevance to the individual's needs and perspective as the proprioceptive perceptions themselves have. A function of proprioceptive perception is to track the animal's movements *as* the animal's own. So whereas, by the point of the previous paragraph, ordinary visual representations are singly indexed, kinesthetic perceptual representations are always *doubly* indexed. In addition to the index, or indexes, for the perspectival origin of the perception, kinesthetic perceptual content associates an ego-centric index with the *perceived* particulars. Proprioceptively perceived bodily movements are marked as privileged with respect to the individual's needs, goals, and perspective.

Third, the proprioceptively perceived body parts and movements are mapped onto a standing proprioceptive body schema. Like all proprioception, this schema is doubly indexed: both representational origin and proprioceptively perceived entities are indexed.

Fourth, for individuals who pass the mirror test, mirror experiences set up an intra-psychological coupling between visual representations of bodily movements and kinesthetic action representations of the same movements. These couplings induce a further coupling between the visual representations of body parts and positions on the body schema.

Fifth, these couplings provide a causal pattern that grounds new representational content. There is a new association of ego-centric indexes with *objects* of *visual* perception. Thus *seen* objects—say, the movement of a body part—are directly and systematically coordinated in the psychology with the individual's needs, goals, and perspective. The movement is seen as ego-privileged. Single-indexed and double-indexed visual perceptions represent the same movement. But the latter are constitutively linked to kinesthetic and body-image representational contents.[15]

this domain over-attributes representation of psychological states to apes and very young children, when attributing representation of teleological states suffices to account for the evidence.

[15] The association of single-indexed and double-indexed visual representations is informative. Such associations can in principle be mistaken. The usual visual representation of the form

[ego_i] that_1 movement

comes to be associated with a visual representation of the form

[ego_i] that_1 [ego_i] movement.

Both represent the same movement. At the level of a psychology that passes the mirror test, the ego-centric indexes may well refer to the whole body rather than merely a body part.

Earlier I said that ego-centric indexes are infallibly referentially successful in their primary function. In associating ego-centric indexes with perceived entities, a perceptual system is fallible

The application in vision of single-indexed and double-indexed representational contents to the same movement is not a propositional identity. It is a coordination between perceptual representations in the psychology. But the visual system functions to treat the different representations as representing the same movement.[16]

Passing the mirror test marks three developmental milestones. They may have been achieved before the test is passed. Passing the test just shows that they have been achieved.

First, by associating an ego-centric index with a purportedly visually perceived physical entity, the psychology invests that entity with the same ego-significance that the origin of the visual perception has. Recall that the second function of ego-centric indexes is to mark a direct psychological relation between the individual's needs, goals, or perspective and what they index. What is new is that this indexing is associated with a *visually* perceived entity. No more in this new step than in the most primitive ego-centric indexing does the individual use psychological attributives that might figure in a *theory* of mind. No psychological states are represented (referred to or indicated) by ego-centric indexes.

Second, the individual's psychology associates occurrences of its ego-centric indexes with the individual's whole, visually perceived body, not just with an indexed anchor *position*. As noted, the proprioceptive body schema constitutively marks a perceptually represented whole body, and does so in a doubly indexed way. It is not part of the *nature* of the visual system to doubly index anything. In coming to associate an ego-centric index with a visually perceived body, the psychological system has an intra-psychological functional connection between the anchor of the visual perceptual representational content and the perceived body. (Of course, perceived body parts can also be associated with an ego-centric index and mapped as parts of the doubly indexed whole body.) As with the first milestone, what is new lies in the linking of visual competence with double-indexing that is already present in another system.

A third step is implicit in the second, and is, I think, the most significant one. Passing the mirror test evinces a step toward *objectification* of self in two respects. One is intermodal. The development of connections among kinesthetic proprioception, the proprioceptive body schema, and visual perception allows the perceiving individual to associate ego-centric indexes with the same individual as

inasmuch as perceptual representation of perceived entities can fail. So association of an ego-centric index with a purportedly perceived entity can participate in a failure to connect with purportedly perceived reality. However, even such an index marks a relation in the psychology of an origin in perception to the individual's needs, goals, and perspective. Occurrences of an ego-centric index cannot fail in representing the origin or in connecting a purported perceptual attribution to the individual's needs, goals, and perspective.

[16] This association ramifies the perceptual constancies that mark all perception. I conjecture that a visual system cannot produce doubly indexed representational contents on its own. A doubly indexed visual representation always depends on some intra-psychological connection to a naturally doubly indexed representation in another perceptual modality—specifically, proprioception—or in a conceptual application. In proprioception, the individual perceives his or her own bodily arrangements or movements, and naturally perceives them as being ego-related.

perceived from the perspectives of different perceptual modalities. The perceived individual is perceived as having ego-significance, in this limited way, from different perspectives; and the perspectives are linked with one another and with actional representations.

The other aspect of objectification concerns visual representation. Couplings among vision, kinesthetic sense, and body-schema representation make possible an informative connection of single-indexed visual perceptions with double-indexed visual perceptions of the same entities. A visual perception of a face—anchored as usual by an ego-centric index—may represent a face as just another face. Then when the intermodal connections are established, the singly indexed perception comes to be informatively associated with a visual perception that attributes an ego-centric index to the same face. So the second perception is double-indexed. The face is seen from two perspectives. (Proprioception is always double-indexed and does not provide two perspectives.) One perspective characterizes one's face, but not as ego-related. The other characterizes the identically appearing face as the individual's own. The case is a primitive, preconceptual ancestor of "Aha, that's me!" It is a primitive ancestor of representing oneself from both first-person and third-person perspectives.

To describe these advances as an emergence of *self*-consciousness or *self*-representation would be misleading. Ego-centric indexes are associated with an object of representation. But no psychological states are attributed to it. So no individual is represented *as* a self. Reflexivity in representation should not be confused with genuine *self*-consciousness or *self*-representation. Still, the reflexivity and coupling of different points of view shown in the mirror test is a *step* toward genuine self-representation.

IV

The psychology evinced in the mirror test is a simple case of intermodal and intramodal perceptual couplings of ego-centric indexes. Such couplings occur in more complex forms. Each form is of interest in itself. Each adds a distinct type of objectification in ego-centric indexing. I will discuss two other couplings—*imitation* and *joint attention.*

Imitation starts from the beginning of human life.[17] Children imitate the expressions of parents from the first few minutes after birth. Purposive social

[17] Andrew N. Meltzoff and M. Keith Moore, 'Imitation of Facial and Manual Gestures by Human Neonates', *Science* 198:4312 (Oct. 7, 1977), 75–78; Alison Gopnik and Meltzoff, 'Minds, Bodies, and Persons: Young Children's Understanding of the Self and Others as Reflected in Imitation and Theory of Mind Research', in Parker, Mitchell, and Boccia, (eds.), *Self-Awareness in Humans and Animals*, 166–186. This copying may be associated with mirror neurons—neurons activated both when an individual observes specific types of activity and when the individual performs the same specific types of activity. Mirror neurons were discovered by Vittorio Gallese et al., 'Action Recognition in the Premotor Cortex', *Brain* 119:2 (April 1996), 593–609. Mirror neurons are popularly construed as

copying plays a major role in the emergence of more refined systematic couplings among an individual's sensory modalities and, of course, in the development of language. *Copying* is matching in which one specific type of behavior in one individual is caused by, and specifically explained by, the sensing—perceptual or not—of the same type of behavior in another individual. Copying divides into several interestingly different subspecies, of which imitation is the most important for the development of a self-concept. *Imitation* is active, purposeful, goal-directed copying of a specific form of behavior. Examples of non-imitative forms of copying are *mimicry*—an automatic sensory-motor type of copying—and *emulation*—purposive copying whose point is to match a behavioral *result*, not necessarily the behavior itself. It is disputed whether non-human animals ever imitate as opposed to emulate.[18] Non-human animals are certainly more oriented toward results than toward coordinating behaviors themselves. From 12 months onward, human children differ dramatically from other primates through greater orientation to imitation, and less to emulation.

Imitation again involves systematic coupling among sensory modalities, and between sensory modalities and the representational actional system. Suppose that the imitating individual has a visual perception, $(V)\underline{[ego_i]that_1F}$. ('(V)' designates the visual modality. '$\underline{F}$' indicates an attributive mode of presentation that is perceptually applied to the token behavior *A* of the imitated individual. I assume that the ego-centric index is spatial. I ignore the inevitable temporal index. The subscripts on 'ego' and 'that' mark occurrent applications of the index and the demonstrative-like capacity, respectively.) The individual adjusts his action—and his psychology adjusts representation of action—so that it is of the same *type* as the imitated action. $(V)\underline{[ego_i]that_1F}$, referring to *A*, is taken as the paradigm model for the individual's imitation. The imitating individual produces a type-similar imitating action A_i in such a way that it satisfies a representation $\underline{F}$ as of the same type of action as the imitated action *A*. So the psychology produces in its actional system and, presumably in its kinesthetic proprioceptive system, doubly indexed representations $(Act)\underline{[ego_i]that_2[ego_i]F}$ and $(K)\underline{[ego_i]that_2[ego_i]\ F}$. ('Act' indicates actional representational mode; 'K' stands for kinesthetic mode.) The actional representation anticipates the imitating act A_i. The kinesthetic representation is caused by A_i. Given that the imitation is successful, acts *A* and A_i are in fact of type F. Imitation is a purposive coupling between the agent's own act A_i and another individual's type-similar act *A*. The coupling of the visual representation $(V)\underline{[ego_i]that_1F}$, which perceptually represents the act *A* that is to be imitated, with

instantiating "theory of mind" or "mind reading". This construal over-intellectualizes the phenomenon. The function of mirror neurons has not been agreed upon. It is natural to think that they are central to copying behavior. But their presence in some primates, macaque monkeys, has not been associated with copying.

[18] The distinction between imitation and emulation is due to Michael Tomasello, 'Do Apes Ape?' in Cecilia M. Hayes and Bennett G. Galef (eds.), *Social Learning in Animals: The Roots of Culture* (New York: Academic Press, 1996), 319–346.

representations (Act)$\underline{[\text{ego}_i]\text{that}_2[\text{ego}_i]\text{F}}$ and (K)$\underline{[\text{ego}_i]\text{that}_2[\text{ego}_i]\text{F}}$ constitutes a social analog of the objectification that emerged in the mirror test.

In imitation, the individual's psychology does not achieve the full reflexivity that it does in passing the mirror test. There is no coupling of a single-indexed visual perception with a doubly indexed actional or proprioceptive perception *of the same token act*. And there is no informative coupling within vision of an ordinary perception of a seen token act with a doubly indexed visual perception *of the same token act*. Only singly and doubly indexed representations of different tokens of the same type of action are coupled.

Although imitation begins earlier in human development, it requires more, in this respect, than passing the mirror test does. Through the mirror, the couplings of seen acts with proprioceptively sensed acts are *given* to individuals that are equipped to register them. Since the individual is tracking his or her own acts visually, what is represented visually is in lock step with what is proprioceptively represented. By contrast, in imitation the individual must adjust his actions to a sequence of actions that he or she does not control. So the type matching must be made through adjustment in imitation. It is not there simply to be apprehended.

A little reflection indicates how fundamental dyadic copying, including imitation, is in infant life. Parent and child enter into games of ritualized turn taking. The range of cases in which coupling between the child's visual representations and its proprioceptive representations, for achievement of sameness of representational type, must be enormous. These couplings are the backbone of early self-objectification.[19]

In human beings, dyadic copying occurs earlier than the triadic coordination commonly involved in *joint attention*. Joint attention is widely regarded as the culmination of coordinated social interaction in the pre-linguistic phases of human development.[20] Joint attention emerges in the middle of the second year of human life. Roughly, joint attention is a capacity to share attention with another, usually toward a third object, with some mutual awareness of the shared attention.

The intermodal couplings in joint attention are complex topics in themselves. But I will not discuss their form or content here. As with passing the mirror test and imitation, joint attention does not in itself involve attribution of psychological states.[21] What is attributed to others is essentially purposive behavior. The purposiveness need not be represented as deriving from psychological states.

[19] Compare. Vasudevi Reddy, 'Before the "Third Element": Understanding Attention to Self', and Sue Leekam, 'Why Do Children with Autism Have a Joint Attention Impairment?' both in N. Eilan et al. (eds.), *Joint Attention: Communication and Other Minds* (New York: Oxford University Press, 2005), 85–109, 205–259.

[20] Malinda Carpenter, Katherine Nagell, and Michael Tomasello, 'Social Cognition, Joint Attention, and Communicative Competence from 9 to 15 Months of Age', *Monographs of the Society for Research in Child Development*, serial no. 255, 63:4 (1998).

[21] Although attending is a psychological act, and although in joint attention individuals represent acts (gazes) that are in fact guided by attention, the acts are first represented as purposive activity, not as psychologically guided purposive activity. See note 14 above.

Both imitation and joint attention do, however, provide further elements in the developmental background for robust *self*-representation.

We mature into a psychological conception of self. But we seem not to start there. Our initial ego-representations are bodily. They involve attribution of purposive activity. The activity is overt physical action. The *psychological* attributions young children come to make are, I believe, primarily built on perceptual representations of bodies, bodily agents, bodily attributes, supplemented with teleological representations of purposive or goal-directed acts. Teleology is not psychology. Our representation of ourselves as psychological beings grows from representation of ourselves and others as purposeful denizens of a physical world.

V

Intermodal perceptual coupling of ego-centric indexes is probably the most primitive representational antecedent of self-representation and self-understanding. A seemingly more advanced antecedent, which I will discuss for the remainder of this lecture, is a certain type of memory. Selves are necessarily extended in time. I think that a point-event self is an incoherent notion. Acting and experiencing take time. A self-concept must be associated with representations of oneself as extended in time. A certain type of memory underlies such representations. This type is systematically associated with a certain type of anticipation.

I will rely on a rather extensive taxonomical background.[22] Please bear with me. The delineation of types of memory is, of course, a matter for empirical psychology. I will present a classification that is a conceptually clarified version of what can be found in empirical work.

I use the term 'memory' only for a representational capacity. I intend 'representational', as usual, in my relatively demanding sense—requiring an explanatory use for appeal to veridicality conditions in type-identifying states and capacities. I do not take retentions of the effects of classical conditioning, or so-called muscle memory, to count as memory, properly so called.

Iconic memory, working memory, and other types of short-term memory have short automatic decay times and, in the latter two cases, fairly strict load capacities. I shall be mainly concerned with a certain type of long-term memory. Long-term memory encodes relatively significant matters, usually tagged through attention. Long-term memory effectively retains representational content indefinitely, with no known automatic decay times. All these types of memory occur in a wide variety of animals, including many that lack propositional attitudes.

Let us distinguish between *experiential* memory and *non-experiential* memory. Experiential memory is remembering a particular entity x in a direct

[22] This taxonomy improves on one I use at the outset of 'Memory and Persons'.

de re way. The particular can be a mental event, a physical event or physical object, or a scene. If I see rain in Salisbury and thereby remember that it rained in Salisbury, without coding the memory perceptually, my memory is not experiential, though it derives from perceptual experience. If I think *cogito* and remember that I did, but do not remember my thinking it, the memory is not experiential.[23]

The division between experiential and non-experiential memory cuts across most of the types already mentioned. Iconic memory is always experiential. But working memory, short-term memory, and long-term memory can be experiential or non-experiential. Propositional thought can occur in any of these latter three types of memory. Experiential memory can be propositional or non-propositional, as long as it is *de re* of a particular or particulars.

I will focus on a type of long-term experiential memory—*episodic memory*. *Episodic* memory is conscious, long-term, experiential memory whose perspective functions to pick out a past particular, representing it as it was at a specific past time.[24] Tulving characterized episodic memory picturesquely, as a type of mental time travel to the past.

Episodic memory evolved more recently than other memory systems. Its neural basis lies in the prefrontal cortex and other neo-cortical regions. It is more vulnerable to disease, injury, and aging than other types of memory. It seems to emerge late in child development, possibly not until 24 months, after imitation and joint attention emerge. Tulving conjectured that episodic memory is unique to humans. This conjecture is controversial. It is hard to test whether long-term animal memory of specific events is *conscious*. Something like episodic memory, not necessarily with consciousness, has been plausibly attributed to birds, dolphins, and primates.

There are two primary distinctions between generic long-term experiential memory and its subspecies, episodic memory. One is that the latter must be conscious, whereas the former need not be.[25] The other is that episodic memory must function to represent a particular *as it was at a specific past time*. Long-term

[23] To be experiential in this sense, a memory need not have the format of *perceptual* experience. It can derive from intellectual events. I am using 'experiential' in a very broad sense.

[24] The past particular could be an object or scene, as well as an event. Discovery and investigation of episodic memory as a distinctive psychological phenomenon derives from the work of Endel Tulving, *Elements of Episodic Memory* (New York: Oxford University Press, 1983); 'Remembering and Knowing the Past', *American Scientist* 77:4 (July–August 1989), 361–367; 'Episodic Memory and Autonoesis: Uniquely Human?' in Herbert S. Terrace and Janet Metcalfe (eds.), *The Missing Link in Cognition: Origins of Self-Reflective Consciousness* (New York: Oxford University Press, 2005), 3–56. Cf. also Alan D. Baddeley, Martin A. Conway, and John P. Aggleton (eds.), *Episodic Memory: New Directions in Research* (New York: Oxford University Press, 2002).

[25] Long-term experiential memory is easier to investigate empirically insofar as one does not require consciousness. There is little agreement about how to determine whether animals' memories are conscious. Often experimenters stipulate that they are studying animal '*episode-like*' memory precisely to avoid the issue of consciousness. Compare Nicola S. Clayton and Anthony Dickinson, 'Episode-like Memory during Cache Recovery by Scrub Jays', *Nature* 395 (Sept. 17, 1998), 272–274; Charles Menzel, 'Progress in the Study of Chimpanzee Recall and Episodic Memory', and Bennett L. Schwartz, 'Do Nonhuman Primates Have Episodic Memory?' both in Terrace and Metcalfe (eds.), *The Missing Link in Cognition*, 188–224 and 225–241 respectively. Tulving's characterizations of

experiential memory can be *de re* of a particular without picking out the particular episodically. A non-episodic long-term perceptual experiential memory could have a perceptual residue that is too generic, or too much a perceptual composite of experiences, to single out the particular as it was at a given time.

Both generic long-term experiential memory and episodic memory are *de re*. *De re representation* is representation that picks out a particular on the basis of a non-inferential, not purely attributional or descriptional representational competence.[26] Non-episodic experiential memory can be *de re* recognition, without representing an entity as it was at any specific time. Episodic memory is not only *de re* reference to past particulars. The *de re* representation retains a past representational state that functioned to represent a particular as it was at a corresponding past time; and the *de re* representation is accurate if and only if it derives from, and accurately represents, the particular as it in fact was, and was represented, at the relevant past time.

The notion of episodic memory, with its evocative metaphor of mental time travel, can perhaps be made vivid by contrasting it with another type of long-term memory. Imagine an animal that buries food and returns to uncover it months later. Whether the unearthing behavior indicates episodic memory—or even non-episodic experiential memory—is so far completely open, even laying aside the issue of consciousness. Consider the following updating mechanism, which is thought to occur in many non-episodic types of memory. The animal perceives the burying. A non-propositional representation with the content, that food in such and such a location now, is laid down. Memory updates this content as time passes: that [remembered] food in such and such a location now. The remembered food is expected to be currently in the location. Although memory derives from past experience, there is no memory of the burying, or of the food as it was at the time of the burying, or of the burial time itself.

Such updating use of memory does not rule out having representational content in past-tense form: that food once buried in such and such location. This non-episodic memory still lacks *de re* reference to the past. Human children younger than 24 months show no signs of episodic memory. Before that time, they can recount past events as being past. But their retellings are generic, script-like.[27] Events are situated within a script, but not on the basis of an individualized

episodic memory confine *representata* of such memory to events. But memory of bodies and scenes can be included, as long as memory represents them as they were experienced at a particular time.

[26] For discussion of *de re* attitudes, see my 'Belief *De Re*', *The Journal of Philosophy* 74:6 (1977), 338–362; reprinted with 'Postscript to "Belief *De Re*"', in *Foundations of Mind*; also 'Five Theses on *De Re* States and Attitudes', in Joseph Almog and Paolo Leonardi (eds.), *The Philosophy of David Kaplan* (New York: Oxford University Press, 2009), 246–316.

[27] Katherine Nelson and Janice Gruendel, 'Children's Scripts', in Nelson (ed.), *Event Knowledge: Structure and Function in Development* (Hillsdale, NJ: Lawrence Erlbaum, 1986), 21–46; Nelson, 'Emerging Levels of Consciousness in Early Human Development', in Terrace and Metcalfe (eds.), *The Missing Link in Cognition*, 116–141; Teresa McCormack and Christoph Hoerl, 'Memory and Temporal Perspective: The Role of Temporal Frameworks in Memory Development', *Developmental Review* 19:1 (1999), 154–182. During ages three to five, there is a leap in richness, detail, and

autobiographical experience. Human adults who have lost episodic memory through neural damage or aging can tell historical, even autobiographical narratives. But the narratives are like scripts learned from hearsay or reading. The narratives rest on memories that do not enable the individual to re-experience past events, or relive the past.

Lacking episodic memory, with its *de re* representation of particulars as they were in the past, an individual would be unable to represent such particulars in the way that the individual experienced them. The form and order of the individual's own past would be lost. For an individual with propositional attitudes, the loss would be a deficit in understanding. The individual's understanding of his or her representational past would be at best indirect.

Earlier I discussed objectification through couplings between different perceptual modalities. Such couplings coordinate representations of the same entity from the perspectives of different perceptual modalities. Such couplings provide a basis for weak ego-objectification, inasmuch as an egocentric index is attached to representation *associated* with a *perceived* entity.

This phenomenon can be replayed in perceptual experiential memory. *Ordinary*, non-proprioceptive, perceptual experiential memory—episodic or not—is analogous to visual perception in being singly indexed. It does not in itself associate an index with any remembered entity. It just carries the ego-centric index of the purported original perceptual experience, or some conceptual analog. Nothing in the remembered scene is represented as having privileged ego-status for the rememberer. There are, however, experiential memories that are the memory-cousins of doubly indexed visual perception that passes the mirror test. Such doubly indexed perceptual experiential memories are autobiographical.[28] I call an experiential memory that associates an ego-centric index, pre-conceptual or conceptual, with the individual rememberer, represented as

specificity in narration. Stories are made up as well as recounted. Temporal indexicals such as yesterday and tomorrow emerge in the third year, later than indexicals like I and you. I disagree with accounts of episodic memory, such as the last one listed, that maintain that episodic memory constitutively involves not only *de re* representation of past entities as they were at past times, but also self-consciousness and a capacity to 'interrelate arbitrary events'. Such requirements are held to be necessary to 'give meaning' to the *de re* elements in episodic memory. I believe such views commit the *philosophical* error of hyper-intellectualizing the phenomenon, along lines analogous to those that I criticize, with respect to perceptual reference, in *Origins of Objectivity*, chapter 6. The *de re* representation of past events as they were at specific past times requires a capacity to distinguish events in different instances of script-like or recurrent phase-like sequences. This capacity requires either a capacity to localize one event in a sequence, or a capacity to distinguish differences in types of events within otherwise similar sequences. But it does not require a use of allocentric, decentered frameworks, or a capacity to refer to arbitrary past events—much less self-consciousness.

[28] Episodic memory need not be conceptualized. It can be purely perceptual. Episodic memory also need not be meta-representational. It need not *attribute* psychological states, beyond the "attribution" involved in associating egocentric indexes with remembered entities, in autobiographical cases. Even when autobiographical, it may make only physical and teleological attributions to entities, marking them as having ego-significance. A common mistake among psychologists who discuss episodic memory is to claim that episodic memory constitutes a meta-perspective on one's perspective. There is no basis in the phenomenon for claiming that it is representation *of* perspective. Episodic memory

an entity in the remembered situation, an '*autobiographical experiential memory*'. An individual's autobiographical perceptual experiential memory not only functions to preserve a previous ego-centric index functioning as perceptual origin. The memory also represents the rememberer, placing the rememberer in the remembered situation; and it marks the remembered rememberer, thus represented in memory, as having ego-significance.

One way in which an autobiographical perceptual experiential memory can place the rememberer in the remembered scene is as an object of perception. An individual can remember and associate an ego-centric index with him- or herself as seen in a mirror—the memory analog of passing the mirror test. But an autobiographical visual experiential memory need not place the rememberer in the remembered scene as an object of perception. One can remember oneself *as* experiencing something other than oneself. Retaining a visual memory is distinct from retaining a visual memory in which one is also remembered as the visual perceiver. Such memory probably requires *conceptual* capacities. Perhaps most autobiographical *visual* experiential memories among selves, and *mature* persons, will be of this sort.[29]

I call a memory that is veridical, and whose veridicality includes preservation of the representational perspective of a representational state that is retained in the memory, 'memory *from the inside*'. I am interested in autobiographical episodic memories from the inside.

In the case of *perceptual* episodic memories, this notion of being from the inside is fairly straightforward. The memory retains the angle of perception—the perceptual perspective—on the subject matter that the original perception had. Not all perceptual memories are from the inside, in this sense. One could have a memory, in perceptual format, as of some subject matter. The memory could get right intrinsic properties and relations of the perceived subject matter. But the memory could be from a perceptual perspective that one never had. The memory is partly veridical, partly not. One might remember the layout of the scene quite correctly, but from a perceptual perspective outside of one's actual past perspective.[30]

retains and enables one to call up perspective. The perspective is *reused, not referred to* in the simpler forms of episodic memory.

[29] All experiential memories that place the rememberer in the remembered situation but do not do so through remembering the rememberer as an object of perception must have a conceptual, or at least post-perceptual, attributional element. Remembering oneself *as* a perceiver requires a representation as of perception. Perception itself cannot represent perception as such.

[30] A common but dramatic way in which a perceptual memory may be partly veridical, but not be from the perceptual perspective of the original experience, is for a memory to have a positional origin that is a rotation from that of the original experience. I have memories of an important childhood event from an impossible perspective twenty feet above the ground. Often a memory nonveridically places the rememberer as a perceived participant in the scene. In the psychological literature, memories from an external observer's position are called 'observer memories'. They are contrasted with 'field memories', those that take up the same positional perspective of the original experience. See Sigmund Freud, 'Screen Memories' (1899), in James Strachey (ed. and trans.), *The Standard Edition of the Complete Psychological Works of Sigmund Freud, Volume 3* (London: Hogarth,

The notion of retaining the perspective of a memory that is non-perceptual—for example, retaining a perspective of a particular occurrent inference or occurrent thought—is similar, even if less phenomenologically vivid. Retaining the cognitive perspective of the original intellectual occurrence—and being an episodic memory from the inside—is veridically retaining the mode and representational content of that occurrence.[31] Inasmuch as the memory is episodic, it must also be a *de re* memory of that occurrence.

Thus to be from the inside, an episodic memory must be veridical, and the veridicality must include retention of the perspective of the original perceptual or intellectual "experience".[32]

I have described autobiographical episodic memory from the inside because I believe that it constitutes another step toward self-objectification. It extends the process into the temporal dimension in ways that passing the mirror test, imitation, and joint attention do not.

There is dispute in the developmental and ethological literatures over how *meta-psychological representation*—representation of psychological kinds and instances as such—emerges. I think that the matter is not well understood, and will not discuss it here. Certainly by age three or four, human children have psychological concepts and apply them to themselves and others. I believe that when married to meta-psychological representation, autobiographical episodic memory from the inside is a momentous type of self-understanding that is constitutively necessary for being a self and for being subject to norms of morality and critical reason.

When autobiographical episodic memory attributes psychological states, it contributes to specifying one's own psychological history. It enables one to revive one's psychological past in some of the same form that it took. It connects one's present and past perspectives. Unlike updating memory, it provides a basis for attributing a contrast of perspectives. When an individual can *refer* to perspectives as such, episodic memory provides a direct, *de re* way for the individual to represent him- or herself as a psychological being with changing commitments.

VI

I conclude by reflecting on how the type of episodic memory that I have been discussing gives point to the Kantian conception of self that I cited at the outset.

1953–74); Georgia Nigro and Ulric Neisser, 'Point of View in Personal Memories', *Cognitive Psychology* 15:4 (1983), 467–482; John A. Robinson and Karen L. Swanson, 'Field and Observer Modes of Remembering', *Memory* 1:3 (1993), 169–184.

[31] There may be more to the original perspective than its cognitive aspects—what Frege called 'coloring', for example. I focus on retaining cognitive aspects of the perspective.

[32] This notion of veridically retaining the original perspective must allow for more or less. Most "veridical" memory involves some omission and some errors. Exact sameness with the contents of the perceptual point of view rarely or never occurs.

Recall that Kant wrote of a being that 'is conscious of its numerical identity, of its self, in different times'. What sort of diachronic consciousness is involved? Why does it matter? I will first make some remarks about consciousness. Then I discuss the relevant consciousness's significance.

Kant is right that consciousness figures prominently in our conception of selves. The relevant consciousness is representational and meta-psychological. The representation is both of the self and of its psychological states and events.[33] It attributes psychological states, and it associates an ego-like representation of them with them.

Consciousness is a distinguishing feature of episodic memory. Meta-psychological autobiographical episodic memory from the inside retains a past perspective in an ego-like way, providing a conscious representation of one's psychological past. Insofar as such memory is consciousness of *self*, it involves consciousness of an individual's *conscious* psychological past.

The consciousness in the remembering is, I think, both phenomenal consciousness and rational-access consciousness.[34] That is, the episodic memories have a what-it-is-like quality to them; and, further, they are accessible to the individual's rational powers—propositional attitudes. Both sorts of consciousness are conditions of the whole individual, as distinguished from a psychological subsystem. Conscious psychological states that the relevant episodic memories are *of* are also both phenomenally conscious and rational-access conscious.

Perhaps not all individuals with psychologies are conscious. Some animals capable of perceptual representation may not be. But selves are conscious in both senses.[35] Phenomenal consciousness is the primitive center of conscious psychological life. Rational-access consciousness either incorporates phenomenal consciousness into propositional attitudes or consists in the individual's direct, occurrent control of propositional attitudes in mental agency.[36]

Imputation of activity or affect *to an individual*, as distinguished from attribution to the individual's psychological subsystems, motivates what has traditionally been taken as the aspect of an individual's psychology that is the individual's *own* in a proprietary sense. This notion applies, I think, to all individuals with a psychology. In higher animals, both sorts of consciousness are sufficient

[33] Kant assumes these points. In the subsequent discussion (*Critique of Pure Reason* A361–362), he takes the consciousness to be not only of self but of 'determinations' (*Bestimmungen*), which are properties of self that are of the sort that help make it a self—psychological properties.

[34] Here I use Ned Block's distinction, with a refinement that I have suggested. See Block, 'On a Confusion about a Function of Consciousness', *Behavioral and Brain Sciences* 18:2 (1995), 227–247; and my 'Two Kinds of Consciousness', in Block, Owen J. Flanagan, and Güven Güzeldere (eds.), *The Nature of Consciousness: Philosophical Debates* (Cambridge, Mass.: MIT, 1997), 427–434, also printed in *Foundations of Mind*, 383–391; 'Reflections on Two Kinds of Consciousness', in *Foundations of Mind*, 392–419.

[35] No type of episodic memory is constitutively necessary for either phenomenal consciousness or rational-access consciousness. There are certainly phenomenally conscious beings that lack episodic memory. There are probably beings, higher animals and/or young human children, that have rational-access consciousness, but lack episodic memory.

[36] 'Two Kinds of Consciousness'; 'Reflections on Two Kinds of Consciousness'.

conditions for this proprietary ownership. They are constitutive of the complex type of psychology that determines what a self is.

I believe that the sort of consciousness that Kant had in mind *and* the sort of episodic consciousness that I have been developing support a more specific point about the consciousness of self over time. The consciousness is direct and non-inferential—hence *de re*. Kant would not have counted conscious theory about one's past, based on *inference*, sufficient for *consciousness of self* in the relevant sense. Nor would he have counted memory that is a mere summation of past experiences *consciousness of self* in the relevant sense.[37]

Meta-psychological, autobiographical *episodic* memory *from the inside* meets this requirement of a *de re* relation to the past—and illuminates it. In meta-psychological uses, episodic memory represents *instances* or *tokens* of past psychological events or states. This sort of memory contrasts with two other kinds of meta-psychological autobiographical memory. One type is generalized memory of types or patterns that is an amalgam or summation of individual past experiences, but that does not retain any one past psychological event or state instance and represent it as it was at a specific past time. Memories with the following contents are examples: my experiences with La Tâche have been thrilling; I have often seen him dominate her; I have learned that she is not to be trusted. The other type is memory that characterizes a specific event, but in a generalized way that does not involve *de re* memory *of* the event. Memories like the following are examples: I have once seen a nut buried there; I have felt this sort of pain once before; at some point, I came to believe that Schubert's string quintet is sublime.

None of these memories is in itself a memory *of* a particular event instance. The memories either summarize several past events without necessarily remembering one of them as it occurred, or they retain *that* a particular event occurred without necessarily remembering that event as it was at a specific past time. Episodic memories are not only non-inferential. They are directly *of* the remembered entity. They are *de re*. I think that meta-psychological, autobiographical, episodic memories from the individual's past perspective provide the sort of *self-consciousness over time* that, on the Kantian conception, is constitutive to being a self.

I turn now to why the diachronic aspect of this self-consciousness matters. What values attach to a self conceived in this way? I think that the conception figures centrally in the applicability and application of norms that are constitutive to the natures of selves and persons.

Locke took persons to be approximately what I call selves.[38] He characterized persons as thinking, reflective, self-conscious beings capable of extending

[37] Kant takes the consciousness to be via inner *sense*, which is non-inferential: *Critique of Pure Reason* A361–363. I agree with Kant that the consciousness is non-inferential. I do not agree that it must be via a *sense*—a capacity that yields only empirical cognition. See note 1.

[38] Locke took persons to be a subspecies of what *he* counted selves. He explained selves thus: '*Self* is that conscious thinking thing, (whatever substance, made up of whether spiritual, or material, simple or compounded, it matters not) which is sensible, or conscious of pleasure and pain, capable of happiness or

consciousness to different times and places.[39] Locke regarded person as a 'forensic' notion, one essentially associated with responsibility.[40] Locke's first characterization of persons probably motivated Kant's focus on diachronic self-consciousness. Locke highlights extending consciousness 'backwards' in time to 'action or thought'. (See notes 39–42.) This emphasis is motivated by a concern that persons be able to understand their responsibility for past misdeeds for which they could be punished.

Locke gives no account of extension of consciousness into the past, although he clearly takes instances of such extension to be memories. His discussion of accountability, combined with his writing of *repeating* past consciousness in extending consciousness into the past,[41] suggests that he may have had episodic memory in mind. I shall assume that he did.[42]

I believe that Locke was right to think that the relation between moral accountability and episodic memory is closely connected to the psychological *kind* that selves instantiate. However, he took the connection to hold at too specific a level. I will discuss these matters in more detail in the next lecture, but I sketch some issues here.

An individual can be morally responsible for deeds that he or she does not remember. An individual can be responsible for deeds that are motivated from a perspective that he or she does not know introspectively at the time the

misery, and so is concern'd for it *self*, as far as that consciousness extends.' *An Essay Concerning Human Understanding* (Oxford: Clarendon Press, 1975), Bk. 2:27: 17. See also 2:27: 25–26. I call 'egos' *roughly* what Locke calls the selves of beings that are not persons. I do not require an ego to be conscious. Egos must be purposive and have some representational competencies, at least perceptual ones. Locke's persons are selves (in his sense of 'self') that are intelligent, have rational reflection, can extend their consciousness to different times and places, and are morally accountable. Locke's persons are *roughly* what I count selves. I say 'roughly', here and above, partly because I do not agree with Locke on the identity conditions for persons, selves, or egos. Some of these differences will emerge shortly.

[39] Locke's fuller characterization of person: '. . . a thinking intelligent Being, that has reason and reflection, and can consider itself as itself, the same thinking thing in different times and places, which it does only by that consciousness which is inseparable from thinking, and as it seems to me essential to it . . . . in this alone consists *personal identity, i.e.* the sameness of a rational being. And as far as this consciousness can be extended backwards to any past action or thought, so far reaches the identity of that *person*; it is the same *self* now it was then; and 'tis by the same *self* with this present one that now reflects on it, that that action was done.' *Essay*, 2:27: 9.

[40] For the claim that 'person' is a forensic term, see *Essay*, 2:27: 26. Locke develops his views on persons and moral responsibility in 2:27: 18–26. For an illuminating discussion, see Patricia Kitcher, 'Two Normative Rules for Self-Consciousness', in Terrace and Metcalfe, (eds.), *The Missing Link in Cognition*, 174–187.

[41] In *An Essay Concerning Human Understanding*, 2:27: 10, Locke writes of persons' *repeating* 'the idea of any past action with the same consciousness it had of it at first'. This passage makes a relatively close fit with meta-psychological autobiographical episodic memory.

[42] It has been disputed whether Locke understands 'extension' of consciousness back in time to be a type of memory. Some have maintained that he meant merely continuity of consciousness. I think that his taking this extension to be something a person does (*Essay*, 2:27: 9–10) and his viewing personal identity and moral accountability as hinging on memory (*Essay*, 2:27: 20–22) suggest that he views extension of consciousness into the past as memory. At least, he takes memory to be a primary *type* of such extension.

motivation is formed. Contrary to Locke, a drunkard can be morally responsible for misdeeds that are neither remembered nor understood at the time.[43] People mired in self-deception or naiveté about their own psychologies are similar cases in point. Moreover, longevity of life can lead to forgetting misdeeds for which one remains responsible. It is simply not true that one must have an autobiographical episodic memory, or indeed any other memory, for each past deed for which one remains morally responsible.

The connection of accountability with meta-psychological, autobiographical episodic memory holds at a more general level. As a first approximation, I think that having a natural competence consciously to revisit some actions and state instances from the inside is a competence that is constitutive to our *status* as morally accountable beings.

Darwin points in the same direction. But like Locke, he underestimates the subtlety of the connection between memory and being subject to moral norms. He writes:

> A moral being is one who is capable of reflecting on his past actions and their motives—of approving of some and disapproving of others; and the fact that man is the one being who certainly deserves this designation, is the greatest of all distinctions between him and the lower animals.[44]

Although Darwin shares Locke's focus on memory, he aims—fruitfully, I think—at a generic connection between memory and moral status, rather than a connection between memory and each morally responsible act. On the other hand, Darwin is less focused on episodic memory than Locke is. Furthermore, approval and disapproval are not enough to make an individual a moral being, even when added to diachronic competencies that are relevant to moral status. An individual must also understand a distinction between right and wrong. Nevertheless, I think Darwin right to connect diachronic competencies with being subject to moral norms.

The constitutive connection between being subject to moral norms and being competent to extend consciousness over time is not that meta-psychological autobiographical episodic memory is the link that enables a person or self to remain the same person or self over time, as Locke seems to have thought. Identity is a necessary condition on primary moral accountability. But episodic memory is not the primary link in transtemporal identity. There are many other

[43] Locke's claims, now widely rejected, occur in *Essay*, 2:27: 22–26. Leibniz, *New Essays on Human Understanding*, ed. and trans. Peter Remnant and Jonathan Francis Bennett (New York: Cambridge University Press, 1981), 2:27: 9, criticizes Locke's views on accountability and personal identity. Leibniz allows for memory gaps in both. He maintains, correctly, that testimony of others can suffice to enable one to understand the justice of punishment for unremembered deeds. Of course, it does not follow that if all that one ever had had—even for very short-term time spans—was testimony from others about one's past, one would be a morally responsible person.

[44] Charles Darwin, *Descent of Man* (1871) (Amherst, NY: Prometheus, 1998), 633.

psychological links, including unconscious ones and ones other than memory, that help make possible the identity of selves and persons over time.

The connection between diachronic competence and moral accountability is more abstract. In lacking a competence for transtemporal 'extensions of self-consciousness'—of which a competence for meta-psychological autobiographical episodic memory is a central example—an individual lacks an essential aspect of a certain *understanding*. A capacity to understand one's actions in this way is necessary for being subject to moral norms, and partly constitutive of being a self. My next task is to explore the kind of understanding that has these features—with special attention to the role that memory plays in this self-understanding.

8 *Self and Self-Understanding*

Lecture II: Self and Constitutive Norms

In this lecture, I pursue a certain methodological strategy. I want to contribute to understanding the nature of selves. I think that certain sorts of self-understanding are among the constitutive features of selves. I explore that type of self-understanding by using clues from requirements on the applicability of two types of norms—those of critical reason and morality. Being subject to norms of critical reason is plausibly constitutive to being a self. Being subject to moral norms is at least arguably constitutive. Having certain psychological competencies, including relevant sorts of self-understanding, is constitutive to being subject to these norms. So the requirements for being subject to these norms provide insight into the nature of the relevant self-understanding, and ultimately into the nature of selves and persons.

I

A *norm* is a standard that governs good or optimal ways of fulfilling some function, or achieving some goal, end, or value.[45] Many norms do not require for their applicability any sort of understanding on the part of individuals that are subject to them. Norms for having a given standard of living or for having a life-sustaining nutrition, for any organism, are applicable even if they are not understood. Norms for the well-functioning of the heart or for formation of veridical visual perceptions require no understanding. They apply whether or not anyone knows about them. Norms are applicable given any function. Norm is a teleological notion. Not all teleology involves psychology, much less understanding.

Even some *rational* norms governing propositional attitudes require neither self-understanding nor understanding the norms. Reasoning by young children is subject to norms of deductive and inductive inference. Being subject to these norms requires no understanding beyond competence to make such inferences.

[45] See my 'Perceptual Entitlement', *Philosophy and Phenomenological Research*, 67:3 (November 2003), 503–548; and 'Primitive Agency and Natural Norms', *Philosophy and Phenomenological Research* 74:2 (September 2009), 251–278.

An individual is subject to these norms if it engages in inference. Inference has the representational function of preserving truth, supporting truth, or supporting some goal. Norms of reasoning apply to any being with propositional attitudes. Organisms that lack propositional attitudes can be evaluated for efficiency, but not reasonability. But young children and probably some non-human animals are subject to norms of reason, even though they cannot understand the norms.

I shall center on moral norms and norms of critical reason. Moral norms are norms for acting morally well, for making good decisions, for reasoning well to moral conclusions, and for having good character. Norms of critical reason perhaps need more explication.

Critical reason is a capacity to recognize and effectively employ reasonable criticism of or support for propositional attitudes and for propositional reasoning, guided by an appreciation, use, and assessment of reasons and reasoning as such. Critical reasoning is the sort of reasoning that evaluates, checks, refines ordinary first-level reasoning, using concepts and standards of reason. Critical reason is constitutively a capacity for meta-reasoning and meta-evaluation. It is commonly reasoning about ordinary subject matters and about reasons at the same time. It evaluates attitudes and reasoning for their reasonability.

Most of our reasoning, even most of our most sophisticated reasoning, proceeds at a first-order level. Most of our reasoning does not refer to representational contents or psychological states, and does not invoke the concept reason or standards for being reasonable. It simply operates under those standards more or less well. Critical reasoning is not in itself any better or more effective as reasoning than first-order reasoning. *Actual use* of critical reasoning is not necessary for making scientific discoveries. But a *capacity* to frame and defend a methodology is constitutive of science. Entering fully into practices that allow us to have our culture and history requires some capacity to explain and defend the practices. One must be able to answer why questions if one is to stand behind—with good or spurious reasons—the values, taboos, and permissions, the institutions and relationships, that make up a civilization. Such ability constitutes a rudimentary capacity for critical reasoning.

We are probably not the only species that reasons. Other animals probably carry out propositional, means–end inferences, subject to norms of reason. Ability to reason is probably not distinctive of selves or persons. But I think that ability to reason *critically* is.

I will not rely on a view about the relation between norms of critical reason and those of morality. Some, like Hume, hold that moral values are a matter of non-rational choice—a rationally optional project.[46] Others, like Kant, hold that moral values are matters of reason. Some think that a person or self could be irretrievably and naturally morally blind or indifferent—without damage, deprivation, illness, or malfunction. Others, following Aristotle and Kant, maintain that

[46] Even Humeans can allow a constitutive role for critical reason in morality—a role for a capacity to defend choices, including moral ones, as reasonable, *relative* to one's moral values.

moral values are associated with constitutive functions of human beings, persons, or selves. I am more sympathetic to the latter in each pair of views. Since the self-understanding that I discuss is parallel for morality and critical reason, little in the discussion will depend on either of the issues cited in this paragraph.[47]

I think it hardly controversial that having a natural capacity for critical reason is constitutive to being a person and being, or having, a self. It is arguable that being subject to moral norms is also constitutive. I will focus on both kinds of norms as clues to the nature of a type of self-understanding that is constitutive to being a person and self.

An individual's instantiating certain psychological kinds—including a kind of self-understanding and an understanding of notions like reason and moral wrong—is constitutively necessary, and arguably sufficient, for the applicability of the norms of critical reason and morality to that individual. Reciprocally, these psychological kinds necessitate the applicability of certain norms to their instantiations. So psychological capacities and normative applicability are interdependent. I shall use the applicability of these norms as clues to the natures of underlying psychological capacities.

The norms of morality and critical reason are not typical norms. Unlike norms for good nutrition, cardiovascular health, and veridical visual-perception formation, these norms apply only to beings with certain psychological capacities for propositional attitudes. The norms' applicability hinges on two other requirements. They require *understanding the norms*—some minimum internalization of them. And they require some *self-understanding*—an understanding of psychological conditions, in one's own case, to which the norms apply.[48]

[47] Locke's claim that person is a forensic notion expresses his view that there is a constitutive connection between being a person and being responsible to moral norms. This view is independent of Kant's more committal view that morality rests on critical reason. Hume differs from Kant in holding that morality does not depend on reason's determining basic values. I do not know whether he thought that a capability for morality is necessary to being a nondamaged fully mature person, or human being.

It is not uncommon to take Locke's claim to imply that the kind *person* does not mark off individuals and is not a basic psychological kind. Locke understood a forensic notion as one having to do with judicial proceedings and law, including divine moral judgments. It is, even now, common to contrast forensic notions with "ontological" notions like substance. Such a contrast is often taken to imply that the kind *person* marks a stance, phase, or contingent property—not a kind of individual. I believe that such contrasts often depend on an outmoded notion of substance. The stance-oriented way of thinking about normative notions also tends to conceive norms as applicable only given desires or choices. There are some norms of this sort. Perhaps norms of table setting apply only if one cares to conform. But norms of critical reason are not like that. And neither, I think, are norms of morality. I believe that these norms apply to individuals if and only if the individual has certain psychological capacities and certain rudimentary understanding. At any rate, I think that counting person and self notions with normative implications does not *entail* that they apply to less ontologically significant kinds.

[48] Tyler Burge, 'Reason and the First Person', in Barry C. Smith, Crispin Wright, and Cynthia Macdonald (eds.), *Knowing Our Own Minds: Essays on Self-Knowledge* (New York: Oxford University Press, 1998), 243–270.

Let us briefly consider the requirement of understanding these norms.[49] Being subject to moral norms requires some understanding of the difference between right and wrong. Both right and wrong are moral, normative notions. Right and wrong centrally concern psychological motivations and psychological initiators of action. The motivations and initiators must involve conceptualization of actions. Application of the norms to actions requires that actions be individuated partly by their psychological antecedents. Helping an old lady across the road mainly motivated by a desire that she arrive at the bank where a scam is planned is a morally different act from helping the old lady across the road mainly out concern for her well-being.[50] Minimal understanding of how moral

[49] Development in children of a capacity to apply norms is not deeply understood. Here is a rough summary. Child development from ages three to six is marked by changes in understanding the role of norms in self-regulation. By age three or four, when children clearly exhibit an ability to attribute beliefs and intentions, they show some awareness that others have different beliefs and goals. Between ages three and six, children exhibit an understanding that social regulation by others is mediated by others' attitudes. Evaluations had been applied to appearance, physical prowess, social behavior. They come to be applied to psychological states. Evaluation is initially accepted from others. It comes to be self-applied. Between ages six and ten, children come to evaluate individuals' psychological conditions and characters not just in terms of whether an outcome matches an antecedent intention or belief. They evaluate individuals in terms of the quality of the antecedent state, somewhat independently of whether an outcome is favorable. This distinction probably grows out of a distinction between talent and effort in physical pursuits. Such evaluations have a three-factor profile. First, there is the contribution of the individual's abilities (those relevant to talent, character, psychological capacity) to whether a standard is met. Second, there is the contribution of the individual's use of these abilities. Third, there is the contribution of external conditions to success or failure. Moral evaluations of individuals and their acts center on psychological capacities, character, and use of them. An individual can act well morally, although things turn out badly. An individual can act badly, although things turn out well. Similarly, evaluations for critical reason center on the individual's contributions, not on extra-psychological matters. Such evaluations are usually independent of the veridicality of an attitude, and of the success of the activity in realizing the individual's goals. An individual can be rational but mistaken, or right but irrational.

For a sampling of literature on development of cognition of norms, see Henrike Moll and Michael Tomasello, 'Level 1 Perspective-Taking at 24 Months of Age', *British Journal of Developmental Psychology* 24:3 (September 2006), 603–613; Charles W. Kalish and Sean M. Shevcrick, 'Children's Reasoning about Norms and Traits as Motives for Behavior', *Cognitive Development* 19:3 (July–September 2004), 401–416; Nancy Eisenberg, Richard A. Fabes, and Tracy L. Spinrad, 'Prosocial Development', in Eisenberg, William Damon, and Richard M. Lerner (eds.), *Handbook of Child Psychology, Volume 3: Social, Emotional, and Personality Development* (New York: Wiley, 1998); Adam Rutland et al., 'Social Norms and Self-Presentation: Children's Implicit and Explicit Intergroup Attitudes', *Child Development* 76:2 (March/April 2005), 451–466; Gergely Cisbra and György Gergely, 'Social Learning and Social Cognition: The Case for Pedagogy', in Yuko Munakata and Mark H. Johnson, (eds.), *Processes of Change in Brain and Cognitive Development: Attention and Performance XXI* (New York: Oxford University Press, 2006), 249–274; David R. Shaffer and Katherine Kipp, *Developmental Psychology: Childhood and Adolescence* (Belmont, CA: Wadsworth, 2009).

[50] I believe that I disagree here with Derek Parfit and Tim Scanlon, both of whom minimize motive in determining the moral goodness of an action. See Derek Parfit, *On What Matters*, vol. 1, ed. Samuel Scheffler (New York: Oxford University Press, 2011); and Thomas M. Scanlon, *Moral Dimensions: Permissibility, Meaning, Blame* (Cambridge Mass.: Harvard University Press, 2008). For a view that I find more congenial, see Barbara Herman, 'A Mismatch of Methods', in Parfit, *On What Matters*, vol. 2, ed. Samuel Scheffler (New York: Oxford University Press, 2011), 83–115.

notions apply requires not only *having* such notions, but being able to use them to evaluate psychological conditions.

Similarly, being subject to norms of critical reasoning requires minimal understanding of norms of critical reasoning—understanding what a reason is. The minimal understanding involved in having the concept reason requires understanding applications of the concept to *instances* of reasoning.[51] Reasons apply centrally to representational contents taken with their mode (belief, suspension of belief, intention). But, as noted, understanding the concept reason requires a capacity to apply it to instances. Understanding conditions under which the concept applies in actual reasoning requires a capacity to apply it to psychological states—instances as well as types—as individuated by their modes and representational contents.

Thus to be subject to either moral norms or norms of critical reason, an individual must have relevant normative concepts and abilities to apply such concepts competently to psychological states and occurrences, and their representational contents, as such. These are *higher-order*, ability-general metapsychological capacities.[52]

My main interest here is the requirement, for the applicability to an individual of norms of morality and critical reason, that the individual have self-understanding. To be subject to norms of morality, one must understand application of concepts of right and wrong to psychological states or to actions individuated in terms of psychological antecedents. This understanding must be applicable to one's own case. One must be able to evaluate one's own psychological states, particularly inasmuch as they determine and help individuate one's bodily actions; and one must be able to act on such evaluation. These abilities require an ability to understand what one's psychological states are.

Similarly, norms of critical reason are applicable to an individual only if the individual has a standing competence, at least in some cases, to understand his or her point of view and to criticize and affirm that point of view by reference to reasons considered *as* reasons. In exploring these requirements, I shall be guided by the idea that there must be a capacity to understand one's point of view and the applicability of normative concepts to it "*from the inside*". Like memory from the inside, understanding from the inside must get the understood state's representational features right, and must do so by preserving those features in certain ways. Preservation is a fundamental connector in any point of view. Selves capitalize on this connector. Conditions for applicability of norms of critical reason and morality center on understanding that makes essential use of preservation.

These points about baseline conditions for the *applicability* of the norms are fairly straightforward, as far as they go. I think that we can gain insight into the specific nature of the required self-understanding by reflecting on conditions for

[51] Indeed, instances of reasoning in one's own case. See my 'Reason and the First Person'.

[52] For explanation of the ability-general/ability-particular distinction, see my 'Five Theses on *De Re* States and Attitudes'.

the *application* of the norms in particular instances. The patterns of evaluations in such applications can be very complex, especially in the moral case. But some simple points about how antecedent psychological states and occurrences figure in such applications will, I think, lead us a step deeper.

II

Let us first look at some applications of moral norms. Moral norms apply most conspicuously to an individual's acts individuated in terms of psychological antecedents. Individual responsibility for such acts is centered on psychological antecedents. An individual can be just as responsible for a bodily act as for psychological antecedents that initiate it. But in such cases full, unmitigated responsibility for the act derives from the act's proceeding from its psychological antecedents in a natural way. The root of responsibility for the act is the individual's responsibility for psychological antecedents—the cognitions, intentions, decisions, and psychological initiations—that are subject to a certain sort of self-understanding. Moral responsibility has its constitutive, explanatory basis in psychological antecedents of acts, antecedents that are subject to a certain sort of self-understanding.

Similarly, an individual is morally responsible for consequences of his or her acts only if the consequences are traceable to something that the individual intentionally did or omitted, where the individual was capable of understanding the intention and initiation of his act, or omission, from the inside, and recognizing the act's moral implications.

On the other hand, an act can be wrong and an individual can be morally responsible for it, even if the individual does not specifically intend, or could not foresee, its harmful consequences. An individual can be morally responsible for acts carried out at times when the individual cannot understand what he or she is doing, in morally relevant terms. I noted the classical case of the drunk last lecture. An individual wild with anger can be morally responsible for injuring another, even if the injury was not specifically intended and even if the perpetrator was not in a state to understand the recklessness of the behavior. The individual might not even realize that he or she was angry or had reckless intentions. The individual could still be morally responsible for the behavior and the injury—not just for not knowing what he was doing and not just for not foreseeing the injury. Similarly, arrogant behavior is often not backed by specific intention or self-understanding. Racist beliefs are often not understood to be racist. Insensitive ingratitude or sexist acts are often not intended or understood as such. Yet an individual can be morally responsible for these states or acts.[53]

[53] Robert Merrihew Adams, 'Involuntary Sins', *The Philosophical Review*, 94:1 (January 1985), 3–31. Several of my examples are drawn from Adams's provocative article. He argues against always tracing responsibility in these cases to earlier culpable willings and understandings. I am not persuaded.

Criticism of the acts and the individual is often *mitigated* in such cases. The drunk driver is not morally as bad as the driver who lies in wait. An inappropriately angry person who would be horrified by his act in a cooler moment is often less bad than one who calculates injuring. Individuals are not strictly morally liable for all causal consequences of their acts. But neither are they always absolved of moral responsibility because they lack the self-understanding to evaluate them as they are formed.

I think that self-understood psychological antecedents in carrying out acts do, however, have a special place in *application*—not just *applicability*—of moral norms. The pattern of mitigation just cited suggests this point.

Responsibility by the drunk person hinges on the relation of the act to a baseline condition in which the individual controls his or her intentional acts and is in a position to understand their psychological antecedents and the acts' moral significance. At some time before the driver became drunk, perhaps in the development of drunken habits, the individual must have had a capacity to foresee risks to others, if the individual is to be counted morally responsible for the act. At some such time, the drunkard was in a position to understand and evaluate psychological antecedents to acts that in some foreseeable way led to being drunk.

Similar points apply to the cases of anger, arrogance, insensitive ingratitude, racism, and sexism. If these characteristics were to have developed in such a way that at *no* point was there *any* capacity to understand, evaluate, and control relevant psychological antecedents, the individual would not be morally responsible for them. It would be as if the states were injected into the individual. We control only little of our psychologies at any given time. But over time, we have opportunities to shape them. Such opportunities make possible moral responsibility for more than we control or understand at any given time.

The mitigation in moral evaluation of the drunk driver in comparison with the calculating murderer derives from a distance between the driver's condition at the time of the accident and the driver's condition when he was in a position to understand what he was doing. The fact that the driver had less understanding of and control over his point of view and its moral consequences at the time of the accident than at an earlier time is *commonly*—I do not say 'always'—a source of mitigation. Moral evaluation of individuals ultimately centers on ways an individual's acts and states relate to uses of the individual's point of view that *are* under intentional control and capable of self-understanding. Moral evaluation takes such uses as a baseline for determining mitigation.

The *application*, as distinguished from *applicability*, of norms of *critical reason*—especially theoretical critical reason—is less complex than application in the moral case. But again, ultimately, certain elements in an individual's point of view are privileged in evaluations. Let me lay out some of these differences and similarities step by step.

Norms of critical reason include norms of first-order rationality—those that apply to any reasoner—together with those rational norms that are specific to critical reasoners. Norms specific to critical reason apply only to propositional states and events that are in principle accessible to an individual's rational powers, immediate self-understanding, and rational evaluation, using the concept reason or some variant. Other aspects of an individual's psychology can be unreasonable, but not specifically *critically* unreasonable. However, elements of a point of view that the individual does not control, access, or understand fall under norms of critical reason. Elements in the Freudian unconscious, elements mired in self-deception or inebriation, elements distorted by entrenched irrationality, and elements camouflaged from self-understanding by anger or depression are subject to norms of critical reason. In this respect, application of moral norms and norms of critical reason are similar.

A difference is that norms of critical reason apply with equal force to all elements of an individual's point of view to which the norms are applicable. In the moral case, intentions and actions not presently understood or available to self-understanding are often *not* just as morally bad as they would be if they had been understood. The norms of critical reason include norms of ordinary non-meta-representational reason. Unreasonable elements in a point of view that are not presently available to self-understanding and self-evaluation are just as unreasonable as they would be if they were. The drunkard's inferences, which are not immediately accessible, but fall under standards of reason, are just as unreasonable as a sober person's making the same inferences under reflectively ideal circumstances. Some psychological states or transitions that would normally be open to self-understanding might be influenced by disease. Their unreasonability need not be explained in terms of something the *individual* did or did not do in allowing them to lapse, or in failing to bring them under the control of critical reason. Unlike immorality, unreason—even specifically critical unreason—can be *purely* the product of disease or malfunction. In assessing *failures* of reason, there seems to be no privileging of elements that are presently open to self-understanding over those that could be open to self-understanding, but are not.[54]

Still, as in the moral case, reasoning that is easily accessible to current self-understanding is privileged in *successes* of critical reason. Successes are constitutively open, readily open, to critical reflection that includes self-understanding. Psychological elements that are not currently easily accessible to self-understanding can fulfill norms of critical reason only via inertia: they were readily open, and have maintained their reasonability.

[54] There is, however, a pattern of mitigation that resembles the moral case. Suppose that an individual's reasoning is periodically clouded by anger. Suppose that the individual tries to change matters for the better, but does not succeed. The unreasonable states are just as unreasonable as if the individual had done nothing to combat them. But the *individual's* critical unreasonability can be mitigated. The *individual* is perhaps less critically unreasonable than if he or she gave in to the unreason.

An individual can use critical reasoning to operate on elements of the point of view that are resistant to critical reason, and—although they may be in principle accessible—may not be currently, easily accessible to self-understanding. We say that the individual *tries to get him- or herself* to think or act more reasonably. The recalcitrant elements are not, however, part of the critical reasoning. They are within the individual's point of view, and are subject matters of higher-order reasoning. But the aspects of the point of view that are the topics of the reasoning are not being used in critical reasoning. They are like objects to be manipulated. The resources of successful critical reasoning constitutively involve elements that are currently, easily accessible to self-understanding.

Meeting the norms of critical reason and morality occurs paradigmatically—I think, only—in cases identical with or derivative from cases in which an individual is currently in a position to readily understand the relevant norms and the application of the norms to relevant psychological states and occurrences. Understanding such application requires that the psychological elements be currently, readily accessible to self-understanding.

Let me summarize my points about applicability and application of the two types of norms. Both types center on psychological elements that are imputable to the individual subject to the norms—to that proprietary part of the individual's psychology in the individual's point of view. What happens in the environment, in intentional bodily actions, or in aspects of the individual's psychology that are irretrievably beyond the individual's control or consciousness, or that do not accord with the individual's self-understanding, have a fundamentally different status in the norms' applicability and application than elements within the individual's point of view that are subject to self-understanding.

III

Animals that are not persons and lack selves have points of view. Even in animals' points of view, there is a distinction between elements that are occurrently conscious, or are immediately or easily accessible to consciousness—whether phenomenal consciousness or rational-access consciousness—and elements that are not. Thoughts or perceptions that are retained in memory, but that would need extensive priming to gain consciousness or be available for rational use, fall in the latter category. Standing beliefs or memories that are available for use and control as soon as the occasion requires fall in the former category.

I call the part of an individual's point of view that is rational-access conscious, or immediately or easily accessible to such consciousness, '*the individual's rationally accessible point of view*'. Elements not immediately or easily accessible but imputable to the individual are part of the individual's point of view, but not part of the individual's rationally accessible point of view. Perceivers that lack propositional attitudes have a point of view, but no rationally accessible point of view.[55] Probably all individuals with rationally accessible points of view have further elements in their points of view, outside the rationally accessible point of view. They probably have memories that need priming to come into play.

For persons and selves, this distinction between point of view and rationally accessible point of view has a higher-level analog. There is a distinction between an individual's representational states and events that are immediately or easily accessible to conscious *self-understanding*, on one hand, and representational states and events that are imputable to the individual, part of the individual's point of view—and perhaps even in principle accessible to self-understanding—but not immediately or easily accessible, on the other. I call the former representational states and occurrences, together with conscious sensations and feelings that are also immediately or easily accessible to self-attribution, the individual's '*apperceptive rationally accessible point of view*'.[56] An individual's unrecognized anger, Freudian unconscious, self-deception, and other states that need extensive priming to be recognized, are outside the individual's apperceptive rationally accessible point of view.[57] The apperceptive rationally accessible point of view is the part of a point of view that is immediately or easily accessible to conscious self-understanding.

Our reflection on applicability and application of moral norms and norms of critical reasoning indicated that apperceptive rationally accessible points of view have a privileged status. Applicability and fulfillment of these norms require a capacity for self-understanding of relevant psychological elements and a capacity to evaluate such elements under the norms. Apperceptive rationally accessible

[55] In individuals with a rationally accessible point of view, there is a further distinction among elements of the point of view that are outside that part of their point of view. There are "outside" states imputable to an individual that are not accessible to consciousness even with priming, as well as those that are. Imputability to the individual is not fundamentally a matter of accessibility-in-principle to either sort of consciousness. Individuals have short-term, unconscious perceptual beliefs and memories that enter into individual-level explanations of an individual's activity, but that are forever beyond conscious control or phenomenal awareness. I think that such beliefs are imputable to the individual, first, because all beliefs are imputable and, second, because they can enter into individual-level explanations of action. This point will have an analog in discussion of critical points of view: some psychological states that are imputable to an individual cannot be claimed by the individual, even in principle, as the individual's own and are not under control of the individual's agency. (Of course, such states tend not to be *morally* imputable.) In other words, psychological imputability and point of view have their roots in committal states and activity not in consciousness. Committal perceptual states and activity are present in the animal world at very low phylogenetic levels. These points are partly terminological. But terminological choice has important substantive implications for how we think about psychology. Whatever the terminology, these distinctions should be drawn.

[56] In persons and selves, anything in the apperceptive rationally accessible point of view is in the rationally accessible point of view. The converse may also be true: anything in the rationally accessible point of view is, in persons and selves, in the apperceptive rationally accessible point of view. I leave this issue open to further investigation.

[57] *Occurrently* conscious elements in apperceptive rationally accessible points of view have a priority in our *sense* of life. I leave open whether they have normative priority. For purposes of the large normative considerations that we are discussing, it is the full rationally accessible point of view that is privileged.

points of view set standards for successful realization of the norms. They form a baseline for mitigation in negative moral evaluation.

I term this fundamental status of apperceptive rationally accessible points of view in explanations of normative evaluation in morality and critical reason their '*buck-stopping status*'. There are no further or deeper grounds for evaluating individuals' contribution to their accountability under norms of morality or critical reasoning. Applications of these norms are rooted in individuals' apperceptive rationally accessible points of view.[58]

The idea behind this point is that elements in a point of view that have buck-stopping status with respect to norms of morality and critical reason are constitutive starting points for expression of an individual's moral being or critically reasonable self. They mark ultimate unmitigated responsibility to the norms. All other elements in an individual's point of view are at best mediate expressions. They get their normative status derivatively—through relations to the apperceptive rationally accessible part of the point of view.

The primary relations here are inferential ones. The buck-stopping status of apperceptive rationally accessible points of view lies largely in their being baseline starting points for inferences. These baseline starting points necessarily are immediately, non-inductively accessible, at least within the conscious apperceptive point of view. 'Easiness' of access is meant to be unspecific. Easiness is relative to the starting points that are immediately accessible to self-understanding. In having easy access to some elements in a point of view by inference, one must self-understand the starting points of the inference.

Although both moral evaluation and evaluation of critical *practical* reason usually center on bodily intentional action, the baseline, buck-stopping starting points for such evaluation are the psychological antecedents of the actions. Let me make this point somewhat more precise. I shall henceforth call intentional bodily action just '*action*'. As noted, one can be equally accountable under the two types of norms for action and for its psychological antecedents. The capacity to reflect on one's actions, especially prospectively, is a factor in relevant normative evaluation. My claim is that the applicability and application of the norms of morality and critical reason to actions goes through their applicability and application to the psychological antecedents of the actions. The individual's contribution to his or her accountability for actions derives from the actions' relations to self-understandable cognitive and motivating psychological antecedents. Full accountability for actions depends on ability to self-understand psychological antecedents.[59] Where understanding actions themselves is a source of

[58] I think that not all self-understanding that grounds accountability under the relevant norms is agential. I discuss relations between rational-access consciousness—the lower-level underpinning of an apperceptive rationally accessible point of view—and individual agency in 'Two Kinds of Consciousness'.

[59] There is some question about whether individuals always know directly the initiation of a bodily action—when and whether the final choice or action initiation is made. Compare Masao Matsuhashi and Mark Hallett, 'The Timing of the Conscious Intention to Move', *European Journal of Neuroscience*

the individual's contribution to his or her accountability, the understanding of the actions must be through understanding their cognitive and motivating psychological antecedents. Basic, full, unmitigated accountability for an action depends on the action's realizing the content of those antecedents, where the antecedents can be self-understood. If the action is not individuated by the self-understandable psychological antecedent—for example, if it does not realize an antecedent intention—the evaluation of the act is mitigated, or at least depends partly on its relation to actions that *are* successful expressions of self-understandable psychological antecedents. The starting point and ultimate source for evaluation of individuals under the two norms is the set of cognitive or motivating psychological antecedents. These help individuate the actions that are evaluated. They obtain their buck-stopping status partly through their potential for being self-understood.

The converse points do not hold. The basis of accountability does not lie in understanding the cognitions and motives through understanding the individual's actions. We often do understand our motives retrospectively, from the actions inward. But such understanding does not ground *fundamental* accountability of individuals under the relevant norms. Motives that are understandable only outside-in are less fundamental expressions of an individual's moral and critically rational being. And actions that derive from such motives are often evaluated as involving mitigated accountability. Basic, unmitigated accountability traces to cognitions and motives that are subject to potential self-understanding in advance of, or during, actions that they produce. The *fundamental* basis for evaluating individuals' contribution to accountability under the two primary norms is the self-understandable cognitive and motivating psychological antecedents, not the self-understandable actions.[60]

28:11 (November 2008), 2344–2351. As long as the individual is capable of understanding and knowing, in a relevantly direct way, the motivating intention, the conditions on self-understanding of an action *in* doing it are met. If the action initiation flows naturally from an intention and decision and is not a surprise, the individual can be fully accountable for his or her action, under the relevant norms, if the intention and decision are open to self-understanding, even if the final occurrent action initiation is not. The latter might well be unconscious.

[60] These points must be understood in the complex context of moral evaluation discussed earlier. Individuals are morally accountable for actions whose outcomes and even natures they did not intend, if those outcomes and natures were foreseeable. Some philosophers have broadened these points by appealing to a notion of moral luck. See Thomas Nagel, 'Moral Luck' (1976), in *Mortal Questions* (New York: Cambridge University Press, 1979). I think that more exact analysis shows that the phenomena that Nagel discusses are not as destructive of our notions of agency and moral evaluation as Nagel advertises. To avoid relying on any particular view of moral luck, I center on full, unmitigated accountability, and on the applicability and application of moral norms and norms of critical reason to the *individual's contribution* to accountability. I believe that in a sense the individual's contribution is what these norms are *all* about. But the notion of the *individual's contribution* must be understood, and applied, in a nuanced way to make out this "sense". In any case, I stand by the view that a competence for self-understanding of antecedent cognitive and motivating psychological elements grounds full, unmitigated individual accountability, and grounds normative evaluation of the *individual's contribution* to his or her accountability for actions, under the relevant norms.

The ground for buck-stopping status of the *psychological antecedents* of action is that an action's nature or consequences may not conform to the individual's motivations, and may not be foreseeable by the individual. The individual's accountability depends on what he or she intends and is in a position to foresee. The specific sorts of accountability that are the focus of the norms of morality and critical reason depend on those aspects of an individual's point of view that are, at some point, accessible to self-understanding.

Self and hence *person* are kinds that are constitutively determined by capacities for self-understanding and for self-applying certain norms. Aspects of a self or person marked by these starting points are *the core self.* Core selves are partly individuated by their rationally accessible points of view. I want to understand the nature and normative implications of the *self-understanding*—distinguished from the self-application of norms—that is constitutive of core selves.

IV

Let us return to the Kant/Locke conception of a person or self as constitutively having *diachronic* powers of self-consciousness. Kant gives little explicit motivation for his conception (see note 1). Locke provides a motivation that claims too much.

In the previous lecture, I discussed meta-psychological autobiographical episodic memory from the inside as a prime instance of a diachronic self-consciousness constitutive to being a self. I want to explain how this type of memory and an analogous type of anticipation figure in competencies constitutive to selves. I discuss three ways in which the self-consciousness constitutive to selves—and to being subject to moral norms and norms of critical reason—has such diachronic implications: in inference, in practical decision making, and in dialectical aspects of critical reasoning and moral thinking.

The first way lies in inference. I make two assumptions. First, I assume that selves must be capable of propositional inference. Second, I assume that they must be able to think consciously *de re* of some of their propositional inferential events, as their own. The apperceptive rationally accessible point of view must reflexively apply *de re* to inferential events in it. Selves not only must be able to be conscious of *steps*, the premises and inferred contents, in inferences and be able to understand steps as reasons for conclusions. They must also be able to be conscious *de re* of events that make up the inference.

The first assumption is straightforward. As critical reasoners, selves must be able to make rational propositional inferences. The second assumption needs elaboration. Its requirement of consciousness is not what needs it. Being able consciously to carry out a self's constitutive representational functions helps define selfhood. The requirement that one think of inferences as one's own is also straightforward. What needs elaboration is the requirement that selves be able to think *de re* of their inferential events. Two considerations underlie this requirement.

One concerns the role of inference in the buck-stopping status of some of a critical reasoner's states. For any states in a critical reasoner to have such status, some inferences must have it. For being rationally accountable for state instances in one's point of view requires ability to support them by rational inference. Inferences, as token events, must be among the buck-stopping elements in a critically rational point of view. As a consequence of a point I made earlier about induction, if an individual could only infer inductively to what his/her inferences are, inferences could never be the starting point for the accountability of a critical reasoner. So one's access to one's inferences must not always be by inductive inference. Inductive inference is the only *inferential* access one *could* have to one's inference events. Moreover, the access must be through canonical—not purely descriptive—representations of the inferences: the access must be to the inferences on their own terms. Purely descriptive access would not allow the inferences, and their premises, to be starting points of accountability. A noninferential, not-purely-descriptive access is *de re*.

The other consideration behind the requirement of *de re* access to one's inferences concerns the role of such access in having a minimal competence with the concept of reason—a requirement on being a critical reasoner. Minimal competence requires understanding implementation of reasons as such—understanding being moved by reasons in particular cases—in the practice of one's own thinking.[61] If one could not implement reasons as such with respect to state instances that one has in mind—in the practice of one's own thinking—one could not understand implementation of reasons as such. If one lacked *de re* access to one's inferences—and knew them, if at all, only purely descriptively or by inductive inference—they would not be *part* of one's own critical thinking: one could not have them in mind. They would at most be topics of appraisal: like unconscious states that one knows one has, but cannot sustain or change in current critical reasoning. Implementation of reasons as such must be *de re*. Implementation requires *de re* access. If an individual knew only in a non-*de re* manner that he or she had engaged in an inference, the individual still could know whether and why he or she came to a conclusion and perhaps could appraise the inference in some way.[62] But the knowledge would not allow implementation of reason appraisals, any more than one could implement reason appraisals in

[61] I have argued this point at some length in 'Reason and the First Person', 250ff. I think that at least the first, and possibly the second, of the two considerations that I am expounding can be modified to show that having *de re* access to one's inferences is necessary to engaging in the minimal reasoning necessary to being morally responsible.

[62] I think that one cannot engage in reason appraisal if one cannot implement reasons as such—be moved by reasons as such to change or sustain psychological elements in one's own thinking. Cf. 'Reason and the First Person'. Both appraisal and implementation are necessary to have the concept reason. It should be noted that the argument of this paragraph could dispense with this premise. I have used it because it fits with the earlier discussion. The minimal point is that critical reasoning requires that one reflexively hold a state instance in mind as one deliberates with it, evaluates it, and determines according to reason evaluation whether to sustain it or change it. Doing so requires *de re* access to the state instance.

others' thinking. The inference that the individual would know about, the topic of the knowledge, would not be an integral part *in* any critical reasoning. The inference could not be held in mind. So in the absence of *de re* access to his or her own inferences, an individual could neither implement reasons as such with respect to inferences in the practice of his or her own critical thinking, nor minimally understand implementation of reasons with respect to inferences—being moved by them as such. Selves are critical reasoners. Critical reasoners must understand implementation of reasons as such in their own inferences. So to be a critical reasoner, hence to be a self, one must have *de re* access to one's own inferences.

The two assumptions that I have just motivated, in two ways, can be used to show that selves must have meta-psychological autobiographical episodic memory from the inside. The first assumption of a capacity for inference entails a capacity for memory. Inference is not a point event. It requires some representational retention. Retention of mode and content through an inference is an instance of purely preservative memory.[63] Purely preservative memory is not autobiographical episodic memory from the inside. But the requirements on being a self in the second assumption entail that having more than purely preservative memory is required for the relevant inferential capacities.

The second assumption, with its requirement of a capacity to think consciously *de re* of inferential events as one's own, immediately entails that the relevant inferential capacities carry a competence with *meta-psychological autobiographical* memory. *Episodic* memory must be long term, conscious, and of occurrences, as they were at a given past time. It is a form of experiential memory.[64] Experiential memory is *de re* memory of some particular. The requirement that there be a capacity for *conscious* memory is embedded in the second assumption. The requirement that there be an ability to remember inferential processes *de re* entails that the memory is experiential. It is *de re* of events in the inferential process.

What of the requirement of long-term memory? Long-term memory is a technical concept. It contrasts with short-term memory—which is, colloquially, *very* short-term. The various types of short-term memory last a few seconds, or less, and have automatic decay times. The memory required of a self in inference—in moral thinking or critical reasoning—is long term in this sense. At least, a being that had a reasonable lifespan and lacked a natural competence to hold together propositional inferences—or practical decisions or deliberative dialectic—longer than a couple of seconds would be rationally deficient in a way that would fall short of having the powers essential to being a person or self. (See note 102 below.)

[63] *Purely preservative memory* is retention of mode and content from an earlier psychological state, retention that does not introduce new subject matter and is not part of a justification for the conclusion of an inference. For discussion of purely preservative memory, see my 'Content Preservation', *The Philosophical Review* 102:4 (October 1993), 457–488; 'Memory and Persons'.

[64] See 'Memory and Persons'. Recall that experiential memory need not be warranted empirically. It is not a type of inner *sense*. See note 37.

Requiring that a capacity for meta-psychological autobiographical episodic memory in inference be *from the inside* entails requiring the memory to be capable of veridically preserving the perspective of the inference. This requirement follows from the second assumption. The relevant sort of *de re* thinking about one's inferential processes involves correctly remembering the mode and content of particular inferential steps through purely preservative memory. When meta-psychological autobiographical episodic memory relies on purely preservative memory in this way, the former type of memory preserves the perspective of the inference and represents it as it was when it occurred.

The two assumptions show that meta-psychological autobiographical episodic memory from the inside, or some short-term analog, is constitutive to selves. Such memory is clearly a type of consciousness of one's past self, plausibly in the sense of Kant and Locke.

I turn more briefly to a second way in which diachronic capacities are constitutive of being a self. I believe that both the special type of memory that I have been discussing and an analogous capacity for anticipation of the future are constitutive to selves, moral beings, and critical reasoners by virtue of their constitutively having *practical* reasoning. Practical reasoning has both forward-looking and backward-looking diachronic dimensions. Intentions have the representational function of being fulfilled by the intender. When they are fulfilled through occurrent decisions to act, and by actions themselves, there is a match with the antecedent intention. Self-conscious selves can recognize these points. Looking forward from an intention, an individual is aware of his or her being the prospective executor of the intention. The individual can think anticipatorily of him- or herself, in the first-person way, as preserving the intention and as realizing it, through a decision to initiate action and through the action.[65] Remembering backward from a decision to initiate action, or from an action, a self-conscious individual can recognize moving on the antecedent intention, and satisfying or failing to satisfy the intention. These diachronic capacities are integral to the self-understanding that is constitutive of being a self and of being subject to moral norms and norms of critical practical reason. They are consciousness of oneself as one will be in the future, as well as consciousness of oneself as one was in the past.

An argument similar to the one given for propositional inference shows that requirements on having the reflexive powers with regard to decision making that are required of selves requires meta-psychological autobiographical episodic memory from the inside—and an analogous *de re* capacity to anticipate particular representational events in one's future.

[65] Framing possibilities for action is an aspect of selves' making choices. Selves have a probably constitutive capacity to think of themselves not only diachronically, but modally. Locke writes of selves as being able to be conscious of themselves in different times *and* places. Cf. *An Essay Concerning Human Understanding*, 2:27: 9. He may have had in mind the relatively primitive freedom of imagination to think of different spatial possibilities in action.

A third source of the requirement that selves have self-conscious diachronic capacities lies in the dialectical reasoning integral to deliberation. A condition on being subject to norms of critical reason and morality is to be capable of dialectical self-criticism.[66] A person or self must be able to stand behind a belief or decision after reflecting on it, to answer self-questioning, and to effect change of mind after self-criticism. Such dialectic requires an ability to think *de re* of one's thoughts, while being committed to them and while criticizing past or prospective commitments. In critically evaluating a present commitment, one must be able to conceive oneself as possibly giving it up. In changing one's mind, one must be able to conceive oneself as having given up an earlier commitment, knowing why. Recognizing actual or possible differences in one's point of view, retrospectively or prospectively, is part of being a *critical* reasoner. Diachronic forms of self-consciousness are part of recognizing what accountability is—a capacity to support or change one's position through rethinking or new information. Such capacities to recognize differences in one's own point of view, as they unfold in dialectical reasoning, are constitutive to being a self.

The diachronic use of self-understanding that is partly constitutive of apperceptive rationally accessible points of view combines *episodic* memory with *purely preservative* memory. Episodic memory gives one diachronic *de re* access, allowing states other than immediately present ones to have buck-stopping status and to be part of the thinking practice that one implements reasons within. Purely preservative memory gives one diachronic access to the relevant psychological elements on their own terms—also a central requirement on their entering into critical reasoning and moral accountability.

To elaborate this latter point: purely preservative memory retains the mode *type* and representational content of psychological state instances. Analogously, for anticipation of the mode and representational content of one's future point of view. Mode and content are the aspects of psychological state instances that are relevant to application of norms of critical reason and morality. If diachronic self-consciousness could not preserve the content and inferential order of past psychological events, one could not review one's inferences and commitments on their own terms. If diachronic self-consciousness could not anticipate the mode and content of instances of one's future points of view, one could not control one's future acts. Reflective review and anticipatory control are basic functions of a self.

I believe that the foregoing considerations provide a deeper characterization and motivation of the relation between selfhood, or psychological personality, and diachronic self-consciousness than we find in Locke or Kant.

[66] Dialectical reasoning is a special case of agency. So the points in this paragraph specialize the points in the previous paragraph. I distinguish practical agency and agency involved in theoretical reasoning. Dialectical reasoning can be either practical or theoretical. The agency involved in reasoning to intentions or to act-initiating decisions is practical.

V

I conclude by summarizing features of the relevant self-understanding. These features will ground discussion of its epistemic warrant. That topic will dominate the third lecture.

The self-understanding must be *meta-psychological*. The norms that I have discussed demand understanding the norms and applying them to one's point of view. To understand application of the norms, it is not enough to have *competence understanding*—the capacity to *use* concepts. One must be able to think about propositional attitudes, and their representational contents, as such. This is meta-psychological understanding.

Second, the meta-representational understanding must be *phenomenally conscious* and *rational-access conscious*. There must be a what-it-is-like quality to the understanding. And the lower-level, "understood" aspects of the point of view must be occurrently accessed by the individual's rational powers. The requirement that a self be phenomenally conscious, some of the time, is so basic that it is hard to motivate. Except in cases of understanding sensations and other states constitutively associated with consciousness, phenomenal consciousness is less important epistemically than in marking off the psychological kind, *self*, that underlies our deepest normative valuations. Those states that are constitutively associated with phenomenal consciousness (pleasure, pain) ground many primary moral valuations. Beings that lack a natural capacity for phenomenal consciousness do not have the moral claim on us that phenomenally conscious beings do. The requirement that a self be able to make accessible to conscious rational control some lower-level aspects of its point of view is just the product of combining the requirement of phenomenal consciousness with the role of self-understanding in applying the norms to one's own case.

Third, the understanding must exercise *self-conscious* meta-representation. A self must have a first-person self-concept and must apply it consciously. It must recognize aspects of its point of view as its own. In thinking I believe that Brahms is greater than Chabrier, I not only refer to myself with I; I take the belief that I represent to be mine. This capacity to represent elements of one's point of view as one's own is, of course, basic to the self-evaluation that grounds being subject to moral norms and norms of critical reason.

An aspect of this self-consciousness is that both the lower-level psychological element and the meta-representational element that constitutes the self-understanding are reflexively *understood*. One is accountable for the meta-representational self-understanding as well as for the understanding's subject matter, the lower-level psychological element.

Fourth, the understanding must include past and future as well as present elements in one's point of view. I have highlighted this constitutive requirement because I believe that it has been less well understood in the philosophical tradition than some of the others.

Fifth, the self-understanding must be *systemic*. This requirement underlies the fourth one. The self-understanding represents not just psychological states and their representational contents. It represents relations among the attitudes and the contents. Reason is essentially relational. An understanding that applies the concept reason must specify relevant relations. Moral evaluation of acts must relate them to beliefs and intentions. Even where moral norms evaluate attitudes, they must do so *relative* to information accessible to the individual.

Sixth, the self-understanding must be *specific*—from the inside. To ground norms of morality and critical reason, the self-understanding must specify psychological states in ways that preserve the point of view to be evaluated. The specifications must enable one to know what the point of view is on its own terms. Since the norms concern the perspective of the individual, an understanding that did not preserve the mode and content of the evaluated psychological elements would be too detached to render the norms applicable. Of course, much is forgotten. But an individual's having a *capacity* to preserve the mode and representational content of evaluated attitudes is constitutive to applicability of the norms.

Seventh, the relevant self-understanding must be *immediate*, non-inferential, in the apperceptive rationally accessible point of view.[67] Access to self-understood psychological elements is not via individual-level inductive inference. Psychological elements immediately accessible to self-understanding are starting points for expression of an individual's moral being or critically reasonable self. They have buck-stopping status for evaluation under the relevant norms. Psychological elements that are not immediately accessible cannot ground applicability of the norms of morality and critical reason. The starting points of, and transitions in, some inference must be capable of non-inferential self-understanding.

There is an existence proof of individual-level non-inferential elements of self-understanding. It is provided by Descartes' pure and impure *cogito* judgments. The structure of these judgments shows clearly that they are not inferential. When I judge I am thinking, or I [hereby] judge that Brahms is greater than Chabrier, or I intend to take the metro, the judgment is reflexive in a way that precludes individual-level inference. The judgment about the psychological state has, as a constituent part, the psychological state instance that the judgment is about. Judgment and subject matter are part of a single thought.

Although *cogito* cases do not involve any individual-level inference, unconscious psychological causal processes connect some relevant types of self-understanding with their psychological subject matters. For example, all self-understanding that includes meta-psychological autobiographical episodic memory relies on sub-individual psychological transitions. Whether these are propositional inferences is an empirical matter. I think that they probably are not, but I rest nothing on this belief.

[67] One can work out inductively things about one's point of view from inside it, as long as one starts one's inferences from a self-understanding that is immediate and from the inside. Wisdom about oneself lies in that direction.

What is important is that these memories are not results of *individual-level* inferences. *Within the apperceptive rationally accessible point of view*, some of them are immediate. Any "easy" inferences to self-understood states in a rationally accessible apperceptive point of view must be open to non-inferential self-understanding of the inferences' starting points and transitions. If an individual comes inductively to a view about a state that is not immediately or easily accessible, and relies entirely on the induction for self-understanding, the individual approaches one aspect of his or her point of view from the perspective of another. The understanding is not from the inside. For it does not preserve that point of view.

Psychological elements can have buck-stopping status only if they can be self-understood from the inside. They must be capable of being self-understood *in* the relevant psychological acts, states, or processes. These points apply to autobiographical memories. To self-understand past states in ways that give those past states buck-stopping status, some of the memories must be from the inside. Hence those states cannot be understood entirely on the basis of induction, within the apperceptive rationally accessible point of view.

An eighth feature of the self-understanding is its *de re particularity*. This particularity has several aspects. Recall that *de re* representation is representation that picks out a particular via a non-inferential, not purely attributional or descriptive competence.

The understanding is *de re* in four ways. First, the application of the self-concept is *de re*. Application of the self-concept is not purely attributional or descriptive. It is effected through a rule that infallibly effects reference through the mere occurrent *use* of the concept.

Second, one's understanding of the *representational content* of one's psychological states is *de re*. Self-understanding uses canonical representation of representational content. Such representation uses the representational content in the very representation of it. When one understands one's belief that Federer will win a match, canonical representation of the belief's representational content (that Federer will win a match) refers to the representational content, and also uses that representational content. Such canonical representation is *de re*.

Third, self-understanding of elements in one's point of view specifies them by applying a temporal index. This indexing is *de re*. One singles out the time non-descriptively and indexically, roughly as *now* or *this time*. In self-understanding that uses episodic memory, there are two *de re* temporal indexes: one for time of the self-understanding, the other for the past time of the understood psychological element.

Fourth, the self-understanding is *de re* with respect to particular state instances, acts, or occurrences. One understands instances of psychological types, not just repeatables. In singling out instances, one uses concepts—the canonical name of the representational content, the self-concept, temporal indexicals, and concepts for psychological modes (like believes, intends, perceives as of). But the use is not purely conceptual. One also occurrently *applies* the concepts to yield *de re* understanding of psychological particulars. In thinking I am wondering if Israel will make peace, I refer to an occurrence of wondering. In thinking, non-inferentially, I believe that Rembrandt is deep, I represent a standing instance of the belief.

It is constitutive to the role of self-understanding in grounding applicability of these norms, and in constraining their application, that it apply *reflexively* to the particular point of view and the particular psychological states of the understander, not just the same type of point of view. The individual must understand applicability of the norms to elements in his or her *own* point of view, *understood as such*.

I have been discussing *representational* aspects of the self-understanding that helps ground selfhood, ground the applicability of moral norms and norms of critical reason, and constrain application of these norms. It is meta-representational; conscious; self-conscious; diachronic as well as synchronic; systemic; specific; immediate, or non-inferential, at the proprietary individual level of the psychology; and, in multiple respects, *de re*. Next lecture, I explore the *epistemic* status of the self-understanding—connecting representational structure and representational competence marked by that structure with their epistemic credentials.

9 *Self and Self-Understanding*

Lecture III: Self-Understanding

I have taken selves to be constitutively self-conscious critical reasoners, and arguably constitutively subject to moral norms. I isolated a type of self-understanding required for applicability of these norms. The norms require an individual to be able to understand some of his or her psychological states from inside having those states.

Norms of critical reason and morality hinge on elements in an apperceptive rationally accessible point of view. Such a point of view is that part of a point of view whose elements are immediately or easily, presently, consciously accessible to self-attribution. Self-understanding helps give such elements *buck-stopping status* with respect to the two norms. Having buck-stopping status is having most fundamental status in explanations of the applicability and applications of a norm.

I

I want to understand the *epistemic warrant* for this self-understanding. To explain any epistemic warrant, one must explain what makes a warranted state an objectively good route to truth, given an individual's background information and competencies. Being a good route implies reliable veridicality in normal conditions. Normal conditions are conditions by reference to which the state's representational nature is constitutively determined.[68]

A warrant must also "make sense of" or rationalize the individual's point of view. This condition is hard to make precise. It is grounded in the idea that

[68] For an argument that normal conditions, in this sense, are the relevant conditions in which the reliability of the underlying warrant is to be considered, see my 'Perceptual Entitlement'. This account shows how the condition of reliability, and warrant for one's beliefs, can remain in place even in conditions—such as brain-in-vat conditions—in which one's beliefs are highly unreliable. Although *de facto* reliability—for example, in abnormal conditions—is not constitutive of being warranted in the empirical case, *de facto* reliability in one's access to some of one's own mental states *is* constitutive of being a self. Individuals that lose all reliability in judgments about any of their mental states, say through pathology, lose their selfhood.

warrant is a norm for psychological competence. There must be something *about the nature of the individual's capacities or point of view* that helps explain the individual's psychological states' being a good route to truth. Brute, "accidental" reliability does not suffice for warrant.

It would be too much to require an individual to be able to explain the warrant. Animals, children, and even adult persons can be warranted in perceptual or other beliefs, and yet be unable to explain why. On the other hand, correct explanation, even if the individual cannot give one, must center on the individual's point of view. For what is being made sense of or rationalized, under epistemic norms, is the individual's competence and perspective.

Epistemic warrant for a belief is sufficient for knowledge, if the belief is true and if its relation to truth is in certain ways not abnormal or adventitious. Warrant concerns the contribution to goodness of route to truth by the *individual's* competencies. Failures of knowledge do not reflect specifically on the individual or the individual's psychology if the individual's states are warranted. In this way, warranted but false beliefs are like good acts with bad consequences. I call a false but well-functioning and warranted propositional attitude, warranted in way *W*, an attitude that incurs *brute error* relative to warrant *W*.[69]

Many epistemic warrants are compatible with brute error. Nearly all *empirical* warrants are compatible.[70] Perceptual beliefs can be warranted but mistaken because of an illusion that one lacks ground to reject. Reasonable inductive empirical explanations can be mistaken. Even most empirical *knowledge* rests on warrants that are compatible with error.

The warrants for self-understanding and self-knowledge that we are exploring show a different pattern. The warrants yield *immunity to brute error*: if the psychological state is warranted, it is veridical. Before developing this point, I shall discuss it intuitively.

Many psychological states are not easily accessible to self-knowledge. Then a belief's being warranted is compatible with brute error. Knowing one's deeper motivations, or character, or the unconscious states and transitions that populate the mind, is hard. Error can derive from the sheer difficulty of the psychological subject matter. Often, mistaken self-ascription is unwarranted. But warranted error can certainly also occur.

By contrast, in many types of cases, if a mature person is wrong about his or her beliefs, intentions, percepts, or sensations, it is common to infer that there is something wrong with the person. Errors reflect on the use or health of the person's cognitive equipment. If a person were wrong about his or her *belief* about whether he or she has a sibling, it would be normal to infer that the person

[69] I introduce the notion of brute error in 'Individualism and Self-Knowledge', *The Journal of Philosophy* 85:11 (November 1988), 649–663; cf. also 'Perceptual Entitlement', section II.

[70] Exceptions are knowledge of one's sensations and perceptions. These are empirical because their warrants depend for their force on *having* of the sensation or perception. I later discuss why at least paradigm cases of these sorts of knowledge are immune to brute error.

was self-deceived, or had allowed emotion to cloud judgment, or was afflicted with some pathology. It is easier to imagine cases in which one has a warranted mistaken belief about whether one *has* a sibling. I am evoking the oddity of a warranted but mistaken belief about whether one *believes* that one has a sibling. Here error commonly reflects on how one formed one's meta-belief. Often, a mistake entails malfunction or other non-standard belief formation, which undermines warrant.

These issues are delicate. The *natures* of *most* psychological states that are accessible to relevant self-understanding do *not* guarantee that they are accessible.[71] Being a belief that one has a sibling does *not* insure susceptibility to self-understanding that is immune to brute error. But I shall argue that, for selves, there are *necessarily some* cases of immunity to brute error, and that such cases ground applicability of norms of critical reason and morality.

I cannot emphasize too strongly that except in *cogito* cases, self-attributions that are warranted in the relevant way *cannot* be determined by the contents and modes of the attributed states. Evidence as to whether one has a sibling could be mixed. One could be unsure of the nature of one's own reaction to the evidence: do I believe that I have a sibling? Using behavioral evidence about oneself, one could be warranted in concluding that one does believe one has a sibling. But one could lack that belief. One could be puzzled, or have a leaning that falls short of belief. Then, the warrant for the mistaken belief that one believes that one has a sibling would not be the warrant that I am reflecting upon. The warrant would not be immune to brute error. Except in *cogito* cases, one cannot generalize apriori about *which* of one's self-attributions can be warranted in the special way. What is apriori determinable, I think, is that *some* psychological states in selves are susceptible to a type of self-understanding that is immune to brute error, relative to the relevant warrant.

Intuitively, we sometimes do understand our own psychological states in a specially warranted way. This intuitive point is supported by theoretical considerations. Let us reflect on the relation between immunity to brute error for the relevant self-understanding and the buck-stopping status of self-understood psychological states in critical reason and morality. Immunity of the self-understanding to brute error is constitutive to having an apperceptive core self. A type of self-understanding that is *liable* to brute error cannot, I think, ground the buck-stopping status of attitudes in applications of these norms. If an individual's self-understanding of a psychological element can be warranted but mistaken, the self-understanding is too detached from that element to give that element buck-stopping status. That is, if one can understand a psychological element only in a way that is liable to brute error, that element cannot be a basic starting point for assessing the individual's contribution to full unmitigated accountability under the norms of critical reason and morality.

[71] Exceptions are pure *cogito* cases and some knowledge of conscious sensations.

Being subject to brute error means that in reviewing one's psychological state or event, one is not fully accountable for getting that state or event right. Error is compatible with one's best use of one's competence for self-understanding. Such reviews do not constitute exercise of a self-understood point of view that one has cognitive control over. They are not exercises of a cognitively unified perspective. The norms of critical reason and morality take the basis for their evaluations to be psychological states that are understood from inside the perspective of those states—preserving that perspective meta-cognitively in the individual's self-review. The norms require the individual to be able to understand what he or she does or undergoes *in* the doing or undergoing, from the inside. If one could not take up the same perspective that those acts or endurings had from the inside—incorporating it into the meta-representational review, by preserving it or making it part of a single meta-representational perspective—one would always be in the position of a privileged outsider. Psychological elements self-understood in ways that allow for brute error lack buck-stopping status in persons and selves.

The idea is that, ultimately as a critical reasoner and as a moral being, an individual's self-understanding must inform, or be capable of informing, some of the individual's cognitions, reasoning, decisions, and other states central to being a self. The self-understanding must inform them in a way that is part of their exercise or part of their being experienced. Working out an inductive, empirical understanding of one's reasoning and decisions, or relying on some other sort of understanding of them that is subject to brute error, is not the kind of self-understanding that forms a basic starting point for critical reasoning or morally responsible decisions. The self-understanding must be from the inside—from a meta-perspective that preserves and is informed by—the psychological states and acts that one is most basically and unmitigatedly responsible for as a critical reasoner and moral being. One must be able to self-understand from the inside those states, acts, and events that are fundamental to being a self.[72]

There are certainly types of self-understanding that rest on warrants that leave one vulnerable to brute error. Immunity to brute error is not an honorific status. It is not invulnerability to all error. It is a normative standing that hinges on possible sources of error, relative to epistemic norms. The deepest wisdom in self-understanding—a result of learning about oneself—*is* vulnerable to brute error. I claim only that the self-understanding required by norms of morality and

[72] I make a version of this point in 'Our Entitlement to Self-Knowledge', *Proceedings of the Aristotelian Society* 96 (1996), 91–116; cf. 109ff. See also the discussion of (CR) below.

One can *get to* a veridical self-understanding that replicates the perspective of a psychological state by inductive inference. If the self-understanding relies on the inference, it would still be subject to brute error, as all induction is. But it would not have buck-stopping status. It could not ground the applicability of norms of critical reason and morality. Only the starting points of the inference could have buck-stopping status, assuming that they are susceptible to the right sort of self-understanding. The starting points of the inference could not have buck-stopping status and be subject to brute error, if they are to be *from inside* the perspective of the understood psychological states, in the relevant sense.

critical reason—and required for there to *be* selves—is immune to brute error: being warranted is incompatible with error. This is a less comprehensive, less insightful self-understanding than wisdom about oneself. But it is the sort that is constitutive to being a self, and to the buck-stopping status of self-understood states in the applicability of norms of critical reason and morality. From here on, I *capitalize* ('Self-Understanding', 'Self-Knowledge') to indicate the self-understanding and self-knowledge that is my focus. *Sans* capitalization, the terms express generic notions that apply to all types.

II

In the last quarter century, self-knowledge has come in for a lot of discussion. I will not canvass the literature. I ignore views that assimilate Self-Knowledge to inner perceptual belief or to empirical hypotheses. I and others have criticized such views elsewhere.[73] I also ignore views that deflate Self-Knowledge, holding that it is trivial or non-substantial. I will, however, say a little to distinguish my view from others.

The psychological competencies that engender Self-Knowledge are not in general infallible. Pure *cogito* thoughts *are* infallible. If one thinks I hereby in this very act think [entertain] the thought that *Figaro* is the greatest opera, that thought about my thinking cannot be mistaken.[74] But pure *cogito* cases comprise only a tiny portion of Self-Knowledge. Immunity to brute error in Self-Knowledge is not immunity to all error. Non-inferential self-attributions of the relevant types can be mistaken because of unwarranted interference or pathology. The epistemic specialness takes subtler forms than infallibility.

The competencies that yield Self-Knowledge are also not to be understood in terms of self-intimation. I doubt that *any* type of psychological state is such that if an individual is in it, it is impossible for the individual not to believe that he or she is in it.

Some traditional accounts of self-knowledge explicate its epistemic specialness in terms of the nature of the subject matter. Conscious mental elements are supposed to render judgments about them epistemically special because of the elements' very nature. Thoughts were said to be transparent, self-evident,

[73] Sydney Shoemaker, 'Self-Reference and Self-Awareness', *The Journal of Philosophy* 65:19 (Oct. 3, 1968), 555–567; Shoemaker, *The First-Person Perspective and Other Essays* (New York: Cambridge University Press, 1996); my 'Individualism and Self-Knowledge' and 'Our Entitlement to Self-Knowledge'; Richard Moran, *Authority and Estrangement: An Essay on Self-Knowledge* (Princeton: Princeton University Press, 2001); Akeel Bilgrami, *Self-Knowledge and Resentment* (Cambridge University Press: Harvard, Mass., 2006), chapter 1; and Lucy O'Brien, *Self-Knowing Agents* (New York: Oxford University Press, 2007), chapter 3.

[74] See my 'Individualism and Self-Knowledge'; Andreas Kemmerling, 'Eine reflexive Deutung des Cogito', in Konrad Cramer et al. (eds.), *Theorie der Subjektivität* (Frankfurt: Suhrkamp, 1987); Kemmerling, *Ideen des Ichs: Studien zu Descartes' Philosophie* (Frankfurt: Suhrkamp, 1996). The original source is, of course, Descartes, *Meditations on First Philosophy*, II.

unmistakable, or self-intimating by their very nature—as if they glowed in the dark. Such accounts are neither explanatory nor accurate about the psychological or epistemic phenomenon.

Some recent accounts have made claims—rhetorically different, but I think ultimately similar—that it is of the nature of 'genuine', 'full-blown' intentionality that mental states be known or normally knowable by any individual who has them.[75] Animals and very young children are attributed a lesser type of intentionality, or none at all. I think that such views are hopeless. Their conceptions of "genuineness" in intentionality lack serious defense. The accounts do not accord with scientific explanations that attribute perceptual, intermodal, and propositional intentional (representational) states to animals and very young children. Many or all of these individuals lack self-knowledge. They have intentional, representational states in the most literal sense of these terms. No argument has shown that their psychological states are any less genuinely representational than those of human adults. There is no difference in kind or degree in their *representationality*. Self-Knowledge, and self-knowledge, cannot be explained in terms of the natures of first-order representational states. There is nothing in their natures that insures that they are self-known or self-knowable.

The natural phylogenetic and developmental order is to have representational states before being able to know them. This natural order is the order of language acquisition. Even if higher-order representational concepts are innate in human beings, which I doubt, no one has given good reason to think that there is a constitutive connection between an individual's having representational states, even having propositional attitudes, and that individual's being in a position to know what those states are.

Another claim is that the subject matters of self-knowledge constitute *reasons* for the knowledge.[76] Both lay-people and philosophers think about reasons in disparate ways. I make just a few points here.

Reasons are necessarily representational. They are considerations that are good routes to truth, in the case of theoretical reasons, or good routes to goodness or aptness of acts, in the case of practical reasons. A thunderbolt or a sea otter cannot be a reason. Only some propositional representational content, with some associated mode—for example, judgment—about the thunderbolt or sea otter can be a reason. Nature *grounds* reasons.[77] It provides entities and situations representation of which can form a reason. But reasons are constitutively propositional

[75] Donald Davidson, 'Rational Animals' (1982), 'The Second Person' (1992), and 'The Emergence of Thought' (1997), all in *Subjective, Intersubjective, Objective* (New York: Oxford University Press, 2001); Bilgrami, *Self-Knowledge and Resentment*, cf. pp. 160ff., 178–182, 205.

[76] Christopher Peacocke, 'Conscious Attitudes, Attention, and Self-Knowledge', in Crispin Wright, Barry C. Smith, and Cynthia Macdonald (eds.) *Knowing Our Own Minds: Essays on Self-Knowledge* (New York: Oxford University Press, 1998).

[77] A major source of the confusion lies in not distinguishing support in reasoning and argument—reasons—from support of other kinds, including physical evidence. For a good account of such distinctions, see Robert Audi, *The Architecture of Reason: The Structure and Substance of Rationality* (New York: Oxford University Press, 2001).

and representational. Nature is not a text and is not made up of propositions or reasons. Since reasons are necessarily *representational*, sensations cannot be reasons unless they are representational. I believe that some sensations are not representational. In such cases, only representation that one has the sensation can be a reason.

Reasons are necessarily *propositional*. Reasons are potential steps in arguments and explanations that, in effect, show why one should believe or intend something.[78] They answer potential why questions. As a matter of logical grammar, non-propositional—and non-representational—entities cannot play these roles. Non-propositional entities, such as perceptions, can contribute to an attitude's being warranted. But reasons are necessarily propositional contents taken with their modes. They are basic units in reasoning. Some Self-Knowledge is of sensations and perceptions, which are not propositional. These entities *could not* be reasons for Self-Knowledge. But it can be shown independently of this point that what one knows in Self-Knowledge is not a reason that supports the knowledge. Let us focus on knowledge of one's own propositional attitudes.

Are first-order beliefs reasons to judge that one has those beliefs? They cannot be. A reason, theoretical or practical, must be about the subject matter of the attitude for which it is a reason, or about some relevantly related subject matter. The reason's representational content must bear on, and in some way count in favor of, the truth or goodness of the conclusion for which it is a reason. First-order attitudes are not about the subject matter of second-order attitudes. They commonly bear no relevant relation to it. They *constitute* the subject matter, but they are not reasons about the subject matter that support the truth or goodness of the higher-order attitudes. If one's first-order attitudes were used as reasons for judgments that one had them, one's reasoning would consist of *non sequiturs*.

Most Self-Knowledge is not supported by reasons. The few exceptions are self-evident *cogito*-like cases. Not being supported by reasons does not make Self-Knowledge any less epistemically warranted or substantial. I shall elaborate this point shortly.

An idea with some currency is to explicate Self-Knowledge by reference to knowledge of one's intentional physical agency.[79] I myself emphasize Self-Knowledge's role in the applicability of norms of morality and critical reason.

[78] I say 'in effect' because the reason need not respond to an actual question, and because a reason is commonly not a meta-level consideration that makes reference to shoulds, beliefs, intentions, or representational contents. Reasons can be about object-level matters. They are reasons by virtue of their potential use in supporting and explaining the "why" of conclusions.

[79] See Moran, *Authority and Engagement*; Matthew Soteriou, 'Mental Action and the Epistemology of Mind', *Noûs* 39:1 (March 2005), 83–105; Bilgrami, *Self-Knowledge on Resentment*; O'Brien, *Self-Knowing Agents*; Peacocke, 'Mental Action and Self-Awareness (I)', in Jonathan D. Cohen and Brian P. McLaughlin (eds.), *Contemporary Debates in the Philosophy of Mind* (Malden, Mass: Blackwell, 2007), 358–376; Peacocke 'Mental Action and Self-Awareness (II): Epistemology', in O'Brien and Soteriou (eds.), *Mental Actions* (New York: Oxford University Press, 2009), 192–214. Some of these approaches are not strictly incompatible with what I say here, but all highlight agency in self-knowledge more than I do.

These norms centrally concern agency. It is certainly true that *some* of what is known in Self-Knowledge comprises mental acts. Persons and selves—the subject matters of Self-Knowledge—are constitutively agents.

The approaches that I have in mind, however, highlight agency in a more specific way. They model Self-Knowledge—or some types of Self-Knowledge—on an individual's knowledge of his own intentional *bodily* action. Knowing one's thoughts and attitudes is taken to be structurally and epistemically like knowing that one is raising one's arm.

I cite two grounds for not following this line. One is that doing so cannot account for Self-Knowledge's buck-stopping role in morality and critical reason. It is not one's bodily actions, but their psychological antecedents that ground these norms. Assessing full, unmitigated accountability for bodily actions depends on assessing their conforming to potentially Self-Understood cognitions, intentions, decisions, and intentional initiations that produce them. A corollary is that warrants for beliefs about one's bodily actions do not yield immunity to brute error.[80] The warrants I seek *do* yield immunity.

A second ground for not modeling Self-Knowledge on knowledge of bodily action is that a lot of Self-Knowledge is not of acts. Selves *are* constitutively agents. Thinking morally or critically requires a self-consciousness and control that insure that the thinking is agential. Critical reasoning requires judgment, an act. Persons and selves constitutively make decisions in the light of values, again acts. All these points bear on constitutive aspects of being a person or self. But persons and selves, *we*, are not *just* agents. We not only act. We undergo and endure. Knowledge of passive elements in one's mind is virtually as important to norms of morality and critical reason as knowledge of active elements.

Some Self-Knowledge centers on sensations and perceptions that we do not bring upon ourselves. Some Self-Known propositional thoughts simply occur to one. Most emotions are not acts. Even many propositional commitments—all standing commitments (standing beliefs or fears) and many occurrent commitments (many occurrent perceptual beliefs)—are not acts. Many do not even derive from acts. Many are acquired through perception and are first stored automatically.[81] Sometimes in Self-Knowledge one reaffirms such states in

[80] See discussion in Lecture II, section III of the buck-stopping status of the psychological antecedents of bodily actions. The immunity to brute error of Self-Knowledge of the psychological antecedents derives from the same fact that gives the psychological antecedents buck-stopping status. Knowing the psychological antecedents depends only on good use of the competence for Self-Understanding. Knowing one's actions through knowing the psychological antecedents is subject to contingencies involved in realizing one's intentions. One can be fully warranted in one's belief about what one's action is, utilizing one's warranted belief about what one is deciding to do. But one can still be mistaken about the action because of occasion-dependent, post-initiation distortion of the bodily action. Then, the individual's contribution to accountability lies fundamentally in the Self-Understood, or Self-Understandable, psychological antecedents. Accountability for the bodily action commonly is mitigated. Mitigation tracks the individual's brute error about what the action is.

[81] Many philosophers emphasize consciousness as a constitutive feature of intentionality or even knowledge. There are, however, many unconscious representational and knowledgeable states. If consciousness makes a type of state more reliable, it is epistemically relevant. This is an empirical

judgments—which *are* acts. But often in Self-Knowledge one simply forms a belief about what beliefs one has long had. The truth and warrant involved in such meta-beliefs do not hinge on reaffirming the standing commitments that did not arise through any agency. Such non-agential states and occurrences are susceptible to Self-Knowledge.

Knowing one's pains, fears, and other aspects of vulnerability is crucial to one's moral status. Knowing one's perceptual perspective and other passive representational states is crucial to critical reasoning. Self-Understood passive elements in a psychology loom large among elements with buck-stopping status for both morality and critical reason. Indeed, Self-Understanding itself can be passive—the product of the triggering of standing competencies.

III

Let us turn from paths not taken to our main route. I noted earlier that most Self-Understanding and Self-Knowledge are not based on reasons, but are nonetheless warranted. I distinguish two types of epistemic warrant. A warrant that is not based on a reason is an *entitlement*. Such warrants need not be accessible, even in principle, to the individual whose thoughts or attitudes are warranted. A warrant that is based on a reason is a *justification*.

Apart from *cogito* cases, I think that the special epistemic warrant underlying Self-Knowledge is an *entitlement*, not a justification. The warranted individual need not have a reason for the warranted belief. On the other hand, the entitlement requires more from an individual than many other entitlements do. A young child can be entitled to perceptual beliefs, even though it has only the competence understanding required to have the beliefs. To be warranted in Self-Knowledge, an individual must be competent in employing meta-representational attitudes, and must think the content of the attitudes that are the subject matter of the Self-Knowledge, while specifying them in the canonical, that-clause-like way.

I believe that warrants for attitudes that constitute Self-Understanding and Self-Knowledge yield *immunity to brute error*. If an attitude *type* is associated with competencies that could ground a warrant W_{ibe} that yields immunity to brute error, any error in an *instance* of an attitude of that type must be explained in one of three ways. First, the attitude instance may be warranted only in another way W that does allow brute error. Second, the instance may derive from misuse of competencies that undermine warrant W_{ibe}. For example, bias, self-deception, or emotion might derail a normally warranted belief formation. Third, the individual might suffer some pathology that undermines W_{ibe}. A lesion might cut off judgment from normal access to intentions or beliefs. In such cases, an

matter. There is probably unconscious self-knowledge that is analogous to Self-Knowledge in being epistemically "special". Any such self-knowledge is not part of the apperceptive core self.

individual lacks the *relevant* epistemic warrant. Having an epistemic warrant that yields immunity to brute error guarantees *getting things right*. Being relevantly entitled to an attitude about one's psychological states ensures that the attitude is true.

Entitlements to Self-Knowledge are not the only warrants that yield immunity to brute error. I shall reflect on three other classes of cases.

The first class includes three types of entitlement associated with first-order deductive inference. In discussing this class, I distinguish first-order inference from such inference *supplemented* by thinking the inference rules that warrant steps in the inference. Individuals can carry out inferences without having the meta-representational concepts, or the ability to generalize, necessary to think the inference rules that justify their inferential transitions. Even if they do have these abilities, the abilities may not be operative in a good deductive inference. Reasoners can be warranted in their reasoning if the reasoning is governed by—correctly explainable in terms of—following correct rules of inference, even if the reasoner lacks resources to think the meta-representational rules, consciously or unconsciously, or does not rely on such resources in carrying out the inference. The *steps* in the inference can be justifications, reasons, for later *steps*. The warrants for the transitions between steps can, however, be entitlements—warrants without reasons. Reasoners can be entitled to rely on the transitions as valid, even though they cannot, or do not, think the inference rules. Entitlement derives from competence with the logical constants in carrying out the reasoning.

The entitlement to make a deductively valid transition yields immunity to brute error in making the transition. One can make a mistake in such transitions. But if one is entitled to a transition in a deductive inference that relies on purely preservative memory, one cannot make a mistake in the reasoning transition.[82] Being entitled to a transition entails that the transition is free of error.

[82] Equivocation between "twin" concepts in "slow-switching cases" requires that this point be understood with some subtlety. I think that errors in reasoning that depend on reasoners' mixing up "twin" concepts that they have unawares rely on mistaken presumptions of equivalence between the twin concepts. In this respect they are like mistaken presuppositions or even mistaken tacit beliefs. The errors do not result from failures of anaphoric, purely preservative memory, since such memory is not relied upon in the inference. They occur when a reasoner relies on sources—such as substantive memory or perception—that introduce content into later premises in an argument *without* relying on purely anaphoric, preservative relations to the content of earlier premises. Nor do the errors result from failures of rational deductive competence. They involve what Luca Struble calls a brute error in exercising a coordination competence, not an error in a reasoning competence *per se*. Although the presumptions of sameness or equivalence of "twin" concepts are not tacit beliefs, since no state puts the two concepts together into a proposition, the presumptions have much the same function. It is clear that possible errors in inference that hinge on equivocation between "twin" concepts are brute errors. They do not reflect on the well-functioning of *memory* or of *reasoning competence*. So the errors do not have their sources in memory or reasoning competence. I introduced slow-switching cases in 'Individualism and Self-Knowledge'. Paul Boghossian thinks that such cases cause problems for anti-individualism in accounting for self-knowledge and inference. See his 'Content and Self-Knowledge', *Philosophical Topics* 17 (1989), 5–26; and 'Externalism and Inference', *Philosophical Issues* 2 (1992), 11–28. I think that there are multiple grounds for finding his arguments unsound. I discuss these issues briefly in

A second entitlement is involved in fulfilling the first. This is entitlement to rely on purely preservative memory. Purely preservative memory is anaphoric retention of representational content and mode type. This entitlement, too, yields immunity to brute error. One cannot be warranted in relying on purely preservative memory and make a mistake in the preservation. Memory errors in reasoning derive from allowing some other type of memory besides purely preservative memory to enter the reasoning, or from some malfunction in purely preservative memory.

A third entitlement is present in those deductive inferences that involve coming to some reason-supported conclusion. Rules for deductively valid transitions are not rules for using deductive inference to provide reasons.[83] The former rules constrain permissible reasons, but do not guarantee that premises that lead validly to a conclusion are reasons for the conclusion. The premise of a deduction could be unwarranted. Then a valid deductive transition from it would provide no reason for the conclusion.

An individual could make an inference between propositions that are in fact connected by a valid rule of inference, without relying on the logically valid connection. The individual might make the inferential connection by rote, for example. Such transitions are not really deductive inferences. They are surely not reason supporting by virtue of deduction. Such inferences are not correctly psychologically *explained* in terms of valid logical inference rules. Inferences that are correctly explained in terms of valid logical inference rules must involve, and must be further explained by reference to, exercise of a competence with the logical constants—not a rote connection, or some other non-logical connection.

For an inferential transition to provide *reason support* for a conclusion by way of a deduction, the transition must be correctly explainable in terms of deductive inference rules and competence with logical constants; the premises of the inference must be warranted; and the reasoning must be relevantly non-circular. For a deductive reason-supporting transition to be correctly explainable in terms of deductive principles, its premise must ground some rational explanation of why the conclusion is worthy of belief *for the individual*. The individual need not be able to give the explanation. But the individual's competence with logical constants must rationalize—ground a potential explanation of—why the conclusion is belief worthy for the inferrer. Reasons are answers to potential why questions. An individual's competence with logical constants must comprise

'Memory and Self-Knowledge', in Peter Ludlow and Norah Martin (eds.) *Externalism and Self-Knowledge* (Stanford: CSLI Publications, 1998). There is further literature on this topic that I will not go into here. The points in the text here supplement my 1998 discussion. For a somewhat different view of reasoning in slow-switching cases, see Mikkel Gerken, 'Conceptual Equivocation and Warrant by Reasoning', *Australasian Journal of Philosophy* 89:3 (2011), 381–400. I regard Gerken's view as broadly congenial. It could be accommodated with minor revisions of the points made here.

[83] I am using Gilbert Harman's point that rules of logic and rules of reasoning are not identical. See Harman, *Thought* (Princeton: Princeton University Press, 1973). One can engage in valid, even sound, deductive inference even if the premises provide no reasons for the conclusions.

some (non-meta-representational) competence understanding of the connection between warranted premise and inferred conclusion—a connection that is belief supporting.

Now, a point analogous to the point about entitlements to transitions as deductively valid applies to entitlements to transitions as reason transmitting or reason supporting. These entitlements again yield immunity to brute error. If a transition in an inference is made with the function of providing a reason—if the individual reasons to commitment to a conclusion that is purportedly supported by the premises, not merely deductively infers the conclusion from the premises—the individual is entitled to the transition step, even lacking an ability to think meta-representationally about it, if and only if the transition is in fact governed by (correctly explainable in terms of) correct principles of reason support.[84]

These entitlements in deductive inference and related ones in inductive inference are, I think, developmentally the first warrants that yield immunity to brute error.

I turn from this first class to a second class of warrants that yield immunity to brute error. This class is comprised of warrants to believe simple, self-evident truths on the basis of understanding them. If one understands I am now thinking, either snow is white or it is not the case that snow is white, or 2 + 2 = 4 well enough to be justified in believing it through understanding it, being warranted insures not being mistaken in the belief.

A third class of cases of immunity to brute error comprises entitlements to *believe* non-inferentially *in simple cases* that one attitude, or content together with a mode type, is a reason for another. *Simple cases* are those in which the individual uses a lower-level reasoning transition—entitled in the way of the third example in the first class—as a basis for the meta-level judgment about the transition relation. This class is the meta-representational analog of entitlements to first-order reason-supporting transitions. I distinguish belief that one thought is a good reason for another from belief about the form of the inference, and from theoretical generalizations about inference. I am concerned here purely with simple, *de re*, evaluative beliefs *about* particular transitions: *P* is good [or bad] reason for *Q* [perhaps relative to background *B*], where the individual can and does reason correctly—*p* so *q*.[85] In simple cases, if one's non-inferential understanding of the reason relation between particular attitudes is good enough to entitle one to the meta-belief that one attitude is a reason for another, the understanding is veridical: the reason-support relation holds. One can be warranted but mistaken in beliefs about reason relations. But such mistakes rest on inference in theorizing, or on complex cases. In simple cases,

[84] Rules for transitions in inductive inference are not well understood. But it is clear that if a step is warranted, it is not inductively fallacious, although it could lead to a false, warranted conclusion. Reliance on the transition rule, not the conclusion, is immune to brute error.

[85] I believe that the points apply to prima facie reasons as well as conclusive reasons.

warranted judgments are true. Errors derive from irrationality or other failures of understanding that undermine warrant.

IV

The key to understanding warrants that yield immunity to brute error does not lie in necessity. There are warrants to believe necessary truths that do not yield immunity—warrants to believe empirical necessities. There are warrants that yield immunity that do not attach to necessary truths—*cogito* cases.

The key to understanding warrants that yield immunity to brute error does not lie in apriority. Some inductively based mathematical conjectures are, I think, apriori warranted but not immune to brute error. Not all warrants that yield immunity to brute error are apriori. Entitlements to beliefs about one's sensations depend for their warranting force on having the sensation. They are not apriori. Such entitlements can still yield immunity to brute error.

What grounds epistemic warrants that yield immunity to brute error? What distinguishes these three classes of competencies? What distinguishes other competencies that yield immunity to brute error? I think that the combination of four features grounds the immunity.

First, all the relevant competencies are, or rest psychologically and epistemically on, one or another type of propositional understanding.

Second, all the relevant types of understanding, including but not limited to competence understanding, are constitutive of having certain explanatorily or normatively significant types of point of view. In order for the type of point of view to be possible at all, these types of understanding must, in their natural undamaged states, be reliable.

Third, fulfilling representational functions of the states that are, or rest on, these types of understanding either forges constitutive connections *within* the relevant type of point of view, or makes commitments that express an understanding of such constitutive connections.

Fourth, representational success in fulfilling representational functions does not depend on anything beyond exercising the understanding in a psychologically well-functioning way. In particular, being representationally successful (being veridical, preserving veridicality, and so on) on any given occasion does not depend on anything outside the psychology on that occasion, and does not depend on having any information beyond that involved in exercising, in a well-functioning way, the understanding that is constitutive to having the relevant type of point of view.

Immunity to brute error is derivable from these four features of a representational competence, together with this constitutive principle governing warrant:

> (E) A propositional state or occurrence is warranted, on a given occasion, if and only if it is the result, on that occasion, of the exercise of a representationally well-functioning propositional competence that provides a representationally

reliable, epistemically good route to veridicality, allowing for the natural limitations of the competence with respect to its subject matter, and limitations of the information available to it.[86]

By the first of the distinctive features, the relevant psychological states or occurrences rest on a reliable propositional competence. By the third and fourth, veridicality on a given occasion does not depend on extra-psychological conditions or on specific information beyond the minimal information necessary to exercise the understanding. By the second, third, and fourth features, being representationally successful (being veridical, preserving veridicality, and so on), on a given occasion, depends only on exercising the understanding of connections within the point of view—the understanding constitutive to having the relevant type of point of view—in a well-functioning way. It follows that a propositional state or occurrence is warranted if it relies only on a well-functioning exercise of a relevant type of understanding. Any malfunction of a relevant competence undermines warrant for a state or occurrence that relies on the competence. By the fourth feature, any error in a state that relies on a relevant type of understanding derives from substandard functioning of the competence—and hence undermines warrant. So any *warranted* state or occurrence that relies only on the understanding is representationally successful. In other words, reliance on any of the relevant types of understanding is immune to brute error.

Let us return to the three classes of warrant that yield immunity to brute error with this grounding in mind.

Take the first class of cases. A reliable competence understanding that consists in making valid transitions in deductive inferences that rely on purely preservative memory is constitutive to any propositional point of view, starting with the simplest empirical ones. To have states with propositional structure, one must be competent to use such structure reliably in inference, including deductive inference. Successful exercises of this competence forge constitutive connections within a propositional point of view. If the competence is exercised in a well-functioning, unhindered way, the transitions are representationally successful. They preserve truth in virtue of form. An individual is entitled to rely on well-functioning exercises of the competence—on the transitions—as deductively valid. So well-functioning exercises of the relevant competence understanding are warranted only if transitions are deductively valid.[87]

[86] For the framework that embeds (E), see my 'Perceptual Entitlement'. The point of the phrase 'allowing . . . ' in (E) is to indicate that as long as the competence is reliably veridical, and functions representationally *well* in using its available information—in a way conducive to getting things right in normal circumstances—exercise of the competence can be warranted. I assume that natural limitations include fallibility. Warranted failure with respect to veridicality derives either from natural limitations—such as natural imperfections in perceptual resolution—or from using *well* information that is in fact misleading.

[87] The rules for deductive transition that meet this description are restricted to simple cases. See the definition of '*simple cases*' above. Norms that are not explanatory of an individual's actual reasoning, because they are too complex, are not relevant to the account.

The same points apply to the competence understanding in exercising purely preservative memory. Having a naturally reliable competence of this kind is constitutive to having any point of view capable of propositional inference, indeed any representational competence at all. The competence makes constitutive connections in a representational point of view. If the competence is exercised without malfunction or hindrance, it is representationally successful in preserving mode and content from earlier states. Individuals are entitled to rely on exercises of a reliable competence that is well-functioning and unhindered. So exercises to which individuals are entitled are representationally successful in preserving mode and content. So the entitlements yield immunity to brute error.

Similarly, for competence understanding that yields reason-supporting transitions in reasoning. A reliable understanding of this sort, applied in simple cases, is constitutive to forging structural connections in any propositional point of view. Almost trivially, individuals capable of reasoning are competent to make moves between propositional steps that, relative to their background information, are in fact reason supporting—as long as their faculties do not malfunction and as long as internal conditions do not hinder natural exercise of the competence. Exercise of the understanding unhindered by disease or bias yields reason-supporting transitions. Being entitled to such transitions guarantees representational success—reason-supporting transitions. So entitlement yields immunity to brute error.

The second class of warrants that yield immunity to brute error concerns epistemic starting points, not transitions. They are otherwise similar. Understanding simple logical and mathematical truths well enough to be warranted through understanding them constitutes an explanatorily and normatively significant type of point of view. That type of point of view is a natural psychological kind. Believing the relevant truths, on the basis of understanding them, is a reliable route to truth. In fact, such beliefs are rational starting points in reasoning to other truths. Any error is a failure of the sort of understanding whose successful exercise constitutes the type of point of view. Warrant for believing simple logical and mathematical truths lies in understanding them.[88] If one is warranted in believing them, one's belief is true. So the acceptance relative to the warrant is immune to brute error.

What of the third class of cases—meta-representational understanding of reason-support relations? Competence to make reliable judgments about reasons *as* reasons is distinctive of critical reason, certainly an explanatorily and normatively significant type of point of view. Recall that *simple cases* are, by definition,

[88] Understanding self-evident logical truths is, I think, an abstraction from deductive transitions in reasoning. Understanding truths of *pure* logic is, I think, a further abstraction from instances of such principles. Understanding simple truths of arithmetic is, I think, an abstraction from counting. Understanding other simple mathematical truths, such as the simplest axioms of set theory (for example, the axiom of extensionality), is not an abstraction from particular cases but is part of a minimum explicational understanding of relevant concepts (here, the concept set). None of these truths is vacuous in the positivists' sense.

those in which the individual uses a warranted lower-level reasoning transition as a basis for the meta-level judgment about the transition relation. In simple cases, being warranted guarantees true judgments. For they function to forge certain constitutive structural connections within a point of view capable of critical reason, by recognizing constitutive structural connections within the lower-level propositional point of view. The meta-representational judgments are grounded in lower-level thinking of the reason relations *in* the meta-representational thinking about them. To get things right, the judgments need use no other information than that involved in the meta-representational understanding that conceptualizes, while relying on the competence with, the warranted lower-level reason-supporting transitions. Forging constitutive structural relations in a critically reasonable point of view just is correctly representing the warranting relations in the transitions in the lower-level reasoning that one evaluates. (As indicated in the first class of cases, the lower-level entitlements themselves yield immunity to brute error.) Any error in using the competence is a malfunction of the competence. Being warranted in using the normally reliable competence in simple cases requires that the competence function well on the occasion of use. So since any error in using the competence in simple cases is a failure in its function to establish constitutive connections within a critically reasonable point of view, any error in simple cases undermines warrant.

In understanding warrants that yield immunity to brute error, it is instructive to reflect on the contrast class. The most primitive warrant that allows brute error is entitlement to perceptual belief. Like all warrants, including those that yield immunity to brute error, perceptual entitlements are explicable only against a background of veridical states. But there are crucial differences in this background.

The first, less basic difference concerns relations between competencies underlying the warrants and having an explanatorily and epistemically significant type of point of view. An individual can have an empirical point of view, even one with warranted beliefs, and yet not be in a position to get any perceptual beliefs right.[89] How could such a thing happen?

[89] I think that something stronger applies. An individual can have perceptual beliefs, but no warranted ones. The nature of perceptual states does not require that they be reliably veridical in their normal content-determining environment. (Of course, most perceptual states *are* thus reliably veridical. See 'Perceptual Entitlement' and note 68 above.) I have been misunderstood to hold, in that very article, that it is apriori and constitutively true that perceptual systems are reliable in normal circumstances—the circumstances in which their contents were formed. I do not hold this view, and consciously wrote around it in that article. An example of the misunderstanding is Anthony Brueckner, 'Content Externalism, Entitlement, and Reasons', in Sanford C. Goldberg (ed.), *Internalism and Externalism in Semantics and Epistemology* (New York: Oxford University Press, 2007), 160–176. I had earlier been guilty of holding, without argument, that the nature of a perceptual state requires reliability. See 'Our Entitlement to Self-Knowledge', 106 n11. But I came to give up the view in the late 1990s. Obviously, if there were a good apriori argument for it, there would be a very direct and comprehensive argument against scepticism.

An individual can inherit perceptual equipment that obtained representational content through patterns of interaction, including veridical applications, in a certain environment. Call this environment '*the normal environment*' for that equipment. (See notes 68 and 89.) Suppose that when given appropriate proximal stimulation, this equipment produces types of perceptual beliefs that would be reliably veridical in the normal environment. However, any individual can be placed in a non-normal environment in which, despite similar proximal stimulation, every perceptual belief lacks the kinds of distal antecedents that gave content to the individual's perceptual equipment. Then the individual has an empirical point of view. The individual's cognitive equipment works as well as it can, and reliably, relative to the warrant-determining normal environment. Yet nearly all the individual's perceptual beliefs, including warranted ones, are false.[90]

An analogous situation is not possible with types of understanding that ground immunity to brute error. Inability to exercise those types of understanding *veridically* would make having the relevant sort of point of view impossible. An individual that could not naturally make valid or reason-supporting deductive inferences would lack a propositional point of view. An individual that lacked an understanding that, when naturally exercised, yielded belief in simple logical or mathematical truths would lack a capacity to understand logic or mathematics.[91] An individual whose natural competence did not yield true beliefs about simple reason-support relations would lack critical reason.

A second, more fundamental difference between perceptual entitlements and entitlements for types of understanding that are immune to brute error helps explain the first. The difference is that well-functioning, unhindered exercise of the perceptual competence underlying perceptual-belief formation does not insure representational success. Success on particular occasions depends on non-psychological, "brute" connections to distal matters, beyond the point of view, in the environment. The distal matters can vary so as to yield either success or failure, even while the psychological competence for forming perceptual beliefs functions well and in an unhindered way.[92]

[90] Clearly, the same sort of point can be made even if the individual is located in its normal environment. Every perceptual belief results from a proximal stimulation with artificial, non-standard distal antecedents. The deeper point is that constitutive explanation of the content of perceptual states in terms of *veridical* cases, which *is* I think apriori necessary, does not necessarily depend on the veridical cases' being the most common cases.

[91] Understanding logic and mathematics, beyond mere competence understanding, contrasts with reasoning logically or mathematically. It involves minimal explicational capacity. I think that such understanding is a stage of development that marks a significant type of point of view.

[92] See my 'Our Entitlement to Self-Knowledge'; 'Perceptual Entitlement'; 'Disjunctivism and Perceptual Psychology', *Philosophical Topics* 33:1 (Spring 2005), 1–78, especially sections VI–VII; "Disjunctivism Again', *Philosophical Explorations* 14:1 (2011), 43–80, especially sections II–IV.

Forming a given kind of perceptual belief depends only on antecedent psychological sets, psychological formation laws, and proximal stimulation.[93] Representational success depends partly, on each occasion, on distal matters—beyond the individual's psychology. Distal matters can vary while producing relevantly similar proximal stimulation. As a result, a reliable competence for forming perceptual belief that functions as well as it can in given circumstances—given the information available to it—can produce warranted but false belief—brute error. No analogous distal, subject-matter variation can occur with respect to well-functioning warranted exercises of the sorts of understanding that are immune to brute error. With such understanding, the constitutive connections—and the connections that representational success depends on—are entirely within the relevant type of point of view. Moreover, exercising the well-functioning understanding that is constitutive of a relevant type of point of view just *is* successfully traversing, preserving, or recognizing the relevant connections within the point of view. Getting things right does not depend on conditions outside the psychology and does not require information beyond that which is necessary to exercise the understanding. Well-functioning, unhindered, warranted reliance on the competence yields representational success on each particular occasion and in any environment. So well-functioning, warranted exercises are immune to brute error.[94]

Of course, default entitlements to perceptual belief are not the only entitlements that allow brute error. Similar points apply, however, to entitlements to empirical substantive memory and to anticipation of one's bodily action or external events. Mistakes can be warranted because they depend on non-normal relations to the environment on given occasions. One can exercise the reliable competence unhindered and be mistaken.

An interestingly different class of warrants that allow brute error comprises the products of various types of inductive reasoning. I have three types of induction in mind: empirical inductive reasoning about the environment, for example, inference to the best explanation from perceptual beliefs; inductive reasoning to conjectures in mathematics; and inductive reasoning about oneself. The conclusions of these types of reasoning can be warranted (justified), but mistaken. For example, in forming well-reasoned judgments about one's

[93] This is a statement of the Proximality Principle, which guides the science of perceptual psychology. See my 'Disjunctivism and Perceptual Psychology', section IV; *Origins of Objectivity*, 385ff.; 'Disjunctivism Again', sections II–III, V. Other issues in this paragraph are discussed in "Perceptual Entitlement."

[94] The fact that a reliable competence represents only internal psychological elements does not suffice to make the representational states that rely on it immune to brute error, relative to warrants for them. Reliable inductive reasoning from warranted premises about one's psychological states can yield warranted but mistaken beliefs. Representational competencies that are basic constitutive elements or basic constitutive connections in a point of view are immune to brute error. Such competencies include purely preservative memory, inferential transitions, using reasons, understanding basic logical and mathematical truths, recognizing simple reason-support relations as such, and the sorts of self-understanding that I will discuss.

character, one can be mistaken. The induction might be based on behavioral evidence, or on types of self-knowledge that *can* be arrived at through Self-Understanding.

Inductive reasoning uses patterns found in a base case, or cases, to form judgments about a further case, or cases—relying on the likelihood that the pattern reapplies. Well-functioning, warranted use of induction never insures veridicality of the conclusion.[95] It is inherent in the method that even if the beliefs that form the induction base are true, the conclusion can be justified but mistaken.

In all these cases of capacities that are vulnerable to brute error, representational success—veridicality—depends, on each occasion, on more than optimal representational functioning of psychological competence, given the information necessary to exercise it at all. Not so, with the competencies that yield immunity to brute error. Although relying on them can yield failure of veridicality, such fault lies in themselves, not in a brute failure to match a subject matter beyond them.

V

By reflecting on other cases of warrants that yield immunity to brute error, I hoped to illuminate Self-Understanding. I now return to that topic. Self-Understanding is constitutive to having an apperceptive rationally accessible point of view. Immunity to brute error in Self-Understanding follows the pattern established by other examples of immunity. To explicate entitlement to Self-Understanding and Self-Knowledge in detail, I must elaborate structural considerations set out at the end of Lecture II.

Self-Understanding and Self-Knowledge involve predication of a psychological element to oneself. I distinguish several aspects of the predication. These elements mark specific competencies that ground specific warrants.

Consider the predication in I believe that Mozart loved Haydn. I refer to myself with a singular application of the self-concept I, and I attributively apply the concept believe that Mozart loved Haydn to myself. This concept consists of three primary components—the verb-concept believe that indicates a psychological mode (belief), the present tense of the verb-concept, and the canonical singular concept that Mozart loved Haydn that denotes a representational content. I shall say that the full concept, believe that Mozart loved Haydn, *indicates* a *type* or *kind* of psychological element—belief that Mozart loved Haydn. It also thereby indicates the mode type, belief.

In veridical predications in Self-Understanding, the attribution establishes a *de re* relation to *instances* of those types—here, an instance of a belief that

95 I exclude mathematical induction that occurs in Peano arithmetic. I think that warrant that derives from relying on such induction does yield immunity to brute error.

Mozart loved Haydn and an instance of the mode, belief. I shall say that, in such cases, the predicate concept that indicates the psychological type or kind *betokens* the instance of the type or kind. So the predicate concept *indicates* the psychological kind and *betokens* the instance of the kind. Both indication and betokening are types of representation.[96]

Predication in Self-Understanding involves attribution as well as indication and betokening. When attribution is veridical, the indicated *kind* is attributed, and the betokened *instance* is also attributed. These are different attribution relations, but I will call them both '*attribution*'. In Self-Understanding, kind and instance are attributed to the individual picked out by application of the singular self-concept. Although all these elements work together in the relevant predications, they are different aspects of the predication and correspond to different aspects of understanding. Thus there is *indication understanding*, *betokening understanding*, and *attribution understanding*.

Betokening is necessarily guided by indications of an attribute—in veridical Self-Understanding, the indicated kind. Indication understanding is always a component of betokening understanding. I will not discuss indication understanding separately.

Betokening understanding is usefully thought of as a *vertical* relation. It is a relation between a higher-level understanding and a lower-level psychological instance that is understood. Both understanding and what is understood are elements in a point of view. The understanding establishes the sort of multi-tier point of view discussed earlier.

My main task is to explain wherein being relevantly warranted in betokening understanding entails *correct* betokening understanding. The explanation should account for the warrant's intuitively yielding immunity to brute error. I begin with Self-Understanding of *propositional* states and occurrences. Later I consider other types of Self-Understanding.

In tracking what follows, one must bear firmly in mind a distinction between lower-level competence understanding and meta-representational understanding. *Lower-level competence understanding* is just a competence to think with propositional contents. *Meta-representational understanding* is a competence to think about the exercise of the lower-level competence, or the modes and contents that are instantiated in the lower-level competence.

In the meta-representational betokening understanding integral to Self-Understanding, a propositional state *instance* or *occurrence* is betokened *de re*. The instance is betokened via its constitutive aspects—its representational content and its mode. In attributing each of these aspects, meta-representational betokening understanding makes essential use of lower-level competence understanding.

[96] Betokening is the predicative analog of syntactically singular *de re* reference. Not all predications involve betokening. The predication 'Some people love others' does not, if the quantifier is not associated with *de re* thoughts about particular instances of loving.

Recall from the previous lecture that in attributing representational content to one's propositional state- and occurrence-instances, one must exercise lower-level competence understanding in the represented thought content as one uses the canonical representation that names that content.[97] An under-recognized aspect of betokening understanding is that it involves an analogous phenomenon in use of the predicate for the *attitude mode* of the propositional content. Application of the concept believes in betokening understanding does not use a *represented* representational content. For believes does not indicate a representational content. But one can preserve in memory, with application of believes, the belief mode; and one can retain dispositions to implement reasons connected to any given belief. That is, in betokening understanding of an attitude instance, one exercises lower-level memory that preserves representational content—of which one has lower-level competence understanding—and preserves sensitivity to its being believed.[98] Further, in understanding the belief, one can prime dispositions to associate it with other attitudes that bear lower-level reason connections to it. In betokening understanding, one uses the lower-level preservation through memory of the attitude mode, along with the content, in picking it out, *de re*. One retains in memory the belief mode as one represents it. I will reflect, first, on aspects of this meta-representational competence and, then, on epistemic warrants for its uses.

Purely preservative memory functions to preserve mode type and representational content, without adding warrant or new subject matter to current psychological transactions. (See note 63.) Fulfilling this function by purely preservative memory is constitutive to *any* representational psychology—propositional or not, meta-representational or not. For example, in propositional inference the mode and content of earlier steps must be preserved for later use in inference. Or in a use of visual perception in action, purely preservative memory retains the mode and representational content of the visual state. Although one is entitled to rely on purely preservative memory, and although such reliance is necessary for the success of the inference or perception-based action, such reliance is not part of the justification or a practical warrant for the action. Warranted reliance on purely preservative memory is a necessary enabling condition for the conclusion's, or action's, being warranted. My account of the epistemic warrants for Self-Understanding and Self-Knowledge in non-*cogito* cases hinges on reflecting on preservational capacities, central to any representational psychology, and

[97] For more on the canonical representation of representational content, see my 'Individualism and Self-Knowledge'; 'Postscript: Frege and the Hierarchy', in *Truth, Thought, Reason: Essays on Frege* (New York: Oxford University Press, 2005); 'Postscript: "Belief *De Re*"', in *Foundations of Mind*; and 'Five Theses on *De Re* States and Attitudes', section v.

[98] I write 'preserves sensitivity' because the lower-level purely preservative memory does not preserve any representation of a belief as such. Lower-level aspects of a psychology are not meta-psychological. They nevertheless track differences in the modes of past steps in inferences. Beliefs function differently in the psychology from suppositions. Purely preservative memory tracks sensitivities to different modes, and in that sense preserves the mode types.

their role in meta-representational understanding—especially betokening understanding.

There are importantly different types of Self-Understanding and Self-Knowledge. Preservation, though not purely preservative memory, figures in all of them. Preservation is necessary to Self-Understanding from the inside, and to the role of Self-Understanding in providing buck-stopping status to psychological elements in critical reason and morality. It is part of the constitutive tissue of selfhood. So understanding the forms that preservation takes between lower-level and meta-psychological elements lies at the heart of my project.

In all cases, preservation between the lower-level understanding and the meta-representational understanding *of the representational content* works in the same way. In propositional cases, the lower-level thought content is always preserved in the meta-representational thought by being *used*—thought—*in* the canonical name for the content. I will concentrate on various types of preservation of *mode type* and *mode instance*.

Preservation between lower-level thought and meta-representation of it occurs trivially in pure *cogito* cases. No memory is involved. In pure—formally self-verifying—*cogito* cases (I [hereby] entertain the thought that writing requires concentration), the mode—entertaining—of the lower-level thought (writing requires concentration) is inevitably preserved in any thinking of the meta-representational thought, no matter what its mode is. Entertaining the lower-level thought content is guaranteed in the meta-representational thought. Even the instance of entertaining is referred to reflexively in the thinking of *cogito*. Pure *cogito* cases are infallible. If one thinks a thought of that form, one thinks a truth.

In impure *cogito* cases (I [hereby] judge that writing requires concentration), the indicated lower-level mode (judgment) is not entailed to be the mode of the content cited by the underlining. In such a *cogito* thought, one could judge that one is judging the lower-level content, and be mistaken about one's making a judgment. The two levels of judgment could come apart. One could judge that one is judging that writing requires concentration, although one does not thereby *judge* that writing requires concentration. The latter content (that writing requires concentration) might pass non-committally through one's mind. Such cases are pathological misuses of the *cogito* form of thought. Normally, the meta-representational judgment coordinates with a performative judgment of the lower-level content. Normally, the lower-level judgment is a constituent of the meta-representational act of judgment. Then the lower-level mode and its instance are preserved in the meta-representational judgment not by entailment, but by coordinated, performative, reflexive agency.[99]

[99] For more on *cogito* cases, see my 'Individualism and Self-Knowledge'; and 'Mental Agency and Authoritative Self-Knowledge: Reply to Kobes', in Martin Hahn and Bjørn Ramberg (eds.), *Reflections and Replies: Essays on the Philosophy of Tyler Burge* (Cambridge, Mass.: MIT, 2003),

Cogito judgments concern present occurrences. Preserving mode type and mode instance is trivial in pure cases, and almost trivial in impure cases. It does not depend on a causal connection between the understood psychological element and the understanding. The connection is contemporaneous. Preservation of mode in judgments that realize Self-Understanding of one's present, continuing, *standing* attitudinal *states* (belief, intention) and certain among one's past psychological states and episodes is more interesting.

How does preservation work in Self-Understanding of standing belief? Memory often figures in a present-tense judgment like I believe that Mozart loved Haydn. Purely preservative memory retains a standing belief—standing from before the judgment. The judgment takes the belief not to have just formed. Purely preservative memory preserves content and mode of the belief for later use. In a meta-representational psychology, the lower-level, purely preservative memory is supplemented by conceptualized representation of both the preserved representational content and the preserved lower-level mode—here, belief. In conceptualizing these matters and *representing* the belief as the individual's own, the individual indicates not only the mode type *belief*; the individual also thinks *de re* of the mode instance. The individual represents his or her own instance of that type of belief in the judgment. When combined with canonical understanding of representational content, such powers yield betokening understanding of instances of standing propositional states.

The memory used in this sort of recognition is meta-psychological autobiographical memory. But it is not episodic memory. The memory is purely preservative memory, supplemented with a meta-representational capacity (a) to conceptualize the tracked preserved mode—type and instance—and representational content of the standing state, and (b) to recognize that it is a state that has been standing from the past.

A similar account applies to Self-Understanding of some past psychological events, especially relatively recent ones. Meta-psychological autobiographical episodic memory again relies on purely preservative memory, supplementing it by conceptualizing the representational content and mode (type and instance) of the past psychological event.[100]

Some Self-Understanding and Self-Knowledge are forward looking. One can be entitled to believe, fallibly but non-empirically, that one will retain a present intention or belief some moments, or longer, into the future. One can anticipate that a specific standing state type and state instance will remain standing into at least the near future, if there is no reason to doubt. Forward tracks of preservation

417–434. In the latter article, I discuss impure performative *cogito* judgments involving modes like intention as well as modes like judgment. See note 74.

[100] Preservation of mode in memory need not maintain commitment involved in the mode. One might have a meta-psychological episodic memory of one's having judged that so and so, even as one currently doubts so and so. Past attitude instances are preserved, inasmuch as betokening understanding relies on lower-level preservational memory to track that commitment.

are as basic to representational psychologies as backward, causally sustained, memory tracks. When utilized by meta-representational powers, including canonical representation of representational content, they figure in *de re* understanding of the mode type and mode instance of future representational states.

Use of these types of preservation—of a present event, of a standing state or a past episode through purely preservative memory, and of a future anticipated state or event—is constitutive to being a self. An individual that lacked any one of these preservational capacities would lack powers integral to applying critical reason in inference or action. I made an analogous point in Lecture II regarding the role of meta-psychological autobiographical episodic memory in critical inference, decision making, and dialectic.

VI

I turn now to *epistemic warrant* for betokening understanding in Self-Understanding. Epistemic norms are grounded in competencies to realize representational functions. Realizing representational functions involves forming and preserving veridical psychological states and occurrences. A propositional attitude is epistemically warranted only if it is the product of a competence that realizes its representational function reliably and well, given the individual's perspective. Powers to produce and preserve veridical representation—representation of mode types and mode instances, and of representational contents—must, in their natural state, be reliably veridical if they are to be constitutive aspects of selves and help ground critical reason. So reliance on these powers by selves—critical reasoners—is epistemically warranted. In particular, in purporting to single out *de re* one's propositional attitude instances, one is default warranted in betokening understanding, if one bases judgments about the instances on the relevant types of preservation.

The judgments based on the preservational routes just discussed are immune to brute error. The meta-representational understanding relies on lower-level preservational capacities in specifying representational content, mode type, and mode instance. The warrants that yield immunity to brute error are all based on this reliance of meta-representational understanding on the lower-level competence understanding, perhaps supplemented by purely preservative memory or anticipation. I will apply these epistemic points to each of the types of Self-Understanding that I have discussed.

Pure *cogito* thinkings are warranted through self-evidence. Individuals are warranted by understanding their thinkings. Such warrants yield immunity to *any* error. Warrant for impure *cogito* judgments lies in *de re* understanding. Understanding coordinates the lower-level performative act with the higher-level *cogito* judgment. Impure *cogito* judgments are not immune to error. Coordination failures that yield error are pathological and undermine warrant. If one judges

I am hereby judging that Mozart loved Haydn and fails to judge that Mozart loved Haydn, one's self-understanding is pathological; warrant lapses. One misuses such judgments if one does not engage in the act that one judges oneself to be engaging in. Such misuses undermine warrant. Impure *cogito* cases are immune to brute error.

Errors in judgments, based on Self-Understanding, about standing propositional attitudes, past or present, derive from malfunction of purely preservative memory or from relying on other capacities besides purely preservative memory. Then the judgment either is unwarranted or is backed by a different warrant from the one that backs Self-Understanding.

Suppose that one applies the concept intend to finish the project or the concept believe that most domesticated dogs are friendly in a purported betokening understanding. Suppose that one lacks the attributed psychological state. The first-level purely preservative memory could have malfunctioned. Or it could have been derailed by competing psychological needs, or other sources of self-deception. Malfunction or relying—however unconsciously—on factors that are irrelevant to supporting a belief undermine the relevant warrant (entitlement). Or one could form a warranted but mistaken judgment with the same content, inferring it from behavioral evidence. Then one's warrant would be different. In all relevant cases of error, either one's default entitlement for the judgment about one's standing propositional attitude lapses, or the judgment does not rest on the *relevant* entitlement. So there is no brute error in judgments warranted in the relevant way. The key route is a conceptualization of mode and content that is preserved by purely preservative memory. The capacity to track one's standing states by way of their mode and content, and to represent them as having been held (not presently formed), is a basic preservational power in a critical reasoner. Minimal exercise of this power yields immunity to brute error.

A similar line applies, I think, to some judgments about past psychological events—judgments based on meta-psychological autobiographical episodic memory that relies on lower-level purely preservative memory. Suppose that one judges I was just then thinking that writing requires concentration. The judgment rests on meta-psychological autobiographical episodic memory. The memory conceptualizes a mode and content preserved by the basic transtemporal psychological connector—purely preservative memory.

I claim that relying on betokening understanding that utilizes purely preservative memory in Self-Understanding of both some standing propositional attitude states and some past propositional attitude events is immune to brute error. I want to elaborate this claim and defend it. First, again, I discuss the warrant for relying on purely preservative memory in these types of Self-Understanding.

Purely preservative memory functions just to retain the mode and representational content of psychological states. It does not represent anything. It only connects. It preserves representation from other states. Relying on it is warranted non-empirically, though, of course, it can preserve empirical content. It is a constitutive precondition of any inference. Its being naturally reliable is a condition on having a functioning representational psychology.

Purely preservative memory is fallible. It can lapse momentarily, or become unreliable, through age or disease. Short-term forms do not operate after their decay times. Longer-term forms, which, as far as is known, lack built-in time limits, not only fail on occasions but deteriorate in reliability. Still, whenever purely preservative memory fails to retain mode or content, the failure is a malfunction. Errors are not attributable to a representationally well-functioning preservative competence. Any malfunction undermines warrant.

Of course, error can derive from other psychological factors' over-riding purely preservative memory, or being used instead of it. Then errors do not derive from purely preservative memory. Since the individual relies on other factors, warranted or not, any such errors do not occur despite warranted reliance on purely preservative memory.

How and whether purely preservative memory interacts with other psychological factors, in general or on any given occasion, is an empirical matter. But when it does not malfunction and is not over-ridden by other psychological factors, it preserves mode and content of past representational states. Where there are errors that are attributable to *its* operations, the errors are not brute errors. They are errors of malfunction that undermine warrant. That is why reliance on purely preservative memory is immune to brute error.

Are other types of memory, besides purely preservative memory, subject to brute error? In a sense, perceptual memory inherits perception's—and perceptual belief's—vulnerability to brute error. A given memory could be brute mistaken because it derives from a perception that is brute mistaken. However, here, the vulnerability is the vulnerability of *perception* to brute error. Errors specifically attributable to memory can seem to result from memory's not making the best cognitive contribution that it can. And it may seem that when perceptual memory undergoes such failure, warrant for relying on it lapses, because memory would on such occasions not operate well, even if it is generally reliable. Such failures would differentiate memory from perception. Perception can operate optimally given its proximal input; but because of abnormal distal conditions, it can still go wrong. Cases in which one is warranted in relying on it are cases in which it operates well, given the information that it has. If inaccuracy is correctly attributable to the operation of the perceptual system in a given case, warrant for relying on perception lapses. It can seem that where error is attributable specifically to perceptual memory, warrant lapses. So it can seem unclear how brute error in *any* memory is possible.

This account is too simple. It is known that some memory—especially propositional, longer-term episodic memory—involves constructing narratives and carrying out implicit (unconscious) quasi-inductions about the past. These constructions are partly based on other things that we believe. They are fairly reliable in providing important general attributes of remembered episodes. But with respect to matters in the relatively distant past, memory is often strikingly unreliable in filling in specific details.[101]

One can make sense of brute error attributable to such memory. To be warranted, the memory type must be reliable with respect to the level of detail remembered in the case in which the memory is employed. One would not be entitled to rely on purported memory of certain types of details that are purportedly remembered in the distant past, if memory in such cases is not reliable. But in cases in which memory is reliable in constructing the general attributes of an episode, it could fall into brute error despite following warranted operations from warranted starting points.

I make three points regarding this conjecture about brute error in episodic memory. First, to count as memory of an episode, a constructionist memory must use not only general background beliefs, but non-constructed, non-quasi-inductively produced representations of the episode. (The non-constructed memories are the singular bases for the quasi-induction.) I doubt that these singular starting points for the memory construction could be subject to brute error. Either they are well formed and accurate, or they are inaccurate by virtue of a non-constructed retention failure by the generally reliable competence. Such retention failures are just the sort of failures in representational function that undermine warrant.

Second, if there were no memories—whether or not these are singular bases for constructionist memories—of one's near-term propositional states and events that were immune to brute error, one's reflexive reviews of such states and events would never give the reviewed states and events buck-stopping status in critical reasoning and moral reasoning—by the arguments I gave in section IV of Lecture II and section I of this lecture. We have good reflective reason to believe that reflective review does give reviewed states buck-stopping status in uses of inference, practical decision making, and dialectic.

Third, the idea that near-term episodic memories always depend on matters beyond the well-functioning of one's memory—and are always subject to brute error—is implausible on its face. The idea that I was just thinking then [five seconds ago] that writing requires concentration or I am carrying out my decision [just made] to help him is always subject to error that could result from memory that functions well in the given circumstances is implausible on its face. Assume that the thought is based on conscious memory of the event, and not some collateral conscious thoughts about the past. To be subject to brute error the memory-based thought must be warranted. So the episodic memory must be generally reliable in that type of case. Error must not derive from contextual failure in the well-functioning operation of memory. For such failure would undermine warrant. Holding that memory of near-term past thinkings is always

[101] See Daniel Schachter, *Searching for Memory: The Brain, the Mind, and the Past* (New York: Basic Books, 1996). Schachter notes that John Dean's, presumably honest, detailed accounts of particular episodes were shown by the Nixon tapes to be massively inaccurate regarding details of the episodes, but accurate on their most important general features.

subject to errors that cast no aspersions on the representational well-functioning of memory is extremely implausible.

I think that our understanding of diachronic reasoning in inference, practical decision making, and dialectic indicates that in such reasoning episodic memory can utilize purely preservative memory—a type of what is called 'semantic memory' in the psychological literature. Such memory is the non-constructionist, anaphoric type of retention required in any diachronic representational operation—including inductive inference and narrative. Our understanding of the epistemic norms governing memory-based meta-psychological thoughts about one's very recent thinkings indicates that errors in such thoughts are attributable to representational malfunctions of memory. Such malfunctions, on given occasions, undermine warrant. Since I think that a natural competence to retain thought content and occurrent thought events over at least short times is constitutive to critical reason in temporal beings, I think that in us, and in any critical reasoner, there are cases of near-term meta-psychological autobiographical episodic memory in which brute error is not possible. I think that any individual that is subject to brute error in all memories of near-term thinkings would lack the cognitive equipment that is constitutive to the use of critical reason in inference, practical decision making, and dialectical thinking.[102]

I have been discussing warrants for relying on purely preservative and episodic memory in Self-Understanding of standing propositional states and past propositional events. Those warrants are aspects of the warrant for betokening understanding in these cases. I want now to focus more specifically on the warrant for betokening understanding in these types of Self-Understanding. In these cases, betokening understanding relies on purely preservative memory in Self-Understanding to retain mode and content of standing or past psychological elements.[103] Betokening understanding supplements this reliance with use of meta-representational competencies to *represent* the preserved representational content in a canonical way and to *represent* the preserved mode instance.

[102] If it is possible that there be beings that carry out critical reason in time periods shorter than those allowed for in current views about long-term memory—of which episodic memory is a subtype—then I think that a competence for an analog of meta-psychological autobiographical episodic memory would have to occur in those beings. The analog would differ only in being short-term memory. It would be memory of reasoning episodes in the very *very* recent—technically, short-term—past. Those analogs would have to be immune to brute error. Any being with representational competencies must, I think, have at least short-term memory. And beings with critical reason must be able to utilize autobiographical meta-psychological memory that retains not only content in reasoning, but retains *de re* memory of episodes of reasoning, individuated by their mode and representational content. Whether the memory is technically long term is, I think, not a constitutive matter. But in all actual critical reasoners that we know of, the relevant memory is surely sometimes long term—a few seconds or longer.

[103] In Self-Understanding of past psychological events, purely preservative memory joins, as we have seen, with episodic memory. In what follows, I focus on the relation between betokening understanding and purely preservative memory. It should be understood that what I say is meant to carry over to the cases of Self-Understanding in which episodic memory is involved as well.

Purely preservative memory preserves mode type and representational content, but *represents* neither as a subject matter. *A fortiori* it does not represent the events or state instances whose mode and content it preserves. Still, it always preserves the mode type and representational content *of a particular antecedent state instance or event*. In Self-Understanding, betokening understanding capitalizes on this fact to represent not only the mode type and representational content, but also the psychological state instance or event—which is always preserved, "there" to be referred to, when purely preservative memory functions well. Betokening understanding *uses* the lower-level purely preservative memory while *representing* what it preserves. This meta-representation is *de re* and is immune to brute error. So, when betokening understanding relies on just these competencies, either it is warranted and veridical, or it goes wrong through malfunction and warrant is undermined. Thus the basic warrant for relying on betokening understanding is immune to brute error.

What are the epistemic bases for my claim that betokening understanding in Self-Understanding of *some* propositional attitude states and events is immune to brute error?

I noted that we know non-empirically, by understanding their reflexive structure, that pure and impure *cogito* judgments involve no inference. Pure *cogito* cases are infallible. By reflecting on the form and content of the judgments, one can recognize that in impure *cogito* cases only a pathological failure to hold meta-representational and lower-level attitude modes together in a judgment could yield error, if one thinks the thoughts in a reflexive, "hereby" way. Errors derive from odd misuses of a form that one can use successfully. Although these are easy cases, they provide a model for understanding a wider range of cases.

What is the epistemic basis for my meta-meta-claim that betokening understanding in Self-Understanding of *some* past propositional events, and *some* standing attitude instances, as standing, is immune to brute error? The betokening understanding consists in a meta-representational conceptualization of the products of purely preservative memory.

The point that relying on *purely preservative memory* is immune to brute error seems secure for reasons already given. The point depends on reflecting on the function of such memory in any representational psychology. A reliable capacity to preserve mode and content is constitutive to having a system that could count as a representational psychology. Purely preservative memory is certainly not a matter of induction. Like deduction, induction *presupposes* a prior non-inferential capacity for purely preservative memory. The retention does not involve representation of what is retained. Error in retention must consist in representational malfunction of the competence, not brute mismatch with a subject matter.

What of the epistemic status of meta-psychological conceptualizations of representational content and mode (type and instance) of the products of purely preservative memory? Canonical meta-representational conceptualizations of *representational content* seem to be exercises of a competence understanding of the content together with competence understanding of the that-clause-like canonical names. Those are competencies that we clearly have. Reflection on them shows at least as clearly as reflection on *cogito* competencies does that they are non-empirical and immune to brute error. The key issue concerns the status of my claim regarding conceptualization of *mode—type* and *instance*—of a state with a specific representational content in meta-psychological memory. How can we know that such conceptualization is immune to brute error?

In Lecture II, section IV, I discussed three ways in which meta-psychological autobiographical episodic memory from the inside figures constitutively in critical reasoning: in inference, in practical decision making, and in dialectic. In such exercises of critical reason, we can recognize conceptualization of a mode—type and instance—of a propositional attitude that is preserved for a rational project (inference, decision, dialectic).

The following principle is a norm of critical reasoning in these enterprises:

> (CR) If, in critical reasoning, one correctly and with warrant judges that a lower-level state is (or is not) reasonable, then it rationally follows directly that one has reason to sustain (or change) the lower-level state.[104]

(CR) is like an inference rule, except that it explicitly concerns transmission of reasons, not preservation of truth. (CR) implies that in the relevant uses of critical reasoning, one's Self-Understanding of the lower-level state instance is immune to brute error. For if an application were subject to brute error, a reason that is attributed, correctly and with warrant, at the meta-psychological level of critical reasoning would not transmit immediately to the lower-level state. The transmission would *always* be subject to the proviso that one's completely warranted meta-level evaluation and self-attribution is not brute mistaken. If brute error were always possible in one's self-understanding, the lower-level attitude would not have buck-stopping status by virtue of being reviewed or reviewable. In fact, *no* attitude, except in *cogito* cases, would have buck-stopping status by virtue of being reviewed or reviewable. In critical reason, the fundamental, buck-stopping states include both the meta-level review and the critically reviewed lower-level states. The buck-stopping status of the lower-level states in critical reasoning partly depends on the potential for meta-level review. They are not separable. If (CR) were not true and applicable in the three cases, critical reason would never occur outside the immediate present of *cogito* cases. Thus I think that the view that understanding our standing propositional attitudes, as having stood, and understanding of some—at least near-term—past propositional occurrences is *always* subject to brute error amounts to a scepticism about our being diachronically extended critical reasoners.

I think that we can know that (CR) governs our reasoning in the way that we know that our reasoning is governed by specific rules of deductive inference—or by specific principles governing simple deductive reason-support relations.

[104] See my 'Our Entitlement to Self-Knowledge', 109ff; 81ff in this volume.

Knowing the principle is based on reflecting on the practice and norms of critical reasoning.

I think that we can know, by reflection, (CR)'s application to some actual particular cases of critical reasoning. That is, I think that in some cases, we can know by reflection that, from a warranted meta-level judgment that a given attitude is reasonable or unreasonable, we immediately have a reason to maintain or change the lower-level attitude. We have a reliable and warranted capacity to recognize the form of transmission of reasons as they occur in some diachronic cases. This capacity is analogous to recognizing in a deductive inference, on a particular occasion, that if a premise is warranted, it provides a reason for the conclusion.

Our ability to recognize immediate reason transmission in given cases is fallible and defeasible. But it seems to me that, as in the case of knowing (CR) itself, the positive warrant for knowing some of its applications is non-empirical. The warrant is for applications of a meta-meta-representational capacity, a capacity to understand our tracking representational content and mode (type and instance) of our diachronic conscious reasoning.

It is true that the unconscious mechanisms that underlie memory in these cases are open only to empirical investigation. But the conscious states are not identical with unconscious counterparts, even those that share *representata*. For the modes of presentation and hence identities of conscious states are different. Consciousness is an aspect of the mode of presentation, hence identity, of the states and processes that I am centering on. I think that the idea that we can know non-empirically the form of reason transmission in particular cases is no less plausible than the idea that we can know non-empirically the form of deductive inferences in particular cases, or that we can know of one type of thought that it is, if warranted, a reason for another. Such knowledge is defeasible.[105] But it is apriori. The idea that we have it seems to me very powerfully plausible. Such knowledge of norms places a prima facie constraint on empirical psychological investigations.

Not all memory in self-knowledge about past psychological states is immune to brute error. Most memories with warrants immune to brute error are nearer term, even if they are technically long-term memories. As time spans lengthen, use of induction about the past and memory that makes essential use of cobbling together different experiential memories become more prevalent. Longer-term memories often rely on factors other than purely preservative memory to access past psychological events—factors subject to brute error. My thesis is that we, as selves and critical reasoners, do sometimes rely on meta-psychological

[105] The claim that we have such apriori knowledge is defeasible. Most apriori warrants for Self-Knowledge are themselves defeasible. Hume was sceptical of reasons, as opposed to natural associations. He tended to assimilate all reasoning to induction, which he construed as blind habitual association. I join the later history of philosophy in rejecting these doubts. I think that some meta-knowledge that we engage in deductive inference is probably not defeasible.

autobiographical episodic memory, supported by purely preservative memory, in calling up past thoughts. I leave open in what cases we do so.

In sum, a betokening understanding that connects *de re* applications to preserved mode instances, and to their representational contents, relies on a lower-level capacity, purely preservative memory, that is constitutive to reason. The betokening understanding itself is a constitutive connector in a psychology capable of *critical* reason.[106] An individual is entitled to such betokening understanding, and to judgments based on it, because (a) he or she is entitled to rely on the lower-level capacity, and (b) he or she is entitled to *de re* meta-conceptualization of the mode and content of attitude instances used in inference, decision making, and dialectic in critical reasoning. Moreover, we have a meta-meta-representational ability to recognize applications of the capacity cited in (b). Judgments that rely on this betokening understanding are immune to brute error, because any error in such judgments derives from failure of purely preservative memory, failure of a *de re* power of understanding constitutive to critical reason, or failure to base the judgment on those capacities.[107]

Analogous points apply to anticipatory betokening understanding. Such understanding of future psychological instances is meta-representational conceptualization that piggybacks on lower-level preservational capacities for control and continuity in states like intentions—for example, in decision making. The default entitlements attach to an understanding that is a basic connector in critical reason. They yield immunity to brute error.

Let me survey whence we have come. I have focused on betokening understanding of propositional states and occurrences. Such understanding is integral to an apperceptive rationally accessible point of view, hence to being a self. Apperception combines betokening understanding with a self's self-attribution of the understood psychological elements. It comprises the multi-tier perspective of critical reason that marks selves. Betokening understanding is also essential to beings that are subject to moral norms. It is a meta-representational capacity that, except in *cogito*-like cases, uses lower-level transtemporal preservational powers to guide *de re* representation of propositional state instances and propositional occurrences. Natural reliability of the lower-level preservational capacities is constitutive to any representational psychology, hence to reason. Natural reliability of the meta-representational betokening understanding that uses those lower-level capacities preserves Self-Understood state instances and events in a way that is constitutive to selves and to psychologies capable of critical reason and morality. Such preservation is basic to understanding one's states from the inside and to the buck-stopping status of Self-Understandable states in critical reason and morality. Epistemic warrants attach to betokening understanding because it is reliable. Successful representation, on particular occasions,

[106] See discussion of meta-psychological autobiographical episodic memory in Lecture II.

[107] I re-emphasize that this failure could involve relying on inference or on considerations that are relevant to supplementing or overriding the basic entitlement.

depends only on its good use. It does not depend on matching a subject matter outside the psychology, or on information beyond the minimum necessary to exercise the understanding. Being warranted follows from using a well-functioning, reliable, epistemically viable competence, relative to information available to it. (See (E) above.) Such uses are certainly epistemically viable and, within their appropriate domain, reliable. And the minimum information needed to use betokening understanding suffices to provide veridical memory. So when the understanding does not malfunction and is not misused, it is warranted. Error derives only from malfunction, from misuse, or from relying on other capacities. So betokening understanding is immune to brute error, relative to this warrant.

Fulfilling epistemic norms governing Self-Knowledge thus entails exercising powers constitutive to *critical reasoning*. They are not norms that are merely necessary to critical reasoning, but constitutively independent of it. Fulfilling them requires using preservational powers that are partly constitutive of critical reason. For betokening understanding of attitudes constitutively uses meta-representation guided by preservational powers that are constitutive to any reasoning.[108] Conversely, such understanding is constitutive to the multi-tier,

[108] These points and the next text paragraphs address an interesting objection by Peacocke, 'Our Entitlement to Self-Knowledge: Entitlement, Self-Knowledge, and Conceptual Re-Deployment', *Proceedings of the Aristotelian Society* 96 (1996), 117–158. Peacocke objected to my account in 'Our Entitlement to Self-Knowledge'. He held that I had successfully shown only that critical reason requires Self-Knowledge, not that critical reason figures constitutively in the Self-Knowledge. He thought that Self-Knowledge is constitutively independent of norms of critical reason, and that mastery of the concept of belief and other attitudes, mastery of concepts of representational contents, and sensitivity to first-order norms of reason "generate" an explanation of Self-Knowledge. My 1996 paper did not focus on this issue; and Peacocke's objection had some resonance, although I believe that my account of the structure of critical reason even then suggested why it was plausible to hold that Self-Knowledge in critical reasoners uses essential preservative routes that are distinctive to critical reasoning. I have maintained that rational capacities that figure in critical reasoning are partly constitutive of Self-Knowledge.

Of course, I do not hold that Self-Knowledge is *based on* critical reasoning. Except in *cogito* cases, it is not based on reasons at all. My view is rather, first, that preservative routes essential to any reasoning, hence critical reasoning, are essential to Self-Knowledge. So Self-Knowledge cannot be constitutively independent of norms of critical reason. Perhaps this point is implicit in Peacocke's own appeal to sensitivity to first-order norms of reason. Second, I think that competence to conceive reasons as such is partly constitutive of mastering concepts of belief and other attitudes. See the discussion, in section VII below, of using reason-support relations as a supplementary factor in betokening understanding. Both points bring out ways in which horizontal and vertical reason-preservation and reason-transmission routes that are constitutive of critical reasoning also figure constitutively in Self-Knowledge. For further discussion, which however does not advance as far as the present account, see my 'A Century of Deflation and a Moment about Self-Knowledge', Presidential Address to the Pacific APA, *Proceedings and Addresses of the American Philosophical Association* 73:2 (November 1999), 25–46.

The account here also answers objections to my 'Individualism and Self-Knowledge', by Sven Bernecker, 'Externalism and the Attitudinal Component of Self-Knowledge', *Noûs* 30:2 (June 1996), 262–275. Bernecker noted that 'Individualism and Self-Knowledge' did not explain warrant in knowing the modes of our attitudes (belief rather than supposition, intention rather than hope). As I have emphasized elsewhere, 'Individualism and Self-Knowledge' was not intended as an epistemic account of Self-Knowledge, much less self-Knowledge.

apperceptive point of view that grounds applicability of norms of critical reason.[109]

VII

I now discuss a further aspect of betokening understanding. Critical reasoners must use understanding of lower-level reason-support relations to guide betokening understanding of instances of propositional attitudes. Betokening understanding must be guided not only by competence with lower-level preservation relations, but also by conceptualizations of reason-support relations among the target attitude and other lower-level attitudes. Such relations place the target attitude instance in a network of reason relations. (See note 108.) Any understanding of attitude instances that could not conceptualize reason-support relations as such could not ground the applicability of norms of morality and critical reason. In betokening an attitude instance, Self-Understanding essentially uses *de re* understanding of specific reason-support relations, practical or theoretical, among that attitude and others. Understanding an attitude's place in a network of reasons, for or against, bears on epistemic entitlement to judgments based on betokening understanding of that attitude. Since the identity of propositional attitudes constitutively depends partly on reason relations to other attitudes, betokening informed by understanding such connections is a better route to veridicality than any would-be betokening not so informed. So this identification through reason support in betokening understanding contributes to warrant for Self-Knowledge.

As with betokening understanding, lower-level competence understanding is used *in* the meta-representational understanding of reason-support relations as such (*R* is a reason for *A*). Understanding reason-support relations requires non-meta-representational competence to *use* one content as reason for another, and some disposition to be moved by, to *implement*, relations of threat or support. To understand a reason as such, for or against an attitude, one must have some lower-level disposition to shift or maintain attitudes in accord with the reason. Recognizing a reason as such essentially relies on the lower-level competence to use the reason to support the conclusion. This competence figures in the conceptual mode of presentation of the reason-support relation between a given betokened attitude and others.[110]

Take the concept, is a reason for believing that someone loved Haydn. Suppose it true of an instance of a belief that Mozart loved Haydn. The

[109] I presented this account in the oral form of these lectures in December 2007 at Columbia University. The general outlines of this account of one's warrant for Self-Knowledge are adumbrated independently by Frank Olav Barel in his Ph.D. dissertation 'Self-Knowledge and Anti-Individualism: A Defence of Neo-Rationalistic Compatibilism', (University of Oslo, 2010).

[110] Cf. 'Reason and the First Person'.

betokening understanding of that belief might be partly guided by an understanding of its supporting a belief that someone loved Haydn. Understanding the instance's supporting and being supported by other possible attitudes helps guide the betokening understanding. Concepts of reason support are meta-representational. In using them, however, a thinker must rely on a lower-level competence understanding of the *represented* propositional attitude types—belief that Mozart loved Haydn and belief that someone loved Haydn—including a lower-level competence to reason from one to the other. One relies on lower-level understanding of the represented contents as part of thinking about those contents, and about reason-support relations among them. The key point is that the guiding conceptualizations of reason-support relations employ lower-level competence understanding of the targeted contents, modes, and reason-support relations.

This point supplements, I think constitutively, the account already given of immunity to brute error yielded by the default warrant for betokening understanding and judgments based on it. Meta-representational understanding of reason-support relations must use lower-level competence understanding. The lower-level competence understanding is warranted through correctly representing the relations of reason support that set norms for epistemic warrant. The meta-representational understanding is warranted by relying on correct lower-level competence understanding of reason-support relations. So a warrant for meta-representational understanding of reason-support relations yields immunity to brute error.

To summarize the idea crudely, meta-representational understanding of reason-support relations depends for its warrant on sucking up warrant, present in the lower-level competence, for engaging in lower-level reasoning. To be warranted, the meta-representational understanding must involve correct understanding, and involve a disposition to be moved by, the lower-level reason-support relations. The meta-representational understanding of reason-support relations helps guide betokening understanding of attitude instances by locating them in a local network of practical and theoretical reasons. Meta-representational understanding of such relations improves one's route to veridical betokening understanding. Meta-representational understanding of reason-support relations contributes warrant to betokening understanding by helping it follow this route, only if it is correct understanding. It must be understanding of relations of genuine reason support. So both the warrant for meta-representational understanding of reason-support relations and the contribution of warrant for such understanding to the warrant for betokening understanding yield immunity to brute error. Clearly, this aspect of the warrant for betokening understanding is constitutive to critical reason—not merely a component of critical reasoning whose warrant comes from outside critical reason itself. (See note 108.)

VIII

I turn to betokening understanding in Self-Understanding and Self-Knowledge of two types of *non-propositional* psychological states and occurrences. One type comprises nonpropositional *representational* empirical states—such as perceptions and perceptually derived images. The other comprises conscious sensory states, such as certain aspects of pains, that are not representational in my sense. I will have to deal with these cases briefly.

The account of betokening understanding of perceptions, perceptual anticipations, perceptual memories, and fictional images broadly follows the account of betokening understanding of propositional states. Capacities to preserve mode and content of perceptions are constitutive to any empirical representational psychology. In a propositional-attitude psychology, such preservational connections obtain between perception and first-order perceptual belief. If a psychology has meta-representational capacities, some of them conceptualize lower-level contents *in* specifying those contents. Just as object-level perceptual concepts use perceptual contents in their modes of presenting environmental entities, so meta-concepts that represent perceptual contents—especially perceptual attributives—employ the object-level perceptual contents in *their* modes of representing those contents. Thinking about perceptions also employs lower-level mechanisms for preserving perceptual mode in conceiving of that mode, type and instance. Meta-representation that uses lower-level capacities is constitutive to the apperceptive core self.

An individual is default entitled to beliefs based on reliably veridical betokening understanding. Reliability is necessary for there to *be* the relevant point of view. Errors derive only from malfunction or misuse of the preservational operations—undermining entitlement—or from failure to base belief on those operations. A relevantly based judgment cannot be mistaken about a current perception, or immediately past perception, or a perceptual anticipation, while all is well with the cognitive apparatus. Brute error—warranted, representationally well-functioning error—is not possible relative to the relevant warrant.

Betokening understanding of *non-representational* conscious sensations, or aspects of them, differs importantly from the other types of betokening understanding. No lower-level capacities preserve representational content. But the basic shape of the account of the warrant for Self-Understanding of conscious sensations is otherwise similar.

Phenomenal consciousness of sensations is constitutive to being a self. Rational access to such consciousness is also constitutive.[111] Here betokening

[111] Higher animals and very young children that lack selves can surely form beliefs about sensations like pain. Such beliefs are as primitive as ordinary perceptual beliefs about the environment. Of course, such beings cannot *self*-attribute having those sensations; for they are not selves. But their beliefs make the sensations ego-relevant through whatever ego-concept they have that precedes a full-fledged self-concept. Traditional theology might allow that God is a self and lacks sensations. I confine my thesis to finite selves that interact with the physical world.

understanding forms concepts of sensations. The concepts constitutively use the sensations, or memories of their types, in representing them. So a lower-level element is again used in conceptual representation of it. A capacity to use such concepts *de re* to pick out instances is not constitutive to reasoning. But it is constitutive to being a self. A self must be able to think about and act on some conscious affect, such as its pains or pleasures. Such a capacity is basic to owning and controlling bedrock aspects of conscious life—the core of consciousness in a point of view. Being subject to moral norms also seems to hinge on this capacity.

A reliable natural capacity to form true beliefs in normal circumstances—those in which the beliefs' representational natures were formed—yields epistemic entitlement to those beliefs. Normally, not all is cognitively well with a person who mistakenly believes he or she is in sharp pain, or that a sharp pain is a tickle. Error can, of course, arise from anticipation, inference, or ideology. But any error that relies on the most direct conscious way of thinking about a sensation is a dislocation of self—a pathology that undermines warrant.[112] Beliefs about one's conscious sensations are immune to brute error, *relative to the entitlement.*[113]

This completes my sketch of applications of betokening understanding, and judgments based on it, to non-propositional cases.

IX

Predications in Self-Understanding are *self-attributions of* betokening understanding. Self-attribution is attribution of the psychological state to oneself via self-conception.

What is the warrant for the transition from warranted *de re* betokening understanding to self-attribution? Wherein is one entitled, in Self-Understanding, to attribute *to oneself*—referred to as oneself, with a self-concept—an instance of an understood state?

[112] There are borderline cases. Perhaps some pains are so close to borderlines with other types of sensation that, even in ordinary circumstances, one could make a brute error about whether one is having one. (Presumably, one could still get right that it is a sensation of a more generic type.) Perhaps one could even be *unreliable* in forming judgments about certain types of sensations that are easily mis-categorized, at least given only short periods of time. Such cases do not, I think, count against the view that many judgments about conscious sensations are immune to brute error relative to certain warrants associated with relevant judgment-forming routes.

Again, I think that whether a warrant yields immunity to brute error is not apriori determinable from the content of the belief. *A fortiori*, I think that immunity to all error is not intrinsic to self-attributions with sensation concepts. I think that because of cognitive malfunction or stoic ideology, an individual can judge even a very intense pain not to be a pain.

[113] I think that *some* judgments about *past* sensations, and perceptions, are immune to brute error relative to certain warrants. The relevant warrants attach to judgments that use purely preservative memory of singular applications of concepts of those sensations or perceptions. Here the account parallels that of betokening understanding of past propositional thought events.

The transition is idealized. There is no step in time. Betokening understanding and self-attribution are aspects of the predication. The relation between the aspects is, however, usefully considered as a transition step, although there is no literal inference. It is *not* an inference from That is a belief that *p* to I believe that *p* (or that belief is mine).

The epistemic norm for the transition step, governing warrant for basic apperception is as follows:

> (WBA) If one is entitled in the relevant way to betokening understanding of an instance of a psychological element, and if the self-attribution relies only on the betokening understanding and on one's competence with the self-concept, one is entitled to attribute that instance to oneself as oneself.

This norm is grounded in the fact that what one is entitled to believe about one's self rests fundamentally on what one is entitled to believe about one's psychological states. Selves are not systems of psychological states. They are the subjects/agents of such systems. But the natures of selves consist in their psychological competencies. As noted in the first lecture, the capacity to form warranted beliefs about one's self through beliefs about one's psychological states has its phylogenetic prehistory in couplings between occurrences of *de se*, or ego-centric, indexes applied to an individual as origin of a representational framework and as subject matter of perception. Such couplings center on perceivable bodily attributes. Selves are not perceivable. But we can know our psychological states *de re*. Such knowledge is a basic route to knowing our selves. Betokening understanding of instances of psychological elements grounds the *Self*-Knowledge that I have been discussing.

A central instance of self-cognition through cognition of one's psychological states is self-tracking in memory. Again, the relevant memory is grounded in more primitive couplings. *De se*, ego-centric indexes, the primitive ancestors of self-concepts, are embedded in all experiential memory.[114] When occurrences of such indexes *that mark an individual* are coupled with occurrences applied to a subject matter, they allow diachronic ego-tracking. The initial tracking is perceptual and body centered. Memories associated with couplings in passing the mirror test are likely examples. When memories become meta-psychological, indexes in them can be used in couplings that track past psychological instances, for example, transitions in reasoning, in meta-psychological autobiographical episodic memory. Such trackings are among the diachronic unities that make selves possible.

These unities are grounded in present-tense couplings—those that attribute current psychological instances, identified through betokening understanding, to a subject/agent that is a central *locus* for them. Descartes' *cogito* focused on performative present-tense couplings. But there are equally important couplings

[114] For more on primitive ancestors of the self-concept, see my 'Memory and Persons'; and '*De Se* Preservation and Personal Identity: Reply to Shoemaker', this volume.

with present standing states. When couplings of ego-centric indexes are incorporated into betokening understanding of psychological states and used in critical reasoning, such indexes support self-concepts. Having a self-concept, indeed being a self, is grounded in reflexive psychological competencies that meet epistemic norms for Self-Understanding.

Fulfilling these epistemic norms is making connections constitutive to selves. Making the connections—engaging in the diachronic and multi-tiered preservations—just is getting things right about one's psychology and one's self. Thus fulfilling the norm set out in (WBA) yields immunity to brute error for self-attributions. The step's starting point—warranted betokening understanding—cannot introduce brute error. If self-attribution relies on that understanding, it veridically self-attributes the Self-Understood state. Self-attributions based in betokening understanding are paradigmatically warranted uses of the self-concept. Errors in self-attribution derive from warrant-undermining pathology or from not relying on the relevant preservational mechanisms. So self-attributions are immune to brute error, relative to the relevant warrants.

Conceiving the psychological elements as one's own via the self-concept is acknowledging ownership of commitments in the psychological elements. Capacity for such acknowledgment helps ground applicability of norms of critical reason and morality.

Many issues arise about this account of warrant for self-attribution. I mention two.

Shoemaker's discussion of quasi-memory may seem to threaten immunity to brute error in Self-Understanding. Through quasi-memory, an individual could identify an attitude instance that is *not* the individual's own in a *de re* way. The individual could make a brute error in self-attributing the quasi-remembered psychological state. Remembering that warrants attach to psychological abilities helps disarm such threats. An ability to preserve one's own attitudes differs, psychologically and epistemically, from any ability to tap into other minds. Brute errors of self-attribution relative to warrants associated with quasi-memory are irrelevant to self-attributions warranted by memory.[115]

Some schizophrenics think that their thoughts are inserted into their minds by others. They deny that their thoughts, which they cite correctly, are their own. Such cases do not threaten the account because they involve pathology in the preservative capacities that ground entitlements to self-attribution. The schizophrenics do not rely purely on betokening understanding to make a self-attribution, in the way specified by the norm. There is some malfunction in connecting self-ascription with betokening understanding. Such malfunction undermines warrant for self-ascription in judgments about state instances. The

[115] Shoemaker, 'Persons and Their Pasts', *American Philosophical Quarterly* 7:4 (October 1970), 269–285, reprinted in his *Identity, Cause, and Mind* (New York: Oxford University Press, 2004). See also my 'Memory and Persons'.

errors are not errors in exercising well-functioning capacities that constitute apperceptive points of view.[116]

My account of warrant for Self-Understanding and Self-Knowledge refines Kant's notion of unity of apperception. Unity lies in an understanding that utilizes transtemporal and inter-level constitutive connections in the structure of selves.

The account highlights a *singular intellectual* power, *de re* understanding of particulars. Descartes emphasized *de re* understanding in contrast to Aquinas's emphasis on reason as a competence with generality. Reason and *de re* intellectual understanding are deeply interrelated. But they are distinct. Even now, the latter is neglected in philosophy.

Kant took up Descartes' notion in his own conception of understanding of particulars. Kant insisted that such understanding must be warranted through sense capacities. I think the insistence mistaken. I think that Descartes was right to hold that we have an understanding of particulars that has a purely intellectual warrant. Descartes over-rated the capacity's power for metaphysical insight. But he was right to highlight the capacity. Self-Understanding provides a central example. Other examples occur in mathematics, certain types of appreciation of intellectual beauty, and perhaps moral thinking.

I hope to have delineated the centrality of Self-Understanding in being a self—and in having the cognitive and valuational capacities, and being subject to the norms, that make us special. Using our intellects to understand the particular, particularly ourselves, is a way in which we are valuable, minded islands in a largely mindless, value-neutral universe.

[116] See Christopher D. Frith, *The Cognitive Neuropsychology of Schizophrenia* (Hove, UK: Lawrence Erlbaum, 1992); Simon Mullins and Sean A. Spence, 'Re-Examining Thought Insertion: Semi-Structured Literature Review and Conceptual Analysis', *The British Journal of Psychiatry* 182 (2003), 293–298.

PART II

INTERLOCUTION

10 *Content Preservation**

Near the beginning of *Rules for the Direction of the Mind* Descartes holds that some things known "with certainty" and "by deduction" are not evident. He notes that in long deductions, we may know that 'the last link is connected with the first, even though we do not take in by means of one and the same act of vision all the intermediate links on which that connection depends, but only remember that we have taken them successively under review. . . .'[1] Though he acknowledges that such knowledge is not evident or purely intuitive, and that long deductions are more subject to error than is intuitive knowledge, Descartes thinks that if the knowledge is deduced from evident mathematical premises, it is certain and demonstrative. Presumably he would not doubt that it is apriori. I lay aside certainty. But the view that the knowledge is demonstrative and apriori seems to me true.

Roderick Chisholm sees matters differently. He defines 'apriori' in such a way that a proposition is apriori (and known apriori) only if it is either evident or follows directly by evident entailment from something that is evident. He explicitly rules out the results of multi-stepped deductions:

> What if S derives a proposition from a set of axioms, not by means of one or two simple steps, but as a result of a complex proof, involving a series of interrelated steps? If the proof is formally valid, then shouldn't we say that S knows the proposition a priori? I think that the answer is no.

He adds:

> [I]f, in the course of a demonstration, we must rely upon memory at various stages, thus using as premisses contingent propositions about what we happen to remember, then, although we might be said to have "demonstrative knowledge" of our conclusion, in a

* I am indebted to Tony Anderson, Hilary Bok, Larry BonJour, Robert Brandom, Michael Bratman, Robin Jeshion, Bill Hart, Bernie Kobes, Ruth Marcus, Stanley Munsat, Christopher Peacocke, W. V. Quine, Corliss Swain, Fred Stoutland, and the editors of *The Philosophical Review* for valuable remarks.

[1] Descartes, *Philosophical Works*, ed. Haldane and Ross (New York: Dover, 1955), vol. 1, p. 8. Locke, in *An Essay Concerning Human Understanding*, 4:2: 7, notes that such knowledge is "less perfect" in the sense of more subject to error than intuitive knowledge.

somewhat broad sense of the expression "demonstrative knowledge," we cannot be said to have an a priori demonstration of the conclusion.[2]

Some of the difference between us derives from different conceptions of apriority. There are many such conceptions. I will be explicit about mine. I understand '*apriori*' to apply to a person's knowledge when that knowledge is underwritten by an apriori justification or entitlement that needs no further justification or entitlement to make it knowledge. A justification or entitlement is *apriori* if its justificational force is in no way constituted or enhanced by reference to or reliance on the specifics of some range of sense experiences or perceptual beliefs.

I take 'apriori' to apply primarily to justifications or entitlements, rather than to truths. There are, of course, conceptual relations between these notions. Justification or entitlement aims at truth since it rationally supports belief. Moreover, the notion of apriori truth is important, though it should probably be explicated in terms of possible apriori knowledge. But in this account, justification and entitlement are fundamental.

The distinction between justification and entitlement is this: Although both have positive force in rationally supporting a propositional attitude or cognitive practice, and in constituting an epistemic right to it, entitlements are epistemic rights or warrants that need not be understood by or even accessible to the subject. We are entitled to rely, other things equal, on perception, memory, deductive and inductive reasoning, and on—I will claim—the word of others. The unsophisticated are entitled to rely on their perceptual beliefs. Philosophers may articulate these entitlements. But being entitled does not require being able to justify reliance on these resources, or even to conceive such a justification. Justifications, in the narrow sense, involve reasons that people have and have access to. These may include self-sufficient premises or more discursive justifications. But they must be available in the cognitive repertoire of the subject. The border between the notions of entitlement and justification may be fuzzy. I shall sometimes use 'justified' and 'justification' broadly, to cover both cases.

A person's knowledge of a proposition might be adequately supported both by an apriori body and by an empirical body of justification or entitlement. Then the person's knowledge would be heterogeneously overdetermined. The person would have both apriori and empirical knowledge of the proposition. To be apriori, the knowledge must be underwritten by an apriori justification or entitlement that needs no further justificatory help, in order for the person to have that knowledge. To be apriori, a person's justification or entitlement must retain its justificational force even if whatever empirical justifications or entitlements the person also has to believe the relevant proposition are ignored.

[2] Roderick M. Chisholm, 'The Truths of Reason', in *Theory of Knowledge*, 2nd edn. (Englewood Cliffs, NJ: Prentice Hall, 1977), reprinted in Paul K. Moser *A Priori Knowledge* (Oxford: Oxford University Press, 1987).

In holding that the justificational force of an apriori justification or entitlement is in no way constituted or enhanced by reliance on the specifics of some range of sense experiences or perceptual beliefs, I do not require that an apriori justification rely on reason or understanding alone—as pre-Kantian rationalists required. A justification or entitlement would count as apriori if it did not rely for its justificational force on sense experience or perceptual belief at all. But it might also count if it depended on entirely general aspects of sense experience or perceptual belief, or on aspects of the structure of the subject's sense capacities and on their function in yielding categories of information.[3]

An individual need not make reference to sense experiences for his justification or entitlement to be empirical. My term 'reliance on', in the explication of apriority, is meant to acknowledge that most perceptual beliefs about physical objects or properties do not refer to sense experiences or their perceptual content. Such beliefs make reference only to physical objects or properties. But the individual is empirically entitled to these perceptual beliefs. The justificational force of the entitlement backing such beliefs partly consists in the individual's having certain sense experiences, or at any rate in the individual's perceptual beliefs' being perceptual.

An apriori justification (entitlement) cannot rely on the specifics of sense experiences or perceptual beliefs for its *justificational force.* An apriori justification will usually depend on sense experiences or perceptual beliefs in some way. They are typically necessary for the acquisition of understanding or belief. But such dependence is not relevant to apriority unless it is essential to justificational force. Distinguishing the genesis of understanding and belief from the rational or normative force behind beliefs is fundamental to any view that takes apriori justification seriously.[4]

No serious conception of apriority has held that all justifications held to be apriori are unrevisable or infallible. Traditionally, the deepest apriori justifica-

[3] Kant thought that all synthetic apriori judgments, except those in his practical philosophy—and perhaps in the critical philosophy as a whole—rested on general ("pure") aspects of the structure or function of sense experience. In fact, he believed that the justificational force of all such judgments depended on one's actually having had sense experiences. My conception of apriori knowledge makes room for Kant's conception. I do not, however, agree with Kant that those apriori justifications whose justificational force is not enhanced at all by sense experience are vacuous, or analytic in the sense of being true independently of any relation to a subject matter. The distinction between reliance on the specifics of a range of sense experiences, or perceptual beliefs, and reliance on the structure or function of one's sense capacities in obtaining categories of information is not sharp. I think it may remain useful.

[4] This explication of apriority applies to justification of *cogito*-type thoughts like *I am thinking*, and of other judgments about intellection. (It does not apply to *I am having an afterimage.*) These thoughts' justification is grounded on understanding, not on sense experience or perceptual belief. I am aware that some traditional conceptions of apriority would exclude *cogito* cases. Some of these conceptions emphasize not justificational independence of sense experience, but justificational independence of any "experience" at all, including intellectual "experience". (I leave open here whether this use of 'experience' is appropriate.) This is one of Leibniz's conceptions (see *New Essays* IV, ix). Of course Leibniz centered on apriori truth rather than on an individual's justification. Frege's conception features justificational independence of any relation to particular

tions were seen to be hard to come by. Putative apriori justifications were traditionally held to be revisable because one could fail to understand in sufficient depth the relevant propositions, or make errors of reasoning or analysis.

Traditional views did tend to overrate the tightness of connection between genuine (as opposed to putative) apriori justifications and truth. First, apriori justification (entitlement) can be non-demonstrative: an apriori justification can be outweighed without being shown to be rationally deficient or based on misunderstanding—without being shown not to have justificational force (not to be a justification). Some mathematical arguments are non-demonstrative, even broadly inductive, yet apriori in my sense. If a principle is accepted because its truth would explain or derive a variety of other accepted mathematical principles, the justification for accepting the principle is non-demonstrative; but it may not derive any of its force from perceptual beliefs. Second, although some apriori justifications or entitlements may be invulnerable to empirical counter-considerations, such invulnerability does not follow from the notion of apriority. As will emerge, I think that some beliefs with genuine apriori justifications or entitlements are vulnerable to empirical overthrow.

In both ways, a belief 's being apriori justified, for a person at a time, does not entail that it is true. There are, I think, some apriori justifications or entitlements that are demonstrative and do entail truth. But they do not do so purely by being apriori. The present conception of apriority fixes on the nature of the positive rational support for a belief. It says nothing about ways in which a belief may be vulnerable to counter-considerations.

Thus apriori justification may be unevident, fallible, non-demonstrative, and not "certain". Beliefs thought to be apriori, and even actually justified apriori, are subject to revision. In these ways, my conception of apriority differs from Chisholm's.

Our differences are not primarily verbal, however. Chisholm regards long deductions as importing memory of particular past mental events into the justification of the deduction.[5] If such memories are a necessary part of the justification

events or facts in time (see Gottlob Frege, *The Foundations of Arithmetic*, section 3). On his conception, only general truths and truths derivable from general truths could be known apriori.

The terminological issues here are complex; but this difference with traditional explications will not affect my argument with Chisholm, which goes through on any of these conceptions. Moreover, the broader argument of the paper does not depend on how one uses the term 'apriori'. I am less interested in the term than in the conception I associate with it. The argument of the paper hinges on the role of perception in justification or entitlement. I do think that there are significant substantive and historical issues regarding these different notions associated with the term 'apriori' that bear on the way the issue between empiricism and rationalism has come to be understood since the work of Kant, Mill, and the positivists. For now, it is enough that the present explication signals my interest in justifications or entitlements whose force is grounded in intellection, reason, or reflection, as distinguished from perception, understood broadly to include feeling.

[5] Descartes' own remark that in deductions we must remember that we have taken the links of the deduction "successively under review" may suggest this view. I find it unclear how he intended the remark.

of the deduction, then—at least where they include memories of empirical beliefs or experiences (memories of reading symbols carefully, for example)—such deductions are not apriori, even on my conception of apriority.

But Chisholm's conception of the *role* of memory in demonstrative reasoning seems to me off the mark. If memory supplied, as part of the demonstration, "contingent propositions about what we happen to remember", the demonstration could not be purely logical or mathematical. But the normal role of memory in demonstrative reasoning is, I think, different. Memory does not supply for the demonstration propositions about memory, the reasoner, or past events.[6] It supplies the propositions that serve as links in the demonstration itself. Or rather, it *preserves* them, together with their judgmental force, and makes them available for use at later times. Normally, the content of the knowledge of a longer demonstration is no more about memory, the reasoner, or contingent events than that of a shorter demonstration. One does not justify the demonstration by appeals to memory. One justifies it by appeals to the steps and the inferential transitions of the demonstration.

Why did Chisholm think otherwise? Long demonstrations are more fallible, and fallible in different ways, than short ones are. As he notes, people make mistakes of haste or incomplete understanding in judgments about relatively obvious propositions. But in longer demonstrations there are not only more opportunities to make these mistakes. One may suffer memory slips, even if one is careful and fully understands each proposition in the deduction. Traditionally, belief that appealed to apriori justification was held to be subject to error. But the sources of error were sometimes limited to failures of understanding and reason. It may seem that failure of memory is a source of error not easily accommodated by the traditional conception.

But relevant differences between short and long demonstrations are at most those between short-term and longer-term memory. Even one-step demonstrations could go bad if the reasoner's short-term memory were defective enough. So if we take vulnerability to memory failure as a sign that a justification of reasoning must *make reference to* memory, no reasoning at all will be independent of premises about memory. This is unacceptable. It is one thing to rely on memory in a demonstration, and another to use premises about memory. Any reasoning in time must rely on memory. But not all reasoning must use premises about memory or the past.

Here as elsewhere, to be justified in a cognitive process, one need not include premises in the justification that rule out all possible sources of error. This is a widely accepted point about perceptual justification. To be entitled to a perceptual belief that there is a bird there, one need not rule out all ways that one could be fooled. The same point applies to reasoning. To be justified in deductive

[6] Chisholm's "thus", in the quoted passage, is clearly a mistake. It does not follow from a deduction's reliance on memory that it, or any justification associated with it, uses "contingent propositions" about memory as premises.

reasoning, one need not include in one's justification propositions that guard against memory lapses, short or long term. Reliance on memory does not even add to the justificational force of the deductive justification.

If a justification depends on valid deductive reasoning from (let us presume) premises that are known apriori, then one's being justified by the justification depends only on one's *actually* understanding the reasoning sufficiently, and on one's reasoning processes *actually* working properly. The justification does not depend on a premise that says that these conditions obtain, a premise that would itself require further justification. (I think that such dependence would involve a vicious regress.) One can presume that they obtain, without needing justification for the presumption, except in special situations in which these presumptions are called reasonably—and perhaps even correctly—into question.

In a deduction, reasoning processes' working properly depends on memory's preserving the results of previous reasoning. But memory's preserving such results does not add to the justificational force of the reasoning. It is rather a background condition for the reasoning's success. Memory is no more intrinsically an empirical faculty than it is a rational faculty. Its function in deductive reasoning is preservative. Its role in justification derives from what it preserves. Our entitlement to rely on memory in long deductions derives from our entitlement to rely on reasoning to carry out its functions. Memory failures that cause demonstrations to fail are failures of background conditions necessary to the proper function of reasoning. Hence the fallibility of memory in deductive reasoning is a source of error that can be countenanced by the traditional conception of apriority—and our conception as well.

Even in empirical reasoning, memory has a purely preservative function that does not contribute to the force of the justification, but simply helps assure the proper working of other cognitive capacities over time. When we perceive events and infer an explanation, memory preserves the perceptual beliefs as we carry out the explanation. But this preservation is not part of the justification of the explanation, nor does it add to it—even though if it were to fail, the explanation would be jeopardized. Rather, memory just holds the results of the perception intact long enough for explanation to be carried through.

Of course, memory sometimes is not purely preservative, but is an independent element in justification. Memory of events, objects, experiences, or attitudes may form a premise in a justification of an empirical belief. The beliefs that such memories support are justified partly *by reference to* the memory. Or else they may partly rely for their entitlement on memory.

Substantive memories of specific events, objects, experiences, or attitudes may play a role in deductive reasoning. They may aid reasoning without being elements in the justification they aid. So, for example, we may draw pictures in a proof, or make use of mnemonic devices to aid understanding and facilitate reasoning, without relying on them to enhance the mathematical justification. Alternatively, substantive memories may be part of an auxiliary, double-checking justification. In such cases, they may play a justificational role, yet be justificationally dispensable.

Substantive memory can even be needed to shore up gaps in a person's deductive reasoning. When a purely preservative instance is reasonably challenged, because memory has proved unreliable, one may have to rely on substantive memory. For example, if one knows one's memory has been slipping, one might have to resort to remembering counting the number of implication signs in a pair of formulas to support one's presumption that one's inference was based on correct memory. In such a case, reliance on the mnemonic devices may be indispensable to the person's justification—not merely a part of an auxiliary double-checking procedure. For the person is no longer entitled to the presumption that memory can be relied upon. I think, however, that the need to make reference to memory in deductions in order to be justified by the deductions is uncommon. In certain cases one might reasonably doubt that one is entitled to rely on one's memory, but be wrong to doubt it.

But the fact that memory can play substantive roles in justification or entitlement should not obscure the distinction between substantive and purely preservative memory. Let me summarize the distinction. Substantive memory is an element in a justification; it imports subject matter or objects into reasoning. Purely preservative memory introduces no subject matter, constitutes no element in a justification, and adds no force to a justification or entitlement. It simply maintains in justificational space a cognitive content with its judgmental force. Like inference, it makes transitions of reason possible, but contributes no propositional content. Unlike inference, it is not a transition or move—so it is not an element in a justification. Hence in deductions, neither reliance on it nor susceptibility to errors that arise from its malfunction prevents the justification associated with the deduction from being apriori.[7]

My discussion of memory is pointed toward exploring analogies between memory and acceptance of the word of others. What is the role of interlocution in the justification of our beliefs?

Relying on others is perhaps not metaphysically necessary for any possible rational being. But it is cognitively fundamental to beings at all like us. Though

[7] The distinction between substantive memory and purely preservative memory roughly parallels a distinction in psychology between "episodic memory" and "semantic memory". There is evidence that these sorts of memory function differently in our psychologies. See E. Tulving, 'Episodic and Semantic Memory', in E. Tulving and W. Donaldson (eds.), *Organization of Memory* (New York: Academic Press, 1972), 382–402.

Another difference between the two types of memory is that purely preservative memory necessarily plays a role in any reasoning in time. The extent to which substantive memory enters into reasoning depends on the psychology of the reasoner, the subject of the argument, and so on. One should not underestimate, however, our dependence on the use of symbols in reasoning. The role of symbols is partly that of providing perceptual objects. Explicating this sort of dependence is a difficult and important matter. Doing so may complicate or blur the distinction between the sometime dependence on substantive memory and the more general rational necessity of depending on purely preservative memory. But I think that the distinction will remain valuable.

ontogenetically later than perception and memory, reliance on others for learning language and acquiring beliefs is deeply ingrained in our evolutionary history. Acquiring beliefs from others seems not only psychologically fundamental, but epistemically justified. We do not as individuals justify this reliance empirically, any more than we justify our use of perception empirically. But we seem entitled to such reliance. Most of the information that we have, and many of the methods we have for evaluating it, depend on interlocution. If we did not acquire a massive number of beliefs from others, our cognitive lives would be little different from the animals'.

What is the epistemic status of beliefs based on interlocution? I will state my view broadly before qualifying and supporting it. The use of perception is a background condition necessary for the acquisition of belief from others. But in many instances, perception and perceptual belief are not indispensable elements in the justification of such beliefs, or in the justificational force of entitlements underwriting such beliefs. The function of perception is often analogous to the function of purely preservative memory in reasoning. Without perception, one could not acquire beliefs from others. But perception plays a triggering and preservative role, in many cases, not a justificatory one. Sometimes, the epistemic status of beliefs acquired from others *is not empirical*. In particular, it is not empirical just by virtue of the fact that the beliefs are acquired from others.[8] Such beliefs are sometimes apriori justified in the sense that they need not rely for justificational force on the specifics of some range of sense experiences or perceptual beliefs.

Thomas Reid insightfully compares acquisition of belief from others to perception as a basic "channel to the mind", with its own functions in acquiring knowledge. Reid also claims that the tendency to rely on others for acquiring beliefs is innate:

> The wise and beneficent Author of nature, who intended that we should be social creatures, and that we should receive the greatest and most important part of our knowledge by the information of others, hath, for these purposes implanted in our natures two principles that tally with each other. The first of these principles is a propensity to speak truth . . . [the

[8] Contrast Chisholm, 'The Truths of Reason', section 5, and James F. Ross, 'Testimonial Evidence', in Keith Lehrer (ed.) *Analysis and Metaphysics* (Dordrecht: D. Reidel, 1975). They assume that belief based on testimony cannot be justified apriori and, if it is knowledge at all, must be empirical.

I think that some of what I am saying here bears on the common assumption that knowledge based on the output of proofs by computers cannot be apriori. Compare Saul Kripke, *Naming and Necessity* (Cambridge, Mass.: Harvard University Press, 1980), 35; also Thomas Tymoczko, 'The Four-Color Problem and its Philosophical Significance', *The Journal of Philosophy* 76 (1979), 57–83. Kripke says that such knowledge is based on the laws of physics. Although such knowledge depends on the functioning of a machine according to the laws of physics, it is not obvious that knowledge of the laws of physics is an indispensable part of our justification for believing in the results of such output. I discuss this issue in 'Computer Proof, Apriori Knowledge', and Other Minds, *Philosophical Perspectives* 12 (1998), 1–37. [This paper was cited in 1993 as in preparation.]

second] is a disposition to confide in the veracity of others, and to believe what they tell us.[9]

Reid notes that credulity, unlike reasoning and experience, is 'strongest in childhood, and limited and restrained by experience'. We restrain credulity by weighing the character and disinterestedness of witnesses, the possibility of collusion, the antecedent likelihood of information. Moreover, our reliance on others is more fallible than our reliance on perception—as Reid also notes. We make perceptual errors, but the errors derive from illusions that often can be explained by reference to natural law. We are led into mistakes by others through lies and emotional interferences that are capricious in comparison to the patterns of nature. Why do these considerations not show that acquisition of beliefs from others is not only necessarily empirical but far more in need of empirical expertise than ordinary perception for its justification?

Justification in acquiring beliefs from others may be glossed, to a first approximation, by this principle: *A person is entitled to accept as true something that is presented as true and that is intelligible to him, unless there are stronger reasons not to do so.* Call this the *Acceptance Principle*. As children and often as adults, we lack reasons not to accept what we are told. We are entitled to acquire information according to the principle—without *using* it as justification—accepting the information instinctively. The justification I develop below is a reflective philosophical account of an epistemic entitlement that comes with being a rational agent.

Justified (entitled) acceptance is the epistemic "default" position. We can strengthen this position with empirical reasons: "she is a famous mathematician". We can acquire empirical reasons *not* to accept what we are told: "he has every reason to lie". But to be entitled, we do not have to have reasons that support the default position, if there is no reasonable ground for doubt. Truth telling is a norm that can be reasonably presumed in the absence of reasons to attribute violations.

It is usually said that to be justified in accepting information from someone else, one must be justified in believing that the source believes the information and is justified in believing it. I think this misleading. A presupposition of the Acceptance Principle is that one is entitled not to bring one's source's sincerity or justification into question, in the absence of reasons to the contrary. This too is an epistemic default position.

The Acceptance Principle is not a statistical point about people's tending to tell the truth more often than not. Falsehoods might conceivably outnumber truths in a society. The principle is also not a point about innateness, though Reid's claim that a disposition to acceptance is innate seems to me correct. The principle is about entitlement, not psychological origin.

[9] Thomas Reid, *An Inquiry into the Human Mind* (Chicago: University of Chicago Press, 1970), chapter 6, section 24.

The epistemic default position articulated by the Acceptance Principle applies at an extremely high level of idealization in most actual communication, especially between sophisticated interlocutors. Social, political, or intellectual context often provides "stronger reasons" that counsel against immediately accepting what one is told. Given life's complexities, this default position is often left far behind in reasoning about whether to rely on a source. One might wonder, with some hyperbole, whether it can ever be the last word in the epistemology of acceptance for anyone over the age of eleven. The primary point—that it is a starting point for reason—would not be undermined if its purest applications were relatively rare. But I think that it has broader application than the hyperbolic conjecture suggests.

Acceptance underlies language acquisition. Lacking language, one could not engage in rational, deliberative activity, much less the primary forms of human social cooperation. (Indeed, this point suggests the line of justification for the principle that I shall begin to develop below.) But unquestioned reliance is also common in adult life. When we ask someone on the street the time, or the direction of some landmark, or when we ask someone to do a simple sum, we rely on the answer. We make use of a presumption of credibility when we read books, signs, or newspapers, or talk to strangers on unloaded topics. We need not engage in reasoning about the person's qualifications to be rational in accepting what he or she says, in the absence of grounds for doubt. Grounds for doubt are absent a lot of the time.

The primary default position, the Acceptance Principle, is not an empirical principle. The general form of justification associated with the principle is: *A person is apriori entitled to accept a proposition that is presented as true and that is intelligible to him, unless there are stronger reasons not to do so, because it is prima facie preserved (received) from a rational source, or resource for reason; reliance on rational sources—or resources for reason—is, other things equal, necessary to the function of reason.* The justificational force of the entitlement described by this justification is not constituted or enhanced by sense experiences or perceptual beliefs.[10] Before filling in this form of justification, I want to make some preliminary points.

I think that I need not show that other rational beings are necessary to the function of one's reason in order for one to have these entitlements. One has a general entitlement to rely on the rationality of rational beings. The Acceptance

[10] Principles narrower than the Acceptance Principle could with luck and context achieve the same utility: rely on the first person one comes across and no one afterward. Such principles are not rational starting points. We are entitled to something more general. In learning a language, one usually need not know the credentials of one's source—beyond the fact that the source is intelligible. Having an apriori entitlement based on the Acceptance Principle is compatible with also having empirical justifications of prima facie acceptance—or of narrower principles, such as "non-aggressive care-givers are more trustworthy than strangers who threaten one". I think that one does not have to have these empirical justifications to be entitled to accept what one is told in particular cases (even though people do have such empirical justifications).

Principle can be apriori instantiated where one has apriori, undefeated, prima facie entitlement to construe something prima facie intelligible as having a rational source. So I think that to maintain that one is apriori entitled to rely upon rational interlocutors, I need not show that a solitary reasoner is impossible.

Our account distinguishes rational sources and resources for reason. Resources for reason—memory and perception, for example—need not themselves be rational beings or capacities to reason. In these senses they need not themselves be rational. Yet they may provide material and services that a rational being is apriori entitled to rely upon. Rational sources are sources that themselves are a capacity to reason or are rational beings.

As with rational sources, I think that to show that we are apriori entitled to rely upon a given resource for reason, I need not show that such a resource is necessary to any possible reasoning. One is entitled to rely upon resources for reason in general—other things equal—even if some particular resource for reason is not indispensable to the function of reason. Such resources may enrich reason without being necessary to every rational activity. This view puts pressure on explicating the notion of a resource for reason. This matter can be postponed, for it is relevant to interlocution only in special cases.

There are deeper questions about rational entitlement that I cannot pursue in depth here. One can ask why one is entitled to rely on rational sources (or resources for reason), in view of the fact that they can be mistaken or misleading. This is tantamount to a traditional sceptical question about how putative rationality or justification is associated with truth. One can apparently imagine systematic misconnections between being justified (entitled), according to ordinary canons, and having true belief. Why then should one ever think that ordinary canons provide ground for belief? I will not take on scepticism here. I will assume that we are rationally entitled to rely on reason, memory, and perception. The Acceptance Principle is an extension of this assumption: we are rationally entitled to rely on interlocution because we may presume that it has a rational source.

Now I turn to filling in the justification for the Acceptance Principle. First, if something is a rational source, it is a prima facie source of truth. For a condition on reasons, rationality, and reason is that they be guides to truth. Explicating this idea is notoriously difficult; but I do not apologize for it. An epistemic reason for believing something would not count as such if it did not provide some reasonable support for accepting it as *true*. The same point applies to rational entitlements for belief. The entitlements that I am discussing are epistemic, not matters of politesse. If one has a reason or entitlement to accept something because it is, prima facie, rationally supported, one has a reason or entitlement to accept it as true. A source is a guide to truth *in* being rational. Rational mistakes are possible. But if there is no reason to think that they are occurring, it is rational to accept the affirmed deliverances of a rational source. For, other things equal, reason can be reasonably followed in seeking truth.

It is not just the rationality of a source that marks an apriori prima facie connection to truth. The very content of an intelligible message presented as true does so as well. For content is constitutively dependent, in the first instance, on patterned connections to a subject matter, connections that ensure in normal circumstances a baseline of true thought presentations. So presentations' having content must have an origin in getting things right. The prima facie rationality of the source intensifies a prima facie connection to truth already present in the prima facie existence of presented content.

The remaining main step in justifying the Acceptance Principle lies in the presumption that the source of a message is a rational source, or a resource for reason. I think that one is apriori prima facie entitled to presume that the interlocutor is a rational source or resource for reason—simply by virtue of the prima facie intelligibility of the message conveyed. That is enough to presume that the interlocutor is rational, or at least a source of information that is rationally underwritten.

The idea is not that we reason thus: "If it looks like a human and makes sounds like a language, it is rational; on inspection it looks human and sounds linguistic; so it is rational." Rather, in understanding language we are entitled to presume what we instinctively do presume about our source's being a source of rationality or reason. We are so entitled because intelligibility is an apriori prima facie sign of rationality.

If something is prima facie intelligible, one is prima facie entitled to rely on one's understanding of it as intelligible. One is entitled to begin with what putative understanding one has. But anything that can intelligibly present something as true can be presumed, prima facie, to be either rational or made according to a rational plan to mimic aspects of rationality. Presentation of propositional content presupposes at least a derivative connection to a system of perceptual, cognitive, and practical interactions with a world, involving beliefs and intentional activity.[11] Belief and intention in turn presuppose operation under norms of reason or rationality—norms governing information acquisition, inference, and practical activity. For propositional attitudes, especially those complex enough to yield articulated presentations of content, are necessarily associated with certain cognitive and practical practices. To be what they are, such practices must—with allowances for some failures—accord with norms of reason or rationality.

To summarize: We are apriori prima facie entitled to accept something that is prima facie intelligible and presented as true. For prima facie intelligible propositional contents prima facie presented as true bear an apriori prima facie conceptual relation to a rational source of true presentations-as-true: Intelligible propositional expressions presuppose rational abilities and entitlements; so intelligible presentations-as-true come prima facie backed by a rational source or

[11] The expression may be derivative in that a non-rational machine might express linguistic content. But such machines are ultimately made by beings who have propositional attitudes.

resource for reason; and both the content of intelligible propositional presentations-as-true and the prima facie rationality of their source indicate a prima facie source of truth.[12] Intelligible affirmation is the face of reason; reason is a guide to truth. We are apriori prima facie entitled to take intelligible affirmation at face value.

We could be apriori entitled to false beliefs. Sounds or shapes could have no source in rationality but seem intelligible. A quantum accidental sequence of sounds could correspond to those of Hamlet's most famous speech.[13] But the fact that we could be mistaken in thinking that something is a message, or in understanding a message conveyed, is compatible with our having an apriori prima facie rational right to rely on our construal of an event as having a certain meaning or intentional content. And where a message has meaning or intentional content, we are entitled to presume apriori that it has a rational source, or is a resource for reason.

Just as the Acceptance Principle does not assume that truth is in a statistical majority, the justification of the Principle does not assume that most people are rational. We could learn empirically that most people are crazy or that all people

[12] I think that the distinction between merely having attitudes with intentional content and being able to understand and present them is deeply significant, and marks a deeper level of rationality than that associated with merely having propositional attitudes and inferential abilities. But I need not explore this point here.

'I have not here argued in depth for the connections between content, propositional attitudes, and rationality because they are a widely accepted theme in much contemporary work. The idea that language is inseparable from propositional attitudes, which are inseparable from assumptions about rationality is present, for example, in the work of Paul Grice, *Studies in the Way of Words* (Cambridge, Mass.: Harvard University Press, 1989), and Donald Davidson, *Essays on Actions and Events* (Oxford: Oxford University Press, 1980) and *Inquiries into Truth and Interpretation* (Oxford: Oxford University Press, 1984). Elsewhere I have sought to show how having linguistic and propositional content is necessarily associated with individuals' having *de re* propositional attitudes to objects of reference and with their interacting practically and perceptually with such objects. See my 'Belief De Re', *Journal of Philosophy* 74 (1977), 338–363, and 'Other Bodies', in Andrew Woodfield (ed.), *Thought and Object* (Oxford: Oxford University Press, 1982), 97–120. The main novelty of the above argument lies in its first step—the claim that we are apriori entitled to rely on our understanding and acceptance of something that is prima facie intelligible—and in its drawing an epistemic consequence from the constitutive, conceptual relations between content and rationality that others have long explored and elaborated.

[13] In *Dialogues Concerning Natural Religion*, part 3, Hume imagines hearing an "articulate voice" from the clouds and asks whether one can avoid attributing to it some design or purpose. He never objects to this inference, though he objects to much of the theological purposes it was put to. He would, however, regard it as a non-apriori causal inference. One of the reasons that he would invoke for thinking that the presumption of a rational source could not be based apriori on prima facie intelligibility is that one could learn empirically that the "voice" was meaningless. This reason is powerless against my conception of the presumption, for I agree that the presumption is empirically defeasible. Apriority has to do with the source of epistemic right; defeasibility is a further matter. For recent criticisms of Hume's view, see C. A. J. Coady, 'Testimony and Observation', *American Philosophical Quarterly* 10 (1973), 149–155; Frederick F. Schmitt, 'Justification, Sociality, and Autonomy', *Synthese* 73 (1987), 43–85. I think that empiricism cannot possibly explain all our justified acceptance of what we read or hear. The idea that we should remain neutral or sceptical of information unless we have empirical grounds for thinking it trustworthy is, I think, a wild revisionary proposal. I also think that empiricism cannot account for norms for children's relying on others in the acquisition of language or knowledge.

have deeply irrational tendencies—not just in their performance but in their basic capacities. Human beings clearly do have some rational entitlements and competencies, even though we have found that they are surprisingly irrational in certain tasks. The justification presupposes that there is a conceptual relation between intelligibility and rational entitlement or justification, between having and articulating propositional attitudes and having rational competencies.

Rational backing is, other things equal, a ground for acceptance of something as true. But in dealing with others, one must often take account of their lies. Why is one *apriori* entitled, except when reasonable doubt arises, to abstract from the possibility that it may be in the interlocutor's rational interest to lie?

This issue is more complex than I can see through now. I will make some general observations, and then sketch one line of reply. (I think there are others.) The Acceptance Principle and its justification are formulated so as to be neutral on whether what is "presented as true" comes from another person. Its application does not depend on an assumption that the source is outside oneself (although further articulation will, I think, give this source a place in the account). Many of the differences between content passing between minds and content processed by a single mind derive from differences in modes of acquisition and in necessary background conditions, that do not enter into the justificational force underwriting an entitlement.

An account of an entitlement that includes, as a special case, relying on the word of others must, however, acknowledge the following issue: The straight-line route from the prima facie intelligibility of a presentation-as-true to prima facie rational characteristics of the source to prima facie acceptability (truth) of the presentation, is threatened by the fact that certain aspects of *rationality* (rational lying) may go *counter* to true presentations. So why should rationality, especially in another person, be a sign of truth? One can have empirical reasons to think someone is not lying. One could have non-rational tendencies to believe, which with luck might get one by. But can one have apriori prima facie rational entitlement to accept what one is told, without considering whether the interlocutor is lying—lacking special reasons to think he is?

Apart from special information about the context or one's interlocutor, neutrality (as well as doubt) is, I think, a rationally unnatural attitude toward an interlocutor's presentation of something as true. (Compare: lying for the fun of it is a form of craziness.) Explaining why, in depth, would involve wrestling with some of the most difficult issues about the relation between "practical" reason and reason. I will broach one line of explanation.

Reason necessarily has a teleological aspect, which can be understood through reflection on rational practice. Understanding the notion of reason in sufficient depth requires understanding its primary functions. One of reason's primary functions is that of presenting truth, independently of special personal interests. Lying is sometimes rational in the sense that it is in the liar's best interests. But lying occasions a disunity among functions of reason. It conflicts with one's reason's transpersonal function of presenting the truth, independently of special personal interests.[14]

The Humean reply that reason functions *only* to serve individual passions or interests is unconvincing. Reason has a function in providing guidance to truth, in presenting and promoting truth without regard to individual interest. This is why epistemic reasons are not relativized to a person or to a desire. It is why someone whose reasoning is distorted by self-deception is in a significant way irrational—even when the self-deception serves the individual's interests. It is why one is rationally entitled to rely on deductive reasoning or memory, in the absence of counter-reasons, even if it conflicts with one's interests. One can presume that a presentation of something as true by a rational being—whether in oneself or by another—has, prima facie, something rationally to be said for it. Unless there is reason to think that a rational source is rationally disunified—in the sense that individual interest is occasioning conflict with the transpersonal function of reason—one is rationally entitled to abstract from individual interest in receiving something presented as true by such a source.

Another consideration pointing in the same direction is this. A condition on an individual's having propositional attitudes is that the content of those attitudes be systematically associated with veridical perceptions and true beliefs:[15] true contents must be presented and accepted as true within some individual; indeed, the very practice of communication depends on preservation of truth. If a rational interlocutor presents intelligible contents as true, one can rationally presume that the contents are associated with a practice of successfully aiming at and presenting truth. Now an inertial principle appears applicable: since the intelligibility of a presentation-as-true indicates a source of both rational and true content presentations, one needs special reason to think there has been deviation from rationally based, true truth-presentation. Other things equal, one can rationally abstract from issues of sincerity or insincerity.

The apriori entitlement described by the Acceptance Principle is, of course, no guarantee of truth. It is often a much weaker sign of truth, from the point of view of certainty, than empirically justified beliefs about the interlocutor. The lines of reasoning I have proposed justify a prima facie rational presumption, a position of non-neutrality—not some source of certainty.

Even if the Acceptance Principle is not an empirical principle, it may seem that particular entitlements sanctioned by it, "applications", must inevitably be empirical. To know what one is being told, one must use perception. One must perceive words as expressing content presented as true. In interlocution, perception does inevitably figure in acquisition of understanding and belief. Perception

[14] Although I think that my claim about this constitutive function of reason is apriori, I do not maintain that it is self-evident. It can be and has been coherently questioned, as I will note. But the claim has substantial initial plausibility, and I believe that this plausibility is deepened through reflection, including reflection on challenges to it.

[15] These true beliefs could fail to be the individual's own, but they must occur somewhere in the development of the content—for example, in the evolution of the cognitive apparatus.

is necessary to minimal understanding; and minimal understanding is essential to belief and justification. But our question concerns perception's role in justification or entitlement. I will first consider its role in justification in our narrow sense, and then turn to its role in entitlement.

One might reason that since the Acceptance Principle counts it rational for a person to accept what is presented as true, and since one can know what is presented as true by another person only through perceiving an event in time, a person must rely for justificational force on perception of particular events to apply the principle.

This reasoning rests on a confusion about the status of the Acceptance Principle and its justification. The Acceptance Principle is not a premise in an argument applied by recipients of information. It is a description of a norm that indicates that recipients are sometimes entitled to accept information from others *immediately* without argument. The justification of the principle is not an argument that need be used by interlocutors, but an account of why the practice of acquiring information from others is rationally justified.[16] It is well known that we do not store the physical properties of sentences we hear or read.[17] The content of the linguistic forms is what is important. We seem normally to understand content in a way whose unconscious details (inferential or otherwise) are not accessible via ordinary reflection. To be entitled to believe what one is told, one need not understand or be able to justify any transition from perceptual beliefs about words to understanding of and belief in the words' content. One can, of course, come to understand certain inferences from words to contents. Such empirical meta-skills do enrich communication. But they are not indispensable to it. To be justified in understanding, we have to reason empirically about what we perceive only when communication runs into trouble, or when special, contextual, non-literal expressive devices are used (see note 21). Other things equal, we are entitled to presume that what seems intelligible is understood. Justification in the narrow sense is not basic to the epistemology of interlocution.

But the question of entitlement is more subtle. In ordinary perception of physical objects and properties we have sense experiences that are not ordinarily the objects of reference or the basis of a justifying inference to perceptual beliefs

[16] Here is a more sophisticated objection along the same line. Suppose that a belief acquired from others may count as knowledge, though one often lacks sufficient grounds, on one's own, to underwrite the belief as knowledge. Suppose that one knows one lacks autonomous grounds for such a belief. Then one's knowledge that the belief was acquired from others would have to be used to enable one's belief to count as knowledge, in view of the known fact that unless the belief had been acquired from others, one's lack of autonomous justification would be insufficient for knowledge. (It is assumed that knowledge that a belief was acquired from others must be empirical. Let us grant the assumption for now.)

This reasoning again rests on a level confusion. If one has acquired one's belief from others in a normal way, and if the others know the proposition, one acquires knowledge. No further reasoning about the practice is needed for the knowledge. No reasoning that does not show that the entitlement has lapsed can undermine the entitlement (though it might mistakenly undermine one's belief that one was entitled).

[17] Kenneth I. Forster, 'Lexical Processing', in D. Osherson and H. Lasnik (eds.), *Language: An Invitation to Cognitive Science*, vol. 1 (Cambridge, Mass.: MIT Press, 1990), 95–113.

to which we are entitled. Yet having such experiences, or having perceptual beliefs, contributes to the justificational force of our empirical beliefs:[18] A perceptual belief's being perceptual is an element in its justificational power. The belief's being causally or constitutively associated with sense perception is part of the force of our entitlement to the belief.

In interlocution, we are also causally dependent on perception. Our entitlements are thus dependent on perception. But in my view, perception contributes nothing to the epistemic force of the fundamental "default" entitlement.

Perceptions or perceptual beliefs about physical objects are constitutively dependent on bearing natural lawlike causal relations to objects of perception—to their subject matter, physical objects. The contents of the beliefs and perceptions are what they are partly because of these relations to specific physical objects or properties. Our entitlement to rely on perception and perceptual beliefs is partly grounded in this causally patterned, content-giving relation which is partly constitutive of perception.

When we receive communication, the situation is different. The objects of cognitive interest—the contents and their subject matters—are not the objects of perception. We do not perceive the contents of attitudes that are conveyed to us; we understand them. We perceive and have perceptual beliefs about word occurrences. We may perceive them as having a certain content and subject matter, but the content is understood, not perceived. The subject matter, word occurrences, of our perceptual experiences and beliefs bears a non-constitutive (quasi-conventional) relation to the content and subject matter of the beliefs to which we are entitled as a result of communication. So the accounts of our non-inferential entitlements to perception and to interlocution must be different.

One might note that the relation between perceived words and their contents or subject matters must involve some sort of explanatory relation. So one might be tempted to think that although one does not typically infer the content from the words explicitly and consciously, the entitlement must somehow be based on this explanatory relation. But it would be a mistake to embrace this temptation without reflecting carefully on the special character of the relation as it occurs in interlocution. The relation between words and their subject matter and content

[18] Davidson and Sellars deny that having sensations plays a role in justifying perceptual beliefs. I am not convinced by their reasons as applied to entitlements to perceptual belief. See Donald Davidson, 'A Coherence Theory of Truth and Knowledge', in Ernest LePore (ed.), *Truth and Interpretation: Perspectives on the Philosophy of Donald Davidson* (Oxford: Basil Blackwell, 1986), 311; and Wilfrid Sellars, 'Empiricism and the Philosophy of Mind', in *Science, Perception, and Reality* (London: Routledge & Kegan Paul, 1963), 164ff. For an alternative to their views, see Steven L. Reynolds, 'Knowing How to Believe with Justification', *Philosophical Studies* 64 (1991), 273–292. My view here does not, however, rest on giving sensations (particularly seen as non-intentional) a role in perceptual entitlement. One need not think of sensations as entities, though I do. It is enough that the perceptual character of perceptual belief contribute to the force of the entitlement. Moreover, I am not convinced that there is an epistemic *transition* from perceptual experience to perceptual belief in the ordinary case. One can, of course, learn to suspend such belief. But perceptual experience seems a constituent element in perceptual belief; and perceptual belief seems to be a default position.

is not an *ordinary*, natural, lawlike causal-explanatory relation. Crudely speaking, it involves a mind.

There are, of course, complex causal-explanatory relations that may be used to infer the content or subject matter of an interlocutor's speech from perceived word occurrences. One could give an account of entitlement centered on possible inferential interpretations, or on reason-giving explanatory connections between words and content. The interpretation might not be accessible to the recipient, but it could represent a reasonable route from the received message to a putative truth. Such an account—broadly familiar in current discussion—would make the entitlement empirical, because it would appeal in the account of justificational force to an inductive connection to perceived word occurrences.

I do *not* doubt that such accounts are true. I doubt that they are fundamental. I think that what is fundamental is not a meta-linguistic connection between word occurrences, taken as objects of perception, and their contents or subject matters. What is fundamental is an apriori prima facie entitlement to rely upon putative understanding, and an apriori prima facie connection between putatively understood contents and rational sources of truths. Understanding is epistemically basic. Traditionally, a justification or entitlement was apriori if it could be derived from conceptual understanding—however experientially dependent the understanding might be. The issue over apriority begins with conceptual understanding and asks whether perceptual experience is needed to supplement the understanding for one to be justified or entitled to one's belief.

The epistemic status of perception in normal communication is like the status it was traditionally thought to have when a diagram is presented that triggers realization of the meaning and truth of a claim of pure geometry or logic. Perception of physical properties triggers realization of something abstract, an intentional content, expressed by the sentence, and (often) already mastered by the recipient. Its role is to call up and facilitate mobilization of conceptual resources that are already in place. It is probably *necessary* that one perceive symbolic expressions to accept logical axioms—just as it is necessary to perceive words in interlocution. But perception of expressions is not part of the justificational force for accepting the contents. In both cases, no reference to a possible meta-inference from expressions to contents is needed in an account of justificational force. The primary entitlement in interlocution derives from prima facie understanding of the messages, and from a presumption about the rational nature of their source—not from the role of perception, however necessary, in the process.[19]

[19] The analogy goes with certain disanalogies. Understanding a simple logical truth yields a justification; understanding a communicated message yields an entitlement. This is because in the logic case justificational force derives from the content itself, whereas in interlocution justificational force derives from one's right to putative understanding and from the presumed status of the source of the message, not (typically) from the content itself. A corollary is that knowledge of a simple logical truth does not depend on anything further than understanding and believing it, whereas knowledge based on interlocution depends on there being knowledge in the chain of sources beyond the recipient. In neither case is correct perception of words or correct understanding of what they express necessary

In interlocution, perception of utterances makes possible the passage of propositional content from one mind to another rather as purely preservative memory makes possible the preservation of propositional content from one time to another. Memory and perception of utterances function similarly, in reasoning and communication respectively. Their correct functioning is necessary for the enterprises they serve. Their failure could undermine those enterprises. They preserve the content of events (past thoughts in proof, word utterances in interlocution)—events that *can* become objects known empirically. But the basic epistemic role of memory and perception in these enterprises is not to present objects of knowledge. They function to preserve and enable—not to justify.

In interlocution, the individual's basic default entitlement normally derives from the presumptive intelligibility of a message understood, not from anything specific in the words perceived. Unless reasonable doubt arises about the reliability or interpretation of the source, the specific perceptions of utterances need not be relied upon in contributing force to the receiver's entitlement to his understanding of or belief in what is communicated.

Perception might be thought part of the justificational force of our entitlement in another way. The justification of the Acceptance Principle says that one is entitled to accept intelligible contents "presented as true". We must perceive a speech act as involving a presentation-as-true to be justified under the principle. Why does it not follow that our entitlement to accept what we are told in particular cases relies for its force on perceptual beliefs?

The issues here are again very complex. But the short answer to the question is that one's intellectually grounded entitlement to one's understanding of content includes an entitlement to understand presentations-as-true. Understanding content presupposes and is interdependent with understanding the force of presentations of content. So entitlement to the former must presuppose entitlement to the latter. In many normal cases the epistemology of our entitlement to understanding assertive force has a default status that is parallel to that of our entitlement to understanding content. Perception is no more basic to understanding assertive force than it is to understanding conceptual content. The default position is that presumed understanding of both content and force is epistemically fundamental. Empirical justification for an interpretation of content or force is demanded only

to the justification (or entitlement). In neither case is correct perception of words necessary even for knowledge. But in the interlocution case (because knowledge depends on inheriting knowledge from a source), correct understanding of what the interlocutor conveys by the words is necessary for knowledge based on interlocution. (Correct understanding of words or interlocutor is not necessary for knowing whatever logical truth one happens to associate with them, if one understands the logical truth sufficiently.) The important analogy between the logic and interlocution cases is that perception of words makes understanding possible, but justificational force can be derived from the individual's understanding without supplementary appeal to perception. I am abstracting, in this discussion of applications, from cases where understanding a particular content itself involves perceiving—for example, perceiving the referents of demonstratives. Such understanding is not purely conceptual; and as a consequence, the relevant entitlement to the particular belief is partly perceptual.

when elements in the context demand reconsideration or supplementation of the default understanding. I find the parallel compelling. But I will sketch in two steps a picture of how default understanding of a presentation-as-true can sometimes be derived from no more than default understanding of propositional content. This picture is not needed, but it may enrich the account.

First, entitlement to one's understanding of a message's content carries with it, indeed rests on, an entitlement to understanding intentional *events* as having specific content. Understanding speech acts or thoughts as they occur is the root of understanding content types. The necessary role of perception in enabling one to follow another's speaking or thinking is not fundamentally different from its role in enabling one to grasp the abstract content of another's sentence. All that I have argued on the latter score applies to the former. Perception's basic role is to make understanding possible and to trigger it on particular occasions. But the justificational force of one's basic default entitlement to understand something as an event with a specific content is not perceptual. It is intellectual in that it resides in one's putative understanding of conceptual content in application or use, in one's ability to think-with.

Second, understanding conceptual content—both abstractly and in contentful events or uses—involves understanding the content's mood. But for contents in the indicative (declarative) mood—as distinguished from interrogative or imperative mood—presentation-as-true is the defeasible default use. The connection between declarative mood and presentations-as-true is conceptual. The justificational force of the entitlement to rely on the connection is correspondingly conceptual, not perceptual.[20]

In the absence of overriding reasons, the default presumption stands. Non-assertive uses (jokes, irony, fiction) that drain declaratives of assertive implications must employ context to make themselves understood. The recipient must infer that the sentence is used non-assertively from empirical information about the context. Although affirmative use of declarative contents must, on occasion, also be inferred from special contextual information, taking a declarative sentence utterance as a presentation-as-true normally requires no such reasoning or empirical interpretation.[21]

[20] Donald Davidson has argued that there is no conventional connection between indicative sentences and assertive use. See 'Moods and Performances' and 'Communication and Convention' in *Inquiries into Truth and Interpretation.* His reason is that one can always use indicative utterances for non-assertive purposes. I find the argument unconvincing. A conventional connection between indicative mood and assertive use could be flouted. I believe that the connection between assertive use and indicative (declarative) mood is deeper and firmer than merely conventional. But it is a contextually defeasible connection.

I use the term 'presentation as true' to cover more than assertions and judgments. Obvious presuppositions, or conventional implicatures, are examples. When someone says to kill the shortest spy, he or she presents it as true that there is a shortest spy. In such cases, as well as the indicative cases, the entitlement to accept what is presented as true can be independent for its justificational force of perceptual connection to context (see note 21).

[21] This point allies with Grice's distinction between conventional and conversational implicature. See Paul Grice, *Studies in the Way of Words,* 28–31. Grice requires that to be "conversational", an

Thus in many instances, one's entitlement to take something as a presentation-as-true in interlocution derives from understanding an event's content, and need not rely for its justificational force on perception of word occurrences. What one is entitled to on intellectual grounds is merely, prima facie, that a given content is presented as true. One gets nothing about the time, form, or circumstances of the assertion. All such information is epistemically grounded in perception of aspects of the context. But the fundamental entitlement to accept something as a presentation-as-true derives from understanding. It can even be derived sometimes from understanding of content (its tokening and the relation of its mood to presentations-as-true). The justificational force of the derivation does not depend on any supplementation from perception. Perception plays its role in making understanding possible and in justifying supplemental information about the form, existence, and context of the assertion.

In appreciating these points, one must distinguish between knowing about the assertion as part of a pattern for explaining the psychology and behavior of the asserter, and using the interlocutor as a source of information. In the former enterprise, perception of an assertion as an action by a particular individual is commonly taken as an element in the justification of an explanation, or an object of interpretation. But in interlocution, perception need not play this role unless some reasonable doubt arises about the informant's message or the recipient's understanding (see note 25).

One can know through memory the events that help recall the previous step in a proof, thereby making those events objects of knowledge. One can know on the basis of perception that a particular person made an assertion at a given time. One can surely construct an empirical meta-justification (or entitlement) for one's belief based on interlocution: "She asserted that p (known empirically); it is prima facie reasonable to rely on others' assertions; so I should rely on her assertion." Such meta-justifications supplement one's epistemic position in interlocution. But they are not, I think, fundamental. Just as remembering events does not enhance the primary object-level justification in deductive argument, so relying on perception does not contribute to the justificational force of one's fundamental entitlement to one's understanding of content, or to one's acceptance of what is presented as true.

Let us return from our entitlement to understanding to our entitlement to believe what we hear, given that we understand it. When we receive a message, we often know a lot about the context of the reception, the biography of the

implicature must be capable of being "worked out" from considerations of the conversational context. Conventional implicatures may be inferred "intuitively" from the meaning of the words. I think that understanding based on conversational implicatures *must* be justified, usually empirically, whereas understanding based on conventional implicatures can rest on apriori entitlement. Analogously, I think that a construal of a sentence or content as ironic *must* be justified, usually empirically, whereas a construal of a sentence as asserted can rest on an apriori entitlement. A parallel story needs to be told about ambiguities. Our ability to understand many ambiguous sentences as they are meant, even apart from context, indicates that certain readings are default readings.

source, the antecedent empirical plausibility of the information. This knowledge is inevitably perceptually grounded. Does this fact make our entitlement to believe what we receive from others inevitably perceptual? I do not think so. Our initial entitlement does not depend on this knowledge for its justificational force.

In areas like politics, where cooperation is not the rule and truth is of little consequence, or philosophy, where questioning is as much at issue as belief, we engage in complex reasoning about whether to accept what we hear or read. Reasonable doubt becomes a norm. But these situations are not paradigmatic. They are parasitic on more ordinary situations where acceptance is a norm.

The default position is justified acceptance. Often we need empirical reasons to defeat reasonable doubts that threaten our right to acceptance. But sometimes empirical reasons simply reinforce and overdetermine the default entitlement. Our being justified does not then rest indispensably on empirical background information.[22]

I turn now from our entitlement to applications of the Acceptance Principle to the role of interlocution in the acquisition of knowledge. In the absence of countervailing considerations, application of the Acceptance Principle often seems to provide sufficient entitlement for knowledge. Most of our knowledge relies essentially on acceptance of beliefs from others—either through talk or through reading. Not only most of our scientific beliefs, but most of our beliefs about history, ourselves, and much of the macro-world, would have insufficient justification to count as knowledge if we were somehow to abstract from all elements of their justification, or entitlement, that depended on communication.

Our entitlement to ordinary perceptual belief is usually sufficient for perceptual knowledge. It is usually sufficient even though we may be unable specifically to rule out various possible defeating conditions. If there is no reason to think that the defeating conditions threaten, one has knowledge despite ignoring them. Something similar holds for acquisition of belief from others. Other things equal, ordinary interlocution suffices for knowledge.[23]

In knowing something through interlocution, the recipient has his own entitlement to accept the word of the interlocutor, together with any supplementary justification the recipient might have that bears on the plausibility of the information. Let this include all the reasons available to the recipient, together with all

[22] The scope for intellection-based justification in interlocution is wider than these remarks may suggest. I think that in certain cases special confidence in an interlocutor can be justified on grounds that are inductive but, with subtle qualifications, intellectual. I discuss these matters further in 'Computer Proof, Apriori Knowledge, and Other Minds'.

[23] The fact that most of our knowledge is dependent on others and has distinctive epistemic status is increasingly widely recognized. See Coady, 'Testimony and Observation'; John Hardwig, 'Epistemic Dependence', *The Journal of Philosophy* 82 (1985), 335–349; Michael Welbourne, *The Community of Knowledge* (Aberdeen: Aberdeen University Press, 1986). For a wildly implausible, individualistic view of the epistemic status of testimony, see John Locke, *An Essay Concerning Human Understanding* 1:3: 24.

the entitlements deriving from his own cognitive resources. Call this body (i) the recipient's *own proprietary justification.*

If the recipient depends on interlocution for knowledge, the recipient's knowledge depends on the source's having knowledge as well. For if the source does not believe the proposition, or if the proposition is not true, or if the source is not justified, the recipient cannot know the proposition. The recipient's own proprietary entitlement to rely on interlocution is insufficient by itself to underwrite the knowledge.[24] In particular, the recipient depends on sources' proprietary justifications and entitlements (through a possible chain of sources). The recipient depends on at least some part of this body of justification and entitlement in the sense that without it, his belief would not be knowledge. The recipient's own justification is incomplete and implicitly refers back, anaphorically, to fuller justification or entitlement. Call the combination of the recipient's own proprietary justification with the proprietary justifications (including entitlements) in his sources on which the recipient's knowledge depends (ii) *the extended body of justification* that underwrites the recipient's knowledge.

At the outset, I explained apriori knowledge in terms of apriori justification or entitlement. The question arises whether apriori knowledge based on interlocution is underwritten by the individual's proprietary justification or by a justification that must include some non-proprietary part of the extended body of justification.

The extended body of justification—the one that reaches beyond the individual—is the relevant one. If I am apriori entitled to accept an interlocutor's word, but the interlocutor provides me with empirically justified information, it would be wrong to characterize my knowledge of the information as apriori. Similarly, if my source knows a proposition apriori, but I must rely on empirical knowledge to justify my acceptance of the source's word, it would be wrong to say that *I* know the proposition apriori—even though I have knowledge that is apriori known by someone. It seems most natural to think that a strand of justification that runs through the extended body into the individual's proprietary body of justification must be apriori for the recipient's knowledge to be apriori. People who depend on interlocution for knowledge of mathematical theorems but do not know the proofs can have apriori knowledge in this sense. The source mathematician knows the theorem apriori and the recipient is entitled apriori to accept the word of the source, in the absence of reasons to doubt. Most of us knew the Pythagorean theorem at some stage in this manner. When apriori knowledge

[24] Because the interlocutor must have knowledge and because of Gettier cases, the interlocutor must have more than true, justified belief if the recipient is to have knowledge. The recipient's dependence for having knowledge on the interlocutor's having knowledge is itself an instance of the Gettier point. The recipient could have true justified belief, but lack knowledge because the interlocutor lacked knowledge.

In requiring that the source have knowledge if the recipient is to have knowledge based on interlocution, I oversimplify. Some chains with more than two links seem to violate this condition. But there must be knowledge in the chain if the recipient is to have knowledge based on interlocution.

is preserved through reports which the recipient is apriori justified in accepting, the receiver's knowledge is apriori.

The Acceptance Principle is clearly similar to what is widely called a "Principle of Charity" for translating or interpreting others. The most obvious difference is that the former applies to situations in which one is not taking another as an object of interpretation, but rather as a source of information presumed to be understood without interpretation. This situation is basic for communication.[25] Radical interpretation is not, I think, the paradigmatic situation for theorizing about linguistic interchange.

We rely on being so formed that we take in information from others without interpretation. Unlike the Principle of Charity, the Acceptance Principle presumes not only that we are like others in being rational. It presumes that we preserve content, other things equal. This presumption works because we share with others around us our cognitive tendencies and means of expressing them, and a common environment. But we do not have to *justify* a claim that these conditions for success are in place to be entitled to rely upon our understanding. (Analogously, we do not have to justify a claim that the environment is normal and we are adapted to it in order to be entitled to rely on perception.) It is enough if we learn how to understand. Once we are in a position to understand, we are entitled to the following presumption apriori, other things equal: We understand what we seem to understand. Or rather, other things equal, we need not use a distinction between understanding and seeming to understand. We need not take what we hear as an *object* of interpretation, unless grounds for doubt arise. Only then do we shift from content preservation to interpretation.

The Acceptance Principle entails a presumption that others' beliefs are justified, that others are sources of rationality or reason. The view that others' beliefs can be presumed to be true is familiar from the Principle of Charity. The presumption that others are reliable indices of truth rests on a presumption that they are rational sources. Their reliability is not some brute correlation between belief and world. We are entitled to treat others as reliable partly *because* we are entitled to presume that they are rationally justified or rationally entitled to their beliefs. We are entitled, most fundamentally, to think of others as sources of rationality or reason not because we take them as objects of interpretation and

[25] The Principle of Charity is illuminatingly used by W. V. Quine, in *Word and Object* (Cambridge, Mass.: MIT Press, 1960), chapter. 2; and Donald Davidson, in 'Radical Interpretation' (1973), in *Inquiries into Truth and Interpretation*. In holding that interpretation is the basic situation for understanding linguistic interchange, Davidson writes, 'The problem of interpretation is domestic as well as foreign: it surfaces for speakers of the same language in the form of the question, how can it be determined that the language is the same?' (Similar passages can be found in Quine.) Davidson presupposes that determining whether we are communicating successfully when we appear to be is a question in place from the beginning. This seems to me mistaken. Such a question arises only when there is some reason to doubt that we are sharing information and preserving content. The default position is that understanding can be presumed until something goes wrong. Incidentally, I do not assume that anything as global as a communal language need be thought of as fundamental. That is a further issue.

explanation, but because prima facie intelligibility is an apriori prima facie sign of rationality.

This focus on others is articulated from a first-person point of view. Each of *us* is justified in presuming that others are justified. But we are possible interlocutors too. The idea that others are prima facie justified in their beliefs makes general sense only if we presume generally: people, including each of us, are reliable rational sources of true justified beliefs. Obviously the conclusion requires qualification and elaboration. But the route to it is, I think, of interest. I arrived at it by arguing that we have intellection-grounded prima facie entitlements to applications of the Acceptance Principle, though they are empirically defeasible. I think that this approach to epistemology may help with some of the traditional problems of philosophy.

11 *Postscript: 'Content Preservation'*

In this postscript I discuss three topics that come up in all my articles on interlocution. I write the postscript to 'Content Preservation', because the topics were broached there. The first topic is the role of knowledge in an antecedent chain of communication in making knowledge by a recipient possible. The second is the Acceptance Principle and its role in articulating a default prima facie entitlement to believe an interlocutor. The third is the epistemic status of the initial comprehension of another's utterance. On the first and third topics, I criticize some of my earlier views. On the second, I elaborate and defend my views.

I

When a person receives information in interlocution, the person frequently acquires knowledge of the subject matter of the report. If a child is told that Magellan was Portuguese but sailed for Spain, the child commonly comes to know that Magellan was Portuguese but sailed for Spain. Often, the recipient lacks any further warrant to believe the report. The recipient commonly gains new knowledge from the report without having warrant that is both specific to what is reported and independent of the report. Students entering a new subject and young children gain knowledge in this way. When we travel, we come to know things about foreign lands simply by asking questions of foreigners. The recipient's background information—about either the topic or the interlocutor—does not suffice to provide any significant independent epistemic support for what the recipient learns. Often the recipient has, at the time of the report, no independent check on the veracity or competence of the reporter. Yet the recipient gains new knowledge through the report itself.

My first topic concerns the epistemic resources in the chain of communication that leads to a reception. The child would not have come to know that Magellan was Portuguese but sailed for Spain if there were no appropriate knowledge in the antecedent chain. What is the role of such antecedent knowledge in making it possible for a recipient to gain knowledge through being told something?

Let us focus on a case in which a recipient's background information makes no significant independent contribution to supporting what the recipient learns.

That is, the recipient does not have, independently of relying on the report, even a nearly sufficient warrant to believe the particular thing that is reported. So, for example, the recipient is not in a position to use whatever epistemic force there is in the reception of the report to tip what was antecedently *close* to being knowledge over the bar. In 'Content Preservation' I assumed that, in such a case, a necessary condition on the recipient's gaining knowledge from a report is that

> (A) knowledge must reside in the chain of communicators that leads up to a report, if the recipient is to obtain knowledge from the report.[1]

This formulation allows that a recipient can gain knowledge even if the recipient's immediate interlocutor lacks knowledge. For example, if the immediate interlocutor tells the recipient something, but just repeats what he or she had heard from a previous interlocutor—without understanding it well enough to know it or without forming a belief about it—then the recipient could gain knowledge—as long as some antecedent interlocutor of the immediate interlocutor had knowledge. In such cases, I call an interlocutor who passes knowledge without possessing it a '*non-knowledgeable conduit*' of knowledge.

(A) indicates that the recipient's knowledge depends on knowledge in the antecedent chain. But (A) does not specify the exact relation that is required. It does not even specify that the recipient's new knowledge be had by some individual in the antecedent chain. As I will explain shortly, this lack of specificity has the advantage of allowing (A) to be flexible about how non-knowledgeable conduits convey knowledge to recipients.

I think that (A) is true, on a natural but liberal understanding of it. However, I also made a more committal assumption in 'Content Preservation'. I wrote,

> (B) If the recipient depends on interlocution for knowledge, the recipient's knowledge depends on the source's having knowledge as well. For if the source does not believe *the proposition*, or if the proposition is not true, or if the source is not justified [warranted], the recipient cannot know *the proposition*.[2]

Unlike (A), (B) requires that knowledge that a recipient gains purely from interlocution be the very knowledge had by someone in the antecedent chain of communication. This assumption is prima facie plausible. But I now think it probably untrue.

[1] 'Content Preservation', *The Philosophical Review* 102 (1993), 457–488, see note 480 n 19; 246 n 19, in this volume. A commitment similar to that in (A) occurs in 'Computer Proof, Apriori Knowledge, and Other Minds', *Philosophical Perspectives* 12 (1998), 1–37; see page 5; 311, 311 n 9, in this volume.

[2] 'Content Preservation', 486; 251, in this volume. Italics added. Earlier in the article, I had indicated that sometimes I would use 'justification' to cover all forms of warrant, as I do here. Thus the brackets.

The assumption, which I shared with several other authors at the time I wrote 'Content Preservation', was challenged by Jennifer Lackey.[3] Lackey's proposed counterexamples do not work against (B). They fail because they do not take account of the provision that the source of knowledge in the antecedent chain need not be the recipient's immediate interlocutor. That provision was quite explicit in my explication of (B).[4] Lackey's mistake seems to have stemmed from insufficiently close reading. But a variant on Lackey's counterexamples has been fashioned by Peter Graham.[5]

Graham describes a teacher of evolutionary biology who does not believe in evolution, but who faithfully and reliably teaches the science anyway. The teacher's students have no other source on evolutionary biology than the teacher. They do not read science books or hear about the science from anyone else. Graham supposes, plausibly, that the students can come to know from the teacher that, for example, humans evolved from non-human primates So far, the case is compatible with (B). The teacher does not know the propositions that the students learn from relying on his or her teaching, because the teacher does not believe those propositions. But the propositions are known by the teacher's own sources. The students gain knowledge from those sources through the teacher, as non-knowledgeable conduit.

Graham adds a twist, however. The teacher discovers a fossil on a solo field trip and infers from the discovery a new proposition that is in fact true and that is strongly supported by evolutionary theory. The teacher does not know the new proposition, because he or she does not believe the science that formed the main basis for the inference.[6] No one else knows the new proposition before the teacher makes a report, because no one else observed the fossil or made the inference. The teacher then tells the students the new proposition, again as conduit of evolutionary biology and its consequences. Graham supposes, plausibly, that the students can come to know the new proposition from the teacher.

Here we have an apparent counterexample to (B). The teacher does not know the proposition, since he or she does not believe it. No one else knew the proposition before the teacher taught it. But the students can come to know it from the teacher's teaching.

[3] Jennifer Lackey, 'Testimonial Knowledge and Transmission', *The Philosophical Quarterly* 49 (1999), 471–490.

[4] In the note attached to (B), I explicitly allow that the immediate interlocutor need not be the *source* of the transmission of knowledge. The source could be farther back in the chain of interlocution. In other words, I allowed for non-knowledgeable conduits. See 'Content Preservation', the note immediately attached to (B), 486 n 24. See also 'Computer Proof, Apriori Knowledge, and Other Minds', 17. Lackey criticizes a range of views, and it may be that her counterexample is successful against some of them.

[5] Peter Graham, 'Can Testimony Generate Knowledge?' *Philosophica* 78 (2006), 105–127.

[6] The teacher knows the science in the sense that he or she is very competent at explaining it. The teacher knows what the science says and implies. But the teacher does not know the subject matter of the science through the science. Evolutionary biology is not part of what the teacher knows about the world, because the teacher does not believe the science. Knowledge in this sense requires belief.

It does not seem to me *obvious* that the students can come to know the proposition under the stipulated circumstances. In believing that they do know, one is easily influenced by the natural assumption (contrary to what is stipulated in the example) that they have other sources. Normally, they would be able to use the teacher's knowledgeable report about the fossil to infer the relevant proposition on their own. Then the students would gain the knowledge needed to carry out the inference from knowledgeable sources in the chain (with the teacher functioning as a non-knowledgeable conduit to those sources). The final piece of knowledge that they gain would not be purely from the report. The propositional bases for the inference would come from knowledegeable sources, and the students would carry out the inference on their own. Such an example would not defeat (B).

One must avoid such natural assumptions in forming intuitions about Graham's example. The teacher must be taken to report the conclusion of the inference without ever having explained its bases to the students. They are not in a position to make the inference themselves. How good are our intuitions at staying clear of such natural assumptions about variant examples? Although I do not find the answer obvious, I believe that Graham is probably right. I take the case to be a counterexample to (B).

The counterexample to (B) is not a counterexample to a liberal, but natural construal of (A). The non-knowledgeable conduit (the teacher) takes an inferential step that combines his or her own knowledge (based on observation of the fossil) with knowledge of the conduit's sources (the knowledge involved in understanding the inferential relation between the belief about the fossil and the conclusion of the inference). True, no one in the antecedent chain knows, or believes, the inferred conclusion that the teacher conveys to the students. But the knowledge that the students acquire is the product of the knowledge of the teacher's sources, the teacher's own observational knowledge, and a good inferential step taken by the teacher. In an abstract sense, the knowledge that the students gain resides, collectively, in the antecedent chain, including the teacher (with his/her knowledge of the fossil). We do talk this way about complex, collective scientific or mathematical work. Knowledge of a theoretical explanation or a proof resides in a group, even though, because of the complexity of the content of the knowledge, no individual has full control of the explanation or the proof. In the students' case, the relevant knowledge—a well-supported, thought-through proposition that is fully available as knowledge in the antecedent chain and that would be known by the teacher apart from the teacher's idiosyncratic irrationality—is conveyed to the students even though no individual knows it before the students do. The knowledge can be regarded as possessed collectively by the individuals in the antecedent chain, even though no individual in the antecedent chain has the knowledge.

The example suggests another issue that interests me more than my (probable) error in (B) and the need to take (A) in a liberal, abstract way. How are we to characterize the students' entitlement that underwrites their knowledge? What exactly is the route to truth and knowledge that they rely upon?

A recipient of interlocution can have different lines of warrant for believing the same proposition at the same time. The students do have and rely on a default warrant that presumes upon the rationality and sincerity of their interlocutor, the teacher.[7] The students do not need to think about rationality or sincerity to have the warrant. Their entitlement is correctly explained by reference to the reliability of interlocutors' being broadly rational and sincere, other things equal. That is the fundamental idea behind the Acceptance Principle. I think that that default warrant is undefeated in the students' psychologies. So they retain it. But *that* warrant cannot underwrite their knowledge. For the presumptions that are associated with the warrant are incorrect. The teacher is neither rational nor sincere. So if the students nevertheless have knowledge, it is because they have a further warrant, beyond the default warrant. I find the example interesting because it forces one to think about different lines of warrant that the students have.

One sceptical way to respond to the example is this. As a conduit for knowledge, the teacher is corrupted by the fact that he or she is insincere. One might support this point by claiming that the teacher is unreliable because he or she would teach whatever was required to have a job. Robert Audi made this point against Lackey's original counterexample.[8] Graham responds by building up the case so that the teacher is reliable in the particular circumstances. Although the teacher is insincere, the reliability of the students' entitlement derives from the school that they attend, not from the teacher's rationality or sincerity. The teacher teaches in a school that has good judgment about its curriculum and requires that its teachers teach its curriculum, which in fact constitutes knowledge.

Although Graham does not develop his point in this way, I take Graham's point to indicate that in addition to the students' default entitlement to rely on the teacher's rationality and sincerity—an entitlement that is undefeated, but that cannot support the students' knowledge—the students have a further entitlement that depends on the teacher's social role. This further entitlement does support the students' knowledge. Graham's idea is that the teacher is reliable because the school's judgment about the curriculum is reliable, and the teacher is as reliable as other teachers in the system. The students' knowledge depends on the school's reliability, although the students need not have any *views* about reliability.[9]

[7] In 'Content Preservation', I argue that a default entitlement to rely on an interlocutor's sincerity is derivative from a default entitlement to rely on an interlocutor's rationality. But I keep track of both entitlements for the sake of expositional clarity. See note 15 below.

[8] Robert Audi, 'Testimony, Credulity, and Veracity', in J. Lackey and E. Sosa (eds.), *The Epistemology of Testimony* (Oxford: Oxford University Press, 2006), 25–49.

[9] One could also stipulate that the teacher's relation to the science is not malleable in the way that Audi suggests. Suppose that the teacher teaches the science faithfully because of its social prestige. I think that neither this response nor the one discussed in the text is the right way to defend the example.

I find this dialect unsatisfying. The school's good judgment and reliability do not seem to be essential to the students' having a warrant that supports knowledge. Suppose that some years before the students arrived, the curriculum had been staffed by an abnormally high percentage of unreliable and worthless teachers teaching worthless subjects—yielding no knowledge. But suppose that the earlier teachers died of a plague and a new set of teachers, more normal in teaching better subjects, arrived to staff the school's curriculum—through no particular good judgment by the school's officials. Then the students would retain a warrant to rely on their insincere teacher that would, I think, still underwrite the students' gaining knowledge of biology from their teacher.[10]

Consider the facts (a) that teachers (or teachers of science) are mostly reliable, at least on large matters about their subjects (the main points in the science); (b) that the particular teacher faithfully teaches the subject (or faithfully teaches science); and (c) what the teacher teaches is a body of knowledge about a subject matter and hence is itself a good route to truth. I think that these are the relevant facts for determining whether the teacher can serve as a knowledge channel for the students. (a) and (b) are social facts that the students' knowledge depends on. (c) is a fact about the science. The school's being run by people with good judgment is not essential. It is the prevalence of knowledge-conveying teachers (or science teachers), and the fact that the students are not in a context in which this prevalence is abnormally and indiscernibly (to the students) overmatched by teacher frauds, that are essential to the students' gaining knowledge.

There remains a question about the nature of the facts in the students' psychologies that ground their entitlement to rely on the teacher in such a way as to yield knowledge for the students. How do they hook into the reliable social resource that connects them to the knowledge in the antecedent chain? In the case of perceptual belief, we can specify the perceptual system and belief-forming system that underlies an individual's entitlements to rely on perceptual beliefs. These are natural psychological kinds that constitute the individual's route to the belief. When they are reliable, they form a natural psychological kind of belief formation that grounds the individual's entitlements to the perceptual beliefs. What are the psychological competencies that ground the students' entitlements to rely on the teacher's reports? What route or routes of belief formation underwrite the students' knowledge?

There are different types of entitlement to rely on the teacher. Relative to one type that they have, the students cannot obtain knowledge, though they have warranted true belief. If they were to rely essentially on the teacher's rationality

[10] If the biology teacher were the only one at the school who taught a respectable subject, capable of conveying knowledge, and all others taught virulently and pervasively post-modernist versions of relativism, no matter what the subject matter, and if the students had no instinct to rely on the biology teacher any more than on the others, then plausibly the students would fail to gain knowledge from the biology teacher. Then the case might be regarded as similar to a case in which an individual is forming perceptual beliefs about a barn in a landscape in which the thing the individual is perceiving is the only real barn among a host of nearby barn-facades.

(and sincerity), they could not obtain knowledge. Relative to another type, or other types, that they also have, the students do obtain knowledge.

Let me back up a bit. I believe that whether an individual is epistemically entitled to a belief depends on the reliability of belief-forming competencies that, in the absence of special information available to the individual, go by way of natural psychological or sociological kinds. For example, an individual is entitled to a perceptual belief if the individual's perceptual system and system of perceptual belief formation (for a particular kind of perception and belief) are well functioning, and are reliable in normal circumstances.[11] The relevant route is specified in terms of natural psychological kinds that the science of perceptual psychology and perceptual belief formation specify.

Let us assume that in the normal environment—the environment in which these competencies were determined to be the kinds of competencies that they are—the perceptual system and perceptual belief-forming system are reliable. Whether an individual happens to be in a psychological experiment with distorting mirrors does not affect the individual's entitlement to rely on perceptual beliefs, assuming that the individual has no information about being in such a situation. Such a situation is not the natural environment in the sense just specified. Being in such a situation may, however, affect whether the individual has knowledge, even if the individual's belief happens to be true. Such situations are the stuff of Gettier cases—true warranted belief that is not knowledge.

Similarly, I think that the sociological kinds that figure in the warrant-giving route that connects an individual's belief in a report to the report's subject matter are general, unless they are made more specific by special information that the individual has. The reliable social channel that grounds a default entitlement for the students is not determined by the particulars of the school that the students attends, unless that school is abnormally infested with unreliable teachers or unreliable subjects. Entitlement hinges, other things equal, on whether the social institutions that occur in the students' society are reliable routes to truth.

I conjecture that the teacher's being a teacher of a body of knowledge is the key fact. This fact is comprised by facts (a) and (c) above. The institution of teaching the subject matter must be reliable in order for the students' relying on it to be entitled to their beliefs. For the students to have knowledge, the *particular* instantiator of that social instituation—here the oddball teacher—must be a reliable route to truth. Fact (b) must be in place for knowledge, though not for entitlement. Conditions (a)–(b)–(c), all external to the students, are necessary conditions on the students' having a route to knowledge. They are satisfied.

There can be hard questions about the social conditions, of course. At what juncture in history did teachers become reliable routes to truth in their science teaching? How should routes be specified synchronically? Perhaps late nineteenth-century teachers in Timbuktu are too unreliable, whereas those in England

[11] See my 'Perceptual Entitlement', *Philosophy and Phenomenological Research* 67 (2003), 503–548.

are reliable. How should the type of teacher be specified? What are the relevant synchronic sociological kinds? For the case of routes for forming individual perceptual belief, we have enough psychological science to answer analogous questions.[12] For the case of relying on others for information, sociology is not far enough along to provide general, well-supported answers. Philosophy must engage with sociology to find the relevant kinds.

Still, the outlines of answers can be reasonably conjectured. It seems to me that, even though reliability varies from teacher to teacher, default entitlement—in the absence of the students' having special information about their situation—cannot rest on local, contextual variations. The reliability relevant to entitlement must center on more general kinds that yield general sociological patterns on which beliefs rely. Such patterns may eventually be specified systematically, as our knowledge of social psychology and sociology grows.

Where it concerns empirical matters, the epistemology of default entitlement must ally itself with reliability of general patterns—the nearest thing to laws that a subject matter will allow. We do not know what the basic kinds are in the sociological world. But it seems clear to me that what particular school a student attends, or what particular teacher they have (assuming that the teacher is a reliable conduit), is at too fine a level of grain to ground scientifically valuable, sociological specifications of relevant *default* entitlement routes to truth—hence relevant types of reliability—that underwrite knowledge. Whether the students happen to attend a school that guides its curriculum in a good way does not affect the default entitlement for their beliefs that hinge on interlocution with their teachers identified as teachers (or teachers of science).

Let us suppose that the social aspects of the students' reliable route to the subject matter concern the institution of teaching science. We have not yet specified the psychological competencies that, along with the social facts, ground the students' entitlements that underwrite knowledge. I think it plausible that the students connect to the social channel by recognizing the teacher as a teacher, or perhaps as a teacher of science. They can recognize the basic social channel as the kind of thing that it is. They need not realize that it is a *basic* social channel. They need not *believe* that it is reliable. They need not even *consider* the teacher's being a teacher as a *reason* to rely on the teacher's reports. It is enough that they

[12] These remarks target the so-called *generality problem*. See Earl Conee and Richard Feldman, 'The Generality Problem for Reliabilism', *Philosophical Studies* 89 (1998), 1–29. An account of the reliable pattern that underlies entitlement must specify relevant kinds that fit into such a pattern, and must specify how the knower's psychology relies on these patterns. The generality problem is taken by some to be a serious difficulty because it is supposed that the kinds relevant to determining the reliability of belief-forming processes are open to endless armchair speculation. Such a view derives, I think, from trying to do epistemology independently of what is known about cognitive processes. In the case of the epistemology of *perceptual belief*, the view derives from failing to think seriously about perceptual psychology. The kinds that are relevant to specifying the relevant reliable process in perceptual belief formation are the most specific, or fundamental, natural psychological kinds of perceptual state and perceptual belief. The kinds are constrained by empirical knowledge of the natures of those states. We know a lot about such natures. Again, see 'Perceptual Entitlement'.

have the social know-how to discriminate the setting and to give some credence to what is taught in that setting. It is enough that they recognize the teacher as a teacher of science, and to have some inclination to allow that fact to be a factor in a tendency to rely upon the teaching. It is enough that they have the know-how to use productions from that social institution as a source of belief. Inasmuch as the institution of teaching science is generally reliable, the students are entitled to rely on it ((a) and (c)). Inasmuch as the particular teacher is a reliable instantiator ((b)) of the reliable institutional route, the students can gain knowledge, not just entitlement in the particular case.[13]

The foregoing account of the students' entitlement is richer than the account that I give in 'Content Preservation' of the most basic default entitlement that individuals have to rely on what they are told. The basic entitlement depends only on the intelligibility of the report, and on there being no sufficient counter-considerations. It does not depend on particular social institutions or on any psychological capacity to recognize them. I believe that the students have and rely on that basic entitlement, as well as the entitlement to rely on the teacher as a science teacher. The basic entitlement depends on the apriori connection between rationality and being a good, reliable route to truth.[14] But they have a supplementary entitlement as well.

[13] I think that if the institution is reliable—if most teachers, or science teachers, are reliable—the students are entitled to their beliefs insofar as they rely on this institutional channel. Relying on the channel involves exercising a psychological capacity to identify the channel appropriately. If a particular teacher were incompetent and unreliable, and students had no ground to realize this fact, the students could be entitled to rely on the teacher. But they could not gain knowledge from the teaching—even if, in a given case, what the teacher taught happened to be true. Here we would have a Gettier-type case.

[14] If a rational being were situated in circumstances in which reasons cannot correct for common, systematically misleading information, the rational being would not be a reliable channel to a subject matter *in those circumstances*. But a recipient could still be warranted in relying on the rational being. (Indeed, the rational being's own beliefs could remain warranted.) For being warranted does not—and in empirical cases, could not—depend on one's route to truth being reliable in all circumstances. Being warranted depends on one's channel's being reliable in normal circumstances—circumstances in which the relevant competence is determined to be what it is. Thus, suppose that a rational being were transported unawares to brain-in-vat circumstances. Then rational procedures might fail to improve on a systematic non-veridicality, and the rational being's empirical reports might be quite unreliable. Suppose that the recipient of the rational being's reports had no reason to think that the rational being had been envatted. The recipient would remain prima facie warranted in relying on the reports. For a discussion of these issues in the case of empirical entitlements, see my 'Perceptual Entitlement'.

A rational being can clearly be unreliable in a particular abnormal circumstance in a way that undermines *knowledge*. Whether, and in what ways a rational being can be unreliable in a normal environment in a way that undermines entitlement seems to me harder issue. Rationality itself is constitutively (and apriori) a good, reliable route to getting things right, in all circumstances. However, some of the resources that rationalitiy relies upon (like perception) are not *constitutively*, hence not necessarily, reliable, even in normal circumstances. (Incidentally, in 'Interlocution, Perception, and Memory', *Philosophical Studies* 86 (1997), 31; 294, in this volume and in 'Comprehension and Interpretation', in L. Hahn (ed.), *The Philosophy of Donald Davidson* (Chicago: Open Court, 1999), 243; 357, in this volume—I make the mistake of claiming that perception *is* constitutively reliable. I have corrected that view in later work, for example in 'Perceptual Entitlement'). Insofar as a rational being relies only on rationality itself, it is, I think,

If the students had *only* the basic entitlement, they would not gain knowledge from the teacher's report. The basic entitlement is an entitlement to rely on the rationality and sincerity of the interlocutor. The students are entitled to rely on the teacher's rationality and sincerity by virtue of understanding the teacher's assertions. Again, they need not think about an interlocutor's rationality or sincerity, or be able to cite them as reasons. Still, their understanding the reports entitles them, prima facie, to believe the reports, because the intelligibility of the assertions is constitutively associated with rationality and because rationality is a good route to getting a subject matter right. The students lack any ground to suspect the teacher's irrationality or insincerity. They lack warrant to believe that the teacher's own attitudes toward evolutionary biology are irrational, or that the reports are not sincere expressions of beliefs. So they have the basic entitlement to rely on the teacher as a rational source. But the particular teacher's irrationality and lack of sincerity prevent the entitlement from yielding knowledge.[15]

clear, that the rationality will necessarily and apriori have the reliability necessary for being entitled to rely upon it. Whether the same can be said for a rational being that relies upon perceptual belief, but can use rationality to correct it, is the complex and deep issue. The issue bears on how best to respond to scepticism. In 'Content Preservation' and 'Perceptual Entitlement', I laid scepticism aside. I *assumed* that not only rationality but its non-rational resources (such as perception and the social institutions that we in fact rely upon) are reliable in their normal environments. I *assumed* that *that* reliability underwrites default entitlements. Whether one can show apriori that one has a right to such assumption is one form of the problem of scepticism about empirical knowledge.

Here also, I assume that perception is reliable. I also assume, for the students' knowledge case, that relevant social institutions are reliable. I also assume that unless special considerations undermine the presumption, one is apriori entitled to rely upon a rational being's presentations as true (cf. (A)). Rationality either supports or protects the presentations-as-true by rational beings.

[15] I continue to think that if one is warranted in finding an individual rational, one is *pro tanto* warranted in taking the individual to be sincere. See 'Content Preservation', 474ff.; 242ff in this volume. Since rationality has an impersonal function of supporting truth, an individual who has shown rationality can be presumed to be sincere, in the absence of evidence for doubt. As I wrote in 'Content Preservation', lying for no reason is irrational. Rational lying needs special reason to go against the impersonal function of supporting truth. Unless there is ground to think that one's source has special interests that would make the individual lie (perhaps rationally), a recipient can presume on a rational individual's sincerity. I leave open whether, if one came across a non-terrestrial blob apparently making English assertions, one would have some reason to doubt sincerity, or doubt whether the blob was actually making assertions. (See the case of a bush apparently talking in 'A Warrant for Belief in Other Minds'.) The blob's form may raise doubts. Still, I think that if one had sufficient reason to think that the blob was speaking English, one should not be neutral or negative in attributing sincerity unless there is some evidence that, from the apparent content or from the context, that counters the presumption of *pro tanto* rationality in sharing apparently harmless information. If one knows nothing about one's interlocutor or about the subject matter except that the interlocutor is an asserter of a not implausible and not specialized or difficult-to-know proposition, one is entitled to rely on the assertion as sincere and belief-worthy. If the recipient lacks ground to believe that a rationally competent individual is acting irrationally, or is lying for special practical reasons, the recipient is entitled to presume on the interlocutor's exercising truth-supporting capacities.

In the teacher's case, the irrational rejection of evolutionary theory falsifies the presumption of the teacher's rationality; and the teacher's not being open with the students about his or her disbelief in the science falsifies the presumption of sincerity. The students have an undefeated default entitlement to rely on the teacher (given no undermining considerations). They are entitled to believe the teacher. But this default entitlement cannot give them knowledge. For the teacher is relevantly irrational and

Here, I think, is a typical case of a recipient's having multiple warrants to believe a report. The entitlement to rely on a teacher does not essentially involve relying on the sincerity of the teacher in expressing individual, personal belief in the what he or she teaches.[16] An entitlement to rely on the teacher as a teacher (or as a teacher of science), given minimum recognition of someone as a teacher, supplements the basic default entitlement. The students' entitlement to rely on the teacher depends on the fact that teachers in the society are reliable in teaching the science and on the fact that the science is reliable. Given that this particular teacher is reliable in conveying the science, acceptance of the reports yields knowledge, where the reports are true.

The socially based entitlement overlays the basic entitlement that derives from the very intelligibility of the report. The students have both entitlements. But only the one grounded in the specific social institution of teaching (or teaching science) yields knowledge. If the students could not recognize the teacher as a teacher, and if they relied only on the intelligibility and internal coherence of the reports, they would not, I think, gain knowledge from them. Yet in many situations, the basic entitlement does suffice to yield knowledge. In the next section, I further discuss the widespread phenomenon of *overlay* among entitlements.

II

In 'Content Preservation' and in my other articles on interlocution, my primary interest lies in the *structure* of epistemic norms that govern interlocution. I want to emphasize both the abstractness of the structure, and its philosophical interest.

I continue to think that the Acceptance Principle (AP) articulates the basic, default warrant in this structure.

insincere. Only their supplementary entitlement to rely on the teacher as science teacher supports their knowledge.

[16] The students' entitlement to rely on the teacher *as a teacher* does not depend on presuming that the teacher's reports express personal beliefs. But, as indicated in note 15, I think that the students are further entitled to rely on the natural presumption—associated with their default entitlement—that the teacher *as a rational source* is sincere. If the students had reason to believe that the teacher was insincere, both entitlements would be threatened. The entitlement to rely on the teacher as a rational, credible source would be directly threatened. The entitlement to rely on the teacher as a teacher would be indirectly threatened, because an insincere teacher is less likely to be a reliable conduit, other things equal. But we are supposing that the students lack any reason to think the teacher insincere. Since their supplementary entitlement (the entitlement to rely on the teacher as teacher) does not depend specifically on sincerity, only reliability, the actual, unsuspected failure of the teacher's sincerity does not prevent the entitlement from yielding knowledge, because the teacher's reliability as a teacher remains intact. These complexities illustrate, I think, the value of getting the form of an entitlement right. Most recipients have entitlements in addition to their default entitlement, as well as many further considerations that bear, pro or con, on entitlements that they have.

(AP) A person is entitled to accept as true something that is presented as true and that is intelligible to him, unless there are stronger reasons not to do so.[17]

The Acceptance Principle is the starting point for an account of warrants to believe what one is told. It is the baseline standard that governs epistemically good belief formation from communication with others. The starting point does not always provide warrant for belief. Sometimes the default warrant is defeated by counter-considerations. Often it is supplemented by further positive considerations, as it is in the case of the students, discussed in section I. But the principle lies at the base of the structure of epistemic norms that govern interlocution.

Some philosophers have doubted this starting point. Some have thought that one must gain information about the trustworthiness of one's interlocutor to be warranted in accepting what one is told. Some have thought that being warranted depends on one's at least trying to assess that trustworthiness. I will not engage specific versions of these doubts. I think that they are all driven by a hyper-intellectualized epistemology. Many confuse *scientia*—the knowledge involved in science and in critical reasoning—with all knowledge. The hyper-intellectualization and this particular confusion are occupational hazards in philosophy.

Such views cannot account for simple cases in which children rightly accept what they are told, without supplementary support and without the minimum sophistication even to try to assess trustworthiness. Information and knowledge simply seem to flow in, without the need for supplementary support or critical scrutiny. I think that if the Acceptance Principle were not true, adults would have too slim a basis to develop the knowledge base that they actually have.

Many philosophers acknowledge these points, but see them as fringe points. They see the epistemology of children's beliefs as a minor part of the whole of epistemology. I think that such a view is very shortsighted. The non-intellectualized starting point of the ontogenetic development of knowledge is a key to understanding the grounds for the epistemology of all interlocution. It is the key to the structure of the relevant epistemic norms.

In 'Content Preservation' I considered, with some hyperbole, whether the Acceptance Principle can ever be the last word in the epistemology of acceptance for anyone over the age of eleven. The most important thing about the principle is that it is the first word. The last word in any given case is what is most salient to epistemic agents. The last word is especially salient to mature, educated adults who commonly scrutinize interlocutors, whose reports they have not developed reasons to trust. Often mature adults weigh pros and cons before coming to a belief, especially with regard to specialized or advanced subject matters. Sometimes an initial ground for doubt prevents the entitlement articulated by the Acceptance Principle from warranting belief. Sometimes a reasonable doubt stands until it is counteracted by further considerations. But since the Acceptance Principle is the first word in the dialectic, it structures nearly everything that

[17] 'Content Preservation', 467; 237, in this volume.

follows. So its role in understanding the epistemic structure of warrant for relying on interlocution is fundamental.

This point is missed if one focuses on the surface of epistemic dialectic or if one does not reflect sufficiently on the structure of defeasible epistemic entitlements and defeasible standards for knowledge.

For example, Elizabeth Fricker writes that even if one were to grant that the Acceptance Principle is correct, it would be 'largely irrelevant to explaining the basis of our right to accept most of the testimony we, mature adults, receive in everyday life'.[18] I believe that two serious errors underlie this claim of virtual irrelevance.

One is a focus on what is most salient to fairly sophisticated adults in their everyday acquisition of knowledge from others. Fricker writes,

[The Acceptance Principle] is largely irrelevant to explaining the basis of our right to accept most of the testimony we, mature adults, receive in everyday life. It is irrelevant because a mature and normally cognitively equipped adult is usually amply furnished with

[18] Elizabeth Fricker, 'Martians and Meetings: Against Burge's Neo-Kantian Apriorism about Testimony', *Philosophica* 78 (2006), 81. Fricker argues against the Acceptance Principle. Some of the argument relies on the hyper-intellectualism mentioned earlier. Some of it relies on the claim that in discussing rational lying, I equivocate between a 'thin' sense of being rational that is constitutive of having propositional attitudes and a presumably thicker sense: 'wholly impartial and disinterested speaker only of the truth', 82. Fricker thinks that my account of the Acceptance Principle depends on conflating these "senses" of being rational. I deny that two such senses of being rational are at issue. I make no use of her supposed thicker sense. I deny that the sense of being rational that is implicit in having propositional attitudes (and being capable of making assertions) is thin. I think that it is the full sense of being generically rational, as distinguished from being critically rational. I think that generic rationality has practical and impersonally theoretical dimensions. I think that, prima facie, when a speaker fails to tell the truth because of special interests, the speaker crosses rationality in one significant dimension, the latter one. Just as a generically rational individual can be irrational on particular occasions, a generically rational individual can be all things considered rational in the particular case, while rationally lying. But the impersonal function of rationality is compromised. So there is some compromise on generic rationality. I think that lacking special reason to suspect that the individual is relying on special needs or special considerations, one can presume that the individual is in a non-special situation and is realizing the impersonal function of rationality. This point is an instance of the more general point that one is entitled to presume that a generically rational individual is in no respect irrational in a particular case, unless there is reason to doubt the presumption. Contrary to Fricker's claim (a wild claim by virtue of the terms 'wholly' and 'only'), I do not need to assume that a recipient relies on a presumption that the interlocutor is a 'wholly impartial and disinterested speaker only of the truth'. The relevant knowledge is defeasible, and it depends on being right about the interlocutor in the particular case—not on presuming that the interlocutor is a perfect rational god. It allows that a given speaker may be rational and may make mistakes, or may, in some cases, not perform up to his or her rational competencies. It allows for different, sometimes conflicting dimensions of rationality. An individual's entitlement can be defeated if there is reason to think that the interlocutor falls short of the default presumption in any of these ways. And an individual who has the entitlement can fail to have knowledge if in fact the interlocutor does fall short in any of these ways. I think that Fricker gives no genuine reason to doubt the explanation in 'Content Preservation' of the constitutive connections that underlie the Acceptance Principle.

Fricker is right (in her section IV) that my account of the Acceptance Principle is not intended to persuade people who doubt that we are entitled to accept what we are told, other things equal. I believe that reflecting on practice and taking care to avoid hyper-intellectualization are the best grounds for a philosopher's coming to accept the principle. My account is intended to articulate an underlying rationale for the principle, granted its truth.

relevant empirical evidence concerning the likely reliability of most of the testimony she encounters in her daily life, and this swamps the significance of any supposed on-no-evidence entitlement to believe what she is told. . . . Whether [the Acceptance Principle] holds is a critical issue for the legitimacy of our everyday acceptance of testimony, only if we are sufficiently often in the position about which it pronounces—that is, we have no defeaters regarding the trustworthiness of the testifier. But more than this: the correctness or otherwise of [the Acceptance Principle] is critical for the propriety of our everyday response to testimony only if we are frequently in a position where we have no positive empirical basis to trust a testifier. If the typical position of a mature adult faced with a piece of testimony is that she has in her cognitive background, and brings to bear, a wealth of empirical knowledge relevant to the assessment of that testimony, then she does not need recourse to a default principle licensing its acceptance in the absence of such relevant empirical information.[19]

This passage illustrates a philosophical focus that fails to engage with the motivations that underlie my proposal and explanation of the Acceptance Principle.

It is true that normal, educated adults usually have lots of evidence that bears, pro or con, on the trustworthiness of a report. I made this point myself. But *irrelevance* of the principle to explaining the entitlements of such adults, much less the entitlements of many less well-equipped persons, hardly follows from this point. Fricker provides no argument that bridges the gap. The relevance of the principle does not depend on the frequency of its operating as the sole factor in recipients' warrant for belief. The principle's relevance does not depend on how often individuals 'need recourse' to it. Recipients frequently have many warrants to rely on the reports of others. They may regularly have a warrant but not *need* to rely on it, since they have others.

My interest in epistemic warrant does not center on the minimum needed to get by in everyday life. It centers on the structure of the relations among the warrants that come into play in the epistemology of interlocution. Frequency, minimum need, salience are not particularly important factors in understanding this structure.

Moreover, I am not focused on all the factors that make communication work in actual human life. It is surely true that the ontogenetically earliest communicative relations are between a child and relatively sympathetic adults, who need to initiate the child into the family or tribe form of communication. The child can rely on the adults not just because they are rational but because they are sympathetic and interested in helping the child learn the language and the environment. I see a deeper epistemic structure that in various ways underlies this practical ground. It underlies this practical ground by applying more broadly and by being a more central component in the epistemic structure of warrant through interlocution.

[19] 'Martians and Meetings: Against Burge's Neo-Kantian Apriorism about Testimony', 81.

Rationality is causally grounded and sustained in mutual need, perhaps. Still, once propositional attitudes and their attendant rational norms and capacities are in place, there are epistemic norms that are apriori fitted to propositional, rational capacities. It is these norms that I am trying to understand.

Fricker remarks, as if in a ridiculing *reductio*, that my abstract explication of the Acceptance Principle 'conjures up the idea of a message beamed in, as it were from Mars, received in abstraction from any social context of real-life everyday communication'.[20] Far from being a rhetorical embarrassment to me, the Mars analogy is close to what I had in mind. I asked whether in abstraction from the trappings of human social context, there is ground to accept what one is told—given only an intelligible presentation-as-true and assuming that no counter-considerations are present.

I continue to maintain an affirmative answer. I believe that the affirmative answer indicates that the minimum source of warrant for receiving communication is more general than human social context. The source lies in something universal to intelligible, propositional presentations-as-true (centrally, assertions). That universality, together with the way dialectical reasoning works with possible defeaters of default entitlements, shows the source to be in a sense basic. All further considerations are more specialized, even if they are more common. As I will try to show, nearly all are in effect reinforcements of or threats on the probity of this basic warrant.

A second mistake that Fricker makes is to misconstrue a fundamental feature of default prima facie (or *pro tanto*) entitlement. She believes that when a recipient has evidential support other than that provided by the Acceptance Principle, that further support 'replaces' the entitlement default entitlement, thereby making it 'irrelevant' in all such cases.[21]

Default, prima facie entitlements do not in general operate that way. The default prima facie (or *pro tanto*) entitlement articulated in the Acceptance Principle certainly does not work that way. Fricker provides no reason to take her non-standard view of default prima facie entitlement.

[20] 'Martians and Meetings: Against Burge's Neo-Kantian Apriorism about Testimony', 82. It is important *not* to allow 'beamed' to suggest that assertive mode is not part of what would be required to be entitled to accept a report in accord with the Acceptance Principle. The principle does not apply unless the recipient is in a position to exercise a reliable competence for taking received content to be asserted.

[21] Elizabeth Fricker, 'Trusting Others in the Sciences: *A Priori* or Empirical Warrant?', *Studies in the History and Philosophy of Science* 33 (2002), 373–383. Peter Graham, 'Testimonial Justification: Inferential or Non-Inferential?', *The Philosophical Quarterly* 56 (2006), 84–95, criticizes unsound argumentation by Fricker in ways I largely agree with. In elaborating on these criticisms, however, Graham appeals to overdetermination of warrant. I think that, in many cases, the default entitlement articulated by the Acceptance Principle and supplementary warrants severally provide warrant. In this sense, Graham is right that there is overdetermination. But most standard cases of overdetermination provide a poor model for the relation between the default entitlement and supplementary warrants. Most standard cases of overdetermination take the overdetermining factors to be independent. As I discuss below, most supplementary warrants are *not independent* of the entitlement articulated by the Acceptance Principle.

A default prima facie entitlement *holds*, unless a further warrant either defeats it, or threatens it in a way that needs to be answered and is not answered. The foregoing claim is the claim made in the Acceptance Principle. That claim is also what is meant by saying that the default warrant holds 'other things equal'. The idea is not that the principle holds only when other considerations are absent or other considerations exactly balance one another. The idea is that it suffices to support belief (and I think, in the absence of further defeaters, supports knowledge), unless there is unanswered, sufficiently strong countervailing warrant. If there is a countervailing threatening warrant and if a supplementary positive consideration defeats or outweighs that warrant, the default entitlement stands. The positive considerations show that the countervailing threats do not, all things considered, undermine the warrant. So the warrant holds. Sometimes the supplementary positive considerations simply show that a threat, which would warrant doubt in the absence of the supplementary considerations, is misleading; the supplementary considerations do not support the belief directly. Then it is completely clear that the initial entitlement retains its force.

Other times, the supplementary considerations support the belief directly, and more strongly than the counter-considerations support doubt. In such a case, the supplementary considerations can be regarded as sufficient to support the belief. But they do not, and cannot do so, in complete independence of the entitlement articulated by the Acceptance Principle. The fact that the report is an intelligible presentation-as-true, with its associated implications spelled out in the Acceptance Principle, must figure in nearly *any* supplementary warrant that directly supports belief. Supplementary considerations show that in the particular case the reporter's rationality and sincerity are to be trusted. Supplementary warrants are specializations and elaborations of the entitlement present in the principle. They are not fully independent voices.

If one were to think of the default warrant and the supplementary considerations that directly support belief as separate units, then it would be hard to see how the the default warrant has any significant role in the final outcome. On this picture, the default warrant and the supplementary warrant are like competing but separate voices. Either voice taken by itself would suffice to support the belief. If the default voice were alone with the counter-consideration, it would not suffice. It would be drowned out. If the supplementary reason were alone with the counter-consideration, it *would* suffice. Given that all three are in competition, the default warrant's voice might seem to be outshouted by the counter-consideration, which is itself outshouted by the supplementary positive consideration.

This picture is misleading about the structure of warranting support. Supplementary positive considerations are normally not independent of the basic default entitlement in the way that the picture of competing voices suggests. As noted, nearly all supplementary warrant presupposes and uses warranting force involved in comprehension of a presentation-as-true. Nearly all supplementary warrant presupposes and uses the warranting force involved in the rationality and sincerity of the recipient's source. Supplementary warrant tends to ensure that this force is in place by supporting the presumption that the recipient's source is likely to be rational and sincere—and consequently worth relying upon—in the particular case. Such support reinforces the warrant already present in the default warrant.

This point is clear from the fact that the same generic kinds of threats that could threaten the general basic default warrant can also threaten these supplementary warrants. For example, if the supplementary warrant is a reason to believe that the individual is a specially competent scientist in the field in which he or she tells one something, that warrant could be threatened by further information that the scientist is an inveterate liar in the type of circumstance in which he or she made the report. Or it could be threatened by further information that the scientist is in a depression and his or her usual rationality has been compromised.

The idea that supplementary positive warrants 'swamp' or 'replace' the basic default prima facie warrant depends on a misconstrual of the nature of the relevant warrants. Nearly all supplementary positive considerations that support the credibility of a reporting incident reinforce the basic warrant. They are not independent of it. They provide special information that tends to indicate that a given case fits the general form of the basic default warrant. They certainly do not emasculate its warranting force, or replace it.

Of course, some supplementary considerations reinforce the credibility of a report not by bearing directly on the credibility of the reporting incident, but by providing independent warrant for the proposition that is reported. For example, the same proposition might be reported by others. Or one might have been an eyewitness to something that adds to the credibility of the reported proposition. If the basic warrant for the initial report is threatened, independent supplementary positive considerations (that do not bear directly on the credibility of the reporting incident) may provide an independent and crucial element in the full warrant for believing the relevant proposition. But such cases certainly do not show that the initial warrant is irrelevant to warrant. They just illustrate what is already obvious—that the basic warrant is prima facie. It is not in general decisive for supporting belief; it can contribute warrant without being decisive.

As noted, nearly all supplementary positive considerations that support the credibility of a reporting incident indicate that the incident is a specific realization of general features that provide warrant in the default case other things equal. The general features that support credibility are those implied by the rationality of presentations-as-true. Reason supports credibility. Slightly less generally, that an individual is prima facie competent in a given case follows from the individual's rationality. A rational person, exercising reason, would not make an assertion on a matter about which he or she is not competent. Also that an individual is prima facie sincere in a given case follows from an individuals rationality. See notes 15 and 16. Sincere, rationally supported or protected, competent presentations-as-true are credible. Further evidence that supports an interlocutor's credibility usually supports specific elaborations on the interlocutor's competence, sincerity, and/or rationality. In nearly all cases, a reporting incident helps ground an entitlement to believe the report at least partly by virtue of the presumption that in presenting the proposition as true, the source exhibits a rationality—including a sincerity and competence—that prima facie supports belief.

I say 'nearly all' because of the complexity in the structure of epistemic warrant illustrated in section I. When the students obtain knowledge from the insincere but reliable teacher, they rely on the source's being a teacher, or a teacher of science. *That* entitlement does *not* rely on the overall rationality or sincerity of the teacher. It relies only on the institution of reliable reporting/ teaching of the science. That warrant is not a specialization of two of the key warranting features that are articulated by the Acceptance Principle—the sincerity of the report and rationality of the interlocutor in backing belief in what is reported. The students' supplementary entitlement bypasses those features. Their supplementary warrant yields knowledge even in the absence of those features.

I think that this supplementary warrant depends on special aspects of social institutions. Such institutions can enlist loyalty and veridicality in carrying on cultural tradition, regardless of the individuals' particular beliefs in what is conveyed. Teaching the subject matter has one function of simply being a conduit of the subject matter. As teacher, the individual can subordinate his or her own belief to the teaching function. Teachers can even function as non-rational but accurate instruments that do not contribute *any* of their own rationality to the process. They can function minimally as telephones or amplifiers that pass on propositions—even knowledge—that resides in their social antecedents.

Such separation of individual belief from accuracy in conveying knowledge is not the normal case. More importantly, the existence of such social institutions—institutions involved in education that reliably conveys truths to recipients—is possible only because of a base of sincere, rational assertions that figured in the development of such institutions. Institutions can go unreliable. Institutions that reliably convey knowledge can be staffed by many non-knowledgeable conduits. However, such institutions are normally dependent on a base of sincere interlocution backed by rationally supported or rationally protected beliefs in the reports. Institutions that lack a such a base will tend to go unreliable and thus will fail to provide warrant, hence knowledge, through interlocution.

Thus, although the epistemic structure of warrant and knowledge transmission is made more complex by the fact that social forms can take over some of the epistemic functions of individual rationality, the role of individual rationality in the epistemic structure of interlocution remains primary. It is the first word in any complex dialectic of pros and cons. And it is normally *present* in the last word, at least in specialized form, if the last word is pro. Most positive considerations that directly support the credibility of a report are specializations and confirmations of the default warrant articulated by the Acceptance Principle.

III

My third topic is the epistemic status of the initial intake of a report—the exercise of a capacity for comprehension. Let me state briefly the background for this topic in my work. Recall the Acceptance Principle:

> (AP) A person is entitled to accept as true something that is presented as true and that is intelligible to him, unless there are stronger reasons not to do so.

I take the warrant for believing the Acceptance Principle to be apriori.

The warrant, which is a justification, for believing the principle is to be distinguished from the warrant that is articulated by the principle. The latter is a default prima facie entitlement that all interlocutors have, if they are reliable comprehenders and if they comprehend an event as having a specific representational content and as a presentation-as-true.[22] To have the entitlement, an individual need not believe the principle, or even have the concepts necessary to believe it. Nevertheless the justification of the principle reveals something about the epistemic status of the entitlement that it articulates.

The justification for believing the principle consists in the argument given in 'Content Preservation'. The argument connects

(1) prima facie intelligibility of a presentation-as-true

with

(2) there being, prima facie, a rational source of the presentation-as-true.

And it connects (2) with

(3) there being, prima facie, a good route to truth.

The argument develops the idea that a presentation-as-true is prima facie backed by reason—either supported by it or protected by it.[23] Reason constitutes a prima

[22] This formulation should not be understood to require that the recipient have the general concepts representational content or presentation-as-true. The idea is that in the particular case, the recipient exercises a comprehension competence. That competence treats the event as a presentation-as-true and assigns a specific representational content to the event. To have the default prima facie entitlement articulated by the Acceptance Principle, the recipient must exercise a reliable competence on the particular occasion.

[23] Not all reports by a rational being are reasoned to or supported by reason. For example, rational beings report their perceptual beliefs. They are entitled to the beliefs. Entitlements are warrants without reason. I say that the rational source's report can be presumed to be supported *or protected* by reason, because I think one can expect a rational being to be able to sustain or protect such warranted beliefs by reasons, even when the beliefs themselves are not supported by reasons. A rational being must have at least a know-how sensitivity to possible defeating considerations or supplementary supporting considerations. And it must be able to think rationally with such considerations when they arise. See 'Perceptual Entitlement'. (Having the know-how sensitivity is *not* to be understood as requiring having epistemic concepts like defeating condition or reason.) The presence of reason enhances the source's reliability even when reason does not directly support the report.

facie good reliable route to getting a subject matter right. Beliefs obtained by such routes are warranted.[24]

There are subsidiary arguments in this account.[25] But the main argument goes through (1), (2), and (3). The links between these steps and the overall argument seem to me to be apriori. So the connection between an event's being an intelligible, propositional presentation-as-true and the proposition's being worthy of belief, other things equal, seems to me to be apriori.

The default prima facie entitlement, articulated by the Acceptance Principle, to believe a proposition that is comprehended, and is presented as true, seems correspondingly apriori. What gives that entitlement epistemic force is the constitutive, apriori connections expounded by the argument that goes through (1)–(3).[26] Those connections are apriori knowable. So I take the entitlement as articulated in the Acceptance Principle to be apriori. I stand by these ideas.

The epistemic status of occurrent exercises of comprehension is a further matter. The entitlement so far discussed is a prima facie epistemic entitlement to form beliefs in presentations-as-true of intelligible propositions. If one is prima facie warranted in one's comprehending an event as a presentation-as-true, one is prima facie warranted in believing the proposition (prima facie) presented as true. But to be epistemically entitled to a particular belief in an intelligible proposition presented as true by another person, one must be epistemically entitled to one's occurrent exercise of one's capacity for comprehension. One must be epistemically entitled to rely on one's (seeming) comprehension of the other person's report as genuine comprehension . Thus the instantiation or 'application' of the entitlement articulated by the Acceptance Principle in a particular case of interlocution requires that the recipient exercise a competence for comprehension. So to be entitled to the belief in the report, the recipient must be entitled to rely on the exercise of comprehension.

In 'Content Preservation', 'Interlocution, Perception, and Memory', and 'Reason and the First Person' (last section), I argued that the warrant for *particular exercises* of our comprehension of others' communication sometimes *is* non-empirical. I relied on this claim, at least for the sake of argument in 'Computer Proof, Apriori Knowledge, and Other Minds'.[27]

[24] 'Content Preservation', 470ff; 239ff. in this volume.

[25] I have in mind, for example, the argument that entitlement to rely on rationality implies entitlement to rely on sincerity. See 'Content Preservation', 474ff., 242ff. in this volume, and notes 15 and 16 above.

[26] The statements of the principle and its epistemic status in 'Content Preservation', 467 and 469 (237, 238 in this volume) respectively go only this far. This point is explicit, 476 (243 in this volume), though it is not as clear on 469 as it should be. The issue of whether in a particular case an individual is warranted in comprehending an event as certain specific prositional content presented as true is a further issue. This is the issue that I shall discuss in this section.

[27] See 'Content Preservation', 476–485 (243–250 in this volume), esp. 478 (245 in this volume); 'Interlocution, Perception, and Memory', esp. sections I and II; 'Reason and the First Person' in Crispin Wright, Barry C. Smith, and Cynthia Macdonald (eds.), *Knowing Our Own Minds: Essays Self-Knowledge* (New York: Oxford University Press, 1998), 264–267; 401–404 in this volume; and 'Computer Proof, Apriori Knowledge, and Other Minds'; all are reprinted in this volume.

I acknowledged that all successful exercises of comprehension of others' assertions depend causally on perception. I maintained, however, that the role of perception in comprehending another's speech or writing is sometimes purely that of an enabling condition. Although perception is necessary for the success of our comprehension, I claimed that perception does not always contribute to the force of one warrant that we have for relying on an exercise comprehension. This claim always seemed to me, as well as to others, counter-intuitive. Still, I thought that I could see through to a way of justifying it. I now think, what most readers have probably long thought, that I was mistaken.

Several authors correctly objected to my claim. But no one explained to my satisfaction why it was mistaken. I will try to sort out rights and, mainly, wrongs in my earlier view.

I begin by reviewing key features of the view. Then I discuss four ideas that seemed to me to support it. Finally, I explain why I think that these ideas, both individually and jointly, fail to justify the claim that in actually comprehending others' assertions, we sometimes have a non-empirical entitlement to rely on the exercise of comprehension in forming beliefs. One's entitlement to rely *essentially* on an exercise of *seeming* comprehension of another individual's report as *genuine* comprehension is never strictly apriori.[28]

I took default prima facie warrant for exercises of comprehension to be an element in default prima facie warrant for forming beliefs in what others present as true. As noted, I stand by my account of the warrant that connects an exercise of comprehension with belief. This is the account that connects propositional presentations-as-true with rational sources, and rational sources with good routes to truth. I believe that the contrast that I draw in 'Interlocution, Perception, and Memory', section II, between using others as sources in communication and using scientific instruments is right. Although one might be equally warranted in relying on a measuring instrument and on an interlocutor in forming beliefs, the explanations of the warrants for belief are significantly different. Both require reliability, relative to appropriate conditions, and goodness of route to truth. But the fact that the latter runs through a rational source makes the forms of the warrant differ in epistemically important ways.

I take exercises of comprehension often to be immediate and non-inferential at the personal-level. Although there are certainly complex, sub-personal, irremediably unconscious transitions in the formation of a (purported or actual) comprehension, the individual often carries out no inference. At the level of processes available to the individual's consciousness, understanding is often as immediate as ordinary perception of the physical environment. Furthermore, individuals need not be capable of giving reasons that justify their comprehension

[28] Note the role of 'essentially'. In mathematics, one could take a piece of communication, find it intrinsically plausible, or find that it fits with other things one knows, and come to know it apriori. Then one's warrant would not rely essentially on the proposition's being presented as true by another. The understanding of the content itself would be the source of warrant.

of others' utterances. So the relevant sort of warrant at issue is an entitlement, not a justification.

Although the comprehension is both a comprehension of content and of mode—presentation-as-true—the comprehension is normally not *based* on a belief that a source asserted (or presented-as-true) that such and such. I believe it likely that children comprehend utterances and are entitled to rely on their comprehension *before* they have any beliefs about assertions as such. They are cognitively sensitive to utterances' being assertions (or other presentations-as-true) and to what the utterances' contents are. But they need not have, and I suspect at certain stages have not, formed conceptions as of content or mode, much less beliefs about them. Understanding here is prior to having beliefs as the "object" of understanding.

I think that comprehending the representational content of another's utterance requires a capacity to understand the primary modes of utterance, or types of speech act. In particular, comprehending such representational content requires responding differentially to assertion, or at least the broader generic mode—presentation-as-true. The epistemic status of comprehending the mode of another's utterance and that of comprehending the representational content of another's utterance are parallel and interlocked.

I did not claim that all exercises of comprehension could be non-empirically warranted.[29] I excluded exercises of comprehension in which the comprehended content itself includes exercises of perceptual representation. For example, if you say 'that is a star-gazer lily', looking at a flower, and I comprehend what you say, part of my comprehension consists in perceiving the flower, or at least perceiving your angle on what you are talking about. My comprehension includes *de re* perception-backed representation. I always took such comprehension to be warranted empirically.

The sort of comprehension that I made my counter-intuitive claim about is purely *conceptual* comprehension—comprehension that utilizes only exercises of ability-general representation.[30] For example, comprehension of another person's utterance of 'the longest strait in the universe is well over 200 miles long', 'the cube root of 27 is 3', and 'blue whales are larger than school buses' are each relevantly conceptual or intellectual. The instances of comprehension do not contain perceptually backed, *de re* applied representational content. The question is whether exercises of conceptual comprehension can sometimes be

[29] Normally, I take epistemic entitlement to apply to an individual's belief, or to an individual's reliance on cognitive capacities that are part of a system that includes belief. I never take it to apply to occurrences of perception. Perceptions *contribute* to entitlement to perceptual beliefs. Exercises of comprehension comprise an intermediate case. I will write of entitlement to such exercises. One could recaste this way of writing, but nothing here hangs on the terminological issue. Comprehension, unlike perception, is propositional. Arguably it can constitute propositional knowledge of the content of another's utterance. Perception cannot constitute propositional knowledge.

[30] For explication of 'ability-general', see 'Disjunctivism and Perceptual Psychology', *Philosophical Topics* 33 (2005), 1–78, esp. section VII; 'Five Theses on *De Re* States and Attitudes', in J. Almog and P. Leonardi (eds.), *The Philosophy of David Kaplan* (New York: Oxford University Press, 2009), section II; and *Origins of Objectivity* (Oxford: Oxford University Press, 2010), 379–396.

intellectually and non-empirically warranted—warranted independently of any contribution by perception to the warrant's positive epistemic force.

I strictly distinguished the nature of the warrant for an exercise of comprehension from the nature of its vulnerability to correction. Although I claimed that some exercises of comprehension are non-empirically warranted, I took all exercises of comprehension of another's utterances to be vulnerable to empirical correction or refutation.[31] The issue concerns sources of positive epistemic force, not sources of epistemic threat.

Because I took perception to be a necessary enabling condition for any comprehension of others' utterances, hence for any *warranted* comprehension of others' utterances, I took malfunctions and other poor exercises of perception to be capable of undermining entitlement to conceptual comprehension. I maintained that it does not follow that perception makes an essential positive contribution to the force of epistemic warrant for comprehension of others' utterances. Malfunctions of the brain can undermine warrant for relying on one's reasoning processes. It does not follow that references to the brain figure in one's entitlement to rely on particular transitions in deductive inference.

I stand by all the foregoing. I will be reconsidering and criticizing only the claim that in some of our exercises of conceptual comprehension, perception contributes nothing to the positive epistemic force of the default prima facie entitlement to rely on the veridicality of an exercise of comprehension of another's occurrent presentation-as-true. Put less technically, I shall criticize the counter-intuitive claim that one's warrant for relying purely on one's comprehension of another's utterance can sometimes be non-empirical.

I discussed four ideas that made me think that I could justify this counter-intuitive claim. The first concerned the intellectual objects of conceptual comprehension and the large role that conceptual mastery plays in comprehension, even acknowledging the necessary genetic role of perception in the process. The second consisted in an analogy between exercises of comprehension and uses of diagrams or concrete symbols in coming to understand mathematics. The third consisted in an analogy between the preservative role of perception in interlocution and the preservative role of memory in deductive inference. The fourth consisted in what I called 'the Injection Argument'.

I regarded no one of the four ideas as decisive in itself. I saw them as mutually reinforcing. I will discuss each idea and try to explain why, even collectively,

[31] Although I agree with the conclusion of Christensen and Kornblith's objection to my earlier view that some of our exercises of comprehension of another's utterances are non-empirically warranted, I continue to think that their reasons were not good ones. A central line of reply in 'Interlocution, Perception, and Memory', section I, esp. 26–30; 290–292 in this volume rests on the distinctions in the text above. It is obvious that empirically warranted considerations can threaten default entitlement—threaten the plausibility of a report or the trustworthiness of the reporter. It does not follow that the default entitlement itself is perceptually warranted. Nearly all the considerations that I mistakenly marshalled to support my view relied on this delicate but important distinction between source of support and source of warranted threat.

they do not suffice to support the counter-intuitive claim that in some cases of being warranted in comprehending others' utterances, perception is just an enabling condition and contributes nothing to the positive epistemic force of the epistemic warrant to rely on the exercise of comprehension.

First, the large role that conceptual mastery plays in comprehension. I take it to be clear that conceptual representational content is an abstract entity that cannot be perceived. Instances or tokens of symbols can be perceived. But individuated as shapes or as acoustical signals in a clearly physical and not content-based way, instances or tokens of symbols are not themselves instances of representational content. They *express* representational content. Expression is not a natural-kind relation, and is not instantiation. It depends on the conventions of a language community, or the idiosyncracies of an idiolect.

As a consequence, language perception is very different from perception of ordinary entities in the physical environment. The latter is known to leave little scope for top-down influence. When we perceive ordinary entities in the physical environment, our background beliefs can affect attention and memory. But the primary processing of perception of basic perceivable attributes—like shape, color, pitch, location, texture, motion, bodihood—is nearly independent of background belief and is broadly common to all mammals.[32]

In language perception, the connection between the physical sounds or shapes that are produced and what sentence is taken to have been uttered by a recipient, let alone what representational content the recipient assigns to the utterance, is vastly more complex. The connection is deeply informed by antecedent linguistic, cognitive capacities. Indeed the acoustical signal in heard speech is not stored or remembered by a hearer of speech.

In 'Content Preservation' and other articles, I emphasized that comprehension is not an inference at the personal-level of psychology. It is fast and unconscious.

[32] A critique of the tendency in twentieth-century philosophy to overrate top-down elements in ordinary, basic perception is one of the main themes of *Origins of Objectivity*. For an introductory explanation of the relative independence of human, non-phonological perceptual systems from higher-level cognitive capacities, see Zenon Pylyshyn, 'Is Vision Continuous with Cognition? The Case for Cognitive Impenetrability of Visual Perception', *Behavioral and Brain Sciences* 22 (1999), 341–365. Pylyshyn's point is among the best supported general facts in perceptual psychology about perception. Many experiments, mainly in social psychology, have tried to show that background beliefs affect perception. But these experiments have never controlled for the psychological source of familiarity effects. They fail to engage with the results in perceptual psychology. Familiarity effects that stem from knowing what sort of thing is perceived have not been shown to be caused by, or part of, perceptual operations as opposed to prior beliefs. A well-entrenched finding in perceptual psychology indicates that formation of perception *per se* does not depend on higher-level cognitive states. For studies that are pointed at poorly controlled experiments alluded to above, and that show that when the sources of familiarity effects are controlled for, the judgment-independence of perception—including human perception—remains intact, see G. Naor-Raz, M. J. Tarr, and D. Kersten, 'Is Color an Intrinsic Property of Object Representation?' *Perception* 32 (2003), 667–680; and M. Olkkonen, T. Hansen, and K. R. Gegenfurtner, 'Color Appearance of Familiar Objects: Effects of Object Shape, Texture, and Illumination Changes', *Journal of Vision* 8 (2008), 1–16. Formation laws for ground-level perception do not involve concepts or linguistic capacities. By contrast, language-perception depends heavily on background linguistic capacities, such as syntactic and semantic competencies, that are not perceptual.

Many comprehenders cannot articulate the inductive and context-dependent transitions that they make from perceiving utterances, as physical tokens, to assigning a representational content and mode to the utterance. To be sure, there is, in principle, a possible person-level inductive, theoretical route from the physical tokens to the assignment of meaning. But the route is not the basic psychological route. The basic route is a sub-personal route. That route begins with perception and proceeds through a complex, context-dependent set of unconscious transitions that are cognitively dominated. The transitions are largely dependent on application of the individual's particular, contextually influenced syntactic and semantic competencies. As I emphasize in 'Comprehension and Interpretation', they are not even a sequence of the individual's unconscious propositional inferences involving beliefs about sounds, speech acts, and the like. The complex processing involved in linguistic comprehension is not a matter of individividual-level propositional inferences at all. Normally, linguistic comprehension involves processing, but not propositional inferences from perceptual beliefs about physical situations. The processing is modular and very likely not even propositional.

I took the role of perception in this process to be more one of triggering the application of the cognitive competencies than one of contributing to the warrant for applying these competencies. I attempted to bolster this first idea with a second.

The second idea is an analogy between the role of perception in triggering comprehension of others' utterances and the role of perception of symbols or diagrams in triggering comprehension of logical or mathematical truths. A traditional rationalist point is that symbols or diagrams simply trigger understanding of representational content. The perception of the symbols is causally and psychologically necessary for understanding. But the epistemic force of the warrant for believing the mathematical content owes nothing to perception. I take this point about the role of perception in coming to understand and be warranted in believing logical and mathematical propositions to be correct. The idea of the analogy was that in the relevant cases of comprehension of an utterance, the role of perception in the *epistemology* of comprehending others' utterances and in believing what one is told is like that of perception in learning mathematics. Perception is necessary to trigger understanding. But the warrant for belief derives from understanding, not from perception.

I acknowledged disanalogies. In understanding mathematics, it does not matter for knowing the proposition whether one correctly understands what proposition is expressed by a symbol, or what proposition the teacher is trying to get across by drawing a diagram. One knows whatever proposition is triggered, if one understands that proposition and if one can know it by understanding it. To know an ordinary proposition by being told it, one must get *right* what proposition is uttered. For, in interlocution, gaining knowledge in the proposition is parasitic on there being knowledge of that proposition (in the sense of section I) in the antecedent chain.

I thought that this disanalogy does not undermine the main claim for two reasons.[33] First, I noted that one can be warranted *and gain knowledge* through interlocution even if one mis-perceives an utterance, as long as one's comprehension is correct.[34] There are many cases in which one mishears what an individual utters and yet correctly understands what the individual says. For example, the interlocutor might mis-speak; but one might correctly understand what the individual intends to convey. Second, I noted that one can be warranted and gain knowledge through interlocution, even if one's perception and perceptual belief about the utterance go wrong because of negligence or carelessness. That point seemed to show that perception of the only relevant thing that one could perceive—the utterance—was not crucial to the warrant for comprehension. Again, I thought, what matters for whether one acquires knowledge is whether one comprehends what the interlocutor asserts—not how perception supports the comprehension.

My idea was that if the default warrant that supports knowledge through interlocution depended for its warranting force on perception or perceptual belief, these two possibilities would not exist—or at least would not be so frequently realized. It seemed to me that perception was just functioning to trigger comprehension. What mattered to warrant and knowledge, I thought, was comprehension, not whether perception was correct or even adept. I thought that these points suggested that perception and perceptual belief sometimes does not provide warranting support for comprehension.

I overrated these two points, which seemed to disarm disanalogies between comprehension in interlocution and coming to understand mathematical/logical truths. They do not show what I thought that they showed. I expound them here to explain how I came to my earlier view, and how I defended it against doubts, including my own doubts. I will come back to why they lack the force that I thought that they had.

Now to the third idea. I thought that the force of the diagram analogy was bolstered by a further analogy. In 'Content Preservation', I argued that purely preservative memory is essential to deductive reasoning, but does not contribute to the warranting force for any premise or inferential transition in the reasoning. Failure of such memory can undermine warrant. But its success, though essential, is not a warranting factor in deductive reasoning. The warrant for believing the conclusion hinges on whatever warrant there is for believing the premises, and whatever warrant there is for relying on one's competence to carry out deductive inferences. Memory does not figure as a positive warranting factor in the

[33] I made these two points in 'Content Preservation', 480 n19; 246 n19 in this volume, and 'Interlocution, Perception, and Memory', 29; 292 in this volume.

[34] One can be *entitled* to rely on one's comprehension and one's belief in what one is being told, even if one does *not* correctly comprehend the utterance. But one is thus entitled only because one is a good, reliable comprehender of others' utterances, in normal circumstances.

epistemology of ordinary deductive inferences. I believe that this argument remains sound.

The argument was in effect an existence proof of a psychological capacity that is psychologically essential to knowledge, but does not contribute to warrant. I thought that, like purely preservative memory in deductive inference, perception could, in interlocution, be psychologically essential, but epistemically inert, unless it failed. Perception seemed to play a role in transduction and preservation of representational content through interlocution that was quite analogous to the role that purely preservative memory plays in keeping steps in an argument in play over the time during which an inference is carried out. The analogy seemed to help explain how perception could be so important to the success of interlocution without contributing to the warranting force for belief or knowledge.

I called a fourth idea 'the Injection Argument'.[35] I tried to evoke the idea that the contents of perception and perceptual belief are inessential to comprehension, by invoking a counterfactual supposition. Suppose that recipients in communication never relied on what they perceived or perceptually believed about utterances in communication in forming their understanding of what was said to them. Suppose that there were a natural, reliable, causal route from interlocutors' intentions in making assertions through some physical medium into the brains of recipients that then reliably caused comprehension. I thought—and still think—that recipients would be warranted and could gain knowledge via such communicative exchanges. If the causal route were part of a natural and reliable psychological competence in understanding another person, then persons could be entitled to rely on it. But the competence would not depend, even causally, on perception or use of one's senses. Perception and sensing would play no role in the account of the positive epistemic force in the warrant. A reliably veridical comprehension competence, supported by a natural causal chain, would suffice.[36]

[35] The idea does surface in 'Comprehension and Interpretation', 244; 358 in this volume. I developed the argument in more detail in the Santayana Lectures at Harvard in 1998 and in an earlier version of 'A Warrant for Belief in Other Minds'. But I never had sufficient confidence in the argument to press it in print. The occurrence in 'Comprehension and Interpretation' broached the idea. I hoped to develop it in more depth later, in the essay that is now titled 'A Warrant for Belief in Other Minds'. The work on that essay halted around 1999. As things turned out, I changed my mind about the Injection Argument. The revised version of that essay is printed in this volume, with the Injection Argument deleted. Of course, the issue over the Injection Argument ran parallel to my pulling back on the larger positions that we could have a warrant for comprehending others' utterances, and that we could have a non-empirical warrant for knowledge of the existence of other minds. I cannot remember when I clarified to myself why I should give up these views, though I began to draw back from them over a decade ago. I did not write down the reasons in detail until around 2010, in preparation for this volume.

[36] One could imagine that the causal route does not run through any of the normal sensory channels. Of course, in a broad sense, the recipient's psycho-biological systems must be sensitive to differences in the causal chain that correspond to differences in content. In this broad sense, comprehension would still depend on a "sensory" route. I would have always been content to concede that interlocution must depend for its entitlement on a "sensory" route in this very broad sense. But I was supposing that no ordinary, and certainly no conscious, sensory capacities and no genuine perceptual representations play any role in the process. The interlocutor's intentions are

I was inclined to think that this counterfactual supposition supported the view that even in actual situations, the nature of perceptual content was inessential to the account of a recipient's warrant to rely on comprehension competence, as long as that competence is in fact reliable in providing comprehension of others' presentations-as-true. What seemed to carry the positive warranting force was the reliable causal chain and the conceptual competence for comprehension, not the specific content of the perception.

I was never comfortable with the position that our default entitlements to rely on exercises of conceptual comprehension of another's utterances were sometimes independent of perception for their epistemic force. The view is clearly counter-intuitive. At first, I thought that some illumination might be gained by developing the position in an exploratory spirit. Then when I failed to see what was wrong with it, and when objections repeatedly seemed not to show it wrong (or to boil down to expressing general dissatisfaction with the explanation, largely in view of the counter-intuitive conclusion), I went with it.

In retrospect, I think that the number of relatively new and abstract initiatives that figured in the discussion concealed, at least from me, the force of the ordinary, intuitive view. The following were all relatively new ideas:[37] (i) the separation of epistemic support from epistemic threat, (ii) the idea of entitlement as a type of warrant that is distinct from justification and that can depend entirely on unconscious sub-personal psychological processes, (iii) the complexities of language perception and its differences from ordinary perception, and (iv) the idea of purely preservative memory. They provided resources for deflecting common sense objections that—at least, as I thought—had not yet been couched in sufficiently sophisticated terms to deal with them. So the common sense objections did not influence me as much as they should have.

In any case, I believe that the four ideas that I mentioned earlier that were supposed to support the thesis that we can have non-empirical warrant for beliefs based on interlocution do not collectively suffice to support that thesis. Taking the four ideas in reverse order, I will explain what I think I missed in each case.

The Injection Argument does, I think, show that exercises of comprehension *could* be warranted independently of perception and of sensing. They could depend only on sensitivity to aspects of a causal chain that affects the intellectual faculties directly. The first, and perhaps even only, *representational* competencies that come into play could be intellectual, and warranted only through the reliability and competence of those intellectual capacities. But our actual comprehension competencies do not work in that way. The key point is this: Epistemic norms—like epistemic warrant—are standards for the good use of

"mainlined" into the central comprehension system. For the distinction between perception and various notions of non-perceptual sensory capacities, see *Origins of Objectivity*, chapters 8 and 9.

[37] I was certainly not alone in emphasizing the first and third points. I mean that the four points were relatively new in the sense that they had not been around for long enough (more than a generation) to be well assimilated and reasonably well understood.

psychological competencies. Epistemic norms are necessarily fitted to underlying psychological competencies. Since our psychological competencies do employ perception, since successful interlocution cannot get by without its being at least broadly reliable (see below), the role of perception in successful exercises of comprehension cannot be avoided *just because* other, psychologically possible beings might not use perception in comprehension.[38]

To put it another way, one's right to rely on comprehension of another's utterances as veridical comprehension depends on not just having perceived some utterances, but on perceiving such utterances reliably enough to warrant taking them as expressing a specific content, which one veridically comprehends. The force of one's right to rely on one's comprehension depends on one's perceiving others' utterances well enough to map them onto the representational content that others in fact utter. If one were systematically unreliable in one's perceptions, one would have no warrant to rely on one's seeming comprehension of others' utterances as genuine comprehension.

So I think that the Injection Argument, though instructive, cannot show that we have apriori warrant to rely on comprehension of others' utterances. We lack the psychological competencies that the relevant warrant would attach to. The argument shows at most that it is in principle possible for someone to have such an apriori warrant.

Now to the third idea. Purely preservative memory does show that a psychological competence can participate in a cognitive process and not contribute to entitlement to rely on the process. There are some interesting parallels between the role of purely preservative memory in inference and the role of perception in enabling content to be preserved through communication. But a key difference is that purely preservative memory does not introduce a subject matter for cognition for an epistemic agent, whereas perception does introduce a subject matter—a communicative event. The epistemic role of purely preservative memory is *purely* preservative, purely anaphoric. The epistemic role of perception in communication is partly preservative; but perception also introduces new, epistemically relevant information into the recipient's cognitive system. How well perception contributes information to the formation of comprehension, and ultimately belief, bears on the epistemic force of the warrant for relying on an exercise of comprehension.

[38] The comprehension postulated in the Injection Argument is analogous to the cognition postulated by Shoemaker in his examples of quasi-memory. See Sydney Shoemaker, 'Persons and Their Pasts', *American Philosophical Quarterly* 7 (1970), 269–285; reprinted in his *Identity, Cause, and Mind* (New York: Oxford University Press, 2004). I emphasize that one must understand an individual's psychological competencies in terms of the *actual* natural psychological kinds that those competencies instantiate in 'Memory and Persons', *The Philosophical Review* 112 (2003), 289–337, reprinted in this volume; and in '*De Se* Preservation and Personal Identity: Reply to Shoemaker', this volume. I emphasize the dependence of epistemic norms on such actual competencies in 'Perceptual Entitlement'.

The second idea—the analogy between communication and use of diagrams or symbols in learning mathematics—still seems to me valuable. The function of perception in contributing to cognition, in both cases, is to put the individual in a position to understand a propositional content. The primary source of warrant to believe the proposition lies in understanding it, not in perceiving anything. Perception does have a triggering function in comprehension of others' utterances. However, as noted, to gain knowledge by relying on communication, one must get right something about what is being communicated. By contrast, in the mathematics case, knowledge does not depend on getting right what proposition is uttered, or what diagram is drawn, by the interlocutor. The warrant attaches to one's understanding of whatever proposition is evoked by the external utterance. The proposition need not have any relation at all to what the interlocutor is trying to communicate. In fact, the triggering events need not have any mathematical relevance at all. There need be no reliable understanding of any event outside one's psychology. The warrant for understanding/believing logical and mathematical truths attaches to the individual understander's understanding of the content of his/her own states.

Recall that I appealed to two reasons to try to disarm disanalogies between warrant for comprehending others' utterances and warrant for comprehending mathematical/logical truths. I tried to discount a role for perception in contributing to the warrant for relying on an exercise of comprehension as veridical comprehension. The two reasons are not compelling.

The first reason was that one could be warranted in relying on comprehension—and even gain knowledge from such comprehension—even if one misperceives an utterance. The second was that one can be warranted in relying on comprehension–and even gain knowledge from such reliance—even if one misuses perception in such a way as to undermine warrant (and knowledge) for perceptual belief. Both of the premise points of these reasons are correct. But the reasons do not suffice to support their purported conclusion—that perception does not contribute to the force of the warrant for relying on comprehension or to the knowledge that depends on that warrant.

Both reasons fail to take account of the immense redundancy of empirical information that is available in a communicative situation. One's comprehension can succeed (and be warranted and yield knowledge) despite a misperception (even if the misperception is not brute error—and involves a misuse of perception), because one has enough other sound, warranting perception to enable one to determine what the individual says. For example, one can negligently misperceive what sentence an interlocutor utters and yet warrantedly (and knowledgeably) comprehend what the interlocutor communicates, despite the interlocutor's mis-speaking. But this is possible only because one relies on correct warranting perception of previous utterances; or one uses perceptually based knowledge of the context to grasp what the individual means. One might fail to realize that one has relied on other empirical sources of understanding, even as one incorrectly and negligently misperceived the individual's (mistaken) actual utterance.

If perception of utterances were not reliably accurate for the most part, we could never have a warrant for our comprehension of what others say, or even for their sayings being propositional. The quality of one's comprehension depends systematically on the veridicality of one's perception, even allowing for much misperception. Warrant depends on a good route to veridicality, and a good route to veridical comprehension requires that there be reliable perception of others' expressions of what they say.

As for the first of my four ideas, it is true that language perception is top-heavy with cognitive supplements. Language perception categorizes in ways that do not line up very well with the acoustical signal or with the shape of written symbols. Memory does not retain perceived elements in the physical world, such as the acoustical signal, that are not categorized in linguistically relevant ways. What is remembered and used in language perception is dominantly categorized in terms fitted to higher-level (syntactical and semantical) cognitive categorization.[39] But regardless of how heavily informed language perception is by higher-level cognitive competencies, perception is not only essential to the success of comprehension. Its operating well, at least on some of its perceived subject matter, and providing a base for the sub-personal transitions that lead to comprehension, are also *part of* having a good, reliable route to comprehension of others' propositional presentations-as-true. And as noted with respect to the Injection Argument, epistemic norms are standards for good use of actual, cognitively relevant competencies. Good use of perception is a part of good exercise of comprehension of others' utterances. If perception were not reliable in getting right (enough of) what an interlocutor uttered, we could not be warranted in comprehending the representational content—or even the propositionality—of utterances by others.

I do think that the role of perception in content preservation through communication is interestingly different from its role in forming beliefs about non-linguistic, non-representational aspects of the physical environment. Language perception is nearly entirely geared toward formation of comprehension. What is comprehended is not what is perceived. In language perception, perceiving is very nearly *merely* causally triggered conceiving. Still, it *is* perceiving. Comprehension involves making good transitions from perception of physical events. Perception's going reliably well, in normal circumstances, contributes to the positive force of warrants for relying on occurrent exercises of conceptual comprehension. Its going reliably well on particular occasions contributes to the force of warrants that are necessary to gaining knowledge through comprehension on those occasions. Knowledge that relies on warrant for comprehension—including knowledge that relies essentially on the default prima facie warrant to believe what another says in particular cases—is always empirical, even if sometimes just barely.

[39] See J. A. Fodor, T. G. Bever, and M. F. Garrett, *The Psychology of Language* (New York: McGraw-Hill, 1974; Alvin M. Liberman, *Speech: A Special Code* (Cambridge, Mass.: MIT Press, 1996).

12 *Interlocution, Perception, and Memory*

David Christensen and Hilary Kornblith raise several interesting issues in their paper, 'Testimony, Memory and the Limits of the *A Priori*'. In response I will clarify and develop some lines in 'Content Preservation'.[1]

I argued that in interlocution we have a general apriori prima facie (*pro tanto*) entitlement to rely on seeming understanding as genuine understanding. And we have a general apriori prima facie (*pro tanto*) entitlement to believe putative assertions that we seem to understand. These are two rational default positions. They can be overridden in particular cases if (possibly empirical) counter-considerations arise. But if one lacks counter-considerations, and if one has a minimal level of conceptual know-how, one is entitled to rely on one's seeming understanding of *particular* putative assertions. And one is further entitled to believe seemingly understood putative assertions—because prima facie they come from a rational source whose backing supports belief (assuming no counter-considerations). If the seeming understanding is intellectual, one can, in the relevant particular cases (no counterconsiderations), be apriori entitled to rely on one's seeming understanding of putative assertions as genuine understanding of genuine assertions. And under the same conditions, one can be apriori entitled to accept particular putative assertions as true. An entitlement (whether it is prima facie or not) is apriori if neither sense experience nor perceptual belief constitute or enhance its justificational force.

What is it for understanding, seeming or actual, to be intellectual? The key idea is that it is understanding whose exercise in particular instances does not require in those instances perceptual warrant for the application of what is understood. A first approximation elaboration is that it is *conceptual* understanding that does not require, in thinking and understanding an intentional content, perceptual warrant for the *de re* application of some aspect of the content. So one could have intellectual understanding of empirical concepts (that of a zebra) and empirical propositional contents, as well as mathematical concepts and propositions.

[1] David Christensen and Hilary Kornblith, 'Testimony, Memory and the Limits of the *A Priori*', *Philosophical Studies* 86 (1997), 1–20. Their paper responds to my 'Content Preservation' *The Philosophical Review* 102 (1993), 457–488.

Intellectual understanding may require *having had* empirical experiences. They are part of learning *how* to use intellectual capacities. With empirical concepts like *red*, intellectual understanding requires being able to apply the concept with perceptual warrant to instances. But one need not perceive instances to exercise such conceptual capacities in particular cases, once one knows how to use them. So one's understanding of what is expressed by an occurrence of "Zebras are larger than red poppies" is intellectual. Thinking the intentional content with understanding does not require on particular occasions an empirically warranted *de re* application of an element in the content. By contrast, the intentional content of "That zebra is running for its life", as applied directly and non-anaphorically to a zebra, cannot be understood just intellectually.[2]

On my view, one's understanding assertive mode is commonly intellectual.[3] One's understanding of asserted content is sometimes at the same time also intellectual. So one's seeming or presumptive understanding of putative assertions is sometimes intellectual.

[2] Thus, given the views I outlined three paragraphs back, I maintain that in interlocution one can on occasion be apriori entitled to rely upon one's intellectual, seeming understanding of (and even apriori entitled to believe) propositions which can only be *known* empirically. The default entitlement to believe propositions one receives in interlocution presumes a more primary epistemic warrant somewhere in the chain of interlocutors. An apriori entitlement carried by the recipient of communication frequently relies upon necessarily empirical primary warrants that others have. I presume that a primary epistemic warrant for belief in the general proposition about zebras can only be empirical. Inasmuch as knowledge gained through interlocution depends on there being primary epistemic warrants that others have, the recipient's knowledge will be empirical even though his own individual entitlement to belief may be apriori—in the sense of "apriori" defined in 'Content Preservation'. Some of the matters just glossed are discussed in more detail in that article.

[3] One of the ways in which the elaborated characterization of intellectual understanding is a first approximation is that it does not elaborate what it means to understand assertive mode intellectually. Getting this right is complex, and not something I will attempt here. The initial explication of "intellectual understanding" will have to suffice in that case, for the present. The assertive mode of an assertion is not part of its content, so the approximate characterization in terms of application of elements of the content is not directly relevant. But the main difficulty is that understanding assertive mode, though commonly what I want to count "intellectual" in the rough intuitive sense I began with, is not entirely conceptual. (Cf. note 12 for an analogous source of complexification in the account of intellectual understanding, even of understanding of intentional content.) So my account of intellectual understanding in terms of conceptual understanding is an oversimplification from the beginning. The claim that understanding assertive mode is commonly intellectual thus invites further elaboration. The situation may appear to be even worse for me, in that the claim is surely one that Christensen and Kornblith would dispute, on any elaboration of it that would serve my purposes. So clarification is needed at this central point. (The approximation that I have given in the text will suffice for the content of empirical generalizations, such as *zebras are larger than poppies*. My claim that understanding in those cases can be intellectual, in the sense given in the text, should be uncontroversial. The controversy in that case centers on whether we can be apriori entitled to rely upon seeming understanding of what an utterance's content is, granted that the understanding is intellectual.) I believe, however, that most of what I will say in favor of the apriority of our entitlement to rely upon our seeming understanding of particular putative assertions is independent of exactly how I characterize intellectual understanding of assertive mode. For I think that the most salient epistemic issues arise about understanding *expressive events*, or uttered presentations of content as opposed to abstract content, regardless of whether the mode or the content of the uttered expressions is at issue. I will, however, have more to say about understanding assertive mode.

A seeming understanding's being intellectual is necessary but not sufficient for one's being apriori entitled to rely upon that understanding. It is necessary because if perceptual warrant is implicated in the understanding, the understanding itself is partly empirically warranted. It is not sufficient because if reasonable empirical doubts arise about one's intellectual seeming-understanding, one's entitlement to rely upon it may require empirical supplement.

The two apriori prima facie entitlements I mentioned at the outset are rational starting points. They function for children at early stages of linguistic competence. They are salient in adults who lack information about their interlocutors, on topics they have no reason to think are problematic. To be entitled to accept what they are told, children or adults need not know that there are no counter-considerations.[4] Children may not conceive of counter-considerations as possible. It is normally enough that there be no available counter reasons. In such cases, they may be apriori entitled to accept what they are told.

I do not hold that we adults are in a high proportion of cases apriori entitled, all things considered, to particular beliefs acquired in interlocution. Perceptual elements are very frequently partly constitutive of our understanding. So understanding is often not purely intellectual. And our apriori prima facie entitlement to accept what we are told commonly needs empirical supplementation to override counter-considerations. But we are sometimes so entitled. The nature of the prima facie entitlements is what interests me.

I

Perception enters into the acquisition of beliefs through interlocution in several ways. Christensen and Kornblith concentrate on the role of perception in

[4] John Biro, in 'Testimony and Apriori Knowledge', *Philosophical Issues* 6 (1995), 301–310, argues that particular applications of the general entitlement to rely upon one's understanding cannot be apriori because one cannot know apriori that there are no (possibly empirical) counter-considerations to one's general entitlement. This argument rests on a misunderstanding of my Acceptance Principle in 'Content Preservation'. The Acceptance Principle is: *A person is [apriori] entitled to accept a proposition that is [taken to be] presented as true and that is [seemingly] intelligible to him, unless there are stronger reasons not to do so.* (I have entered the bracketed clarifications. All are explicit in my original explanations of the principle, though I failed to make them explicit in the statement of the principle itself. None of them is relevant to Biro's argument.) The force of 'unless there are stronger reasons not to do so' is to indicate that the person's entitlement is prima facie. The principle says that the entitlement holds unless there are stronger reasons (available to the person) that override it. It does not say that the person must know there are no stronger reasons; the individual may lack the concept of a stronger reason, and in any case need not rule out the existence of defeaters in advance. It is enough for the individual's being warranted that there *are* no defeaters; defeaters of the entitlement must be available to him. As Glenn Branch pointed out to me, an individual with the right conceptual abilities *might* know through non-empirical self-knowledge that there are no relevant counter-considerations. On the other hand, it might be that some candidate counter-considerations could be known not to be stronger only through empirical considerations. But the main point is that the individual does not have to *know* that there are no stronger reasons counting against the apriori entitlement in order for the apriori prima facie entitlement to entitle him to belief. It is enough that there *be* none that are available to him.

understanding assertions. I acknowledged that perception is necessary for the acquisition of belief through interlocution. But I maintained that with respect to the default entitlements that we have in interlocution, and some of their applications in particular cases, perception functions only as a causal enabling condition, not as a contributor to the justificational force of our entitlement. I appealed to two analogies.

One is an analogy with the traditional rationalist view of the role of perception of symbols and diagrams in triggering understanding of (and belief in) a mathematical truth. On this view (with which I sympathize), perception is necessary to being justified in believing the theorem. But it need not contribute to the justificational force of one's justification. Even though perception is necessary to obtain such understanding, the justification flows from intellectual sources, from one's understanding and reason, not from perception.

The other analogy is with the role of memory in preserving thoughts over time in a deductive proof. One must rely on memory of earlier steps, and their assertive mode, to carry out the proof. Memory is necessary for reasoning in time. But appeals to memory are not normally part of our justification for believing the conclusion. The justification is the argument itself. *We* are justified by our having thought it through and understood it. But memory, though essential to our being justified, does not contribute to the epistemic force of our reasoning; it simply preserves it. I claimed that in communication, perception plays a connecting and preservative role analogous to the role memory plays in deductive reasoning. Perception is necessary to understanding—even to being entitled to rely upon one's understanding and to being entitled to one's belief in what the interlocutor says. But it need not enhance the justificational force of one's basic prima facie entitlements.[5]

Christensen and Kornblith target the first analogy. They emphasize a certain disanalogy between the case in which perception is used to trigger understanding of a mathematical truth and the case in which perception is used to understand a piece of interlocution. They point out that in the mathematical case, the understanding that the perception triggers is an understanding of propositional content. The justification flows from understanding the content alone. In interlocution, entitlement derives from understanding a putative assertion. They hold that the

[5] Although Christensen and Kornblith acknowledge this distinction between genetic and epistemic roles for a cognitive capacity, they often state the dispute in ways that ignore it. For example, they write, 'The root intuition we are trying to capture [in deciding whether something is apriori] involves a distinction between learning about the world through our senses and learning about the world through thought' (p. 9). On my view, and most rationalist views, one's learning about the world through the senses, as a genetic matter, need not compromise the apriority of a justification, as long as the senses play a merely enabling role rather than a justificatory one. Similarly, they write, apparently in criticism of my view, 'Propositional content does not pass directly from one mind to another; rather the passage is mediated by perception of an utterance . . . ' (p. 12). As a description of the causal process there is of course no disagreement here. My metaphor of passage of propositional content from one mind to another concerned the account of justification, and explicitly acknowledged the essential role of perception in mediating communication.

role of perception in understanding a putative assertion is inevitably justificatory and empirical.

The disanalogy can be sharpened. The classical account of understanding mathematical truths is not an account of one's being justified in understanding the *presentation* of content. It is not central to the account whether the content understood is the content presented. What matters is that some appropriately understood mathematical truth is triggered and understood, regardless of its provenance. In interlocution, the understanding that perception triggers and to which the entitlement attaches is an understanding of an event, a presentation of content, not merely understanding of a content abstracted from any instantiations. We are further entitled to rely on understanding of the mode of the presentation—centrally an understanding of the event as an assertion. The second prima facie entitlement—to believe what one seemingly understands—derives from understanding an event as an assertion. At issue is understanding both the presentation of content and the presentation's mode.

A further disanalogy is that the entitlements in interlocution derive from seeming understanding of the putative assertion. They do not require genuine understanding of a genuine assertion. In the classical account, the justification derives from genuine understanding of an abstract content.

As they acknowledge, I noted the disanalogy that they cite too.[6] But I maintained that a prima facie entitlement to believe a putative assertion still rests on putative understanding alone. I took the disanalogy to force the epistemic warrants in interlocution to be prima facie entitlements rather than decisive justifications. I held that it did not prevent them from being apriori. I maintained that putative understanding alone yields an apriori prima facie connection to a rational source that is, other things equal, to be believed.[7]

In elaborating this view I want first to discuss understanding of assertive mode of a presentation event. Then I turn to understanding the content of a presentation event (cf. note 3). Why are we entitled to rely upon seeming understanding of presentations of content as *asserted*? I claimed that understanding the mode of instantiations of content is a corollary of, indeed a necessary condition for, understanding instantiations (including presentations) of content. So entitlement

[6] Compare 'Content Preservation', p. 480, note 19; 246 n19 in this volume. There are also, of course, differences in what is required for knowledge. Although one has a prima facie entitlement in interlocution to belief based on seeming understanding of a putative assertion (presuming that one's seeming understanding involves a genuine content that one is thinking), one has knowledge only if one does understand a genuine assertion. In the mathematical case, one can have knowledge as well as justification regardless of whether one has understood an assertion, or indeed any communication, as long as one genuinely understands a relevantly simple logical or mathematical truth.

[7] The argument for the apriori connections between seeming understanding of putative assertion and prima facie rationality of the source, and between such rationality and the truth of the putative assertion, is given in 'Content Preservation'. There is much to be said about this argument. But Christensen and Kornblith do not discuss it. They concentrate on my claim that we have apriori prima facie entitlement to rely upon seeming understanding of an assertion. So I will not defend the larger argument here.

to rely upon seeming understanding of instantiations of content requires entitlement to rely upon seeming understanding of presentations-as-true, including assertions.

Here is a fuller explanation of this point. Assume for the sake of argument that we are apriori prima facie entitled to rely upon our intellectual, seeming understanding of instantiations (including presentations) of content as genuine understanding. It is necessary and apriori knowable that the entitlement for understanding instantiations of content gets its epistemic force partly from the presumption that we normally understand the mode of instantiations. The central sort of mode for these purposes is presentation-as-true, which includes assertion. Epistemic support for understanding presented content must normally include epistemic support for understanding presentations as assertions. So one is apriori prima facie entitled to rely upon one's seeming understanding of putative assertions as genuine understanding of genuine assertions.

This point then connects with the rest of the argument: there is a necessary and apriori knowable prima facie relation between a content's being asserted and its having rational backing, and between its having rational backing and being true. So assuming that one's seeming understanding is intellectual, one's seeming understanding of putative assertions provides one with an apriori prima facie entitlement to accept such contents as true.

The foregoing concentrated on the nature of our entitlement to rely on seeming understanding of assertive mode. It assumed for the sake of argument that we could be entitled apriori to rely upon seeming understanding of presentations or particular expressions of intentional content. We turn now to reflect on this assumption.

When we understand events that are expressions of content in interlocution, we must do this through perception. I think, however, that the justificational force of the relevant prima facie epistemic entitlement to rely upon seeming understanding of the content of presentations as genuine understanding, derives from the seeming understanding alone. Perception or perceptual belief can provide supplemental justificational force for one's presumption of understanding. But it need not contribute to the force of the epistemic warrant associated with the basic presumption. Given that one *has* the ability to understand, one is entitled to rely upon applications of seeming-understanding, including intellectual applications, *per se*. The prima facie entitlement is—in intellectual, undefeated applications—apriori.

How do Christensen and Kornblith argue that perception must contribute to the force of one's entitlement to rely upon seeming understanding of putative assertions? They give two arguments.

One continues the contrast with the mathematical case. If in understanding a mathematical truth, one were shown that one had hallucinated the symbols that called it to mind, one's justification for belief would not be undermined. But if one were given sufficient reason to believe that one had hallucinated one's interlocutor's telling one something, one's entitlement to believe what one had putatively been told would be undermined. They conclude that in the former case, perception plays (or may play) a triggering role, whereas in the latter, perception plays (or must play) a justificatory one.

The conclusion of the argument does not follow from the premises. In fact, the argument has no force at all. I grant the premises, but deny the conclusion. I maintain that one's entitlement derives from seeming understanding of a putative assertion (or more broadly, presentation-as-true). Other things equal, one has an epistemic entitlement to rely upon one's putative understanding of an asserted content, as genuine understanding. If one is given reason to believe that one has not understood any such assertive occurrence, that reason can override one's prima facie entitlement to rely upon one's seeming understanding (and this in turn would override one's prima facie entitlement to believe the putative assertion). If one were shown that one had hallucinated, that would show that one had not understood an assertion. But it would not show that the initial entitlement gets its force from perception in addition to understanding—only that understanding depends on perception, which I already maintain.

Moreover, I emphasized in 'Content Preservation' that reason or evidence that overrides one's apriori prima facie entitlement can be empirical. The fact that overriding considerations are empirical does not show that the entitlement to one's presumption of understanding, or to one's belief, is empirical. The nature of the initial positive justificational force is one matter. The nature of overriding considerations is another. Perception is a necessary condition for understanding an assertion. So evidence that one has not perceived anything will be evidence that one has not understood an assertion. Having such evidence can override one's default entitlements. But this fact has no tendency to show that the entitlements do not derive their justificational force purely from understanding, but must draw justificational force from perception as well. It shows only that the entitlements are empirically defeasible, which I already maintain.

Evidence that one has hallucinated is not the only evidence about perception failures that can override one's apriori prima facie (*pro tanto*) entitlement to rely upon one's seeming understanding. Suppose that one is subject to brute perceptual error that leads one seemingly to understand an assertion, when no assertion has occurred. Then one could later obtain empirical evidence that shows one had misperceived. Such evidence could override one's entitlement. Or suppose that one's understanding is veridical—that is, there is an assertion that one does understand; but suppose that one has non-veridical but excellent empirical evidence that one has misperceived. Again one's entitlement is in principle overridable—since the prima facie entitlement rests on the seeming-intelligibility of a putative assertion, not on actual understanding of an actual assertion. These cases do not show that the relevant entitlements in interlocution are empirical. They show only, what I already maintain, that they are prima facie, and subject to empirical overriders.

We are not primarily concerned with whether or how the relevant entitlements are overridable. We are concerned with the nature of the justificational force underlying those entitlements. I think that seeming understanding of a putative assertion can itself provide prima facie justificational force. One can override such force with an empirical reason to believe that one has not actually understood an assertion. But reliance on one's understanding is a rational starting point that need not be epistemologically grounded in anything about the perceptual conditions that make understanding possible.

Christensen and Kornblith's first argument does not touch this position. It begs the question by assuming that if empirical evidence can override a justification or entitlement, the justification or entitlement is itself empirical.

The role that perception plays in enabling understanding to take place makes it hard to separate out the distinctive epistemic entitlement associated with understanding. Let us consider hallucination again, going further in raising threats to my position. Not only can *evidence* of hallucination outweigh our prima facie entitlement to rely upon our understanding. But—whether or not we have evidence that we have hallucinated—actual hallucination can in particular cases undermine (in the sense of prevent our having) that entitlement. For the entitlement presupposes the actual normal functioning of faculties on which it depends. If the perceptual faculty malfunctions, seeming-understanding, which relies upon perception, can fail to carry epistemic warrant. This point might seem to show that the entitlement to rely upon seeming understanding is empirical. But it does not show this. There remains the question of whether perception contributes essentially to the epistemic force of the entitlement, as opposed to merely being a necessary condition on which the entitlement depends.

We have an entitlement to rely (at least prima facie) upon our intuitive mathematical judgments of validity, or of self-evidence, or of plausibility (given sufficient mathematical understanding and expertise), or upon our deductive reasoning. But our intuitive judgments depend on the proper functioning of our brains; and our deductive reasoning depends as well (in a different way) on the proper functioning of preservative memory. If brain-functioning or memory-functioning were to fail, our intuitions or reasoning could seem compelling yet fail to carry any warrant. This fact hardly shows that the epistemic entitlement to rely upon mathematical judgment or deductive justifications, gets part of its justificational force from something about our brains or memories.

Are there cases where one is entitled to rely on one's seeming understanding even where one is not entitled to rely upon the perceptions that purport to put one on to what one's interlocutor uttered? Prima facie entitlement to rely upon seeming understanding can certainly survive brute perceptual errors. But brute perceptual errors are those that occur despite one's being entitled to rely upon one's perception. Most failures of perceptual entitlement that are not matters of malfunctioning (as hallucination is) involve some conceptual interference or prejudice. But these failures are rational failures and would involve misuse or malfunctioning of one's rational powers. These would tend to undermine one's entitlement to seeming-understanding as well as one's entitlement to one's perception.

I think, however, that one can sometimes be entitled to rely on one's seeming-understanding of another's discourse even though one has been negligent (even negligently mistaken) in one's perception of what the other person uttered. Let us suppose that the negligence undermines the perceptual entitlement. For example, suppose that one's interlocutor mis-speaks and one fails to notice the tongue slip by being perceptually careless. Suppose, that one's understanding of what the other person said is governed by considerations of intellectual plausibility (imagine it to be a simple mathematical truth); so that one's understanding of the person's meaning is warranted and veridical. Then one could be entitled to rely upon one's seeming understanding of what the other person said, even though one might lack entitlement to, and be mistaken about, one's perception of what the other person uttered. (It does not matter whether what the person means is what the person "said". Perhaps there is a sense in which it is; perhaps not. What matters is that one is prima facie entitled to one's seeming understanding of what the person said—even though one lacks perceptual entitlement.) This point suggests, though it does not entail, that one's reliance on one's understanding—in particular cases and in general—can be a source of epistemic warrant, independent of perceptual contribution to the warrant's force.[8]

But even if one's entitlement to rely upon one's seeming understanding always lapsed when one lacks entitlement to rely upon one's (apparent) perception of what the other person uttered, it would not follow that one's entitlement to rely upon one's seeming understanding is empirical. Take any piece of reasoning in which one relies upon preservative memory but (say, because it malfunctions) in which one is not entitled to do so. Such a piece of reasoning will fail to warrant acceptance of its conclusion. But preservative memory does not contribute to the justificational force of reasoning. It makes reasoning possible without contributing to its force. I think that the perception normally involved in understanding putative assertions plays a similar role in the epistemology of interlocution.

Perception is one source of epistemic warrant. Understanding is another. We are all familiar with epistemologies that take (implicit or explicit) inference from perception to be at the root of understanding others' speech. Such epistemologies do not seem to me plausible. They make understanding in communication a matter of interpretation or translation, when it seems, in normal cases, to be epistemically immediate, once the capacity for understanding is in place. Inference from perception in normal interlocution seems more characteristic of cases

[8] It does not entail it, of course, because one might claim that the entitlement to rely on the understanding derives from other previous perceptual entitlements. I do not think that this is so. I do not think that any reasonable inductive or perceptual inference can explain our entitlement to rely upon our seeming understanding of strangers in unusual contexts. And I do not think that perceptual experience that precedes acquiring understanding plays an indispensable role in justifying applications of our intellectual faculties. But the point in the text is not meant to settle the issue about the nature of our entitlement to rely upon understanding. It is just to explore the relation between understanding and perception, showing that our entitlement to rely upon understanding can be independent of particular perceptual entitlements in the context of that understanding.

when there is some breakdown or failure of communication, or some special, non-normal use of language.

One might claim that understanding just seems to be epistemically immediate. The inference from perceptual belief is unconscious and implicit. But this appears to me to be analogous to views that build the epistemology of perception on implicit inferences from sense data or from registrations of stimulation of the retina. The position seems to confuse psychology with epistemology.

There must always be a psychological story in individual cases that explains a route from sounds and shapes through perception to understanding. Perhaps such a story will enter into some science. But in interlocution understanding seems normally to be epistemically immediate, a defeasible, epistemically warranted starting point. Our reliance on our understanding is in itself no less epistemically fundamental than our reliance on our perception. Other things equal, our seeming-understanding needs no justification by reference to perception, once perceptual and conceptual know-how are in place. It can be given supplementary support by appeal to perception, rather as perceptual belief can be given supplementary support by appeal to beliefs about appearings or perceptual sensations. But the primary entitlement to rely upon understanding stands on its own.

Strictly speaking, we do not perceive the assertive mode, or the conceptual content, of utterances. We understand them. These are exercises of intellectual capacities. We understand events as assertions by perceiving other aspects of assertions. We understand the concepts in assertions, by perceiving expressions of them. But here perception is part of the condition for exercising the intellectual capacity, not—or not normally—part of the warrant for the individual's relying on his understanding. It is a necessary triggering mechanism, but it is not the understanding itself. In the order of epistemic warrant, seeming understanding is a rational starting point. Insofar as seeming understanding is normally an epistemic starting point with its own justificational force—not the *epistemic* product of an inference (explicit or implicit)—and insofar as such understanding is sometimes purely intellectual, it must carry an apriori entitlement (cf. note 7).

Intellectual seeming-understanding is subject to empirical correction. But we can sometimes be warranted in relying upon our conceptual know-how in understanding assertions, without the warrant's being bolstered by perception or perceptual belief. It is enough that one have a capacity to understand—have some degree of reliability in understanding—and know how to use the capacity. I think that being reliable in normal circumstances can be shown to be necessary to a capacity for seeming-understanding, in something like the way that a capacity to register perceptual appearances can be shown to be necessarily reliable in normal circumstances (in order to count as a capacity for perceptual appearances). I believe, however, that I have not gotten to the bottom of this matter. I hope to have suggested some grounds for thinking my view worth exploring further.

II

Christensen and Kornblith's second argument consists in a challenge to distinguish the entitlement involved in interlocution, which I claim to be apriori, from entitlements involved in certain clear cases of empirical justifications or entitlements. They compare interlocution to our acquiring beliefs about the time from looking at a clock. The clock example is representative of a range of examples of belief acquisition from observing scientific instruments, especially instruments with conventionally established, linguistic ways of reading their indications. Beliefs acquired from these instruments are normally empirically justified, or carry empirical entitlements. The challenge is to say what differentiates entitlements to the relevant beliefs derived from interlocution.

Here I think that Christensen and Kornblith raise an interesting issue. I had deleted a section of a draft of 'Content Preservation' which sought to say something about it—postponing the attempt. I hope to make some progress here.

I agree, of course, that the conventional mode of expression on a clock's dial, which introduces a rational element into the causal chain relating our beliefs to the time, does not by itself give one a non-empirical entitlement to beliefs acquired from looking at the clock. I also agree that whether one makes subconscious inferences from a perceptual belief about the clock's face, or simply forms a non-inferential perceptual belief about the time from looking at the clock's face, is not an issue that is crucial to whether one's justification or entitlement is empirical.

The issue has to do with the nature of the understanding in the two cases (interlocutor and clock) and with the nature of the connection between understanding and subject matter. I did not deal with the closely related matter of the relation between perceptual entitlement and interlocutional entitlement in a satisfactory way in 'Content Preservation'. But I think that I was nevertheless on to an important difference in the passage my critics quote:

> The relation between words and their subject matter and content [in interlocution] is not an *ordinary*, natural, lawlike causal-explanatory relation. Crudely speaking, it involves a mind.[9]

Their discussion of this passage gets off track. I do not hold that minds are 'set apart from the same causal order that encompasses clocks'. My emphasis on "ordinary" was meant to indicate that mind-involving relations may be special cases of lawlike, causal-explanatory relations. I took no position on this. Moreover, this challenge is fruitless: 'If minds are part of the causal order, we have been given no reason for thinking that a mind-mediated connection cannot be ordinary lawlike or causal-explanatory.' I took it that in involving a mind, the connection is *ipso facto* not "ordinary". My concern was not to suggest that the connections are not lawlike or causal-explanatory, but to suggest that they are sometimes not

[9] 'Content Preservation', 479; 245–246 in this volume.

empirical connections, from the point of view of understanding epistemic warrant, because of the role of mind in the connection.

I thought that the role of mind in the relation between words, their subject matter, and their intentional content is relevant this way: it makes possible an apriori, conceptual, connection between seeming understanding of a putative assertion and being rationally entitled to form beliefs. A seemingly understood putative assertion is a prima facie sign of rational backing for the assertion and rational commitment to truth; and rational backing together with rational commitment constitute a prima facie sign of truth. Even if this epistemic connection must be associated with a lawlike, causal explanatory relation (one that relates one's understanding through perception to utterances and to a subject matter), the epistemic account of interlocution will be different from that of the epistemic connection between belief, clock dial, and subject matter—which is of course itself associated with a lawlike, causal explanatory relation. In the latter case, the reliability of the causal relation carries most of the epistemology, without reference to reason.

It is true that minds figured in the relation between one's beliefs about the time and the subject matter of the clock dial. Minds gave the clock dial meaning—and played a role in making the clock. In the case of more sophisticated scientific instruments, understanding the empirical theory behind them is needed to understand their outputs. But these roles are different in epistemically relevant ways from the role of minds in interlocution. Let me develop these metaphorical suggestions in a more concrete way.

Reading a clock or other instrument differs from understanding interlocution in two ways. What one understands is different. And the relation between what one understands and the way the instrument or interlocutor relate to the subject matter is different.

In reading a clock, one does not understand what one reads as an assertion or presentation-as-true. There is no intentional intellectual act on the part of the instrument. More fundamentally, we do not rely on rationality in the instrument. The instrument is not understood as carrying through reasoning about a subject matter. (An appropriately complex talking robot might be different.)

Although the presence of linguistically intelligible signs on the instrument would bespeak the presence of rationality in the chain linking one with the subject matter, we understand that the clock has been set up to provide merely a natural sign of, or natural meaning in Grice's sense about, the subject matter. More precisely, we do not—and are not entitled to—understand its outputs as intelligible acts of assertion, or as any other expression of reason.

The entitlement in interlocution rests on a prima facie conceptual relation between assertions, reason, and truth about a subject matter. One does not understand clocks as asserting anything or as otherwise expressing rational processes. And the chain relating clocks' outputs to their subject matter does not support a conceptual relation featuring only understanding and reason. The relation is not guided by reason. Thus putative understanding of the clock's output does not entitle us to rely upon apriori prima facie conceptual connections between understanding, reason, and truth (connections which we need not have mastered to have the entitlement, but which can be understood apriori). Our understanding of a clock's output does not entitle us to rely prima facie on the source's being rational, as our understanding of an interlocutor does.

In learning to understand scientific instruments, we understand these matters—or at least lack an understanding of instruments that would allow us to rely on them as expressions of reasoning. Simple instruments are mere amplifications of our perceptual powers. More sophisticated empirical instruments may incorporate checking devices. But even in these cases, with possible exceptions like the one already noted, we do not rely on their carrying out rational processes and being committed to truth. We rely on them as non-rational reliable indicators. Our warrant for relying on them is empirical.[10]

Of course, we can often infer from the fact that the instrument is an artifact that reason went into its construction. We can infer that reason was involved in setting up the non-rational mechanism for providing reliable indications about the subject matter. But this is an inference from empirically justified belief about the artifact as object. One's warrant does not derive purely from seeming understanding of putative expressions of reason. What is operating in one's justification is a substantive thesis *about* the thing's being an artifact, not a preservation of instances of reasoning. The instrument itself does no such reasoning, nor does it express any. The meta-reasoning about the artifact introduces empirical elements that simple preservation through understanding does not.

In understanding others we are in a position to rely on the fact that seemingly understood putative assertions are a prima facie sign of rationality or reasoning. Insofar as we are relying alone on the rationality of our source, our entitlement is apriori.

Our not understanding empirical instruments as we would rational beings is probably the most central reason why our warrant for relying upon them is not apriori. But there are other considerations. I am not in a position to develop these in detail because there is much I do not understand. But I want to broach one of them.

Even in understanding human interlocutors, our entitlement to accept what they tell us often fails to be apriori because our understanding is not purely intellectual. Often understanding what another person says involves seeing what the person is pointing to. In such cases, our warrant for believing what the interlocutor asserts depends partly on warrants backing our perception.

It is not clear to me how far this empirical *de re* element in understanding extends. But one sort of extension is to a case like this: The interlocutor provides

[10] As empirical scientific instruments approach the sophistication of androids in the way they express and process information, they become more nearly like interlocutors. Whether there is a *sharp* line between a rational source and a sophisticated empirical instrument (with lots of the functions of reason) is a deep question in the philosophy of mind.

a context-dependent indicator of a particular object, where the indicator is a component in what he tells us. Suppose that the indicator is understood to indicate and track some individual or event that the interlocutor can see, but we cannot. In at least some cases of this sort, we seem to be engaging in deferred ostension: we see the indicator and take it as our means of "demonstrating" an individual or event that we cannot perceive. We seem to use the indicator as an amplification of our perceptual powers, since we understand it to be connected to the referent that it indicates through deferred perceptual means.

The indicator is not an expression of a general descriptive concept of the individual. It tracks the presence, absence, or changes of the indicated individual or event. Perhaps it changes in some conventionally understood way in accord with how the object changes. The indicator contributes a perceptually backed *de re* element to our understanding of what the interlocutor tells us about the individual. And our epistemic warrant for understanding the *de re* element is empirical. Our understanding is not purely intellectual; and our entitlement to accept what we are told is not apriori.

Now some such *de re* empirical element seems to be part of much understanding of scientific instruments. Normally we understand the instruments' readings as tracking, in a context-dependent manner, particular state or property instantiations, or aspects of particular events or individuals. The relation of the instruments' indicators to the particulars that they indicate often seems broadly analogous to a deferred demonstrative perceptual relation.[11] The instruments are amplifications of our perceptual powers in relating us *de re* to particulars. I do not claim that all instruments for yielding information about the empirical world provide us with an empirically backed *de re* relation to particulars. But I think that most do. The instruments' role is normally to help track individual events or property instantiations by means that are not purely conceptual. In most cases understanding contents that include such tracking seem to require empirical warrant through a kind of deferred empirically backed ostension.[12] But

[11] This is, I think, one important difference between computers that we rely on to solve mathematical problems and most scientific instruments. I argue elsewhere—in 'Computer Proof, Apriori Knowledge, and Other Minds', *Philosophical Perspectives* 12 (1998), 1–37—that we can obtain apriori mathematical knowledge through computer-aided proofs. It is not just that the computers rely on rational processes, and that we can be entitled to understand this by understanding their output. It is that the computers do not put us into *de re* empirical relations to a subject matter. Our understanding of their output is purely intellectual.

[12] I think that all *de re* relations to times or particulars in time involve non-conceptual elements in the mode of reference. Compare my 'Belief *De Re*', *The Journal of Philosophy* 74 (1977), 338–363. But I do not think that *understanding* contents that involve such *de re* elements always involves *empirical* warrants, or perceptual application. I believe that certain *de re* contents (for example, those involving *I* and certain uses of *now* and perhaps *here*) can be understood without being backed by a warrant whose epistemic force derives (even partly) from sense experience. One can understand such *de re* contents through mastering a conceptual rule, and applying the rule with such understanding in a context, where the application is not guided by the senses, only by thought. In effect, one allows the context to provide a *re* given the intellectual application of the rule. The application need not be guided by perception. Such understanding seems to me to be intellectual in a way that would not preclude an apriori justification or entitlement. So it appears that applicational elements in thought can

interlocutors can provide us with non-empirical information; and they can provide empirical information in purely conceptual form. So our understanding of content in interlocution can be intellectual in a way that allows our entitlement to the understanding—and beliefs based upon it—to be apriori.

How deep a role empirically backed *de re* understanding plays in the empiricality of our warrants for relying on scientific instruments is a matter that I leave open. I hope that I have broached the subject in a way that will prove fruitful.

III

The last part of Christensen and Kornblith's paper deals with purely preservative memory, which was the key to the second analogy I offered—the analogy between perception in interlocution and preservative memory in deductive argument. They raise objections to my discussion of preservative memory. They begin by misdescribing my views about it. Perhaps my exposition was at fault. But I do not invoke 'the distinction between explicit invocation of memory as a premise and implicit reliance on memory to divide the empirical from the apriori' (p. 12). Preservative and substantive memory are distinguished by their function, not their degree of explicitness; and they do not mark the divide between the apriori and the empirical.

Substantive memory frequently functions even if it is not invoked explicitly. And preservative memory can be conscious as well as unconscious. Moreover, substantive memory can participate in an apriori justification, if its object is a past intellectual event (say if one remembers thinking the *cogito*). And preservative memory can, frequently does, preserve empirical thoughts and warrants.

The difference has to do with the function of the two sorts of memory. Preservative memory preserves thoughts and their assertive mode, and does not contribute new elements in a justification, or add to justificational force. Substantive memory refers to events or objects and provides elements in a justification, whether or not the justification is explicit (or conscious).

I have shown, I think, that preservative memory—memory that preserves beliefs with their justifications, but contributes no independent source of justification—is epistemically necessary if we are to understand any argument as justifying beliefs through the steps of the argument. I think it unquestionable that preservative memory *normally* preserves not only beliefs but warrants (at least as entitlements, often as justifications) for the beliefs it preserves. This can

be intellectual, without themselves being wholly conceptual. I think that understanding instances of the *cogito* is intellectual but not purely conceptual. The intentional content of such instances involves *de re* non-conceptual elements of application, but the understanding of the instances is not normally guided by sense experience. There are analogs in understanding *de re* elements (e.g., tensed elements) in thoughts about physical objects. These matters obviously complicate the relation between *de re* contents and the nature of one's epistemic entitlement to rely on one's understanding of such contents.

be argued on apriori grounds, assuming only that we reason in time and are sometimes entitled to beliefs that we argue for.

Christensen and Kornblith provide two reasons 'for doubting that the distinction between substantive and purely preservative memory can mark the border of the *a priori*' (p. 13). Since I do not mark the border this way, we will have to consider these reasons carefully.

The first reason rests on the example of someone's remembering a theorem, but not remembering how he or she acquired the belief. Christensen and Kornblith appear to assume that on my view, as long as she does not invoke memory explicitly, her justification is apriori. They cite against this the possibility that she learned the theorem in an a posteriori way. But the objection rests on a mistake about my view. If she relies upon preservative memory (consciously or not) and that memory preserves in an appropriate way an acquisition that rests on an apriori entitlement or justification, then her current belief (or its warrant) is apriori.[13] If preservative memory appropriately preserves an empirically entitled belief, then her current belief is empirical. It does not matter whether she remembers how she learned the theorem. What matters is how she in fact learned it and whether her memory preserves those beliefs from those acquisitions in an appropriate way. So their first objection is irrelevant.

They produce a supplementary objection to a position that is closer to mine. They consider the view that whether a justification of a memory-based belief is apriori or not depends on historical factors, such as how the person acquired the belief in the first place. They object to this view because they think

> it entails that two people could remember the same fact—indeed, they could share all the same present beliefs, reasons, cognitive abilities, and memories—and one of them be justified *a priori* and the other justified *a posteriori*, because one of them had done the proof 20 years ago, while the other had taken a student's word for it after self-consciously considering the possibility that he was lying [where each has forgotten the original grounds for the belief].

If a person remembers the theorem she has proved, but has no access to the proof, then the person is not *justified* (has no justification, in my sense of justification) for believing the theorem. In the case described, neither person has a *justification* for belief in the theorem. But what are we to say about their *entitlements*? Insofar as they are entitled to believe the theorem, their entitlements are surely different.

Although nothing in 'Content Preservation' commits me on this matter, I believe that a person clearly *can be* entitled to believe a theorem she believes because of preservative memory even if she cannot remember the proof she gave long ago, and even if she cannot remember that she gave a proof. Most of what

[13] I will return to "appropriate way". Preservative memory does not add to one's justification or entitlement. But there may be conditions on preservation of justification or entitlement that go beyond merely preservation of the belief. I am sceptical about such conditions' looming large, but I leave room for their possibility.

one is entitled to believe from past reading, past interlocution, past reasoning, or past empirical learning, derives from sources and warrants that one has forgotten.

Why is the person entitled to believe the theorem when she has forgotten the proof? By hypothesis, the person has a properly functioning preservative memory of the belief; and the belief derives from the acquisition or reinforcement of the belief through proof. Preservative memory does not contribute to the force of the justification or entitlement. The force of the entitlement derives from the nature of the warrant for the acquisition of the belief, a warrant that, via preservative memory, entitles the person still to believe the theorem. The entitlement to believe the theorem is apriori because the entitlement derives from the original apriori justification. Similarly, the person who originally acquired the belief empirically, would retain an empirical entitlement. Given the plausible view that we are entitled (at least normally) to beliefs preserved in memory even if we do not remember the grounds for these beliefs, I find these results unexceptionable.

The objection was that it is unintuitive that two people 'could share all the same present beliefs, reasons, cognitive abilities, and memories—and one of them be justified *a priori* and the other justified *a posteriori*'. The view just stated does not entail this result. As I have noted, it makes a claim about entitlement, not justification. Further, although the two relevant people remember the same belief, it seems to me very doubtful that their memories are properly individuated as the same, given that they preserve a belief that in the two cases was associated with very different types of warrant. It is doubtful that a preservative memory can be individuated independently of the warrant that it preserves. In either case, I find the idea that the two people have different warrants (here, entitlements) for their belief fairly obvious.

In special cases (say, when memory is reasonably called into question), one's entitlement to maintain a belief when one has lost access to one's original grounds may come to depend on meta-reasoning about the reliability of one's memory. Such warrant might be empirical. But a generalization of this view would be preposterously hyper-intellectualized. An individual's warrant for relying on memory-preserved beliefs when he forgets his grounds is not normally an empirically justified meta-belief in the reliability of his memory. The entitlement to rely, without justification, on beliefs retained by preservative memory greatly outruns the ability to call up the warrants for the beliefs that are preserved. If it did not, very little of what we count as knowledge would be knowledge.

The entitlement to the remembered belief seems to derive from the warrant for the original belief. For we would not be entitled to the belief if it were preserved from *unwarranted* acquisitions that we had forgotten: we cannot, I think, become warranted by forgetting the poor grounds originally had, and then relying on the remembered belief.

The second objection to my conception of preservative memory centers on observations about how memory works. Christensen and Kornblith emphasize the degree to which background beliefs affect memory. I accept this emphasis.

But I do not think that my proposals about epistemic warrant are affected. We may rely on background empirical beliefs in maintaining a view. These beliefs may mix with and reinforce our apriori grounds for the view. We may have empirical and apriori warrants for the same conclusion. The question in judging whether a warrant is apriori is always whether there is a line of justification or entitlement to whose justificational force sense experience makes no contribution. Arguing that empirical background beliefs affect the epistemic status of beliefs based on memory does not show that those beliefs do not have apriori warrant. It shows only that they have empirical warrant. One can have both.

Christensen and Kornblith discuss a case in which someone acquired the belief that the Vikings preceded Columbus "for good reasons", remembers that the Vikings preceded Columbus, but does not remember how she acquired the belief. They think mistakenly that the fact that she does not explicitly think about the fact that she is relying on her memory makes it the case that her belief hinges on purely preservative memory. But let us suppose that the belief does hinge on purely preservative memory, in my sense, going back to the original acquisition. They further suppose that she is able to retrieve the belief from memory only because it is "inferentially connected" with other beliefs about the Vikings. They maintain that whether the belief is justified depends on whether these other beliefs are justified. So if the belief were inferentially integrated with irrational beliefs, the belief would not be justified.

They consider an objection that the initial warrant 'provided by . . . purely preservative memory is present, but is *undermined* by the belief's inferential connections with other beliefs' (p. 17). They reject the view that there is any initial prima facie warrant because irrational beliefs with which the Viking belief is integrated call into question neither the Viking belief nor the workings of preservative memory. They maintain that the problem lies with the memory. They claim that the justificatory power of a memory cannot be divorced from the justificatory status of beliefs whose inferential connections sustain it.

They appear to see this example as refuting the view that preservative and substantive uses of memory correspond, respectively, to apriori and empirical justifications. But this, as I have noted, is not my view. A belief that the Vikings preceded Columbus adopted for good *reasons* will be *empirically* justified, and the belief retained by preservative memory would remain empirically warranted.[14]

[14] There is perhaps an even more serious mistake in the discussion. It was part of my characterization of preservative memory that it does *not* confer or enhance warrant (neither apriori nor empirical warrant). It preserves beliefs and, normally, their warrant. We are entitled to rely on it. But it is not itself an element in an individual's justification: it provides no justification, and adds nothing to the force of justification or entitlement. They appear to be asking whether the warrant provided by preservative memory is genuine, overridden, or dependent on other matters. This would be a misunderstanding, unless "provided" simply means "preserved". Later (p. 18) they speak of the 'justificatory power of memory', again suggesting this basic misunderstanding, insofar as preservative memory is at issue.

Let us imagine that the individual acquired the belief without reasons but with an apriori prima facie entitlement to accept what she is told. Suppose that the belief is retained from this acquisition by preservative memory. Suppose that memory is sustained by the belief's being inferentially integrated with empirical beliefs. Suppose that apart from this causal sustenance, the memory would have lapsed.

Here again we face issues about how justifications or entitlements (as distinguished from beliefs) are retained by preservative memory. In 'Content Preservation' I held only that in normal argumentation, one's justification was retained over time through argument. What I said about the analogy between perception in interlocution and preservative memory in argument is independent of how one glosses cases in which one's initial grounds are not only forgotten but inaccessible to reflective memory.

Still, what are we to say? One must distinguish between the causal role of the inferential background beliefs and their role in justification. Lots of beliefs and experiences might function mnemonically to help sustain preservative memory. Some might have no justificatory connection to the preserved belief. The fact that a psychological condition keeps preservative memory going does not show that the condition bears on the epistemic status of the preserved belief. It seems clear that we are commonly entitled to beliefs retained by preservative memory even though we have forgotten the circumstances of their acquisition and our warrant for them.

For the sake of argument I will adopt the possibly over-simple view (cf. note 13) that as long as preservative memory works to preserve a warranted belief from its initial acquisition, one has at least a prima facie entitlement to the belief, which derives from entitlements or justifications originally associated with the belief. Suppose one's initial warrant is apriori. Then, of course, one could acquire additional empirical grounds, good or bad, for the same belief. And it might happen (though I think this by itself is epistemically irrelevant) that these additional beliefs causally sustain the memory that goes back to the acquisition. Without them, the memory would have lapsed. Then one has different sources of possible warrant—those preserved by the original memory and those associated with the subsequently acquired empirical beliefs whose inferential connections to the belief help support it epistemically.

Christensen and Kornblith believe that these suppositions are mistaken. They think that there is not even a prima facie entitlement to the belief that derives from the belief's being preserved from its initially warranted acquisition, in cases where the belief is "inferentially integrated" with other beliefs. I disagree, *at least if (as they seem to stipulate) there remains a preservative memory connection to the times the belief was held with the original good warrant*. The individual need not remember the acquisition of the belief as an event, of course. It is enough that the belief is preserved in memory from that acquisition. If the inferentially connected background beliefs were given up; and if the relevant preservative memory were still sustained by some other (say, epistemically irrelevant) means;

and if the individual still remembered that the Vikings preceded Columbus; then the individual would surely be entitled to that belief, even if the individual could not justify it or remember how or when she acquired it.

I can imagine positions that would accept this point but hold what while the individual had the irrational beliefs supporting the same conclusion, the individual would lack even prima facie entitlement to the belief. But I find such positions implausible. The prima facie entitlement seems to be present all the way through because of the individual's memory connection to an epistemically warranted acquisition of the belief. Of course, if there remains no causal memory-connection to the period of the original warrant, and the only preservative memory relations are to times when the belief was supported by irrational grounds, then one has no entitlement preserved from one's original warrant.

Where one has a prima facie entitlement deriving from a warrant preserved by preservative memory and one also has inferential empirical justifications (good or bad), one has two sources of warrant. These should be epistemically distinguished, regardless of how complex and overgrown the psychological situation is. One source might be apriori and the other empirical. I see no difficulty here.

What are we to say about whether the individual is entitled to the belief *all things considered*, in the case where his background empirical inferential support is irrational, but his memory derives from a belief acquisition that was rational? We continue to suppose that the individual cannot remember the initial rational warrant, but he has a good prima facie entitlement for the belief. As Christensen and Kornblith in effect point out, the bad reasons do not impugn the good entitlement. It is not obvious to me that they *must* undermine the entitlement. In many cases they can be seen as purporting to reinforce a warrant that the individual already has: it is just that the reinforcement that they purport to add is worthless.

Again, if the irrational beliefs were given up and some other element causally sustained the preservative memory deriving from the initial acquisition, the individual would still believe (apparently with entitlement) the proposition. Perhaps there are cases where the entitlement is overridden by the bad reasons for the same belief. Since the individual cannot cite the prima facie entitlements, perhaps the bad reasons she can cite sometimes dominate. But if she is not so entitled, it is because she relies on bad reasons, even though she *has* a source of rational entitlement. What I think mistaken is the view that the defect lies in the warrant (entitlement) preserved by the memory that the Vikings preceded Columbus. Preservative memory tracing back to the time of the original acquisition carries, I think, a prima facie entitlement to maintain the belief.

Obviously in such a case, there will be various memory routes which the individual cannot sort out: those tracing to the original acquisition, and those tracing to instances of inferential empirical support (good or bad) for the same belief that came along later. All routes preserve the same belief but do so in epistemically different ways. Epistemic warrants stem from these routes in ways that are largely independent of whether the individual can sort out the routes or warrants. The route to the original acquisition is, we are assuming, causally and epistemically traceable. There is no defect there. Normally where the memory route is intact, some warrant, at least prima facie entitlement, would seem to be preserved.

It would be a mistake to underrate the normative strength of preservative memory mechanisms that tie present to past practice, or the independence of warrant preserving mechanisms from the individual's ability to call them up. Like intentional content, much epistemic warrant depends on normatively loaded causal chains which need not be fully available to the individual.

13 *Computer Proof, Apriori Knowledge, and Other Minds**

Almost two decades ago there was a flurry about the significance of the computer-proved Four-Color Theorem. Humans had programmed the computer, of course. But it had run a proof that was too long for any human to carry out or check. The full proof could be checked only by other computers. Nevertheless, nearly all mathematicians conceded that the theorem had been proved.[1] The status of the theorem has not changed substantially since then. I will be interested in the situation only as thus described. I want to discuss a philosophical question that arises from the supposition that we are ineliminably dependent on computers for some of our mathematical knowledge.

The question concerns the justification for a person's believing, indeed knowing, a theorem on the basis of a computer proof. I want to ask whether reliance on a computer must in itself prevent one's warrant for believing a theorem from being apriori.

The dominant philosophical view is that reliance on a computer in itself makes a person's justification empirical—prevents it from being apriori. Discussing an analogous case (without reference to whether the computer's work could be replicated by humans), Saul Kripke argues that such beliefs are a posteriori:

> No one has calculated or proved that the number is prime; but the machine has given the answer: this number is prime. We, then, if we believe that the number is prime, believe it on the basis of our knowledge of the laws of physics, the construction of the machine, and so on.[2]

* I want to thank Joseph Almog, Tony Anderson, John Bartholdi, Jon Barwise, John Carriero, David Kaplan, Bill Hart, Paul Hovda, Tony Martin, Chris Peacocke, Tony Roy, Jim Tomberlin, the UCLA Philosophy of Mathematics workshop—which scrutinized a presentation of the paper over six all-afternoon sessions—and audiences which heard versions of the paper at various times from 1991 to 1997 at CUNY, MIT, Oxford University, Santiago de Compostela, University of Southern California, University of Barcelona, University of California at San Diego, University of California at Northridge, and University of Illinois at Chicago.

[1] There are also cases in which computers provide overwhelming inductive evidence for a conclusion. I think that much of what I say will carry over to these cases, but I will not assume they are similar in every way. Cf. M. O. Rabin, 'Probabilistic Algorithms', in J. F. Traub (ed.), *Algorithms and Complexity: New Directions and Recent Results* (New York: Academic Press, 1976).

[2] Saul Kripke, *Naming and Necessity* (Cambridge, Mass., Harvard University Press, 1980), 35.

Thomas Tymoczko writes:

This appeal to computer, whether we count it strictly a part of a proof or as a part of some explicitly non-proof-theoretic component of mathematical knowledge, is ultimately a report on a successful experiment. It helps establish 4CT . . . on grounds that are in part empirical.[3]

These views are plausible. But the issue is more complicated than they indicate. I think that their suggestion that knowledge of these results *must* rest partly on an empirical justification is problematic and probably mistaken. I want to explore the situation in a socratic spirit, aiming less to settle the matter than to identify and broach some epistemological questions.

The case is worth discussing for its own sake. Issues regarding unsurveyable proofs and issues regarding the reliance on computers or on interpersonal communication in mathematics are interesting in themselves. But a co-equal motive for discussing the case is that it provides a context for exploring a distinction that is fundamental to understanding rationalism as an epistemological position.

For most of this century, various forms of deflationary empiricism have dominated thinking about knowledge. In my view, this dominance is poorly grounded and will collapse. For all empiricism's success in criticizing a procession of unfortunate metaphysical views and in championing the methods of natural science, it fails to yield a credible, complete account of our knowledge of logic, mathematics, self-knowledge, ethics, and several other parts of philosophy. The dominance of empiricism has led to complacency regarding empiricism's vulnerabilities and ignorance regarding rationalism's resources.

It is still too little recognized that rationalism is not about unrevisability, infallibility, indubitability, or innateness. It is about the force of epistemic warrant. It claims that in some cases the force of our warrant for believing (or knowing) certain propositions or for relying on certain practices or capacities derives from intellectual understanding or reason, not from sense experience. The phrase "force of our warrant" should be highlighted. Rationalists need not (and many classical rationalists did not) claim that any of our beliefs are independent of sense experience *tout court*. Perhaps all of our thinking depends on having had sense experience. Rationalism claims rather that sense experience does not contribute to the normative or justificational force carried by some warrants. So arguing that a belief "depends on" sense experience does nothing in itself to support an empiricist epistemology.

The issue centers on distinguishing epistemic or justificational dependence from various other sorts of dependence—causal or psychological dependence, or dependence for grasping intentional content or for learning. If one does not recognize the prima facie possibility of drawing this distinction, one cannot even begin to reflect on the issues between rationalism and empiricism, or to

[3] Thomas Tymoczko, 'The Four-Color Problem and Its Philosophical Significance', *The Journal of Philosophy* 76 (1979), 63.

see what rationalism is driving at. The point has been emphasized by nearly every major rationalist.[4] But it is repeatedly forgotten in current discussions. I shall explore the distinction between dependence on the senses for acquiring putative information and dependence on the senses for warranting our understanding and acceptance of the information thus acquired. I will be developing a rationalist line. But a more fundamental aim is to make progress on understanding the issues surrounding computer proof, with special reference to this distinction. Understanding the distinction is more important than the position on the special case that forms the topic of this paper.

I

I begin by explaining what I mean by "apriori". I take apriority to be a feature primarily of justification or entitlement. A warrant (either a justification or an entitlement), is *apriori* if neither sense experiences nor sense-perceptual beliefs are referred to or relied upon to contribute to the justificational force particular to that warrant.[5] A person's knowledge is apriori if the knowledge is supported by an apriori warrant that needs no further warrant for the knowledge to be knowledge.

Let me highlight some features of this conception of apriority.[6] The distinction between justification and entitlement concerns two types of epistemic warrant. Both justifications and entitlements are epistemic warrants with rational justificational force. But entitlements need not be understood by or even accessible to the individual subject, whereas justifications, in my narrow sense, involve reasons that individuals have and have access to. Entitlement is my partially externalist

[4] I mention three cases. Leibniz emphasizes the point repeatedly. Compare, for example, at the very beginning of *New Essays on Human Understanding*, Preface, 48–49. Kant makes the point at the beginning of both editions of *Critique of Pure Reason*, A1, B1. Frege writes, 'If we call a proposition empirical on the ground that we must have made observations in order to have become conscious of its content, then we are not using the word "empirical" in the sense in which it is opposed to "a priori". We are making a psychological statement, which concerns solely the content of the proposition; the question of its truth is not touched.' *Foundations of Arithmetic*, trans. J. L. Austin (Evanston: Northwestern University Press, 1968), section 8. For me, it is not so much truth as warrant that is at issue. I will speak of the *warrants for believing* propositions (intentional propositional contents), rather than the propositions themselves, as apriori or empirical. But the fundamental point is the same.

[5] Although justifications or entitlements are what are primarily apriori, I count a justificational method or resource apriori if someone who used that method or had that resource could have apriori justification or entitlement through it, or if our entitlement to rely upon it does not rely for its force essentially on sense experience.

A different application of "apriori" is to truths. I am sympathetic with this usage. I leave as an open problem the relation between my application (to warrants, methods, capacities, uses of concepts) and this application (to truths). I am inclined to think that the apriority of truths is to be explained in terms of the apriority of possible, canonical warrants for believing those truths. But the issues are complex.

[6] Compare my 'Content Preservation', *Philosophical Review* 102 (1993), 457–488, for more detailed exposition of this conception, as well as for some defense of the fourth assumption set out below. I shall say more to support this assumption in the course of this paper.

analog of the internalist notion, justification. I will leave open exactly what may count as an entitlement, although I do not believe that *any* entitlements are *mere* matters of reliability. What is important is that with an entitlement, a *full* reason, or a full expression of the warrant associated with a state or capacity need not be available to the individual.

We are entitled to rely on our capacities for perception, memory, interlocution, deductive and inductive reasoning. Children are entitled to rely on particular perceptual beliefs even though they could not understand why. Perhaps only philosophers can explain why. Justifications, in my narrow sense, are available on reflection to the justified individuals. Justifications may be self-sufficient premises, or reasons that a person could "in principle" call up.

That a warrant for a belief is apriori does not entail that the belief is self-evident, infallible, indubitable, innate, unrevisable, or even unrevisable by *empirical* considerations. Apriority concerns the nature of the rational support for an attitude, not the nature of its vulnerability to criticism.

A non-demonstrative justification can be apriori. Arguably, much non-demonstrative reasoning in mathematics is apriori: neither sense experiences nor sense-perceptual beliefs need contribute to the justificational force involved in such reasoning. The reasoning explains or proceeds inductively from judgments that are fundamentally mathematical or intellectual, not sense-perceptual, in their justificational underpinnings.

Sense experience may be psychologically necessary for an acquisition or warrant of a belief, without contributing to the force of the warrant—hence without making it empirical. The role of sense experience in the psychology and acquisition of belief must be distinguished from its role in contributing to the normative force associated with the belief's warrant.

In deciding whether a belief is apriori warranted, it does not suffice to determine whether it is empirically warranted. A belief can have empirical as well as apriori justification or entitlement. The issue is always whether there is justification or entitlement that has justificational force independently of any force contributed by reliance on sense experiences or sense-perceptual beliefs.

I shall let these remarks suffice for now to provide a working understanding of my conception of apriority. But my stage-setting is not complete. I shall make four large, controversial assumptions. I stand by these assumptions. But I make them not primarily to support a conclusion but to isolate the issues I want to discuss. I hope that the discussion will be of some interest even to those who do not share the assumptions. Here I make them for the sake of argument.

First, I assume that individuals' knowledge of pure mathematics, resting on specifically mathematical understanding or reasoning, is ordinarily apriori.[7]

[7] As I will indicate in discussing my third assumption, ordinary mathematical reasoning by working mathematicians often involves taking for granted theorems passed on by other mathematicians, even if the recipient does not know the proof. Since I think that it is plausible that the mathematical enterprise as a whole, in its communal form, is apriori warranted—grounded in intellection—I believe that there is

Although this assumption is larger than the thesis under discussion, its role here is just to help us focus on whether knowledge of computer-proved theorems *must in itself* add an empirical element to the warrant for mathematical knowledge.

Second, I assume that the computers that are used in carrying out the relevant proofs do not have minds and are not autonomous thinkers. They can do too little to relate their operations to reality to count as having autonomous meanings or thought contents. For example, they have no perceptual abilities and do nothing for themselves. Their manipulation of mathematical symbols is meaningful mathematical activity only insofar as human beings have programmed the machines to manipulate contentful mathematical symbols. Whether robots could have autonomous thought is a question I leave open.

Third, I assume that we, including mathematicians among us, can obtain mathematical knowledge from others, even when we do not know a proof. There are positivistic views, still voiced sometimes, according to which one cannot know (or even, on some views, understand) a mathematical proposition unless one knows a proof of it. This seems to me quite unbelievable as applied to knowledge in an ordinary unstressed sense. Many people know the Pythagorean theorem who have never seen a proof of it. Mathematicians know theorems from other mathematicians, even though they have only the vaguest idea of how the proofs go. These are cases where someone among the individual's sources knows a proof. But even this does not seem necessary. Newton knew simple truths of the calculus, even though they were not self-evident and even though no one, including Newton, knew a proof at the time. The positivist, and intuitionist, insistence on proof for mathematical knowledge is simply a mistake.

Fourth, I assume, what I have argued elsewhere, that although commonly empirical, knowledge that an individual obtains by being told a proposition by another person, where the individual's warrant resides in the interlocution, *can be apriori*.[8] If one's source knows a proposition apriori and there is no reason to

some force to arguing from the dependence of mathematicians on one another for their knowledge, to the possibility of preserving apriori knowledge through communication—that is, to the truth of my fourth assumption below. In other words, since mathematical practice is apriori warranted, and since mathematical practice depends on knowledge being preserved through communication, individuals can and do obtain apriori knowledge through communication (even without going through all the reasoning that underwrites the knowledge). I will not argue in this way here. I am fully aware that philosophers with a Cartesian bent will not find this route to my fourth assumption persuasive. Moreover, I think that the fourth assumption can be argued for independently of assumptions about mathematical knowledge. Since the present discussion will use these assumptions as starting points, however, I will not here argue for them directly at all. To divide the labor among my assumptions as cleanly as possible, I will apply this first assumption, in the course of this discussion, only to mathematical knowledge for which the *individual's* own thinking and understanding provides sufficient warrant for his or her knowledge. So for the sake of my argument, the assumption is to be taken to mean that when an individual obtains knowledge of propositions of pure mathematics through his or her own mathematical understanding or reasoning, the knowledge is ordinarily apriori. Thus insofar as the warrant for the reasoning and understanding is the individual's own, and is specifically mathematical, the warrant is apriori.

[8] 'Content Preservation', and 'Interlocution, Perception, and Memory', *Philosophical Studies* 86 (1997), 21–47.

doubt the source, one can in certain cases know it apriori by accepting the source's word.

This is the least familiar of my assumptions. Again, I make it to isolate distinctive aspects of the epistemology of reliance on computer proof that do not arise in ordinary interlocution. Because of its unfamiliarity, I want to elaborate it at least briefly.

The basic idea is that rational acceptance of an interlocutor's word is an apriori default position. The seeming intelligibility of another's talk gives one apriori prima facie warrant to presume that the other is a resource for rationality. Resources for rationality are, other things equal, to be believed. The presumption of another's credibility can be defeated by special considerations, including empirical considerations. But in the absence of reasonable doubt it is a rational starting point—independent of empirical support. Sometimes, acting in accord with this default entitlement, together with the fact that one's source is knowledgeable regarding the information being passed on, suffices to provide one with apriori knowledge. In such cases, the recipient's apriori prima facie warrant can reside fundamentally in the interlocution, not in an independent and sufficient warrant that the recipient has for the communicated knowledge.

Of course, most knowledge preserved through interlocution is empirical. When a recipient is told something, there are a multitude of possible empirical sources of warrant. The original knowledge passed along might itself be empirical. The recipient may need empirical grounds to supplement his apriori default warrant—because there may be reason to doubt the interlocutor's competence or veracity. The recipient may need empirical observation or empirical reasoning to justify an understanding of the interlocutor's utterance. The recipient may have reason to believe that the topic (like politics or philosophy or many specialized subjects) is one on which one cannot accept just anyone's word; and empirical reasons may be needed to bolster the credentials of the interlocutor or the plausibility of what he says. It is only in special cases that apriori knowledge can be preserved through interlocution. What I am assuming, with some background of argument, is the mere possibility of such preservation. The explication of apriori knowledge that I gave requires that the knowledge be supported by an apriori justification or entitlement that needs no further justification or entitlement for it to be knowledge. Knowledge gained by interlocution raises special problems. When one depends on an interlocutor for knowledge, one's knowledge depends partly on one's own entitlement for accepting the interlocutor's word and partly on the knowledgeability of the interlocutor.[9] More particularly, the

[9] I am oversimplifying. The conditions of dependence on previous members in the chain are complex. One's immediate interlocutor might perhaps not be knowledgeable. But the interlocutor must either pass on knowledge that reposes somewhere in the chain or, as a computer might, otherwise indicate the existence of grounds for knowledge.

There are other sources of Gettier cases in which the recipients's true, warranted belief fails to constitute knowledge—beyond those that depend on lack of knowledge-ability in the chain of sources. For example, a recipient may be told something true by a knowledgeable source, may believe it, and

recipient depends on there being in the chain of sources sufficient justification or entitlement to underwrite knowledge—or else indication of the existence of a method of justification sufficient to underwrite knowledge. For the recipient's entitlement to rely on interlocution is never by itself sufficient to underwrite the recipient's knowledge.[10]

So in interlocution we distinguish two bodies of epistemic warrant: (i) the recipient's *proprietary warrant* for a belief—that is, the reasons available to him together with his epistemic entitlements for holding the belief; and (ii) the *extended body of warrant* for a belief—which includes not only the recipient's proprietary warrant, but those warrants for the belief that are possessed or indicated by interlocutors on whom the recipient depends for his knowledge (though not for his proprietary warrant). The recipient "depends" on interlocutors' justificational resources for his knowledge in the sense that if the interlocutors had lacked or failed to indicate the existence of these justificational resources, the recipient's warranted true belief would not have been knowledge. A person's knowledge based on interlocution is apriori only if some sufficient part of the extended body of warrant (including also an appropriate part of the subject's proprietary warrant) for the knowledge is apriori. There must be a body of warrant or justificational resources within the extended body of warrant that suffices for knowledge and that is itself apriori.

may be warranted in believing it. But conditions in the context may be such that if the recipient knew them, he would lose his warrant to believe the interlocutor. To give a slightly fanciful but provocative example, suppose that Fermat had proved (hence known) his last theorem along the lines that it was actually proved; and suppose that a contemporaneous, fellow mathematician, knowing Fermat's general reliability as a mathematician, and thinking reasonably that the proof was relatively easy, had believed Fermat. Then the recipient would have been warranted, let us suppose, in believing Fermat. If the theorem had been no harder than some ordinary equation in three variables, perhaps the recipient could have obtained knowledge of the theorem through the interlocution. But in view of the fact that the proof is vastly harder than anyone had realized (other than Fermat, on our supposition), it seems plausible that the mathematician's warranted true belief would not have constituted knowledge—even if Fermat had produced the proof in the privacy of his study. If the mathematician had known how hard the proof was, he would have needed supplementary reasons to be warranted in believing Fermat's assertion—reasons beyond knowing that Fermat was a good enough mathematician to have solved the ordinarily difficult problems of his day. For Fermat's proof of Fermat's last theorem would have required substantially more mathematics than any of the other problems that Fermat had solved.

The problem of Fermat's last theorem is notorious for having yielded false reports of proofs by excellent mathematicians. In fact, of course, the eventual prover, Michael Wiles, first reported erroneously that *he* had proved it. Within a couple of months an error was found. It was fifteen months from the time Wiles made his highly publicized announcement to the time when he closed the gap in his erroneous proof and produced a correct one. It is widely assumed that the proof is much too hard for Fermat to have discovered.

[10] The recipient might hear a statement from a source who is not knowledgeable, but find the statement self-evident, or otherwise mathematically convincing, or might immediately see a proof, or might find justification in the fact that the statement coheres with other things he knows. But in such cases the recipient is relying not on interlocution but on other resources in his or her proprietary justification for knowledge. The individual's warrant for the knowledge does not reside in the interlocution.

II

I turn now to accounts of why reliance on computers for proof of a theorem must be empirically justified. Tymoczko writes,

> Some people might be tempted to accept appeal to computers on the ground that it involves a harmless extension of human powers. On their view, the computer merely traces out the steps of a complicated formal proof that is really out there. In fact, our only evidence for the existence of that formal proof presupposes the reliability of computers.[11]

Tymoczko adds that our knowledge that computers are reliable can only be justified empirically.

One difficulty with this argument is that the key word "presupposes" is unclear. The question is in what form an assumption about the reliability of the computers is an indispensable part of the warrant for believing in the existence of a proof. In a certain sense our ordinary formal arguments for mathematical results "presuppose" the reliability of our brain processes. If our brains were relevantly defective, a faulty proof could seem to be sound. We can know that our brains are reliable only by empirical means. This hardly shows that all our formal arguments in mathematics are empirical. Although we depend on the reliability of our brains, the assumption of their reliability need not be a part of the warrant for our mathematical result. The question is how our reliance on a computer's reliability is different from our reliance on our brains' reliability. I believe that the reliance *is* different. But do the differences force the warrant for relying on a computer to be empirical?

A similar question arises about Kripke's claim that one believes a computer's saying that a number is prime only "on the basis of" knowledge of the laws of physics and the construction of the machine. It is true that we could have gotten the wrong answer if the physical world had been less regular or if the machine had been physically defective. And it is true that we may and do support our belief with empirical information about the machine. But it does not follow that a justification of our belief must rely for its justificational force on these empirically known physical facts. To be justified in believing the machine's mathematical results, do we *have to* include reference to the physical mechanism underlying the proof, when we do not need to do so in giving ordinary proofs? If so, why?

An obvious answer is that when we have a proof, the proof suffices for our justification. Unless there is some actual defect in the proof, there is no defect in the justification. So our knowledge that our reasoning depends on our brains is not needed in carrying out an apriori mathematical justification. But in the situation we are considering, we do not have a proof—at least not one of our own that we can check fully. We rely on the computer's having carried out the details. So the justification might appear to have to make reference to the computer's physical reliability. Perhaps this is the relevant difference between the two cases.

[11] Tymoczko, "The Four-Color Problem and Its Philosophical Significance', 72.

The difference is relevant. But although we do not have a proof "of our own", this does not suffice to explain why our belief in the computer-produced theorem must be empirically justified. For in accepting information from other people, we often lack our own justification for the acquired beliefs. We presume that the source has, or indicates the existence of, justificational resources; we need not have those resources ourselves. Yet we do not always need to know that our source is reliable in order to be warranted in accepting what the source says. It is enough if we have no reason to think the source unreliable. Acceptance is a default position, a rational starting point that needs no empirical support in the absence of reasonable doubt. One's fundamental entitlement to accept what one is told is, in the absence of reasonable doubt, apriori. This was our fourth assumption above.

Thus on our assumptions, one can have apriori knowledge of a proposition one has received from another source, even though one cannot oneself provide a justification sufficient in itself to make one's belief knowledge. One relies on there being justificational resources in one's chain of sources. The extended body of warrant that enables one to have knowledge might be apriori. People who accept the Pythagorean theorem (or comparable theorems in arithmetic) on the basis of a diagram and the word of another, but who themselves cannot produce a proof of it, are often in this position.

The case I cited in which one could have such knowledge is one in which one lacks reasonable doubt about what one's source says. So our questions have become: Why must there be reasonable doubt about the computer, so that its outputs cannot fall under one's general right to accept the word of others? And why must such doubt force empirical considerations to underwrite belief in the computer's offering? I will consider several answers to these questions.

To be warranted in believing the computer's result, we do need reasons to supplement our entitlement to believe what others tell us. The theorem's difficulty and the fact that no one has checked the proof force some supplement. But I want to postpone discussing this issue, to focus first on this narrower question: Is the unsurveyability of the computer's deduction sufficient ground in itself for taking belief in the computer's result to be empirical?[12] The answer is "no".[13] An

[12] Tymoczko holds that neither the argument run through by the computer, nor the argument given by the programmers, is a proof in the traditional sense, because both are unsurveyable. He thinks that counting either argument a proof would be to change the traditional meaning of "proof". I agree with Paul Teller in 'Computer Proof', *The Journal of Philosophy* 77 (1980), 797–803, that the computer has completed a proof in the traditional sense, and that the requirement that a proof be surveyable by humans is not part of the sense of the word "proof". My argument does not, however, hinge on the outcome of this dispute. Teller sees Tymoczko as arguing directly from the unsurveyability of the computer's work to the empirical character of the human argument. There is some suggestion of this argument—for example, Tymoczko, 'The Four-Colour Problem and Its Philosophic Significance', 74. But I am not sure that this is an accurate interpretation.

[13] It is a delicate question whether to count the computer programmers' argument a proof—an elliptical proof that involves appeal to the computer runs—or as merely a non-demonstrative argument that the computer has completed a proof. I think, however, that the answer to this question does not matter for our purposes.

unsurveyable deductive argument is no *more* inherently empirical than a non-demonstrative argument in mathematics.

Much justified mathematical belief is underwritten by non-demonstrative reasoning: Newton's belief in, indeed knowledge of, the elementary truths of the calculus, before they were given mathematical explications, is an example. Our belief in the consistency of arithmetic seems thoroughly warranted; in fact I think it constitutes knowledge. But no proof of it adds significantly to the ground for our belief. Zermelo's belief in the axiom of choice, new axioms in descriptive set theory, and Church's thesis also count as examples of warranted belief, though the latter two cases may not amount to knowledge. It was reasonable to believe these propositions in the absence of proof, and in the absence of self-evidence. The relevant considerations were mathematical. There appears to be no indispensable reliance for justification on sense experience, or on the role of the arguments in physical theory.[14] The form or length of an argument does not determine whether its justification must depend, in the relevant way, on sense experience.

Let us try a new tack. What role does the source's being a computer, rather than a person, play in our evaluation of the argument? Is this a source of reasonable doubt about the computer's output? Is it a necessary source of empirical elements in our justification? These issues are complicated. In accepting the word of others in normal discourse, we presume that the seeming-intelligibility of others' talk is an apriori prima face sign of the others' rationality. When we accept the word of another, we treat the other as a rational source. Lacking reasons to the contrary, we are apriori prima facie entitled to accept what others tell us. But by our third assumption, the relevant computers are not people. They do not have minds. They do not have warrants, strictly, because they lack beliefs. They are not autonomously rational.

To investigate whether our source's being a computer is a ground for reasonable doubt about its output—and a source of empirical elements in our

[14] Imre Lakatos has highlighted the variety of types of non-demonstrative reasoning in mathematics. Compare Lakatos, 'What Does a Mathematical Proof Prove?', and other papers in *Mathematics, Science and Epistemology*, ed. John Worrall and Gregory Currie (Cambridge: Cambridge University Press, 1978). Compare also George Polya, *Induction and Analogy in Mathematics* (Princeton: Princeton University Press, 1954). As I have been emphasizing, the reliance on the calculus before it received its foundational explications in the nineteenth and twentieth centuries seems to have required non-demonstrative reasoning. But I think that competent mathematicians had apriori knowledge of the calculus. The basic principles were not (and were not seen as) axioms, or as theorems derived demonstratively from axioms. It seems to me very implausible to think that the role of the calculus in physical theory was the only sufficient ground for accepting it as a part of mathematics, or as true, before the calculus received its modern formulations. Mathematical fruitfulness seems to have been the most significant ground. My suggestion is that the ability to see fairly deeply into an unsurveyable proof might give one mathematical but non-demonstrative grounds for believing its conclusion, especially given that one has reason to think that the computer has carried through the full proof. The epistemic status of this reason still needs discussion. But the fact that the relevant proof is unsurveyable, and the fact that the warrant that a mathematician has is non-demonstrative, clearly do not *themselves* force the mathematician's warrant to be empirical.

justification—I will consider what about the computer can be seen as possible sources of error.

The programming language and the programming of the computer constitute possible sources of error—by far the most likely sources. But these are themselves versions of a mathematical language and a strategy for mathematical argument. The problems they introduce are not in principle different from problems inherent in any mathematical language or mathematical form of reasoning. Since we are assuming that ordinary mathematical reasoning is apriori, I shall ignore these aspects of the problem of relying on a computer to give us mathematical information. The language and program are expressions of the mathematical rationality of the programmers.

Of course, there are physical transitions in putting the program into the computer and in running the program. These are possible sources of error. The source might lie in some defect in the machine, or in some error of programming execution. Let us consider errors of programming execution. Errors in entering the program into the machine seem analogous to errors of expression in ordinary mathematical argumentation. Proofs can go wrong because the author writes down a different symbol from the one intended. Although such slips are sources of error that can be checked for only by empirical means (one has to look at the written symbols to verify that they are the ones intended), this does not show that mathematical argumentation that uses written symbols, diagrams, and so on, is empirical. Since we are making the large assumption that ordinary mathematical reasoning is not empirical, I lay this matter aside here.[15]

Because of some physical defect, the computer may fail to carry out correctly the mathematical program that has been entered into it. As noted, an analogous point applies to human beings. Defects in others' brains may distort their sense for correctness. But we may presume that they are rational sources, or resources for rationality, absent reasonable doubt, because we may consider the content of their output without knowing or needing to know anything about the physical mechanisms that produce the output. The same point applies to interaction with computers. We take up their language, their mathematical offerings, and need not consider their physique.

There is this difference: Machines must be made. The designers must make them carry out rational procedures (make them computers) before anything can

[15] Michael Detlefson and Mark Luker, 'The Four-Color Theorem and Mathematical Proof', *The Journal of Philosophy* 77 (1980), 803–820, think that the fact that ordinary mathematical proofs involve calculation that may fail to correctly realize a mathematically sound algorithm by itself shows that proofs are empirically justified. This view seems to me quite mistaken. I think that it involves one or more of the following conflations: empirical reasoning with fallible reasoning; aids that are relevant to, or necessary to, the psychology of proof with aspects constitutive of the justification associated with giving or accepting the proof; meta-knowledge about the circumstances in which proofs are given with the reasons actually given in a proof; procedures for checking against possible doubts with reasoning procedures that may justifiably be relied upon in the absence of reasonable doubt. I shall not pursue these issues here. But I think that this sort of empiricism about mathematical epistemology shows the effects of decades of philosophers' not taking rationalism seriously enough to understand it.

be presumed about their reliability as interlocutors. The computers we are discussing are not autonomously rational.[16] To make a computer, the maker had first to view it as an object of empirical engineering. The maker could not simply presume that the machine carried out instructions according to the rational principles that the maker intended it to realize. The maker had to check and revise, through empirical experimentation, to make the machine into a computer capable of carrying out its intended functions. In the designing process, it does seem appropriate to see the machine's output as the result of an empirical experiment. But so far we have seen no reason to think that we must always join the designers in requiring empirical assurance that the machine is a reliable physical realizer of the relevant design intentions, as long as the machine is producing seemingly intelligible and credible information.

Empirical knowledge is available about the physical reliability of the computer. Having this knowledge may increase our confidence in the computer's offerings. But given that we have rationally intelligible and credible mathematical presentations from the machine, we can presume that we are being confronted with a resource for rationality, in the absence of specific reasons for doubt. Following a large part of the machine's proof can indicate the existence, form, and nature of a justification (proof) for a theorem. This remains true even though the machine does not itself "have" a justification.

The meaning of the machine's activities, and indeed the nature of its rational powers, are derivative from intentionality and rationality of rational beings. Such activities provide an amplification of the designers' and programmers' rationality. In understanding the machine's offerings, we have an apriori entitlement to rely on the rationality of these offerings, and the justifications backing these offerings. If the justification indicated by the machine's output contains no premises or rules whose only justification is empirical, and supposing that we are apriori entitled to accept the machine's propositional offerings and proof-fragments as backed by rational procedures and rational sources, our extended body of warrant can be apriori. Again I invoke my assumptions about interlocution and about the apriority of the relevant mathematics.

III

The central epistemic issue is not that we are dealing with computers, but that the reasoning that we depend on is not only not ours but is known to be difficult. To be warranted in relying on the computer here, we need more assurance regarding its capabilities than can be gotten from knowing that it is a source of plausible

[16] The notion of autonomous rationality has been a matter of dispute. I take it seriously. Of course nature "fashioned" us. But nature does not have a mind.

mathematical presentations. Could this assurance be gained non-empirically? Let me approach this question through a parable.

Imagine that a mathematician (M) had computing power and speed equal to that of the relevant computer. Suppose that M had as much access to the details of the unconscious computations going on in her as the access that the programmers have to the computer's computations in the actual situation. Such computations are too long, fast, and complex to be brought fully to M's conscious mind and checked. But parts of the computations can be brought to mind. Suppose that M understands the principles of the unconscious computations that she carries out exactly as well as the programmers understand the nature of the program that they have entered into the machine. So M creates a proof strategy that is the same as the one the programmers create.[17] M knows that she can carry out correct proofs unconsciously, since she has often checked the details of shorter, but still difficult proofs that she first carried out unconsciously. M's ability to reach sound conclusions unconsciously from her premises by way of consciously understood proof strategies is well-established. M's unconscious mathematical powers and track record exactly parallel those of the relevant computers.

To keep the parallels going, suppose that other mathematicians have computational powers similar to M's. Such mathematicians check each others' proofs by running fast unconscious calculations—in the way one computer checks another. Sometimes these checks can be consciously replicated by those mathematicians and by lesser humans, and they are found to be very reliable, though not infallible.[18]

Suppose that M tries a proof that is not consciously checkable. Any portion can be consciously checked. The strategy can be consciously checked. But the

[17] I am relying here on the fact that the programmers of the computer that proved the Four-Color Theorem had considerable insight into the specifics of the proof strategy that the computer carried out, even though they could not go through all the details of the proof. The proof was what is known as a "computer assisted" proof. There are cases in which one can know that a proof has been carried out but have virtually no insight into how. There has recently grown up an area of mathematics which studies the properties of what are known as "Zero-Knowledge" proofs. Compare Shaffi Goldwasser, Silvio Micali, and Charles Rackoff, 'The Knowledge Complexity of Interactive Proof Systems' *SIAM Journal of Computing* 18 (1989), 186–208; Oded Goldrei Silvio Micali, and Avi Wigderson, 'Proofs that Yield Nothing But Their Validity or All Languages in NP Have Zero-Knowledge Proof Systems', *Journal of the Association for Computing Machinery* 38 (1991), 691–729; Manuel Blum, Alfredo De Santis, Silvio Micale, and Giusepp Persiano, 'Non-Interactive Zero Knowledge', *SIAM Journal of Computing* 20 (1991), 1084–1118. I leave open wherein the epistemology of zero-knowledge proofs is different.

[18] I am assuming that these mathematicians can know the proof strategies of their unconscious calculations because they can bring to consciousness fragments of those proofs as they are being carried through, just as many human fast calculators have some insight into the methods of their mostly unconscious calculations. So the hypothesized mathematicians can know that the conscious checks are checks of (parts of) the same proofs that they had carried out unconsciously. We could imagine cases in which the mathematicians lacked this insight into their unconscious methods. Then their conscious checks would be checks only of the results of the unconscious calculations. These checks would still lend credence to the unconscious calculations, even if it were not known what sort of unconscious proof had been carried out, or even whether the result had been established through a proof.

proof is too long to be fully gone over. M reaches a result and reports it. After the proof has been checked by M and similarly endowed mathematicians, who also cannot check all details consciously, the result is found to hold. Clearly, M and her colleagues would have reason to conclude that the theorem had been proved—much as actual mathematicians have reason to conclude that computers have produced proofs that cannot be checked by humans.

I think that M's justification for believing her conclusion remains apriori. The full proof demonstration is not available to her, as a conscious, deliberative being. Her conscious justification for believing that the result has been proved contains a non-demonstrative element. She relies not only on her understanding of the premises and proof strategy, but on her unconscious mathematical powers. But, as I have noted, non-demonstrative elements in mathematical reasoning do not themselves render the reasoning empirical. The mathematical powers that M relies upon are an authoritative resource for rationality, analogous to other reasoning that M relies upon.

What is M's warrant for relying on her partly inaccessible, unconscious mathematical powers? I think that she has two warrants. One is an entitlement; one is an inductive justification.

She is apriori *entitled* to rely on those powers even lacking an articulateable justification. *She is entitled because those powers are in fact rational and she knows how to use them with sufficient understanding*. This is, I think, the form of our entitlement for all reliance on our rational faculties or our other resources for reason.

M's second type of warrant is an inductive *justification* for relying upon the relevant rational faculty. The induction is roughly, "I unconsciously proved t_1, $t_2, \ldots t_m$; so I proved tn." I think that the knowledge of the induction base and the relevant type of inductive reasoning are intellectual, non-sense-perceptual, and apriori.

How is the induction base warranted? There are the provings of $t_1 \ldots t_m$. These are apriori warranted on our first assumption. Then there is the knowledge that she has proved them. This knowledge rests first on self-knowledge: a knowledge that one is thinking through the proof, together with knowledge that it is a proof. This self-knowledge is, I think, intellectual and does not rest for its justificational force on sense experience. Then, second, there is M's memory of her thinkings-through of these proofs. One function of such memory is the same as that of preservative memory, which makes possible thinking through any argument over time. The memory at issue here preserves not only the content of steps in arguments, but the past events of having reasoned through those steps and having recognized them as a proof. Such memory is fundamentally intellectual. Like the relevant self-knowledge, it does not rest on sense experience for its justificational force. It is a preservation of intellectual events and their content.

The memory of one's thinkings-through of past proofs also functions to support a continuing use of the first-person. I think that one has an intellectual, apriori warrant through direct memory to attribute past intellectual acts to oneself via the first-person concept. Of course, such warrant is defeasible, even empirically defeasible. But if such memory of one's past intellectual acts—which in the normal case seems not to involve any reliance on sense experience—were not warranted, one could not hold a rational deliberative point of view together over time.[19]

So the warrant for the combination of the mathematical knowledge of the proof together with the self-knowledge of one's proving the theorem and the memory preserving that self-knowledge, remains apriori. The preservative memory and relevant sorts of self-knowledge are underwritten by apriori, non-sense-perceptual entitlements; and the original mathematical knowledge is by hypothesis apriori.[20]

But it is also plausible that the transition rules or norms of passage in the inductive inference that enables M to infer from her induction base regarding past proofs to the conclusion that she has proved the new theorem are also apriori. Such induction is as constitutive of being rational, to rationally processing information (including empirical information), as deductive inference is. So I believe that the induction as a whole is warranted apriori.

Defending all this in detail is beyond my scope here. I present it as plausible. It is certainly clear that in knowing that a proof has been carried out, M need not rely on knowledge of the physical laws governing her brain, or any measurements of the reliability of her *physical* capacities. I think that, at the very least, the empiricist has a tough road in showing that empirical sources of warrant *must* play a role in underwriting M's knowledge of the induction.

IV

Of course, computers are outside us in a way that M's powers are not outside her. We cannot access their results from a first-person point of view, as M could access her results. This remark raises three issues. First, we have to perceive physical events to know the results the computers express. Second, M has learned to use her mathematical powers, and has them in her, whereas we take over the computer's reasoning second-hand, regarding it from the third-person point of view. Third, to know that a computer can deal with difficult mathematical problems, to have some sense of the *degree* of its mathematical power, we must have used our sense capacities. Do these points show that our justification

[19] This is of course a complex and disputed issue. It is the subject of Kant's third paralogism, and Kant seems to take a position opposed to mine. I do not hope to do justice to this matter here, but I shall discuss the matter somewhat further below—compare note 27.

[20] See 'Content Preservation'; 'Our Entitlement to Self-Knowledge', *Proceedings of the Aristotelian Society* 1996, 1–26; and 'Interlocution, Perception, and Memory', for discussions of the apriority of self-knowledge and preservative memory.

for believing that a computer has carried out an unsurveyable difficult proof is inevitably empirical?

All these issues have analogs in the epistemology of communicating with other people. I shall deal with the first two quickly, so as to concentrate on the third. As to our reliance on perception in knowing what a computer's results are: We do have to hear the words of others, just as we have to read the print-outs of computers. But perceiving the words need not be, and normally is not, part of the fundamental entitlement for understanding what intentional content they convey, or for understanding them as presentations-as-true. The words are like diagrams or symbols that help one see the point of a mathematical claim: They call to the recipient's mind the thoughts of the interlocutor that they express. There is no other way. But perception of them does not normally enter into the recipient's entitlement for relying on understanding of them, or their assertive force. These issues are complex, but they are covered by my assumption that we can have apriori knowledge through interlocution.[21]

Second, although M has mastered the forms of her unconscious reasoning, that reasoning is no more available to her as a justification than the computer's transformations are available to us. We have to access through print-outs the computer's results and the individual passages of the proofs that we check, whereas M knows her results and passages of her reasoning "directly". But this reduces to the previous point about the role of perception in interlocution.

I turn now to the third issue, which I want to dwell on for the remainder of the paper. To have some sense of whether the computer has mathematical power sufficient to solve a difficult mathematical problem, we seem to have to rely on perception in a different way than that involved in simply understanding the print-outs of results. We have apriori prima facie entitlement to accept intelligible presentations-as-true, expressed by the print-outs. But the entitlement can be overridden if there is reasonable ground for doubting what one's interlocutor presents. The known difficulty of the relevant mathematical theorems seems to be prima facie ground for doubt or suspension of belief, in the absence of supplemental knowledge about the power and reliability of the interlocutor.

Suppose that I know that some proposition (say, Fermat's Last Theorem or the Four-Color Theorem) has been alleged to be proved many times before, always mistakenly. If a stranger tells me that he or she has proved it, I have reasonable

[21] 'Content Preservation', 'Interlocution, Perception, and Memory'. Tymoczko, 'The Four-Color Theory and its Philosophical Significance', and Detlefson and Luker, 'The Four-Colour Theory and Mathematical Proof', place heavy emphasis on the view that knowing that the computer has given certain results is empirical, as does Michael Resnick, 'Computation and Mathematical Empiricism', *Philosophical Topics* 17 (1989), 129–144. Much of Resnick's paper is devoted to defending the modest (and I think correct) conclusion that sound argumentation for mathematical conclusions *can* involve empirical premises (e.g. premises about the empirically known reliability of a computer or human). See especially ibid. 133–137. But Resnick infers from his arguments that these empirical methods are for us "largely ineliminable"—presumably ineliminable from the warrant for the mathematical results (141). I think that he has not established this stronger conclusion or even seriously argued for it.

ground for withholding belief. As things stand, I am not justified in believing what I am told. The initial generalized apriori entitlement to accept intelligible assertions needs supplementation. An analogous point applies to computers' computation of theorems of hard, unsurveyable proofs.

Our question is whether the extended body of warrant of a belief in the report of a difficult theorem can ever yield apriori knowledge. More precisely, I ask whether one can ever abstract from empirical elements in the extended body of warrant and retain sufficient apriori justification in that extended body to have apriori knowledge. An affirmative answer requires that the sources' knowledge, or indicated justification, be apriori. It further requires that one's entitlement to accept the report and one's supplementary justification to maintain acceptance, despite reasonable doubts based on the theorem's difficulty, also be apriori.

We have transformed our problem into a question about the nature of the supplementary justification. Clearly, much of what we know that might supplement our basic entitlement is empirical. Knowledge of the psychology, biography, statistical reliability, or physical capacities of the mathematician or computer would be empirical. If I depended on an interlocutor that depends on such information for his or her warrant, the extended body of warrant underwriting my belief would also be empirical, even if my proprietary warrant to accept the interlocutor's assertions were apriori. Most supplemental historical knowledge of the sort needed to lend credence to the claim of a difficult theorem will inevitably be empirical. Is there a relevant type of supplemental knowledge that might rest on apriori, non-sense-perceptual justification?

One thing to notice is that we can rely on others to provide warrant for accepting the mathematician or computer's word, without depending in the extended body of warrant *indispensably* on empirical facts about psychology, biography, or physique. A simple sort of thing that would be relevant is if several sources said that there were near proofs of the theorem, and a proof would not be a major mathematical surprise. Then one's reliance on the mathematical authority of the mathematician or computer (call her or it "Q") would be distributed through several intermediaries; and the reliance would derive from mathematical considerations that backed, for our sources, belief in the theorem. These might buttress accepting Q's word.

It remains likely, however, that in the cases of hard, uncheckable proofs, the supplemental information needed to believe what Q says would have to include something about Q's authority or competence as an individual—or at least about the authority of our interlocutors, reporting on Q. Let us assume this to be true.

Now one could be impressed not so much by Q's notoriety or physical characteristics, as by the force of Q's reasoning. One (or one's interlocutors) might have checked prior mathematical successes of the individual. The mathematical *results* are by hypothesis apriori warranted. But to support believing in the authority of an individual, the results must be attributed *to the individual, in third-person form.* Does the reliance on cumulative mathematical successes in another person or computer require empirical justification?

Recall that M had two sorts of non-empirical epistemic warrant—an induction and an entitlement. I want to discuss the third-person analogs of both sorts of warrant, for the case of relying upon Q. Let us begin with the analog to M's inductive justification.

Inductive form does not by itself make the justification empirical in either the first- or third-person case. We must consider the basis steps of the induction. Any such induction about Q has a basis in attributions to an individual in time. The induction might be of the form, "Q produced proofs p, p_1, p_2; these proofs indicate that Q is a source of great rational (mathematical) power; so Q is worthy of special epistemic reliance".

One might hold that any conclusion based on beliefs about particular events in time is *ipso facto* based on experience, and hence empirical. So any justification that bolsters Q's authority as a mathematician or computer will inevitably be empirical.

This line of thought is based on a conception of empirical experience and apriori justification that differs from mine. Frege held that an apriori truth just *is* one that has to be derived and justified from fundamental general laws rather than from assertions about particulars. Leibniz, from whom Frege got his conception of apriority, maintained that judgments about one's present thoughts (*cogito* thoughts) are judgments of experience, hence not apriori. These philosophers saw the divide between the apriori and the empirical as lying between justifications that are founded on general rational principles, and justifications that are founded on judgments about particulars, especially events or objects in time. The latter judgments counted as judgments of empirical experience, regardless of whether the "experience" is sensory.[22] On this view, neither M's induction about herself nor our induction about Q could be apriori.

By contrast, my conception of apriori warrant features independence of justificational force from sense experience or sense-perceptual belief—not independence of any intellectual "experience". With Kant—and the empiricisms of Mill, the positivists, and Quine—empirical experience has come to mean sense experience. Thus apriori justificational force is force that derives from non-sense-perceptual sources—sources of reason, understanding, or perhaps the form of experience. An induction based on judgments about particular intellectual events is not *ipso facto* empirical.

It is clear that in attributing intellectual successes to another individual, we must utilize perception. The question is whether such perception must contribute to the force of our justification for relying on another individual as specially authoritative. I think that there are relevant apriori warrants that derive from intellectual understanding.

A conception of understanding that is congenial with the Leibniz–Frege conception of apriority is that intellectual understanding operates only on

[22] Frege, *Foundations of Arithmetic*, section 3; G. W. Leibniz, *New Essays*, Book IV, Chapter IX.

intentional content *types*: perception is always needed for warranting the application of understanding to particulars. I believe that this is a mistaken view of intellectual understanding. I think that intellectual understanding operates originally on token events with intentional content. One understands one's own token judgments—the content and judgmental mode of intellectual (particular) acts. One understands them as judgements with certain conceptual content, and such content is an abstract type. One also understands the *mode or judgmental force of the particular tokening* of the type (understands its judgmental or assertive force). One cannot understand one's own thinking or the assertions of another without grasping intellectually the judgmental or assertive force of events with intentional propositional content.

In some cases, one understands what I call intellectual *applications* of the conceptual types. For example, in understanding the first-person concept in application to oneself (or, I think, another) on a particular occasion, or in understanding some other indexical or demonstrative as applied to some intellectual act, one understands a token event. Such understanding of token events always requires conceptual guidance, but it is not reducible to the understanding of abstract conceptual (or intentional-content) types. These are cases of intellectually understanding a token event; our warrant for relying upon such understanding is non-empirical.[23]

To gain conceptual understanding of one's own thought as such, one must understand token instances. Conceptual understanding of content through language develops in tandem with understanding token judgments or assertions by others. To understand intentional propositional content types, one must normally understand intentional content tokens that are presentations-as-true, or more narrowly, token assertions or judgments. More generally, understanding content types requires understanding content tokens. I believe that such understanding is commonly intellectual. Individuals with the relevant competence are apriori prima facie entitled to rely upon it. This view is embedded in our fourth assumption.

We have an apriori warrant to rely on our putative understanding of token assertions, or other presentations-as-true. In the first-person induction, M relied on this warrant in her self-knowledge. She was able to identify and (through intellectual memory) reidentify herself via the first-person concept. I maintained that she could be apriori prima facie warranted in doing so. To carry out the analogous third-person induction, one must attribute successes to the relevant source, Q. Normally we identify such a source perceptually. Our warrant for such identification is empirical. But to be apriori warranted in the induction that we are considering, we cannot lean on this form of identification. We have available, however, an alternative: we can fix the source as the source of the putatively

[23] For more on applications and intellectual understanding of them, see my 'Belief *De Re*', *The Journal of Philosophy* 74 (1977), 338–363; and 'Interlocution, Perception, Memory', e.g. notes 3 and 12.

understood event with intentional content—*the source or author of this presentation of intentional content*. I believe that we have an apriori warrant to presume that an instance of intentional, propositional content has a source with propositional powers.[24]

In any given case, on our assumption about interlocution, we have apriori prima facie entitlement to accept what the source presents as true. And we (or at least some among us) can supplement this entitlement by thinking through an argument that the source gives us. If what we reason through lends credence to what the individual tells us, the individual's assertion gains credibility; and our defeasible default entitlement is supplemented with a justification. Our own going through some or all of the mathematical reasoning is by hypothesis apriori. So on our assumptions, *in given cases* we can obtain apriori justification for supplementing our apriori entitlement to accept the intentional contents that our source presents as true.

To yield an inductive justification for taking Q to be more than ordinarily authoritative, however, it seems, at least at first blush, that we must attribute presentations to a single source not only from the third-person point of view, but *more than once*. Must such attributions be warranted empirically?

V

In discussing an apriori warrant for third-person attributions, we have been implicitly advancing a position on an aspect of the problem of understanding knowledge of other minds. This issue demands much fuller development. But I want to sketch more explicitly the line that has been guiding my thinking.[25] Clearly, perceptual belief is a necessary enabling condition for having any warrant regarding another being's intellectual accomplishments, indeed for understanding any given accomplishment. It is a standard view that it must also enter into the justification of any knowledge of another mind. Most would hold that one must infer such knowledge from the behavior, broadly and liberally construed, of an individual. But I do not find the matter so straightforward.

I think that one can know of the existence of another mind non-inferentially, via apriori entitlement associated with understanding intelligible assertions, or other acts with intentional content. If a reliable linguistic understander finds immediately intelligible any seemingly propositional (seeming) event, that

[24] The powers can be either autonomous or derivative for these purposes.

[25] I want to emphasize immediately that what I am about to say is not meant as an answer to scepticism about other minds, although it may play a role in such an answer. I intend to describe one of our actual warrants for our knowledge of other minds. But I do not discuss what the legitimate sceptical questions are and whether the approach I am taking avoids begging such questions. These are further tasks.

individual has an apriori prima facie entitlement to presume that it has a rational source.

In reception from a computer (lacking careful observation!) one might mistakenly but justifiably presume that one's immediate source is a rational mind. This possibility is compatible with the relevant apriori entitlement. First, the entitlement is prima facie, and thus compatible with being mistaken. But second, the entitlement is not a warrant for the presumption that one's *immediate* source is rational. It is a warrant for presuming that there is *some* rational source in the chain of sources behind the seemingly intelligible, seemingly propositional, seeming occurrence. Although this presumption too is defeasible, it would remain not only warranted but true in the case of reception from a computer. The presumption does not depend on recognizing the nature of one's immediate source, other than as a locus and causal source of events with intentional content. The immediate source need not be a mind. Thus seeming understanding provides apriori prima facie entitlement to, and potential apriori knowledge of, the conclusion that a particular (seemingly) propositional (seeming-) event has some (ultimate) rational source, in some mind.

There are two necessary conditions on an individual's having the relevant entitlement that need to be highlighted here. The first is that the individual be a reliable linguistic understander. That is, the individual must be capable of understanding, from a third-person point of view, events with propositional intentional content according to some pattern that distinguishes, reliably, what is ("literally" or objectively) expressed from the momentary intended contents or implicatures of any one speaker. The second is that the individual must, using his linguistic capacity, find an event *immediately* (that is, non-inferentially) intelligible as an instantiation of understood propositional intentional content.

Some have wondered whether the foregoing view will justify finding rational agency behind the regularities of nature. I think not. The argument is importantly different from traditional arguments from design. The warrant attaches to the immediate application of a linguistic capacity by someone with a reliable ability to understand linguistic expression of intentional content. The physical world is not a text. It is made up of objects and properties, not (except perhaps in the case of the products of minds) intentional contents—not of entities with intentionality and some claim on truth. So perceptions of regularities in the world are normally not non-inferential applications of a linguistic capacity. In special circumstances, one might be warranted in non-inferentially taking what are in fact mere random physical events as propositional expressions. But perceptual experience of most non-propositional events does not involve application of intellectual capacities associated with propositional understanding. The warrant that I am discussing attaches specifically to seeming understanding by reliable understanders of seemingly propositional events. Understanding *intentional, propositional* events—whether issuing from computers or people—lies at the basis of the apriori warrant for presuming the existence of a rational source.

Of course, my view about apriori knowledge of *other* minds depends not only on my assumptions about interlocution. It also depends on the claim that one can be apriori entitled to presume that the rational source is not oneself. I cannot defend this claim here. I sketch it only to indicate a direction of reasoning. In my view, one can have a rational, non-sense-perceptual, knowledge-yielding entitlement to believe that the source is another mind, by having non-perceptual entitlements to presumptions about one's own agency and responsibility, or lack thereof, in the reception of information.

One thinks an intentional content but takes it as received from an event that one did not produce and that has the same intentional content. That *is* third-person (putative) understanding. By recognizing that the propositional act that one is considering for evaluation is not one's own, one can rationally presume the existence of agency from *another* rational source. Again, the warrant underlying this recognition is defeasible. But I hold that the ability to recognize whether or not one is the source of understood propositional acts is fundamental to being rational. The warrant for the exercise of the ability is not grounded in sense-perception. Thus I think that one can know the existence of another mind without resting the knowledge's justificatory elements on inference from observation of behavior or on other particular sense-perceptual beliefs.

There is a sense in which the possibility of knowing apriori a computer-proved theorem does not depend on whether the rational source behind the computer's work is *another* mind. The mathematician that created the program might be the recipient of the computer's offerings. The relevance of issues about other minds is not that the computer's offerings *in fact* stem from another mind (though often they do). The relevant aspect of the parallel is that the recipient of the computer's offerings must confront them *from the third-person point of view*. Even if the computer is in some sense an amplification of the mathematician-programmer's mind, and even if the recipient is the mathematician-programmer, the recipient must consider the computer's offerings as presentations to which the recipient is not necessarily committed.

So the *epistemic orientation* of the recipient toward the rational source (the programmer-*cum*-computer) is the same as it would be toward another mind. We are investigating the third-person orientation toward evaluating the computer's offerings and determining the scope of its power and reliability. This orientation is like a third-person viewpoint on one's own unconscious, except that one is not in a position to access the source as one's own, in the way that M accessed her unconscious as her own. Nevertheless, it seems to me that using this third-person orientation, one can know, attribute, and evaluate the contents of the computer's offerings without resting the knowledge's justificatory elements on inference from observation of behavior or from other sense-perceptual beliefs.[26]

[26] Given the conceptual tie between intentional events and there being an ultimate rational source of those events, and given an apriori reliable capacity for determining whether the events are one's own acts or another's, one can be apriori defeasibly warranted in presuming that the ultimate source of

VI

Even assuming that what I have said about knowledge of other minds is correct, our problem of finding a supplementary apriori ground for believing the computer's claim to the proof of a difficult theorem is not solved. For the inductive warrant that we are considering requires an induction on the *same* source. To know inductively that some source is specially authoritative in mathematical matters, one must accumulate grounds to find the same source authoritative. So there appear to be issues about reidentification.

Many philosophers influenced by Kant, notably Strawson, have emphasized the role of perception of physical objects in reidentifying other minds, and even one's own. I value this tradition and find its basic line broadly plausible. What gives me pause is reflection on the fine line between the genetic role of perception in enabling one to learn to use a rational source and the justificatory role of perception in warranting acceptance of the deliverances of the source.

Let us return to the first-person case. I held that M could do an induction from apriori warranted basis steps, over her past proofs, to ground a conclusion that she was specially trustworthy in carrying out a new difficult argument not all of whose steps could be consciously checked. She used the first-person concept to attribute the proofs to a single source. She might have used a name whose reference and intentional content is conceptually tied to some use of *I*. But the argument was fundamentally first-personal.

Now as Kant pointed out, any use of *I* could in principle be tied to apparent memories that in fact connect not to one's own past exploits, but to those of another being.[27] So the use of *I* in these inductions does not *guarantee* a single source for the induction. But a guarantee is too much to ask even for an apriori warrant. The lack of a guarantee does not show that one's warrant to presume a single source requires empirical grounding. What normally provides a prima facie apriori warrant for the presumption of a single source is not a direct self-reidentification. It is, as I noted earlier, an apriori connection between uses of *I*, on one hand, and certain reliable uses of preservative memory (or more broadly, intellectual uses of memory), on the other. The ability of the individual to use intellectual memory, especially to carry on arguments over time, grounds a prima facie apriori warrant for the individual's presumption that he or she has remained

the relevant intentional acts is another mind. My point is just that in order to know the computer's offerings, one need not *know* whether the ultimate source is *another* mind. One could even be warranted in believing that it was, and be mistaken. The induction that we are discussing—the one needed to lend credence to the computer's special competence in presentation of a difficult theorem—can be carried out with a third-person attribution of relevant intellectual accomplishments, without taking a position on whether the source has autonomous or derivative rationality, or whether the ultimate rational (autonomous) source is oneself or another. Compare note 24.

[27] Immanuel Kant, *Critique of Pure Reason* A361–365. The notion of quasi-memory was later introduced by Sydney Shoemaker, 'Persons and Their Pasts', *American Philosophical Quarterly* 7 (1970), 269–285, reprinted in Shoemaker, *Identity, Cause, and Mind* (Cambridge: Cambridge University Press, 1984).

constant through an induction that depends on his or her identity. Note that this ability itself carries unguaranteed but non-empirical warrants for presumptions of sameness of content, and continuity among the steps of the inference. By extension, memory of past argument steps and past intellectual acts provides defeasible, but non-empirical entitlement to a presumption, in first-person form, that one is the author of those past acts.

Is the third-person case relevantly different? We can construct a justification of the form: Q proved t_1; Q proved t_2; . . . Q proved tm; Q has carried out a large and difficult part of the proof of t_n; so Q is an unusually reliable source whose claim to have proved t_n can be trusted. I have maintained that if Q is identified only as the source of a given intellectual act which one presumes to understand, one can be apriori warranted in knowing and attributing any given basis step in the induction. To carry out the induction, however, one seemingly must hold Q constant. One can, of course, reidentify Q empirically. But can one justifiably presume with *non-empirical* warrant that it is the same rational source through the induction?

Before beginning to answer this question, I want to point out that it does not seem necessary that one reidentify an *individual* as rational source. It would not matter to one's warrant or one's knowledge, I think, if someone had substituted for one's computer another indistinguishable one that was made to be identical in its methods, power, and output. (It would not matter, that is, if one were not committed to its being an individual, but was rather open to there being some "source" that might take a variety of forms.) It would not matter if a committee of sources masqueraded as a single individual, but cooperated in contributing different elements in a proof. As long as the source is a single *type* of power and the individual components of the type (either members of a committee or members of a series of "replacement" individuals) are non-accidentally and rationally related, the relevant warrant and knowledge will, I think, be substantially unaffected. What is important is identity in the type of source of information, not identity of a concrete individual. I shall assume that the induction depends on reidentification of a rational source in this broad sense. But this assumption will not be critical to my argument. The points I make will, I think, carry over to defeasible attributions to an individual.

What resources do we have for non-empirically warranted re-attributions to a rational source from the third-person point of view? The role of intellectual memory in the first-person case provides a model. One might be entitled to re-attributions to the rational source through continuities in understanding the source's output, where these continuities are held together through preservative memory. That is, one uses intellectual continuities as basis for an entitlement to presume that the source is the same, rather than empirical observations to justify premises of reidentification. In turning on one's computer, one is not confined to checking the look and position of the computer. One understands what remains from the previous day as providing continuity. One goes on with the manuscript, presuming defeasibly that the resource is the same. One is not confined to reading a single screen. One can check through the document and check other documents as ways of justifying the presumption that one is dealing with the same source.

To be more concrete about the case at hand, in understanding the output of a person or computer, one has various continuities in the discourse to rely upon. One can not only continue with the proof one left off with. One can check the person or computer's abilities and memory by having it re-prove theorems whose proving gave it special authority. One can note anaphoric connections or other memory connections both within a single argument and across arguments. One can find definitions recurring that might remain constant both within and across arguments. Such checks would, of course, be subject to error. But they could be justificationally independent of sense experience, on our fourth assumption about interlocution, even though the ability to gain access to those powers requires sense experience. The relevant continuities are intellectual. The *warrant* for relying on them does not seem to depend essentially on continuities in sense perception of the computer's body. The presumption of sameness of source seems to be a justifiable prima facie default position, as long as one's preservative memory indicates intellectual continuities from one session and proof to another. The continuities can be relevantly like those in thinking through a discourse or in reading a novel.

Although I have represented the induction as being over a series of past successful arguments, the induction need not always take this form. What matters for establishing the special mathematical authority of one's source is not that the source prove several separate theorems or make several separate calculations. What matters is that the source somehow show relevant mathematical power and reliability. This could be shown in a fragment of a *single* argument if the fragment were complex and hard enough. In fact, there may be only one argument on which to check actual computers which have been programmed to deal with a problem as hard as the Four-Color Theorem. Still, by thinking through a large portion of the proof, with its various lemmas and sub-proofs, a competent mathematician will be in a position to determine on intellectual grounds that this source has considerable power and special authority.

The fact that the induction can be carried out from the case of a single proof affects our understanding of the reidentification problem. We tend—rightly, I think—to assume that we are entitled on intellectual grounds to rely on presumptions that we are following a single argument, no matter how complex. If a source is credited with special authority through its producing a single, sufficiently difficult argument fragment, the induction need not depend on multiple attributions at all. One need identify it only as the source of the relevant argumentation. One can induce from its virtuosity with the argument fragment that it is worthy of special credence.

Of course, what is to be counted as a specious present for these purposes is subject to possible dispute. Most arguments of the requisite complexity cannot be held in mind in one session (on any reasonable conception of "session"). This is true of the Four-Color Theorem. Thinking through a substantial fragment of that proof would involve going home, sleeping, returning for more. Here, some philosophers might hold that the problem of reidentifying a source for a single argument, in these cases, is not substantially different from the problem of reidentification across different arguments.

I am inclined to think that special presumptions of continuity do attach to the thinking through of arguments (or their fragments), no matter what their (finite) length. The ability to follow and carry through arguments is so fundamental to being able to understand intentional content, and to being a single subject, that I think that this ability has a special place. But waiving this point, it seems obvious that it is in principle possible for a mathematician or computer to establish special authority through an argument that can be held in mind, by sufficiently competent recipients and with sufficient background knowledge, in a single "session". So reidentification is not essential, in every possible case, to carrying out inductions that support a source's special authority.

Even if one lays aside appeals to single sessions, the continuities in a single argument—no matter how long—provide a defeasible intellectual basis for presuming a single source. As I have indicated, there are further intellectual continuities that hold regardless of whether one is dealing with a single argument. These continuities seem to me sufficient to solve, on intellectual grounds, the problem of reidentification for purposes of carrying out a successful induction—regardless of whether the induction takes as its basis a single intellectual achievement or a series of them.

So far I have explored how giving the source special credence in its claims to have proved a difficult theorem rests on an apriori-based induction on the source's virtuosity in solving mathematically difficult problems. But there is a dimension of difficulty different from mathematical difficulty that I have not discussed—namely, sheer length. One might think that a substantial element in possible doubt as to whether the machine has proved the Four-Color Theorem is the length of the proof. One seems to need a warrant to believe that the machine is "big enough" to have carried out the proof. And this warrant might seem to be obtainable only from empirical information about the storage capacity of the machine. This (one might think) could not be gained from induction on past proofs that had been checked. For by hypothesis, this proof is too long to check. The whole issue might seem to be whether the machine has the storage capacity to deal with a proof that requires x bits of storage capacity, where x is much larger than the length of any checked proof. How can one be *non-empirically* warranted in presuming that the machine has carried out a proof that is longer than any proof that one could think through?

I believe that this problem is less difficult than it may first appear. How do we know how long the proof in fact is? The machine is (or might be) equipped with a counter that counts steps in the proof. By understanding the content of the output of the machine, one could know inductively that the counter is reliable in counting steps. So an induction that projects beyond the actual lengths one has checked could be warranted. Such an induction would proceed from mathematical, apriori basis steps. Similarly, one could have recognized that the machine announces a proof (stops and produces the turnstile) only when it has one—to an extremely high degree of reliability. Its announcing a proof in the case of a long proof could be deemed credible on inductive grounds. Again, the basis steps for the induction seem to be warranted non-empirically. We can recognize assertion of theorems through understanding. This is a corollary of our fourth assumption. And the recognition of the mathematical probity of the individual proof-, or sub-proof-, announcements is mathematical, hence apriori by our first assumption. The reidentification problem is substantially the same in cases of determining mathematical endurance as it is in the cases of determining mathematical virtuosity.

In summary, one can be apriori warranted in a third-person inductive argument of the form: Q (identified as the rational source of the proofs) has carried out the proofs p_1, $p_2 \ldots p_m$; these provings indicate that Q is a source of great mathematical power; Q claims a proof p_n and has produced a large relevant fragment of such a proof; so Q can be given special credence in Q's claim to have proved p_n. This is how the apriori warrant goes: One is apriori entitled to one's understanding of a sequence of contents presented as a proof. One can presume apriori that that sequence has a rational source. Call the source "Q". One can be apriori warranted in believing that the sequence in fact constitutes a proof by thinking through the sequence and realizing that it is a proof. Call the sequence "p_1". Using intellectual memory to hold the initial understanding together with the mathematical replication, one is apriori warranted in believing that Q has carried out the proof p_1. Similarly, for $p_2 \ldots p_m$, and the fragment of p_n. One can hold this sequence of proofs together by intellectual memory. And one can be warranted in thinking that the proofs have a single source, Q, by noting intellectual continuities among the proofs, and by rechecking Q's ability to carry out proofs again. One is in a position to appreciate, on mathematical grounds, the special mathematical ability of Q as indicated in these proofs by appreciating the mathematical difficulty of finding and carrying out the proofs. One's use of the inductive form of argument, as well as one's beliefs in the inductive base (just sketched), are apriori warranted. So the whole induction is apriori warranted.

I conclude that, given our assumptions, one can be in a position, from the third-person point of view, to be apriori warranted in believing, in fact knowing, on defeasible, inductive grounds, that the theorem has been proved. One can know this even if one cannot fully replicate the proof. The problem of reidentification in relevant inductions can in principle be solved through apriori warrants. I do not presume to have discussed the reidentification problem in appropriate depth. But I think that I have sketched a view that is plausible and deserves development.

VII

We have been discussing third-person, apriori, inductive warrants for granting special authority to an interlocutor who asserts a difficult theorem. Recall that M, our first-person reasoner, had two forms of non-empirical warrant—an inductive form and an entitlement that does not rest for its warrant on an induction. I want to turn to the third-person analog of the non-inductive entitlement.

I took M to be *non-inductively and non-empirically entitled* to rely on her mathematical powers because they are in fact rational and she knows how to use them with sufficient understanding. Is there a third-person, non-inductive, apriori entitlement to accept the offerings of some other person or computer, Q, as specially authoritative?

Through use of her unconscious mathematical powers and through understanding their methods and products, M acquired an entitlement to rely on those powers as more than normally powerful. In being an entitlement, the warrant does not involve a meta-induction on past successes. The warrant is the counterpart, for a rational faculty, of an individual's entitlement to rely on the sharpness of a very sharp perceptual faculty, without doing a sophisticated induction on its comparative merit. I want to consider whether there is a third-person analog of M's reliance on her own unconscious mathematical powers. Such an analog would be an entitlement to rely on Q as a powerful amplification of one's own rational faculties. The delicate matter is to articulate an apriori entitlement that meets two conditions. It must reside in knowing how to use a cognitive faculty or source, not in an induction on products of the source. But it must be in third-person form—an entitlement to rely on the products of a rational resource that is not taken to be one's own.

For an individual that has the appropriate relation to Q, the entitlement would take the form: *the powers one relies upon (Q's) are rational and are more than normally powerful, and one knows how to use those powers and understands their methods and products to a degree sufficient to appreciate their power*. I think that there are two primary aspects of *knowing how* to use Q's powers in a way that entitles an individual to rely on those powers more than one might rely on those of any arbitrary rational being.

The first aspect is a kind of "know how" that is not to be distinguished from tracking some of Q's intellection—thinking through reasoning that one correctly (and with warrant) attributes to Q. For example, one might follow some of Q's proofs or sub-proofs. One need not have any idea how Q targets potential theorems, or how Q discovers the proofs. It is enough that one think through Q's justifications for theorems, understanding them well enough to recognize their soundness and appreciate the power and difficulty involved in coming up with justifications for the relevant results. This appreciation need not be meta. One need not think *about* a series of past proofs. Nor need it involve comparisons between Q's intellection and that of other mathematicians whose accomplishments one thinks about from the third-person standpoint. It is enough that one understand the justifications sufficiently to realize *through understanding them* that they involve special powers (perhaps more power than one has), and sufficiently to understand their ability to solve problems whose difficulty one appreciates. One gains an appreciation of Q's power by being expanded by it, or at least by internalizing fragments of it.

To this degree, one's appreciation of Q's power seems similar to a great mathematician's appreciation of his or her own power. Such a mathematician *might* do an induction on his or her past successes or a comparison with the abilities of others. But to be entitled to rely on the power, the mathematician need not take such a meta point of view. The mathematician's basic entitlement derives from using the power in a way that enables appreciation "from the inside" of the type and difficulty of the problems that he or she can solve.

For contrast, consider a case in which one has no insight into the reasoning of one's source. One simply finds that he, she, or it gets answers to difficult problems that when checked are found to be sound. Assuming that one has some ground to recognize the other's methods as rational, one can be warranted in trusting the other's assertions to a greater than normal degree. A historical approximation to this sort of case is Hardy's increasingly warranted confidence in Ramanujan's mathematical results, despite at best sketchy justifications of them by Ramanujan.

I believe that one can have a non-empirical *inductive* warrant, for relying on such a source to a greater than normal degree. But it seems more questionable that in such cases, where one has no insight into how one's source solves its problems, one can have a *non-inductive* apriori entitlement to rely on the exceptional rational powers of one's source. If one cannot incorporate any part of the source's ratiocination, there is no clear sense in which one is using the other's powers as an amplification of one's own. One is not in a position to appreciate one's source's powers of inferential justification "from the inside", by reproducing them in one's own cognitive processes. If one can replicate none of the source's inferences, one does not know how to use the power—except in the weak sense that one can recognize, check, and come to rely on its *results*. I doubt that this weak type of know-how suffices to provide a non-inductive entitlement in this case. It seems to me that in such a case one is forced to treat the source as an object whose special rational reliability can be established only through induction on its past successes. The non-inductive entitlement depends, I think, on an ability to "incorporate" the ability (at least partly) into one's own point of view. So I will assume that the first aspect of the relevant sort of know-how with *mathematical* powers involves being able to follow some of the justifying reasoning, and to appreciate, from the reasoning itself, its degree of power and insight, as well as its reliability.[28]

[28] The issue about whether one could have a non-inductive non-empirical warrant for relying on another being as a rational source of mathematical information in the absence of insight into the source's justifications bears comparison to the issue over whether one could be entitled non-inductively to rely on a capacity in oneself to have correct answers to difficult mathematical

Of course, insofar as this first aspect of knowing how to use the source's rational powers involves simply mathematical reasoning, it contributes, given our first assumption, no empirical element to the entitlement to rely on the source as specially authoritative. The question of whether the putative non-inductive warrant can be apriori hinges on a second aspect of knowing how to use another's rational power in some domain. This second aspect concerns knowing how to *access* the power, indeed how to access the same power (or relevantly type-identical power).

With first-person appreciation of a mathematical power as special, one's access to the power's products can be through self-knowledge. I maintained that this access is apriori warranted. But in accessing the power of another source, one does not have first-person access. According to our fourth assumption, one can have apriori warrant for one's understanding of what another being says. But can one have third-person apriori warrant for a reliance that depends on *re-using* the same power over time in a way that does not involve meta-attributions of the results to the power, yet yields a non-inductive, third-person apriori entitlement to the use of that power?

Let us consider another parable. Suppose that to access mathematical reasoning in oneself, one had to see a red light. Suppose that to access higher mathematical reasoning, say analysis, one had to see a red light of a certain type juxtaposed with a green one of another type. Suppose that one is very good at analysis. As one solves more difficult problems, one develops, through the practice of one's reasoning, a warranted confidence in one's ability to solve difficult problems. Part of one's competence lies in an ability to determine whether a given problem is within the range of one's powers. To access and develop one's competence in analysis, one would have to use one's senses to reidentify the types and colors of light. But such use would clearly not, I think, be

problems simply pop into one's mind, without the slightest understanding of why they are true, or of how to use these results in further reasoning. One could learn, by checking through normal inferential means, that the answers one gave were very reliable. One could support one's acceptance of future answers through induction on past successes. But I doubt that one could be non-inductively entitled to rely on this manifestly inferential, non-perceptual capacity without justification simply because it *was* reliable. I am supposing that one lacks the slightest insight into why the answers were true, how one got them, or how to use them further. I believe that an individual's free-standing (non-interlocutionary) warrant with respect to mathematical results involves some understanding of something to be said in favor of them or something about their further mathematical uses.

This is a familiar issue in debates between externalists and internalists about epistemic warrant. Although I endorse a qualified externalism as regards both interlocution and perception, I am doubtful about an unqualified externalism that would hold that if one had a reliable capacity to be smitten with answers to difficult mathematical problems, then one would be rationally entitled to rely on that capacity purely because of its *de facto* reliability. The case that I considered in the text is not quite so unqualified. The recipient can check the source's individual answers to difficult problems, to determine that each is correct. But it is hard for me to see how the recipient can be rationally entitled to rely on the special power of the source in the absence of an induction on these successes. I am inclined to think that the recipient's "knowing how" to use the source is too thin, too much like the uncomprehending idiot savant's power, to count as a rational entitlement (in this domain). I want to leave the issue open, however. It deserves deeper exploration.

a part of a basic warrant for relying on one's reasoning or one's special abilities in analysis. It would merely be part of knowing how to start the reasoning on any given occasion. Empirical cues that go into accessing one's rational powers are not part of one's entitlement for relying on those powers to whatever degree one is entitled to rely on them. They are part of knowing how to use them.

One could, of course, construct an empirical induction: because one had gotten reliable answers to difficult problems after seeing the relevant lights, one could rely on one's answers to difficult problems in analysis after seeing the lights. But such an induction would, I think, be an unnecessary superimposition on a non-inductive entitlement that would reside in knowing how to access and use one's mathematical powers.

In such a case, one would be entitled to more-than-ordinary reliance on those powers because they are rational and more than ordinarily powerful, and because one knows how to use them, in the sense that one understands their methods and products to a degree sufficient to appreciate their power. The fact that one has to see certain types of lights to use one's powers bears only on one's knowing how to access the powers, from the standpoint of this entitlement. The perceptual experience, though a necessary condition for using one's mathematical powers, does not contribute to the force of the warrant, however psychologically essential it is to the exercise of one's powers. The warrant is a non-inductive apriori entitlement.

Learning a language and learning certain types of mathematical reasoning both require use of one's senses. One must hear the words, see the symbols, reflect on the diagrams, to acquire the abilities to reason mathematically. But such reliance on perceptual belief is normally not part of one's warrant for accepting the propositions that one comes to understand in this way. Similarly, in accessing and coming to exercise one's mathematical powers on any given occasion, perceptual belief could have been necessary. But even if it were, it need not function justificationally in one's non-inductive entitlement to rely on the rational capacity to a degree appropriate to its rational power.

In the third-person case, it is not obvious to me why perception of a computer, like perception of its symbols, cannot be a cue for drawing one's attention to a resource for one's rationality, rather as seeing red and green lights might be a prerequisite for using one's higher mathematical powers, but no part of warranting belief in their offerings. The perceptual experiences needed to tap into the resource would be part of *knowing how* to gain entry to the resource, but no indispensable part of a warrant for relying on it to whatever degree one's actual mathematical use and understanding of it allows. Using Q as a rational resource would require empirical recognition. But the recognition would only be a matter of knowing how to access the source, from the standpoint of a non-inductive entitlement to rely on its mathematical power. It would no more contribute to the force of the relevant non-inductive warrant for relying on the source than seeing the lights would contribute to the force of a warrant for relying on one's own rational powers. The entitlement would derive from one's employment and understanding of products and procedures of the mathematical resource, given that the resource was accessed.

Once one has accessed the source, an appreciation of its exceptional mathematical powers can be gained, as we have seen, by reasoning through its solutions to difficult problems (even perhaps *one* difficult problem), and understanding "from the inside" the level of problem that the power can deal with. The warrant for resting more-than-ordinary confidence in the source need not be inductively based. One's warrant resides in one's understanding the source's solutions to difficult problems. The warrant resides in one's ability to reason through, and appreciate the reliability of, the source's reasoning with respect to a range of difficult problems. By incorporating Q's mathematical work, and understanding it, one's warrant to rely on the source expands to fit the level of power that the source has exhibited.

The third-person, non-inductive apriori entitlement is analogous to the non-inductive entitlements we have discussed before. *One is entitled to more than ordinary reliance on Q's powers because they are rational and are more than normally powerful, and because one knows how to use those powers and understands their methods and products to a degree sufficient to appreciate their power.*

What happens to the problem of reidentification, with respect to the non-inductive entitlement to rest more-than-ordinary reliance in this source? Part of what it is to be entitled to rely on Q is not to need to justify the reliance in terms of attributions to Q, of the sort that formed the induction base of the inductive justification for relying on Q. So to be apriori entitled to such reliance, the recipient need not, in beliefs about Q and Q's offerings, *refer to* Q and Q's offerings as objects. The recipient need not refer to Q even in the form "the source of these [intellectually understood] presentations". The recipient must access Q's intentional contents through perceiving Q's symbols, and perhaps Q as well, to trigger intellectual understanding of the products of a single, coherent resource for rationality. But the "God's eye" articulation of the recipient's entitlement to rely repeatedly on Q can specify Q only as the source of the offerings that Q understands. And among these offerings, there is sufficient continuity, of the sort discussed in connection with the inductive justification, to entitle the recipient to presume on the sameness of the source.

No analog of the reidentification problem arises, of course, if one's appreciation of exceptional intellectual power derives from one session with Q (thinking through a large sub-part of a single proof, or, given enough background knowledge, appreciating the power and difficulty of a single, short, creative justification). But the single-session case is not essential to the point. One's warrant can be non-inductive in the normal multi-session case. The "God's eye" articulation of the entitlement need not refer to the empirical characteristics of the source. It can specify Q as the source of a series of events with intentional content that the recipient understands. Where there are appropriate continuities among the intentional contents and proofs (continuities of anaphora, definition, level of ability, and so on), the recipient can be entitled to rely on the sameness of the source even if the recipient cannot conceptualize the continuities as such.

M needed to *know how* to access and use her powers. But to have warranted confidence in them, she did not need to reason about them, beyond understanding and checking their products. The account of the warrant to rely on one's own reasoning clearly does not require that one reidentify one's reasoning capacity in that domain whenever one accesses it. Even if one must see lights in order to access one's capacity, those perceptual pre-conditions need not be specified in an induction that depends on one's reidentifying one's reasoning capacity each time one accesses it. The capacity need not figure as an object in one's reasoning. Nor need the recipient's dependence on perception in accessing the source be specified in the non-inductive entitlement which explains the recipient's rational right to rely repeatedly on the source's reasoning power. It is enough that there be actual identity of the power and perhaps some simple continuities among its products. One's ability to regularly reaccess the capacity suffices to enable one to be entitled to rely upon it.

I am hypothesizing that certain third-person cases can be similar in these respects. In reaccessing the source, one perceives its symbols and its empirical characteristics; but they play the role of triggers for access, for the immediate application of intellectual understanding, and for reliance on the source's intellectual continuities. But one need not *reidentify* it as part of an induction. One need not refer to it or its products as *objects* in one's reasoning at all, anymore than one need refer to oneself or one's own thoughts in being entitled to carry out one's own reasoning. One learns how to use the source's rational power and to rely upon it through understanding its reasoning. The entitlement is based on the rationality of the resource and on one's understanding its activity and knowing how to use it as an extension of one's point of view.

Despite the fact that, as recipient, one need not reidentify or refer to the source (or its intentional content) as an object in justificatory reasoning, one does bear an objective, third-person relation to the source. What does this third-person relation consist in? I think that there are three relevant features. First, there is a rational gap between its asserting something, or presenting it as true, and one's own acceptance of the same thing. Its presentations are not part of one's point of view unless one incorporates them. Second, there can be failure of access and failure of (non-explicational) understanding—failure of comprehension—without there being any rational failure or malfunction in one's own cognitive system. For example, one can mistakenly presume, with warrant, that a putative source is offering propositional contents, when there is really only random noise. Or one can access a source that appears to be the same as the source last accessed, but it is not. These failures need not signal any malfunction or rational failure in one's own point of view. One is subject to *brute error* with respect to one's presumptions. Third, any rational being with appropriate abilities of understanding could have been equally well placed to understand and make use of the relevant source.

And such understanding could have had the same epistemic status or warrant with respect to that source.[29]

But the *objective*, third-person relation to the source does not entail that the individual recipient need make reference to the source as *object* in some justification or warrant. The recipient must perceive the source and expressions of its intentional content. But this perceptual identification need not be specified in the account of the individual's entitlement. The entitlement for relying on the source does specify the source. But it need not specify the object's empirically discernible characteristics, or the empirical background conditions that enable the recipient to access and rely on the source. It can specify the source in the non-empirical way that the apriori inductive justification does. With respect to this non-inductive entitlement, perceptual access is as much a matter of know-how and as little a matter of object reidentification by the recipient as it is in the first-person red-light green-light cases.

It is a delicate and unresolved matter (in my mind) how to distinguish the cases in which *warrant* for continuing reliance on a source Q requires an empirical induction, or even an empirical entitlement, from the cases in which empirical recognition can be submerged into knowing how to access a rational resource. The latter cases seem to me to require that the perceivable properties of a computer or person that one uses as rational resource be relatively simple. I think that they must be incorporated into a nearly automatic routine. It is important that the recipient need not engage in context-dependent empirical (or non-empirical) tracking exercises, or complex theorizing, to reidentify the resource (type or token) through its possibly changing physical characteristics. We rely on empirical constancy for access, without having to refer to it in our warrant.[30]

If the world were such that we did have to engage in complex theorizing to access and reidentify a rational source, then perhaps we could not avoid relying on empirical reasoning to justify a presumption of repeated reliance on the authority of the same person or computer. Access-know-how would not simply be the occasion for triggering intellectual understanding. The application of the understanding would require an empirical justification.[31]

[29] These aspects, or notions, of objectivity are discussed in my 'Individualism and Self-Knowledge', *The Journal of Philosophy* 85 (1988), esp. 657ff.; and 'Reason and the First-Person', in Crispin Wright, Barry C. Smith, and Cynthia Macdonald (eds.), *Knowing Our Own Minds: Essays on Self-Knowledge* (New York: Oxford University Press, 1998), 243–270.

[30] George Boolos reminded me of Frege's apt ridicule of Schroeder's Axiom of Symbolic Stability, the idea that our justification in understanding a proof must invoke an empirical postulate about the physical stability of symbols used to express the proof. Compare Frege, *Foundations of Arithmetic*, Introduction, viii. Frege was fully aware that use of one's senses and the physical stability of symbols might well be psychologically necessary conditions for understanding mathematical reasoning over time.

[31] A similar point would apply to our understanding of what is said. If reidentifying words required a complex empirical inference, then understanding would be interpretation, not comprehension. It would depend on reasoning *about* the words and their relation to what they expressed. They would be objects of reference in one's understanding. In such cases, I think understanding itself might rest for its

As things stand, the perceptual properties of a source seem capable of functioning as triggering codes for accessing rational processes, rather as familiar words are ways of accessing meaning. Perceptual experience of the words or of the body of the source need play no role in justifying one's understanding of, or intellectual uses of, the content of the words or the presentations of the source. Perception is submerged into know-how: the conditions that make possible the intellectual understanding and continued use of a resource for rationality. Perception is only the mode of access, an enabling condition which makes no contribution to the epistemic force of the warrant.

The distinction here between background enabling conditions and justificational force is fundamental to understanding the contributions of reason and sense-perceptual experience in providing warrant for our beliefs. What is difficult is separating the role of perceptual reidentification in justifying an inference from its role in enabling one to know how to use a resource for rationality on which one is entitled to rely. Although I know of no recipe for applying it, the distinction is fundamental to developing a reasonable and nuanced epistemology. Such an epistemology must recognize the ways in which reason and understanding underwrite many of our cognitive practices, even though being able to reason and understand depend—both in general and often in particular cases—on an enabling background of sense experience. I hope that the foregoing discussion will contribute to a deeper understanding of the distinction.

I conjecture that if perception is used only to access the mathematical offerings from a single source which are then checked for validity or plausibility and found to be more-than-ordinarily reliable and powerful, then the perceptual features of the object need not play a role in one's inductive justification or non-inductive entitlement for relying on it. Insofar as one's reliance on the computer as a powerful mathematical source is the product simply of an association of perceptions of a given look, name, or model tag with a series of powerful pieces of understood mathematical reasoning, I conjecture that perception is merely a non-justificatory pre-requisite to special reliance on an authoritative source for intellection. Perception is merely a condition that enables one to make use of a resource for reason and understanding. I believe that one can, in many cases, treat recognitional aspects of another person or computer as keys to tapping a resource for rationality, in the way that one might use empirical triggers for accessing or starting-up one's own rational faculties in some domain. I am less sure of my grip on the apriori non-inductive entitlement that I have discussed than on my grip on the apriori inductive justification. But I think that both are types of apriori warrant for accepting difficult mathematical results from another being.

warrant on empirical reasoning rather than the direct application of an intellectual ability, given perceptual triggering. In such a case, one could not have an apriori entitlement to rely upon one's linguistic comprehension of particular events with intentional content.

VIII

As one comes to learn from the computer, to understand and rely on its arguments and results, it becomes analogous to one of one's own rational faculties. One's empirical activity comes to be submerged into one's knowing how to use the computer as an extension of one's own rationality. At least for those who understand the mathematics and rely on other mathematically adept persons or on computers, the knowledge that they obtain from understanding, checking, and appreciating the relevant mathematical work seems sometimes to rest on entitlements or justifications underwriting use of resources for reason. For those people, perception appears to be playing only an enabling, not a justificatory role with respect to some of their warrants. They will normally also have empirical justifications for their reliance on the computer or person. But their knowledge that the theorem has been proved can be underwritten by an extended body of warrant that is apriori. In these cases, other mathematicians, or computers, can be sources for apriori justification of one's knowledge of the source's result. Apriori knowledge can be transmitted through communication—even when the recipient cannot alone justify his knowledge, and even when the source must be accorded special authority if reliance on it is to be warranted.

I offer these reflections on the line between apriori justification or entitlement and non-justificatory empirical background conditions in a conjectural spirit. There are numerous difficult issues that I have not gone into. I commend the subject to reflection.

14 *Comprehension and Interpretation*

I will explore some relations between my account of content preservation and the accounts of radical translation and interpretation developed by W. V. Quine and Donald Davidson.[1] I center on Davidson's account, and concentrate on epistemic issues.

Quine and Davidson claim that theorizing about understanding a foreigner's words from scratch is a good model for theorizing about understanding one's communal fellow's words, and indeed all linguistic understanding. Quine writes, 'radical translation begins at home'. Davidson writes, 'All understanding of the speech of another involves radical interpretation.'[2] Both can see differences in the two cases as well as anyone. Still, they maintain that similarities are what are important in principle, or for theoretical purposes. We need to get straight about the principles and purposes.

Quine and Davidson motivate using the radical situation as a model for understanding all speech by considering coming to understand a language. Quine writes,

> We have been beaten into an outward conformity to an outward standard; and thus it is that when I correlate your sentences with mine by the simple rule of phonetic correspondence, I find that the public circumstances of your affirmations and denials agree pretty well with those of my own. . . . The case of the linguist and his newly discovered heathen . . . differs simply in that the linguist has to grope for a general sentence-to-sentence correlation that will make the public circumstances of the heathen's affirmations and denials match up tolerably with the circumstances of the linguist's own.[3]

I gave an earlier version of this essay at Rutgers University in 1993 on a week-long panel with W. V. Quine and Donald Davidson. I am grateful to both for comments that improved the paper, and to Davidson for pointing me to work of his that improved my understanding of his views. I have benefited from giving the paper at Brandeis, and at Harvard as the first of two Santayana Lectures, in the Fall of 1998. Bob Nozick, Hilary Putnam, and David Wong made comments on these two occasions that led to improvements.

[1] Compare my 'Content Preservation', *The Philosophical Review* 102 (1993), 457–488; W. V. Quine, *Word and Object* (Cambridge, Mass.: MIT Press, 1960), chapter 2; Donald Davidson, *Inquiries into Truth and Interpretation* (Oxford: Oxford University Press, 1984), chapters 9–12.

[2] W. V. Quine, *Ontological Relativity and other Essays* (New York: Columbia University Press, 1969), 46; Donald Davidson, 'Radical Interpretation' in *Inquiries*, 125.

[3] Quine, 'Speaking of Objects', *Ontological Relativity*, 5.

Davidson writes, 'what evidence plausibly available to a potential interpreter would support the theory to a reasonable degree?' He continues, 'To deal with the general case, the evidence must be of a sort that would be available to someone who does not already know how to interpret utterances the theory is designed to cover'.[4]

I think that the line of reasoning is this. People understand what others say. Since they must start from a position of not knowing whatever language they learn, and since it is possible to understand a foreigner from scratch, this understanding must rest ultimately on non-linguistic evidence—evidence that does not already involve a construal of words. One must use a person's relations to an environment and the person's behavior as evidence for what the person's words mean. So a general theory of linguistic understanding should explain how one could get from non-linguistic evidence to linguistic understanding in any given case—even the hardest case. Communication among people who know one another, or who speak the same language, uses a mixture of contextual evidence and entrenched assumptions. But these assumptions go back to past, publicly available non-linguistic evidence. So fundamentally, the domestic and radical situations are the same.

Since Quine (*in absentia*) and Davidson (*viva voce*) were my most influential teachers, this original and fruitful line was basic to my education in philosophy. It is clearly on to something. But it passes over important distinctions. Use of the term "evidence" to cover the innate background, the training, and the contextual bases for understanding is a case in point. I believe that one must distinguish, more sharply than the radical interpretation model does, differences between the psychology and epistemology of understanding. Quine blurs the distinction on principle. I am less certain how Davidson views it.

Davidson's aim in developing a theory of interpretation is different from my aim in investigating the epistemology of linguistic understanding. Davidson presents his account as a rational reconstruction, explaining neither the actual epistemology nor the actual psychology of communication. The account is meant to explain one way that someone *could* understand another.[5] It thus answers to the Kant-like question 'how is linguistic understanding possible?' where the answer may give one of various possible ways. Although it is supposed to represent what an interpreter knows (presumably the truth conditions of sentences understood), it is not meant to explain how he knows it. Thus it is not assumed that an interpreter actually has, even tacitly, the truth theory that the model outlines.

Davidson's account of the relevance of his theory to actual psychology is sometimes elusive. He counts interpretation theory a 'model of the interpreter's linguistic competence', adding that 'claims about what would constitute a

[4] Davidson, 'Radical Interpretation', *Inquiries*, 125, 128.

[5] Davidson, 'Radical Interpretation', *Inquiries*, 125; 'The Structure and Content of Truth', *The Journal of Philosophy* 87 (1990), 324f., esp. note 67. Compare also 'Belief and the Basis of Meaning', *Inquiries*, 141.

satisfactory theory are not . . . claims about the propositional knowledge of an interpreter, nor are they claims about the details of the inner workings of some part of the brain'.[6] There follows this difficult sequence of statements:

> They [claims about what would constitute a satisfactory theory] are rather claims about what must be said to give a satisfactory description of the competence of the interpreter. *We* cannot describe what an interpreter can do except by appeal to a recursive theory of a certain sort. It does not add anything to this thesis to say that if the theory does correctly describe the competence of an interpreter, some mechanism in the interpreter must correspond to the theory.

If the radical interpretation model is a claim about what *must* be said to give a satisfactory description of competence, it is not clear wherein the model concerns only *our* descriptions of interpreters' competencies rather than the structure and content of the competencies themselves. Perhaps Davidson is being non-realist about psychological descriptions generally. But if he is not (as I am certainly not), it seems to me that anything that must be said to give a satisfactory description of the competence of the interpreter is a true description of part of the competence of an interpreter.

I see no difficulty in principle in taking truth-theoretic interpretation systems to correspond to actual psychological structures. The competence described is certainly not explicit, conscious, knowledge. Whether the theory would describe a *basic* psychological system for understanding others seems open. It could describe a system superimposed on a basic system only when an individual develops concepts of truth and reference. I believe that communicators understand utterances from the beginning by associating them with truth conditions—with intentional content that sets conditions on what it would take to make them true. It is an open question whether the semantical concepts are ground-floor concepts (either innate, or present as soon as communication is), or whether they are later conceptualizations of non-conceptualized operations that assign truth conditions.[7]

As long as the best psychological accounts continue to treat the competence as a computational mechanism, Davidson's approach may well have a place in psychological theory. My epistemic picture will be broadly compatible with truth-theoretic interpretation systems taken as structures of a psychological competence, possible or actual.[8]

[6] Donald Davidson, 'A Nice Derangement of Epitaphs', in *Truth and Interpretation* (Oxford: Basil Blackwell, 1986), 438.

[7] I do not accept Davidson's claim that to have beliefs one must *understand* (in the sense of conceptualize) the possibility of being mistaken, hence that to have a belief one must have the concepts of truth and falsity. Compare 'Thought and Talk' in *Inquiries*, 170.

[8] This point suggests that Quine's account of translation, which shares many basic features with Davidson's account of interpretation, may also be a contribution to psychology. The pleasant irony does not escape me.

Even where Davidson considers epistemic warrant, his project is orthogonal to mine. He asks what evidence would suffice to make linguistic understanding possible. I ask what warrants underwrite actual understanding. For all its value for other purposes, the interpretation model is likely to mislead as regards this question.

Davidson's rational reconstruction emphasizes similarities between domestic and radical cases. I believe that the domestic and radical cases are epistemically very different. The differences can be usefully developed by first considering differences between standing and contextual elements within *domestic* understanding.

There is a traditional and worthwhile distinction between elements of linguistic understanding that are "standing" or "constant" within an individual's repertoire and elements that depend on "context". Logical constants, common nouns, verbs, prepositions, adjectives are items in one's language or idiolect whose understanding does not depend in a linguistically systematic way on the context in which they are used. Indexicals, tense, demonstratives, proper names depend for a primary aspect of their construal on a context of use. There are constant elements in the context-dependent devices, and contextual elements in the constant ones. But at our level of description, these reminders will suffice.

Context can figure in another way in understanding, as Davidson has emphasized. One can use constant aspects of one's language to yield meaning in context that is at variance with or supplemental to one's ordinary understanding. Irony, humor, metaphor, conversational implicature, and speaker meaning are cases in point.

Whereas understanding contextual devices commonly depends essentially on background knowledge and generalized intelligence, standing linguistic understanding is commonly a specifically linguistic competence. Of course, in understanding even standing aspects of language, one relies on a background of belief about non-linguistic matters. And in acquiring linguistic understanding, one must rely on attitudes about non-linguistic matters. But I assume, with backing from empirical psycholinguistics, that linguistic understanding is in part a distinctive ability.

As a way of developing epistemic implications of this distinction between standing linguistic understanding and contextual understanding, I want to go over some familiar ground in the epistemology of perceptual belief. One *can* justify perceptual beliefs. One can justify one's perceptual belief that there is a brown lectern, by saying that one knows that one has a perceptual experience as of a brown lectern (or alternatively that one has stimulations commonly caused by brown lecterns), that one has checked one's perceptual apparatus and insured that no sources of illusion are in play, and that having such experiences under such conditions makes it reasonable to believe that there is a brown lectern in front of one. But this routine is not routine. The individual need not even be able to go through it to be warranted in perceptual beliefs in ordinary circumstances.

To be warranted in one's perceptual belief, one normally need not refer to—or be able to refer to—experiences, seemings, stimulations, ordinary conditions, possible sources of error, or the like. One need not be able to *justify* one's perceptual belief at all. One is normally warranted in one's perceptual belief straightway, without reason, evidence, or justification. In my terminology, one is *entitled* to the belief without raising questions about it, unless specific contextual grounds for doubting it arise. Entitlement is an epistemic warrant that need not be fully available to the warranted individual. The individual need not even have the concepts to explain the warrant. Entitlement thus contrasts with justification, including unarticulated but operative justification, the kind of warrant that involves evidence or reason accessible to the individual on reflection. One can be warranted in one's perceptual beliefs, without being able to justify those beliefs, and even without having the concepts to do so. Such warrant requires having perceptual experience, but it requires no reference to it. Experience does not function as evidence which the individual need think about.

Ordinary understanding of constant aspects of words used by another person is, I think, in this way analogous to ordinary perception. In neither case is evidence or justification needed for warrant. The point is not that justification is unconscious. It is that justification need not even be conceptually available to the individual on reflection. The warrant for the understanding does not rest on an unconscious transition from a reason or evidence. Nor is the warrant a matter purely of coherence with other beliefs. The reliability of the competence (in perception or understanding) is the main source of the individual's warrant. One develops a competence to take in what other people say, when they use words one shares with them. Unless specific contextual grounds arise for doubting one's understanding, one is entitled to it.

Talk of evidence and theory is misleading here in the way that it is in talking about ordinary perception. Perceiving the words and behavior of the interlocutor need be no more *evidence* for understanding, in ordinary cases of smooth domestic communication, than stimulations or sense impressions and the background conditions for perception are, ordinarily, evidence for perceptual beliefs. In both cases, what would mistakenly be called evidence is part of the causal enabling conditions for having and exercising the competence. The sense-perceptual presentations or experiences contribute to the individual's warrant. But they do not do so by being reasons or evidence that the individual needs to have conceptualized as such, or needs to refer to. Sophisticates can think about experiences as reliable signs of the physical objects the perceptual beliefs are about. But such meta-inferences are supplementary justifications, at best. The fundamental perceptual warrant is an entitlement that does not go through such inferences.

Similarly, with understanding. We can reconstruct an inference from perceptual beliefs about words and behavior to our understanding. But such an inference would be supplementary. The basic warrant for distinctively linguistic understanding is an entitlement that rests on the content and reliability of our non-inferential understanding.

Of course, the psychology of understanding depends on perceiving words and having some mechanism that assigns them meaning. The psychology of perception depends on having sense impressions and having some complex mechanism that assigns perceptual content to those impressions. In both cases, psychology may describe the mechanism in computational terms. But it is a mistake to take psychology as itself an account of our warrant for (entitlement to) perceptual belief or standing linguistic understanding. For the points about an individual's subsystems do not track the *individual's* reasons.

Wherein lie one's entitlements to perception and ordinary linguistic understanding? The general *pro tanto* entitlement to perceptual belief resides in one's status as a perceiver, and one's default entitlement to presume on that status. Being a perceiver *necessarily* involves certain reliabilities in perceiving normal perceptual objects in normal circumstances. The entitlement derives from reliabilities coded in the causal, transformational, and presentational conditions that are constitutive of being a perceiver, and in the workings of the perceptual system. The same point applies to linguistic understanding. Being an understander necessarily involves certain automatic reliabilities in understanding expressions. Warrant resides in the reliabilities encoded in this cognitive competence and in the exercise of the competence.

All the preceding pertains to *standing* linguistic abilities applied in normal domestic cases. To understand many *context-dependent* devices one must exercise different abilities. To understand the application of the demonstrative in "That is beautiful", one must normally find the physical referent and note the angle of perception on it. One's standing linguistic competence will not suffice for understanding the utterance. In this case, exercise of non-linguistic perceptual abilities seems integral to the understanding.[9]

There is the second class of contextual elements to consider. Here also, understanding involves assignments of meaning that commonly go beyond standing linguistic competence. In understanding conversational implicature, for example, standing competence is supplemented with reasoning about the speaker's intentions.[10] This case is analogous to cases in which one has

[9] The analog to such context-dependence in the perceptual case is not immediately salient. Here is a case. A certain type of visual perception might never in itself distinguish for the individual between two sorts of object, only one of which is what the individual's perceptual apparatus and reaction system has evolved to discern. The individual is then not entitled to rely on visual perception alone to yield a warranted justified perceptual belief. Perhaps vision must be supplemented by touch, or the word of an adult, to yield a warranted perceptual belief. Young monkeys cannot distinguish different types of eagles, as seen from below, only some of which are predators. They give predator calls for all of them. But their escaping action is often contextually dependent on the further predator call of an adult (which can make the distinctions). The adult in effect provides needed confirmation of the youth's perception of a given bird as a member of a predator species. The example is suggestive regardless of whether one thinks, as I do, that monkeys have warranted perceptual beliefs, with (roughly) the content of "there is a predator up there". Compare D. Cheney and R. Seyfarth, *How Monkeys See the World* (Chicago: University of Chicago Press, 1990), 129ff.

[10] Grice identified the key epistemic feature of conversational implicatures: one's understanding has to be "worked out" from contextual considerations, rather than inferred "intuitively" from

contextual reason to doubt one's perceptual presentation of an object, and one must invoke further evidence to be justified in forming a belief.

So I distinguish two sources of justificational force or warrant. First, there are entitlements deriving just from applying standing cognitive abilities. One is immediately though defeasibly entitled to a perceptual belief or to understanding aspects of an utterance. Then, second, there are warrants that derive partly from contextual, non-linguistic supplementation of such entitlements.

How does Davidson's position on evidence relate to this picture of the epistemology of domestic communication? Since Davidson intends his remarks as an idealized account of psychological input and psychological structure, my epistemic points might be seen as entirely compatible with this approach.

But even on this irenic construal, I think that the interpretation model underplays differences between radical and domestic cases. The radical case is a matter of theory development until the foreign language is learned. We use an already mastered language to interpret forms over which we have no competence, by thinking about words as objects to be theorized about. In domestic cases, we are competent with the same words our interlocutors use. The vast preponderance of the time, we presume on understanding the speaker's words as he or she does.

In the domestic case, we do not use evidence much to understand what another says, except to correct or supplement the standing mastery that we already have. Normally we need not scrutinize the behavior of a stranger who uses words familiar to us. When we talk on the phone, or when we read, we have no new world–word relations and little or no non-linguistic behavior to rely upon. The standing mechanism yields an enormous amount of understanding. It is so reliable that the unsophisticated are slow to recognize verbal disputes. In treating the radical and domestic cases alike, interpretation theory passes over epistemic differences associated with the reliability of linguistic competence.

The interpretation model's portrayal of word–world, behavioral, and psychological facts that go into fixing content as *evidence* does have application to idiolect variations, tongue slips, irony, conversational implicature. Davidson fixes on malapropism in domestic communication because it maximizes analogies between the domestic and foreign cases.[11] But malapropism is not the norm.

meaning or constant understanding. Paul Grice, *Studies in the Way of Words* (Cambridge, Mass.: Harvard University Press, 1989), 28–31.

[11] 'A Nice Derangement of Epitaphs'. I believe that Davidson underrates the role of shared linguistic framework in communication. He calls it 'an enormous convenience' (p. 438) that many people speak in similar ways. He maintains that in principle communication 'does not demand that any two people speak the same language'. I think that this latter is true. But it does not follow that an account of the epistemology of communication can dispense with presumptions about overlapping linguistic frameworks and competencies. Davidson points out that the linguistic elements that are shared are not sufficient to enable one to understand all that is said, and that one can expect variations in what is shared from one speaker to another (p. 444). He seems to infer from these points that shared linguistic frameworks do not characterize a communicator's linguistic competence. But no one ever thought that such frameworks did the whole job of characterizing linguistic competence, and variations are easy to allow on the presumption of substantial overlap.

The norm in normal linguistic communities is taking one's homophonic understanding for granted. Things go wrong often, but not significantly so in comparison to their going right.

No one need disagree with this. But there is danger, even for an account of competence, in taking the abnormal as a paradigm of the normal. Since I am not centering on psychology, I will not ask how far assimilating ordinary communication to interpretation might distort psychological theory.

Let us pursue the epistemological issue. On a natural extrapolation of the interpretation model, domestic communication is like radical interpretation except that the evidence used in standing understanding is held over, unconsciously, from past cases. In past cases, one is justified because one rested understanding on behavioral and environmental evidence. What grounded past understanding grounds understanding of new interlocutors who sound similar, unless there are reasons to call this background evidence into question. So in communication that smoothly treads well-worn paths, one need not attend to contextual evidence about what standing elements in speech mean.

One doubt about this empirical-theorizing model derives from the unlikelihood that many have grounds that support the inferential justification that the theory postulates. Competent neophyte understanders have no inductive-explanatory reason to think that if a stranger makes sounds that they have previously understood, the stranger is likely to mean what others meant by those sounds. Yet competent, neophyte understanders seem warranted in relying on unreasoned understanding of strangers, until things go wrong. There is no doubt sociological evidence that underwrites such an inference. But many do not have it, and they do not seem to need it to be warranted in presuming on understanding.

One could avoid appeal to inductive-explanatory evidence and invoke the idea of justification through coherence with other things one believes. But we seem to be warranted in understanding single utterances by strangers out of the blue. Here, coherence carries little weight. In *most* cases, any of a wide range of interpretations is equally coherent with everything we know.

Some of one's standing understanding derives from training during a period in which one lacked a language or theory. Assimilation of this background to evidence used in the radical situation is especially problematic for epistemology. The point applies not merely to past behavioral and word–world evidence. It extends to perception of the words themselves. All our exposure to what the empirical-theory account treats as "evidence" goes into the formation or exercise of our competence. But not all of what goes into how we gained the competence enters into the justificational basis for our understanding of utterances. This was a lesson of the comparison of understanding with perception. The fact that sensations, or irritations of our nerve endings, play a necessary role in mediating our perception of objects should not lead one to conclude that they play the role of evidence for our perceptual beliefs. Much of what the empirical-theorizing model counts as "background" evidence in the interpretation of strangers is merely part of the formation conditions or enabling conditions for our linguistic competence.

I would like to distinguish two kinds of understanding.[12] Let us label them "*comprehension*" and "*interpretation*". Comprehension is understanding that is epistemically immediate, unreasoned, and non-inferential. First-person comprehension is the minimal understanding presupposed in any thinking, in beings that understand their thoughts at all. (I believe that animals think but lack comprehension of what they think, since they lack a third-person perspective on thought content.) One thinks a given intentional content, and comprehends that content and its force in thinking it. What one comprehends is not ordinarily thought *about*. Of course, one can comprehend meta-thoughts that take thoughts or thought contents as their referential objects. But normally one comprehends thought contents that apply to other things.

I include words, in a derivative sense, as things one can comprehend in the first-person way. One comprehends the words in one's idiolect as one uses them. The comprehended words are the *direct* expression of thoughts one comprehends. They express one's thoughts without mediation of further words or thoughts.

Comprehension in the third-person way is understanding that is epistemically immediate, unreasoned, and non-inferential and that carries no presumption that the comprehended material is one's own. It may *be* one's own. But it is comprehended without relying on taking it as one's own immediate or remembered product. I shall shortly discuss whether one can have third-person comprehension of others' thoughts and words.

Interpretation involves taking what is interpreted as the primary object of epistemic interest. Interpretation arises out of there being a question or issue about how to understand a candidate object of interpretation. Interpretation is always from the third-person point of view. I conjecture that it is always epistemically inferential. What one interprets may be a thought content, or it may be words. It may be directed toward one's own products as well as those of others. But in interpreting, one necessarily carries out the interpretation of the object of cognitive interest in terms of thoughts one presumes—and is entitled to presume—one comprehends.[13]

The distinction is a functional one. It concerns one's epistemic starting point. One must comprehend something to interpret something else. One cannot

[12] Wittgenstein famously, though with characteristic obscurity, insisted on a related distinction in his dictum 'There is a way of grasping a rule which is not an interpretation.' Compare *Investigations*, 201. Michael Dummett, 'Comments on Davidson and Hacking', in *Truth and Interpretation*, cites Wittgenstein and criticizes Davidson's treatment of communication as involving interpretation. Dummett does not note an important difference between Davidson's conception of interpretation and Wittgenstein's. Wittgenstein sees an interpretation as a substitution of one linguistic expression for another. Davidson's conception of interpretation is decidedly not one of relating words to words. My conception of interpretation makes no use of this point of Wittgenstein's. I do not assume that my conception of interpretation is Davidson's, but I think that it is strongly suggested by his interpretation model.

[13] Interpreting something for someone else may involve conveying an interpretation that one could oneself simply comprehend. This is a different use of "interpret" than the one I focus upon.

comprehend something that is interpreted until it no longer functions as object of interpretation. What one comprehends is a content one can take for granted.

In saying that the distinction is functional, I am allowing that it is relative. What one comprehends could become an object of interpretation if some question arises about the comprehension. But to pursue such a question, one must use other contents that one presumes to be comprehended. However fluid, the distinction is fundamental and inevitable. A critical reasoner *must* presume comprehension of some intentional content to engage in any thought, including any interpretation. Interpretation, by contrast, does not inform all critical thinking. So the understanding that I call comprehension is functionally basic.

First-person comprehension can be criticized as more or less good. One can worry about whether one's understanding is flawed or incomplete. This is a constant theme in philosophy. But it occurs in the progress of science and in ordinary life. Learning more is not always sharply separable from enriching or even correcting one's comprehension. Our thought contents are fixed by a web of inferential and applicational connections. The inferentially associated thoughts which help fix the meaning or content of our thought contents make substantive and fallible commitments. Although no one capable of any comprehension at all can fail to comprehend their own thoughts—in the sense that minimal, unreasoned understanding of the thoughts is necessary to thinking them—the quality and depth of the comprehension can vary. Minimal comprehension can be infected with erroneous commitments. Improvement of comprehension can utilize self-interpretation. It can involve regarding the comprehension as an object to be reasoned about from the point of view of an outsider. Or it can utilize the sort of critical, elaborative reflection on conceptual connections that stays within a comprehended point of view.

My primary claims do not depend on holding that first-person comprehension can be better or worse, though the argument I have just given shows that it can be. What I want to insist on here is that even minimal first-person comprehension falls within the broad domain of norms and warrant. This point derives from two sources. One is the insight—variously developed by Kant, Frege, Wittgenstein, Quine, and Davidson—that understanding is essentially associated with inferential, applicational, explicational abilities. These abilities involve propositional commitments that may be true or false. The other source is Quine's insight, richly developed by Putnam, that even the most fundamental propositional summaries of meaning or understanding are about a subject matter, can be true or false, and are thus subject to epistemic norms.

What is the nature of our warrant—our entitlement—to first-person comprehension of our thoughts and words? The entitlement is non-sense-perceptual—apriori. To be apriori a warrant's justificational force must be independent from contributions to that force from sense experience or sense-perceptual belief. Apriority concerns the force of a warrant, not its vulnerability to criticism. Some beliefs that are apriori warranted are subject to empirical correction. As traditional rationalists noted, a warrant can be apriori or non-sense-perceptual even if sense experience is necessary for the acquisition of the understanding involved in the warranted cognitive abilities. For a warrant to be apriori it is enough that sense experience and sense-perceptual belief not contribute to its force.

I believe that we have an apriori *pro tanto* entitlement to our minimal comprehension of our own thoughts and words—the sort of comprehension necessary to think or use them at all. Unless some specific reason arises to question comprehension, we are warranted in relying on it; the force of the warrant does not derive from sense experience. Let me motivate and provide an argument for this claim that the entitlement is non-sense-perceptual. The issue of rational warrant that interests me arises for non-modular attitudes and capacities of beings with minimal understanding. The relevant attitudes and capacities require minimal understanding—comprehension. In the relevant cases, such beings must presuppose comprehension of an intentional content (and words directly expressing the content) in taking anything as an object of cognitive interest. So all rational warrant for the relevant attitudes requires as a necessary condition that such minimal understanding is in place. Such rational beings cannot be warranted, in relevant cases, in an attitude whose content they do not minimally understand (comprehend), at least for any warrants and attitudes that depend on understanding. Any interpretation or questioning of one's thoughts or words must presume a comprehension of content in terms of which those activities of questioning are carried out. Raising questions without ground is rationally capricious. So such beings have a rational entitlement to unquestioned reliance on comprehension of their thoughts and words, unless there is reason to question the nature or quality of that comprehension.

This account of our entitlement derives purely from conditions for thought of certain rational beings. Sense-perceptual experience is not epistemically relevant. The entitlement to the presumption of this comprehension is non-sense-perceptual. All this, notwithstanding the fact that comprehension depends on sense experience for its genesis, and may be subject to questioning, interpretation, and even revision.[14]

Traditionally, the view that apriori warrants derive from understanding was applied to understanding of thought contents. I will assume that comprehension of one's idiolect is subject to substantially the same points about entitlement that I have just made.

I conjecture that interpretation requires justification, not merely entitlement. The justification need not be conscious. But to be warranted, an interpretation must be backed by grounds or evidence available on reflection to the interpreter. The warrant for most interpretation is empirical. As I shall later note, some such warrant can be apriori.

[14] I have discussed these matters further in 'Content Preservation', and 'Interlocution, Perception, and Memory', *Philosophical Studies* 86 (1997), 21–47.

I have argued that questions of warrant for attitudes of certain rational beings require presuming a starting point in comprehension. One is apriori *pro tanto* entitled to rely upon comprehension of one's own thoughts and language, unless reason for doubt arises.

I turn now to warrants for understanding others. I want to discuss three questions. First, in domestic communication that depends on constant linguistic understanding and that does not require contextual supplementation or correction, is the understanding of *another's* speech always interpretation? Or is it often comprehension? I will take the answer to hinge on a second, closely related question: Are we ever entitled to rely on our presumptive understanding of another person without evidence or justification? If the answer is affirmative, our understanding of others' speech is like our perceptual beliefs and our understanding of our own thoughts and speech, in that we have a prima facie entitlement that requires of us no justificatory inference or evidence, however unconscious. I have already signaled my answers to these two questions. But I wish to develop these answers more fully. Third, if understanding another is sometimes comprehension, is the entitlement always empirical, as our entitlement to our perceptual beliefs is? Or is the entitlement sometimes apriori or non-sense-perceptual, as our entitlement to rely on first-person comprehension is? I do not hope to answer these questions decisively here. But I want to engage them. Let me begin with the second.

Can one ever reasonably presume, as a defeasible default position, that one comprehends what others say?[15] Is the warrant for such understanding ever an entitlement that requires no evidence or justification? I assume that there is always some unconscious psychological transformation from perception of words to understanding. I question whether, if the understanding is to be warranted, the perceptual basis for the transformation must provide evidence for that understanding. Similarly, I assume that in learning words, one's understanding is formed through associating behavior, words, and objects in the physical world. I question whether the understanding's being warranted requires that these associations justify the understanding through evidential relations, or inferences to the best explanation, or coherence with antecedent beliefs.

There seems little intuitive ground to think that evidence of these sorts is needed to support untrammeled domestic understanding. Not only are we not conscious of inferences. In most cases, explicating the justifications is well-nigh impossible—not only for ordinary speakers, but even for philosophers. Barring

[15] Davidson writes, 'Speakers of the same language can go on the assumption that for them the same expressions are to be interpreted in the same way, but this does not indicate what justifies the assumption. All understanding of the speech of another involves radical interpretation', 'Radical Interpretation', *Inquiries*, 125. Davidson seems to suggest here that what justifies the assumption is the evidence for radical interpretation. On my view, the assumption is warranted by an entitlement that need not be supported by any evidence available to the individual. The "evidence" postulated by the theory of radical interpretation need not function as evidence in the individual's warrant for understanding standing aspects of another's speech.

anomaly, understanding an interlocutor's speech, on ordinary topics, seems no less immediate than understanding one's own speech in thinking out loud.

These points are not decisive. One could postulate fast, unconscious inferences. There are psychologically relevant transformations. But these are mostly not person-level inferences. Moreover, the postulation would have to show some advantage in accounting for epistemic warrant. For it has no intuitive plausibility for ordinary understanding.

I am impressed with analogies to epistemically non-inferential epistemic abilities. As with perception, and comprehension of one's own speech, the processes underlying domestic understanding of others are fast, unconscious, difficult to articulate, nearly automatic, almost modular, and very reliable in ordinary contexts. Inferences relevant to epistemology are acts in the central cognitive system, though often unconscious ones. They are acts by the individual, not processes in subsystems. The transformations present in these three cases normally do not involve acts by individuals.

One reason why it is unreasonable to see ordinary perception as warranted by an inference from evidence about sensations or perceptual presentations to a conclusion about physical properties derives from the degree to which sensations are not an object of cognitive focus. The function of the central perceptual systems (especially vision, touch, and hearing) is to facilitate the individual's interaction with the physical world. For higher animals it is to enable the individual to form reliable beliefs about physical objects and properties. Probably only critically rational beings conceptualize sensations or perceptual representations involved in perception of physical properties—conceptualize them as distinct from the physical properties they indicate. Such beings come to do so long after they have warranted perceptual beliefs about physical properties. Moreover, memory of the properties of sensations or perceptual representations, in ordinary perception of physical objects or properties, is less reliable and more ephemeral than memory of the physical items perceived. These points tell against taking the primary warrant for perceptual beliefs as an inference from properties of sensations. The sensations or perceptions are psychological stepping stones which once negotiated sink out of sight. They and beliefs about them are not ground-level evidential bases for perceptual beliefs.

I think that similar considerations tell against treating ordinary domestic understanding of others' speech as an inference starting with observations about words or behavior. Understanding depends on perceptual awareness and implicit memory of words in something like the way that perception depends on utilization of sensations or perceptual representations in forming perceptual beliefs. But forming beliefs about the properties of words is no more the aim of understanding than forming beliefs about sensations or perceptions is the aim of perception. Understanding of speech may well precede an ability to conceptualize and form beliefs about a distinction between words and the objects they indicate or the meanings they convey. It is well known that memory of the properties of words is less reliable and more ephemeral than memory of what one understands through the words. These considerations suggest that our warrant for understanding others' words in domestic cases does not rest on inferences involving beliefs about words as such.

A further ground for doubting the empirical theorizing model for domestic understanding rests on how we reason about justification in this domain. When communication runs smoothly, the question of justifying one's understanding does not seem to arise. It is no more in place to ask someone who is a perfectly competent language user to support his or her presumed understanding of someone who says "push-button telephones are more common than rotary ones" than it is to ask a normal perceiver how he or she justifies a perceptual belief that that is a brown lectern, when he or she is looking at one in a good light. These questions about warrant are philosophers' questions. Addressing them well requires giving weight to the fact that they do not arise in that form in ordinary life.

Normal perception and perceptual belief involves a default warrant (an entitlement) that requires no justification from the individual. Justification is required only when specific difficulties arise. If the light is bad, one may need to justify one's attribution of a color. If holograms are afoot, one may need to justify a belief that it is a lectern rather than a hologram. But then specific alternatives are relevant and threatening.[16] In the absence of such alternatives, one's warrant resides in one's being a well-functioning perceiver and having the perceptual experiences. Being a perceiver involves having a perceptual system formed through reliable connection with entities categorized by the perceptual representations of the system. To be warranted in perceptual beliefs, one need not be able fully to conceptualize the warrant. One needs no justification or evidence in any space of reasons available to reflection (or even "unconsciously").

Similar points apply to domestic understanding. Linguistic training gives one a reliable understanding of what others say. Status as a competent understander and normal use of associated conceptual apparatus yields a defeasible warrant that obviates the need for evidence or justification. Justification is needed only when anomalies arise, or when one cannot rely on the transformations afforded by presumptive overlap with one's own idiolect. It is not a from-the-beginning open question what someone else says, if a reliable understander presumes on seeming immediate understanding. The understander is prima facie entitled to immediate presumptive understanding.[17]

[16] For an introduction to the relevant possibilities idea in theory of perception, see Fred Dretske, 'Epistemic Operators', *The Journal of Philosophy* 67 (1970), 1007–1023; 'The Pragmatic Dimension of Knowledge', *Philosophical Studies* 40 (1981), 363–378; *Knowledge and the Flow of Information* (Cambridge, Mass.: MIT Press, 1981).

[17] These claims may be compatible with Davidson's theory, in view of its rational-reconstruction methodology. But I think that Davidson's emphasis on seeking evidence that enables us to deal with "the general case" (cf. the passage cited at notes 4 and 15) tends to ignore the restrictions on relevant possibilities and the obviation of the need for evidence that come with an individual's being contingently embedded in a relatively homogeneous speech community. Similarly, Davidson's emphasis on cases, such as malapropisms, in which standing understanding becomes problematic tends to suggest that because of the possibility of non-standard cases, evidence is always needed. Cf.

Our usage does diverge from that of our fellows. People need not use words in the same way in order to communicate. But in normal communities, except where divergences appear to occur, there is no rational ground for departing from, or having to justify, the presumption of comprehension. A person's use of words could diverge from others' so completely as to leave no overlap. Then the person would be unable to comprehend others' speech and would have to rely on interpretation. But this no more shows that the epistemology of our actual situation centers on interpretation, as opposed to comprehension, than the fact that an individual could be placed in a world where perceptual inclinations could not be trusted, and had always to be double checked, shows that actual perception is epistemically dependent on inference and checking. Warrant for comprehension in the domestic case derives from contingent facts about reliability in a given social environment.

So my answer to the second question is that one's warrant for understanding others' speech and writing is commonly an entitlement, not based on evidence, inference, or coherence. My answer to the first question is that understanding others in domestic cases is often comprehension, not interpretation.

What of the third question? Is all such entitlement empirical, as our entitlement to perceptual beliefs is? Or is one sometimes entitled through some non-empirical warrant to understanding of another's speech, as one is entitled to rely on comprehension of one's own thoughts and words? I think that the basic entitlement is sometimes non-sense-perceptual, non-empirical.[18] The entitlement derives from one's status as a competent understander, and from the conceptual aspects of understanding. Sense perception is necessary for this understanding; but the warrant need not lean on it for justificational force.

I believe that we have a *general* apriori *pro tanto* entitlement to rely on understanding of others' words that are immediately prima facie intelligible to us. In certain circumstances, we can be apriori defeasibly entitled to comprehension of what others say *in particular cases* when what they say is immediately prima facie intelligible to us. The justificational force of the entitlement derives from the role of understanding in intellection. It rests on no *epistemic* underpinning from perception.

One reason for my view is that comprehending standing, conceptual aspects of one's own thought and idiolect is itself, as a matter of psychological and sociological fact, normally dependent on having comprehended thoughts (one's own) that were shaped and expressed through the words of others. Even innate non-linguistic concepts are commonly associated with one's words only via

'A Nice Derangement of Epitaphs' in *Truth and Interpretation*, especially 442–443. But we need evidence only when such cases threaten the presumption, warranted for competent understanders, that we understand what we seem to understand.

[18] I have argued for these claims in 'Content Preservation', and in 'Interlocution, Perception, and Memory'. In cases where we have a non-empirical entitlement to comprehension of others' speech, we may simultaneously have other entitlements or justifications that are empirical.

understanding them as expressed by others' words. So homophonic comprehension of one's own words is normally interwoven with homophonic understanding of others'. So it cannot in general provide a prior framework for empirically grounded third-person understanding. This is a primary source of reliability in understanding others. Our reliability derives from the fact that understanding others to a large degree constitutes the nature of comprehension of our own idiolects, the entitlement to which is clearly apriori.

This is not a traditional private language argument. It does not depend on a claim that we metaphysically must begin with others' speech. The point is that that is how we do begin—and probably psychologically must begin. Epistemic norms arise out of actual situations of reliability and interdependence—rather as the norms for perceptual warrant center on facts about what is reliable in the environment that the perceiver is embedded in. There is no metaphysical necessity that the perceiver *be* in an environment that makes perception reliable (though one's perceptual system must have been formed in such an environment). Entitlements to comprehension rest on contingent, external reliabilities outside the control and perhaps even ken of the individual.

Here is not the place to explore differences between my and Davidson's views on the role of the social in linguistic understanding. It is, however, worth indicating where the chief difference lies. Davidson thinks that to have the concept of belief, and even to have beliefs at all, an individual must actually enter into a communication-interpretation relation with another language user. Actual social relations are necessary to thinking—not merely psychologically necessary but metaphysically necessary, or necessary in whatever way that goes with Davidson's apriori arguments for his view.[19] I do not accept these claims. I think that it is metaphysically possible for an individual to have beliefs without language (in fact, higher animals are actual examples), to have language without social relations, and to have the concept of belief without social relations.

I think it metaphysically necessary that certain attitudes, given certain facts about the individual, depend for their intentional content on relations that they bear to others' uses of language. The individual need not be in those relations to think. But to think thoughts he does think, given his actual intentions, relations to others, and limits on background knowledge, the individual (metaphysically) must have been in those relations. I locate the social dependence of mental states at a less global level than Davidson does. The necessities in my account depend on contingent parameters' being fixed.[20]

[19] Donald Davidson, 'Thought and Talk', in *Inquiries*, p. 170; 'Three Varieties of Knowledge', in A. Phillips Griffiths (ed) *A. J. Ayer Memorial Essays*. (Cambridge: Cambridge University Press, 1991), 153–166; 'Rational Animals', *Dialectica* 36 (1982), 317–327; 'The Social Aspect of Language', in *The Philosophy of Michael Dummett* (Dordrecht: Kluwer Academic Publishers, 1994), 1–16.

[20] Tyler Burge, 'Individualism and the Mental', *Midwest Studies in Philosophy* 4 (1979), 73–121; 'Intellectual Norms and Foundations of Mind', *Journal of Philosophy* 83 (1986), 697–720; 'Wherein is Language Social?' in A. George (ed.), *Reflections on Chomsky* (Oxford: Blackwell, 1989),

Suppose that we could not perceive words others speak. Suppose that the stimulus effects of the words nevertheless affected us by some natural causal process in such a way that we reliably understand their sense and their being used assertively, interrogatively, and so on, as received—rather than as initiated. Suppose that we could not directly know or even reliably guess anything about the words whose effects were thus injected. Suppose that the word sounds in the relevant contexts called up understanding of conceptual content and assertive force by bypassing the perceptual system, but triggering the same central mechanisms by which we understand our own speech. Thus I comprehend the interlocutor's conceptual content and assertive mode without perceiving words, as long as the communicator is using words in the ways I use them. The words might become perceptible when but only when something in the context provides grounds to doubt the standing comprehension of what the interlocutor is saying. Understanding, however, remains as good as ever.

In such a case, much of what we know in communication would, of course, be lost. But I think that the *basic* entitlement underlying understanding communication of intentional content and assertive mode, in ordinary domestic cases, would in its fundamentals be unchanged. Our starting point in domestic understanding is a non-sense-perceptual entitlement to rely on understanding of what one's system takes as immediately or non-inferentially intelligible. The warrant for relying on the apparent intelligibility can stand on its own. It need not lean on sense experience of words or behavior, once a system of language-comprehension is developed.

In the injection case, we would lose access to empirical evidence for inference and interpretation. But such interpretation is not epistemically basic. Unreasoned putative comprehension of the interlocutor's thought content and force is basic. The role of words and intonation is to call up the conceptual mechanisms that make comprehension possible. In basic cases, they are not objects of cognition which serve as evidence, or even indispensable sources of warrant for comprehending the intentional content and basic illocutionary force of others' intentional acts.[21]

175–192. Davidson criticizes some of my arguments in 'Knowing One's Own Mind', *Proceedings and Addresses of the American Philosophical Association* 60 (1987), 441–458; and 'Epistemology Externalized', *Dialectica* 45 (1991), 191–202. He makes some important points, but I am not persuaded by the criticisms. I will not answer them here. Davidson is mistaken in attributing to me the view that '"speaking in the socially accepted" way is essential to verbal communication'. Compare 'The Social Basis of Meaning'. Moreover, I do not take social anti-individualism to explain the essence of communication.

[21] Clearly the argument for the apriority of first-person comprehension is different from the argument for the apriority of third-person comprehension, as applied to the speech of others. The former is a transcendental argument. The non-empiricality of the warrant for first-person comprehension is a necessary condition for the possibility of any warranted comprehension, or comprehended thought, at all. The argument for the apriority of third-person comprehension yields only a prima facie warrant, allows for brute error, and adverts to contingent, reliable causal processes between the understanding and the event that is understood. Although they attribute different subspecies of apriority, the arguments agree in attributing a warrant whose justificational force is

A presupposition of this argument is that non-inferential sense-perceptual beliefs are warranted through a distinctively empirical, sense-perceptual entitlement. The entitlement derives its force not merely from the reliability of the beliefs' connection to their subject matters, but also from the beliefs' depending systematically for their content and application on the way perceptual objects are presented through sense perception. Some might deny this. They might maintain that all beliefs that depend at all for their warrant on reliable causal "tracking" connections to their subject matters have the same epistemic status. I believe that such a position would be seriously mistaken. It ignores what is distinctive about empirical warrant and runs together the very different contributions to the internal aspects of warrant made respectively by the systems of sense perception and understanding. These are, however, topics for another occasion.

I have not discussed context-dependent aspects of linguistic understanding except as a foil to standing aspects. The warrant for understanding perceptually guided demonstratives would seem to be empirical. Indeed, most of the warrant for fully comprehending utterances about objects on display, in the here and now, will be empirical all the way down. But not all context-dependent elements in an intentional content require empirical elements in the understanding. Establishing certain tense parameters can normally rely on a rule working directly off the comprehension of the intentional act. The referent of the first-person pronoun, as used by someone else, is fixed as the author of the comprehended act. Similarly, certain context-dependent parameters surely have default readings that are fixed by comprehension of the language. The language is context-sensitive in being systematically open to having the default reading overturned by background knowledge. (The background knowledge could be non-linguistic and empirical. But it could also be associated purely with comprehension of prior elements of the discourse.) Lacking special information, the contextually vulnerable default readings stand. I understand "red hair" normally to apply only to hair that in its natural state, in normal light, looks more like certain orange objects than most red objects. But empirical background knowledge in certain contexts could require that one understand that the phrase applies to hair that looks blood-red in normal light because it is painted, and not to hair normally counted "red". So although much contextual resolution of contextual-dependent elements in discourse is empirical, some depends on default settings associated with comprehension. Some depends on reasoning from comprehension of prior discourse.

Similar points apply to comprehending context-dependent elements of language whose understanding turns on the speaker's intended force. Understanding conversational implicature requires reasoning that goes beyond linguistic comprehension. The supplemental considerations will commonly be empirical. Certain simple types of irony can, however, be comprehended using standing linguistic competence: a certain tone might mark irony, and the marker might

independent of the contributions of sense experience or sense-perceptual beliefs. That is the sense in which both first- and third-person comprehension can be apriori warranted.

be comprehended through a standing linguistic competence—very much like the way that normal cases of serious assertion can be comprehended. Then a simple, standing transformation on the standard meaning might yield understanding (even comprehension) of the irony. In other cases, understanding that irony is in play might require subtle empirical background knowledge; and the reinterpretation of the sentence might require ingenuity that goes well beyond automatic application of standing linguistic competences.

Although much *interpretation* is justified through perceptual beliefs, not all justificational force behind interpretation is sense-perceptual. Some interpretation rests on making a message as coherent as possible, or on prior comprehension. Insofar as such interpretation does not rely on empirical inference or on perceived elements in the context, its justification might be non-sense-perceptual.

Thus four combinations seem possible: apriori, non-sense-perceptual entitlement to comprehension; entitlement to comprehension that is partly sense-perceptual; apriori, non-sense-perceptual justification of interpretation; and justification of interpretation that is partly sense-perceptual. As noted, I am inclined to believe that interpretation must rest upon justification rather than entitlement.

I have claimed a fundamental symmetry in epistemic status between comprehension of standing aspects of our own thought contents and comprehension of similar aspects of our interlocutors'. Yet Davidson is clearly right in maintaining that there are significant asymmetries. What asymmetries do I acknowledge?

The question leads into fascinating issues about self-knowledge and knowledge of other minds. The issues are complex. But some simple points of agreement can be cited. First, whenever we use a word with a definite meaning, we can meaningfully but trivially disquote. We cannot misidentify the meaning of our words or the intentional content of our current thoughts. We can get them wrong only in the sense that our explications of them, or inferential associations with them, can be mistaken. But although we are apriori prima facie entitled to our seeming-homophonic comprehension of another person's words in the absence of reasonable doubt, we *can* in any given case question our homophonic comprehension of another's speech, without doubting that the speech has a definite meaning.

Second, we can utilize self-interpretation to doubt commitments associated with the comprehension of our words or thoughts; but we cannot treat all our words or thoughts and their associated commitments as objects of interpretation. By contrast, we can take all of another's words as objects of interpretation. We can view another person as a foreigner, or even a black box. But we cannot view ourselves purely that way.

But the fact that we can, and sometimes have to, view others as foreigners to be interpreted does not indicate that we always must, or always do, or even rationally can, given our actual situations, view others that way all of the time. I think that fundamental to the social character of language is the fact that the norm within domestic communication is content preservation. We comprehend another's content without interpretation. We are entitled to take others as linguistic fellows rather than as objects of interpretation unless reasons arise for doubting our comprehension. For all their substantial philosophical virtues in other respects, radical interpretation and radical translation are misleading models for the epistemology of communication. Linguistic community provides epistemic norms in communication despite the wonderfully multifarious ways—including linguistic ways—in which we are individuals.

15 *A Warrant for Belief in Other Minds*[*]

What warrants do we have for believing that there are other minds? We believe with great confidence that there are other beings that think and/or that are conscious. We are as confident that there are other minds as that there are physical objects. At least, I am. But as in the case of knowledge of the existence of physical objects, the nature of the justifications or entitlements for our knowledge of other minds has sometimes seemed difficult to articulate, particularly in the face of sceptical doubts. I will not try to answer scepticism here. I want to reflect on the nature of one type of warrant for belief in other minds, laying aside its relevance to scepticism. I think that my reflections *are* relevant to scepticism, but not by being directly convertible into an anti-sceptical argument. I will discuss a type of warrant that I think we have.

Mill, followed by Russell, gave an early statement of an argument from analogy: We know in each of our own cases that we have thoughts. We can observe that these thoughts produce behavior of certain sorts in us. We observe similar behavior in others. We infer by analogy that similar behavior has a similar cause.[1]

* [Added 2011:] A parent of this essay was given at New York University in April, 1997; at Harvard University, October, 1998; and at Florida State University, February, 1999. I benefited from comments from the audiences on those occasions, and from a conversation with Amelie Rorty in the Fall of 1998. Parts of this essay overlap 'Reason and the First Person', reprinted in this volume. I never submitted the present essay for publication. I have edited the essay lightly, except for one significant substantive change. I altered step (1) in accord with my giving up the view that our entitlement to comprehension of another's utterances on particular occasions can be apriori. See 'Postscript: Content Preservation', section III, this volume.

[1] J. S. Mill, *An Examination of Sir William Hamilton's Philosophy and of the Principal Philosophical Questions Discussed in His Writings*, 6th edn. (London: Longman, Green, Longman, Roberts, and Green, 1889), 243–244; and Bertrand Russell, *Human Knowledge: Its Scope and Limits* (London: George Allen & Unwin, 1948), 482–486. For more recent developments of the approach, see Stuart Hampshire, 'Feeling and Expression', in his *Freedom of Mind and Other Essays* (Princeton: Princeton University Press, 1971), 143–159; and Christopher S. Hill, 'On Getting to Know Others', *Philosophical Topics* 13 (1985), 257–266. More recently yet, several philosophers have used the notion of simulation to yield variants on the argument from analogy. I think that these latter may avoid the charge of hyper-intellectualization that I present below against the classical arguments from analogy. I think that improvements on Mill and Russell may contribute to understanding one warrant that we have in knowing that there are other minds.

The issues over arguments from analogy are complex. I am not persuaded by some of the traditional criticisms. However, the mode of knowledge championed by these arguments does not do very well with minds, bodies, or behaviors quite different from our own. It seems to me that we can know that non-human animals, perhaps even Martians, have minds. Furthermore, insofar as the *classical* arguments from analogy are presented as our only way of knowing that there are other minds, the view tends to hyper-intellectualization. The classical arguments from analogy place each person's observation of his own behavior, and reflection on how his or her own mental states cause it, in too prominent a place to account for the primitiveness and solidity of our knowledge of other minds. Inference from self-observation and inference from self-knowledge are rather sophisticated abilities. I believe that our initial and fundamental warrants for belief in other minds are more primitive and less complex.

Strawson claims that the idea that mental predicates apply only to oneself is incoherent. He holds that since (as he assumes) all predicates must be significantly applicable to a range of individuals, one must be prepared to ascribe mentalistic predicates on appropriate occasions to other individuals. The condition of being so prepared is that one operate with a conception of an individual to which both mental and physical predicates are applicable. One must be able to identify subjects of mental predicates by identifying them as subjects of physical predicates. Strawson holds that one must ascribe mental predicates to others on the strength of observation of their physically identifiable behavior, where the behavioral criteria one goes on are "logically adequate" for the ascription of mental predicates.[2]

I doubt several parts of his argument. I doubt that it is impossible to have concepts like pain or belief unless one can apply such concepts on the basis of behavioral observation. We *do* apply these concepts in these ways. But I think that Strawson fails to show that it would be impossible to have such concepts and not be able to relate them to behavioral manifestations. It seems obvious that one could have the concept of pain simply by conceptualizing, and fitting into a framework of propositional inference, one's experiences of pain. The concept of belief is more difficult. However, I see no essential need to recognize expressions of belief in others in order to have the concept of belief. Reflection on one's own committal attitudes seems prima facie enough. Although in human development reflection develops in concert with observation of others, Strawson does not show that it is conceptually impossible to develop such reflection independently. Moreover, I think that Strawson is clearly wrong to hold that we have 'logically adequate' behavioral criteria for the existence of other minds. And he gives no convincing reason to think that the third-person application of mentalistic

[2] P. F. Strawson, *Individuals* (1959) (London: Routlege, 2002), chapter 3. Strawson thinks that scepticism about other minds presupposes but silently and incoherently repudiates the conceptual scheme that entails the existence of other minds—given that behavioral criteria are observed to be fulfilled.

concepts must entail application to more than one individual. Finally, as is well-known from Stroud's criticism, his account does not explain why our standard modes of third-person attribution are warranted.[3]

Descartes' explanation of our warrant for believing that there are other minds seems to have been a form of inference to the best explanation. On his view, nothing but thoughts could explain an observed activity that is as complex as human behavior, particularly linguistic behavior. This is the upshot and method of explanation in cognitive psychology—without the exaggerated claim that resides in 'nothing'. Descartes' explanatory approach has the advantage over classical versions of the argument from analogy of being extendable in natural ways to animals, to the insane, and to aliens whose bodies and behavior might be substantially different from ours. Partly because of this extendability, it seems to me at least a part of a general account of how we know other minds.

For all that, if it were the only answer, I think that it would suggest too theoretical a basis for our knowledge of other minds. It seems doubtful that our knowledge of other minds rests entirely on a complex, even if entrenched and tacitly inferred, explanatory theory—any more than knowledge of physical objects does. Knowing other minds seems epistemically more direct and straight-forward than such an account can allow.

I believe that one insufficiently developed approach centers on innate perceptual or close-to-perceptual ways that we have of perceiving other individuals as being in pain, or as having certain emotions. I think that such knowledge derives from either a perceptual or an intermodal, perceptually based, pre-propositional capacity. It is a misleading metaphor to say that we see another's pain, much less feel it. But we can see certain bodily or facial configurations as expressing pain. And similarly, for other feelings and emotions.

There is much more to be said about all the foregoing approaches. I will, however, take a different approach. In tackling our problem, it would be well to step back a bit. What is our primary source of knowledge of other minds? It is our understanding through the spoken and written word. Among our early exposures to other minds is our hearing the instructive, encouraging, or admonitory talk of adults. All of the most famous traditional philosophical views are striking in placing little weight on this central fact. They all fail to center on what is clearly a fountain of knowledge of other minds—understanding communication.

No doubt, Mill, Russell, Strawson, and Descartes would claim that they intended to include linguistic activity in the behavior that they cite. But none of them explicitly features linguistic communication as a central case. Descartes does highlight the complexity of language use. But his account features *explaining* linguistic behavior, rather than *understanding* speech. It is as if only when we rise to the level of taking others' activities as objects of explanation are we in a position to know other minds. But in the normal course of things, understanding

[3] Barry Stroud, 'Transcendental Arguments', *The Journal of Philosophy* 95 (1968), 241–256.

speech appears to be more basic than explaining it. The translation theory of Quine and the interpretation theory of Davidson are applications of the best-explanation approach that does concentrate on speech.[4] But they, too, take interpreting and explaining speech to be central. I believe that comprehension is fundamental.

I suggested that Descartes' view, taken as a full account of our knowledge of other minds, is redolent of hyper-intellectualization. Perhaps one can defend Descartes' view from this criticism by invoking the implicitness of our early theorizing. We get onto meaning by utilizing some tacit theory about the sounds and behavior that we perceive. This view may be correct. But it seems to me more likely that the formation of semantic competency is a lower-level process than that of theory formation. I conjecture that explanation and interpretation are not basic, even at some unconscious level. Explanation and interpretation by the whole person, as opposed to computations by sub-systems, seems certainly to be a later development than comprehension of speech. At the very least, Descartes' view glosses over the centrality of *understanding*—more specifically, of *comprehension*—in our knowledge of others. Most of our explanation of others' behavior presupposes comprehension of what they say. We resort to interpretation and explanation primarily when comprehension fails.

Descartes, Quine, Davidson, Mill, Russell, and Strawson share a picture of our knowledge of other minds. For them, knowledge of other minds is epistemically grounded in observation of behavior. Most of these views also agree in drawing a certain contrast between knowledge of other minds and certain paradigmatic sorts of self-knowledge. Such types of self-knowledge are taken, correctly I think, to be immediate and not to depend on ordinary perception. Mill appealed to inner observation. Descartes and Strawson, more plausibly, held that central sorts of self-knowledge are non-observational. Descartes maintained that they are underwritten by an intellectual warrant deriving from understanding our own thinking.

I believe that, in important respects, a strand of self-knowledge and a strand of knowledge of other minds are more similar than these views suppose. I will present an argument that outlines a warrant for belief in other minds that rests on linguistic comprehension. Much of the argument's interest derives from its eliciting the fact that certain relevant questions about epistemic warrant are very subtle and difficult.

The relevant warrant is defeasible and, at psychological levels that are available to introspection, non-inferential. Although the warrant is strictly speaking empirical, it is not otherwise dramatically different from knowledge our own mental states. The main difference lies in mastery of the relation between

[4] W. V. Quine, *Word and Object* (Cambridge, Mass.: MIT Press, 1960), chapter 2; Donald Davidson, 'Radical Interpretation' (1973) in *Inquiries into Truth and Interpretation* (Oxford: Oxford University Press, 2001); and my 'Comprehension and Interpretation' in L. Hahn (ed.) *The Philosophy of Donald Davidson* (Chicago: Open Court Publishers, 1999). The contrast between comprehension and interpretation/explanation that figures in what follows has its roots in this latter article.

first- and third-person points of view, and in an intellectually warranted awareness of the source of rational agency.

Comprehension is a type of epistemically immediate, unreasoned, non-inferential understanding. Although comprehension inevitably involves complex perceptual and cognitive processing, it is not—at the level of conscious, or even individual-level, psychology—inferential. I use the term 'comprehension' to cover both comprehension of meaning or thought content (and associated mode or force) and comprehension of words as expressing such meaning or content (and mode). The latter usage will be primary in the main line of my argument.

There is first-person and third-person comprehension. *First-person comprehension* is comprehension of a thought that carries a presumption that the thought is one's own. It is the minimal understanding involved in the ability to think one's own thoughts. The ability to think thoughts—competence understanding—does not count as comprehension unless it is accompanied by third-person comprehension. Third-person comprehension is understanding that operates with no presumption that the words or thoughts are one's own. Third-person comprehension does *not* entail the existence of another thinker. It allows the possibility of brute error in assignment of meaning or content and in determining whether the source of the content is oneself or another. One can make mistakes about these matters while performing at one's epistemic best. Correct third-person comprehension involves (a) correctly assigning a thought content of one's own thinking to an expression or expressive event that causes that thinking and (b) correctly assigning a force or mode (which need not be one's own attitude to the content) to that same expression or expressive event.

A normal case of third-person comprehension would be immediate unreasoned understanding of what is expressed by another person—mode and content. The other asserts, 'All healthy adult blue whales eat more than one hundred pounds of foodstuff a day', and one understands immediately.

Third-person comprehension is, I think, essential to understanding language. Understanding language requires having an ability to understand a content or meaning that one might be mistaken about. That ability is additional to an ability to think with and express a meaning or content (first-person comprehension). Any understanding of language requires both types of comprehension. For example, one must be able to hear or read one's words regardless of who produces them, and still understand them. For representational content expressed in language is necessarily shareable, since it is fixed through objective relations to the individual's environment. It is thinkable through conceptualizing patterns associated with those relations. Any genuine linguistic understanding requires a competence to comprehend that content (and its mode) as something objective, independently of one's immediately thinking or meaning it.

Thinking thoughts does not itself entail comprehending them—in the sense of having a capacity for a third-person perspective on them. Animals think thoughts, but do not comprehend them. They have minimal competence understanding, but not comprehension. Regarding one's thoughts from a third-person point of view requires an ability to evaluate them from a point of view that abstracts from one's own commitment to the representational content.

Comprehension requires a competence not only to think one's thoughts, but to abstract from one's own commitments (or attitudes) toward the representational content that one nevertheless occurrently thinks. Thus third-person understanding presupposes a minimal competence to think the thoughts (and to have one's own attitudes toward them). But the latter competence does not count as comprehension unless it is coupled with a competence for third-person understanding.

I have explicated third-person comprehension so as not to entail that it is necessarily comprehension of the others' utterances or thoughts. I assume only that it could possibly include comprehension of intentional acts of others.

Now, for some background about what I mean by 'empirical'. I use the term as a contrary for 'apriori'. Apriori warrants derive from reason or understanding, or the nature of a capacity that functions to contribute to reason or understanding. I take apriority to be a feature primarily of warrant—justification or entitlement. Justifications are reasons in the repertoire of the justified individuals. The reasons may be self-sufficient (as in self-evident premises), or they may be antecedent steps in inferences. Entitlements are warrants that are not reasons. An entitlement always has its warranting force whether or not the individual with the entitlement has the capacity to think it. So for example, a very young child can be entitled to its perceptual beliefs, even though it may not yet have the meta-representational concepts needed to explain why. The individual is entitled to the belief even though nothing in his or her psychology suffices to provide a justification—a propositional reason for the belief. Epistemic norms for epistemic goodness or badness of the individual's cognitive processes and activity remain applicable.

I believe that entitlements always make reference to the way an individual's cognitive (or practical) psychological competencies operate. So they are partly internal. They are never merely statements that the individual's beliefs are produced by just any old process that is reliable in producing true beliefs.[5] They are certain norms governing operation of the individual's internal states. For example, the norms may make reference to the belief's dependence on a reliably veridical type of perceptual representation. But the individual need not be able to explain why the perceptual belief is warranted. To have a warranted perceptual belief, an individual need not even have the concepts to think (entertain) such an explanation. Individuals are entitled to their perceptual beliefs, and they are entitled to rely on their memory, their deductive and inductive reasoning, and their comprehension, even though they may not be able to explain—or even understand—why.

[5] I do think that any warrant must apply to a competence that is reliable in producing veridical cognitive states in certain relevant normal circumstances—the circumstances by reference to which the nature of the competence is explained. See my 'Perceptual Entitlement', *Philosophy and Phenomenological Research* 67 (2003), 503–548.

A justification or entitlement is apriori, or non-empirical, if its justificational force is neither constituted nor enhanced by reliance on specific sense perceptions or sensory registrations. In deciding whether a belief is non-empirically warranted, it does not suffice to determine whether it is empirically warranted. A belief can have empirical as well as apriori warrant. The apriority or non-empiricality of a warrant does not entail that it is self-evident, infallible, unrevisable, or even unrevisable by *empirical* considerations. One can be apriori defeasibly warranted in believing something that is false. Sense perceptions or perceptual beliefs can be psychologically necessary for an acquisition or even a justification of a belief, without contributing to the force of the warrant, hence without preventing it from being apriori. Our beliefs in simple logical truths, which are surely warranted apriori, may require having had sense experiences of particular types of configurations of symbols, or of physical objects. The role of sense experience in the psychology and acquisition of belief must be distinguished from its role in contributing to the belief 's warrant.

A central feature of this conception of non-empiricality is that it does not apply merely to warrants for believing eternal propositions. We can be apriori warranted in believing in the existence of certain thought *occurrences*.[6] Our warrant for knowledge of instances of *cogito* (I am thinking) is normally apriori. The warrant's epistemic force resides in understanding the temporally occurrent thought. Understanding the occurrence suffices to warrant (justify) one in believing it. Sense perception is not a constitutive part of the understanding or an element in the warrant for believing an instance of *cogito*. The understanding suffices for the belief 's being warranted.[7] So a judgment that a given thought is occurring is not *ipso facto* empirical.

I will now state the main lines of an argument that we have a barely empirical warrant for believing in other minds—a warrant that centers on comprehension of utterances. The point of the argument is to elicit the structure of what I take to be one fundamental entitlement that we have for our belief in other minds. Here is the argument.

(1) An individual is prima facie entitled, in some particular instances, to rely on seeming, third-person comprehension of apparent active instantiations of propositional representational content, as being *genuine* comprehension of genuine, active instantiations of propositional representational content.

The entitlement to rely on particular exercises of a capacity for comprehension is empirical. I think it barely empirical. I will explain this point shortly.

[6] [Added in 2011:] See my 'Frege on Apriority' in C. Peacocke and P. Boghossian (eds.), *New Essays on the A Priori* (Oxford: Oxford University Press, 2000); reprinted in my *Truth, Thought, Reason: Essays on Frege* (Oxford: Clarendon Press, 2005).

[7] I believe that we are entitled to accept many other self-attributions of present mental states and events, even those whose truth is not evident from understanding their content. In these cases, entitlement can also derive from our understanding. See my 'Individualism and Self-Knowledge', *The Journal of Philosophy* 85 (1988), 649–663; and 'Our Entitlement to Self-Knowledge', *Proceedings of the Aristotelian Society* 96 (1996), 91–116.

A presupposition of this step is that the individual *has* a linguistic competence for third-person comprehension. I presuppose that this competence is reliable in circumstances in which its nature was formed. (See note 5.) The individual need not assume this presupposition, or be able to understand it, in order to have the warrant stated in (1). It is enough that the individual in fact have the competence.

(2) An individual is prima facie entitled to presume that genuine active instantiations of propositional representational content must have a rational source—must be the acts of a mind or mental agent.

(3) So, both in general and in some particular instances, an individual is prima facie entitled to rely on seeming third-person comprehension of apparent instantiations of propositional representational content as being genuine comprehension of the acts of a rational source—a mind or mental agent.

(4) An individual is prima facie entitled, both in general and in some particular instances (including some of the same ones that are covered by (1)) to a belief as to whether the source of such acts that one seemingly comprehends is oneself or another.

A presupposition of this step is that the individual has the first- and third-person concepts necessary to have such beliefs.

(5) So in particular instances in which an individual seemingly comprehends apparent active instantiations of propositional representational content, the individual can be prima facie entitled to a belief that the relevant rational source is another mind, a different rational agent.

The cases in which an individual has the entitlement are cases in which the individual exercises seeming third-person comprehension of an instantiation of an active mode of propositional representational content, and where the individual takes the activity not to be the individual's own.

Whereas I believe that the entitlement articulated in step (1) is empirical—but only barely empirical—I think that the entitlements articulated in steps (2) and (4) are strictly apriori. So the whole entitlement to believe in the existence of other minds, which is articulated by the argument, is barely empirical. I think that reflection on the entitlements articulated by steps (1), (2), and (4) is philosophically worthwhile—as is reflection on the entitlement articulated by the whole argument.

Before defending the steps, I want to begin by explicating (1):

> (1) An individual is prima facie entitled, both in general and in some particular instances, to rely on seeming, third-person comprehension of apparent active instantiations of propositional representational content. An individual is prima facie entitled to rely on such seeming comprehension as being *genuine veridical* comprehension of genuine, active instantiations of propositional representational content.

The entitlement that I believe we have to rely on comprehension—both in general and in given particular cases—is a very abstract default entitlement. It holds only prima facie, or *pro tanto*. It can be overridden by reasons that show that one's seeming comprehension is *merely* seeming. One's seeming comprehension can be, and often is, shown by empirical means not to be genuine comprehension. For example, one might be given empirical reason to think that one has hallucinated an utterance, or that someone was not using words in familiar senses. So one's entitlement to rely on seeming comprehension is vulnerable to overriding empirical reasons. In themselves, these points do not bear on whether the default prima facie entitlement to rely on one's seeming comprehension, in particular cases, is empirical.[8] That issue hinges on the nature of the positive force of the warrant, not on the nature of possible threats to it.

In saying that the warrant resides in *seeming* comprehension, I mean only to indicate that in given cases, it is left open, as a matter of fact, whether or not exercise of the competence is veridical. One can exercise the competence, take an event as having certain content and force, and be mistaken. Such exercises, as well as exercises of the competence that assign the right mode and content, count as seeming comprehension. I do *not* assume that seeming comprehension involves a belief *about* comprehension, or is of the form: that seems intelligible. Similarly, I do not assume that the reliance on exercises of seeming comprehension as veridical comprehension involves a meta-representational belief about veridicality or about comprehension. The point is that the individual is epistemically entitled to the comprehension in the same way that an individual might be epistemically entitled to a perceptual belief. Such entitlements do not require having cognitive states that make reference to psychological, semantical, or epistemological matters.

It is crucial to the development of the argument that (1) claims that our entitlement applies to reliance on seeming comprehension not only of tokened or expressed intentional content, but of tokened or expressed activities. The comprehension includes understanding mode or force of certain public acts. Thus utterances are assertive, imperative, interrogative, and so on. Comprehending representational content is impossible without normally comprehending at least basic modes or forces associated with the content of an utterance.[9]

[8] See my 'Content Preservation', *The Philosophical Review* 102 (1993), 457–488; and 'Interlocution, Perception, and Memory', *Philosophical Studies* 86 (1997), 21–47. These papers discuss the prima facie nature of a similar entitlement, spelled out in what I call the Acceptance Principle. This principle entitles recipients to accept what they are told. The present claim is weaker: it entitles recipients of interlocution to rely on their understanding what is said.

[9] See 'Content Preservation'. I think that the most basic linguistic mode is presentation-as-true. Assertion is a central species of presentation-as-true. I credit Quine with centering on assertion as basic, and approximating the present view. See his use of the principle of charity in *Word and Object*, chapter 2. To begin to translate a language one must be able to discern presentations-as-true. The point applies not only when translation is at issue, but also in the lower-level cases of ordinary comprehension in communication. Whatever entitlement attaches to understanding of content must attach to comprehension of some basic modes of activity associated with the content. So what is comprehended includes the representational content's being presented as true, in some cases.

Since I am not discussing scepticism, I will just assume that we are dealing with individuals who are competent in comprehending the meaning, or content, and mode of utterances in normal circumstances. Comprehenders can seemingly comprehend only because they do in fact sometimes comprehend when they exercise their competence. I will assume that relevant comprehenders reliably comprehend in normal circumstances. (See note 5.) Seeming comprehension provides a defeasible epistemic entitlement to presume genuine understanding. Clearly the entitlement can gain force as conversation or reading goes smoothly on. Such cumulative warrant internal to the understanding is worth remembering when one considers the strength of one's default, unthreatened entitlement.

The minimal default warrant that I cite in (1) can co-exist with many supplementary or independent warrants for relying on comprehension. If one has moderate sophistication about language, one knows that words are being uttered, that representational content is expressed through sounds and shapes. One knows that a person with a certain look is making the sounds. One often has memories of the same person saying similar things that might contribute to our warrant for relying on current comprehension. We have empirical warrant for trusting our understanding of familiar interlocutors. It is uncontroversial that this knowledge is warranted not merely by immediate occurrent comprehension but by further considerations.

What an individual is entitled to according to (1) is not that a content has been expressed in words—not that an utterance has occurred that expresses such and such content—not that a person with a certain look is talking. I am interested in an entitlement just to presuming on the veridicality of a comprehension of a certain representational content as being instantiated with a certain mode or force.

The claim in step (1) is closely associated with the stronger Acceptance Principle, articulated in 'Content Preservation'. The Acceptance Principle underwrites entitlement not merely to comprehension, but to acceptance of what one is told, other things equal. This entitlement is possible only because the individual has an entitlement to rely on comprehension of utterances as veridical—as getting the content of the utterance right.

For purposes of the present argument, one can take the entitlement cited in (1) to be less committal in both respects. It is not an entitlement to believe what is uttered, but to a certain comprehension. And although I think that we do have a entitlement to rely on seeming comprehension of the specific representational content and mode (such as assertion) of what is uttered, the argument works equally well with an entitlement to much less rich aspect of seeming comprehension. The minimum entitlement required by the argument is merely an entitlement to rely on the veridicality of a comprehension of an event as an active propositional event. Thus one need not get right the specific representational content of the utterance, as long as one gets right its propositionality. Similarly, one need not get right the specific mode or force—whether it be a warning, an imperative, a question, an assertion, or what not—as long as one gets right that its mode is active. What the individual has to be warranted in is merely reliance on a comprehension of an event as tokening or expressing propositional agency.

So much for explication of (1). I turn to considerations that favor it. An epistemic norm is a standard for exercising a competence in a way that contributes to good routes to veridicality—here, ultimately, true belief. I will reflect on individuation of third-person comprehension competence. I think that reflecting on the function of third-person comprehension is a start toward understanding the nature of the competence. The function of third-person comprehension is to yield understanding of instantiations of representational content.

An exercise of comprehension competence is warranted if is a non-pathological exercise and the competence is reliable. Since, in accord with my methodology of ignoring scepticism, I am assuming that the competence is reliably veridical in the circumstances by reference to which its nature is explained, it is relatively trivial that exercises of the competence are warranted. Similarly, in accord with the methodology, I am assuming that there are no hard issues about whether we are in abnormal circumstances. So it is relatively trivial that the relevant exercises of comprehension can lead to knowledge. The point of the argument is not to convince the sceptic, but to outline the nature of a warrant and route to knowledge that we in fact have.

The interesting issues about (1) lie in the nature of the relevant warrant. I think it strongly plausible that the warrant that attaches to exercises of comprehension of an event as an instance or expression of propositional agency is an entitlement, not a justification. Good comprehension is certainly not veridical by virtue of self-evidence. At the level of conscious, individual psychology, comprehension is not inferential. It seems epistemically and consciously immediate. Although it could in principle derive from propositional inference that is not available to the individual, but is modular, what we know about comprehension indicates that it is brought about by perceptual processing, not what would ordinarily be counted reason-transmitting propositional inference. Even though propositional elements enter into the processing (certainly as products, possibly earlier), there is no reason to think of these elements as providing reasons for—explanations of the acceptability of—any further propositions. To count as a justification, a warrant must be grounded in use of reasons.

I mentioned earlier that I thought that the warrant articulated by (1) is empirical, but barely empirical. It is empirical because comprehension must rely on the specifics of sense perception. Comprehension of specific content and mode is by way of a capacity to discriminate certain events perceptually and to systematically associate certain types of perception with comprehension of certain contents and modes. That is what the competence consists in. Even the mere comprehension of an event as tokening or expressing some active propositional event or other (regardless of the event's specific content and mode) consists in systematically transitioning from certain types of perception to comprehension as of active propositionality. Inasmuch as this type of comprehension has a constitutive perceptual component, epistemic warrant for relying on the comprehension depends on meeting standards for the well-functioning and reliability of the perception, and for transitioning from the perception to the comprehension. So the warrant is empirical.[10]

So the warrant is empirical. Why is it barely empirical? The type of perception involved in language perception is cognitively top-heavy. The perceptual representation is triggered by sensory input. But perceptual attributives are molded not so much by physical patterns that are sensed, as by coordination between minds. The role of intellection is disproportionately large in the account of the warrant for the exercise of the comprehension competence.

Language perception is a very special type of perception. The perceptual attributives that figure in ordinary perception—for example, visual perception of shapes, colors, motions, locations, types of body—are stamped into the perceptual systems of animals through patterned, pre-perceptual, sensory interaction with instances of those physical attributes.[11] These are attributes described by the natural sciences. Propositional structures, whether syntactical or semantical, are not structures in the ordinary physical environment, as described by the natural sciences. Predication, subject–verb agreement, propositional structure are not kinds described in the physical sciences. They cannot be stamped into an individual through pre-perceptual sensory interaction with them. Physical instantiations or expressions of these kinds must be *comprehended* if the individual is to even have perceptual attributives for them. The physical particulars that we perceive and their physical characteristics, however, played little role in molding the perceptual attributives used in language perception. Cognitive capacities largely molded the nature of the perceptual attributives.

In meeting standards for good routes to veridical comprehension, one must rely on perception. Perceiving well is a constitutive part of comprehending another's utterance, as I have acknowledged. But since the perception itself is constituted primarily by a cognitive super-structure, perception seems to play the role of triggering the recipient's use of this cognitive super-structure and enabling it to align with the cognitive super-structure of the interlocutor. The empirical element in language perception functions primarily as a connector between two cognitive systems and systems of understanding. It is only secondarily a source of information about the perceivable world.

I believe that all the remaining steps in the argument—steps that explain the structure of the warrant to believe in the existence of other minds—are strictly apriori. Let me turn to (2):

[10] [Added in 2011:] These points revise views expressed in 'Content Preservation'; 'Interlocution, Perception, and Memory'; and 'Reason and the First Person', in Crispin Wright, Barry C. Smith, and Cynthia Macdonald (eds.), *Knowing Our Own Minds: Essays Self-Knowledge* (Oxford: Oxford University Press, 1998), reprinted in this volume. See 'Postscript: "Content Preservation"', this volume, section III.

[11] [Added in 2011:] The anti-individualism about perception that is expressed here is developed in some depth in *Origins of Objectivity* (Oxford: Oxford University Press, 2010).

(2) An individual is entitled to presume that genuine active instantiations of propositional representational content must have a rational source—must be the acts of a mind or mental agent.

I use the word 'presume' just to signal that the individual need not think, or be able to think, this connection. The presumption is supportable by apriori reflection. It constitutes a rationally supportable connection that individuals are entitled to rely on, even if they do not understand the connection. Expression of propositional representational content necessarily presupposes some source in a system of cognitive and practical interactions with the world (including belief and intention)—a system that involves the use of reasons under rational norms.

The propositional *expression* of content bears a particularly marked relation to a higher sort of practical and cognitive intentional *agency* than can be assumed in non-linguistic higher animals, which I also freely assume to have propositional attitudes. In any case, a rational source of the expression of propositional representational content must make use of reasons in the formation of beliefs and intentions.[12]

In discussing an argument for knowledge of other minds by Price, Malcolm in effect challenges the idea that there is an apriori connection between seeming propositional intelligibility and there being a rational source.[13] Malcolm claims that no amount of intelligible sounds coming from a tree or bush, or even a computer, would provide evidence that the sounds were understood by the tree or bush. So he concludes that prima facie intelligible sounds provide no ground at all, in themselves, to think that they come from a rational source.

Malcolm's claims rest on two points. The first has to do with the necessity of empirical application for understanding. He claims that unless one perceives the initiator of the sounds as applying them to things that they are plausibly true of, one has no ground for attributing understanding to the initiator. He extends this point, plausibly, to utterances of sounds that seem to express purely mathematical content: '... suppose that there was a remarkable "calculating boy" who could give right answers to arithmetical problems but could not apply numerals to

[12] For work that stresses relations between having language and using reasons, see H. P. Grice, *Studies in the Ways of Words* (Cambridge, Mass.: Harvard University Press, 1989); Davidson, *Inquiries into Truth and Interpretation*; and 'Content Preservation'.

[13] H. H. Price, 'Our Evidence for the Existence of Other Minds', *Philosophy* 13 (1955), 425–456; Norman Malcolm, 'Knowledge of Other Minds', *The Journal of Philosophy* 55 (1958), 969–978. Both are reprinted in David M. Rosenthal (ed.), *The Nature of Mind* (Oxford: Oxford University Press, 1991). Price's argument combines Cartesian inference-to-the-best-explanation and analogical arguments. Price's argument is unique, to my knowledge, in placing central emphasis on language. I developed my views in ignorance of Price's paper. But I have come to admire it as a stimulating and broadly congenial anticipation of the line of this paper. Price advocates an old-fashioned empiricism about comprehension. He thinks that we build up evidence by reckoning likelihoods or best explanations on the basis of experience. The structure of the warrant that I am laying out is somewhat different. I think it supplements and indeed underlies the warrants that Price focuses on.

reality in empirical propositions. . . . I believe that everyone would be reluctant to say that he understood the mathematical signs and truths that he produced.'[14]

I agree that understanding and even representationality itself presupposes *de re* applications.[15] I agree that if one had reason to think than an initiator of seemingly intelligible sounds could not carry out such applications, one would have reason to think that the initiator did not understand. But it does not, even remotely, follow that perceiving applications of the instances is necessary for one to be warranted in attributing understanding. It seems to me obvious that if we hear intelligible sounds, we are rationally entitled in the absence of countervailing considerations to presume that they have a rational source. It is not rational to remain neutral about whether they have a rational source until and unless we can check to determine independently that there is understanding backing the apparently intelligible sounds. Malcolm's points show only that our entitlement to rely on our seeming understanding is prima facie, not that it does not exist.

I need not depend for my warrant on the presumption that the initiator of the sounds is itself the rational source. One may be hearing a recording or the outputs of a computer, neither of which would be rational or would have any understanding. The entitlement is to presume that there is an ultimate source of seemingly intelligible content that is rational. Here I can leave open how to think about chains of transmission. When we regard a table computer as unable to initiate rational activity, we still presume prima facie that the computer manipulates material whose intelligibility derives from a rational source, perhaps the programmer or the last person to use the computer. In the case of the bush, perhaps we suspend belief about whether the bush is an initiator or even a transmitter of rational activity. If we heard seemingly intelligible sounds apparently coming from a bush, we would look into the matter further. But the initial presumption that there is a rational source is warranted and robust. Nothing in Malcolm's points shows that we lack the apriori prima facie entitlement to move from warranted seeming intelligibility to warranted belief that the occurrences have a rational source.

Malcolm's second point is that when something lacks a human face or body, it *makes no sense* to say of it that it understands: 'things which do not have the human form, or anything like it, not merely do not but *cannot* satisfy the criteria for thinking'.[16] I think that this claim is patently mistaken. It is a sign of philosophy really running amok. Although our knowledge of some objects, like bushes, does suffice to show that *they* could not understand, we are not so provincial in our use of mentalistic concepts as to require that only things that have something like human form could understand.

[14] Malcolm, 'Knowledge of Other Minds', 974.

[15] See my 'Belief *De Re*', *The Journal of Philosophy* 74 (1977), 338–362, section II; reprinted in my *Foundations of Mind: Philosophical Essays Volume II* (Oxford: Clarendon Press, 2007).

[16] Malcolm, 'Knowledge of Other Minds', 974.

So far, the argument is that in particular cases, seeming comprehension provides warrant for belief in there being a rational source. The argument still must show that one is prima facie entitled to take the rational source to be *another* mind.

I turn to (4), the next substantive step.

> (4) An individual is prima facie entitled, both in general and in some particular instances (including some of the same ones that are covered by (1)) to a belief as to whether the source of such acts that one seemingly comprehends is oneself or another.

Any initiation of an utterance with propositional content is an act. The formation of a predication for a propositional utterance is an intellectual act. So comprehension of an instantiation of propositional content can presume that the source (however far back in a chain of transmission) of the instantiation *acted* in producing it.

To be entitled to presume that the rational source of an event that one comprehends in the third-person way is another agent, one must be entitled to presume that it is not oneself. Knowledge of other minds is distinguished from self-knowledge not by being necessarily inferential or just by being necessarily grounded in perception, but by being in some known contrast with taking a comprehended representaitonal event to be one's own.

I believe that any individual with representational powers has a natural capacity to discriminate its own activity from activity of other individuals. Such a capacity is marked *de se* by an ego-centric index. In individuals capable of comprehending utterances with propositional content, this capacity is associated with a capacity to be moved by reason. In persons, individuals with the first-person concept and a capacity to use a concept of reason, the capacity takes a more complex form. The key feature of the first-person concept, as I have argued elsewhere, is that it marks acknowledgment of the immediate rational relevance of reasons to intellectual practice.[17] In a person, the capacity to distinguish the person's own acts from those of others must be associated with a natural competence to reliably discriminate acts for which one is responsible from acts for which one is not responsible, and to acknowledge responsibility to implement reasons that support or count against such acts.[18] In understanding utterances in interlocution, one exercises this capacity—taking the act not to be one's own and recognizing that one cannot directly sustain it in response to reasons that favor it, or directly alter the act in response to reasons that count against it. To be rational in the employment of first- and third-person points of view, one

[17] See my 'Reason and the First Person', in Smith, Wright, and Macdonald (eds.), *Knowing Our Own Minds*. Substantially the argument given in the present paper is outlined in there, and also in 'Computer Proof, Apriori Knowledge, and Other Minds', *Philosophical Perspectives* 12 (1998), 1–37.

[18] This ability is one aspect of what Kant called 'the unity of apperception'. *Critique of Pure Reason*, B131ff.

must have, and be apriori entitled to, a competence for fallible awareness of whether an act associated with seeming comprehension of an instantiation of propositional content is one's own.

To be a person, or any individual with capacities for reason, one must have and be entitled to rely on a natural capacity to be aware of the agent of comprehended propositional acts. A person must be able to distinguish cases in which he or she is committed under rational norms from cases where one is not. A person must—in a natural, mature, undamaged state—be able to reliably determine the applicability of rational norms—in particular, norms of epistemic or practical critical rationality—to a given comprehended intentional act in a rationally immediate way, on pain of regress. So entitlement to it is apriori. This is to say that one must have some apriori entitled awareness for one's *not* being the agent of relevant comprehended active instantiations or expressions of content, and for one's thereby *not* being rationally committed under rational norms governing the relevant agency.

The conceptual space between understanding and acceptance of comprehended actual assertion is one to which a rational agent must be sensitive—and be entitled to be sensitive—if he or she is to be subject to rational norms governing acceptance. To be subject to such norms, a rational agent must be able to discriminate the sorts of requirements on warrant and defense that apply to autonomously held beliefs from those that apply to beliefs that are gained from interlocution. One must have an entitlement to presume that an interlocutor is not oneself, but another rational agent.

One most naturally thinks of cases in which the rational source whose assertions one comprehends is one's immediate interlocutor. But the warrant that I have outlined does not differentiate between immediate sources that are rational and immediate sources that are non-rational conduits for rational sources. I see that there is a person standing there in front of me, holding forth. But insofar as I am relying only on my comprehension of an utterance, that comprehension is compatible with the possibility that the rational source of the intelligible propositional act bears a more remote causal relation to the event that I comprehend. The entitlement that I am delineating tends to be an entitlement only for the belief that there is a rational source in the chain, not that it is immediately present.

One might hear a recording or read a computer's print-outs, or hear a voice from the sky, whose intelligibility warrants presumption of a rational source. But the source may be further back in a causal chain. Of course, this opens the possibility that the source is not another mind, but one's own. Perhaps I am understanding a computer that I myself programmed. Or perhaps my voice got recorded a while ago, and I do not recognize my phonological characteristics. Then I comprehend my own productions in the third-person way.

Does one need to rule out such possibilities in order to be warranted in presuming that they are not in play? Can one be prima facie warranted in presuming that they are not in play without invoking evidence that they are not?

For all the byzantine complexity of such possibilities, I stand by the claim that one is prima facie entitled to rely on one's taking a comprehended propositional act not to be one's own. If one loses confidence about the provenance of an act, then one loses one's warrant, since one is not exercising the rational discriminative capacity that the warrant is built upon. One can certainly be mistaken in taking something not to be one's own. But such mistakes, possible or actual, do not in general undermine warrant. And it seems to me that they do not do so here.

What the cases bring out is that the relevant capacity to discern the source of rational agency involves a reliance on memory. One must be entitled to rely on a capacity to be aware not only that the comprehended act is not one's present intentional act. One needs to be entitled to the view that it is not an act that constituted a past representational act. The reliance on memory here is a reliance on memory of past intellectual acts, or on a capacity to recognize an intellectual act as not in one's past. There is nothing essentially empirical about such memory.

Price said that news signals otherness.[19] It does, but news is not essential. What is essential is recognition that an intentional or other representational act is not one's own. Price thought that if an utterance does not express a belief that one does not already hold, and is simply 'intrusive' into one's own thought process, one could not rule out the possibility that the occurrence resulted from one's own unconscious. Also a new utterance that expresses a belief that we are not consciously aware of, or that contradicts our self-construed beliefs, might in principle issue from *unconscious* cognitive dissonance. The warrant and knowledge that we in fact have is never strong enough to infallibly rule out the in-principle possibility that one has been fooled—either in comprehending an utterance or in taking it to come from another. We have defeasible warrant and knowledge in any case.

Price is perhaps right that our taking in *news* through comprehension strengthens our warrant for our belief that the comprehended utterance comes from another. But a lot of the content of what we comprehend is not news. Only the occurrence of the utterance is news. We have a reliable and warranted, even if fallible, capacity to determine whether a propositional act is our own or comes from another.

If one thinks—from a sceptical standpoint or from the standpoint of a quest for philosophical certainty—about the possible ways in which one *could* go wrong about who did what, one can be misled into philosophical hopelessness or solomonic neutrality. But in fact, we are reliable at telling whether a thought is one we have thought before. We are good at growing uncertain in problematic cases. We mostly remember what we have done or have not done, and what acts we are responsible for or not responsible for. Loss of reliability in these respects is a form of craziness. I believe that being rationally entitled to one's views about

[19] Price, 'Our Evidence for the Existence of Other Minds', 431.

what commitments one has made is necessary to being a rational agent—and certainly a self or person.

The title of the essay refers to minds, plural. So far I have argued only for an entitlement to believe in the existence of another mind. Considerations regarding differences in styles, capabilities, and views among different sources can be developed internally, from reasoning about the contents of utterances that one comprehends. Such considerations can justify differentiations among other minds without using more empirical resources than I have already laid out. To be warranted in believing that there is a plurality of other minds, it is not metaphysically necessary to rely on the look or sound of a speaker, beyond simply comprehending what the speaker utters.[20]

My topic has been warrant. What of knowledge of other minds? I assume that the relevant warrant can yield knowledge if and only if the warrant is not overridden and the belief it supports is true and not entangled in Gettier problems. The warrant that I have argued we have can support knowledge if sufficient stretches of new information are comprehended and held together in preservative memory. I leave open whether the stretches must go beyond a single propositional act. However, I doubt it.

Knowledge of the existence of other minds is not knowledge of just one more fact. It is for most people a tacit, framework knowledge. It is rarely articulated. But it can be derived from particular non-inferential instances of comprehension. In this respect, the knowledge is like knowledge of the existence of physical objects, which is a tacit, framework knowledge derivable from particular non-inferential beliefs grounded in ordinary, non-linguistic perception.

The argument that I have given seems to me to yield a new angle on old issues.

[20] For discussion of these issues, see the last sections of 'Computer Proof, Apriori Knowledge, and Other Minds'.

PART III

REASONING AND THE INDIVIDUALITY OF PERSONS

16 *Reason and the First Person*

A small but persistent tradition in philosophy insists that there is a large divide between knowledge of one's thoughts and attitudes, and knowledge of one's thoughts and attitudes *as* one's own. The introduction of the *I* concept (please allow this convenient barbarism) has been characterized as a misleading, or at any rate momentous, step in need of special argument.[1] Hume complained that he could not find a self when he introspected.[2] He wondered whether 'the self' was simply an evolving bundle of sensations and ideas, which he thought he *could* find in introspection. Lichtenberg suggested that Descartes' *cogito* is less certain, or more objectionable, than an impersonal substitute: He recommended that one substitute *a thinking that there are physical objects is occurring* for *I am thinking that there are physical objects.* He wrote:

We are acquainted only with the existence of our sensations, imaginations, and thoughts. 'Thinking is going on' (*Es denkt*) is what one should say, just as one says, 'Lightning is

[1] By the *I* concept or (ignoring the plural *we* for now) the first-person concept, I intend an indexical concept shared by fully mature language-users by virtue of their mastery and understanding of the term 'I', or exact translations thereof. This is only a rough reference-fixing explication. I do not assume (though I think it may be true) that only language-users have the relevant full first-person concept. The main argument of the essay does not depend on any very exact understanding of what is essential to having the concept. But I assume that having what I call the full first-person concept involves having other concepts and conceptual abilities that go beyond mere ego-centric awareness—for example, concepts of thought and agency and some reidentification or self-tracking abilities. I believe that autonomous use of the full first-person concept is possible only for persons, and that it applies to entities of a certain important kind—persons or selves, which I take to be by nature (in part) critical reasoners. But the argument of the essay does not depend on, or establish, this view either. Nor does it depend on distinguishing this concept from lower-level ego-centric sensitivities or modes of reference (even perhaps ego-centric concepts) utilized by animals that are not persons. The argument I shall give only supports the view that necessarily when critical reasoners use the full first-person concept, it fulfils certain functions. I want to start with a notion that is relatively non-committal from a theoretical point of view and assume that it is familiar. I think it would be a mistake to get into deep issues about ontology of persons, selves, and concepts, or fine-grained issues about concept-individuation, in advance of considering the argument I will offer as applied to a recognizable element in intentional thought contents that is commonly expressed with the word 'I'. Concepts are elements in intentional thought contents. If one wants to avoid calling intentional indexical elements in intentional thought contents 'concepts', one can find a different terminology. The key assumption is that there is a structural intentional element or aspect of thought that is shared by all thoughts properly expressed using the first-person singular pronoun. I am interested in the role and epistemic status of this element or aspect.

[2] David Hume, *A Treatise of Human Nature*, I. iv. 6.

occurring' (*Es blitzt*). Saying '*Cogito*' is too much, as soon as one translates it as 'I am thinking'. Accepting, postulating, the *I* is a practical requirement.[3]

Some have extrapolated these suggestions to the point of holding that there is something suspect about the use of the *I* concept to indicate an individual. A few have held that it is epistemically and metaphysically appropriate to dispense with the *I* concept altogether.

Lichtenberg's epigrammatic remarks provide a text for my discussion. Let me begin by taking up his emphasis on acquaintance. Lichtenberg is surely right, as was Hume before him, in claiming that what yields a usage for the *I* concept is not an acquaintance with something. We do not seem to 'introspect' a self. A view loosely associated with Hume maintains that since we cannot introspect a self, we should not regard *I* as having a referent. I mention this view only to set it aside. It stems from empiricist dogma so crude as not to merit serious consideration. There is no reason to accord such weight to the notions of acquaintance and looking-within in arbitrating an issue about reference or self-knowledge.

One could advance a less of dogmatic point along similar lines, however. Lichtenberg's and Hume's observation that we are not directly acquainted with a self might be combined with the view that we do 'introspect' our thoughts, or at any rate have more immediate access to them. Then 'postulating' an agent (to echo Lichtenberg's words)—an agent in addition to the thought itself—may seem like a significant *step* that might be doubted. The result of foregoing the *I* concept, and making do with impersonal reference to thinking's going on, may seem less subject to doubt than the *cogito* itself.

It is not clear in what sense we 'introspect' thoughts, any more than we do a self. Thoughts present no inner-perceptual resistance (as perceptions of a physical object do); they commonly have no phenomenology. Moreover, the notions of acquaintance and introspection are elusive. They can hardly be taken as firm tools for understanding these matters. Still, we do, sometimes, 'run through' thoughts. In such cases, we seem to have some occurrent grasp or understanding of them. There is, as far as I can see, no analogous occurrent grasp of a self. If one were impressed with this difference, one might sympathize with the view that the move from awareness of a current thought to the assumption of a self involves a *step* that is problematic in a way that the awareness of the thought is not.

But there is something misleading about this reasoning. It is entirely external to actual uses of *cogito*-like thoughts. For someone who has the *I* concept, there is no step from recognition of the occurrence of a thought to the conclusion that there must be a self. There is no step, inference, or postulation at all. There is no identification of a self based on awareness or based on anything. Normally one simply applies the first-person concept immediately, not in *response* to anything. Such applications fall under the rule that the referent is the author of the thought.

[3] G. C. Lichtenberg, *Schriften und Briefe*, II (Munich: Carl Hanser Verlag, 1971), 412, §76.

Given that the first-person concept is applied, there is no possibility of reference failure. And if one's ascription of the thought to oneself is immediate and non-inferential in this way, there is no possibility of mis-attribution or misidentification of the thinker of the thought.

The claim of differential certainty based on considerations of relative closeness to introspection seems uninteresting. The epistemic issues do not concern missteps within one's cognitive economy. Moreover, the character of the rule that governs reference with the first-person concept suggests that the introspectionist or perceptual model is mistaken. Mastering the first-person concept is sufficient to guarantee that applications will be successful. This suggests that the epistemic warrant associated with applications of the concept comes with mastery of the concept—and is non-empirical. It does not derive from experiences associated with particular applications of the concept. I shall return to this point.

Insofar as we are to find a philosophically interesting challenge in Lichtenberg's remarks, I think that we must associate them with issues about the point and commitments of the first-person concept. The challenge is that acquisition of the *I* concept contains some error, or at least is dispensable for cognitive purposes.

Some have held that the first-person concept carries an objectionable commitment to mental substance separable from physical entities. I think this a mistake caused by overreaction to Descartes' claims to derive dualism from mere reflection on the *cogito*. Uses of the *I* concept make no obvious commitment regarding the metaphysical nature of its referent, other than that it be an author of thoughts. Deriving metaphysical implications from this commitment would require further argument, which would have to be evaluated on its merits.

But Lichtenberg seems not to be raising a question about the nature of thinkers. He is questioning whether there are thinkers—referents of applications of the first-person concept—at all. Or at any rate, he is questioning whether belief in their existence, via judgments involving application of the first-person concept, stands on an epistemic par with knowledge of the existence of thoughts.

An issue often raised about Descartes' use of the *cogito* is whether one could refer to oneself in the first-person way if one did not have various perceptual experiences that enabled one to individuate oneself, or at least reidentify oneself over time. This question was raised sharply by Kant and has been pressed by Strawson and others in modern times. Sometimes it is inferred that one could not have purely intellectual knowledge of oneself or of one's thoughts as one's own; for self-knowledge inevitably depends on perceptual experience. But to know that certain thoughts are occurring, we seem only to have to think about the matter.

It is surely true that self-knowledge and the mastery of the first-person concept depend on perceptual experience. But it does not follow that reference with the first-person concept, or knowledge of oneself through employment of the first-person concept, rests for its justification on sense experiences. It is important here to distinguish dependence on sense experience for the mastery of concepts—ability dependence—from dependence on sense experience for fixing a reference or for being justified in or entitled to one's judgments. Perhaps understanding any concept—including logical ones—depends on having sense experiences of stable objects. But it does not follow that the reference of all concepts is fixed through sense experience.

The referent of a use of the *I* concept is *not* fixed by sensory experience. It is fixed purely by the rule: the referent is the author of the occurrence of thought containing application of the *I* concept. No perceptual ability to track that author enters into fixing the referent in any given instances.

The role of sensory experience in justification of *cogito*-like judgments is equally indirect. Although the very thinking of the thoughts depends on having had certain types of sensory experiences, one's epistemic right to accept such judgments does not rest on such experiences. The relevant judgments are not reactive. One does not find oneself in introspection and then make a judgment about what one is thinking. One does not, or need not, connect oneself with some body that one tracks through time and base one's first-person judgment about one's own thoughts on this connection. One simply makes the judgment. One's epistemic right to make it is, at least prima facie, purely intellectual. It does not rest on any warrant given by sense experiences epistemically associated with the judgment.

So the dependence of the conceptualized first-person perspective on having some third-person perspective on oneself or on other stable objects is not one that enters into the account of one's epistemic warrant for making such judgments. I think that Descartes was entirely right in his view that many first-person judgments are warranted through no more than their being understood. His being right about this in no way shows that it is coherent to conceive of someone with the *I* concept who takes only the first-person perspective. So Lichtenberg's claim that thoughts involving the first-person concept are epistemically less basic than thoughts (about thoughts) that lack that concept cannot be usefully developed by reflecting on the role of third-person perspectives in enabling us to think about ourselves.

There is, I think, a point about conceptual priority that one *can* usefully associate with Lichtenberg's remarks. They suggest the question of whether the *I* concept could be 'dispensed with'. On this line, one would employ only propositional attitude concepts impersonally attributed in Lichtenberg's format. Lichtenberg compares the thought *that thinking is going on (es denkt)* to the thought *that lightning is striking (es blitzt)*. A closer grammatical equivalent in English to the German *es denkt* would be *it is thundering (es donnert)*. What would be lost if one followed Lichtenberg in using only these conceptions?

For the sake of argument I will not take a position on whether it is possible to have the concept of propositional attitudes, or even to reason critically, yet lack the full first-person concept. (To reason critically in my sense, one must correct, suspend, change attitudes, conceived as such, on the basis of reasons acknowledged as such.) But I think that such beings would be conceptually deficient.

They would lack a full conceptual perspective on themselves and their acts. I want to explain the deficiency and indicate what epistemic rights attach to self-attributions of thoughts containing the full first-person concept. I would like to do this without begging questions against Lichtenberg's position. I will develop an answer to Lichtenberg that assumes only what he is surely committed to: that reasoning occurs, and that it is a worthwhile theoretical enterprise to understand reason and reasoning.

Given this objective, I will neglect other answers to Lichtenberg that I think obvious and sufficient in themselves. For example, I think that the idea of mental states and events without an individual *subject* is incoherent. Thinking requires an agent that thinks. For *persons* who think, the first-person concept makes possible reference to themselves from the perspective most basic to their thinking.

One can take the dependence of mental states on a subject further back ontogenetically. Consider subjects that have phenomenal or intentional states, but that by their nature lack critical reason—and hence, in my view, are not persons and are not (or lack) selves. The very existence of perceptual states or sensations—even in the absence of propositional ability—requires a subject, an individual with subjectivity or consciousness. Perceptual systems of lower animals require a subject; and it is clear that those systems have some sort of non-conceptual ego-centric sensitivity. Similarly, animals that think but by their nature lack critical reason lack selves; I think that they lack a full first-person concept. Their thinking too requires an individual subject. Animals with propositional attitudes certainly have non-conceptual ego-centric sensitivity; perhaps they also have some indexical concept that applies to themselves and that is an ontogenetic predecessor of the full first-person concept. All these beings' mental states require a subject, whose subjectivity is a necessary aspect of their sensations, perceptions, or propositional attitudes (cf. n. 1).

Lichtenberg's format ignores the conceptual requirement that such states and events presuppose an individual subject with a subjective perspective. Ego-centric sensitivities or concepts mark this perspective. I think that these truisms are decisive. Pursuing them might carry us further into the nature of persons or selves and into the ontological and ontogenetic roots of the first-person concept.

But my project here is not primarily to determine the nature or ontology of persons or selves, or the range and variety of ego-centric sensitivity and conceptualization. It is to answer Lichtenberg's epigrammatic challenge to explicate the cognitive role and epistemic status of the first-person perspective, assuming only things about reason that Lichtenberg is committed to. Thus my argument will not depend on how one views the relation between persons and animals, or between selves and mere subjects. It does not even depend on my view that only beings whose natures make them capable of critical reason can have the full first-person concept. It depends only on an argument that that concept has a certain necessary and unique role in fully understanding reasoning.

So what does Lichtenberg's format leave out?[4] One deficiency is articulated by Bernard Williams. Williams points out that Lichtenberg's formulation, 'Thinking is going on', needs 'relativization'-intuitively, to a thinker or point of view. For there is a distinction between cases in which we regard thinkings of mutually contradictory propositions as indicative of a violation of a law of logic and cases in which we regard them as indicative of disagreement. Similarly, there is a distinction between cases in which a thinking that *p* and a thinking that *q* indicate some normative pressure in the direction of a thinking that *p* and *q*—and cases in which there is no such pressure. The first case in each pair intuitively involves thoughts by a single thinker (at roughly the same time). The second case in each pair involves thoughts by different thinkers, or within different points of view.[5]

These points do force some sort of 'relativization'. But it is not evident from them alone what the relativization should be. Lichtenberg might still resist use of the *I* concept. He might maintain the impersonality of formulation that he began with. Derek Parfit has tried to remain true to Lichtenberg's spirit by providing a substitute for the *cogito* that makes explicit use of the notion of a point of view: *In the point of view or life to which this thought belongs, thinking is, in this very thought, going on.*[6]

I will assume that the key element in Lichtenberg's position is captured by this proposal. The key element is a claim that full understanding of reason or cognition can dispense with the first-person concept: the concept has no special epistemic status or cognitive value. It has at most merely 'practical' uses.

I think that this position is untenable. To understand fully the fundamental notions associated with reason, including the notions of reasoning, judgment, change of mind, propositional attitude, point of view, one must have and employ a first-person concept. Indeed, understanding the notion of reason itself—epistemic

[4] One relatively minor intuitive deficiency is that there is no self-referentiality or self-verification in Lichtenberg's purported analogies to the *cogito*. Even laying aside issues about the first person concept, the realization that thinking is going on is different from the realization that it is thundering. The former realization is, or will become on reflection, self-referential, and not subject to illusion or error. This difference could be admitted by Lichtenberg. He could simply understand *thinking is going on* as *thinking is in this very thought going on*.

[5] Bernard Williams, *Descartes: The Project of Pure Enquiry* (Harmondsworth: Penguin, 1978), 95–100. Williams does not pursue the question whether impersonal (third-person) specifications might replace the first-person way of specifying a subject that thinks, or even whether the reference must be to an agent. Williams accuses both Lichtenberg and Descartes of failing to provide a basis for individuating minds, and claims that some reference to physical bodies is necessary. In this, he follows Strawson, *Individuals* (1959) (London: Routledge, 2002), 93–100.

[6] Derek Parfit, *Reasons and Persons* (Oxford: Oxford University Press, 1984), Sections 81, 88. Parfit goes beyond Williams in developing the questions whether the 'relativization' to a mind must specify a person in unreduced terms, and whether the specification must be with the *I* concept. He suggests a negative answer to both questions. He hopes to provide a reductive explanation of what a person is by specifying various sorts of continuity among mental states and events. And he purports to express the truth of the *cogito* by dispensing with the *I* concept in favour of self-referential demonstratives. The project of giving a reductive description of what persons are is not presently at issue. But the proposal to de-personalize the *cogito* will be the subject of what follows.

or practical—requires the first-person concept. I will not prejudge whether one must *have* the *I* concept in order to have these other concepts. Here I will argue that any being that had concepts of propositional attitude, reason, change of mind, and so on, but lacked an *I* concept, would be conceptually deficient in the sense that it would lack the conceptual resources to understand fully the most basic necessary and apriori knowable features of the relevant notions. The notions of reason and first-personhood are, at the deepest levels, necessarily and apriori involved in understanding one another.

Reasoning is necessarily governed by evaluative norms that provide standards that count reasoning good or bad—reasonable or unreasonable. But to understand reasons and reasoning fully, it is not enough to understand abstractly that some purported reasons are good and others are bad. For reasons necessarily not only evaluate but have force in forming, changing, confirming attitudes in accord with the reasons. All reasons that thinkers have are *reasons-to,* not merely rational appraisals. But to understand reasons and reasoning, it is also not enough that one understand that rational evaluations should be, and normally (in thinkers) are, associated with some motive or impulse to think or act in accord with the reason or rational evaluation. One must, further, have and understand this motive or impulse in one's own case, and actually apply reasons as rational evaluations to affect judgment and action—to support a judgment, change an attitude, or engage in action. In other words, fully understanding the concept of reason involves not merely mastering an evaluative system for appraising attitudes or relations between thoughts, and not merely realizing abstractly that in any reasoning such evaluations must be (somehow) associated with a motivating impulse to implement them. It requires mastering and conceptualizing the application of reasons in actual reasoning. And this requires being immediately moved by reasons in reasoning and understanding what it is to be so moved. There are thus applicational, or implementational, and motivational elements in understanding reasons.

These motivational elements are intrinsic to a broad notion of agency. I do not mean by 'motivational' to imply some interposition of desire or motive or volition. I mean that to understand reasons one must know how to use reasons, and indeed actually use them, to support or change one's own attitudes in one's own thinking practice. To understand the notion of reason, one must be susceptible to reasons. Reasons must have force for one, and one must be able to appreciate that force. Considerations seen as reasons must have some tendency to affect one's judgments and inferences according to the norms associated with the reasons. And one must recognize that this is so.

Having reasons and having some capacity to be moved by them—to think or otherwise act on account of them—are necessarily connected. The connection is not that everyone who has reasons must at every moment have some tendency to be moved by them. One can perhaps imagine schizophrenics or mystics or quietists lacking such a tendency some of the time. But to have reasons one must, I think, have had some tendency to have one's thoughts and attitudes be affected by them. Beings who have reasons must sometimes be in continuing, uncoopted control of some events, in the sense that the events are a direct guided product of the reasoner's central rational powers. Events guided by reasons issuing from a thinker's uncoopted central rational powers (from the thinker *qua* individual) are acts, as are the guiding events.

So in reasoning, no thinker can be a mere observer of reasons and their effects on reasoning. For having reason requires at some point having some tendency to be affected by reason's power in motivating reasoning. *Understanding* what a reason is, is partly understanding its motive force, as well as its evaluative norms. To understand reason and reasoning, this force must be operative in one's own case; and one must conceptualize its implementation. That is, one must be susceptible to the force and implement normative evaluations in guiding thought and other acts that fall under those evaluations; and (to understand reasoning), one must regard reasons as effective in one's judgments, inferences, and other activity. Doing so amounts to an acknowledgment of one's agency. If one conceptualizes this fully, one recognizes oneself as an agent. Here we see a point about agency that Lichtenberg missed in comparing thinking to lightning's occurring. Thinking is necessarily associated with reasoning—thinking guided by reasons—and reasoning cannot in general be a mere 'going on'. In making inferences, a being is *ipso facto* an agent.

Let me depart from the main line of argument to elaborate these remarks about agency. The relevant effects of reasons are effects on one's judgments, inferences, and other acts. In recognizing the effect of reasons on one's judgments and inferences, one cannot reasonably think of oneself as powerless. Reasons give one reason to make, change, or confirm a judgment or inference. Recognition of a contradiction in one's attitudes gives one reason to change them. Recognition that one's means will not suffice for one's end gives one reason to change one's means or end. To understand reasons, one must understand their force and application in one's reasoning. To understand their force and application one must have some tendency normally to make them effective in forming, changing, or confirming one's attitudes or inferences.

An instance of this sort of point is commonly associated with a view about moral reasons—the view that reasons that are associated with obligation or with a good must, at least in normal cases and given that the person understands the reasons, be associated with some sort of motivation. This view is shared by many who differ over the relation between reason and motivation (whether, for example, the motivation must reside in an independent desire and is a prior condition on a reason, or derives from understanding the reason itself). The point is normally applied to what are commonly called practical reasons. I think that it is embedded in the broader, less restrictive notion of reason, and applies no more to practical reasons and practical agency than to epistemic reasons and epistemic agency. The notions of agency and practice that I am explicating are broader, and I think more fundamental, than the standard notions of action and practical reason.

I return to the main line of argument. Reasons must sometimes provide immediate reason-to—must sometimes be rationally applicable to affect an attitude or action—*immediately*. On pain of regress, in actual reasoning one cannot require a premiss or further reason for applying reasons, for implementing rational evaluations. In reasoning, reasons must have force in a way that is obvious and straightaway. The rational relevance of reasons to their first implementation within one's thought must be rationally necessary and rationally immediate.

A fully explicit understanding of reason must be capable of marking conceptually the cases in reasoning where evaluating or appraising attitudes or activity under rational norms rationally motivates *immediate* implementation of the evaluations in shaping the attitudes or activity being evaluated. One can evaluate a system of attitudes (in another person or in the abstract) as unreasonable without its being immediately rational for one to change those particular attitudes, or even immediately rational that those attitudes be changed from the perspective in which implementation has to occur. To understand reason one must distinguish conceptually from such cases those cases where particular evaluations immediately rationally require being moved to affect the attitudes or activities being evaluated in accord with the evaluations.

These distinctions are knowable apriori. We can know apriori not only the distinction between evaluation and implementation. We can also know apriori how to conceptualize and recognize instances where implementation is immediately incumbent, and understand wherein these instances are relevantly different from cases where an evaluation of attitudes does not rationally demand immediate implementation of the evaluation on the attitudes being evaluated.

Many thinkers with reasons—many animals, I think—cannot mark the distinction. They lack full understanding of reason. They have not conceptualized what is fundamentally involved in reasoning. Full understanding of reasoning requires a form of thought that marks conceptually those particular attitudes where implementation on those attitudes of a rational evaluation of those attitudes is rendered immediately rationally incumbent by the evaluation.

The first-person concept fills this function. Its association with a thought ('I think...', 'I judge...', 'I infer...') marks, makes explicit, the immediate rational relevance of invocation of reasons to rational application, or implementation, and motivation. It both designates the agent of thought and marks the acts and attitudes where a rational evaluation of the act or attitude immediately rationally requires using that evaluation to change or maintain the attitude. Acknowledgment of a reason for or against an act or attitude to which one attaches, or can attach, one of these forms of 'I think' makes it immediately rationally incumbent on one to give the reason weight in making the act or attitude accord with it.

Acknowledging, with the *I* concept, that an attitude or act is one's own is acknowledging that rational evaluations of it which one also acknowledges provide immediate (possibly defeasible) reason and rationally immediate motivation to shape the attitude or act in accordance with the evaluation. Unless further evaluations of the attitude must be taken into account, there need be no further intervening reasoning involved for it to be rational to have the reason affect the attitude or act. The first-person concept fixes the locus of responsibility and marks the immediate rational relevance of a rational evaluation to rational implementation on the attitude being evaluated—to epistemic or practical agency.[7]

First-person concepts, of which the singular is paradigmatic, are, I think, the only ones that fill this function. (I lay aside the plural 'we', though I think this notion deserves reflection.) Let me try to make this claim plausible by considering alternatives.

One can attribute irrationality to a judgment of the form 'It is judged that . . . '. But such an assessment makes explicit no immediate reason to change the commitment being evaluated, for the judgment is not attached to anyone who makes the judgment. The assessment marks no locus of responsibility or power associated with the judgment. One can conclude only that someone has reason to change the judgment.

Judgments in third-person form—like 'She judges that . . . ' and 'Burge judges that . . . '—do identify an author of the judgment. So they do identify a locus of power, responsibility to norms of reason, and rational motive. But these forms cannot mark the immediate rational relevance of a rational assessment to modifying or standing by the judgment. Here the notion of *immediacy* is significant. I want to clarify the role of this notion in the account.

As I have noted, reasons enjoin thinking or acting in accordance with them. And anyone who has a reason normally has some motive force for implementing it in thought or action. But there is a further point. Anyone who has a reason that evaluates any act or attitude, no matter who is actor or subject of the attitude, has some rational motive—however attenuated—to affect the act or attitude in accordance with the reason. That is, reason has the *transpersonal function* of presenting true thoughts and guiding thought to truth, regardless of individual perspective or interest. This function is valid for any rational agent. But such a function operates only through the reasoning of individuals. So an individual's assessment of some judgment as irrational carries with it some prima facie ground not only that it be altered—but some prima facie ground to alter it, regardless of who the source of judgment is. But when the source is not understood to be oneself, the reason to implement the evaluation cannot be *immediate,* in at least two respects.

One respect has to do with the person- or system-dependence that attaches to the having of reasons. What may be a reasonable evaluation by person (or system) A of an attitude held by a person (or system) B may not be a reasonable evaluation for B. For example, if A knows something on which the reason is

[7] John Perry, 'The Problem of the Essential Indexical', *Nous* 13 (1979), 3–21, repr. in *The Problem of the Essential Indexical* (Oxford: Oxford University Press, 1993), insightfully makes the point that attribution of beliefs involving the first-person indexical is essential to the explanation of certain actions. Perry does not connect the point to fundamental features of reason, or to the broader notion of agency, that includes mental agency, that I have highlighted.

based that B does not know (and has no reflective access to), then A's reasonable evaluation cannot be immediately rationally applicable for B. B would have to acquire the additional background knowledge. Similarly, if A's all-things-considered reasonable evaluation of B's attitude were based on information that B had but which was superseded by knowledge that B had but A lacked, then A's all-things-considered reasonable evaluation of B's attitude could provide no all-things-considered rational motivation for B. Again, the rational applicability of A's rational evaluation of B's attitude would not be immediate. This is a variant of Williams's point, discussed earlier. The fact that it is reasonable for A to make an inference with premises for which A has good reasons does not immediately imply that it is reasonable for B to make the inference, since B may lack reason to believe one or more of the premises. Since mismatches in information on which reasons can be based are always possible, no rational evaluation that is not universally self-evident, however reasonable, has rationally immediate application, with consequences for immediate implementation, across persons or across points of view. As long as the attitude is not taken to be one's own, there is always the possibility of a gap, and filling that gap involves a rational step.

The second respect in which rational evaluations of attitudes not understood to be one's own are necessarily non-immediate in their implementation has to do with means. When the subject of the evaluated attitude is not understood to be oneself, one can propose to affect the attitude in accordance with the evaluation only non-immediately, by some *means*. One can propose to do so only by force or persuasion. One's power over, and responsibility for, the attitude (or activity) are not direct. So the question of how one is to bring about any alteration must inevitably arise. One cannot simply alter the thought immediately, with no intervening practical premises.

In one's own case, these questions do not normally arise. One may ask what element in one's point of view to modify in the face of reasons that count against a thought or an attitude. But, except in special cases, the rational relevance of reasons to implementation is direct, and does not pass through premises about means.

The special cases are cases in which an attitude is psychologically immovable in the ordinary way, or those in which one sees one's own attitudes as objects, rather than as parts of one's critically rational point of view. One may then have to reason about one's attitudes as if they were those of another person, perhaps even using methods of manipulation on oneself. But then there must be other attitudes and thoughts over which one has immediate power. If there were no such attitudes and thoughts, one would not be a reasoner at all.

So third-person attributions do not mark the immediate rational relevance of rational evaluation to implementation of the evaluation. Even when a third-person attribution is to oneself, the relevance is not rationally immediate. For one could fail to know that the third-person attribution applied to oneself. I could fail to know that I am Burge. And although I do know, the *rational* relevance of reasons to their affecting my attitudes is not conceptually immediate. It must pass through the assumption that I am Burge.

Even third-person attributions that draw on the epistemology of first-person authority do not mark the immediate relevance of reasons to reasoning. For example, the Lichtenberg-like formulation—'in the point of view or life to which this thought belongs, it is being judged, in this very thought, that . . . '—does not do so. Such a specification constitutes no acknowledgment of proprietary power over, or responsibility for, the thought, much less a locus of power and responsibility. There is nothing in the content of 'this very thought' that ensures that it is one's own and makes for immediacy of rational evaluation to rational implementation. We tend to presume that *all and only* thoughts referred to that way, and that can be known non-inferentially, *are* one's own.

But there is no rational necessity that this be so. Even if there were, understanding the necessity would require that one make explicit that such thoughts are necessarily one's own. And doing this would require use of the first-person concept. So any presumption of immediacy associated with such conceptualizations relies on an implicit premise identifying the thoughts as one's own. Lacking such a premise, the rational relevance of reasons to implementation is not immediate.

Similarly, specifications of oneself like 'the thinker of this very thought judges that . . . ' or 'the agent of the point of view that contains this very thought thinks that . . . ' do not do so. They do specify a locus of power. But they do not acknowledge proprietary power over, and responsibility for, the thought. They are simply objectively descriptive of the thought's owner. Such specifications express a point of view on oneself from the outside.

The relevance of third-person self-descriptions, and of the Lichtenbergean description of a 'point of view', to implementation of rational evaluations is not rationally immediate. They depend on connection to the first-person conception. The premiss that one is the relevant thinker—or that one is the author of the relevant point of view—is necessary for making the description immediately rationally relevant to connecting reasons to their application in reasoning.

Only the acknowledgment of authorship or ownership for thoughts or attitudes makes conceptually explicit the immediate rational connection between rational assessment of those thoughts and the affecting of the attitudes according to the norms of the assessment. Any way of thinking of oneself, or of one's point of view, that does not carry this acknowledgment conceptualizes associated attributed attitudes as objects of thought, but not necessarily and immediately ones to reason with in accordance with the evaluations.

Recognition that a thought is one's own—taking up the subjectivity and proprietary ownership expressed in the first-person concept—is the only basis for conceptually expressing having a rationally immediate and necessary reason to tend a point of view, to make the reasons effective on the attitudes they evaluate. Attributions of attitudes in first-person form instantiate recognition of ownership and power of agency, and of the rationally immediate motive force and implementational encumbency of reasons. Rational activity presupposes a distinctive rational role for the first-person singular concept.

Much of the content of science and mathematics includes no first-person elements. Scientific writing leaves out such elements on principle. Such omission acknowledges the transpersonal function of reason. It also acknowledges the fact that theory and evidence in these disciplines are perspective-independent, in the sense that anyone could have made the same observations or come to the same theory. But the application of reasons within such theorizing—indeed, the very notion of reason—nevertheless presupposes the first-person concept. Understanding reason and the objective point of view of science and mathematics is inseparable from taking on and acknowledging explicitly a first-person way of thinking.

The reason why this is so is that reason has an essential relation to reasoning, to the *practice* of being moved by reasons. The practice of reason, not just the form and content of reasons, is inseparable from the nature of reason. Having reason and having a reason are essentially associated with some impetus to think or otherwise act in accordance with reason. Understanding reason requires being inclined to be affected or motivated by reasons—to form, change, or confirm beliefs or other attitudes in accordance with them—when those reasons apply to one's own attitudes. So understanding reason entails some optimism and commitment regarding the possibility and effect of reason in one's thinking. Here Lichtenberg was on to something deep in the last remark of the passage we began with: 'Accepting, postulating, the *I* is a practical requirement.' Despite the misleading point about postulation, and despite the fact that Lichtenberg was wrongly thinking of a practical requirement as in some opposition to epistemic or theoretical requirements, the linkage of the first-person concept with practice is on to a fundamental point.

Let me summarize the main line of argument. To fully understand basic features of the concept of reason, it is not enough to understand the concept in the abstract. It is not enough to understand the evaluation of attitudes or thoughts as being reasonable or unreasonable. And it is not enough to understand, in the abstract, that reasons enjoin and normally motivate thinking or acting in accordance with the normative standards that they set. Fully understanding the concept of reason also requires engaging in reasoning, and understanding basic features of such reasoning. Engaging in reasoning requires implementing reasons or rational evaluations immediately on the attitudes to which the reasons or rational evaluations apply—being moved to think in accordance with one's reasons. Understanding basic features of such reasoning requires understanding such implementation. Fully conceptualizing and understanding such implementation requires an ability to mark conceptually, in actual particular instances, the attitudes or acts for which it is rationally immediate that one's all-things-considered reason or rational evaluation of the attitude or act enjoins shaping it in accord with the reason or rational evaluation. Such understanding requires being able to distinguish those attitudes from attitudes in which one's all-things-considered evaluation of the attitude indicates (as always) that the attitude should be shaped in accordance with the evaluation, but in which this indication does not presume to be all-things-considered in the point of view from which the implementation must be carried out. That is, the implementational relevance is not rationally immediate: it is subject to further possible rational considerations that bear on the rational appropriateness of its implementation. The first-person concept marks the former set of attitudes. Its use marks those attitudes where the individual's rational evaluation of them carries a rationally immediate incumbency to shape the attitude in accord with the evaluation. Acknowledging them as one's own is acknowleding such responsibility. The first-person concept is the only concept that fills this function in the actual practice of reasoning. So fully understanding the concept of reason, and engaging in reasoning in the most reflective and articulated way, require having the *I* concept and being able to apply it for this purpose.

I have summarized this argument in a way that brings out that it does not beg the question against Lichtenberg. It assumes only that Lichtenberg is committed to understanding reason and reasoning. The argument shows that the first-person concept is indispensable to a full understanding of reason, including theoretical reason. Given the understanding of agency expressed earlier, and given the fact that thinking presupposes reasoning, the argument yields a corollary—thinking presupposes agency. Each of these points is incompatible with the view I have associated with Lichtenberg.

The argument also undermines the view that the first-person concept is of *merely* practical significance. As I noted, Lichtenberg holds that accepting the first-person concept is a 'practical requirement'. The context suggests that practical requirements are to be distinguished from more 'substantive' requirements that might be relevant to knowledge or reality. But the first-person concept is essential to understanding reasoning of any sort—theoretical or practical. The understanding involved in marking conceptually, through the first-person concept, individual cases where rational evaluation of attitudes rationally requires immediate implementation of the evaluation on the evaluated attitudes is no less theorietical than practical. In fact, a sharp distinction between the theoretical and the practical makes no sense at this level of reflection. Any reasoning necessarily involves agency. Fully understanding all reason and all reasoning requires the first-person concept. So the first-person concept is as relevant to metaphysics and scientific reasoning as it is to 'merely practical' matters.

Thus the role of the first-person concept in understanding reason cannot be taken as 'merely practical' in a way that would undermine the natural idea that uses of the concept *refer*. I have in effect provided an argument, as if one were needed, that such uses do refer: True accounts of subject matters of theoretical importance are committed to referents for their irreducible singular terms. True accounts of the nature of reasoning are theoretically important and are irreducibly committed to uses of the first-person concept. Uses of the first-person concept constitute uses of a singular term. So in being committed to such accounts, we are committed to referents for uses of the first-person concept.

It is not my purpose to rebut attempts to show that the first-person concept is non-singular, or attempts to challenge the standard view of referential commitment just sketched. I know of no interesting, clear-headed challenges of these sorts to the ordinary view that uses of the first-person concept refer. To this extent, Descartes and common sense are confirmed.[8] My main purpose, however, has not been to argue reference, but to establish the role of the first-person concept in understanding reason and reasoning.

The first-person concept plays a central role in apriori understanding of reason, agency, and ourselves. I want to say a little about the place I have given understanding in this account. I have not argued that to reason, in the weak sense of making good inferences, one must *have* the first-person concept. I think that animals engage in rudimentary thinking, which (given that it *is* thinking) constitutively occurs in normal cases according to norms of reason. Inferential thinking is caused or guided by reasons, and is explained by their being reasons. But animals lack the first-person concept that interests me. They have some sensitivity to their own points of view, but I think that they lack the conceptualized self-attributions necessary to employ a full-blown first-person concept.

I have not even argued that engaging in *critical* reasoning—the sort that evaluates attitudes as reasonable or unreasonable, and that shapes attitudes according to such evaluations—requires, by necessity, having a first-person concept. I have not argued this because I think the relevant issues need further clarification. In our actual social development, it is of course true that one acquires the first-person concept before or during the development of critical reasoning. The hard issue is whether this order is necessary and knowable by apriori reflection. On the other hand, we can certainly imagine critical reasoning proceeding without explicit *linguistic* expression of a first-person point of view. Whether it is necessary and knowable by apriori reflection that the first-person point of view be *implicitly* conceptualized whenever critical reasoning occurs is the delicate matter that I have left open.

So I have allowed, for the sake of the present argument, that a critical reasoner might lack the full first-person concept. Such a reasoner would conceptualize reasons and attitudes as such, and would be sensitive to cases where attitudes had

[8] As I noted earlier, I do not think that Cartesian dualism can be inferred from applications of the first-person concept. But I do think that the concept's cognitive role is relevant to metaphysics and epistemology. The argument just sketched helps show why it is mistaken to embrace the strange idea that Lichtenberg's remarks have sometimes inspired—that thinking is best seen (perhaps best seen for 'metaphysical' purposes) as going on without a thinker, or that the first-person concept never literally has a reference. Note that the earlier argument that thinking requires agency also tends to undermine this view, in so far as it is especially hard to conceive of agency without an agent. It is not an accident that Lichtenberg's formulations gravitate to locutions that do not attribute agency. In so far as the first-person concept is necessary to a full understanding of any sort of reason, including theoretical reason, there is no room to see its implications as dispensable or merely practical.

to be shaped immediately by reasons. But the reasoner could not mark those cases conceptually in the implementation of reasoning.

Critical reasoning is the sort of reasoning that we associated with the nobility of being a person, with science, mathematics, art, practical reasoning, and with rational inquiry of all kinds. So supposing for the sake of argument that such reasoning does not require having the first-person concept, what philosophical significance is there in the argument that one cannot *understand* reason (*a fortiori* critical reason) without that concept?[9]

I want to highlight two types of significance. One stems from the sort of understanding that is involved. The understanding is apriori, and can be derived from reflection on fundamental aspects of the nature and functions of reason. I maintain that every step of the argument which established the role of the first-person concept in fully understanding reason is apriori.

Such understanding can be derived from reflection—on concepts and on actual reasoning. And it is not part of some esoteric theorizing about reason. It normally arises from the most elementary cognitive development in a social setting.[10] Uses of the first-person concept in claiming acts or attitudes as one's own are normal acknowledgments of authorship and responsibility in critical reasoning. They are part of a full expression of what it is to be reasonable. A being that reasoned but lacked a first-person concept would not have conceptualized or rationally expressed a *fundamental* function of reason. Being able to conceptualize, for implementation in reasoning, the cases where there is a rational demand and motivation immediately to shape evaluated attitudes in accordance with the evaluation is placing under conceptual control one of the most basic functions of critical reasoning. Use of the first-person concept is a conceptual expression of one of the central functions of reason.

As a consequence, use of the concept is underwritten by reason. We are entitled to first-person concepts in judgments partly because they are necessary to the fully articulated exercise (as well as understanding) of reason. The first-person concept earns its place in the general non-empirical entitlement to self-attributions of thoughts partly through its constitutive association with a particular fundamental feature of critical reasoning.

Thus I believe that I have provided a rational 'deduction', in Kant's sense, of the first-person concept. I have shown that we have a right to use the concept, a right that is grounded in reason. The steps of this exposition of right are warranted apriori. Moreover, the points that I have made about the dependence of our understanding of reason on practice—actual applications—and on understanding practice suggest a sense in which our apriori understanding of the concept of reason, and of the first-person concept, is not purely 'analytic', in the sense

[9] I owe this question to Barry Stroud.

[10] Thus, although it is a delicate question whether critical reasoners metaphysically *must* have the first-person concept, it is certainly normal for critical reasoners to have it; and the concept enters into ordinary understanding of those critical activities that mark their nobility.

of being grounded in abstract conceptual analysis.[11] Fully understanding the concept of reason requires understanding reasoning. Understanding reasoning requires use and understanding of the first-person concept. The relevant use and understanding resides in conceptualizing an awareness of the rationally immediate applicability of rational evaluations to affecting attitudes *in the actual practice of reason*. Such awareness must be an understanding of actual applications of reasoning. It cannot be obtained from conceptual analysis alone. So the 'deduction' is synthetic by any measure.

The second type of significance bears on the role of the first-person concept in conceptualizing rational agency. Part of being a fully rational agent is, in Kant's phrase, to act under an idea or concept of that agency. A being that lacked the first-person concept could be sensitive to the norms of reason, and might (I am conceding for the sake of argument) even sensitively shape its attitudes according to a *conception* of good and bad reasons and reasoning. But the agent would lack full conceptualization of what it is doing.

More specifically, it could not conceptualize cases in which reasons had immediate rational relevance to implementation of the reasons on the acts or attitudes that they bear on. It could not fully conceptualize its agency and acknowledge its responsibility to rational norms. It would not be 'acting under the idea' of its responsibility or agency. Insofar as *full* intellectual (or any other) responsibility requires the capacity to understand the way norms govern agency and the capacity to acknowledge the responsibility, a being that lacked the first-person concept would not be fully responsible intellectually. It would not have a fully realized rational agency. Conceptualized self-consciousness seems a necessary condition for fully responsible agency. Using the first-person concept is necessary to being a fully realized person.

* * *

I want to step back now and consider briefly how this discussion of the role of the first-person concept in reasoning bears on self-knowledge and knowledge of

[11] I reject any conception of analyticity that claims truth independent of the way 'the world' (or a subject matter) is. The notion of analyticity that applies simply to truths of logic plus definitions seems to me harmless if one does not build bad theory into one's understanding of logic or definitions. I am not hostile on principle to the third notion of analyticity—the one associated with analysis of concepts. But I am agnostic about how fruitful or important the notion is. There may be broader and narrower conceptions of such analysis. On the narrower, traditional conception, analysis must take the form of decomposition. On a broader conception, analysis might include any constitutive account of the nature of a concept partly or purely in terms of its relations to others. I am doubtful that there is any clear historical basis for calling truths that are products of analysis in the broader sense 'analytic'. Conceptual truths that 'go beyond'—or depend on conceptual relations beyond—the putative components of a concept are, I think, traditionally counted synthetic. On the interpretation in terms of the narrower notion of analysis, I think that the conception of analyticity has nowhere near the importance accorded to it by Leibniz or even by Kant. I presume that analyses of either sort, like the truths or logic, are true not only in virtue of the nature of concepts, but in virtue of (presumably necessary) features of the world. Cf. my 'Philosophy of Language and Mind: 1950–1990', *The Philosophical Review* 101 (1992), 3–51, especially 3–11.

other minds. This is a subject that needs fuller development on another occasion. But a brief sketch may place in a sharper light the preceding discussion.

Elsewhere I have maintained that self-knowledge has a special epistemic status by virtue of its role in critical reasoning. I argued that the nature of critical reasoning requires that some self-knowledge, that which is essential to rational review, must be epistemically different from observation of objects. I maintained that our epistemic entitlement to relevant self-attributions derives, in one sense, from the essential role of such judgments in critical reasoning.[12] The relevant self-knowledge is non-inferential and intellectually grounded. Whatever the details of this account, it is natural to think of self-knowledge as independent of perception of objects for its epistemic warrant.

How does self-knowledge differ from knowledge of other minds? A natural answer contrasts the intellectually grounded character of the relevant self-knowledge with the observationally based character of knowledge of other minds. The relevant self-knowledge is epistemically warranted by an immediate intellectual entitlement, one sanctioned by reason and present in a being with the right conceptual equipment as a consequence of his simply thinking normally. By contrast, according to this natural answer, knowledge of other minds is indirect in that it requires an empirical inference from the perceived behavior of another being—or else it is drawn from complex criteria applied to observed behavior. In any event, its epistemic warrant rests on perception of behavior.

It may be that self-knowledge requires as a psychological condition that one have or have had knowledge of other minds. It might even be (though I doubt it) that it is impossible in some more metaphysical sense to know one's own mind without knowing another mind, or the existence of another mind. But, runs this natural reasoning, self-knowledge has an immediacy and non-empirical intellectual epistemic warrant that is not shared by knowledge of others' minds. Its warrant derives from intellection, whereas knowledge of other minds rests on sense-perceptual observation.

I think that the situation is more complicated. Both self-knowledge and knowledge of other minds can, of course, be inferential or perceptually grounded. But in my view both self-knowledge and knowledge of other minds can be epistemically immediate and epistemically grounded in intellectual, non-empirical entitlements. The fundamental epistemic differences between self-knowledge and knowledge of other minds are more subtle. I do not have the space to elaborate and defend my view that knowledge of other minds can be non-inferential and can rest on an intellectual, non-perceptual entitlement. But I will sketch the main line of reasoning.[13]

[12] See my 'Our Entitlement to Self-Knowledge', *Proceedings of the Aristotelian Society* 96 (1996), 91–116.

[13] Much of the reasoning that immediately follows is layed out in my 'Content Preservation', *The Philosophical Review* 102 (1993), 457–488.

This sketch is necessary to motivate the point of this concluding section. The point will be this: The role of the first-person concept in reasoning illuminates a common source, as well as a key difference between self-knowledge and knowledge of other minds.

I think that we can have a non-empirical, apriori epistemic entitlement to knowledge of other minds through our intellectually grounded entitlement to accept our seeming understanding of speech as genuine understanding. We have an apriori entitlement to prima facie reliance on our seeming understanding of an apparent utterance of content as genuine understanding.

A justification or entitlement is apriori if neither sense experiences nor sense-perceptual beliefs are referred to or relied upon to contribute to the justificational force particular to that justification or entitlement. So, roughly, *justifications or entitlements* are apriori if their force derives from intellection, understanding, or the nature of other cognitive or practical capacities. *Knowledge* is apriori if it is grounded in an apriori justification or entitlement that suffices to make the knowledge knowledge. This conception of apriority allows that one can know apriori of the existence of particulars—for example, particular mental events—if one's justification or entitlement is intellectual, not sense-perceptual. For example, I think that one knows apriori, in this sense, *cogito*-like thoughts. The argument I will sketch supports the view that one can know with apriori (defeasible) entitlement of the existence of other minds.[14]

Let me emphasize that the issues here have to do with the nature of the epistemic warrant, not the mechanism that makes the knowledge possible. Of course, we need perception to hear or see words. So we need perception to understand speech emanating from another mind. That is *how* we do it. This *is* a difference between knowledge of other minds and knowledge of one's own. For one normally does not need perception to know one's own thoughts. But these points concern the mechanism of knowledge acquisition, not, in my view, the nature of our epistemic warrant—justification or entitlement. I believe that our epistemic entitlement to our understanding of content need not have, and sometimes lacks, a perceptually based element.

We can apprehend the presentation of propositional content in speech by simply understanding it, by thinking the content and understanding it as being

[14] This conception of apriority is discussed at greater length in 'Content Preservation'. I hope to show elsewhere that the conception is a traditional one, rooted in Kant, despite the fact that apriority was traditionally not associated with defeasibility, and was often not applied to knowledge of events in time (even sometimes the *cogito*). (Kant refused to apply his conception in any of these ways, but I think that this was the upshot of ancillary doctrine, not a direct consequence of his conception of apriority.) I think users of the conception did not always see possible consequences of its use, or were blocked from accepting such consequences by other doctrines. Earlier in the essay I spoke of apriori reflection. Reflection or understanding is apriori if it rests on an apriori justification or entitlement.

presented. This understanding depends causally and psychologically on perception. But that dependence need not be justificational.[15]

The epistemic entitlement has its force in abstraction from the background dependence on perception. Perception of words, of utterance events, commonly plays an enabling role but not a justificatory role, in our understanding and, indeed, acceptance of intelligible, expressed contents.

I see the matter on partial analogy with the way in which traditional rationalists saw the role of diagrams or symbols in enabling one to apprehend and see the truth of geometrical or mathematical contents.[16] The fact that perceiving something (symbols or utterances) is psychologically necessary to understanding the content is fully compatible with the epistemic warrant's deriving from understanding and being non-empirical, in the sense that the justificational force of the warrant does not derive from perception. In the mathematical case, one's warrant for believing the content derives from genuine understanding of the content alone. In the interlocution case, one's warrant for presuming that one understands derives from one's seeming understanding of an apparent *instantiation*, or *token occurrence*, of content.[17]

Understanding content requires (in normal cases) understanding the attitudinal (e.g. assertive) mode of the content. And understanding attitudinal mode is further inseparable from understanding instantiations, or token occurrences, of content. One's entitlement to rely on one's seeming understanding is fundamentally an entitlement to rely on seeming understanding of instantiations. This is, other things being equal, an intellectual or apriori, defeasible entitlement. Its probity or justificational force as a rational starting point derives not from experience, but from conceptual understanding. In my view, where perception of physical events functions to provide access to an instantiation (utterance) of

[15] When the understanding is not purely intellectual, it may involve perceptual elements. For example, if to understand what someone is saying in pointing to some observed object, I have to see the object, or have some perceptual or imaginative image of how they are thinking of an object, then the understanding is not purely intellectual. One's general prima facie entitlement to rely on seeming understanding of apparent utterances of content is always apriori. But instantiation of this entitlement to (seeming) understanding of a particular (apparent) utterance of content is apriori only if the understanding in the particular instance is intellectual. I take it that although such perceptually infected *de re* cases are very widespread, they are not ubiquitous. Utterances in pure mathematics and some empirical generalization provide examples. What interests me is the very possiblity of apriori prima facie entitlements to believe in the existence of other minds.

[16] In 'Interlocution, Perception, and Memory', *Philosophical Studies* 86 (1997), 21–47, I discuss this analogy, and its partialness, in some detail.

[17] For the sake of my argument about knowledge of other minds, I do not need the claim, which I defend in 'Content Preservation' and will allude to later, that we have an apriori prima facie default entitlement to *accept as true* (particular) seemingly understood apparent *assertions*. All I need for present purposes is that one has an apriori prima facie entitlement to accept one's seeming understanding of an apparent *utterance*, as genuine understanding of genuine utterance. Such seeming understanding is to include seeming understanding of the content and mode of use of the utterance (for example, understanding the instantiated content as asserted). One needs seeming understandings of the form: 'It is asserted that *p*'. More qualifications to this argument are needed in a full statement.

content with its attitudinal mode, not to provide information about objects, perception is no more an element in the justification of the understanding (and of beliefs based on the understanding) than memory is an element in the justification of deductive reasoning.[18] The role of perception is to make understanding possible. But the seeming understanding carries justificational force in itself, in abstraction from its genetic reliance on perception.

So seeming understanding provides an apriori prima facie entitlement to presume genuine understanding of an instantiation of content. But the presumption of the existence of an instantiation (for example, an assertive utterance) of content in explicit propositional form provides an apriori prima facie entitlement to presume that the event has a rational source. For instantiation of content can be known apriori to be constitutively dependent on a system of rational practices for belief formation and content formation.

There are many difficult issues about the points just made. I will have to leave them in undeveloped form for present purposes. So seeming to understand an instantiation of content, together with its mode, gives one apriori prima facie ground to presume that it ultimately has a rational source. That is enough to give one apriori prima facie ground to presume the existence of a rational agent or mind. It seems to me that if the presumption is undefeated and veridical, one will have knowledge of the existence of a mind on the basis of seeming understanding of what is prima facie intelligible.

This presumption need not be the product of an inference, any more than there need be an inference to the existence of oneself in the thinking of *cogito*-like thoughts. Anyone with the requisite conceptual equipment (concepts of thoughts, and first- and third-person pronouns) will be apriori entitled to the presumption of a rational agent both from first-person thinking of one's thoughts and from understanding of thoughts articulated by others. Indeed, anyone unable to immediately associate an instantiated propositional content with the existence of a rational source—a rational author, agent, or locus of power—would be conceptually deficient in something like the way that someone confined to Lichtenberg's formulations would be conceptually deficient. For a reflective understanding of propositional instantiation of content entails understanding that rational norms associated with uses of content apply to agents, loci of rational power. In the first-person case, one indicates rational agency with the *I* concept. As I have argued in the main part of this essay, application of that concept marks acknowledgment of intellectual responsibility and agency. Since this acknowledgment expresses a fundamental function of reasoning, we are rationally entitled to the application of the concept. What sort of epistemic entitlement do we have for attribution of authorship to *others* when we understand their utterances?

I have argued that when one seemingly understands an utterance in interlocution, one is apriori prima facie entitled to a belief in the existence of a *rational*

[18] Compare 'Content Preservation', 476–484; 243–249 in this volume. The points in the next paragraph are also argued for in that article. All of these points require more development and support than I have given them.

source—some agent capable of producing utterances with propositional content and attitudinal force, and responsible for acting under rational norms. For one to be entitled to presume that such a source is *another* agent, one with *another* mind, one must be entitled to presume that it is not oneself. So knowledge of other minds is distinguished from self-knowledge not by being necessarily inferential or by being necessarily grounded in perception, but by being in some known contrast with acknowledgment of an understood instantiation of content as one's own.[19]

The key feature of the first-person concept is that it marks acknowledgment of the immediate relevance of reasons to intellectual practice. In understanding utterances in interlocution, one lacks ground for this acknowledgment. I think that to be critically rational, one must have, and be apriori entitled to, a capacity for a fallible sensitivity as to whether an act associated with a seemingly understood instantiation of content is one's own.

One is also apriori prima facie entitled to rely on particular applications of this capacity. To be critically rational, one must have, in normal cases, sufficient awareness as to when and whether one is the agent of propositional acts to distinguish instances in which one is committed under rational norms governing thoughts with the relevant attitudinal modality from instances in which one is not. This sensitivity is necessary for the ability to apply reasons straightaway. Indeed, it is, as we have seen, a constitutive part of reasoning and understanding reason. So entitlement to it is apriori. If norms of critical reason that indicate how one ought to reason (or otherwise act reasonably) are to apply to one's mental states, one must have, and be rationally entitled to, awareness of instances where they apply and where they do not. This is to say that one must have some apriori entitled awareness for one's *not* being the agent of relevant instantiations of content, and for one's thereby not being rationally committed under rational norms governing the relevant agency.

To know the author of an instantiation of content to be oneself or another, one needs to apply concepts in accordance with the sensitivities discussed above. As indicated before, to be fully responsible to the relevant norms, one needs to be able to act under the idea of the norms. One needs to be able to know and acknowledge one's responsibility. Thus conceptualization of the sensitivities is necessary for being fully responsible to the norms of critical rationality.

These remarks apply to any utterances in interlocution that fall under rational norms—to assertions, to suppositions, to promises, perhaps even to story-tellings. Any understood utterance might be such that one is apriori entitled not to see oneself as its responsible author, relative to whatever rational norms are

[19] I do not claim that one develops this other-attribution only *after* one makes self-attributions. The issues here concern the relation between the entitlements. My point does not even entail any priority of entitlement to take agency as one's own over entitlement to take agency as coming from another.

I have moved freely from talk about prima facie epistemic entitlement to talk about knowledge. I think that entitlement or warrant is the main philosophical issue in a philosophical account of the relevant knowledge. But there are separate issues about knowledge that a full account should address.

relevant. Ability to apply the rational norms entails an awareness of differences between reception and initiation. This sort of awareness is fundamental to being a rational agent. Given that one has first- and third-person concepts and the concept of agency, and given that one understands—and is entitled to understand—some particular content instantiations which one is aware of as not being one's own, one's entitlement to this awareness gives one apriori prima facie entitlement to presume that there is a rational agent other than oneself.[20] One's entitlement to believe in other minds can depend for its justificational force on intellectual understanding of instantiations of intentional content—intellectual 'experience'—rather than sense-perceptual experience.

Let me illustrate these ideas for the case of understanding assertions in interlocution. Suppose that we are apriori prima facie entitled to rely on seeming understanding of events as presentations-as-true, more particularly as assertions (compare note 17). Of course, one's understanding is compatible with one's not accepting what is asserted. Unless one accepts an assertion, one is not rationally committed to there being rational support for the assertion, much less rationally committed to defend it. This conceptual space between understanding and acceptance of an actual assertion is one to which a rational agent must be sensitive—and be apriori entitled to be sensitive—if he is to be subject to rational norms governing acceptance. So to be subject to such norms, one must be apriori entitled to a sensitivity that differentiates merely understanding assertions from making assertions. But this is equivalent to a sensitivity to whenther the source of an assertion is another or oneself.

To articulate another side of this same point: a critically rational being must be able to—and be apriori entitled to—discriminate the sorts of rational warrant that are relevant to acceptance of understood propositional content. In the case of one's own judgments, one must be able to advert to grounds, accessible to one, that would provide some justification. Or else one must (as in perceptual judgments) have access to some mark of one's entitlement (for example, one's experience). But the norms of reason governing *interlocution* allow that one be rational in one's acceptance of an assertion and lack independent epistemic warrant for the proposition accepted. One is not rationally responsible for defense of these beliefs in the same way as one is for defense of one's autonomous beliefs. One must rely on rational entitlements or justifications (in others) that one lacks. One's acceptance presumes justifications, or entitlements, that one may not oneself have. To be subject to the epistemic norms governing interlocution, one must have and be apriori entitled to awareness of this dependence. This awareness yields apriori entitlement to presume that the agent of an assertion is not oneself.

So one is apriori entitled to awareness of whether or not a commitment associated with a putative assertion is one's own. For to be subject to epistemic norms one must be able to discriminate cases in which one is committed to

[20] I should say that I think that entitlement to an awareness of the type of rational commitments one has obtains in cases other than interlocution—in inference, perception, memory, and so on.

rational support of the commitment from cases in which one is not. To be rational one must have, and be apriori entitled to, some sense for one's *not* accepting actual assertions. Where one's seeming understanding of an apparent assertion is accompanied by an awareness that one is not the agent of the assertion, one has an apriori prima facie ground for presuming that there is another mind, another rational agent.

These entitlements to an understanding of the type of rational commitments that our intellectual activities fall under underwrite a non-inferential ability to discern one's authorship or non-authorship of intellectual (or practical) acts or commitments. We need not *infer* that a rational source of interlocution is another mind. We believe it through understanding an assertion in the third-person attributive way. If one has the requisite conceptual equipment to make explicit third-person attributions of propositional content, one can know immediately in understanding an utterance its being a sign of another person, just as in using the first-person concept in *cogito*-like thoughts, one knows non-inferentially a thought as one's own.

Thus, at the base of rational practice is an awareness of the source of rational agency. We are entitled to a non-inferential belief that there is another agent through the very understanding of utterances in interlocution. Third-person attributions have a source in a rationally required and rationally entitled ability to distinguish, at least in normal cases, our own acts and commitments from acts and commitments that are not our own.

We can be mistaken. Something that appears to have a rational source or to be endowed with mind can be random. Something that appears to come from another mind might have its well-spring in our own unconscious. But infallibility is too much to hope for. Our entitlements in these matters are inevitably defeasible.

What Lichtenbeg missed is the role of the first-person concept both in designating a source of rational agency and in acknowledging subjection to epistemic norms and power to act under them. The reverse side of this ability to acknowledge the commitments of one's rational agency is an ability to acknowledge sources of commitments other than one's own. One can sometimes do this non-inferentially, on the basis of intellectual understanding of utterances of content. When this is so, one's ability to recognize and understand other minds is not epistemically grounded in sense experience. It is grounded in understanding content in interlocution, and in an entitlement, underwritten by apriori requirements of rational agency, to recognize one's liabilities and entitlements as a rational agent.[21]

[21] An earlier version of the main part of this essay was given as the fourth of six Locke Lectures at Oxford in 1993 and as the second of two Whitehead Lectures at Harvard in 1994. I have benefited from audience comments on those occasions. I have subsequently benefited from discussions when drafts of the whole paper were given at St Andrews, Berkeley, and New York University, where Tom Nagel presented valuable comments.

17 *Memory and Persons*

I want to reflect on some functions of memory and their relations to traditional issues about personal identity. I try to elicit ways in which having memory, with its presupposition of agent identity over time, is integral to being a person, indeed to having a representational mind.

I

Three types of memory figure in the discussion. The first is *remembering x*, where *x* can be a particular thing, event, property instance, experience, state, or act. Call this type "*experiential memory*". The second is vividly exemplified as *remembering that p*, though as will emerge I conceive of it more broadly. Call this second type "*substantive content memory*".

I construe remembering that p as involving a belief that p, retained from earlier acquisition of the belief.[1] One might have acquired the belief retained in substantive content memory by various means: perception, self-knowledge, interlocution, reasoning. Experiential memory places greater restriction on the antecedent cognitive state from which the memory derives. To remember that Napoleon was defeated at Waterloo, one can have acquired the information in many ways. To remember Napoleon's defeat at Waterloo, one must have been there.

Well, perhaps I exaggerate. Linguistic usage is loose and varied. Here I just assume that experiential memory, as the term suggests, requires previous perception

I have benefited from comments by many people, including Richard Gale, Alan Gibbard, Mark Greenberg, Harold Hodes, Terry Irwin, Mark Johnston, Hans Kamp, David Kaplan, Stephen Neale, Lucy O'Brien, John O'Leary-Hawthorne, Gideon Rosen, Abe Roth, Seana Shiffrin, Sydney Shoemaker, Ted Sider, Zoltan Szabo, and David Wiggins. I have also learned from audiences in San Marino, Paris, University of Pittsburgh, Cornell University, Princeton University, Syracuse University, University of Toronto, University of California at San Diego, and the Australian National University.

[1] One can, of course, have a memory presentation that p, but not believe it. My construal of "remembering that p" simply clarifies how I use the term. I believe that the less committal type of remembering-that and non-veridical memories are ultimately to be explained in terms of *reliance* on *veridical* memories.

or experience of the remembered object, or some other fairly direct awareness, such as introspection.[2]

The third type, *purely preservative memory*, is distinguished by the representational and epistemic roles played by the retained content. Experiential memory and substantive content memory introduce new subject matter into current thinking. They yield or require new warrant. Purely preservative memory merely retains representational content for further use. It introduces no content or warrant into an argument. When I recall a general fact, I use substantive content memory. When I recall an event, I use experiential memory. When I use an earlier-instantiated step in an argument to combine with an inference rule, I rely on purely preservative memory.

I introduced substantive content memory and purely preservative memory in propositional form. I allow for non-propositional analogs. A lower animal's memory of a color type that attributively guides its activity can be a non-propositional substantive content memory. Such memory does not retain a representation of an instance of the color. The memory is only the residue of general aspects of perception of instances. So the distinction between experiential memory and substantive content memory is not that between non-propositional and propositional memory. All experiential memory is *de re* and must go back to a previous experience of a particular. (It must, of course, still involve general, categorizational elements.) Substantive content memory consists either in purely general, non-*de re* presentations, or in *de re* propositional memories that do not go back to a perception or experience of a particular.

Purely preservative memory need not be propositional either. It can retain perceptions or representational action sets for further use, even in animals that lack propositional abilities. Purely preservative memory is not distinctive in what it preserves—general or singular, propositional or non-propositional. It can retain any sort of representational content. It is distinguished purely by its role—preserving content and attitude for later use, while introducing no new subject matter or new warrant into a current representational transaction.[3]

[2] I leave open whether video, television, or radio transmissions provide the relevant sort of direct awareness. I incline toward liberality. Grice in effect suggested that "I remember Napoleon's being defeated" more clearly requires experiencing the defeat first-hand. I find that intuitions vary even on this point. Some think that receiving direct reports and participating in the thrill of the event would warrant the gerund construction.

[3] Tulving distinguishes episodic memories from semantic memories in 'Episodic and Semantic Memory', in E. Tulving and W. Donaldson (ed.) *Organization of Memory* (New York: Academic Press, 1972); *Elements of Episodic Memory* (Oxford: Oxford University Press, 1983); 'Multiple Memory Systems and Consciousness', *Human Neurobiology* 6 (1987), 67–80. All *experiential* memories are instances of Tulving's episodic memories, with one caveat. Tulving explicates episodic memories as memories of events. I intend experiential memories to include a wider range. Some *substantive content memories* are Tulving's semantical memories. Those substantive content memories that do not involve *de re* constructions are semantical memories. Tulving does not distinguish substantive content memories from purely preservative memories. I also differ in allowing both to be nonpropositional. For recent discussion, see A. Baddeley, M. Conway, and J. Aggleton (eds.), *Episodic Memory: New Directions in Research* (Oxford: Oxford University

All experiential memories, and many memories of the other types, have *de se* form. To have *de se* form, to be *ego-centrically indexed*, a memory's representational content must meet two conditions. It must include an element that indicates the rememberer in such a way as to mark other entities referred to by the content as being in relation to the rememberer's position or perspective. And the element must mark other entities so referred to by the content, as being of immediate relevance to the rememberer's needs, aims, or perspective.

Understanding *de se* elements may be facilitated by considering their role in perception. Perception is prototypically *de se*. *De se* elements in vision indicate the perceiver as at the origin of vision's spatial and temporal representational frameworks. Thus, an ego-centric index might mark the origin of a spatial framework in which a perceived object is represented in relation to the perceiver (for example, to the left). To be *de se*, the perception must also be linked with an immediate sensitivity to ego-related implications. If a large object moves speedily toward the perceiver's position, the representation is immediately associated with motivation to get out of the way. Experiential memories retain *de se* elements of the representations from which they derive.

Ego-centric indexes figure in the perceptual representations even of lower animals such as insects. They are the phylogenetic ancestors of the first-person concept. The first-person concept bears more complex relations to indications of the person's spatiotemporal position, but inherits fundamentally the same framework-fixing and motivational implications of *de se* representations. It differs in being conceptual, in being essentially associated with certain background propositional and inferential capacities, and in being used in *acts* of reference.

The relation between the mature first-person concept and its *de se* ancestors will not be crucial here. A few further remarks might nevertheless be helpful. I can associate the first-person concept with my memory of having gone to a concert in which Glenn Gould played a Beethoven piano concerto. I leave open whether my memory representation of "having gone" contains the mature first-person concept, or whether it contains some lower-level *de se* element. In either case, on apriori reflection, acceptance of the memory commits me, given that I have the mature concept, to applying it: If the memory is veridical, *I* went to such a concert.

Higher non-linguistic animals have conceptualized memories, I believe, but lack a mature first-person concept. Many lower animals have perception and

Press, 2002). The generic types of memory that I discuss are further subdivided into working memory, shorter-term memory, and longer-term memory.

I use "purely preservative memory" here as I do in 'Content Preservation' (*Philosophical Review* 102 (1993), 457–88, reprinted in E. Villanueva (ed.) *Content*, (Atascadero: Ridgeview, 1995) and in A. Casullo (ed.), *Apriori Knowledge* (Aldershot: Ashgate Publishers, 1999)), except that I am more explicit in applying the term to non-propositional content preservations. In the earlier article I used "*substantive memory*" to cover both experiential memory and what I here call "substantive content memory." *Substantive memory* is distinguished by its role—and directly contrasts with purely preservative memory. It introduces subject matter and carries new warrant.

perceptual memory, but lack concepts and propositional attitudes altogether. There is a *de se* element in the form and function of both sorts of animals' perceptual memory. That is, such animal perceptual memories meet both requirements on *de se* markers. Suppose that a dog has *de se* experiential memories of burying a bone in a certain place, and wants the bone. It will not only navigate to that place and dig. It will act toward the territory of the bone in proprietary ways—let us suppose, differently from how it would act if it remembered a rival's burying the bone. The perceptions and memories are constitutively associated with immediate use for the needs of the animal itself.

This case illustrates the presence of two grades of *de se* involvement in an experiential memory. There is an ego-centric index in the memory connecting it with perceiving the burying. That is part of the explanation of the dog's returning to the scene. This dimension of ego-centric indexing is common to all experiential memories. In this particular case the memory also marks the rememberer *de se* as the agent of the burying that was the *referent* of the memory. This is part of what explains the dog's proprietary activity. The dog's memory is doubly ego-centrically indexed. One or both of these indexes will be associated with spatial co-ordinates of the burying.[4]

Some memories have a third grade of *de se* marking. These are experiential memories *from the inside*. First, the notion of being from the inside; then, the three grades. An experiential memory from the inside is one that retains the perspective of one's remembered agency, state, or experience, as it occurred.[5] If an experiential memory is not from the inside, it is *from the outside*.

Experiential memories of entities other than acts, mental states, or experiences (of entities that lack a perspective to get inside of) and experiential memories of *others*' acts, mental states, and experiences are from the outside. One can remember a statue, or M. L. King's 'I Have a Dream' speech event, only from the outside. By contrast, in remembering drinking a Clos Vougeot at a restaurant, I might remember how the room looked from where I sat, how the glass tilted to obstruct the view, how the wine looked, smelled, tasted as I drank it. That would be to remember the act from the inside. I can also remember my actions from the

[4] A *de se* form in memory must fix the framework of the content of the remembering. It will commonly retain a spatiotemporal framework from the remembered context, but may not. To be veridical, it must coordinate with the agent of the remembered context; and to be a *de se* element in memory, it must normally function to further the needs, aims, perspective of the agent of the remembered context. There is evidence that animals that cache food also retain the time of the act of caching. Compare N. S. Clayton, D. P. Griffiths, N. J. Emery, and A. Dickinson, 'Elements of Episodic-like Memory in Animals', in Baddeley, Conway, and Aggleton (eds.), *Episodic Memory*. Animals like birds and dogs seem to have experiential memories with the two grades of *de se* involvement. It is less clear whether they remember acts from the inside (see below). This is a topic of conjecture and methodological debate in psychology.

[5] This distinction comes from Shoemaker, 'Persons and Their Pasts', in his *Identity, Cause, and Mind* (Cambridge: Cambridge University Press, 1984), 27, originally published in *American Philosophical Quarterly* 7 (1970), 269–285. I believe that the distinction was somewhat misformulated by Shoemaker, but this did not affect his primary points. Compare Gareth Evans, *The Varieties of Reference*, ed. John McDowell (Oxford: Oxford University Press, 1982), chapter 7.

outside. If I remember drinking the wine by way of an image of myself in a mirror, perhaps not even realizing that the image is of me, I remember it from the outside.

When I remember my act or event from the inside, the memory is indexed to mark all three *de se* grades. It is indexed to my having experienced the act or event, to my having been the agent or subject of it, and to my perspective as agent in the past act or event.[6] All experiential memories have *de se* form along the first dimension. Some are *de se* along the second dimension. These are experiential memories of having done, thought, or experienced something. A proper sub-class of these are *de se* along the second dimension. These are experiential memories of having done, thought, or experienced something. A proper sub-class of these are *de se* along a third dimension—those that are remembered from the inside.

I assume that experiential memories, and *de se* memories generally, need not themselves apply the first-person concept. They do carry first-person presumptions, or would-be commitments for an informed, reflective, conceptually mature person.

Presumption is not a propositional attitude in the individual whose states carry the presumption. A *presumption that p is associated with* an individual's being in a representational state if and only if veridical recognition that p would rationally derive from fully informed, conceptually mature reflection on the conditions that make that state possible, *from the would-be perspective of the individual in that state*. Thus, the presumption of application of the first-person concept is associated not only with mature persons' experiential, *de se*, memories. The presumption is also associated with lower animals' *de se* memory states, even though such animals lack propositional attitudes and the first-person concept. Conceptually mature, informed reflection, that elaborates the animal's perspective, would yield first-person memory beliefs.

Substantive content memories and purely preservative memories need not have *de se* form. But all of them have *de se* and first-person presumptions.[7] To be any kind of memory is constitutively and necessarily to preserve a past representational state that is the rememberer's own. Elaborating the rememberer's perspective yields *de se* and first-person presumptions.

[6] As regards *experiences*, I think that we can (experientially) remember only our own, and only from the inside. So here the three grades are always filled. I think that we can remember another person's *having* an experience only from the outside. I remember your having pain, from the outside, because I remember your expressions of physical suffering. I cannot experience your pain, and I cannot remember from your perspective, as sufferer, on the pain, no matter how much I empathize and talk projectively about sharing your pain. I can remember my *having* an experience, say a pain, either from the inside—by remembering the pain—or from the outside. In the latter case my memory of having pain may derive from seeing myself in a mirror reacting to the pain. I can experientially remember an act as either my act or another's and either from the inside or from the outside.

[7] A *de se* element in a representational content is distinct from a first-person concept, for reasons given above. The deep distinction between the two sorts of representational content and psychological capacity will surface only occasionally in this paper.

Presumption is linked to *presupposition*. Being in a representational state *presupposes that p*, if and only if a metaphysically necessary condition on being in that state is that p and this condition can be arrived at by mature reflection on the conditions that make that state possible. Presupposition relaxes the requirement on presumption that reflection be from the individual's perspective. Having a memory with a certain representational content presupposes that the rememberer was in a representational state with that content, or a content that implies it.

II

With these distinctions in hand, let us consider relations between memory and intentional agency. Experiential memories have *de se* form. If I remember an explosion, the memory retains a (first-grade) *de se* mark of its derivation from *my* perception of the explosion. There is immediate psychological relevance to its deriving from the *de se* marked perception. If I remember hearing a symphony, the memory is indexed at least at the second grade—to my being the agent of the remembered event. Something analogous is true of common intentions. They are *de se* at the second grade. If I intend to listen to all of Bruckner's symphonies, I am committed to my doing the listening. Someone else could not carry out my intention. If I remember listening to the symphonies pursuant to the intention, then I can take satisfaction in fulfilling my plan.

I think that an individual could not have intentions in *de se* form unless he or she could maintain them in memory. There are two ways in which this connection is necessary.

The primary way is that retention of intentions in memory is necessary to guiding and controlling intentional activity. Intentions are not point events. Some intentional activity may be instantaneous. Most—I would say necessarily some—requires working out over time. Acts take time and a sequence of sub-acts to be realized. Memory must retain the intention from the time it is formed to the time the action is carried out, if the intention is to control the intended act. Memory must connect the intention to a representation of the act, to record that the act has satisfied the intention. Memory must maintain the *de se* form that the intention has. And the representation of the act must share *de se* form to record a match. If an animal could not record a match between the act and the intention that it is supposed to fulfill, the animal would have no psychological basis for not repeating an act that it had already carried out.[8]

There are non-psychological ways of initiating and ceasing activity to meet need. But an explanation in terms of intentions must invoke this role for memory.

[8] For discussion of indexes in the explanation of action, including pre-intentional action, see Marc Jeannerod, *The Cognitive Neuroscience of Action* (Cambridge, Mass.: Blackwell, 1997). The attribution of ego-centric indexes is ubiquitous in perceptual and animal psychology.

The idea that a being could form intentions, constitutively lack a capacity to retain them in memory in carrying out the action, and cease activity when the intention is fulfilled, always through some non-psychological process, is, I think, incoherent. Ability to maintain continuity in *de se* perspective and presumptions between intention and act is necessary to control in practical agency.

Such retention must reside in purely preservative memory, although it could also be exercised in the other types of memory. Thus, the mere retention of the force and content of an intention from its onset to the time when it is acted upon is an exercise of purely preservative memory. Here it must be recognized that a retentive, memory element is involved in the very having of representational *states* that last over time—inasmuch as these states are capable of having psychological effects beyond the time intervals when they are first formed. The capacity to retain a representational state over time is a memory capacity, which requires continuity of content and force, and retention of representational content and force, and retention of representational content and force until they are called upon in later psychological transactions. In the case of intentions, the capacity includes a capacity to retain both referents and representational contents of *de se* markers. And this retention must link up with *de se* markers in act-representations so as to, again, retain both referents and representational contents of the markers as they occur in the intentions. Whether an individual has an additional experiential memory of the onset of the intention is, I think, not crucial. Similarly, the individual's having a substantive content memory (retaining a belief) that he has an intention with the relevant representational content is not crucial. Each of these latter two types of memory retention enriches intentional agency. But what is fundamental is the preservative capacity to maintain an intention, together with its representational content, and to call upon it, over time.

The second way intentional agency and memory are necessarily connected involves the retention of sequencing of acts in the fulfillment of intentions. If I intend to listen, in order, to Bruckner's ten symphonies in a month, I have to fulfill this intention serially. To record a match between my activity and my intention, I must record listening to the first of the symphonies, then the second, and so on. Each memory of a listening must have *de se* form. For at the end of the month I must have some *de se* memory of *listening* to all the symphonies if a match is to be recorded with my intention. As I have argued, this intention too must be retained in *de se* form. In the case of such sequencing, the essential type of memory is purely preservative memory, not experiential memory. In this case, the purely preservative memory must retain a temporal sequencing and content-specification of each of the acts of listening. But it is not apriori necessary that the individual remember each episode in the sequence via experiential memory.

I believe that these points state necessary conditions on having *intentions-to*. Having an intention to do something requires having (as a matter of normal psychological constitution) a capacity for memories that retain the *de se* element from the intention. *Intentions-to* constitute the basic kind of intention. They are the analogues of *de re* belief, and have the same priority.[9] Having other kinds requires having this kind. A dying person may intend that his or her worldly goods be distributed in a certain way. This is *intending that*, not *intending to*. Intendings-that (like impersonal desires) depend on a background of intendings-to (wantings), which are essentially *de se*. The infinitival, *de se* forms are basic to agency. The *de se* forms must be retained in memory.

None of the memories just discussed need refer to intentions. *Normally* memory simply preserves the intention—its content and conative force. *Persons*, however, have memories that conceptualize and refer to past intentions. Understanding one's life (as distinguished from simply having a life) requires having experiential memories of past intentions. Such memories in first-person form are basic to reflecting on how well one's projects have been satisfied.

III

Agency relies upon memory, which stems from perception. Perceptual representation, in turn, constitutively depends on use. Commonly this use lies in overt agency.

When we perceive a physical object, we perceive it as located relative to our spatiotemporal viewing position. Ego-centric indexes carry psychological implications in addition to indexing positions. They mark not only the origin of a framework within which the information can be used. They also mark the immediate psychological relevance of perceptual information to the perceiver's own use—to the perceiver's own needs, aims, or activities. I believe that the representational contents of all perceptions are ego-centrically indexed. All perception is certainly dependent on perception that represents objects in an ego-centrically indexed framework.

For a perceptual state to have content, some content must be of possible immediate use for the perceiver.[10] So an object must be represented not just in relation to what is in fact the viewer's viewing position, but as in relation to the viewer's perspective and needs *de se*. *De se* elements, and only such elements, mark the inevitable connection between perception and use.[11]

The most bare-bones notion of use is perceptual use. Having a perception requires having repeatable perceptual abilities—abilities to perceptually

[9] Cf. my 'Belief *De Re*', *Journal of Philosophy* 74 (1977), 338–362, section 2.

[10] For an individual's states to have any representational content, some of this content must be associated with uses by and for the individual. This is one reason why thermometers and computers do not have autonomous representational content. Of course, they have a derivative kind of content in that they can be used for expressing and processing representational content.

[11] For an argument on this sort of point that focuses on the mature first-person concept, but that is transferable with relatively obvious modifications to the contexts of this paper, see my 'Reason and the First-Person', in Crispin Wright, Barry C. Smith and Cynthia Macdonald (eds.) *Knowing Our Own Minds: Essays on Self-Knowledge*, (Oxford: Oxford University Press, 1998).

discriminate particulars (including property instances) in terms of perceivable kinds or types that they instantiate. This includes an ability to apply repeatable perceptual representational content to different instances of these kinds or types. Such repeatable abilities include a systematic ability to connect, from moment to moment, successive perceptions to one another and to the standpoint from which they represent. This is what any perception's being *de se* partly consists in. The retentive aspects of these abilities are the basis for *de se*, or *de se*-presuming, purely preservative memory of perceptual content.

I think that the bare-bones notion of use is necessarily swathed in more flesh. A richer relation between perceptual content and use emerges in the nature and individuation of perceptual states. For any property that a perceiver can perceive, there are patterns of interventing causal factors (such as light arrays) and patterns of stimulations on and in the perceiver that make perception of that property possible. Perceptual sub-systems are sensitive to those patterns, and as a matter of physical law must be, if the perceiver is to perceive the property. Why is the perceptual state a perception of that property rather than a perception of a pattern of intervening causal factors or sensory stimulations? Because the property (or associated ones) enters into explanation of animal agency—eating, navigating, mating, fleeing. The perceptual content of perceptual states is individuated not only by what the animal or its sub-systems can discriminate, but by entities figuring in the animal's needs and activity. These are properties and objects that the animal, not just its sub-systems, engages with in using perception. At the most primitive levels of perception and agency, the relevant animal needs, aims, and activities are biological.

So perceptual states with representational content are correctly individuated by reference to relations to objects and kinds in the environment that perceivers use perception to engage with in fulfilling their needs and goals. The capacity to perceive and act in the immediate service of the individual perceiver's own perspective, needs, and goals is a fundamental topic of psychological explanation. It must be marked in the representational content that type-individuates the perceptions and acts. *De se* or ego-centric indexes fulfill this role. Perceptual contents must be preserved in memory if they are to be used. Reapplicability of such contents is essential if perceptions are to be useful. So representational contents of perception must be preserved in memory if they are to be of use for the perceiver—hence if they are to have any intentional content at all. To have a repeatable use, perception must be associated with a memory capacity to preserve its contents over time, for use by the individual. This capacity is *de se*.[12]

For my main purposes, I need not explicate the richer notion of use. I want to pause over it, however. How are we to understand it? Broadly. Some perceptual contents might be useless, evolutionarily accidental companions in a system of other contents that *are* usable. A perceptual category may have evolved as useful

[12] For a discussion of these matters in somewhat greater depth, see my 'Perceptual Entitlement', *Philosophy and Phenomenological Research* 67, 503–548.

in one species and be retained in another that no longer has a use for it. A dodo's visual system might have retained representations of properties relevant to flight although dodos could not fly. Perhaps perception could yield a fear reaction in a species after the species lost any ability to flee. Should one count the fear reaction a use? I do not know.

In actual animal life, use for perception lies in guiding agency—nearly always agency in the species in which the perceptual system and content were fashioned. I am tempted by a more committal view. The relevant use for perceptual content of a perceptual system *must* be for agency—allowing for the by-product and inter-species points made in the previous paragraph. This "must" can surely be constructed as an evolutionary "must". Is it a deeper "must"? Is it metaphysically possible that there is a perceiver of physical features where the perceptual representational content cannot be associated with any function or use in animal agency?

Perhaps a helpless sea creature watches large forms pass by and has survived because few enough of them are predators. Perhaps the creature feels fear; but the fear has never been associated with any avoidance movement, even in evolutionary ancestors with the creature's visual system. Perhaps there never were any active uses for the perceptual system. Even granted that this never in fact happens, is it not possible? One can forget about evolution and try to conceive a being that takes an interest in the passing show but never could do anything about it. Again one has to add that there is nowhere in the individuation of the perceptual system any connection to agency—including no mental agency, no inference. These ideas tempt some people.

I think that they mislead. One can imagine a human paralyzed and without mental agency, who can still perceive. Such cases take over a perceptual system that had use in agency. One can imagine God's creating a perceiver with no evolutionary history who has not yet acted. I believe that here again one implicitly relies on a function for agency in God's designing and God's design. Some insist that no such function need be in the wings. Perception is conceived as purely passive—connected to no pursuits, either in God's design or in the system's history.

I think this position untenable. I think that it involves magic thinking about perceptual content. Reapplication of perceptual categories to serve mere perceptual interest in the passing show is too insubstantial a "use" to make perceptual content possible. The position involves magic thinking inasmuch as it cannot non-arbitrarily explain why the animal's perceptual content specifies physical kinds and particulars rather than any number of other entities that are causally relevant to its perception and systematically correlated with those kinds and particulars. Better, why does the animal have *perception* at all?[13] For vision,

[13] Sensation is an indicator that conduces to survival but is not genuinely perceptual. I discuss the distinction between sensation and perception in 'Perception', *International Journal of Psychoanalysis* 84 (2003), 157–167. I intend to elaborate the distinction elsewhere. The argument that immediately follows in the text is also given in 'Perceptual Entitlement'.

there are the patterns of light that are correlated in lawful ways with kinds and particulars in the environment. There is the array of proximal stimulations. There is the array of internal stimulations at various stages along the optic nerve. Reliable and regular discrimination is not enough for perception, even if it correlates counterfactually with the physical properties that are in the environment. For these other patterns also correlate counterfactually. Perceptual ability is individuated in terms of the uses in agency for the biologically relevant needs and goals of individual animals, including humans.

Perception must categorize in ways that are immediately relevant to use by the perceiver. *De se* elements, and only *de se* elements, mark the immediate relevance of content to animal perspective, needs, and aims, and mark immediate sensitivity to use of content in agency that serves them. *De se* elements have this explanatorily relevant function. To be of use, perception must be linked with a capacity to preserve its contents *in memory* over time. The memories must have *de se* forms and presumptions.

Whether or not the connection of perception to agency is metaphysically necessary, the role of *de se* memory in mediating perception and agency is necessary *in individuals that use perception in agency*. The simplest connections between perception, memory, and agency involve ego-centrically marked short-term or working memory. Think of quick reactions to perceived situations, such as fleeing or catching prey. In any being with *intentional* or other complex agency (like birds' burying caches of food), there must be a capacity for retention of projects over longer periods. So any view that attempts to understand beings with relatively complex agency must attribute long-term memories with *de se* form or presumptions.

IV

So far I have discussed dependence of intentional agency and perception on memory. I want now to discuss the role of memory in inference. Norms of inference are sensitive to the identity of the reasoner. If I rationally believe that p and I rationally believe that q, then I am rationally committed to believing that p and q (other things equal). But if I rationally believe that p and *you* rationally believe that q, there is no rational pressure for either of us to believe the conjunction. I might even believe rationally that you rationally *ought* to believe that p. It does not follow that you should. The information available to you may be sufficiently different that it is not rational for you to believe that p. From points about what it is rational for one person to believe, nothing follows—in many cases—about what it is rational for another person to believe.

De se forms of memory constitute a special case of the applicability of rational norms of inference to an individual's perspective. When I remember listening to a Mozart Piano Quartet in Boston, I remember an event that is indexed to me. The logical form of the memory indexes me *de se*. Given that I have a mature first-person concept, I am committed by acceptance of the memory to believing that I listened to music then. Experiential memories combine under logical norms that reflect these facts. If I remember listening then and I remember enjoying a Botticelli painting the same afternoon, then if I accept my memories, I am logically committed to accepting that I both listened to the music and enjoyed the painting on that day.

Normative dependence of inference on memory and on the identity of the reasoner is pervasive. It does not merely occur in inference with *de se* contents. Purely preservative memory is anaphoric. It preserves the representational content and attitudinal force of an antecedent state. The representational content of a purely preservative memory is the content preserved, which can be any content. Inferences, whether deductive or inductive, by scientists and mathematicians frequently contain no *de se* elements. All such inferences rely on purely preservative memory. Purely preservative memory contributes no force of its own to justifications or entitlements in an argument. Still, if purely preservative memory did not function properly, an individual's warrant for taking a step in an inference or argument would lapse.

If an argument is to support its conclusion, one must rely on purely preservative memory to preserve past steps with warrant unchanged. A step and its warrant must be held constant if it is to combine with another step or just an inference rule to take a further step. The presupposed constancy of content and of warrant over time depends on purely preservative memory.

Suppose that a perceptual belief's representational content and warrant are maintained over time by purely preservative memory in an argument. They are maintained between the initial instantiation of the belief and the time when it is reinvoked to be combined with other contents and rules of inference, or simply with a rule, to take a further step. If the content, as preserved at the later time, had derived from an earlier instantiation of the content in another person (a person who had the perceptual experiences), the warrant for the later instantiation could not be the same as the warrant for the earlier one (the warrant for the perceptual belief). For the recipient, the putative agent of inference, did not have the perceptions. So the recipient cannot have the same warrant. Transference across persons would not preserve warrant for a step in an inference. I believe that this case illustrates a general principle: Inference and argument, as psychological processes that function to transfer and provide warrant for a conclusion, must occur within the psychology of the agent of the inference or argument.

Consider a mathematical argument. Suppose that an individual is warranted in believing a premise by thinking it through and understanding it. The individual's later reinvoking the premise to combine it with another content, or just with an inference rule, to take a new step, presupposes that the warrant for the reinvocation is the same as the original warrant. Otherwise, the individual would have to begin the argument again by establishing the premise. If purely preservative memory cannot be relied upon to preserve not only the content but the warrant for the earlier instantiation of the step, the individual cannot proceed with the argument. If the individual only had a capacity that preserved content from some previous person's thought, warrant could not be preserved from the earlier instantiation. For the content had been warranted by the other person's thinking through the premise, not by any thinking-through by the agent of the inference. Any warrant the agent may have would not be a preservation of the warrant for the original instantiation of content. It would be new warrant. So he would be starting the argument anew.

The warrant for re-invoking a content in an inference must be presupposed to be the same as the warrant for establishing that content in the argument. This is a constitutive feature of inference inasmuch as it functions in support of a conclusion. In supporting a conclusion, an agent can apply a rule to a step only on the presupposition that the step has not suffered some shift in epistemic status. Being a step in an inference in support of a conclusion constitutively presupposes constancy of content and warrant through the argument. Constancy is possible only through use of purely preservative memory with its presupposition of identity of the agent of the inference through the argument. I will elaborate this point in more depth in section X.

Of course, mathematicians accept lemmas from others, even if the recipient has not thought through the proof. And there is simple acceptance of the word of others in less mathematical domains.[14] In both cases the recipient's warrant for acceptance will never be the same as the original prover's or informant's warrant for the lemma, if the recipient relies on the source. Dependence on another forces a difference in the warrant had by the dependent recipient(s) from that had by the ultimate source.[15] The recipient is warranted through interlocution. The source is not. Transitions across persons through communication do not have the same epistemic status and thus cannot be preservations of steps in an inference or argument.

The point applies to practical inferences. Suppose that one wants to see the Vermeer paintings again, but wants time to cross the park and hear some Beethoven quartets at Lincoln Center. The reasoning weighs the importance of

[14] I have discussed interlocution or testimony in 'Content Preservation' and in 'Computer Proof, Apriori Knowledge, and Other Minds', *Philosophical Perspectives* 12 (1998), 1–37.

[15] If the informant is not the ultimate source, there is still a difference in warrant. Although both informant and recipient are warranted by interlocution, the particular reliances are different, since each relies on different informants with different degrees of reliability in the causal chain. But the main point is that the ultimate source of warrant for the original instantiation of the step is different from the warrant of any given recipient through interlocution.

There are, of course, arguments carried through jointly by many scientists, where no one scientist independently confirms all the steps. Such arguments are found convincing because they can be seen to be logically valid and because an individual can have meta-warrants that the steps he or she has not autonomously confirmed are warranted. For example, one is warranted in thinking that the scientists responsible for step 3 are reliable, hence that their advocacy is a sign of the step's being warranted. The meta-warrants must remain constant for a step from establishment of the step to its re-invocations, if the overall argument is to support its conclusion for an individual. Further, I think that one could not use meta-warrants unless one were capable of carrying out autonomous arguments that follow the scheme I am outlining.

the two aims, reviews information on how long things take to do, and so on. One's argument must preserve over time—through the reasoning—the contents of and warrants for one's goals, values, and beliefs. If one lacked a capacity to rely on earlier instantiations of content as one's own—and as backed by warrants for one's own goals, values, and beliefs—one could not carry out practical reasoning. The reasoning is carried out to meet the agent's needs, aims, and values. If the contents that the agent preserves may or may not be from someone else's psychology, norms governing practical reasoning will make no sense. The autonomy of the agent in making a decision will be undermined. Practical norms are geared to fulfilling the agent's needs, goals, and values, as his own. The inferential norms presuppose that the agent of practical reasoning is the same through the reasoning.

To summarize our sketch: Theoretical and practical inference necessarily depend on a presupposition that a step invoked earlier can be reinvoked with unchanged warrant when it is to be combined with another step or just with an inference rule. Inference aimed at supporting a conclusion constitutively presupposes that the step's epistemic or practical warrant has not changed between initial instantiation and re-invocations. Purely preservative memory carries this presupposition. Its presupposition that the agent of an earlier inferential step is the same as that of the later use of the step is necessary to its role in explaining inference. If there were no explanatorily relevant presupposition that the author of the previous instantiation of a content is the same as the agent of inference, the later invocation would be a new beginning, since its warrant could not be presupposed to be the same as the warrant that backed the initial instantiation. In the practical case, it could not be presupposed that the agent was acting on his own needs, aims, and values. From the point of view of carrying out the inference, the agent would, in effect, have to (re-) claim the needs, aims, and values for himself. The new instantiation would not preserve the step. It would require its own warrant. In beings whose representational transactions take time, purely preservative memory is necessary for these presuppositions. It is necessary to the explanation of inference, in both theoretical and practical domains, through its presupposition of a single agent of inference over time.[16]

V

I have argued that *de se* forms and applications must be preserved in memory as a condition on the possibility of intentional practical agency and perception. *De se*

[16] As noted, section X elaborates and defends this line of reasoning. The argument does not claim that the agent of inference must have *de se* memories. But I think that any representational activity does require such memories in the psychological system. The argument's claim that the agent must have purely preservative memory yields the presupposition that the agent is the same through the inference, as well as the *de se presumption* that earlier instantiations of steps are the agent's own.

presumptions are clearly derivable from the presuppositions that I have argued are associated with inference.

De se forms and presumptions mark a competence in the individual. Use and competence entails a capacity for veridical application and presumption in appropriate conditions. *Veridical application of a de se memory-form* is successful coordination with a past veridical application of a *de se* form: the memory application connects with a past application by the same agent in such a way as to be immediately sensitive to the agent's needs, and aims. An individual could not have *de se* forms in memory if they did not mark an explanatorily relevant ability of the individual to maintain (or index) his perspective—in the relevant non-meta-theoretic way—through intentional agency, perceptual use, or inference. Similarly, an individual's exercising any sort of memory, including purely preservative memory, requires a core of veridical *de se* presumptions. *Veridical de se presumption* in exercises of memory is connection of a memory with a past psychological state of the rememberer in the service of the individual's perspective, needs, and aims. So the presence of memory competencies with *de se* form and presumption entails competence for veridical applications of *de se* forms and veridical *de se* presumptions.

What are the veridicality requirements on having a memory competence with *de se* forms or presumptions? The adage that error is intelligible only against a background of veridicality seems to me correct, but too vague to be more than a general guide. Most representational competencies are normally reliable. But there are exceptions. Some states' representational content is fixed by relations between the individual and what are in fact the content's referents, even though the competence associated with the state is *not* reliable in normal circumstances, that is, in circumstances in which it was individuated. A rabbit may represent a moving figure as a predator, even though *normally* such representations are nonveridical. The rabbit's panicky trips down the hole are normally unnecessary. It remains possible that the rabbit's representational state has the content *predator*. The content is fixed partly by its role in explaining rabbits' relatively few but crucial successes with respect to predators.

De se or ego-centric indexes are very different from the representation *predator*. *Predator* is applied in perceptually identificatory ways. It is subject to brute errors of illusion. *De se* elements do not identify an entity via perceptual categories at all. Applications of *de se* elements do not always depend on being guided by perceptions and are not always subject to identificatory error. They depend merely on the individual's capacity to act out of immediate sensitivity to its own perspective, needs, and aims. That is part of what it is to be an individual agent.

An individual's *having de se* memory competencies requires having an explanatorily relevant competence to coordinate his perceptions, needs, aims, acts over time. This competence is individuated by reference to at least an evolutionary background of veridical applications.

An individual's *exercising* such competencies requires more. Such exercise is fallible. But unlike applying a representation like *predator*, exercise of these competencies requires a core of veridical applications *in the individual's own life*. To have any career as a person or an agent with a psychology, the individual must hold together over time representational transactions in such a way as to make *de se* applications or presumptions veridical. And he must do so in a way that is relevant to explaining his psychology. If there were no core of explanatorily relevant instances of an individual's holding his *de se* perceptions *veridically* in *de se* memory for his use, controlling his agency through veridically tracking his *de se* intention to its putative exercise, preserving steps in inferences through memories whose *de se* presumptions are veridical, the individual could not exercise these competencies. The individual could not exercise memory with *de se* form and presumptions. Hence, the individual would not be an agent with a career of agency. Veridical *de se* applications and presumptions entail individual identity over time.

The fact that the exercise of *de se* memory presupposes sameness of individual over time is grounded in the fact that exercise of memory with *de se* forms and presumptions is *individuated* by reference to the individual who has the memories. The individual's role in the individuation of exercising memory (and of other psychologically fundamental competencies) underlies the fact that exercising it requires veridical *de se* application and presumption. Veridical *de se* application and presumption require retention of past psychological material from the individual who exercises memory. The center of this system of individuation lies in individual agency deriving from the individual's perspective, functions, needs, and aims. I touched on this point in section 3 and will return to it in section 11.

Veridical, explanatorily relevant applications are essential to exercising memory with *de se* forms. But numerical dominance of veridical memories is never fundamental in determining the presence or absence of a competence, or whether it is exercised. An individual might be born with innate capacities for perception and inference, but be obliterated before exercising memory in their service. Dementias make memory unreliable; *de se* forms might remain. Pathologies of loss can be laid aside. They presuppose a competence that is undermined. We can imagine a competence intact but statistically unreliable on other grounds. A being could have nonveridical experiential memories that outnumber veridical ones. Still, there must be *explanatorily relevant* veridical applications of *de se* memories as the individual lives life, if he is to engage in intentional agency, perception, or inference. To be an agent with a psychology—whether a pre-intentional, intentional, or inferring agent—the individual must veridically connect *de se* applications at different times to his own past, and must exercise purely preservative memory in ways that veridically fulfill *de se* presumptions and same-agent presuppositions of such memory.

Let me summarize the reasoning that I have been developing: A necessary condition for being a person, or other agent with a psychology, is to have one or more of the following competencies: intentional agency, pre-intentional agency involved in use of perception, inference. These competencies constitutively involve memory. The relevant types of memory must have *de se* forms or presumptions. *Exercising* one or more of these competencies, and attendant types of memory, is necessary for living the life of a person, or other agent with a psychology. Exercise of memory with *de se* forms or presumptions necessarily requires an explanatorily relevant core of *veridical* applications of *de se* memory forms and *veridical de se* presumptions. That is, applications of such forms must coordinate instantiations of psychological states within the same individual to preserve the individual's perspective, needs, aims, over time. To live a life as a person or other agent with a psychology—to exercise any of these three types of agency—the individual must successfully coordinate his acts and veridically coordinate his act-representations in ways that are immediately sensitive to his perspective, needs, and aims. This coordination must extend over time and be explanatorily relevant. So to exercise any of these types of agency, the individual must engage in an explanatorily relevant core of veridical *de se* applications of psychological contents, and must carry veridical *de se* presumptions associated with purely preservative memories. An individual cannot be an agent with a psychology unless he acts from his perceptual perspective, with immediate sensitivity to his needs and aims. Agents with propositional attitudes must veridically retain intentions, particularly their *de se* forms, from their formation to their realization in activity, and must preserve and transfer warrant of inferential steps in purely preservative memory. These veridical exercises of memory must include veridical applications of *de se* forms, or must carry veridical *de se* presumptions. Veridical applications of *de se* forms and veridical *de se* presumptions require sameness of individual over time. They require that the memory retain past psychological material from the same individual as the individual who exercises the memory. The nature and individuation conditions for the exercise of the relevant memories, then, requires an explanatorily relevant core of veridical applications of *de se* memory-forms and veridical *de se* presumptions. Individuation is the most basic form of explication of the nature and exercise of a psychological competence. So the most basic (correct) explication of what it is to exercise these sorts of agency and memory must make reference to the individual agent as identical over time. More materially, the nature of the exercise of these sorts of agency and memory presupposes agent identity over time. So there is a reciprocal relation between the individuation and nature of individual agents with psychologies, including persons, and the individuation and nature of psychological activities that are essential to being such agents.

Memory, with its *de se* presumptions and its presuppositions of transtemporal agent identity—deriving from the individuation of exercises of basic sorts of agency—is a condition on the possibility of an individual's having a representational mind.

VI

Now to personal identity. Locke is commonly read as having attempted to analyze the concept of personal identity partly in terms of the concept of memory.[17] Butler complained that such an analysis is circular, because memory presupposes personal identity and is not intelligible independently of it.[18] Shoemaker proposed a way of answering Butler. Parfit followed this proposal with a similar one. They appealed to a notion, *quasi-memory*, that applies to an ability that is like memory in certain respects but that does not presuppose personal identity. To anticipate my conclusion, I think that this notion cannot contribute to an analysis of personal identity. I think that the notion of quasi-memory is coherent, and that Shoemaker's introduction of it was valuable. The notion cannot, however, serve its intended purpose. This is because quasi-memory is both explanatorily and metaphysically parasitic on memory after all. So Butler cannot be answered by appealing to the notion. I think that by reflecting on why the reductions fail, one can gain insight into important features of the nature of persons.

Shoemaker explains quasi-memory as a kind of knowledge of a past event involving a causal correspondence between an individual's present cognitive state and a past cognitive state about the event. He writes: 'although otherwise just like that which exists in memory, [the correspondence] does not necessarily involve that past state's having been a state of the very same person who subsequently has the knowledge'. Thus, a quasi-memory is a state that represents an event. It is causally derived from and retains information from an earlier representation, say a perception of the event. The earlier perception might have been a perception by the same person. Shoemaker holds that then there would be both a memory and a quasi-memory. The earlier perception might have been another person's perception. Then there would be quasi-memory but no memory. The causal derivation in the latter case is to be as close to that involved in ordinary memory as is possible, consistent with its being between different people. Shoemaker holds that memory is a "special case" of quasi-memory.[19]

Parfit introduces what may seem to be the same concept as follows:

I have an accurate quasi-memory of a past experience if

(1) I seem to remember having an experience,
(2) *someone* did have this experience,
and
(3) my apparent memory is causally dependent, in the right kind of way, on that past experience.

[17] John Locke, *An Essay Concerning Human Understanding*, ed. Peter H. Nidditch (Oxford: Clarendon Press, 1975), 2.27. 9. It is not clear to me that this passage shows that Locke regarded memory as part of a definition or analysis of the concept of a person.

[18] Joseph Butler, 'Of Personal Identity', First Dissertation to the *Analogy of Religion*.

[19] Shoemaker, 'Persons and Their Pasts', 24. I shall ignore the view that quasi-memory is a type of knowledge and regard it as a putative ability.

Parfit remarks, 'ordinary memories are a sub-class of quasi-memories. They are quasi-memories of our own past experience'.[20]

What is "the right kind of way"? Parfit imagines that a surgeon changes brain states that underlie memory traces. The surgeon connects these changes with brain states in another person that underlie the other person's experience. Causal continuity is supposed to preserve information and phenomenal vividness from the prior experience, by the grace of the surgeon. This causal route corresponds to no natural competence. It is friendlier to Parfit's purposes to imagine more natural ways in which persons might be connected. I shall return to some of these.

Shoemaker and Parfit's explications of "quasi-memory" differ. Shoemaker seems to represent a person's quasi-memory as neutral on whether the quasi-remembered experience is the person's own. He writes,

> One way of characterizing the difference between quasi-remembering and remembering is by saying that the former is subject to a weaker previous awareness condition than the latter. Whereas someone's claim to remember a past event implies that he himself was aware of the event at the time of its occurrence, the claim to quasi-remember a past event implies only that someone or other was aware of it.[21]

This characterization seems to entail (by "only") that in having a quasi-memory that derives from another's experience, one does not *thereby* err in taking the quasi-remembered experience as one's own. He writes,

> Suppose that at time t_1, a person, call him A, does action X and has while doing it a quasi-memory from the inside of the immediately previous occurrence of the doing of action X. A's having this quasi-memory of the doing of X is of course compatible with X's having been done by someone other than himself.[22]

Shoemaker writes of A's having a quasi-memory of *the doing* of X, not of *doing* X.

Shoemaker claims that memory is a special case of quasi-memory in that it requires that the previous experience be that of the same person, whereas quasi-memory lacks this requirement. This claim is not ideally specific. It surely means that to be a memory the memory must connect to a state of the same person, whereas to be a quasi-memory the quasi-memory can connect to a state of either the same or a different person. Shoemaker's remarks quoted earlier seem to entail a further difference. Quasi-memories lack the commitment, present in the *de se*

[20] Derek Parfit, *Reasons and Persons* (Oxford: Oxford University Press, 1984), 220. Parfit sees himself as following Shoemaker in this characterization. In addition to the difference that I will highlight in this section, there are two other differences in their characterizations. One is Shoemaker's characterization of quasi-memory as a type of knowledge. The other is that Shoemaker writes of quasi-memory as being of an event that was experienced, whereas Parfit writes of quasi-memory as being of an experience. I think neither usage is sufficiently general to cover all types of memory and quasi-memory. This will not matter in what follows.

[21] Shoemaker, 'Persons and Their Pasts', 24.

[22] Shoemaker, 'Persons and Their Pasts', 32.

representational content of experiential memories, that the previous states were the individual's own.

By contrast, Parfit's condition (1) takes a person's quasi-memory to involve a seeming memory of having the prior experience. This entails that in having a quasi-memory that derives from another person's experience—lacking knowledge of one's condition, and taking the quasi-memory at face value—one will be mistaken in taking the experience to have been one's own.

On Parfit's characterization, memory is a special case of quasi-memory in a different sense. Both memory and quasi-memory involve first-person presumptions—they both *present the past experience as one's own.* The difference is merely that to be a memory, the past experience must *be* one's own, whereas to be a quasi-memory, it need not be.

The difference between the two characterizations turns on a difference in veridicality condition marked by a difference in representational form.[23] Parfit's characterization builds *de se* form into having a quasi-memory. Shoemaker's does not. Neither philosopher gives serious attention to the representational form of what they characterize as "quasi-memory". I believe that understanding such form is crucial to making progress on the philosophical issues.

VII

The concept of quasi-memory was introduced to aid a reductive analysis of the concept of personal identity. The analysis must represent quasi-memory as an ability that is explanatorily independent from memory. Attributions of memory presuppose that the remembered material comes from the same individual as the rememberer. If quasi-memory were metaphysically or explanatorily parasitic on memory, with its presuppositions about individual identity over time, use of a concept of quasi-memory to analyze the concept of personal identity would be circular.

Such circularity need not be definitional. There are definitionally non-circular concepts of quasi-memory that seem to apply to possible cases. *Definitional circularity* does not even follow from the presence of *de se* elements in the concept of quasi-memory. For if, as on Parfit's definition, applications of such elements need not be veridical, attributions of quasi-memory would not definitionally rely on the truth of propositions about individual identity.

But if the notion of quasi-memory could figure in *explanation* only by presupposing an explanatory function for memory, or only by otherwise presupposing facts about individual identity, the notion could yield no reduction. Call

[23] I am not sure how aware either philosopher was of the features that I take to differentiate their characterizations. Parfit does not remark on the *de se* character of his characterization. Shoemaker never provides a completely sharp and straightforward characterization. Whether the distinguishing features of the characterizations were recognized will not matter.

these sorts of presupposition *explanatory circularity*. I will argue that reliance on the notion of quasi-memory in psychological explanation or in reductive analyses of personal identity is involved in explanatory circularity. Thus, I will object to the putative explanatory independence of the concept of quasi-memory from the concept of memory—hence to reductive explanation of personal identity in terms of quasi-memory, or in terms of causal sequences of any agent-neutral psychological states.

I shall argue in this section that reductive use of Parfit's conception of quasi-memory, having a *de se* element in its logical form, is involved in explanatory circularity. Then in sections VIII–X I turn to invocations of quasi-memory conceived as lacking *de se* elements.

I begin with background assumptions. I remarked that Parfit illustrates quasi-memory by reference to surgical implants. Such cases seem less forceful for the reductionist case than naturally occuring quasi-memories. They cannot yield knowledge;[24] they disrupt functional continuities; and so on. I assume quasi-memories that retain the pasts of other individuals, with etiologies that are normal for the species. For simplicity, I assume that only one other person is involved. I suppose that the person who has quasi-memories cannot distinguish cases that go back to the other person from cases in which the quasi-memory goes back to his own past. At least, the person cannot do so internally and immediately, as part of the quasi-memory competence. If the two cases were differently marked, memory could provide continuity, and quasi-memory (where it differed) might be merely a source of further information. Supposing immediate distinguishability would be at odds with the aims of the original proposals.

Let us assume with Parfit that quasi-memories have *de se* representational form. When they derive from the person's own past, the *de se* form is applied veridically. These quasi-memories, we may suppose, are also memories. When they derive from the other person's past, the *de se* form occurs nonveridically. These quasi-memories are not memories.

The representational form of a psychological state does not get to be what it is in the abstract. Representational forms reflect competencies whose presence grounds representational norms. Such competencies are not explanatorily idle. *De se* memory-forms mark an ability to hold together psychological states as immediately relevant to the perspective, needs, and aims of the individual. Psychological explanation keys on psychological kinds typed in terms of such forms. The form of a quasi-memory can be *de se* only inasmuch as the individual is competent in connecting present representations specifically to his own past. Only relative to such a capacity could *de se* applications be non-veridical. Error presupposes a background of norm-setting competence for veridicality.[25]

[24] Evans, *The Varieties of Reference*; Parfit, *Reasons and Persons*, 220–222, 516 n. 15.

[25] Let me caution against a common mistake. Some infer from the possibility of systematic error in memory that systematic error is possible from the ground up. People do lose all accuracy in memory. Someone could also be systematically fooled in abnormal circumstances. These cases presuppose a

As argued in section V, a psychological state's having *de se* form marks it as being part of an explanatory core of competencies that involve veridical *de se* applications and presumptions. An exercised state cannot be individuated as having *de se* form unless there is a core of veridical applications and presumptions by the individual that figure in core explanations of the individual's activity in co-ordinating events and activities with his own perspective, needs, and aims.

Relevant exercises of *de se* competencies form a core for *explaining* specifically the unity of an individual's psychological life, *including unity over time. De se* indexes mark the practical, perceptual, and spatio-temporal origin of the subject's perspective, needs, aims, and activities over time. If the relevant states played no such explanatory role, there would be no sense in which the individual is making an error in exercising quasi-memory. And there would be no basis for attributing *de se* form to his experiential quasi-memories.

We are assuming that the individual cannot through the competence coordinate quasi-memories with his own past any better than with the other's past. So the states do not count as *de se* through any competence for immediate discrimination of his past from another's. As argued in section V, statistical considerations are not basic. Veridical applications and presumptions need not predominate at a given time. The states count as *de se* through providing a basis for explanation of individual agency and for grounding norms for such agency. That is what *de se* markings of psychological states are *for* in scientific and ordinary explanations. Attributing such states presupposes that the individual can coordinate present states with his own past. The exercise of the competence is *individuated* by reference to its relations to the individual.[26]

Consequently, a reductive explanation of personal identity that appealed to quasi-memory with *de se* form would presuppose, in the individuation of exercises of quasi-memory, facts about identity of individual agent over time. So it would be involved in explanatory circularity.[27]

memory competence that is subverted. It is impossible for an individual to have representational content, but never have an explanatorily fundamental memory capacity to preserve his own past representational states or events. In the case of many representational capacities, successful application may have occurred not in the life of the individual but only in the formation of the species' capacity. But purely preservative memory and *de se* aspects of memories are special. An individual whose activity was never correctly explainable in terms of successful preservation of its own contents from one moment to the next, or in terms of holding agency together with perception and aim over time, could not have a mind at all. Having attitudes with representational content is constitutively dependent on having *de se* and purely preservation memory competencies that issue in explanatorily relevant, *successful* holdings together, over time, of perceptions, needs, aims, and activities. In the absence of an explanatory core of *veridical* applications *by the individual*, there would be no representation.

26 The individual could learn inductively to distinguish presentations that connect to his own past. This would not suffice to give the presentations *de se* form. For that form necessarily yields motivations, norms, and acts that are immediate from the presentation itself.

27 Thus, I think that Parfit has a right to his *de se* conception of quasi-memory only if quasi-memory is, in explanatorily relevant ways, parasitic on normal memory. His conception is not strictly incoherent. We can imagine that a person connects to other people and cannot through any immediate phenomenological marker distinguish those cases from veridical memory. The competence could

This argument leaves open the possibility that in a reductive explanation of personal identity one could appeal to a notion of quasi-memory that does not involve *de se* logical form. Then one might hope to explain *de se* attributions in more basic terms that do not presuppose facts about individual identity in the individuation conditions of the psychological states that are attributed. I turn now to the task of showing that this possibility is merely apparent.

VIII

Is it coherent to take quasi-memory as explanatorily fundamental, conceived as a capacity without *de se* form or presuppositions about agent identity over time, and attributed without any explanatory reliance on memory, with its same-agent condition? Presentations of the past, on such a view, do not support *de se* or first-person presumptions because they do not connect to the individual's own past through any psychological competence that has an explanatory role. Such a being would lack memory properly so called. The quasi-memories stemming from his own past do not ground norms for correctness and are not a distinctive explanatory kind.

Let us begin with intentional practical agency. Such intentional agency is impossible if a being has only quasi-memories in the sense just discussed. International practical agency is necessarily dependent on having intentions-to and wants-to. As argued in section 2, these states are necessarily *de se*. Having them necessarily depends on having memories that track their content *de se*. Hence having them presupposes agent identity over time. To act as an intentional agent is to have intentions whose point and efficacy are to be explained in terms of the agent's coordinating through memory his intentions with acts flowing from those very intentions, while being sensitive to the immediate relevance of this coordination to the agent's own aims. The intentions, memories, and act-representations are *de se* inasmuch as they involve a pattern of acts and competencies that track relevance to agent aims in the sequence from intention to execution. The *de se* forms must be veridically applied in normal exercises of intentional practical agency. In conceptually mature persons, informed reflection on these competencies, and exercises of them, yields *de se* and first-person attributions. So they carry *de se* and first-person presumptions. Intentional practical agency would be impossible if a being had only quasi-memories lacking *de se* forms and lacking any explanatorily relevant, veridical application of *de se* forms in memory.[28]

have a *de se* form (mistakenly applied in the cases of connections to others' past) only if fallible but successful coordination with the individual's own past states were the explanatorily fundamental case.

[28] Several fine papers have cited practical agency as a problem for appeals to quasi-memory. Compare Susan Wolf, 'Self-Interest and Interest in Selves', *Ethics* 96 (1986), 704–720; Christine Korsgaard, 'Personal Identity and the Unity of Agency: A Kantian Response to Parfit', *Philosophy and*

IX

I return to perception and use. In each perceiver at least some perceptions are necessarily *de se*. The representational content of a perception is fixed not only by patterns of causal relations between the perceptual system and types of objects in the world. The content necessarily also depends on repeatable perceptual abilities' being associated with use.

A minimal notion of use is repeated exercise of perceptual competencies. This capacity requires retention of innate and learned representations and a capacity to relate instances of such representations at different times to one another in such a way as to mark perceiver-perspective and serve the perceiver's own needs. This amounts to a minimal competence in *de se* memory.

As argued in section III, more substantial notions of use are required to account for the individuation of perceptual content. Use must include agency in fulfilling the needs and aims of the perceiver.[29] Such agency requires an explanatorily relevant competence to connect perceptions to subsequent acts in ways that are immediately sensitive to the individual's perceptual perspective, needs, aims, and activity. Again, the connections must have *de se* form.

Explanation of a psychological system in terms of quasi-memories that *lacked de se* form would be incompatible with both perception and agency that uses perception. An explanation that did not attribute, as explanatorily central, immediate sensitivities that coordinate the individual's needs, aims, and activity with its perspective in perception could not attribute even non-intentional individual agency. Individual agency deriving from perception requires a capacity to coordinate perceptions with acts that use the individual's perceptual perspective and that maintain, over at least short periods of time, the individual's motivational, and action-representational orientation through the act. So primitive individual agency deriving from perception requires veridical applications of *de se* memory-forms—memory that presupposes agent identity over time. In higher animals the agency that uses perceptual content includes intentional agency. But the basic pattern is necessary for even the most primitive sorts of pre-intentional agency in perceptual use.

To simplify the argument: Having perceptual representational content requires having uses for it. The connection between content and use must be forged

Public Affairs 18 (1989), 101–132; John McDowell, 'Reductionism and the First-Person', in Jonathan Dancy *Reading Parfit* (Oxford: Blackwell, 1997). These papers do not give the present arguments. Each contains large points that I do not accept. Yet each offers insights into what I consider to be sound and fruitful directions for thinking about the issues.

[29] I always allow the use associated with aspects of psychological state-types corresponding to perceptual representations (contrast *de se* elements) to be established in an animal's conspecifics or species' ancestors (Compare note 25). One must, of course, not require that every representation have a use special to it. For example, representations of certain shapes that are of no use to an animal species might be individuated through their being constructed from principles governing representations of shapes that are of great interest or use.

through a capacity to preserve content over time for reuse *for* the perceiver. Such use requires that the perceptual base of representational content, the psychological states that figure in use of it, and the exercises of memory that connect them, all be *de se*. The states must be immediately sensitive to the individual's motivations and perspective; and the *de se* elements must be systematically connected over time in the psychology of the perceiver. Individuation of such uses requires veridical applications of *de se* memory-forms. Even those memories that are not themselves *de se* must be associated with memories with veridical *de se* and first-person *presumptions* and with *presuppositions* of perceiver-user identity over time. An explanation that relied on quasi-memory with no explanatory reliance on veridical *de se* applications and presumptions in exercises of memory could attribute no perceptions or uses. Perception and perceptual use would be impossible without veridical, explanatorily relevant applications of *de se* memory-forms that presuppose agent identity over time. For persons, they presuppose personal identity over time.

X

In this section I elaborate the argument sketched in section IV—the argument that a psychological system that does not presuppose agent identity cannot carry out inference.

I assume, first, that inference requires an ability to invoke representational contents as steps, relying over time on the same warrant that backed these steps when they were first instantiated. This is what it is to be a step in an inference or argument construed as psychological process, where the process is aimed at providing warrant for a conclusion.[30] Part of what it is to carry out an inference is to be able to rely on representational content, logical form, and logical inference rules to make transitions, without having to acquire new warrant for

[30] An argument can be considered in the abstract, regardless of who, if anyone, put forward the steps, and regardless of what the warrants for the premises or steps are. We can construct (or abstractly, there is) an argument made of pieces extracted from the psychologies of different people. In some cases, warrants for the steps can perhaps be considered under impersonal standards of rationality. Moreover, the validity and soundness of *any* argument are person-independent. But we are considering argument and inference not as abstract sequences of propositions, but as psychological processes—as reasoning whose point is to support a conclusion. Such inferential reasoning is individuated not only through the logic of the argument but through the warrant of the steps. Deductive reasoning, considered as an inferential process that supports a conclusion, is meant not only to preserve truth and stay within the rules of deductive logic. It is meant to support or transfer warrant to a conclusion. A step in such reasoning must be considered to include its warrant. Reinvoking a step involves relying on its warrant. What I call "*inferences*" in this paper are norm-governed transitions in such reasoning. What I call "*arguments*" are chains of such inferences aimed at supporting a conclusion. The narrower, more formal conceptions of inference and argument presuppose these notions, I believe, in this sense: Any being that carries out inferences and arguments in any sense must be capable of carrying out inferences and arguments whose function is to provide warrant for a conclusion.

steps already taken when they are invoked later. Warrants may change, but then one is not simply relying on an earlier step. One is reconstituting the inference or argument.

I assume, second, that epistemic norms and warrant attach to the agent of the inference. Psychological states are also spoken of as warranted, but they are warranted for individuals. The individual that carries out a step in an inference is the agent who is warranted or unwarranted in relying on a step or in making a transition between steps.

I assume, third, that epistemic norms for inference, and the warrant that an agent has in a step in an inference, must be explained in terms of epistemically relevant capacities, acts, experiences, or states of the agent of the inference. It is his agency that is evaluated under the norms. So his capacities, acts, and states are the relevant ones for explaining fulfillment of them.[31]

Consider an individual engaging in a putative argument. Suppose that his activity is to be explained without reference to purely preservative memory. The individual's retentive activity is to be explained only in terms of quasi-memory, which lacks presuppositions about individual identity over time. Consider a point in the putative argument at which the individual must rely on a step taken earlier, with its warrant, in order to use the step's content in an inference.

Suppose that the step was earlier warranted for another individual by his having perceptions. By the second premise, we ask what warrant the agent of the inference has for relying on the earlier-instantiated content. The earlier instantiation was warranted through the having of perceptions. The agent of inference did not have the perceptions. So by the third premise, whatever warrant the individual using quasi-memory might have for relying on the content cannot be the same as the warrant for the original instantiation of the content. The agent of the inference cannot be normatively evaluated in terms of experiences that he did not have. His warrant must derive from his own states or capacities. By the first premise, since the individual's re-instantiation of the content cannot preserve the original warrant for relying on the content, the individual lacks an ability to rely on the earlier instantiation of the content as a step in the putative inference. The individual needs new warrant and needs to start an argument anew.

Now suppose that the individual relying on quasi-memory to take the putative step *did* have the relevant perceptions. By hypothesis, the appeal to quasi-memory in explaining the putative inference by the individual does not presuppose that the individual is the same. The explanation does not attribute exercise of a competence that presupposes individual identity. So the epistemic and psychological functions of quasi-memory cannot differ from the case in which the

[31] Epistemic internalists hold that warrants must be fully accessible to the warranted individuals, as reasons or justifications. I believe, by contrast, that not all warrants need be accessible to the warranted individual. Warrants are nevertheless always explained in terms of states and capacities of individuals, supplemented in some cases by relations to a subject-matter environment. I believe that epistemically externalist views join internalist views in either accepting the premise or yielding an analogous and equally effective premise that accords with their terminology and theory.

individual who had the relevant perceptions is different. By hypothesis, the individual's quasi-memory is, from the point of view of explanation, neutral as to whether its presentations derive from the same individual's past or another's past. So by the same argument as before, even if the quasi-memory happens to derive from the past of the individual who exercises quasi-memory, the individual cannot be treated, in the psychological explanation, as preserving the warrant that derives from the perceptions. Quasi-memory may yield a warrant for relying on the content. But the individual's psychology cannot be explained as preserving the perceptual warrant. So the individual's activity cannot be explained, in these terms, as carrying out an inference.

The argument claims that taking the warrant of a step to remain constant between the time it is first instantiated and the time it is relied upon in an inference presupposes that the same individual relies on the step at both times.

The reductionist might respond as follows. The claim is plausible when only ordinary situations are considered. It is less plausible when applied to fission examples conceived to occur naturally. Suppose two cases. In one, *A*'s brain is divided so that two people, *B* and *C*, result. (We grant this fiction.) In the other, a physical duplicate of *A* (called "*D*") loses one brain hemisphere; so there is no fission. *D*'s brain activity after the loss duplicates *B*'s. Just before the loss, *D* starts a process of deductive reasoning and completes it soon after the loss. *A* begins a parallel process of reasoning. After fission, *B* thinks thoughts analogous to the thoughts *D* uses to complete his own (*D*'s) deduction. The beliefs that occur in the two sequences (*A*–*B*, before and after fission, and *D*, before and after the hemisphere loss) and how they are causally related in the brain are the same. It would be strange to hold that *D* is warranted in the conclusion he draws and that *B* is not. It would be strange to hold that *D* carried out an inference but *B* did not. So the claim that argument or inference presupposes sameness of person must be wrong.

Before answering this line, I make two points. First, if *B* is a person, *B* has perception and agency. By the arguments of sections II–III and VII–IX, in counting *B* a person the reductionist is involved in explanatory circularity. Second, as far as the argument regarding inference goes, *B* can be warranted in inferring conclusions from contents initially warranted by quasi-memory. Such inferences would be intrapersonal. They would rely on purely preservative memory proper, preserving the content and warrant of a step initially warranted by quasi-memory. Quasi-memory plays no role in explaining the *transitions* in such inferences. (Cf. also note 30.)

The questions at issue are precisely these: Can *B* make the same deductive inference that *D* makes after hemisphere loss, with the same type of warrants at each step? Can appeal to quasi-memory explain *B*'s making a "cross-personal" inference, inferring a conclusion from a premise supported only by the warrant that *A* had for initially instantiating the premise? Underlying these: Can an individual's making inferences that take time be explained without relying on purely preservative memory, with its presuppositions about agent identity over time?

The answers to these questions remain negative. Suppose that *D* had a perceptual belief before hemisphere loss. Afterwards, *D* makes a deductive inference from the content of that belief. *A* is warranted in a parallel perceptual belief before fission. Suppose that *B* acquires type-identical content through quasi-memory retaining *A*'s perceptual belief. Can *B* rely on a re-instantiation of *A*'s content, without new warrant, to infer the conclusion?

To review the argument that *B* cannot, by using quasi-memory instead of memory, draw an inference purely from a step instantiated by *A*: Like *D*, *A*'s belief in the content is warranted through having perceptions. *D* has purely preservative memory and so can use it to preserve his perceptual warrant for the initial step when later he uses it in inference. Suppose that *B* is to carry out an inference from the step that *A* instantiated. *B* is the agent of the putative inference. So by the second premise, relevant epistemic norms apply to *B*. *B*'s warrant is what is relevant to assessing *B*'s inference and determining what counts as a step. *A*'s belief (like *D*'s) is warranted through his having perceptions. By the third premise, since *B* did not have the perceptons, *B*'s warrant, if any, must be explained in terms of some other act, experience, capacity, or state of *B*'s. Warrant could come from *B*'s quasi-memory of *A*'s perceptions. *B*'s warrant cannot come *purely* from *A*'s perceptions—since *B* did not have them. By the first premise, *B* cannot infer from the step established by *A* and so cannot infer with the same warrant as *D*. *B*'s quasi-memory constitutes a new epistemic start in his reasoning (even if the beginning is warranted).

The argument applies even if quasi-memory happens to provide information from *B*'s past—since explanations in terms of quasi-memory cannot rely on this fact. If the explanation of *B*'s mental processes relies on agent-neutral quasi-memory while excluding memory proper, no transtemporal inference is attributable to *B*.

What of the three premises? The first premise is an evident point about what it is to be a step in an inference, where inference is construed as a psychological process that provides or transfers warrant. I see nothing in the reductionist programs to motivate objection to it. As for the second premise, given that there is an agent of the inference, there is an epistemically fundamental question whether the agent is warranted at stages of the inference. One might deny that individuals are agents of inference, holding that some composite is the "agent". I make no objection here to allowing larger units, such as corporations or even swarms, to be agents of inference. To block the argument along this line, however, one must *deny* that individuals are agents of inferences. Such a denial would undermine a reductive account of personal identity. Reductionism must preserve fundamental features of personhood, including agency in inference, in its account. Such a denial is in any case refuted empirically.

The third premise seems as solid as the other two. An agent can fulfill norms of inference and be epistemically warranted only through his own states and capacities, together perhaps with their relations to a subject matter. Internalist and externalist epistemologies accept this principle, or relevantly similar principles that could be used in analogous arguments.

The argument rests on independently motivated principles about individuation of inference steps, individuation of warrant, and the relation between norms and abilities to fulfill them.

Resistance can easily feed on unclarity about what is at issue. Let me address this matter.

Whether *B* is warranted in accepting the same content that *A* is warranted in accepting is not at issue. If *B* is a person, *B* can, as far as the argument goes, be warranted in accepting type-identical content through quasi-memory that connects to the same events that *A* perceived. *A*'s warrant is having perceptions. *B*'s is having quasi-memory. If quasi-memory were counted a type of perception, *A* and *B*'s warrants would still differ. All perceptions have different causal relations to a subject matter; these also have different temporal distances and degrees of reliability.

Whether *B* can deduce from the same content the same conclusion that *D* deduces is not at issue. Perhaps *B* can deduce a conclusion type-identical with *D*'s from content that is type-identical with *D*'s. *B* can be warranted in establishing the content by quasi-memory. *B* can use ordinary purely preservative memory to draw the same conclusion that *D* does. Such an inference is not cross-personal and has a different warrant for its initial step from *D*'s warrant, or *A*'s.

Whether *B*'s quasi-memory can *"preserve" A*'s *content* is also not at issue. There is a sense in which *A*'s content can be "preserved" in *B*'s quasi-memory. Quasi-memory can preserve the content of *B*'s perceptions in the sense that it can produce a causally dependent re-instantiation of type-identical content.[32] Quasi-memory can depend on *A*'s being warranted for its being warranted. What it cannot preserve is *A*'s warrant for the step. Being warranted in a belief or step lies in fulfilling norms. One fulfills norms by doing certain things, having certain abilities, being in certain states. *A* has warrant by having certain perceptions. Not having had the perceptions, *B* must fulfill norms of epistemic warrant in some other way—perhaps by having quasi-memory. If quasi-memory is warranted, it must *produce* warrant for *B*. In fact, quasi-memory cannot preserve the warrant associated with the having of perceptions even if *B* did happen to have them, since quasi-memory is neutral as to who had them. Quasi-memory is, from the point of view of fulfilling epistemic norms by epistemic agents, a source of new information and new warrant—not a genuinely preservative capacity at all. I shall return to this point.

What is at issue is whether quasi-memory can effect and explain cross-personal warrant preservation without relying on purely preservative memory with its same-agent condition. I have argued that it cannot, and hence that it

[32] Insofar as there are *de se* elements in the content, of course, the referents of the indexes will shift. But the content can be type-identical.

cannot explain inference. An agent that had quasi-memory but lacked memory proper could not carry out inferences.

The argument generalizes from perception to the other types of warrant—thinking through a content with understanding, interlocution, conclusions inferred from empirical or mathematical premises, commitment to practical goals based on practical reasoning from values, and so on.[33]

Genuine purely preservative memory, like all memory, presupposes that its content came from the individual who relies on purely preservative memory.[34] If an explanation tried to get by with quasi-memory, all presentations of content in quasi-memory would have to be treated as explanatorily neutral as to whether they preserved a content from the individual's own past or from another individual's past. Then the individual cannot be treated as reinvoking a content and proceeding on the same warrant that supported it earlier. From the point of view of explanation, this is to treat each quasi-remembered instantiation of content as new to an argument. Every putative re-invocation of a step at a later time would have the formal and epistemic status of a new premise, not a re-invocation or preservation of a step already warranted. No instantiation of content could be treated as the preservation of warrant earned earlier. So no inference in an argument meant to provide warrant for a conclusion would be possible.[35]

[33] The supposition that the content was warranted earlier through an exercise of quasi-memory will not change the situation. Each chain will have different warrant that depends on the nature of the chain from the quasi-remembered event to the quasi-memory. (Compare note 15.) In having different warrants, instances of quasi-memory are like instances of perception. Quasi-memory is formally a new source of information. It is not purely preservative in my sense. Even supposing that an individual begins life with a set of reliable quasi-memories, these are sources of information and warrant for him. Although they derive from the states of another individual, and depend for their warrant on those earlier states' being warranted, they do not preserve warrant; and they do not preserve information within the quasi-remembering individual.

I should note that inference in practical reasoning, which is discussed near the end of section IV, is subject to the same-agent condition for *two* reasons. One has to do with the nature of steps in inference—by the argument just given. The other has specially to do with the fact that practical reasoning must preserve the practical commitments of the reasoner. In theoretical reasoning the commitment (to truth) is in a sense common to all inferrers. In practical reasoning the role of preserving the individual's own commitments—motivational elements—is additional. Individual practical reasoning would be doubly incoherent in the absence of the same-agent condition on purely preservative memory.

[34] I believe that purely preservative memory clearly presupposes the presence of a *de se* memory competence in the same individual (as well as vice versa). But I do not rely on this belief in the present argument. It is enough that purely preservative memory constitutively involves an explanatorily relevant competence to preserve content from one's past. Purely preservative memory thus *presupposes* agent-identity and *presumes* the veridicality of *de se* attributions—whether or not the purely preservative memories are themselves *de se*.

[35] I caution here about a special case of the mistake discussed in note 25. One might think that since we rely on others in interlocution, it does not matter whether a step of any argument comes from someone else. The broadbrush answer is that our reliance on others presupposes that our representational content and inferential abilities are already in place. If a recipient could not make intrapersonal inferences, taking lemmas or other propositional information from others would be impossible. Further diagnosis of this mistake can be derived from the discussion in this section.

To summarize: It is essential to inference, as a psychological process that contributes to providing warrant for a conclusion, that steps maintain the same warrant through the argument or inference over time. Warrant in the activity of inference necessarily attaches to the agent of inference. Warrant that attaches to an agent must be explained in terms of the agent's acts or states. A putative explanation of inference that confined itself to quasi-memory with no presuppositions about individual identity over time cannot treat warrant as preserved in steps used in inference after the steps are first established. For quasi-memory cannot presuppose that the acts or states that first warranted a step were those of the agent of the inference. So quasi-memory cannot help explain an individual's activity as involving argument or inference.

Descartes featured maintaining an argument in a single insight.[36] Can "immediate insight" replace temporally sequential reasoning in explanation? I think not. To be capable of immediate insight, one must have a standing competence for inference, which involves an ability to re-invoke steps and reapply inference rules in reasoning. We are necessarily discursive reasoners.[37]

The point about inference has an analogue that applies to pre-inferential representational processes. All use of representational content requires purely preservative memory. This is because purely preservative memory is necessary for explaining the re-applicability of any representational competence. Such explanation must appeal to representations and information as already present—as not new for the individual.[38] Quasi-memory cannot do this. An individual that lacked preservative competencies—with their *de se* presumptions—would have no mind at all. Each putative representational event would be a new beginning for the individual. In a being with only quasi-memories, without explanatorily relevant presuppositions about individual identity over time, there would be no repository of

[36] For criticism of Descartes on memory see my 'Content Preservation'. Descartes was, of course, right to emphasize a role for non-inferential comprehension in carrying out a proof. For a proof to be effective, there must be component steps that can be comprehended without inference.

[37] This is the great lesson of Kant's emphasis on the role of inference in making judgment possible. Compare *Critique of Pure Reason*, for example A 76ff./B 102ff., B 130–131. It is also a deep theme in twentieth-century philosophy-stemming from work of Frege. I have in mind his assumption that meaning is to be understood only by understanding inferential structure, an assumption that suffuses his method in *Foundations of Arithmetic*, trans. J. L. Austin (Evanston: Northwestern University Press, 1968), and 'On Sense and Reference', in *Translations from the Philosophical Writings of Gottlob Frege*, ed. P. Geach and M. Black (Oxford: Blackwell, 1966). Compare my paper 'Frege on Sense and Linguistic Meaning', in David Bell and Neil Cooper (eds.), *The Analytic Tradition* (Oxford: Blackwell, 1991). Compare also Wittgenstein's *Investigations*, trans. G. E. M. Anscombe (New York: Macmillan, 1968). The picture of understanding as simply an immediate flash of insight has been repeatedly and convincingly undermined.

[38] The point applies to norms as well as psychological competence, since relevant norms are grounded in such competencies and their functions. Normative standards for operating well in fulfilling the function of veridical representation, given the perspectival and operational limits of the individual, are the predecessors of epistemic norms. Applicability of such norms depends on certain processes' functioning to preserve fulfillment of them over time. A framework for these remarks is developed in 'Perceptual Entitlement', section 1.

information or repeatable representational competencies. Not only would such a being not be a person. It could not represent anything at all.

Quasi-memory's inability to help explain inference, or lower-level preservation of representational competencies, lies in its nature. In psychological explanation, quasi-memory is more like perception than memory.[39] Quasi-memory is formally a source of *new information* to the individual. It yields non-inferential information about the past. Even when information happens to come from the individual's own past, it is psychologically new for the individual—since it is not associated with an explanatorily relevant competence to retrieve information that comes immediately from the individual's own past. As with perception, each instance of quasi-memory carries *new warrant.* So quasi-memory cannot preserve warrant for the individual who relies upon it. In the absence of purely preservative memory proper, each instantiation of content would be a new beginning. Preservation of warrant and information over time is essential for inference, indeed for content. So in the absence of explanatorily relevant purely preservative memory proper, an individual cannot make inferences or have representational content.

Why is newness of information *relative to an individual's perspective* significant? The psychology and epistemology of inference concerns the agency of individuals. One must consider what the agency of inference is *for*. Since it is the agent's perspective, functions, needs, aims that govern use of representational content, what counts as new or preserved is dependent on the individual that is informed. Similarly, explanations of and norms for use of content are grounded in the activities, experiences, and competencies of the individual who is the agent of use.

These points about the role of memory in preserving information and warrant for an individual reflect on Shoemaker and Parfit's assumption that memory is a special case of quasi-memory. Quasi-memory is indeed defined by relaxing a condition on memory—the condition that the content is preserved from the rememberer's past. So in a sense the assumption is trivially true. The condition that is relaxed is, however, fundamental. Memory generally is not an original source of information for the individual. It preserves information. Purely preservative memory is not a new source of warrant and not even a *contextually* new source of information for the individual. Quasi-memory, construed as lacking presuppositions about individual identity, is formally a source of new information and warrant. Memory's function is to preserve.[40] Preservation for use *for* the

[39] Quasi-memories are like memories and interlocution in conveying information from prior representational states. They also resemble memories in that (presumably) one could reverse temporal order among them. Unlike all memories, they are formally or explanatorily sources of new information. Unlike purely preservative memories they are sources of new warrant.

[40] Of course, experiential memory and substantive content memory often embellish what was originally taken in. Some of this embellishment is distortion, and does not fulfill memory's representational function (though it may fulfill some biological or practical function). There is, however, the phenomenon of ordering, summarizing, drawing inferences, so as to extend the

perspective, functions, needs, aims, activities of individuals is fundamental in the individuation of psychological kinds and in the aims of psychological explanation. The claim that memory is a special case of quasi-memory is like a claim that fruits are special cases of animals, except that they are plants and their DNA is different.

XI

I have argued that one cannot explain fundamental aspects of individual psychology if one invokes quasi-memory without explanatory reliance on memory with its same-agent condition. Absent explanatory appeal to memory, individuated in terms of success in holding an individual's past with his present, one cannot reasonably attribute intentional practical agency, perceptual content and its use, or inference. In this section I consider reductionist views more concretely.

The central tenet of Shoemaker's and Parfit's reductionisms is that the basic explanatory psychological notions do not presuppose individual identity over time. The basic explanatory notions are supposed to be person- and agent-neutral. The notion of an individual person or agent is to be explained in terms of continuities of states characterized agent-neutrally.

I believe that the arguments in earlier sections undermine this strategy. Being an agent with a psychology—whether a person or not—requires exercising psychological competencies supported by *de se* memories and memories with *de se* presumptions. The notion of an agent with a psychology is partly individuated in terms of such exercises. Reciprocally, such competencies and their exercises are partly individuated by reference to relations to their agents. Individuation of *de se* aspects of basic psychological acts and states is not agent-neutral.

An explanatory scheme that takes agent-neutral psychological notions as basic cannot recover the *de se* notions. The notion of being a non-branching continuity or of beginning at a branch (in a fission case, for example) can *correspond to* distinctions among individuals. But such notions lack psychological explanatory power. If they had such power and if psychological states were individuated and explained in terms of them, explanatory circularity would be reinstated.[41]

material of the original belief or experience. Cf. Daniel L. Schacter, *Searching for Memory* (New York: Basic Books, 1996). This may constitute an epistemic function of memory. I maintain, however, that insofar as it is an epistemic or representational function of memory, it must be a drawing out of elements implicit in material already present in memory.

[41] Shoemaker and Parfit seem aware of this threat. Each suggests that the first-personal elements in memory attributions derive from a "trivial" linguistic point (Shoemaker, 'Persons and Their Pasts', 24; Parfit, *Reasons and Persons*, 222–223). I have tried to show that *de se*-and first-person-presuming aspects of mental states are explanatorily fundamental. Fundamental psychological capacities are *individuated* in terms of *de se*-presuming competencies, which are in turn individuated partly in terms of the individual who is the agent of the competencies. Individuals are extended in time; their perspectives, needs, aims, and activities are extended in time.

Let us reconsider fission cases. Such cases are often taken to be the reductionist's strongest example. *A*'s brain is divided to allow two equally worthy continuants, *B* and *C*. *B* and *C* are in separate bodies closely matching *A*'s. Neither body more closely continues *A*'s than the other. I shall grant the fiction that two halves of *A*'s brain are each sufficient to guide and continue on from *A*'s life, equally well. Let us suppose that *A* ceases to exist at the time of the branching. *B* and *C* begin new lives. (*B* and *C* are not identical; *A* therefore cannot be identical to both; each has equal claim to be a good continuation of *A*; so *A* is identical to neither.) In the first moments of the lives of *B* and *C*, their quasi-memories of *A*'s past will vastly outnumber any memories they have of their own pasts. Their mental lives are largely derivative from *A*'s.

Inasmuch as *B* and *C* are regarded as individual agents with psychologies, the case silently presupposes that they exercise genuine memory capacities with *de se* forms or presumptions. These exercises must play an explanatory and norm-grounding role in understanding their agency. The quasi-memories, though initially in a majority, cannot play these explanatory or norm-grounding roles. So we have here the case discussed in section VII. The reductionist has not escaped explanatory circularity.[42] I believe that fission cases constitute no ground at all to think that quasi-memory, as opposed to memory, is explanatorily fundamental.

Although they rarely are, fission cases could be elaborated to avoid presumption of *de se* explanatory memories, and presupposition of agent-identity. In such cases, there would be no individual agency. So they could not illustrate a reduction of individual agency to more fundamental terms. Since such cases correspond to no actual psychological kinds identified in psychological explanation—all of which presuppose kinds with *de se* elements—they cannot help an eliminationist who wishes to hold that individual agency, as a psychologically relevant phenomenon, is dispensable. Such a view of animal or human psychology is refuted empirically.

It might be tempting to infer that since the individual cannot distinguish cases where his memory-like presentations go back to his own past from cases where they do not, the individual must lack an explanatorily relevant memory competence. This inference might be especially tempting in examples where instances of the latter sort outnumber instances of the former.

The temptation must be resisted. Most competencies are subject to illusion that cannot be avoided purely from the inside. Statistical dominance is, as argued in section V, not determinative in individuating a competence. Moreover, the

[42] An alternative is that quasi-memory of another individual's past is *not* a non-veridical exercise of *de se* memory (that is, not the case I discussed in section VII), but a separate faculty, like perception. Such a faculty is still explanatorily and individuatively parasitic on genuine memory in the same psychological system. Explanation of its use, like explanation of perception, agency, and other representational capacities, requires attribution of memory proper. For being an individual agent requires being able to act out of an immediate sensitivity to one's own perspective, needs, and aims. This ability requires *de se*-presuming memories.

temptation leads to incoherence. Being an agent with a psychology requires acting on a perspective to serve one's needs and aims. Such agency requires coordinating perspective, needs, aims, and acts over time by means of veridical *de se* applications of memory forms or veridical *de se* memory presumptions. Such coordination is individuated partly in terms of agent identity. Any *agent* in a fission case will engage in such coordination and agency. Imagine short-term memories serving acts on immediate needs. These will grow over time. Quasi-memories may be a rich source of partially misleading background information. They cannot substitute for *de se*-presuming memories in explanations of agency.

One might press the reductionist case as follows: Imagine creatures that reproduce by fission. Quasi-memories of these creatures may go back indefinitely far in the chain of splittings. Or they may include just the history of the creature since it split off and the recent history of its most recent ancestor. Without auxiliary information, a creature cannot tell whether a quasi-memory presentation goes back to its own history or its ancestor's. These creatures do not care about the distinction. They have a term "remember*" translated "quasi-remember", but no separate term for remembering. Similarly, they have terms "I*" and "*de se**". *De se** contents are not *de se* contents. Yet they do much the same work. We can suppose that the cognitive dynamics in a sequence of mental states that consists of the mental career of the immediate ancestor of one of these creatures plus the mental career of that creature itself is the same as the sequence of mental states in the history of a single individual person or other agent.

I believe that this sort of response lives on under-description. The claim that *de se**-marked capacities enter into the same cognitive dynamics is mistaken. The role of *de se* form and presumption in memory goes deeper than language. Lack of a word for "I" or "remember" is not significant. Animals with *de se* capacities lack language. The role of *de se* elements goes deeper than caring. The issue is whether the individuals act from their own perspective coordinating with their own needs and aims. If they do, they exercise explanatorily relevant *de se* memories. Then reductionism falls into explanatory circularity in the way indicated in section VII. All the standard cases of fission are of this sort. They assume that the individuals involved are people. If there is no such set of activities, then these are not creatures with individual psychologies. They are incapable of psychologically based agency or perception, as opposed to reaction and sensation. Even the attribution of quasi-memory to them is incoherent. For these creatures lack the retentive and coordinating powers to have representational content. (Compare notes 10 and 42, and section III.)

I do not deny the possibility of cases in which the notion of an individual agent with a psychology is inapplicable. Such cases are not relevant to understanding actual psychologies. They are not relevant to giving an account of the identities of persons.

Explanatory kinds in psychology are entangled, in their individuation conditions, with the existence of individual agents. Reductionisms typically associated with Locke and Hume systematically underestimate this point. Exercises of *de se*-

presuming competencies are individuated partly by reference to particular individuals, not merely to general tendencies and causal sequences of their instantiations. An account of cognitive dynamics that takes fundamental psychological explanation to make reference only to states that have forms and norms that are agent-neutral underestimates how deeply psychological kinds are bound up for their individuation with agent identity—with the perspectives, functions, needs, aims of individuals.

Full appreciation of this point requires detail about the way perceptual content and primitive, pre-intentional individual agency are (I think must be) explained and individuated in psychology. Space is insufficient for such detail here. I have, however, tried to evoke some of the ways in which individuation of primitive psychological states is bound up with the biologically basic needs and activities of individuals—eating, fighting, fleeing, navigating, reproducing, parenting. Perceptual content and primitive uses of it are individuated in terms of a system whose functions are grounded in biological needs and activities of individuals.

In the fission case, *B* and *C* have different bodies, needs, perceptual perspectives, and agency-ends from one another. It is easy to assume that they have common interests (and so on) with their ancestor *A*. Largely, they will. However, since they have different bodies from *A*, they have different perceptual perspectives, different defensive and nutritional needs, possibly different reproductive needs, from their common ancestor. Being an individual agent with a psychology requires having explanatorily relevant capacities to act on one's own functions, needs, aims, and perspective. These capacities must include memories with *de se* forms or presumptions. At the most primitive level of psychological explanation, the reductionisms underestimate the role of basic biological needs and functions of animal individuals in individuating psychological states.

Analogous points apply to higher-level psychological explanation. As argued in section II, intentional agency requires explanatorily relevant capacities individuated in *de se* ways. Activity that is not explained in terms of psychological kinds that privilege the individual's perspective, needs, aims, values, would not be that of an individual agent, much less a person.

The idea that in persons there might be no explanatory role for psychological states individuated in such a way as to presume *de se* and first-person capacities—so that *de se** and *I** forms take their place—is unacceptable on its face. Being a person requires having states that are explainable as having a place for an *I*-concept.[43] A psychological capacity can be correctly individuated as involving a place for an *I*-concept only in terms of a psychological unity that is explanatorily grounded in the individual's agency. Such unity is in turn partly explainable in terms of memory with its presuppositions about individual identity.

The point that individuals figure in the individuation of psychological states that are central to explanation of perception, agency, and inference applies

[43] Cf. my 'Reason and the First-Person', 247ff; 386ff. in this volume.

whether or not agent-neutral accounts attempt a further functionalist reduction. Individual agents enter into the individuation conditions of such psychological competencies and their exercises. Individuation is the *basic* level of kind-individuation. I believe that the basic kinds in psychology are representational kinds, not some further nonintentional, functionally specified kinds. I reject functionalism. But to be adequate, even functionalism would have to include, among the anchor points in its account of individuation, the individual as well as aspects of the individual's environment.

General intuitions that facts about psychological states must be fixed by continuities in neural causation do not confront basic forms of empirical explanation of agency, perception, and inference—or the way those forms are individuated. The role of individual agents in individuating psychological states is just as central as the role of kinds in the referred-to environment.[44] Here the relevant part of the "environment" is the individual agent, not the kinds or properties that the individual interacts with. But the reasons why sequences of neural states considered in isolation cannot fix psychological states are similar. Individuation of representational states must take account of the "environmental" context of neural states, including what needs they serve. Insofar as the reductionisms rest on intuitions about neural or causal continuity *per se*, their mistake is analogous to the mistake made by individualism about psychological states.

The point that individual agents figure in the individuation of psychological states applies whether the account is associated with an attempt to downgrade the importance of individuality in psychology—as it is for Parfit—or with an attempt to rework ordinary individual psychology with agent-neutral psychological kinds—as it is for Shoemaker. Both approaches neglect either the central role of psychological states with *de se* forms and *de se* presumptions in psychological explanation or (inclusive) the fundamental role of individuals in individuating exercises of *de se* psychological states. As regards persons, they in effect relegate to a trivial "special case" (cf. note 41) the central role of taking one's own agency as a source of epistemic, practical, and moral norms. Reductionism misses how individuation, explanation, logical form, and representational or epistemic norms are grounded in the perspectives, needs, aims, and values of individuals.[45]

[44] Cf. my 'Individualism and the Mental', *Midwest Studies in Philosophy* 4 (1979), 73–121, and 'Other Bodies', in Andrew Woodfield (ed.), *Thought and Object* (Oxford: Clarendon Press, 1982), 97–120.

[45] Parfit's line is especially affected by the empirical success of individual psychology. One cannot eliminate the notions of individual agency, or of the individual, in giving an empirical psychological account of transactions in our world. Epistemic and practical norms are grounded in these empirical facts. Shoemaker's line is especially affected by the fact that the notions of agency, inference, and personhood are apriori connected to the explanatory centrality of *de se* presumptions, and their individuative presuppositions about individual identity. Memory is in on the explanatory ground floor of any account of the representational content of an individual agent.

I have been emphasizing individuation and explanation in this section. Earlier sections, especially sections IV and X, also featured the role of representational norms (norms of veridicality) and epistemic norms in our understanding of presuppositions of agent identity in psychological states. Both

I see no threat from the fission case to any of my arguments. Explanatory circularity emerges in using such cases for reductionist ends as soon as *A* and *B* are considered agents with psychologies. Reductionsim is much further from intuition, from individuation of psychological natures, from empirical psychological explanation, and from a reasonable epistemology than it might have first appeared to be.

XII

The foregoing reflections tell against a prominent class of theories of personal identity, indeed of mind, commonly associated with Locke and Hume. The common assumption of these theories is that psychological states are more basic than personal (or agent) identity.

This assumption has sometimes been combined with attempted reduction of the personal to the impersonal. Examples are Hume's bundle theory, Lichtenberg's attempt to eliminate the first-person from the *cogito*, Parfit's impersonalist theses, and various philosophers' hopes to accord explanatory hegemony to the third-person stand-point.[46] Since quasi-memories can exist only in a system that depends for its representational functions on memories, with *de se* presumptions, quasi-memories cannot support an impersonalist reduction.[47]

More modest reductionist approaches popularly associated with Locke are also affected. Attempts to reduce personal identity to a sequence of psychological states, or a sequence of arbitrarily small time slices each of which has the relevant states seem to me to fail.[48] Psychological states that are essential to being a person presuppose a core of veridical applications of *de se* memory forms and veridical *de se* presumptions in purely preservative memory. These presuppose,

reductionisms are vulnerable to criticisms based on normative considerations, but I will not undertake a separate diagnosis of these vulnerabilities here.

[46] Compare Hume, *A Treatise of Human Nature*, 1.46; G. C. Lichtenberg, *Schriften und Briefe* (Munich: Carl Hanser Verlag, 1971), 2.412.76; Parfit, *Reasons and Persons*, 224–226. From the point of view of this project, the presence of a *de se* notion in the concept of quasi-memory would make that concept illicit. It is notable that Parfit's conception does not exclude *de se* elements from the concept of quasi-memory. Parfit tends to conflate the issue of reduction of the personal to the impersonal with two other issues. One is whether thinkers are "pure egos" that are "separately existing entities". The other is whether one can correctly and completely specify experiences and connections among them without presuming the existence of a subject that has these experiences. Materialists and many non-materialists would join Parfit in inclining toward a negative answer to the first question—pending clarification of its meaning. Parfit, the Hume before the appendix, Lichtenberg, Mach, and James maintain an affirmative answer to the second question. Our discussion of *de se* elements supports a negative answer.

[47] Cf. my 'Reason and the First Person'.

[48] It is disputable whether the view is Locke's. Examples of this view are Shoemaker, 'Self and Substance', *Philosophical Perspectives* 11 (1997), 283–304 (and references therein); David Lewis, 'Survival and Identity', in Amélie Rorty (ed.), *The Identities of Persons* (Berkeley: University of California Press, 1976); John Perry, 'The Importance of Being Identical', in Rorty (ed.), *The Identities of Persons;* and H. P. Grice, 'Personal Identity', *Mind* 50 (1941), 330–351.

and are individuated in terms of, the unity of a representing individual over time. So correct explications of the possibility of these psychological states and abilities, and norms governing their exercise, entail individual identity over time.

One can conceive of beings that lack perception, intentional agency, and inference, but that get along in the world. The most primitive layer of the animal kingdom is replete with such beings. They lack representational minds. One can conceive of beings that have only an analog of inference—a state transition statistically favorable to the organism, described in "information theoretic" terms but falling under no inferential norms.[49] If such beings have biological needs and either perception or any sort of agency, their psychological states have representational elements. Individuation of exercises of such states depends on individual identity over time. Neither sort of being would be a person. The idea that we might be either sort of being is absurd. Among the most solid things we know are that we perceive objects, intend to do things, draw inferences.

I have not faced various puzzles about personal identity: transplantation, transportation, multiple personalities, dementia, fetuses. I hope to have found insight into necessary conditions for being a person.[50] Memory is essential to

[49] Hume's scepticism about inference suggests this approach (cf. *A Treatise of Human Nature*, 1.4.1, 1.3.13). Another *ersatz* for inference might be regarded as warrant purely through credulity. If X seems to quasi-remember that p, X is warranted in believing p. Each surrogate for a step could have this sort of warrant. The view is incoherent if it omits purely preservative memory. A being that lacked a capacity for deductive inference could not have logical form for its representational states, hence no propositional attitudes. Such a view would also not account for the role of purely preservative memory in perception and agency. It would not account for *use* in representational transactions. So I think that this imagined case collapses into the case in which there is no inference and no representational content.

[50] This paper has an obviously Kantian flavor. It investigates necessary conditions on the possibility of having certain competencies. It also claims that psychological states presuppose a certain unity reflected in the form of the content of psychological states. Kant called his version of such unity "the transcendental unity of apperception". He saw such unity as necessary to applications of "I think" (*Critique of Pure Reason*, B 131–142). By grounding his account in self-consciousness, Kant tends to ignore lower-level *de se* elements in mentality, present even in many lower animals. Such animals have perception but no thought, much less a first-person concept.

At a more specific level, my arguments regarding inference are kin to Kantian suggestions about what is involved in holding a propositional thought together over time (Compare *Critique of Pure Reason*, A 98–110). I do not see in his work the specific argument I give. My arguments regarding perception are certainly congenial to Kant's view of empirical intuition (compare A 98ff.). I know of no cousin of the argument regarding intentional agency in Kant. I did not develop these arguments by reflecting on Kant, but no doubt I was somehow influenced.

Kant sees his account of the unity of apperception as "purely formal", with no immediate ontological significance for being in time. He bases this view on what I regard as an untenably restrictive epistemology. In the third Paralogism, which contains what must be the first appeal to quasi-memory, he imagines as an empirical possibility a series of persons who use "I think" in purported memories about their pasts, each having memory-like presentations that derive from the previous person (also applying "I think"). Each person mistakenly construes the previous one's past as his or her own. The case is primarily directed against the idea that we can through mere reflection have a certain scientific cognition that we are substances in the Cartesian sense. I accept Kant's rejection of this idea. But Kant also holds that only empirical experience could rule out such a series of erroneous self-attributions in an actual case. (Compare *Critique of Pure Reason*, A 361–367 (esp. A 363 and A 363 n.), B 408.) I believe that this claim is mistaken. I think that such a case can be shown by apriori reflection alone to be impossible, or at least incompletely described. I think that Kant's target,

representation because use and competence make representation possible. Uses and competencies must be reapplicable by and for the representing agent. The *de se* forms and presumptions of memory—and attendant presuppositions about agent identity—are necessary for reapplicability. Memory is at the root of representation.

I believe that most of the points I have been making are apriori. I believe that their apriority has important implications for method in philosophy.

"rational psychology" (which relies on certain features of Descartes's method of reasoning from the *cogito*) has more to be said for it than Kant allows. I do not accept Kant's restrictive epistemology of self-attribution, or his views about the *cogito*. Being fair to Kant's subtle position requires a much richer historical account, one I hope eventually to give.

18 De Se *Preservation and Personal Identity: Reply to Shoemaker*

Sydney Shoemaker replies to my criticisms in 'Memory and Persons' of his and Derek Parfit's attempts to reduce personal identity to continuities among psychological states.[1] Locke had seemed to invoke memory to analyze personal identity. Butler claimed that such analysis is circular because the concept of memory presupposes connection to past states of the same person. Imaginatively, Shoemaker and Parfit introduced a concept of quasi-memory that avoids the presupposition. I argued that although they avoid Butler's charge of definitional circularity, they incur explanatory circularity.

The circle takes this form. Shoemaker and Parfit attempt to explain constitutively what it is to be a person in terms of a certain preservational power, what they call 'quasi-memory'. The point of the explanation is to avoid appeal to memory in the explanation. For memory constitutively involves a presumption of sameness of person (between the rememberer and the individual whose psychological states—perceptions, thoughts, and so on—are remembered). But if quasi-memory is to function at all in an individual, it must do so in an individual that also uses memory. And it is memory that makes it possible for there to be an individual who exercises quasi-memory. Memory plays a role in constitutive explanation of the psychological individuality of the individual that is necessarily prior to any (putative) role for quasi-memory. So quasi-memory cannot ground a reductive explanation of personal identity.[2]

Shoemaker thinks that I beg the question against his and Parfit's views. I believe that his reply is mistaken.

Quasi-memory of a past event is 'representing it in a way that is as much like remembering it as is compatible with the person having the quasi-memory not being identical with the person whose experience of the past event caused it' (87).

[1] Sydney Shoemaker, 'Careers and Quareers: A Reply to Burge', *The Philosophical Review* 118 (2009), 87–102. Citations of passages by page numbers will occur in parentheses in the text.

[2] I am not proposing a reductive constitutive explanation of personal identity in terms of memory. I think that memory and psychological/personal identity are constitutively inter-dependent. One could not be an individual with a psychology and lack memory; indeed, memory is part of what constitutively marks a psychology as an individual's psychology. On the other hand, memory, as Butler noted, is constitutively what it is by virtue of preserving psychological states of a given individual.

Memory is supposed to be a special case of quasi-memory. Shoemaker summarizes his strategy in this passage:

> Someone who wants to define personal identity in terms of memory continuity . . . can hold that the persistence of a person over time consists in the occurrence of a series of events linked by quasi-memory, where the series is nonbranching. Because of the nonbranching provision, the quasi-memories will turn out to satisfy the identity requirement on memory, and so will be memories. But the notion of memory is not invoked in the analysis, so there is no circularity. If one has the more promising project of construing the persistence of persons as consisting in psychological continuity, of which memory continuity is just a part, one can use the same strategy to avoid circularity. (87)

Shoemaker describes imaginary persons, whom he calls 'Parfit people'. He notes that these beings may be metaphysically impossible.[3] He presents his fiction, however, as conceptually possible. These persons' lives begin as ours do. At age twenty-one, there occurs an episode of fission. In fission, 'a person's body divides into two exactly similar bodies, and each of the bodies is the body of a person psychologically continuous with the original person' (88). If both products of a fission lived on, each would grow increasingly different psychologically. But in this society, one fission product is killed shortly after fission occurs. The surviving creature takes the name of the original person, and for all practical purposes lives a life continuous with the original person's. Shoemaker postulates that the surviving person is a distinct person who has quasi-memories from the original person's life, and lives out the remainder of what Shoemaker calls a '*quareer*'. A quareer is as much like a person's career as is compatible with a later quasi-rememberer's not being identical with the person whose past is quasi-remembered.[4]

[3] Parfit does not clarify his attitude toward the two-person continuants. It is certainly natural to take them as entities with representational powers. I will track this issue in my discussion.

[4] Shoemaker believes that persons after the fission event are not identical with pre-fission persons—even persons that they share quareers with. I think that given how he describes the case, this belief is not obviously true. Shoemaker argues, 'Because identity is transitive, both products of the fission cannot be identical with the original person, and there is nothing that can pick out one of them rather than the other as identical with him or her—the killing of one of them cannot retroactively make the other someone that he or she earlier was not' (89). This is not a decisive argument. Shoemaker assumes that once fission occurs, the two products of fission are not identical with one another, and that there is no ground for identifying one rather than the other with the person that underwent the fission. So by transitivity, neither product of fission is identical with the original person. Shoemaker argues that since neither of products of fission is identical with the original person, the one that continues to live cannot be identical with the original person "retrospectively", once the short-lived product of fission is gone. But the assumption that neither product of fission is identical with the original person is not obviously true. Identity is timeless. Looked at timelessly, the fact that one product of fission continues the original person's life, and the other does not, and is killed shortly after the fission event, seems some ground to identify the surviving product with the original person. The eliminated fission product might be considered a mutant outgrowth on the career of the original person. The fact that it is chosen randomly does not seem to me to matter. On metaphysical issues that deal with cases as far from actual cases—and probably possible cases—as this one is, I think that there is little point in arguing. So I will accept Shoemaker's view of the matter for the sake of argument. It is

On Shoemaker's analysis, persons are beings with non-branching psychological continuities established by such capacities as quasi-memory. Shoemaker uses his fiction to try to rebut my argument that his and Parfit's attempts to explain personal identity in terms of such capacities as quasi-memory are explanatorily circular. He thinks that since quasi-memory does not involve first-person *de se* elements, as memory does, the explanation is not circular.

The gist of my argument was as follows. Being a person requires capacities to have and successfully realize intentions, to use his or her own perceptual experience, and to make inferences in argument.[5] All these capacities either involve or presume *de se* capacities that function intra-individually to preserve contents of the person's psychological states over time. An individual's intention to do something has a content that is made successful only by the individual's carrying out the intention. An individual cannot use his or her perceptual experiences unless the use is an application of the individual's own retained experience. An individual cannot draw an inference, unless the reinvoked premises preserve the individual's warrants for attitudes with the representational content of the premises, from when the premises are first used in the inference. These competencies, and norms governing their exercise, require preservation in memory of the individual's own states. Such preservation is either marked *de se* or is apriori associated with *de se* attitudes. *De se* markers apply only if the individuals' preservational competencies are keyed to the individual's own history or future. A person's having quasi-memory and other such quasi-capacities is explainable only by reference to the person's also having *de se* preservational powers. I wrote, '. . . quasi-memories can exist only in a system that depends for its representational functions on memories, with *de se* presumptions . . . '.[6] Since *de se* preservational powers *in persons* entail assumptions of personal identity, and since explaining persons' having quasi-powers requires appeal to the persons' having such *de se* powers, explanation of what a person is in terms of quasi-powers is circular.

The circularity does not lie in the point that in defining 'quasi-memory', Shoemaker makes reference to memory. (Quasi-memory is like memory except that. . . .) The circularity lies in the fact that being an individual that exercises quasi-memory depends on being an individual that exercises memory. Memory is constitutively the basic psychological kind. Constitutively, to have quasi-memory at all and to be an individual with a psychology at all, an individual must utilize memory. The contrary is clearly not true: to be an individual with

obvious that one can allow the two products of fission to live out full lives. In that case, one would have strong ground to claim that there are three people, not one or two.

[5] Indeed, an individual's having a representational mind requires having lower-level analogs of the first two of these capacities; and having propositional attitudes requires the third.

[6] 'Memory and Persons', *The Philosophical Review* 112 (2003), 327; 444 in this volume.

memory it is not necessary at all, much less constitutively necessary, to have quasi-memory.

Shoemaker's reply does not engage with the argument. Some of the reply imputes to me positions that I do not hold. Some of it fails to confront the argument's key claim—that an individual's having quasi-memory-type powers is explainable only by reference to the individual's having *de se* preservational powers.

I

I begin by disclaiming some positions that Shoemaker's reply imputes to me.

First he holds that I believe that 'there can't be quareers that are not careers' (99, 100). I believe that the quareers that begin with the original person and continue with a different person after a fission event are not careers. I think that quareers are necessarily made up of one or more careers,[7] but quareers that cross the relevant fission events are not careers.

Second, he imputes to me the view that his fictional beings are not conceivable (101). I find them quite conceivable, and nowhere intimate that I do not.

Third, he holds that to make my case, I would have to argue 'that the differences between the Parfit people and us are such that they lack representational minds, or at least that they could not preserve knowledge and warranted belief over periods of time that span episodes of fission . . . ' (93). All the claims here are mistaken. I am quite willing to take pre- and post-fission persons in Shoemaker's scenario to have representational minds. I think that the two-person continuants in the scenario probably also have representational minds. I also believe that, in a perfectly good sense, the persons and the two-person continuants preserve knowledge and warranted belief across fission episodes. Except in cases in which knowledge or belief has *de se* content, each post-fission person can have, through the postulated causal connections, knowledge and warranted belief with the same content that the pre-fission person's states had. Two-person continuants can also retain knowledge and warranted belief across fission episodes.

My view is that the representational minds of pre- and post-fission persons can be coherently understood only by taking them also to have the *de se* preservational competencies cited earlier. Their cross-personal competencies—for example, quasi-memories—augment a core of *de se* intra-personal competencies. It is these latter that are constitutive of the relevant individuals, as psychological beings. Specifically, each pre-fission and post-fission person has *de se* intentions

[7] One might be inclined to think that this is all that Shoemaker means by his attribution to me that all quareers are careers. However, he himself makes the distinction, and mistakenly attributes to me the view that 'there cannot be quareers that are not careers' (100), while holding that 'quareers that are not themselves careers are made up of careers' (101).

and capacities to preserve them in actional representations, *de se* perceptions and *de se* capacities to preserve the perceptions in memory for using the perceptions, and competencies to engage in inference that presuppose *de se* preservation in memory of premises that are reinvoked in the inference. These powers are fundamental to understanding the persons as persons, indeed any individuals with propositional attitudes. In fact, the powers that allow cross-personal preservations can function as representational powers at all, only by being employed by individuals with these *de se* preservational powers. Shoemaker's reductive explanations of what it is to be these individuals is explanatorily circular, because they tacitly presuppose for their coherence that the individuals employ the ordinary *de se* preservational powers—ordinary memory.

Let us say that quasi-memories, and other similar psychological capacities that allow cross-individual realization, are *de se** preservational capacities. In Shoemaker's scenario, the function and content of *de se** preservational capacities are unspecific as to whether states that they connect to are the person's own states or those of a person on the other side of a fission.

In addition to the core intra-personal *de se* competencies, persons in Shoemaker's scenario have *de se** cross-personal powers—quasi-intentions-to or quasi-memories, for example—that can extend to another person across a fission event. Having these *de se** powers cannot prevent persons from having *de se* competencies that are constitutive of their being persons.

Similarly, I think that the continuants consist in the pre- and post-fission-event persons are conceivable. But they can be coherently understood as having representational minds only because they are understood either as being individuals with *de se* powers, or as being group minds—two-person minds—consisting in the individual persons that have *de se* preservational capacities as well as cross-personal *de se** capacities.[8]

Having intentions, using perception, and engaging in inference are constitutive of persons. These powers are associated with diachronic, intra-personal *de se* capacities to preserve representational contents of the person's own

[8] I leave open whether the continuants have *de se* powers. Nothing in my discussion hinges on how one regards the continuants. I take no position on this metaphysical issue, partly because the issue does not even concern a clear possibility.

If the two-person continuants are persons or other individuals with representationl minds, they have *de se* powers. They would then also have cross-individual *de se** powers, because they would have powers to connect directly to the minds of other individuals—in particular the minds of the shorter-lived persons that are co-terminus with periods in the continuant's existence on the other side of a fission event. If the continuants are not individuals, they cannot strictly have cross-individual capacities. But they can have capacities that connect directly to persons with whom the continuants are not identical. Such capacities enable them to have "internal" psychological connections, analogous to quasi-memory, into the psychologies of persons—the shorter-lived persons which quareers join—with whom they are not identical. If the continuants are not individuals, then it is less clear *exactly* how to describe and explain their minds, granting that they have minds. But I suppose that their minds would be explained in terms of relations among the psychological powers of the shorter-lived persons—in something like the way we describe the socially constituted minds of organizations or groups.

psychological states. Persons in Shoemaker's scenario have *de se** powers only because they have such *de se* preservational powers.

I make three claims, then, about the Shoemaker's scenario. First, I claim that the individual persons in the scenario have these *de se* preservational competencies. Second, I claim that these *de se* competencies ground explanation of the cross-personal "quasi", *de se** powers that the persons also have. Third, I claim that the *de se* competencies are partly constitutive of the persons that occur in the scenario—and that whatever competencies the continuants have are ultimately explainable in terms of *de se* competencies of either the continuants themselves or of the shorter-lived persons.

I will develop these three claims together. But I will lead with the first. Consider a post-fission person who is continuous with but distinct from a pre-fission person. As an individual person, this individual has his or her own point of view. The person can preserve this point of view by retaining recent perceptions, intending to do things, and carrying out inferences. The person can and does use short-term *de se* memory indexed to his or her own needs and perceptions, from the first moments of the person's career. Part of what makes the person an individual is having needs, perceptions, and preservational capacities that figure in the individual's actions—actual and potential. Perceptions are used, via *de se* preservational memory, to serve the person's needs through action. The person has intentions that bear on his own future; those intentions are indexed to that person. Exercise of those intentions utilizes *de se* preservational powers. The person can engage in inferences that reinvoke premises that that person established earlier in an argument. The preservational powers that operate in reasoning are typically relatively short-term, since inferences typically do not extend over long periods of time (though they may). Their use presumes *de se* preservational powers. Analogous points apply to the pre-fission-event person. In each case, being an individual representer requires having *de se* capacities to preserve the individual's point of view for the individual's own needs and activities.

In addition to their *de se* psychological states, and *de se* preservational capacities, the persons in Shoemaker's scenario also have *de se** capacities (like quasi-memories) that allow cross-personal quasi-memories, quasi-intentions, and the like. Our ability to conceive a fiction in which such capacities operate cross-personally provides no ground at all for thinking that the persons in the fiction *lack* the *de se* psychological states and intra-individual preservative powers that lie at the heart of psychological understanding of persons as individuals.[9]

[9] *De se* attributions are basic in the various psychological sciences, not just in ordinary common-sense discussions. See the discussion of ego-centric indexes in my *Origins of Objectivity* (Oxford: Oxford University Press, 2010), in chapters 9 and 10. Ego-centric indexes are non-conceptual instances of *de se* representation. They are representational contents that occur for example in the perceptual and actional systems of pre-conceptual animals and in pre-conceptual (pre-propositional) aspects of the psychologies of beings with propositional attitudes.

Shoemaker's scenario description tends to distract from these straightforward points. He describes the persons in the scenario as theorists that

> regard the surviving fission products as "as good as identical" with the original persons and treat them as such in all their social practices. Among other things, they regard the beliefs of the fission products as inheriting the warrant those beliefs acquired from the reasoning and experiences of the original person . . . and assign the same status to memories, or quasi-memories, of things done or experienced prior to the fission as they do to memories not separated from the remembered events by episodes of fission. (89)

The persons on either side of a fission event are by hypothesis *not* identical—whatever 'as good as identical' is taken to mean. Shoemaker sometimes writes as if their not caring about the difference between persons on either side of fission events has some effect on whether they have *de se* capacities (89). But the fact that the persons in his scenario do not care much about the non-identity does not cancel it. Members of groups may subordinate their interests to the group in their 'social practices'. It does not follow that they lack individual needs, interests, goals, and motivations. By hypothesis, the persons on either side of a fission event are distinct. So they have their own needs, interests, and points of view. They have *de se* preservational powers that serve those needs, interests, goals, and motivations.[10]

Given that there are persons in a scenario, our understanding of them uses concepts for actual natural psychological kinds, including *de se* kinds. A sound methodology for considering non-actual, possible situations (or as in Shoemaker's case, conceivable situations that may not be possible), requires determining whether the competencies that underlie actual psychological kinds are present. *De se* kinds are easily found in Shoemaker's scenario.

De se aspects of psychological powers are an absolutely fundamental natural psychological kind. No individual with a representational psychology can lack them. Such powers realize fundamental functions in any individual's representational psychology—including a person's. Powers to realize these functions are part of what marks an individual with a mind *as* an individual. I emphasize here two such functions.

One is to exercise sensitivity to the individual's own needs, goals, and perspective as privileged in a way that is distinct from those of others. Every individual's psychology functions to privilege its own needs and perspectives. A second function is to provide the representational basis for serving the individual's own needs—doing things for him- or herself—from the individual's own

[10] Shoemaker's scenario is set up to obviate conflicting interests among beings in the scenario as far as possible. By killing one of the post-fission beings, Shoemaker eliminates actual post-fission conflict of interests (at least once the post-fission being is dead!). But since, by hypothesis, there are at least two persons in each quareer—not to mention the continuant—there is the potential for conflict in needs and interests. There is even potential for conflict in needs and interests between a shorter-lived person and the continuant that exists at the same time. There are certainly *differences* in needs and interests.

motivations or other representational powers. The unity of persons consists partly in having primitive representational powers that unify and distinguish their individual psychologies by fulfilling these two functions. Having *de se* functions, *de se* representational states, and *de se* preservational powers is partly constitutive of being an individual with a representational psychology, hence of being a person.[11]

These points do not derive from introspecting a property of mineness.[12] Individuals need not be able to discriminate introspectively the intra-individual operation of *de se* preservational capacities from that of *de se** preservational capacities. Where there are individuals with minds, however, there are sensitivities to and privileging of individual needs; and there are powers to serve those needs. Representational states and processes with these sensitivites and powers have *de se* representational content, which marks those aspects of the psychological states and processes. Similarly, as I argued, individuals with minds must have *de se* powers that function to preserve those sensitivities and privilegings in serving the individual's needs, projects, and functions. What makes persons in the scenario individual persons is their having these powers.

The two functions that I have emphasized, and the *de se* states and *de se* preservational powers that realize them, can be expected to be present in the individuals in Shoemaker's scenario, since those individuals are (plausibly) specified as individuals with representational psychologies. And in fact, there is not the slightest ground for thinking that the individuals in the scenario lack such functions and powers. Like any other individuals, they have powers to be sensitive to their own needs and projects, and to use representational capacities to serve them.

The pre-fission persons have many memories and no quasi-memories with respect to the post-fission person.[13] They have *de se** intentions that are unspecific as to whether they are to be fulfilled by themselves or by the post-fission counterpart. But these powers are vastly outnumbered, and are certainly undergirded, by ordinary *de se* intentions to do things. Such intentions, especially short-term intentions that fix *de re* on immediate actions and do not span the fission events, hold their daily, individual lives together. Analogous points apply to the post-fission persons. They have many *de se* intentions, and no *de se** intentions that could be fulfilled by their pre-fission counterpart. Their

[11] See 'Some Origins of Self', this volume.

[12] Shoemaker sometimes writes as if the opposition is committed to such a view. See (98).

[13] It might be argued that if continuants are individuals, a shorter-lived pre-fission person can have cross-individual *de se** intentions that would be fulfilled before fission by the simultaneously existing continuants. I do not object to such a view. But it depends on the *de se** preservations riding piggyback on simultaneously occurring *de se* preservations that occur in the shorter-lived pre-fission person or in the pre-fission continuant. I think that Shoemaker is not thinking of *de se** capacities of these kinds. Unclarity on these issues derives partly from the oddities of the fiction, and from the lack of obviousness that there really are two pre- and post-fission persons in the scenario. (See note 2.) There are certainly differences in needs and interests between the longer-lived continuant and the shorter-lived persons.

quasi-memories are additional to a large core of *de se* intra-personal memories, especially short-term *de re* memories that function to retain particular near-term events in the rememberer's history, events that as a matter of fact, could not have been experienced by anyone else.

Trying to *dispense* with person-specific *de se* concepts, in understanding the persons in the scenario, in favor of concepts that have no actual explanatory use and that apply to no known psychological kinds, would be a serious conceptual and methodological mistake. The capacities that ground *de se* notions are clearly present in the persons in the scenario. The idea that the persons in the scenario lack short-term memories and short-term *de se* intentions has no justification. We can intelligibly attribute *de se** powers to them. But it would be a mistake to think that such attributions block them from also having *de se* powers, states, and processes.

Shoemaker makes this mistake.[14] He takes the persons in his scenario to lack *de se* psychological states and to have *only de se** psychological states (94, 92, 101). He does not argue for, or even highlight, this view, even though it is essential to his position.[15]

Shoemaker does write,

> Functionally *de se** forms are just as efficacious as *de se* forms in satisfying needs and furthering the achievement of goals—bearing in mind that in the case of the Parfit people the satisfaction conditions of needs and goals have to do with effects on their quareers rather than with effects on their careers. (93)

This claim may seem pertinent to whether *de se** psychological states can be regarded as replacing all *de se* psychological states in the scenario. The non-individualized description of needs and goals fudges the issue, however. The *de se** psychological states can indeed further needs and goals held in common between the two persons whose lives make up a quareer. They can further needs and goals that are unspecific as to whether they are those of an individual person, those of the other person in the continuant quareer, or those of the two person continuant. But these are cross-personal needs and goals. They are analogous to one individual's having the goal that a philanthropic foundation be established, where the intention is unspecific as to whether that individual, his son, or some group of which they are members establishes the foundation. *De se** states cannot have the function of serving needs and goals specific to individual persons.

[14] Shoemaker's assumption that only *de se** powers occur in his scenario is an illusion born of focusing on novel psychological elements in the case and neglecting the familiar elements. In fact, having a novel power that crosses a single temporal divide for each of the persons in the scenario provides no ground for thinking that the standard natural psychological kinds—the *de se* kinds—are inapplicable in the scenario.

[15] Shoemaker writes, generically, of *de se** states 'doing the work' of *de se* states (94). But they cannot do all the work, for the reasons given earlier. They cannot mark capacities that realize the two basic functions of *de se* aspects of competencies—functions to mark off and serve the individual's needs, goals, and perspective.

A being is not an individual person unless he or she has such needs and goals, and can serve them.

The following passage also bears on whether scenario persons have *de se* powers:

> Except for personal beliefs and other present tense mental state ascriptions that carry no implications about the past or future, attitudes whose verbal expression would involve their "I" will not strictly speaking be *de se* attitudes. But they will be very much like *de se* attitudes. We can call them *de se** attitudes. Instead of being indexed to the possessor of the attitude, or to the possessor's career, they are indexed to the possessor's quareer. So, for example, the intention to do X will be a *de se** attitude whose satisfaction condition will be a *de se** attitude whose satisfaction condition is the doing of X at some future point in the quareer of the intender (90).[16]

This passage contains a number of unargued claims that evince serious underestimation of the role of *de se* attitudes in persons' psychologies, including the psychologies in the scenario.

It is not clear what Shoemaker means by 'personal beliefs'. I take him to mean ordinary first-person (hence *de se*) self-ascriptions of mental states. He concedes attitudes involving such *de se* self-ascriptions, but claims that they have no implications for the past or the future. But it is unclear why he thinks the persons cannot retain those self-ascriptions in at least short-term memory, and why he thinks that they cannot use them in forming *de se* intentions to do things. I believe that the view that they can have such *de se* self-ascriptions, but not retain them in memory or use them to set up future use—in action for example—is incoherent. Having the first-person conceptual *de se* representational capacity is not a point-event. Having such a capacity requires abilities to preserve such *de se* capacities for use. These are *de se*, diachronic abilities.

This issue regarding implications of present-tensed mental-state self-ascriptions is not the basic one. Present-tensed mental-state self-ascriptions are constitutive to persons, and persons must have *de se* powers to preserve them. But *de se* powers that are constitutive to persons—indeed to all individuals with representational psychologies—go well beyond the self-ascriptions that Shoemaker cites.

[16] Just prior to this passage, Shoemaker holds that the analogs of first-person pronouns in the language spoken in his scenario cannot be translated as our "I", for they count as true the judgment 'I did *A*', when *A* was done not by the speaker but by the "ancestral self" of the speaker, that is, the person whose fission had the speaker as its surviving fission product. (90)
Although Shoemaker disclaims relying on points about linguistic usage (94), he commonly appeals to such points, both in the present article and in the articles that I discuss in 'Memory and Persons'. I think that psychological issues, not linguistic issues, are basic; and he seems to agree. In any case, I believe that the claim just quoted is mistaken. I think that in natural readings of the scenario, utterances of sentences containing the word-form 'I' cannot be translated uniformly in the way that Shoemaker suggests. In utterances of 'I did *A*' that express short-term realizations of *de se* intentions—and there will be many such realizations—occurrences of 'I' should be translated as our 'I'. Some occurrences can, I agree, be translated differently.

'Memory and Persons' elicited ways in which *de se* indexing enters into sub-conceptual capacities that are the ancestral psychological core of conceptual self-consciousness. For example, all perceptions and perceptual beliefs are indexed *de se*. *De se* psychological states index the possessor of those states in a way that makes them ego-relevant—by linking them directly to the needs, goals, and intra-individual representational functions of the states' possessor. To have *de se* status, the states must be preservable in memory, for at least short periods of time. That is a condition on their having any content at all. Shoemaker has given no reason to believe that the persons in his scenario lack *de se* indexes on their perceptions, perceptual beliefs, and intentions, and on retentions in memory of such states.

Whether or not the individual persons (or the two-person continuants) in the scenario notice or care about the point, there are many individual proprietary needs and goals—most obviously short-term ones—that are not shared between the two persons in a quareer. For example, given that the person before a fission event does not exist any longer, that person does not need the breakfast that the person after the event needs. The pre-fission person does not have the goal of eating it, and may never have foreseen it. (Suppose that the breakfast occurs in the ripe old age of the second person.) The post-fission person certainly has a *de se* indexed goal. The goal can be fulfilled only by the post-fission person who has the goal.[17]

Let us look at intentions of *pre*-fission persons. Consider again Shoemaker's remark,

> So, for example, the intention to do X will be a *de se** attitude whose satisfaction condition will be a *de se** attitude whose satisfaction condition is the doing of X at some future point in the quareer of the intender. (90)[18]

Shoemaker assimilates the intention-like attitudes of pre-fission persons to *de se** attitudes whose satisfaction condition is 'the doing of X at some future point' in the existence of either that person or the person on the other side of the fission event. But few intentions are so indefinite. The overwhelming majority of the intention-like attitudes of pre-fission persons will have definite, *de re*, short-term timing conditions. These attitudes function specifically to serve the needs and goals of the pre-fission person. They are constitutively *de se*.

Consider an example. If at age 14 a pre-fission shorter-lived person had no such intentions as the intention to raise his or her arm with a relatively immediate,

[17] It is acceptable to claim that the two-person continuant needs the breakfast. But the continuant's needing the breakfast and the second person's needing the breakfast are different matters. They are different beings, assuming that the continuant is a being; and their needs are indexed to the different beings that have the need. (See note 5.)

[18] Shoemaker's remark, 'the intention to do X will be a *de se** attitude' is a clear error. An intention to do X—by the plain meaning of the locution 'intention to do X'—is not fulfilled unless the intender fulfills it. It would have been more accurate to have written 'intentions* to do X'. Intentions to do X are *de se*.

de re, short-term, future timing indication, then there would be no pre-fission person to reach the age of 21 and go out of existence through fission. Most intentions have *de re* timing constraints in their contents. Intentions to move body parts, like the 14-year-old's intention to raise his arm, nearly always have a short-term timing specification in their contents. Thus, many pre-fission intentions will not cross the fission time-line. They will function specifically to serve the needs of the pre-fission person, not just the needs of one or other of the beings in a quareer. Without such intra-personal, *de se* intentions, there can be no persons.

Analogous points apply to the preservational capacities involved in uses of perceptual beliefs. A core of *de se* memories and intentions, most of them short-term, underlies those *de se** capacities that function to span a fission event, or that function in such a way as not to be specific as to whether their fulfillment spans a fission event.

Where there is a difference in individuals with representational minds, there is a potential for difference in need, goal, and function of attitude. Whether or not there is an actual difference, the difference in the being that has a perception, need, or goal is indexed—for a core of psychological states—in the being's psychologies. Perceptions, needs, and goals of a being are indexed to that being. Given that the individual persons are different, their having shared needs or goals does not prevent *even those shared needs and goals* from being indexed severally to each of the shorter-lived persons. Each being has *de se* preservational capacities that function to preserve the contents and modes of *de se* states. (See note 5.)

De se psychological states function to index and preserve perceptions, needs, goals, intentions, and other aspects of the perspectives of individuals. *De se** psychological states lack these functions. *De se** notions are blind to states and preservational capacities that function intra-individually. Given that there are distinct individual persons, as there are in Shoemaker's scenario, there are intra-personal perceptions, needs, goals, perspectives, and intra-individual preservational capacities. These are *de se* capacities. The individual persons must have *de se* states and preservational capacities to be individuals. Where the individuals are persons, the *de se* psychological states are functionally connected to capacities to use a first-person concept, and to self-ascribe mental states.

I believe that I have said enough to establish my first claim—that the persons in Shoemaker's scenario have *de se* psychological states and *de se* preservational capacities.

Shoemaker appeals to the psychological and epistemic beliefs of persons in his scenario. He presents these beliefs as supporting his own view that these persons lack *de se* preservational capacities. He says that the persons in the scenario treat the different persons before and after fission events as, for practical purposes, 'as good as identical'. On Shoemaker's own account, they are not identical. We must evaluate their beliefs in light of that fact. Their beliefs have no special authority. The issue is whether the beliefs are true.

I have already argued for my first claim—that the persons in the scenario do have *de se* preservational powers, as well as *de se** states. I will sharpen it by evaluating the beliefs of the individuals in Shoemaker's scenario (which are really Shoemaker's own beliefs).

Shoemaker gives two descriptions of the self-conception of the persons. I quote both.

> They regard the beliefs of the fission products as inheriting the warrant those beliefs acquired from the reasoning and experiences of the original person, regard them as knowing the things the original persons learned before the fission and which the fission products believe, and assign the same status to memories, or quasi-memories, of things done or experienced prior to the fission as they do to memories not separated from the remembered events by episodes of fission. (89)

> [They think] that the warrant of beliefs is preserved across episodes of fission, that there is preservative memory across such episodes, that seeming memories from the inside of a thing done or experienced prior to an episode of fission can have as good a right to count as knowledge as seeming memories from the inside that are not separated from the remembered events by episodes of fission, that actions done after an episode of fission can count as executions of intentions formed prior to it. Burge is committed to holding that if my imaginary creatures are possible at all, they are radically mistaken in thinking these things. (91)

Let us go through these views one by one.

> (1) They regard themselves as knowing and being warranted in believing things across fission events that their predecessors also knew.

Via quasi-memory, post-fission persons can both know and be warranted in believing propositional contents known and believed by pre-fission persons, except for any *de* se elements in the propositional contents. For example, they can have cross-personal singular *de re* knowledge (based on quasi-memory) of the things that the pre-fission person experienced.[19] Post-fission persons can meet

[19] Shoemaker thinks (96ff.) that I may deny [the possibility of] singular *de re* quasi-memories. I do not. The locutions in 'Memory and Persons' that misled him were highlighting the fact that in ordinary specifications of quasi-memory it is left open which and how many individuals one has quasi-memory relations to. I take it that in any given use each such relation is singular and *de re*.

There is a minor disagreement over interpretation of Parfit's account of quasi-memory. In the passages in *Reasons and Persons* (Oxford: Oxford University Press, 1984) that I focused on, Parfit writes of individual subjects' seeming to remember having an experience. This locution *entails*, in plain English, that the individual subjects' seem to remember *their* (*de se*) having the experience. Shoemaker points out (97 n5) that, in an essay thirteen years before, Parfit states that [on his interpretation of his own usage] it does not follow from my seeming to remember having an experience that part of what I seem to remember about the experience is that "I, the person who now seems to remember it, am the person who had this experience". I think that Shoemaker is right that this is a denial that quasi-memories have *de se* content. But although Parfit states that having *de se* content does not follow from seeming to remember having an experience, in fact it *does* follow. So in 1971 his locution does not capture the meaning that he intended. I concede that in 1971 Parfit did not believe that quasi-memories have *de se* content. But Parfit does not repeat this gloss on his use of words in *Reasons and Persons*. Parfit's account is what he writes down. Since the words mean what they mean, Parfit's account in 1984 is committed to quasi-memories' having *de se* content, even if

standards of epistemic good-use and reliability, and thus be epistemically warranted, in using their *de se** cross-personal powers.

> (2) They assign the same status to quasi-memories of things done or experienced prior to the fission as they do to memories not separated from remembered events by episodes of fission.

It is not clear what 'assigning the same status' amounts to. The scenario beings may not care about the difference. But memory and quasi-memory are different types of psychological capacity. Memory has the representational function of preserving the contents of psychological states within an individual's psychology. Quasi-memory does not. Its function allows cross-individual operations. Memory is a *de se* preservational capacity. Quasi-memory is not. That is, since memory is an intra-personal capacity, it can preserve both the referent and the representational content of *de se* occurrent applications; quasi-memory cannot. Trivially, a person cannot preserve another person's referent and representational content, when the other person uses the first-person concept I̲ (or other *de se* elements) to self-refer. Moreover, even laying the *de se* elements aside, knowledge that rests on an initial quasi-memory of what a pre-fisson person experienced is inevitably new knowledge for the quasi-rememberer. By contrast, knowledge associated with initial memories of what an individual previously experienced is commonly a reactivation of knowledge that the rememberer already had.

Let me illustrate some of these points with an example. Suppose that a pre-fission person at age 18 has a perception. All perceptions are *de se*. Suppose that the person forms a warranted *de se* perceptual belief from that perception. The belief inherits the *de se* framework of the perception. That person can preserve the *de se* belief in memory. (Pre-fission persons, in the scenario that Shoemaker sets out, do not even have quasi-memory, though one could alter the case so that they do.) Preserving *de se* belief in memory requires preserving the reference and *de se* representational content of the *de se* element. The post-fission person is a distinct person. This person (necessarily) has no power to preserve both the referent and the *de se* representation content of the pre-fission person's belief. In this sense, trivially, the post-fission person cannot have the *de se* content that the pre-fission person had in beliefs acquired through quasi-memory.

Through quasi-memory, the post-fission person *can* obtain the non-*de se* aspects of the content of the pre-fission person's perception and perceptual belief. These aspects are preserved through a causal chain. But they are not preservations within the post-fission person's psychology. For the post-fission person, quasi-memory of events that the pre-fission person experienced is acquisition of new content and new knowledge, although it is the same non-*de se* content that

Parfit himself (and perhaps Parfit's account) in 1971 were not committed to that view. What Parfit himself believed in 1984 is unclear.

the pre-fission person had. So even the non-*de se* aspects of content constitute new belief and new knowledge for the post-fission person. In this respect, as I suggested in 'Memory and Persons', quasi-memories are more like perceptions than memories.[20] They are like perceptions of past events through the instrument of another person's perceptual experience.[21] The post-fission person's quasi-memories, and beliefs based on quasi-memories, are anchored in the post-fission person's own *de se* perspective. The post-fission person has *de se* memory capacities to preserve states, including quasi-memories, within that perspective.

> (3) They believe that beliefs by post-fission persons inherit the warrant that those beliefs acquired from reasoning and experiences of the pre-fission person.

This belief is mistaken. Types of warrant are fitted to types of psychological states and processes. Warrants are standards for epistemic good use of the psychological powers. Different types of powers are subject to different standards, different warrants. Warrants specific to memories that preserve *de se* elements concern how well the memories retain *de se* states intra-individually. Persons that meet those standards have those warrants. A different person cannot preserve *those* warrants in using different powers, because that person cannot preserve intra-personally the contents of *de se* psychological states of another person. A post-fission person cannot have the same warrants for relying on *de se* memories of the pre-fission person's perceptual beliefs, because those warrants are standards for intra-personal preservation of the reference and content of those *de se* states. Indeed warrants for quasi-memory are standards for the acquisition of new information, not for intra-individual preservation. As noted, however, beliefs based on quasi-memory can be just as warranted as beliefs based on memory.

> (4) They think that actions after a fission episode can count as executions of intentions formed prior to it.

Some intentions are intentions that something be done, or that something be done by one or more specific executors. Such intentions of pre-fission persons can be executed by different post-fission persons. But actions by post-fission persons cannot count as executions of pre-fission persons' intentions to do something. Pre-fission persons that intend to do something and do not get around to doing it

[20] Although quasi-memories are not *de se* preservational capacities—so that with respect to preservation they are *de se**—they themselves have *de se* markers, just as perceptions do. They are psychological states that mark the perspective of the quasi-rememberer, and that are functionally and directly connected to that person's needs and goals.

[21] Despite its similarities to memory, quasi-memory is representationally more like perception than like memory. Quasi-memory passes through other persons, but is not intra-personally preservational. It differs from ordinary perception in that it goes through another person. But it is like perception and unlike memory in that it is a source of new information for the user. Note that perception too can reach well back into the past (think of perception of distant stars).

before they go out of existence cannot have *those* intentions executed by other persons. Intentions-to are constitutively *de se* intentions.

To be persons, both pre- and post-fission persons must intend (*de se*) to do things. Such intentions can be fulfilled only by the intender. It should be remembered that to be an intention to do something, the intender need not specifically *formulate* or *consider* the intention as *de se*. It is *de se* by virtue of its role in the person's psychology, not by virtue of the person's meta-attitudes toward the intention or its fulfillment. Insofar as the timing on a pre-fission intention is unspecific both about whether it be done before or after the fission event and about whether it be done by the intender or the post-fission person (or by the continuant), such intentions can be executed by others. But as noted in the example of the 14-year-old, the core intentions of both pre- and post-fission persons are *de se* intentions with specific *de re* timings to do things conceived of *de re*. These intentions function to be fulfilled by the intender. To be persons, the pre- and post-fission persons must be able to form such intentions to do things themselves.

I turn now to my second claim—that *de se* powers ground constitutive understanding of cross-personal *de se** powers of the persons in the scenario. I think this claim fairly obvious, once it is recognized why persons in the scenario must have *de se* preservational powers.

Here again is Shoemaker's introduction of the notion of quasi-memory: 'quasi-remembering something is representing it in a way that is as much like remembering it as is compatible with the person having the quasi-memory not being identical with the person whose experience of the past event caused it' (87).

I note two points about this introduction. First, quasi-remembering is explained in terms of remembering. Our grip on the notion starts with memory and then relaxes its intra-personal *de se* aspect. I do not say that 'quasi-memory' is *defined* in terms of memory. I say that explaining quasi-memory makes essential use of the notion of memory. Second, the introduction assumes that *persons* have quasi-memory. Understanding the quasi-, *de se** cases rests on assuming that persons are up and running. We understand persons as having *de se* states, like perception, and *de se* preservational memory (which preserves perception, intentions, and premises for use). The 14-year-old again illustrates the point.

We use this base of understanding to add special additional *de se** powers that enable one person to connect to the psychology of another person. We can also attribute such powers to the two-person continuants. The supposition that some persons in the scenario are fission products and are causally continuous with a pre-fission person should not obscure the fact that we utilize the ordinary notion person with its presupposed *de se* notions as basis for explaining the quasi-notions and the continuant two-person beings. We augment persons' intra-personal powers with cross-personal powers. Again, focusing on relatively short-terms needs, memories, uses of perception brings out the prevalence of *de se* preservational functions and powers in our understanding of the individuality of the persons involved in cross-personal preservations. Each person has *many* more intra-personal *de se* memories, intentions, and so on, than cross-personal or person-unspecific memories, intentions and so on. More importantly, these intra-personal states and powers ground understanding the persons as persons. The existence of a single fission event at age 21 of a person hardly undermines the centrality of standard intra-personal intentions or memories in grounding our understanding of the persons—and continuants—involved.

The explanatory dependence of *de se** notions on *de se* notions is not just an artifact of Shoemaker's explanation. Explanation of *de se** powers is ineliminably parasitic on prior constitutive understanding of *de se* powers. I think it incoherent to postulate a representational mind that neither has intra-individual *de se* preservational capacities nor is explained in terms of constituent beings that have *de se* capacities. Consideration of the timing of intentions, the short-term character of preservation of perceptions for use, and the short-term nature of most inference brings out that attempting to explain an individual mind purely in terms of preservational powers that are functionally unspecific as to whose content is preserved is hopeless. The unity of such a mind would be impossible without capacities with intra-individual functions to preserve content for use. A being that lacked such *de se* capacities would lack the capacities to represent for its own use and do things for itself. It would lack the capacities integral to being a locus of representation. In understanding representational mind, we perforce rely on assuming that intra-individual *de se* capacities preserve the mode and representational content of basic representational powers—and to provide the unity that is constitutive of representational mind.

To understand cross-individual *de se** capacities, we must assume, as explanatory base, a person with the intra-individual *de se* preservational powers that are involved in use of perception to meet the individual's needs, in realization of intentions-to, and in exercise of inference. One cannot make sense of a power that operates across minds (including persons' minds) unless one can make sense of the individual minds. Making sense of individual minds requires invoking *de se* notions—notions that apply to intra-individual needs, preservational capacities, and functions.

In 'Memory and Persons', I claimed, 'Memory with its *de se* presumptions and presuppositions of transtemporal agent identity—deriving from the individuation of basic sorts of agency—is a condition on the possibility of an individual's having a representational mind'.[22] I think that this claim stands. Any attempt to explain personal identity, or the identity of any individual with a representational mind, in terms of *de se** powers—powers that are not specific to serving and preserving intra-individually the individual's perceptions, needs and goals—is doomed to explanatory circularity.

[22] 'Memory and Persons', 306; 423 in this volume.

The foregoing points justify my third claim. The third claim is that the *de se* capacities involved in using perception, applying intentions, and engaging in inference are constitutive of persons, including persons in Shoemaker's scenario. I leave open whether *de se** powers or *de se* powers are constitutive of the two-person continuants that occur in the scenario. That depends on whether the continuants are assimilated to individuals or organized groups. I think it coherent to take the continuants as just as important as the pre- and post-fission persons, for social and other practical purposes. Some might even take them as individuals, or as persons. But explanation of the nature of *de se** powers is parasitic on explanation that invokes *de se* powers in individuals with representational powers. Shoemaker's fiction contains such individuals—at least the shorter-lived persons. The fiction is imaginative, thought provoking, and worthy of reflection. But the idea that it provides independent explanatory ground that is more basic than the natural *de se* kinds in terms of which we understand the nature of persons is an illusion.

II

I attended especially to inference in 'Memory and Persons'. In section VI of his essay, Shoemaker discusses my views on inference. The issues are special cases of those just discussed. They deserve attention anyway.

I argued that a psychological system that does not presuppose intra-individual agent identity cannot carry out inference. I argued this conclusion from three premises:

(1) Inference requires an ability to (re-)invoke representational contents as steps, relying over time on the same warrant that backed these steps when they were first instantiated.
(2) Epistemic norms and warrant attach to the agent of the inference.
(3) Epistemic norms for inference, and the warrant an agent has in a step in an inference, must be explained in terms of epistemologically relevant capacities, acts, experiences, or states of the agent of the inference.[23]

Suppose that a pre-fission person has a perceptual belief warranted through that person's having perceptions. A post-fission person cannot use purely preservative memory to reinvoke that belief's content as a previously established premise, relying on the same warrant that supported the other person's belief, because the post-fission person did not have those perceptions. So although the post-fission person can be warranted through quasi-memory to rely on a belief based on quasi-memory, the belief would be a new premise. It could not be warranted by reinvoking a past premise. I argued that even if the perceptual belief occurred

[23] 'Memory and Persons', 314–315; 431–432 in this volume.

in the career of the post-fission person, a quasi-memory capacity that was not specifically an intra-individual competence could not support preservation of warrant from the earlier belief so as to allow a reinvocation of the originally warranted premise. The individual could be warranted in the belief through quasi-memory, but not through the original perceptions. For quasi-memory is not an intra-individual competence, and cannot support intra-individual warrants involved in inference.

Shoemaker responds by objecting to (3).[24] He says that his scenario people accept a 'slight revision' of (3): 'the norms and warrant are explained in terms of epistemologically relevant capacities and so forth of the agent *and of earlier occupants of the agent's quareer*' (96).

Now this is not a *slight* revision. It evinces the tendency of persons in the scenario to ignore differences between intra-individual capacities and cross-individual capacities (and the attendant epistemological standards or norms for good use of them) that I criticized in section I above. In the first place, the phrase 'the norms and warrant' is unspecific on the point at issue. I explicitly avoided doubting that uses of quasi-memory can be warranted. Beliefs that rest on quasi-memory of experiences of other persons in a quareer can be warranted. That point is not in question. What is in question is whether warrants that derive from quasi-memory can support reinvocation of a premise in an inference. My argument that they cannot do so stands. Shoemaker does not address that argument. He simply asserts something that is not in question.

In the second place, although quasi-memory may be 'very much like inference', it is not inference. Inference, as I argued, requires intra-individual preservation of warrant. Shoemaker envisions explaining mentation that occurs in the scenario purely in terms of quasi-memory rather than *de se* preservational memory. This move requires that we give up standard natural psychological kind notions in understanding the scenario, and that we explain the situation entirely in terms of kinds that are not even clearly possible. I think that this move has no merit.

Third, and crucially, attributing quasi-memory to persons in the scenario presupposes that the persons who use such cross-personal capacities have intra-personal preservational capacities. In particular, the persons that use quasi-memories also make ordinary inferences that rely on intra-personal preservational memory, with its *de se* presumptions. In fact, an overwhelming

[24] Shoemaker claims that the premises are 'unclear' (96). He maintains that they must be understood 'as saying that a warrant, and conformance to a norm, attaches to an agent in virtue of what has occurred during some relevant period in the career of that agent' (as distinct from the *quareer* of the agent). But the premises are not unclear. I do not use or need the 'in virtue of' locution in premise (2), for premise (3) connects norms and agents to capacities, acts, experiences, or states of the agent of the inference. Shoemaker's response depends on rejecting premise (3), which he does. There is no need for clarification of either premise, and he suggests none for premise (3). Premise (3) connects the reinvocation of steps in inference to past states of the individual—hence to states in the individual's career, not quareer.

number of clear cases of ordinary intra-individual inference occur in, and function to occur in, their lives. Nothing that Shoemaker writes comes to grips with these points. A capacity for inference is constitutive of being a person. A person's having additional cross-personal capacities that in certain fictional (but perhaps impossible) cases can seem 'very much like inference' presupposes, for its intelligibility, thinking of the person as having the intra-personal preservational powers, including those used in inference, that persons constitutively have.

III

I turn now to Shoemaker's appeal to the Ramsey procedure to vindicate his view. I find this invocation unappealing on both general and specific grounds. First, the general grounds.

I think that the Ramsey procedure is an overused philosophical crutch. It is an inappropriate tool for illuminating explanatory notions. It purports to replace theoretical explanatory notions by a massively complex, and usually not fully explicit, existentially quantified description. Theoretical explanatory notions are natural kind notions—here notions for psychological natural kinds. Most theoretical explanatory notions are name-like. They are like common nouns that name the kinds. Descriptions are in general not explanatorily equivalent to natural-kind common nouns that they purport to explicate or be equivalent with.[25]

There are two general grounds implicit in what I just wrote, for rejecting Shoemaker's appeal to a Ramsey-description to replace notions like memory and *de se* intention. First, it is too unspecific to be explanatory. It is a mere handwave toward explanation. An unspecified unity or continuity relation is said to be as much like the ordinary relations as is compatible with the relations' being cross-personal. This is not an explanatory notion comparable to the standard ones. Shoemaker's use is worse than the original uses of Ramsey's procedure, since the original uses putatively operated on actual scientific theories—quantifying out the theoretical terms. Shoemaker offers only a handwave toward an actual theory, since the relevant unity/continuity relations are not specified. No one knows how to specify them. Second, Ramsey-descriptions make the old mistake of treating descriptive (in Shoemaker's case, functional) notions as explanatorily equivalent to the name-like theoretical notions that they are supposed to replace.

The more specific ground for rejecting Shoemaker's appeal to Ramsey's method to replace notions like *de se* memory derives from the points made previously about Shoemaker's scenario. We understand and apply cross-personal *de se** notions only through presupposing the application of the constitutively more basic intra-personal (or more broadly intra-individual) *de se* notions. Any

[25] I am relying here on the work of Saul Kripke, *Naming and Necessity* (Cambridge, Mass.: Harvard University Press, 1980); and Hilary Putnam, *Philosophical Papers*, vol. II: *Mind, Language and Reality* (Cambridge: Cambridge University Press, 1975).

Ramsey-description that contains *de se** notions—or functionalist analogs of *de se** notions—is explanatorily posterior to a specification that contains *de se* notions. So the constitutive explanation of the nature of personal identity through the Ramsey form of reduction remains circular.

IV

I summarize my main points from a more general perspective. The basic problem for the Shoemaker–Parfit reductionist explanation of personal identity is that it cannot account for the way a person's being an individual with a mind depends on capacities for self-conception and other forms of *de se* indexing that are constitutive to being a person. I elaborated this point in section XI of 'Memory and Persons'. I began that section by stating the point in general form:

> The central tenet of Shoemaker's and Parfit's reductionisms is that the basic explanatory psychological notions do not presuppose individual identity over time. The basic explanatory notions are supposed to be person- and agent-neutral. The notion of an individual person- or agent is to be explained in terms of continuities of states characterized agent-neutrally. . . . Being an agent with a psychology—whether a person or not—requires exercising psychological competencies supported by *de se* memories and memories with *de se* presumptions. The notion of an agent with a psychology is partly individuated in terms of such exercises. Reciprocally, such competencies and their exercises are partly individuated by reference to relations to their agents. Individuation of *de se* aspects of basic psychological acts and states is not agent-neutral.[26]

Shoemaker replies,

> Much the same could be said on my view, although I would replace "*de se*" with "*de se**". Mine involves a holism about the individuation of mental states, which includes a reciprocal relation between the individuation of the states and what individuals have them. Burge does not say what is meant by a state's being "agent neutral"—but what the term suggests is something incompatible with such a holism. (101)

This reply vividly illustrates the basic difficulty with the Shoemaker–Parfit reductionist view. *De se** indexes do not refer to the agent that has the states in whose content the *de se** indexes figure. They index states that realize cross-individual powers that serve either the agent's own perceptions, needs, and goals or another (perhaps specific) agent's perceptions, needs, and goals. They are in that sense agent neutral: they are not specific to a given agent, or a given agent's perceptions, needs, goals, premises. They are not specific to which agent's needs, goals, and perspective the relevant psychological states function to serve.

[26] 'Memory and Persons', 322; 439 in this volume.

In Shoemaker's scenario, *de se** indexes are unspecific at least as between indexing the needs and goals of the pre- or post-fission persons. (Nothing in the notion of *de se** indexes limits the individuals connected by quasi-powers to two.) Being an individual with a mind—hence being a person—requires having *de se* indexes that mark the individual, or at least that individual's perspective—the individual's perceptions, needs, goals, and so on. To be individuals with minds, individuals must have capacities indexed to their own perceptions, needs, goals; and they must have preservational capacities that function to serve those needs and goals. These capacities are marked by *de se* indexes that single out the individual or that perspective.[27]

Shoemaker's approach does invoke a holism that connects states with unity relations. But the unity relations do not coincide with individual agents that he takes to have representational minds. *De se** unities are individual-neutral in that they allow cross-individual preservation and do not specify connection to individual's perceptions, needs, and goals. His approach fails to account for the individual pre-fission and post-fission persons that the *de se** powers cut across.[28] I think that his approach does not even account for the two-person continuants. Suppose, first, that the continuants are themselves persons, or at any rate some type of individual with a representational mind.[29] Then the *de se** aspects of their states do not mark them off as individuals. For their *de se** states mark unspecifically both the continuants and the shorter-lived persons that their existence partially overlaps with. To be persons, or other individuals with representational minds, the continuants must be explained in terms of *de se* states and *de se* preservational capacities that mark their proprietary needs, goals, perspective, and functions.

Suppose, second, that the continuants are not individuals, but are organizations of individuals—organizations that consist in relations between the individual persons that make them up. Then they are not individual minds at all. Perhaps then they need not themselves have *de se* states. But their representational states are, again, to be explained in terms of the states of participant individual minds that have *de se* states and *de se* preservational capacities.

Any kind of individual with a mind is that kind of individual partly because of its natural psychological powers. Among these psychological powers are representational powers. Underlying representational contents of states are the representational powers or competencies that those contents help type-individuate. The psychology of any individual includes representational psychological powers that

[27] In the case of persons, there must be a self-concept that specifically indicates the person-agent him- or herself, not just a privileged set of needs and goals that are the agent's, as might be the case for *de se* indexes in animals that are not persons. For a discussion of this matter, see the first of my Dewey Lectures, 'Self and Self-Understanding', Lecture I: 'Some Origins of Self'.

[28] In fact, I believe that in discussions of *de se** preservational powers, Shoemaker's writing systematically invites confusion as to whether such powers are unity relations for the shorter-lived persons or the continuants. This point comes out clearly in his invocation of his holism.

[29] See note 5.

function to mark off itself and to serve its own needs, goals, and representational perspective. These powers are type-individuated partly by *de se* representational contents. To serve its own needs, goals and representational perspective, an individual must have powers that function to retain its representational take on things over time, so that that take can be *used*. Such preservational powers are also *de se*. The core of pre-conceptual *de se* states and preservational powers in *persons* is systematically related to conceptualizations—representational powers that occur as elements in propositional attitudes and underlie inference. In persons, these conceptualizations include the first-person self concept. Thus, at both pre-conceptual and conceptual representational levels, persons have powers with representational contents that mark underlying functions and powers of persons to distinguish themselves and their needs, goals, and perspectives from those of other individuals. They also have powers to preserve states that make such distinctions in uses that serve those needs, goals, and perspectives. *De se* representation is fundamental to having a psychology.

Shoemaker attempts to individuate persons as individuals in terms of (a) psychological *de se** "unity relations" that are not specific to individual persons, and (b) a non-branching condition. The unity relation must not be specific to an individual, if it is to avoid Butler's circularity objection. But one cannot capture the individuality of *persons* via the non-branching condition. Persons are individuated in terms of their representational competencies. The concept non-branching marks too generic a point about representational competencies to do the work in grounding psychological explanation that *de se* notions do. *De se* notions mark competencies to fulfill fundamental functions in any individual representational psychology: the functions of marking off one's needs, goals, and perspectives, in contradistinction to those of others, and the functions of doing things to meet those needs and goals that are connected to one's perspective. The unity of persons consists partly in representational competencies that unify and distinguish their minds by fulfilling functions to serve specifically the persons' own needs, goals, and perspective. They are individuals with representational capacities partly because they have *de se* representational competencies that distinguish them and unify them as individuals, privileging their own needs, goals, and perspectives. The concept non-branching is not reflexive. It cannot mark the basic psychological representational competencies that individuals (with representational psychologies) constitutively have to mark off their proprietary needs, goals, and perspectives, from those of others. It cannot anchor spatiotemporal representational frameworks, or, more especially, the *de se* ego-related psychological functions and preservational competencies that are necessarily associated with such anchoring.

The Shoemaker–Parfit approach systematically underestimates the centrality of *de se* representation in marking (constitutively, type-individuating) the unity and privileged status of individuals with representational powers—and for each individual, the privileged status of that individual's needs, functions, and goals. At the level of explaining the natures of individuals with representational powers, that approach makes fundamentally the same mistake that Butler thought Locke made: thinking that unity relations can be understood in a way that is independent of the *de se* reflexive representation by individuals whose minds are unified. Psychological unity relations that cut across the identities of individuals with representational minds cannot lie at the heart of individuating such individuals. For such relations presuppose the presence of ordinary psychological kinds that constitutively are internal to an individual's psychology.

19 *Modest Dualism*[1]

I am grateful to Bernard Kobes for his sympathetic and insightful survey, 'Burge's Dualism'. He presents concise sketches of my methodology, my views on mind–body causation, and my argument from anti-individualism against materialist token identity theories. I will comment briefly on each of these three topics before centering on his primary contribution—his discussion of compositional materialism. Then I will present a new line of thought on the mind–body problem.

My methodology requires metaphysical *claims*, as distinguished from metaphysical speculation and heuristic guidance, to be grounded as closely as possible in specific knowledge that resides in explanations and judgments in science and common sense. The history of philosophy has shown that a metaphysics that either goes it on its own or takes off from insufficiently specific features of science and common sense tends to be wayward.

Our best understanding of causation comes from reflecting on good instances of causal explanation and causal attribution in the context of explanatory theories. Similarly, our best understanding of what sorts of things exist comes from reflecting on ontological commitments of explanations in science, or clear-cut judgments in common sense. Ontological commitments are what must be the case if such explanations or judgments are to come out true. Of course, both scientific explanations and common-sense judgments are fallible. And sometimes they are reduced to other forms of explanations or judgments. Science and common sense remain our best epistemic routes to understanding both causation and ontology.

Understanding mind–body and body–mind causation is philosophically important and difficult. But such understanding has regularly been distorted by antecedent metaphysical commitments. Thus Descartes' account of mind–body causation as a relation between two self-sufficient substances over-dramatized the problem, to the detriment of nearly all subsequent discussions. The claim of some modern materialists that mental events, properties, and kinds must be associated with a physical mechanism if they are to have causal upshot raises the concern that mental attributes are in themselves irrelevant to mind–body or body–mind causation. Neither approach sufficiently grounds its metaphysics in what is known from psycho-physical explanation. We should start with our knowledge that there are causal interactions among psychological events and

[1] I thank Ned Block for several astute critical comments on an earlier draft. Kobes' paper is printed in *The Waning of Materialism*, R. C. Koons and G. Bealer (eds.) (Oxford: Oxford University Press, 2010).

between psychological events and physical events, and try to understand the interactions in light of scientific explanations and common-sense judgments.[2]

I continue to think that my old argument against the materialist token identity theory is sound. Indeed, it seems to me decisive. Any given token physical neural event that is a candidate for being identical with a mental event could have been associated with different causal antecedents in the distal environment from those it in fact had. The different causal history could have been part of a pattern of physical or social environmental relations that constitutively determined a different kind of mental event, with different representational content. Mental events are different events if they have different representational contents. Such differences in distal causal history are not constitutively determinative of the identity and kind of the neural event. The same type and token neural event could have derived from various causal histories, of the sort that determine different mental events, type and token. So the given neural event could exist in a situation in which the mental event with which it is supposed to be identified did not occur, and in which some other mental event (with a different representational content) occurs instead. So the neural event is not identical with the original mental event. The argument is general. So token neural events are not identical with token (representational) mental events.[3]

Kobes expounds a closely related argument: token physical events in the brain are not constitutively dependent for having the natures that they have on relations to specific attributes in the physical or social environments. Representational mental events are constitutively dependent for having the natures that they have on such relations. Token events are different if their natures are. So token physical events in the brain and representational mental events are not identical.

I think that this argument is implicit in the argument that I gave against token identity theories in 'Individualism and the Mental'. The modal points made in the first argument were grounded in my view of the natures—or basic explanatory attributes—cited in the second. Kobes in effect emphasizes the different forms of individuation in different sciences, and emphasizes that the modal argument that I gave has its underpinnings in the different identity conditions of basic explanatory kinds referred to in the sciences.

I focus now on Kobes's interesting and, I think, successful attempt to give some substance to my doubts about a materialist view that does not maintain *identities* between mental and physical states or events, but holds that mental states and events are *composed* of physical entities. Call this view *compositional materialism*. This view is less committal than the materialist token identity theory. A statue could be composed of marble. The statue is not identical with

[2] For elaboration of these points see my 'Mind–Body Causation and Explanatory Practice' and 'Postscript to "Mind–Body Causation and Explanatory Practice"', both in Burge, *Foundations of Mind: Philosophical Essays, Volume* (Oxford: Oxford University Press, 2007).

[3] The argument was first given in 'Individualism and the Mental', section IV, in Burge, *Foundations of Mind*. The argument is further defended and elaborated in 'Mind–Body Causation and Explanatory Practice'.

the marble. The same marble could have composed a different statue, or no statue at all. Then the marble would have existed but the statue would not have existed. The statue is materially composed of the marble. So identity is a stronger relation than material composition.

The natural sciences are built on the view that more complex material entities, such as planets, crystals, and plants are materially composed of material parts. Chemistry and physics were conceived in terms of material composition from the beginning. Some biologists once thought that explaining life would be a problem for compositional materialism in biology. But that view was never a majority position. The idea that plants are to be understood in terms of some soul or entelechy within them that is not composed of material entities has never had a serious foothold in western thinking, at least since early-modern science; nor should it have.

Because the framework of material composition has been explanatorily successful in the natural sciences, it is prima facie reasonable as a heuristic strategy to explore whether relations of material composition hold between psychological entities and neural or chemical entities. But *belief* in compositional materialism at this stage of inquiry seems to me not to be reasonable. The difference is between heuristic worth-trying and having reason for belief.

In the first place, the entities that psychology theorizes about (psychological states and events) are not perceived as material in the way that planets, salt crystals, and plants are. Although much of the *evidence* for psychology is physical movement conceived functionally (more specifically, as behavior), what psychology tries to understand and explain, and what it theorizes about, is not perceived or introspected as material.

In the second place, there are prima facie differences between psychology (sociology, economics, and so on), on one hand, and the natural sciences, on the other, that ground caution about *assuming* that material composition is the right relation between psychological events and brain events, merely on the basis of the success of the framework of material composition in the natural sciences. There are widely articulated problems in understanding how consciousness and qualitative states can be understood as materially composed. And there are attributions of reason in parts of psychology that have no analog in the natural sciences. Material composition is not a relation that grounds theorizing in mathematics or logic. As will emerge later in this chapter, I think that because psychology makes essential reference to structures from these disciplines, there is some question whether a framework of material composition applies within relevant areas of psychology. In sum, the prima difficulties in construing psychology in terms of a material-compositional framework—difficulties that have concerned philosophers for centuries—are much more substantial than they ever were in biology.[4]

[4] There have been massive attempts in mainstream philosophy since the 1950s to show that worries about materialism are just mistaken. I think that in some cases the attempts succeed in showing that

In the third place, there are large areas of psychology in which compositional materialism has literally no positive support in the explanations or methods of the science. Although there are limited correlations between neural events and low-level sensory states, which give some hope to a material-compositionalist view about them, the view has no positive support in science as applied to propositional thought, and nearly none as applied to consciousness.

I believe that there is no good reason to believe compositional materialism. I think, however, that it is the least implausible form of materialism about the mind. I will pursue critical discussion of the view as applied to *propositional thought*, an area where it seems to me to be particularly doubtful. Kobes concentrates his attention on this area, and I join him in this focus.

I like Kobes's approach. He explores ways the world could empirically turn out to be, avoiding strong commitment to any of the ways. Unlike most authors who have written on this topic, Kobes does not claim more than he or anyone has good and specific reason to believe.

Kobes introduces a highly knowledgeable mathematical archangel as a heuristic device. He proposes two scenarios. On both, the archangel can use facts about fundamental physical entities over all of space–time to compute chemical and biological facts, including neural facts, regarding a community of psychological subjects. On both scenarios, Kobes stipulates that assuming global supervenience of the representational aspects of psychology on the physical, and bracketing issues regarding phenomenal character, the archangel can also compute the representational facts over the community of subjects. The two scenarios differ over what neural states and events that systematically correlate with the representational facts are computed *from*.

On the first, the archangel can discern at the neural level, prior to attributing propositional attitudes to the subjects, a network of states and events that interact causally, and that are candidates for composing the intentional [representational] states and events to be attributed at a later stage of the computation.

The idea is that the archangel computes certain salient, discrete neural states and events in a causal network. At a later stage of the computation—a stage that invokes psychological kinds and principles—the already distinguished neural states and events can be seen as specifically correlating with representational states and events that are individuated anti-individualistically. And at this later stage the correlated neural states and events can be seen as materially composing the representational states and events.

On the second scenario, the archangel cannot compute, without using principles from representational psychology, neural events that form patterns that naturally correlate with the representational states and events. The archangel

certain apriori arguments in favor of dualism are unsound. And in some cases, the attempts set out prima facie empirically possible materialist pictures regarding *certain* psychologically relevant phenomena. The attempts have, in my view, failed to give good reasons for believing any form of materialism about the mind.

cannot, independently of psychological concepts and principles, identify a pattern of neural events that correlates structurally with states and events in the psychologically identified structure.

Kobe proposes, 'compositional materialism is false just in case the archangel would have to first recapitulate . . . intentional psychology, and only then seek neural event correlations'. In the second scenario, 'if neural correlates can be found at all for token mental events, the direction of metaphysical explanation [would be] from mental events and their causal patterns to the correlated neural events'. In the first scenario, just as the archangel first identifies 'chemical and psychological kinds and their instances by their physical components, distinctively arranged', so the archangel identifies, without using psychological kinds or principles, distinctively arranged material items that can later be seen to correlate with and compose instances of psychological kinds.

The key issue for Kobes is whether there are principles, other than psychological ones, for independently demarcating boundaries of neural events and states that are the units out of which psychological events and states are (and are later seen to be) composed. Kobes holds that if demarcation of neural correlates is possible only with help of psychological principles, compositional materialism is false. I believe that his idea is that if ideal non-psychological explanations in science do not independently indentify the material complexes that correlate with (and compose) psychological states and events, there is reason to believe that compositional materialism is not true. He holds that the explanatory question is empirically open.

A second issue, which Kobes takes to be independent, is this. Even if the archangel 'computes principled and determined correlations between [representational] and neural events', would the correlations do explanatory work analogous to that of compositions in familiar material sciences? If the correlations did not illuminate diachronic causal relations in a way analogous to diachronic relations among materially composed things, then correlation would not suffice to establish material composition. Kobes insightfully supposes, for example, that the neural correlates of psychological causes might be *causally relevant* to the neural correlates of a psychological effect of that cause, without being a neural *cause* of the neural correlate of the psychological effect. In such a case, neural causation would not line up with psychological causation in the way that would be required by compositional materialism.

Again Kobes holds that the facts are empirically unknown. Our actual neural and psychological theories do not provide us with relevant correlations or with parallel causal structures—neural and psychological. Perhaps the situation is simply a product of our ignorance. Kobes's point is that nothing that we now know makes the situation postulated by compositional materialism an explanatorily illuminating or distinctively likely situation.

Kobes's suggestions are insightful, intriguing, and provocative. I do not see my way to the bottom of these issues. I will, however, hazard some tentative comments.

One doubt starts very far back. Kobes assumes that the archangel can compute biological facts from a starting point of 'only the complete facts about fundamental physical objects, events, fields, laws, and causes over all of space and time'. The archangel can identify 'chemical and physiological kinds and their instances by their physical components, distinctively arranged'.

In contrast to Kobes, I doubt that it is possible to identify the material components of biological kinds and compute the behavior of the components that causally correlates with the behavior of instances of biological kinds, relying entirely on entities' *distinctive arrangement and behavior as described by physics alone*. If, for example, the physical components of biological species or the physical mechanisms of phenotypic expression of genotypes could be identified purely by distinctive physical arrangement and physics-described behavior of physical components, then biological kinds and laws would be reducible to physical kinds and laws much more simply than they seem to be.[5] I think that to identify the physico-chemical components of biological kinds and compute the physico-chemical behavior of the components of biological kinds, one must make use of biological principles.

I am not certain that these points are correct. The issues are very abstract, and Kobes grants the archangel large powers. What seems to me hard to believe is that biological kinds in general coincide with independently identifiable, 'distinctive' patterns of arrangement and movement of physical particles. I find it doubtful that generalizations of physics that apply specifically to the physical components of distinctively biological transactions, can be derived from independent physical observations and independently identifiable principles of physics that govern 'distinctive' movement of those components. Biological kinds are individuated by theoretical notions in biology. I doubt that their components can be picked out as 'discrete' or salient patterns among entities identified in physics.

Like specifications of all kinds and principles, specifications of biological kinds and principles are inevitably intensional. They are not reducible to specifications of finite groupings of particles, for example. I doubt that an effective procedure can compute the component physical behavior of instances of the kinds governed by those principles—even relative to an intensional base of physics-kinds and physics-principles. I think that the relevantly distinct component kinds and movements are distinctive only in light of biological categorization and explanation, not—or at least not in general—independently. The supervenience, and even the compositional dependence, of the subject matter of biology on the subject matter of physics, do not entail that one can *compute* the physical behavior distinctively associated with biological entities from the behavior of entities described in physics together with principles of physics.

Let us suppose that these points are correct. Then it would *not* be a sufficient, condition for rejecting compositional materialism that the archangel could not,

[5] I know of no clear sense in which all of biology is reducible to physics.

prior to using psychological kinds and principles, identify physical components of psychological kinds. It would *not* be a sufficient condition for rejecting compositional materialism that the archangel could not compute distinctive patterns of neural states and events that could be slotted into a psychological network, once representational psychological kinds were individuated in an anti-individualistic manner. For instances of biological kinds are *composed* of entities identified in physics, even though (by hypothesis) the distinctive components of biological kinds cannot be identified as salient and distinctive types through concepts and principles available only in physics.[6]

A second doubt concerns the alleged independence of Kobes's two worries about compositional materialist theories. Kobes takes the question whether the archangel can establish representational–neural correlations that show how neural events 'asymmetrically and synchronically sustain representational events' to be independent of a second question. The second question is whether the correlations illuminate diachronic causal relations involving representational events. I do not see how synchronic correlations could be of any ontological interest or integrity if they did not track diachronic causal relations at the level of representational events. So I think that Kobes's two worries are not independent.

Of course, as Kobes recognizes, the key issue regarding compositional materialism is not merely a generic correlation and coincidence, even between causal sequences of neural events and causal sequences involving propositional representational events. It is whether the correlated neural events explain their effects in a way that illuminates causation at the psychological level 'in the manner of familiar sciences of materially constituted things'. As I have long emphasized, composition is a *specific* theoretical relation. The issue is not only whether the two levels of causal structure can be correlated, but also—as Kobes rightly emphasizes—whether they can be correlated *in a manner familiar from sciences that make use of causal aspects of material components to illumine causal aspects of higher level composed kinds.*

Are there the relevant event-by-event, structure-preserving correlations? Do they ground explanations that use the causal powers of the candidate neural and chemical components to explain the causal behavior of the psychological propositional events in a compositional manner?

Many philosophers will be inclined to ask, dismissively, how *could* answers to these questions not be affirmative. I believe that such philosophers allow ideology to turn them dogmatic about a very complex and not very deeply explored empirical matter.

Psycho-physics gives us no such correlations. There are promising correlations in pre-perceptual aspects of the perceptual system. For example, firings of banks of neurons in the retina correlate with registered information regarding the spatial distribution of proximal stimulation from light arrays. This registered

[6] Ned Block noted that the same point can be made about computations between different levels within physics.

information forms the first input into the psychological mechanisms that yield visual perception. There are even some nice correlations between aspects of genuine visual *perceptual* representation (such as edge representations) and firings of lines of neurons. Moreover, there are correlations between the timing of neural occurrences in areas of the brain and the timing of certain stages in the formation of visual perceptions. There is certainly the beginning of a massive scientific effort to correlate processes in perceptual systems with neural processes. Correlation and perhaps even composition present a natural paradigm for research in these areas. Still, correlations are currently quite generic, except at the periphery of psychological processes. More importantly, correlation is far from explaining all psychological causation in the compositional manner of the material sciences.

With regard to *propositional* psychological events, there is not even a serious beginning in establishing the correlations needed to support compositional materialism. Propositional thinking is known to have a variable correlation with brain processes, both over time in given individuals and across individuals. This situation is compatible with compositional materialism. But if there is little stability or uniformity in the types of physical events that underlie types of propositional thoughts, there will be limited scope for explanations of patterns of psychological causation in terms of (putatively composing) patterns of neural causation.

In the natural sciences, relative stability and uniformity in relations between kinds of components and kinds of things that they compose facilitates part–whole causal explanation. Of course, science can appeal to many levels of kind specifications that might be used to carry out compositional explanations of psychological states. And we are in the earliest stages of understanding the brain, not to say its relation to psychology. So it would be premature to take the lack of established correlations between neural events and propositional attitude events to show that compositional materialism is false. It seems possible, however, that the plasticity of the brain and the genetic and developmental differences among individuals may prevent compositional forms of explanation from being viable for psychological science. More crucially, neural occurrences may not match up well with propositional attitude occurrences, *even in individual cases*. There might turn out to be no clear correlations, even in individuals at specific times, between particular propositional occurrences (among the propositional occurrences at any given time) and particular neural occurrences (among the neural occurrences at the same time) that could provide the beginning of a compositional materialist account. If science does not provide correlations, there will remain no reason to believe in compositional materialism.

Kobes delineates one way in which a failure of correlation could emerge. As noted earlier, he holds that psychology could provide explanations that feature psychological events as causes of a given physical or psychological event, and could cite further psychological states and events as part of the causal *enabling* conditions for the transaction. At the same time, neural science could cite a complex of neural and chemical causes of the physical event, and a background of causal *enabling* neural and chemical events. There could be a general correlation between causally *relevant* events at the two levels, where causally relevant events include both causing and enabling events. But there could turn out to be no illuminating correlation of any subset of the *causing* neural events with the *causing* psychological events. Yet each theory, psychological and neuro-chemical, could provide a true and illuminating account of why certain effects, including physically specified effects, occur. It seems to Kobes, and to me, an open question whether the two types of causal explanation will line up so that causation at neural/chemical levels is correlated with causation at the psychological level, even in given contexts, in the way required by compositional materialism.

The burden on compositional materialism is heavy. It must correlate neural causes and their effects with psychological causes and their effects. And it must illuminate *psychological* causation, of both physical and psychological effects, in ways familiar from the material sciences. I will not try to formulate a precise assumption about physical causation entailed by compositional materialism. However, causation at the level of wholes must be a physical composite of causation at the level of material parts, for some natural division of the material parts. Of course, causation at the level of the parts can capitalize on physical *relations* among the parts. For the psychological causing event to be composed materially, *psychological* causation must depend on the causation of the material parts in one of the ways familiar from causation in the natural (material compositional) sciences. To *know* that such causation occurs, we must have explanations that take psychological causation to operate in such ways.

For example, can one explain the psychological causation of an occurrent thought that is the conclusion and causal effect of a piece of reasoning in such a way that the inference to that thought, including the inference's causally relevant *rational* aspects, are illuminated by the composite causation of the components of an antecedent chemical or neural event? As I elaborate shortly, it is hard to see how the rational aspects of psychological causation can be illuminatingly explained as a material composite of the causal operations of putative neural or chemical components of the inferential process, even taking into account the physical relations among those components in their causal operation.

Some philosophers claim that appeal to a language of thought is assumed in the psychology of propositional attitudes.[7] The idea is that there is a level of psychological activity that is 'purely syntactical'. Representational content is taken to be a further matter attributed at a different stage of explanation. It is frequently also assumed that the tokens of the syntactical, 'linguistic' items are brain events. (Here one often hears intoned the portentous but utterly misleading slogan that the brain is a syntactic engine.) It is concluded that a language of

[7] The picture that I will criticize is substantially that of Jerry Fodor *The Language of Thought* (Cambridge, Mass.: Harvard University Press, 1979).

thought illuminates how occurrences of thought—inner 'linguistic' episodes—could be composed of brain processes. The language of thought picture is sometimes said to indicate that compositional materialism is already in place in psycho-physical science.

Such reasoning skates too fast at each turn. In the first place, except perhaps in psycho-linguistics, there is no autonomous account of syntactical processing in cognitive psychology. Theories of perception and perceptual belief, theories of natural inference and practical reasoning, do not attribute a syntax *except* as a structure *in* the representational contents of psychological states. There is no purely syntactical level of explanation in most of cognitive psychology. Even in the syntactical part of psycho-linguistics, syntax appears to be an abstraction from—and the structure of a capacity embedded in—capacities to understand meaningful sentences.

Thus the view that psychology contains a theory of an autonomous 'purely syntactical' level of processing—one that operates independently of a representational capacity—is misleading even as applied to syntactical aspects of psycho-linguistics. It is without *any* solid grounding as applied to perceptual or propositional-attitude psychology, indeed the whole representational part of psychology. The idea that all of propositional thinking is, at some level, a processing of syntactical symbols that in themselves are neutral as regards representational content has no scientific basis. Propositional psychology is about thought. Thought has a structure. One can abstract that structure and study its properties. Some thinking—for example, certain deductive proofs—but probably not very much thinking, hinges purely on that structure. Even then, the structural elements are not content-neutral. In fact, the structural elements have representational content that is relevant to rationality-based explanations. The category of logical constant or predicate, for example, gets its content from roles in reasoning with representational contents. Even in purely deductive thought, the logical constants have representational content.[8] The structure has no psychological status apart from its association with representational thought.

In the second place, since syntactical elements in psychology are associated with attributions of contentful representational states, they have no specific association, in current psychological theory, with brain states or brain events. Correlating brain states with syntactical states is just as much an open empirical problem as correlating brain states with representational states. In fact, these are different specifications of substantially the same problem, insofar as we have any clarity about what the syntactical states are (to wit, aspects of representational states). So appeal to a language of thought gets us *no* closer to material correlates that are supposed to constitute the compositional material of the syntactical states, or the representational states. It is just part of the materialist ideology,

[8] Hilbertean proof theory is an abstraction from reasoning that uses the representational contents of logical constants.

not part of scientific theory, that 'syntactical' states or events are instantiated by specific neural states or events.

Thus in our present state of knowledge, the language of thought hypothesis cannot make compositional materialism more plausible. The hypothesis is not even a plausible gloss on psychology unless standard presentations of it are severely qualified.

One important difference between representational psychology, on one hand, and neuro-physiology and chemistry, on the other, is brought out by anti-individualism. Representational kinds are partly individuated through patterns of causal relations that they bear to entities in an environment that lies well beyond them. Neural and chemical kinds are not individuated in ways that rely on these long-range patterns.

A further difference between representational psychology, particularly the psychology of proportional attitudes, and the neural bio-chemical sciences is that the part–whole relation of material composition plays no evident role in psychological theorizing. Propositional psychological structures are compositional in a different way. They are broadly rational structures. They include predication and structures of deductive inference. These structures are notoriously not assimilable to physical structures. The bonds between elements in rational structures are not physical but propositional. Since the seventeenth century, it has been evident—in fact, virtually axiomatic in the natural sciences—that the physical world, as described by the natural sciences, does not have the form of content of a text. It is not made up of rational structures at all. It is not a direct expression of reason. The brute nature of physical relations—the difference, for example, as Kant put it, between resisting force and logical negation,[9] or between property inherence and predication—should, I think encourage puzzlement about how physical structures *per se* could compose instances of propositional attitudes.

There have been ways of trying to blunt this concern, insofar as it is ever raised. One might, for example, say that rational structures are relevant only to *norms* governing psychological transactions; material composition is relevant to the ontology and the causal transactions among psychological events that may or may not fulfill the norms. This line is not plausible. Psychological explanation takes the propositional structure of propositional attitudes to be fundamental to *what they are*. Psychological explanation gives the rational, propositional aspects of psychological states a causal role. It is not that we are infallibly guided by reason. Obviously we often fall short. Rather, both common sense and scientific explanation indicate that rational, propositional aspects of psychological states and events figure both in the ontological individuation and in the causal powers of psychological occurrences. Indeed, I think that any psychological science that

[9] Immanuel Kant, 'Attempt to Introduce the Concept of Negative Magnitudes into Philosophy' (1763), in *Kants Werke*, A.K. II (Berlin: Walter de Gruyter, 1968), 175–176.

did not acknowledge a role for rational elements in psychological identity and psychological causation could be reasonably counted inadequate.[10]

Let us reflect on examples involving the causal powers of propositional attitudes with rational structure. The thought occurrence that is the conclusion of an individual's deductive inference is caused by transitions involving premise thoughts that incorporate competence with the logical structures of the premises. The rational-structural aspects of the premise attitudes figure causally in drawing the conclusion. An individual's predicating a concept of a perceived particular, in a perceptual judgment, is part of the cause of the individual's practical reasoning about how to deal with the particular. These rational aspects of psychological causation—deductive inference and predication—are not construed as summations of material forces.

Similar examples support a constitutive role for rational structure in the ontology of propositional attitudes. What it is to be an occurrent thinking *since all humans are mortal, if that human is Socrates, then Socrates is mortal* depends constitutively on the propositional structure of the thought. Any event occurrence that lacked that structure would not be the same event.

I have two concerns about compositional materialism, beyond concern about lack of evidence for it. First, it is hard to see how material compositional structures could ground causation by propositional psychological states or events. Second, it is hard to see how material compositional structures are consistent with the nature of propositional psychological states or events. I shall elaborate that causal point first.

Rational, propositional structures are fundamental aspects of psychological causation by propositional states and events.[11] Rational, propositional structures do not appear to be identifiable with structures of material composition. Everything we know about causation by material composites indicates that such causation is not rationally structured. In fact, as noted, this point is a virtual axiom in the natural sciences (physics, chemistry, biology, including neurophysiology). Moreover, scientific reduction of rational, propositional structures to material compositional structures has little prospect of success. For such a reduction to succeed, natural science would have to show that a *constitutive* aspect of causation by propositional psychological states and events—their rational structure—is fully *explainable* in terms of the causal properties of brute material composition. There is not the slightest reason to think that such a reduction can succeed. These points derive as much from fundamental commitments of the natural sciences as from fundamental commitments of psychology.

Earlier, I did not attempt a precise formulation of the sort of physical causation produced by material composites. But any such causation must be a physical

[10] For a discussion of this point, see the end of 'Mind–Body Causation and Explanatory Practice', in *Foundations of Mind*.

[11] The psychological cause could have a psychological effect. Or it could have a physical effect, for example a physical activity motivated by propositional reasoning.

composite of causation by the material parts (on some natural division of those parts), where the parts operate through their physical relations to one another. Since nature is not a text, such causation does not have rational or propositional structure. And as noted, rational, propositional structures do not appear to be explainable in terms of brute material compositional structures. Thus causation that depends on rational, propositional structure appears not to be identifiable with or reducible to causation by material composites, or by material components of material composites operating through their physical relations to one another. So material composites, such as chemical and neural composites, appear not to exhaustively constitute causation by rational, propositional states or events. Causation by states or events that are material composites is purely causation of material composites. So rational, propositional states or events appear not to be material composites.

More simply: reason is a constitutive structural feature of causation by propositional psychological states and events. According to the natural sciences, reason is not a structural feature of material composites. The causation by material parts of material composites, operating in their physical relations to one another, must suffice to alone compose causation by material composites. It is hard to see how the causal structure of material components could alone compose the causal powers and causal structure of propositional states and events. So it appears that rational, propositional, psychological causation is not the causation of a material composite. Propositional psychological states and events are material composites only if their causation is purely that of a material composite. So it appears that propositional psychological states and events are not material composites.

The second concern about compositional materialism is similar, but does not feature causation. Here it is: the physical structure of material composites consists in physical bonds among the parts. According to modern natural science, there is no place in the physical structure of material composites for rational, propositional bonds. The structure of propositional psychological states and events constitutively includes propositional, rational structure. So propositional states and events are not material composites.[12]

Both arguments depend on a structural contrast between material composites and propositional psychological states and events. Psychological causation hinges often, but then constitutively, on rational propositional structure. And

[12] There is a distant kinship between these arguments and the argument for the simplicity of the soul that Kant criticizes in the second Paralogism. Compare Immanuel Kant, *Critique of Pure Reason*, A351–361. The argument that Kant criticizes does appeal to the propositional unity of thought, in effect predication, which is one of the rational, propositional structures that my arguments appeal to. But the argument Kant criticizes aims to establish the simplicity of thinkers, and by (alleged) extension, their exemption from dissolution (A356). My arguments are not for simplicity, only against *material* compositeness. And my arguments center not on the thinker but on psychological states and events. They also center on a basic feature of modern natural science—that natural physical relations do not include rational, propositional structures. I believe that my arguments are not subject to any of the objections that Kant raises.

the nature of propositional psychological states and events constitutively involves rational, propositional structure. Causation associated with material composites is, to all appearances *constitutively*, *not* causation that involves rational, propositional structure. And it is a principle of physical nature that physical structures of material composites are constitutively *not* rational, propositional structures. So it appears that psychological causation by propositional states and events is constitutively not causation by material composites. And it appears that propositional psychological states and events are constitutively not material composites.[13]

These arguments are, of course, very abstract.[14] Perhaps developments in empirical science will show how to overcome them. But the developments would have to be fundamental. At present, the arguments seem to me to support provisional rejection of compositional materialism—independently of doubt that derives from the absence of the correlations and of the explanations of psychological causation that compositional materialism requires.

What are we to say about demands for a *mechanism* for psychological causation? Such demands are often just question-begging insistence on physical mechanism, specified by the natural sciences. Postulating such mechanisms is warranted if and only if they enhance empirical explanation. We want to understand relations between respective explanations and subject matters in representational psychology and the biological and chemical sciences. It does no good to insist that empirical explanation in psychology conform to explanations in very different sciences. There may be no deeper way to explain *how* psychological events cause psychological or physical events than to specify the law-like patterns by which they do so, and the neural or chemical patterns some of which seem necessary to those psychological patterns' occurring.

The most popular way of indicating how psychological causation could be composed of physical causation has been to appeal to computers. Computers are claimed to be physical machines that 'instantiate' reason. This claim does not solve the problem. Computers' processes express reason only insofar as we give them programs and interpret their processes in accord with those programs. There is nothing in the computer's physical processes *per se* that makes the computer reasonable, or explains whatever rational causation might occur in computers.

[13] These arguments differ in two respects from the arguments, discussed early in this chapter, against the materialist token identity thesis. First, the earlier arguments depend essentially on anti-individualism and on the view that the identity of neural/chemical events in the brain does not depend on the sorts of long-range patterns of relations to the distal environment that the identities of representational states do. The arguments against compositional materialism do not depend on anti-individualism or on denying that the identities or neural/chemical events in the brain depend on the same long-range causal patterns that the identities of representational states do. Second, the earlier arguments do not center on the causal or constitutive roles of rational, propositional structures, whereas the later arguments do.

[14] It should be obvious that both of the arguments could be modified into additional arguments against materialist token *identity* theories. Such arguments would supplement the two arguments against such theories that I discussed near the beginning of this chapter.

Actual computers do not reason autonomously. They go through a sequence of states that were fashioned to express and amplify our rational states. In them, we simply mimic physical symbolization of our own reasoning, and amplify our reasoning, by relying on the computers' processing of those symbols.

I leave open whether more sophisticated robots could reason autonomously. It would, however, not follow from the assumption that a robot reasons autonomously that rational causation in the robot is explained by material composition, even on the further assumption that there are known correlations between its physical states and its propositional states. One still needs to explain rational causation in terms of a composition of physical causal relations. That is what we have some reason to believe cannot be done. Supposing that a robot could reason autonomously does not entail supposing that its reasoning events are *composed* of material processes. The robot's reasoning events would depend for their existence on material processes, just as ours appear to. Its supposed rational propositional events would appear not to be materially composed any more than our appear to be. Only a cartoonish view about what rejecting material composition amounts to—a view that would see the rejection as postulating immaterial soul stuff in the robot—would have to be embarrassed by a reasoning robot.[15]

The language of thought hypothesis is often conjoined with the computer analogy to try to support materialism. The idea is that reasoning in computers hinges causally on the shape, size, and physical configuration of symbols. In addition to the difficulties with its hypothesis that I catalogued earlier, there is a further one. The argument I gave regarding rational causation applies to any attempt to use the language of thought hypothesis to support compositional materialism. Psychological causation depends on the rational, hence representational, aspects of psychological states. Insofar as the language of thought hypothesis tries to account for psychological causation in terms of the shapes, sizes, and configurations of symbols, it fails to connect with the type of causation that is referred to by common sense and psychological explanation in science. Psychological causation that hinges on the rational, propositional properties of psychological states is not independent of representational content in the way that causal processes that hinge on the physical properties of symbols are. As I have indicated, there is no scientific basis for an explanation of psychological causation in terms of an autonomous syntax realized in neural or chemical entities. Even if there were, the account of causation provided by such explanation would fail to explain psychological causation that hinges on the rational aspects of the form and content of psychological events. The language of thought hypothesis is not only ungrounded in scientific explanation. It is irrelevant to

[15] I am abstracting from issues about consciousness. I think that robots that have the *sorts* of material bases that they commonly have would not and could not be conscious. I am doubtful about counting such beings autonomous reasoners, without serious qualifications. I claim that even if one *lays aside* issues about consciousness, supposition of autonomously reasoning robots does not threaten objections to compositional materialism about propositional reasoning.

explaining the central feature of psychological causation by propositional psychological states and events. Kobes cites my interest in Descartes' apparent view that psychological states and events are distinctive in that their being or nature is grounded in consciousness, activity, power, and point of view, not in substance, soul stuff, or composition. Let us reflect on that list.

With Kobes, I have bracketed consciousness. I think that consciousness is constitutively associated with our psychological being. Constitutive relations between consciousness and specific types of psychological states are complex. I will not take on these issues here.

Not all representational psychological states or events, even propositional ones are active. But all propositional states and events are *constitutively associated* with activity. For all propositional psychological states and events are constitutively associated with inference—the exercise of the capacity to make use of propositional structure. Inference is activity.

Propositional psychological states and events seem to be constitutively associated with at least generic causal vulnerabilities and powers, including some active powers. Those states and events seem to be constitutively associated with points of view that mark them—representational contents. Unlike material entities, propositional psychological states and events are not identifiable through their material substance, stuffings, or compositions. This conception of the natures of propositional psychological states and events accords with what we now know about them. I see no good reason to believe that such psychological states and events are materially composed. I think it reasonable to think that they are not.

Kobes highlights elements in my dualism that are at odds with traditional dualisms. First, it is not a substance dualism, in the early-modern sense of 'substance'. I have no reason to believe, alas, that psychological events can exist *self-sufficiently*, or *independently* of physical material. Representational psychological events seem to depend for their existence on physical events and material that 'sustain' them. A lot of physical states do not depend on psychological states, but all psychological states seem to depend on physical states. (It may be that the particular underlying physical states could not be what they are if they did not sustain, underlie, or otherwise associate with psychological states.) Second, a strengthening of this first point: although I am not committed to a belief in global supervenience of the psychological on the physical, I incline toward such a belief, pending better reasons against it than any that I know of. Third, the physical world is without gaps in physical causation, modulo quantum indeterminacy. It is *approximately* causally deterministic. Most or all of these points are uncongenial with most early-modern forms of dualism. My dualism is not only conjectural; it is modest and undramatic. It does not encourage belief that our souls can soar out of the material world.

On the other hand, it is not merely a dualism of concepts. Nor is it merely what is often called a property dualism. It is a dualism of occurrent events as well as states, kinds, and properties. For now, I leave open how to think of psychological *agents*. I am not convinced by Strawsonian claims that one *must* conceive of *every* such agent as having material properties, at least if material properties are not comprised of *relations* of dependence on matter.

Kobes speculates that my modest dualism might be counted an 'extremely weak' form of physicalism. I am inclined to resist this speculation. Psychological states and events depend on the physical. But I see no intellectual substance in counting them physical. They are unlike numbers and logical forms in that they have causal powers and vulnerabilities, and in that they occur in time. They are like numbers and logical forms in seeming to lack material composition, mass, physical force, and physical energy. In fact, propositional psychological states and events are what they are through their having logical forms. None of the primary attributes that we cite in theorizing about them—including logical forms—are cited as physical structures in the natural sciences. I see no clear sense in which propositional psychological states or events are physical. Supervenience is consistent with dualism. To be materialist (or physicalist), a view must claim that psychological entities are themselves material (or physical), not merely that they vary with or depend on material (or physical) entities.

So I do not think that counting my view a weak form of physicalism accounts for the diffidence in my position in recommending—in the passage that Kobes quotes—that philosophers be more open and relaxed about whether some form of materialism (or physicalism) is true.[16] I am no type of physicalist or materialist. But I recommend—not sloth or indifference, but—disinterested, open reflection on the issue.

On relaxation: I think that the question whether something like my form of dualism or some non-reductive form of materialism is true is not momentous. At least, it is not momentous for traditional reasons for caring intensely about whether dualism is true. Traditional issues of life after death, the existence of freedom and moral responsibility, and the explanatory powers of natural science within its own domain, do not seem to hinge on the answer to the question.

On openness: I think that we do not know enough about the relation between psychology and the natural sciences to take hard lines for or against materialism. I believe that at this point a modest dualism is clearly more reasonable than materialism. But I think that materialistically motivated empirical research is heuristically tenable. The key to any view in this area is openness to empirical exploration and philosophical reflection. Both positions should be developed in a dialectically open spirit.

A modest dualism, however, cleaves more closely to what we know. It does not make warrantless claims. It is more reasonable than compositional

[16] From 'Mind–Body Causation and Explanatory Practice', in Burge (*Foundations of Mind*, 360), 'It seems to me that philosophers should be more relaxed about whether or not some form of materialism is true. I think it a thoroughly open—and not very momentous—question whether there is any point in insisting that mental events are, in any clear sense, physical . . . What matters is that our mentalistic explanations work and that they do not conflict with out physcialistic explanations. But it serves no purpose to over-dramatize the conflict between different ontological approaches.'

materialism, or other sorts of materialism of physicalism, partly through abnegation, and partly through appreciation of the deep differences between rational structures and physical structures. Psychological events have not been shown to have any attributes that are distincitive of physical events. Their primary attributes are not those cited in the natural sciences.

I confess to a more psychological influence on my dualism. I was educated in philosophy in a climate in which materialism had become smug ideological dogma. It dripped with the more supercilious aspects of blind religious orthodoxy. It was not just that herd instinct in philosophy was itself a danger signal. I thought that there was an implicit hypocrisy in the climate. What concerned me was that many philosophers saw any doubt about materialism as *ipso facto* irrational and unscientific. Many still do. Many philosophers exuded a certainty that was out of line with the speculativeness and lack of force in the grounds supporting their positions. Many still do. Such philosophers assume the mantle of science while contravening its letter and spirit. I hope that new generations of philosophers will do better.

20 *Epistemic Warrant: Humans and Computers**

I begin with some remarks about the epistemology of human cognition. Then I compare human cognition, taken by itself, with human cognition enhanced by relying on computers. Finally, I discuss some uses of computers in pure mathematics and in empirical science.

Knowing a proposition requires that the proposition be true, that one believe it, and that one's belief be epistemically warranted. These three conditions are not jointly sufficient for knowledge. But they are necessary. I will focus on the third condition—that one's belief be epistemically warranted.

Being epistemically warranted in having a belief is having the belief in a way that is good for having *true* beliefs, given limitations on one's information and cognitive capacities. One can be warranted but mistaken. But if one is warranted, one's belief is held through a natural competence that is epistemically good—conducive to the belief's being true and, usually, to the belief's constituting knowledge. To constitute warrant, this "good" route to truth must meet a certain minimum standard. It must be *reliable* in yielding true beliefs, in normal circumstances.

These remarks do not constitute a definition, analysis, or reduction of warrant. They are intended as orientation. They hide complexities. There are hard questions about limitations on information and capacities, about reliability, and about the normal circumstances in which reliability is required. I will not discuss the complexities.[1] I assume that we have an intuitive grip on the notion of epistemic warrant. I want to develop certain aspects of that notion.

Being warranted in having a belief is in the same ballpark as being justified. I count warrant the genus and justification a primary subspecies. I call the other main subspecies '*entitlement*'. What differentiates justification and entitlement? Being *justified* is having a reason that figures in an appropriate way in arriving at or in sustaining the relevant psychological state. (For *epistemic* justification, the primary relevant psychological state is belief.) *Having* a reason requires that the

* This essay was written for an interdisciplinary conference on applications of computers in empirical science, in Paris 2011.

[1] See my, 'Perceptual Entitlement', *Philosophy and Phenomenological Research* 67 (2003), 503–548.

reason is in one's psychology, or could easily be brought into one's psychology by simple, easy, obvious inferential transitions from what is already there.[2] So having a reason requires having the capacity to think the reason; and it requires being able to connect the reason with what it is a reason for. One can have a reason without the reason's being operative. A reason that one has for an attitude is *operative* if and only if the reason figures in a cognitively relevant causal way in forming or sustaining the attitude. Thus being epistemically warranted in having a belief, in the sense of being justified, requires having in one's psychology a reason that is operative.

I assume that reasons are propositional contents together with modes (like belief or intention). Reasons have the same structures as sentences, even though they need not be linguistic or even symbolic. Some think of perceptions or pictorial images, which I assume are not propositional, as reasons. I do not. Perceptions and images can figure in supporting an attitude, but they are not reasons for it. Reasons are explanatory as well as justificatory. Reasons are answers to potential 'why' questions. In effect, they provide a kind of explanation of the credibility of what they are reasons for.[3] Explanations and answers to questions are propositional. Perceptions and pictures, except as *glossed* by propositions, cannot explain—cannot answer a 'why' question or complete a 'because' clause. They are not themselves reasons. The point is a conceptual/grammatical one. Perceptions and pictures can certainly figure in entitlement and thus in support of belief.

Being *entitled* to a belief is being warranted in holding it, without depending for being warranted on having an operative reason for it. Entitlement is warrant without reason.

[2] This formulation allows modular justifications—justifications that occur in an individual's psychology, but that cannot, even in principle, be brought to consciousness by the individual. On this formulation, being justified hinges not on availability to consciousness, but on having a rational structure in one's psychology that functions to support and explain the relevant attitude. We might label being justified in this sense '*justified*$_{p}$', where the subscript indicates that an individual is justified in holding a propositional attitude if and only there is an operative (undefeated) reason for the attitude in the individual's *psychology*—whether or not the reason is in-principle accessible to the individual. A narrower conception of being justified would require in-principle accessibility to consciousness. We might label being justified in this sense '*justified*$_{cp}$', where the subscript indicates that an individiual is justified in holding a propositional attitude if and only if there is an operative (undefeated) reason for the attitude in the individual's *conscious psychology*, or at least that part of the psychology that could be brought to consciousness through introspection, prompting, or the like. One might use this distinction to produce correspondingly different notions of entitlement. I will return to this distinction later in the essay.

I take reasons to be the abstract representational contents, marked with a certain mode, of psychological states. Strictly, the abstract reasons do not cause anything. Only the psychological states do. Strictly, an operative reason is one that is the mode-content of a psychological state or occurrence that has that content and that figures causally in forming or sustaining the attitude whose mode-content the reason is a reason for.

[3] The explanation need not be in meta-representational terms. It is not essentially *about* belief or truth. It is fundamentally at the same level as the belief: *p* because *r*, where *p* is the content of the belief and *r* is the reason.

An individual can have both a justification and an entitlement for the same belief. Epistemic warrants derive from meeting standards for having epistemically good propositional attitudes, or for making epistemically good transitions among such attitudes. One can simultaneously meet different standards.

In the history of philosophy, epistemology featured justification, partly because of a focus on science—which aspires to justification. I focus first on entitlement.

A simple fact about human cognition is that we often have knowledge for which we lack a justification—a reason. Often an individual knows a proposition, and hence is warranted in believing it. Yet the individual may be unable to think a reason for the proposition. Or whatever reason that the individual does have, or could come to have, may not be operative in the individual's believing the proposition. The belief is not caused or sustained by a reason. Yet the individual is warranted in the belief, and the belief may even constitute knowledge.

A vivid example is a very young child's perceptual belief that red sphere is there. The child may lack the concepts necessary to having a reason for the belief. The simplest reason might be: I am having a perception as of a red sphere there. There are developmental reasons to believe that children have beliefs about colors, shapes, bodies, and locations before they can think about psychological states like *perceiving*. Even if they have psychological concepts (like perceiving) innately, their having meta-beliefs about perceivings is certainly not the primary warranting basis for perceptual beliefs about spheres. Children first form perceptual beliefs about spheres without any *thought* about perceivings. They form beliefs directly from their perceptions. They are warranted in doing so. Operative *reasons* come later.

An analogous point applies for perceptual beliefs in mature adults, including scientists. They may be able to cite their perceptions in rationalizing their perceptual beliefs. They may develop more articulated reasons. But they are first warranted in their perceptual beliefs because the beliefs were formed via a good, reliable, truth-conducive psychological competence, not because the believers can provide a reason that explains the beliefs' belief-worthiness. Their justifications are posterior to their entitlements. The justifications are not needed for the entitlements to hold.

A similar point applies to *transitions* in a deductive inference. To be warranted in believing the conclusion of an inference, one must be warranted in believing the premises and in relying on the inferential transitions. Children and some adults make warranted propositional inferences even when they are not in a position to think the inference rules that help explain their inferential transitions. One can make an inference that relies on a transition that is correctly explained as an instance of *modus ponens*. One can be entitled to rely on the transition in coming to a warranted conclusion, without *being able to* think the rule as a justification for a transition. Inference rules are meta-representational schematic generalizations about propositional contents. They hinge on isolating and representing the logical constants (here the conditional) on which the transition depends. Being competent to make a *modus ponens* inference does not require a capacity to think schematic generalizations or to think about propositional contents or logical constants—much less a capacity to think the rule and use it to justify a transition. Entitlement to rely on such transitions does not require *justifications* that cite the rules that codify and explain the transitions. They do not even require a *capacity* to cite the premises *as a reason* for the conclusion.

So at the most elementary level of empirical belief and of inferential transitions, entitlement (warrant without reason) precedes and is independent of justification (warrant with reason). Warrant without reason is warrant without full understanding. One can do well cognitively without being able to explain and rationalize what one is doing.

Many pieces of good reasoning mix entitlement and justification. Perceptual beliefs that one is entitled to can form premises for inferences. As premises, they can be reasons for an inference's conclusion. As just noted, one can be entitled to make inferential transitions without thinking the inference rules as justification for the transitions, even unconsciously. Thinking correct rules is not easy, even for mature thinkers. Even when one can think the rules, this ability is often not operative in a transition. Except for when an individual deliberately carries out an explicit proof, individuals who do think an inference rule as justification usually provide a further warrant, after the fact. Warrant for the conclusion of an inference is a combination of warrant for the premises and warrant for the transition-inferences. Even where the premises constitute reasons, justifications, for the conclusion, the reasons are often mediated by inferential transitions which the inferrer is entitled to, but not justified in. The transitions are warranted, but not justified by operative representation of the rule (consciously or unconsciously). So the full warrant for the conclusion is a mix of justification and entitlement.

In all inferences, a warrant for believing a conclusion depends not only on the reason-giving powers of the premises, but also on warrants for relying on the transitions. If the warrant for relying on a transition is an entitlement, the warrant for believing the conclusion will be a mix of justification (from antecedent steps for the propositional steps, including the conclusion) and entitlement (to the inferential transitions). Entitlement resides in an actual competence to make the relevant deductive transitions, not in an ability to understand and represent the rule governing the competence.[4]

When one is warranted in believing a conclusion because of an inference to it from premises, one commonly has some justification for the conclusion. It is natural to hold that the premises justify—provide reason for—the conclusion. I think that it is natural and correct to hold this even when one is just entitled to the transition steps, as long as the premises, together with the rules governing the

[4] Of course, although the premises in empirical inferences are often reasons for the conclusions, the premises themselves may be warranted by entitlement, not reason. As I have indicated, the basic warrants for perceptual beliefs are entitlements, not justifications.

transition steps, constitute a rationalizing explanation of the belief- worthiness of the conclusion.

I conjecture that, in *deductive* inference, when one is warranted in accepting a conclusion by virtue of the inference, the essential premises of the inference are always reasons for the conclusion. Often premises in good inductive inferences are reasons—justifications—for the conclusion. The premises constitute reasons, justifications, for the conclusion, by way of the inference, if but also only if those premises, when combined with the rule of inference (whether or not the inferrer understands and is justified in relying on the rule), yield some rationalizing explanation of the belief-worthiness of the conclusion. But as I will soon explain, this condition is probably not always met, even when one is warranted through the inference in believing the conclusion.

I have mostly concentrated on deductive inference. We do not understand induction very well. We do not have a theory of induction comparable to deductive logical theory. Many inductions—including what we call 'inferences to the best explanation' (abductions)—are complex, hard to articulate, and partly unconscious. There have been many attempts to codify induction into a logic. The best of these is probably Bayesian subjective probability theory. But what we call 'inductive inference' is probably a motley of significantly different kinds of transitions.

Some of what we call 'inductive inference' may not be genuine, reason-giving inference, even when it is warranted. Of course, any transition from one or more propositional attitudes to another one, according to some warranting transition pattern, is propositional inference. If inference is to provide a reason for a conclusion, the premises, together with the inferential transition, must constitute some sort of explanation of the acceptability of the conclusion. They must provide some answer to a 'why' question. They must be of the form: *p* because *q*, *r*, and *s*. The conclusion may be taken as only likely, relative to the inference from the premises. Or it may be taken as more reasonable than not, or as enhanced in credibility, relative to the inference. But if the inference is reason-giving for the inferrer, the premises and the transition rules must combine to go some way toward providing an (object-level) explanation for the individual inferrer of why the conclusion is credible. That is a minimum necessary condition on the premises' functioning as reasons for the inferrer.

As noted, even with respect to deductive inference, one need not be able to think the inferential rule that 'because' stands in for. One can be entitled to a transition without being able to explain the connection. One need not have a complete or fully satisfying explanation of the conclusion's being made credible by the premises, even at a non-meta-representational-, object-level of thinking. But if the premises are to provide reasons for the conclusion, they must provide some sort of rationalizing, explanatory support for the belief-worthiness, through the nature of the inferential transition, of the conclusion for the individual.

Spelling out this requirement is difficult. I do not know how to do so. Perhaps an example will help. Suppose that a human or higher animal has a perceptual belief. Suppose that some aspect of the belief triggers, by a natural psychological competence, an inference to a non-perceptual belief, such as a belief that it would be dangerous to move. Suppose that the connection between the truth of the perceptual belief and the truth of the triggered belief is a good, probabilistic one. Suppose, however, that nothing in the individual's experience could explain the connection. Perhaps the individual has no evidence that supports the inference. Perhaps the connection was innate, selected through evolution. It is well known that intuitively *very* unobvious probabilistic connections between properties can be significant and valuable psychological transitions.[5]

Of course, there is a biological/psychological explanation of why the perceptual belief is connected to belief about danger. There is some causal or statistical explanation of the relation in the environment between the perceived property and danger, that grounds the psychological connection. But the perceptual belief that is the premise may not provide the slightest explanation of the belief-worthiness of the conclusion for the individual. The individual and the individual's psychology cannot use the premise to rationalize, explain, or make sense of the belief-worthiness of the conclusion—even unconsciously. For the individual, it constitutes no *reason* for the conclusion.

In an abstract sense, one might claim that the premise *is* a reason for the conclusion. It is a small part of an explanation that could, in principle, be filled in by evolutionary theory or by an ideal epistemology. I think, however, that the premise would not function as a reason in the individual's psychology. The transition rule in the psychology may provide no insight into the connection, even for a theorist who knows the rule. The rule may be as simple as: if perceptual belief has concept F̲ in it, belief about danger is to be formed. Only an explanation of why the rule is in place explains and rationalizes the belief-worthiness of the conclusion, given the premise. Such an explanation would have to show why the environmental property correlates with danger, and why the connection between the properties in the environment grounds the psychological connection.

Not only does the believer not understand a reason connecting premise and conclusion. The contents of the premise states that are in the (possibly unconscious) psychology combined with the rules of inductive transition do not add up to what even a super-psychologist would count as providing the slightest hint of a reason-explanation of the conclusion. As noted, a full account of why the rule came to be in place could show the premise as reason for the conclusion. From a God's eye point of view, one could reasonably say that the premise is a reason for the conclusion. But it is not a reason for the individual. It is not a rationalizing,

[5] W. S. Geisler, 'Visual Perception and the Statistical Properties of Natural Scenes', *Annual Review of Psychology* 59 (2008), 10.1–10.26; J. Burge, C. C. Fowlkes, M. S. Banks, 'Natural Scene Statistics Predict How the Figure-ground Cue of Convexity Affects Human Depth Perception', *Journal of Neuroscience*, 30 (2010), 7269–7280; W.S. Geisler, 'Contributions of Ideal Observer Theory to Vision Research' Vision Research 51 (2011), 771–781; Tyler Burge, *Origins of Objectivity* (Oxford: Oxford University Press, 2010), 359–366.

explanatory premise within the individual's psychology. I see too little in such an inductive inference to count it as reason-giving for the individual.

The premise and inferential transition still entitle the individual to believe the conclusion. For accepting the conclusion and relying on the inferential transitions are certainly as warranted as forming perceptual beliefs. In both cases, one uses a natural psychological competence that in fact constitutes a good, reliable route to getting things right.

What matters here is not whether or not one calls such inferences 'reason-giving'. What matters is to fix ideas on how dumb and non-rationalizing inductive transitions can be, while still providing entitlement to believe their conclusions.

It seems to me that more of our elementary induction may take this form than we philosophers are inclined to think. Patterns of inference are hammered into us by evolution. The principles that they instantiate and that explain them may yield little insight into why the conclusion is rationalized, made sense of, or explained by the premises. So, even knowledgeable reflection on the principle or rule according to which the inference is carried out may not show the premises to be reasons for the conclusion for the individual who carries out the inference. We do not know enough about how we actually carry out inductions to know whether the foregoing is so. But I think that we should be open to the possibility, indeed, I think, likelihood.

For some decades, computers have solved problems that are so complex that human beings lack the time and cognitive power to check the solutions. This situation emerged dramatically with the proof of the Four-Color Theorem by Appel and Haken in 1976.[6] Although any given step can be checked, the length of the proof prevents humans from checking the whole proof. Other computers checked the proof, however; and the theorem is considered proved.

This situation has grown more complex in the succeeding thirty-five years. Other important conjectures have apparently been proved with the help of computers. Checking them has sometimes been harder than checking the proof of the Four-Color Theorem. In 1998 Thomas Hales announced a proof of Kepler's conjecture.[7] The proof combined traditional geometrical analysis and

[6] K. Appel and W. Haken, 'Every Four Color Map is Colorable, Part I: Discharging', *Illinois Journal of Mathematics* 21 (1977), 429–490; 'Every Four Color Map is Colorable, Part II: Reducibility', *Illinois Journal of Mathematics* 21 (1977), 491–567. The theorem states that given any separation of a plane into contiguous regions, producing a figure called a map, no more than four colors are required to color the regions of the map so that no two regions that share a border segment have the same color.

[7] Thomas C. Hales, 'A Proof of the Kepler Conjecture', *Annals of Mathematics, 2nd Series* 162 (2005), 1065–1185; 'Historical Overview of the Kepler Conjecture', *Discrete & Computational Geometry: An International Journal of Mathematics and Computer Science* 36 (2006), 5–20; Thomas C. Hales and Samuel P. Ferguson, 'A Formulation of the Kepler Conjecture', *Discrete & Computational Geometry: An International Journal of Mathematics and Computer Science* 36 (2006), 21–69. The Kepler Conjecture states that the highest density that can be achieved by filling a three-dimensional space with equal-sized spheres is *pi* divided by the square root of 18, or about 74%. This is the density of spheres stacked in a regular pyramid.

computer proof, with extensive descriptions of the computer portions. Several years of trying to check the proof by a committee of referees ended in failure. Again, the proof was too long to be checked by humans. But the computer program was not written with a view to being checked by computers. The referees regarded the proof as 99% *likely* to be correct. Subsequently, Hales launched an attempt to produce a fully formal proof that can be verified by computers. Hales regards it as a multi-year project. It is currently in progress.

Since computers are programmed to run *understood* mathematics the inferential transitions that they model in proofs, taken one by one, certainly give reasons, to humans who understand the mathematics and rely on the computers, for believing later inferential steps. Mathematicians understand steps as *rationalizing* later steps—providing some explanation of why they are belief-worthy. So the inferential transitions provide them with reason for the conclusions of the transitions. The problem, relative to classical conceptions of proof, lies simply in the uncheckably large number of transitions in the more complex proofs.

In earlier work, I gave an account of the epistemic status of relying on computers to complete and check a proof.[8] The account compared gaining knowledge from computers with gaining knowledge from communicating with other people. Computers are not people. However, their outputs have language-like character; and they produce propositional knowledge in a receiver much as human communicators do.

In exchanges with other people, I think that we have a default prima facie entitlement to accept what they assert unless there is reason not to. The entitlement is grounded in the prima facie rationality of a being that makes propositional assertions. Prima facie rationality implies prima facie competence and openness, at least on ordinary topics. As a constitutive point about rationality, rational individuals tend not to make incompetent assertions on such topics, although of course they sometimes do so. And rational beings tend not to lie without special reason. Lacking evidence that one's interlocutor is irrational, and that the subject matter is one on which competence and openness cannot be assumed, and lacking reason to believe that one's interlocutor has a special reason to lie, one can rely on one's interlocutor's prima facie rationality. I believe that recognizing such a default entitlement is the right way to account for childrens' gaining knowledge by being told things that they cannot evaluate, for adults' gaining knowledge by asking for unproblematic information from strangers, and for students' gaining knowledge when they are introduced to a new subject. This default entitlement structures most further warrants that can support or undermine an interlocutor's credibility on given occasions.[9]

[8] Tyler Burge, 'Computer Proof, Apriori Knowledge, and Other Minds', *Philosophical Perspectives* 12 (1998), 1–37; reprinted in this volume, section II. The discussion over the next few paragraphs summarizes some of the points in this article.

[9] See my 'Postscript: Content Preservation', this volume, section II.

In using computers to solve difficult problems, however, this default prima facie entitlement never suffices to yield knowledge, or even warranted belief. For the extreme difficulty of the subject matter grounds the need for supplementary warrant to believe that the computer is specially competent to solve the problems that mathematicians rely on it to solve.

Various types of warrant can support relying on a computer as credible. One division is between those that concern a machine's physical make-up and those that concern the content of its outputs. Knowing that a computer is made of stable materials, that its workings are predictable by laws of physics, that its power source does not die quickly, and that it is physically complex can help support believing that it could solve a difficult problem.

Such physical evidence is inevitably indirect. I think that it is not essential to warranting belief in a computer's outputs. Although the computer's success depends on the stability, reliability, and complexity of its physical operations, we do not *have to* know anything about such operations to have reason to believe that a computer's outputs are credible. Consider relying on a human mathematician or empirical scientist for information. Although the human must have a stable, reliable, and complex brain to solve a hard problem in mathematics or science, one need not know anything about biology or physiology to learn to trust his or her statements. One's warrant can derive entirely from considering the content of the statements.

The second, and primary, category of warrants that support relying on a computer depend on evaluating the content of the computer's outputs. For example, a mathematician can check shorter difficult proofs, or difficult parts of a long proof. One can study its program. Continuities in a proof and similarities of argument structure among different proofs enable one to ascribe highly competent processing to a single source, even apart from identifying the source as a single physical machine. One can check one computer by relying on another. The second computer can use different methods to solve the same problem. Thinking through such outputs and performing an induction on cases can provide warrant to rely on a computer to solve hard problems, even when one cannot check the full proof. Warrants that rely on understanding the content of a computer's output are the primary warrants for believing what a computer produces.

Warrant to rely on computers in solving hard problems bears comparison to warrant that a gifted mathematician has to rely on his or her own powers in arriving at mathematical beliefs without proof. Imagine a Ramanujan-type genius, who can do proofs, but who often proposes unproved mathematical results. Repeatedly, but not infallibly, the results are proved later. The individual has only a sketchy sense of how he or she arrives at the results. The individual firmly believes the results, however, even before they are proved.

I believe that a reliable individual, like Ramanujan, is often *warranted* in believing such results, before they are proved. The individual has exceptional mathematical powers and uses them well in forming true beliefs.[10] Lacking knowledge of Ramanujan's psychology, we cannot know exactly what type of warrant he had. I presume that no proof occurred in his psychology. If unconscious proofs were carried out in this psychology, he would have a justification at the unconscious-but-personal-level.[11] Perhaps he had inductive reasons to believe the theorems—unconscious but in principle available to consciousness. If his routes were non-propositional (say, pictorial), or if they were propositional but did not constitute an operative, explanation of the conclusion, he would have an entitlement. Then he would be like a person who reliably forms perceptual beliefs, but lacks a reason *anywhere* in his psychology that provides some explanation of the conclusion's belief-worthiness.[12]

A mathematician like Ramanujan is likely to have another warrant. The mathematician can develop an inductive *meta-representational* justification for relying on the powerful competence. He or she can reason: I have reliably come up with answers that have later been proved; so I have a reliable competence. The mathematician can strengthen the primary warrant or this meta-representational warrant by showing that a given result of the unconscious competence coheres well with other proved propositions, or by producing parts of a proof for it.

Thus the gifted mathematician can have both unconscious entitlement for belief in particular propositions and conscious, reasoned justification that falls short of proof. By contrast, a person who relies on a computer to solve difficult problems can, I think, have only inductive justification for believing the computer's results. Since the mathematician cannot produce solutions to the problems in a reliable way, the mathematician must have some inductive reason to accept the computer's offerings. The mathematician cannot rely on a general default entitlement to accept what one is told, other things equal. Other things are not equal. The known difficulty of the problems demands a reason to justify reliance on the computer's outputs. The justifications can be inductive. They have the defeasible, prima facie character of all inductive reasoning.

Even though reliance on computers to carry out proofs inevitably involves an inductive element, the way in which this element compromises the deductive heart of mathematics does not go very deep. Relying on computers to carry out proofs seems in some ways not substantially different from one mathematician's relying on another, as a source about what has been proved.

A deeper reason for rejecting the view that mathematics concerns only proof has been available for eighty years. Gödel's results show that for any given system of consistent axioms of sufficient power (including axioms as weak as

[10] I am assuming that the individual understands the relevant mathematics, can explain its significance, and can relate it insightfully to other mathematics. Thus the individual is not an idiot *savant* with marvelous calculating powers, but no broad mathematical competence. I will not take a position here on the epistemology of the beliefs of idiot *savants*.

[11] He would be justified$_p$ but not justified$_{cp}$. See note 2. In any case, he would still be warranted in his beliefs.

[12] I think it unlikely but possible that Ramanujan's warrants were mainly entitlements.

those for arithmetic), there are truths expressible in that system that are not provable from those axioms.[13]

Other mathematical practices also show that proof from compelling starting points is not always expected in mathematics. Many proofs in advanced descriptive set theory, for example, begin with unproved propositions that are postulated as plausible but that are far from compelling. The value of the postulations and proofs from them lies in unifying and helping to explain other results in set theory.[14] Of course, computers can be used in these proofs, as well as in proofs that begin with compelling assumptions.

Moreover, much work in descriptive set theory involves inductive support for conjectures that may frame further work—with no immediate expectation of proof for the conjectures.

Some uses of computers in mathematics go well beyond producing and checking proofs. A significant departure from traditional uses of computers in mathematics consists in computer-driven *probabilistic* tests for the truth of a hypothesis. There are, for example, probabilistic ways of determining whether a number is prime or for solving combinatoric problems. The computing itself is algorithmic and deductive. However, its starting point is a sample of possibilities. The result is argued to be true on probabilistic grounds, often assigning specific probabilities.[15]

Although expanded probabilistic uses of computers in pure mathematics is inevitable, there is some inertial resistance to such expansion in the mathematical community. Some of this resistance is simply a natural unease, born of the worry that any movement away from the gold standard of mathematical argument—rational compulsion through proof—may ultimately compromise mathematical standards. I think that no one doubts that proof from conceptually compelling

[13] It is sometimes said that the view that mathematical truth must be distinguished from mathematical theoremhood depends on specialized, doubtful philosophical views, such as Gödel's platonism. For an example of such confused writing, see Brian Davies, 'Wither Mathematics?', *Notices of the American Mathematical Society* 52 (2005), 1350–1356. The evidence of the truth of the relevant unprovable sentences does not depend at all on philosophical views. Contrary to Davies' assertions, the point depends neither on Gödel's platonism nor on reliance on an allegedly specialized notion of truth, such as Tarski's. I do regard ontological platonism about central mathematics as the natural and correct view. Of course, some mathematical problems are, inevitably, afflicted by vagueness in their key concepts. Then one cannot expect determinately true or false answers regarding determinate abstract structures. I think that we have no recipe—in particular, no general strictures (such as that vagueness is present when no proof is possible)—that indicate when and where there is vagueness in our mathematical concepts.

[14] Donald M. Martin, 'Mathematical Evidence', in H. G. Dales and G. Oliveri (eds.), *Truth in Mathematics* (Oxford: Clarendon Press, 1998), 215–232.

[15] David H. Bailey and Jonathan M. Borwein, 'Future Prospects for Computer-Assisted Mathematics', *Notes of the Canadian Mathematical Society* 37 (2005), 2–6; Leonard M. Adelman, 'Molecular Computation of Solutions to Combinatorial Problems', *Science* 266 (1994), 1021–1024; C. W. H. Lam, 'The Search for a Finite Projective Plane of Order 10', *American Mathematical Monthly* 98 (1991), 305–318; Carl Pomerance, 'The Search for Prime Numbers', *Scientific American* (December 1982), 136–147; Michael Rabin 'Probabilistic Algorithm for Testing Primality', *Journal of Number Theory* 12 (1980), 128–138.

starting points is to be preferred where it can be obtained. One simply has to maintain perspective on the relevant epistemic statuses of different methods. Sometimes one can use less traditional methods to put one in a better position to use traditional ones. But mathematics is very large, and mathematicians are comparatively small. Some mathematical problems simply do not submit to traditional methods. They can still be worth investigating.

I shall assume that non-traditional, probabilistic uses of computers in pure mathematics are intellectually worthwhile. I shall assume that they yield at least knowledge of the high probability of the truth of conclusions. I ask whether there are epistemically principled differences between probabilistic uses of computers and traditional methods of proof that include computer-assisted proof. I think that there are.[16]

The differences are not matters of rational certainty.[17] Although many proofs in elementary mathematics are rationally certain, use of inductive methods in set theory and other areas of pure mathematics is standard. Such methods cannot yield rational certainty. Computer proofs and long proofs that involve many mathematicians are also not rationally certain: One could, in principle, have reasonable grounds to doubt them, even when they are correct.

The differences are not matters of apriority. Arguments from self-evident axioms, broadly inductive-explanatory arguments, and probabilistic arguments can be all be apriori warranted. The force of the warrant for believing them can owe nothing to sense perception. Moreover, reliance on other mathematicians or on computers in carrying out deductive proofs[18] and reliance on computers to carry out probabilistic inferences are both warranted empirically.[19]

[16] In what follows, I disagree with a main conclusion of Don Fallis, 'The Epistemic Status of Probabilistic Proof', *The Journal of Philosophy* 94 (1997), 165–186. I agree that there are no good epistemic grounds for mathematicians to *reject* non-traditional probabilistic methods in pure mathematics. I do not agree that 'there is no important qualitative difference between [probabilistic uses of computers in pure mathematics] and the [more deductive, traditional methods] acceptable to mathematicians' (166). Although I disagree with the letter of the argument, I find its spirit largely congenial: probabilistic methods in mathematics are epistemically warranted and entirely legitimate. Fallis focuses exclusively on the epistemic value—*supporting the truth of mathematical claims*. As he says, probabilistic methods can make a mathematical truth just as credible as a very long human or computer-assisted deductive proof can. I think, however, that his article underplays epistemic values that reside in *how* a belief is warranted and what sort of understanding a given method provides. Probabilistic methods do not yield a direct understanding of mathematically necessary structures and relations.

[17] A belief is *rationally* certain if no possible rational consideration can justify rational doubt. *Psychological* certainty is just unshakeable total confidence. Such certainty can be irrational. Moreover a belief can be rationally certain without being psychologically certain, if the believer is overly cautious or timid, or does not sufficiently understand the power of his or her reasons. Although a lot of discussion of probabilistic methods focuses on their (high) degree of psychological certainty, psychological certainty is of no particular interest to epistemology.

[18] I note that I have given up my earlier view that reliance on computers or other human beings' reports can be strictly apriori warranted. See 'Postscript: "Content Preservation"', this volume, section III.

[19] A striking instance of empirical computation is Adelman's use of a computational interpretation of experiments involving DNA to solve a mathematical problem. See Adelman, 'Molecular

I believe that the epistemic difference lies in the sort of understanding associated with the different methods. Deductive inference yields understanding of why a mathematical truth must be true, at least relative to the premises. Purely probabilistic inference does not.

Understanding is not a precise notion. There are some ways of caricaturing the distinction that I am trying to get at. I want to resist such caricatures.

First, one should not think in terms of full understanding. Deductive mathematical practice allows many forms of incomplete understanding obtained from a deduction from self-evident starting points. A computer-assisted proof, like the proof of the Four-Color Theorem, does not give anyone full understanding of the proof. The mathematician knows much of the proof, understands the principles used in it, and has inductive reason to think that the computer has carried out a proof. Understanding is partial. It is partial understanding of how the proof goes, backed by inductive ground to believe that the proof has been completed. It is partial, idealized, but genuine understanding of the necessity of the conclusion relative to the premises.

The computer-assisted deductive proof is similar to a mathematician's sketching a proof and recognizing that the proof is completeable.[20] It is also similar to a mathematician's having in mind a proof of a closely related proposition and recognizing that a variant proof for the different proposition is viable—all, without actually going through the proof. Part of being a good mathematician is being able to recognize and understand, in a rough how-it-would-go way, the provability of a proposition. Such recognition provides understanding, even if not rigorous complete understanding, of the necessity of the conclusion relative to the premises.

It is common among writers on these subjects to point out cases in which individuals make mistakes in purporting to recognize that a proposition is provable. But centering on these cases tends, I think, to miss the most important point. The most important point is that in these cases, there is commonly a genuine mathematical competence that grounds epistemic warrant. The recognition competence is inevitably fallible. It varies in power and reliability with different mathematicians.

However, the competence is a reliable capacity in trained mathematicians that commonly warrants their belief in unproved but provable propositions. Assuming that a worked-out, operative proof is not present in the mathematician's unconscious psychology, there remain reasons—proof sketches or systematic analogies

Computation of Solutions to Combinatorial Problems'; and Keith Devlin, 'Test Tube Computing with DNA', *Math Horizons* 2 (1995), 14–21. The mathematical computation clearly is warranted partly through warrants to believe biological theory about DNA.

[20] This kind of situation is *very* common in mathematics. Gödel only *sketched* his second incompleteness theorem in 1931. The proof was not written out, or probably even thought through in full detail, by anyone until 1939. The mathematical community recognized the theorem as "proved" much sooner. See John W. Dawson, 'The Reception of Gödel's Incompleteness Theorems', *Philosophy of Science Association 1984* 2 (1985), 253–271.

to other proofs—that are available to the mathematician. These reasons can combine with an entitlement to rely on a capacity to transition from these reasons to belief. Thought, even unconscious thought, about the specific rule that governs these transitions may not (indeed, I am sure, often does not) figure operatively in the transition. So the warrant for relying on the transition is an entitlement, not a justification.

We do not know enough about the psychology of mathematicians to articulate these warrants. It seems likely, however, that the transition rules are often not represented and operative in the mathematicians' psychologies. And it seems even more likely that the whole process, including both the reason sketches and the transitions, often provides warranted understanding. Such understanding is *incomplete* understanding of why a mathematical proposition is a necessary consequence of the inference's premises.

In sum, there are different kinds of understanding of the necessities associated with deduction. The ideal is full deduction from first-principles. However, the starting points of deduction need not be compelling. They can be acceptable because they help explain other mathematical truths. And the understanding can be sketchy and incomplete, while still pointing toward how the explanatory reasons of a deduction would go.

The understanding need not be general. Brute force proofs in finite domains determine all relevant possibilities, though the truths are individual to the problem. (I am thinking, for example, of finite, brute force solutions to versions of the traveling salesman problem.) Understanding does not derive from general principles, and may provide little general insight. Still, such deductions yield singular understanding of necessary truths and necessary relations between truths.

Further, one should not caricature my point to mean that *probabilistic* proofs provide no understanding at all. Probabilistic proofs can yield insight into the structure of a problem, or inductive insight into mathematical relationships. They are part of an experimental approach to pure mathematics that is similar to the method of hypothesis and testing in empirical science.[21]

Deductive proof gives grounds not just for believing a proposition, but for understanding the necessity of its truth and its necessary relations to premises. It is not surprising that mathematical practice counts such understanding—even in idealized, partial form—epistemically distinctive. Mathematics aims to obtain such understanding, where possible.

I turn now from computers in mathematics to computers in empirical science. Empirical science cannot aspire to deductive proof from propositions that are, when understood, compelling. It must reach its conclusions through induction from perceptual belief.

Uses of computers in empirical science are more varied than those in pure mathematics. The variety of types of mathematical applications to empirical

[21] J. M. Borwein and D. H. Bailey, *Mathematics by Experiment: Plausible Reasoning in the 21st Century* (Wellesley, Mass.: A. K. Peters Ltd, 2004; 2nd edition 2008).

problems seems nearly endless. I discuss briefly just three such applications. I relate each to issues in human cognition.

A common computer application in biology, psychology, and engineering is computation of an optimal solution to a problem. In some such applications, the program is governed by a fixed, although possibly multiply re-applied, mathematical formula. In biology and psychology, such computations are often carried through to determine a baseline for testing actual performance—for example, by a colony of ants in foraging for a food supply, or by a visual system in encoding proximal stimulation for accurately representing some environmental condition.[22]

These computations consist in applying well-understood—often Bayesian—mathematical operations to huge data sets—perhaps repeatedly altered by Monte Carlo randomization—at very fast rates. These uses differ from human reasoning mainly with regard to length and complexity. Given a task with a clear goal, given a data set, given prior probabilities and limiting conditions, the computation provides an optimal solution to the problem, considering costs and benefits.

There are complications. Being certain that a solution is optimal often depends on a separate proof that there are no local maxima. Often one does without certainty and takes a computation of a local optimality result to probably apply globally. Even when one has a proof, the putative certainty lies in the machine's proof, not in the scientist's conclusion. The warrant for the scientist's conclusion consists in an inductive justification based on understanding the machine's program and reliability. The computer run is, however, often a proxy for reasoning that except for its complexity could be carried out by the scientist.

The Bayesian transitions in a computer run are commonly in a form that makes them propositional proxies for human reasoning *about* non-rational processes. Of course, the non-rational processes lack propositional structure. The subject matter of the programs—for example, ants' foraging behavior or processing in a visual system—are not pieces of reasoning. Even where the scientist has not fully thought through the computer's transitions, the scientist has a general

[22] Peter Nonacs and Joanne L. Soriano, 'Patch Sampling Behaviour and Future Foraging Expectations in Argentine Ants, *Linepithema humile*', *Animal Behavior* 55 (1998), 519–527; Sasha R. X. Dalla, Luc–Alain Giraldeaub, Ola Olssonc, John M. McNamarad, and David W. Stephense, 'Information and its Use by Animals in Evolutionary Ecology', *Trends in Ecology and Evolution* 20 (2005), 187–193; Thomas J. Valone, 'Are Animals Capable of Bayesian Updating? An Empirical Review', *Oikos* 112 (2006), 252–259; Yoram Buraka, Uri Roknia, Markus Meistera, and Haim Sompolinskya, 'Bayesian Model of Dynamic Image Stabilization in the Visual System', *Proceedings of the National Academy of Sciences of the United States of America* 107 (2010), 19525–19530; Wilson S. Geisler, Jiri Najemnik, and Almon D. Ing, 'Optimal Stimulus Encoders for Natural Tasks', *Journal of Vision* 9 (2009), 1–16.

Sometimes optimization algorithms are developed from observation of frequencies in actual behavior, for example, the swarm behavior of ants or bees. Such algorithms are often carried over to apply to different domains, including pure mathematics. See M. Dorigo and T. Stützle, *Ant Colony Optimization* (Cambridge, Mass.: MIT Press, 2004). These cases are rather like the DNA computing mentioned in note 16, in that they use natural empirical phenomena to help solve mathematical problems, as well as problems in natural science.

understanding of the explanatory relation between premises and conclusion of the computer run. The scientist can, commonly, understand any given step as a reason for a later step. Even where the scientist's understanding is partial, the computer run can be interpreted as a realization of reasoning. The scientist can enter into the reasoning at any point. The computer's transitions, supplemented by the scientist's inductive reasoning about the computer's performance, yield justification—warrant by reason—for the conclusion.

The transitions effected by the computer are not operative in the scientist's psychology. The scientist makes an induction about the reliability of the machine's outputs and is justified in relying on them. The scientist is *better* off than in many cases in which the scientist relies on his or her own inductive reasoning. Transitions in ordinary inductions are, to be sure, operative in the scientist's psychology, whereas transitions effected by the machine are not operative in the scientist's psychology. Principles governing transitions in human inductions are, however, often less well understood than the principles governing transitions effected by the machine.

A second type of computation in empirical science raises more interesting epistemic issues. I have in mind [what are broadly called] '*genetic algorithms*'. Genetic algorithms are search and optimization techniques based on principles governing mechanisms of evolution. The algorithms require individuals in a population to be represented as solutions to a problem, and they require a fitness function that maps solutions to a quality-of-solution evaluation. An initial population of solutions is generated randomly. A proportion of this population is selected through fitness evaluation to begin breeding a new population of individual solutions. Selection is biased toward better solutions. A new population is produced by genetic operators—the most common being mutation and crossover (or recombination). *Mutation* applies a probability that a random aspect of a solution is to be modified in some usually random way to produce an individual solution for the next population. *Crossover* is a genus of operations that combine aspects of the solutions of two or more "parent" individuals from a given population to produce a new individual solution for the next population. Different parents are chosen from the selected sub-population to produce new individuals. When a new population of a certain size is produced, the process of fitness evaluation, selection, and genetic operation is reapplied. The process is usually terminated when a given optimization level is reached, though it can be terminated after a set number of generations.[23]

Such algorithms are very good at producing optimal solutions in large, complex domains. An optimal solution can be used in science as a baseline for generating empirical hypotheses about actual domains and for empirically testing the nature of such domains.

[23] Melanie Mitchell, *An Introduction to Genetic Algorithms* (Cambridge, Mass.: MIT Press, 1996); J. R. Koza, *Genetic Programming: On the Programming of Computers by Means of Natural Selection* (Cambridge, Mass.: MIT Press, 1992).

What is the epistemic status of a scientist's belief that a certain population of solutions is optimal, when that belief rests on the result of a computer's running a long, complex genetic algorithm? As before, the belief can be epistemically justified through inductive reasoning about the program and the computer's reliability in implementing it. It seems to me, however, that there is a difference with the case in which a scientist bases a belief on the generation of an optimality result that derives from Bayesian updating.

As noted, like genetic algorithms, Bayesian updating processes often yield results about non-rational processes. But Bayesian updating processes are, at least on some formulations, proxies for *reasoning* in Bayesian probability theory. Each stage in the algorithm is naturally construed as a reason for the next stage. Genetic algorithms are not proxies for reasoning. The random mutations and relatively random recombinations are not analogs of reasoning. They lack the explanatory structure of reason transmission. The computer's transitions do not transmit reasons from previous steps to later steps. Earlier processing stages are not reasons that help constitute answers to why-questions about the credibility of later steps. The computer does help the scientist reason about a non-rational process. Its runs provides evidence for scientific inductions about the relevant subject matter. The computer's processing is *not*, however, to be construed as an instantiation or amplification for reasoning.

Given our understanding of evolution, we understand why the non-rational process yields an optimality result. We are justified—in principle *just as justified*—as in any other inductive case. But the computer's processing does not serve as a proxy and amplification for our own reasoning—as it does in carrying out mathematical proofs and in many construals of Bayesian algorithms. Of course, the evolutionary process itself is not an exercise of reasoning. Computer processing that serves as a proxy for that process is not a simulation of reasoning.

I emphasize that these reflections do not bear on the power of our warrants for believing the computer's results. Inductive warrants can be equally strong in the different cases. The reflections bear on the relations between natural construals of particular transitions in the computer's processing, on one hand, and steps in our reasoning, on the other.

I mentioned earlier that some warranted inductive propositional transitions in our own psychologies may not always constitute reasoning. Previous steps may enhance credibility of later steps without providing explanatory rationalization of the later steps—without being reasons for them. Some propositional inductive processing in our own psychologies may be as non-rational as evolution is, while still being a good route to truth. Given that our minds reflect nature in many other respects, it would surprising if they did not do so in this one. Any such warrants to believe conclusions from inferences would be purely entitlements, not justifications. Even the premises of the warranted inferences would not be justifications—reasons—for the conclusions.

There are Hume-inspired conceptions of inductive inference that treat *all* inferences as non-reason-giving associative transitions. Some psychologists who take connectionist programs to model human reasoning tend in this direction. I think that such models are not adequate to model deductive reasoning or most verbally articulated induction. Much of the reasoning in mathematics and natural science *is* reason-giving. What I am noting is that there may well be inductive inferential transitions, even in science, that support their conclusions by making them reliably more credible, but without providing even a partial explanation of the sort that reasons provide. Such inferences yield entitlements to their conclusions, all the way down, not justifications.

I turn briefly to a third example of uses of computers in empirical science. Recently, some striking results have been obtained in producing algorithms for discovery of scientific explanations. A computer was given data on certain types of motion, such as that of a double pendulum. It was given basic arithmetic, geometric, and trigonometric operations, and some basic parameters. It was programmed with a genetic, symbolic regression algorithm—one that looked simultaneously both for parameters to plug into equations and for equations that simulate conservation principles. It was also programmed to balance simplicity of equation against accuracy in fitting the data. The algorithm returned a small number of equations (on the order of ten). Although some were of no scientific interest, some were classical laws of physics—such as Newton's second law of motion and Lagrangian equations that apply to the double pendulum.[24]

Here the genetic algorithm works on formulae, weeding out the "less fit" ones in favor of equations that fit the data and are simple. Again, particular transitions in the computer processing do not yield reasons for subsequent stages. And again, the scientist has only general insight into the computer's operations. The algorithm does simulate the scientific method of posing a question, offering a hypothesis, testing the hypothesis, adjusting the hypothesis until one finds one that tests better, and so on. But the generation and adjustment of hypotheses have the random character of all genetic algorithms.[25] The process produces better hypotheses over time because only the better fitting equations are selected for each new cycle of testing.

In scientific discovery of new laws, we know very little about human inferences to the best explanation. It is clear that hypothesis revision in our reasoning is often more directed and more based on rational considerations than it is in any genetic algorithm. Such algorithms can afford to search for and produce new hypotheses by more nearly random methods.

Human discoveries are often described loosely as leaps of intuition. In such cases, the scientist has little conscious recognition of how he or she finds a

[24] M. Schmidt and H. Lipson, 'Distilling Free-Form Natural Laws from Experimental Data', *Science* 324 (2009), 81–85. There are important issues here, which I shall not go into, about the scientist's understanding non-basic parameters that the computer comes up with and about the scientist's role in recognizing which among the equations that the computer comes up with are scientifically interesting and genuinely explanatory.

[25] By contrast, as far as I can see, the testing phase of the algorithm *does* operate as proxy for rational testing in scientific reasoning.

hypothesis. But discovery often derives from weighing reasons at various levels of generality—from observations of evidence to attempts to fit in mathematically with high-level generalizations that are already known. Normally, a lot of reasoning goes into the inductive discovery of new explanatory principles.

Even so, we do not know the full psychology of such inductions. We cover our ignorance with the all-purpose, but uninformative epithets 'intuition' and 'insight'. Although human induction to scientific laws commonly involves reasoning, it may not consist purely in reasoning. There may be non-rational, evolutionarily drummed-in associations that are not backed by any explanatory reasons that are operative in our psychologies. Reasoning may sometimes figure mainly in testing hypotheses.

I think that this Popperean picture of scientific discovery is at most one factor in the right overall account of inference to best explanations. I emphasize, however, that although much human induction (and all deduction) surely differs from the genetic algorithms used in the case just described, we know very little about human induction. We are not in a position to say that our unconscious psychologies never use the brute, non-rational, "natural selection" methods that occur in genetic algorithms. Thus our sense of not understanding the computer's processing as *reasoning* should not be allowed to obscure the fact that we do not understand our own inductive processing very well—especially in the case of inferences to new scientific principles.

What we should insist upon is that science *looks for* rational explanations that yield reasons where it can find them. Fundamentally, science attempts to produce reasoned explanations. Both the non-rational processing in computers that simulates non-rational selectional processes in nature and whatever elements of non-rational processing there are in our own unconscious psychologies must be used to find reasoned explanations.

Empirical science, like mathematics, aims at reasons, explanation, and understanding. Science is a particular form of understanding. It strives not just for knowledge, but for *scientia*—systematic knowledge backed by reasoned, explanatory understanding of why its conclusions are to be believed. Relying on computers underscores that understanding is often partial, even in science. Reason is not itself inadequate. But our capacities to reason are small in comparison to the vast complexities of our mathematical, physical, and psychological subject matters. We must rely on computers, and perhaps on our own non-rational but warranted inductive processing, for all the help we can get, even though such help guarantees that our understanding of some aspects of science will remain partial.

PART IV

REFLECTION

21 *Reasoning about Reasoning*[*]

Mutual expectations cast reasoning into an interesting mold. When you and I reflect on evidence we believe to be shared, we may come to reason about each other's expectations. I develop expectations about your expectations, expecting them to be expectations about mine. The double-mirror aspect of such situations suggests that there may in some sense be no stopping. Of course, we do run out of expectations. But to some this seems to result merely from inability to understand sufficiently complicated nestings, rather than from lack of *reason* to acquire each of infinitely many nested beliefs.

Limitless nesting of psychological "contents" has played an important role in accounts of communication, convention, and related topics. But there have been few attempts to analyze precisely how the infinity of contents might be generated. An exception is David Lewis's account in *Convention*, pp. 52–57. This account contains significant errors which are easily overlooked because of the complexity and subtlety of the subject matter. Fortunately, they are also easily corrected, at least within the general framework of the account. The purpose of this note is to identify and correct the oversights.

A key to the discussion is the term 'have reason to believe'. The term is used by Lewis in such a way that if a person has reason to believe something and does not already believe it, he would be extremely irrational to disbelieve it and he should come to believe it with only a little reflection (pp. 53, 59, 63). Having reason is understood to vary with one's inductive standards and background information.

The reasons normally "had" in the situations we shall be discussing are inductive reasons. They therefore admit of uncertainty. In what follows I shall sometime speak of one's reason to believe φ as making φ probable for one to a certain specified degree. It is unimportant to my purpose whether inductive reasoning be actually, or even ideally, explicable in terms of numerical probabilities. The appeal to probability is merely a way of making applications of the following point picturesque: When one has reason to be less than fully certain about each of a plural number of epistemically independent theses the assumption of each of which is needed to justify a conclusion, one's certainty about the conclusion (relative to those assumptions) should be less than one's certainty

* I am indebted to Gregory Kavka for remarks on an earlier version.

about any one of the assumptions. It is common to assume that if numerical probabilities are assigned, the probability of the conclusion (relative to the reasons for believing it) is no more than the product of the probabilities assigned to those reasons, assuming the reasons independent.[1]

Lewis illustrates his theory of interpersonal reasoning by reference to a particular case. We imagine that *state of affairs A* is any sort of evidence for the conclusion that we will meet at a given spot tomorrow—for example, a signal of some kind. I shall call the state of affairs of our meeting at that spot and time *state of affairs B*. Lewis makes the following assumptions, and I shall adopt them:

(1) You and I have reason to believe that some state of affairs A holds.
(2) Each of us is such that if he had reason to believe that A held, then he would thereby have reason to believe that you and I have reason to believe that A held.
(3) Each of us is such that if he had reason to believe that A held, then he would thereby have reason to believe that B would hold.
(B) Each of us has reason to believe that if he would have reason to believe B held granted that he has reason to believe A held, then the other would have reason to believe B held granted that *he* had reason to believe A held.

Lewis claims that these four premises imply

(4) Each of us is such that if he had reason to believe that A held, then he would thereby have reason to believe that each of us has reason to believe that B would hold.

He also claims that they imply

(5) Each of us is such that if he had reason to believe that A held, then he would thereby have reason to believe that each of us would have reason to believe that the other has reason to believe that B would hold

and so on, with indefinitely many iterations in the consequent.[2]

These claims are mistaken. We can most easily bring out why by representing simplified versions of the relevant sentences in symbolic, abbreviated form:

[1] The assumption of the independence of justifications must be seen as an idealization. It has been argued that it is an unfruitful and fundamentally misleading idealization. Since our appeal to probability strength is purely heuristic, this issue can be set aside.

[2] (1)–(5) are formed from like-numbered sentences in *Convention* (Cambridge, Mass.: Harvard University Press, 1969), 52–53, by replacing 'indicates' in the latter sentences by Lewis's definition of the term. A *indicates* to someone x that – – – if and only if, if x had reason to believe that A held, x would thereby have reason to believe that – – –. (B) is the relevant instance of Lewis's assumption that you and I have reason to believe that we share the same background information and inductive standards, at least nearly enough so that A will indicate the same things to both of us. Actually Lewis claims that instances of this assumption would be needed to get (5) and more complicated iterations. Thus it is (1)–(3) and relevant *analogs of (B)* which are alleged to imply (5). Ignoring this point will not be crucial to our criticism.

(1') R(I,A)
(2') R(I,A) → R(I,[R(I,A)]) & R(I,[R(Y,A)])
(3') R(I,A) →R(I,B)
(B') R(I, [R(I,A) → R(I,B). →. R(Y,A) → R(Y,B)])
(4') R(I,A) →. R(I [R(I,B)]) & R(I, [R(Y,B)])

'R' stands for 'have reason to believe'; 'I', for 'I'; 'Y', for 'you'. The brackets are intended for ease of reading. The arrow stands for 'if... then (thereby)... '—except in its occurrence as main connective in the content clause of (B'), where it stands for ordinary 'if... then... ': the difference in the meaning here can be slurred without harm. I assume that the analog of *modus ponens* holds for the arrow. We do not need to give a semantics for the notation. Whereas the argument form (1') through (4') represents only the first person point of view, (1)–(4) incorporate an analogous argument from the second person viewpoint. Our remarks will apply (by analogy) to both arguments.

Using (1') and (3') we can detach

(a) R(I,B).

To get the first conjunct in the consequent of (4'), we need the iteration principle

(6') R(I,C) → R(I, [R(I,C)])

where 'C' stands for any content clause. Let us grant (6') for the present. This together with (a) gives us

(b) R(I, [R(I,B)]).

Using (1') and (2') we can detach and simplify to

(c) R(I,[R(Y,A)]).

Now my reason described in (c) is inductive. Suppose it has a probability strength of 65%. My reason described in (B') is also inductive. Suppose it has a probability strength of 66%.

Using (B') and (3'), one is tempted to try a sort of detachment that would yield

(d) R(I, [R(Y,A) → R(Y,B)]).

But such a detachment would be within the scope of 'R'—within the first person point of view. And so far, we have assumed only (3'), not that I have reason to believe (3').

Let us grant the following for the sake of the argument:

(7') R(I,C) & (R(I,C) → R(I,D)). → R(I,[R(I,C) → R(I,D)])

where 'C' and 'D' stand for any content clauses. The idea is that if I have reason to believe some state of affairs obtains, and that state of affairs gives me reason to believe that another state of affairs obtains, then I have reason to believe that this evidential or reason-given relationship holds for me.

With (7'), (1') and (3') we can derive

(e) R(I, [R(I,A) → R(I,B)]).

The status of the reason described in (e) is a deep problem. Hume argued that such reasons were clearly not deductive and that if they were inductive, then induction could not be justified. I do not want to embroil myself in this matter. But I shall suppose, what seems plausible, that my reason described in (e) has a probability strength of less that 100%.

We are now in a position to diagnose the sources of Lewis's mistake. My reason in (B') has a probability strength of only 66%, and any premises I conjoin with that reason may carry their own degrees of uncertainty. Unless such premises are 100% probable (or epistemically derivative), the conjunction will have less strength than any of the conjuncts. Thus we may derive (d) from (B') and (e) *only if* the product of the probability strengths of my reasons—those reasons described in (B') and (e) is sufficiently high. If it is, we obtain (d). The same point applies in conjoining (d) and (c) to get

(f) R(I, [R(Y,B)]).

(This derivation is more clearly problematic than that of (d) because its interpretation does not hinge on the Humean issues surrounding (e).) In fact, on our assumptions about probability strength in (c) and (B'), we *cannot* derive (f). For the reason described in (d) has a strength of less than 66%; that in (c) has a strength of 65%; the reasons are (reasonably) independent epistemically; so their product is less that 50%. In ordinary circumstances, this situation might be regarded as one in which I lack the reason described in (f). If my inductive support were strong enough to justify (f), of course, we could conjoin (f) with (b), and thereby have (4').

The moral is that the truth of (4) depends on the combined *strength* of our reasons regarding three matters: the other's appreciation of the evidence ((2')), the other's inductive standards and background information ((B')), and our own inductive reasons ((7') and actually (6') as well).

How will iteration affect matters? (1), (2), (4) and (B) (or analogs, cf. note 2) do not entail (5) for the same reason that the previous alleged entailment failed. But if (B'), (7') and (c) are strong enough to produce (4'), then by relying on (4') (and an analog of (B') which is of at least equal probability strength), one can derive the appropriate representation of (5)—on one condition. The reasons generated by (6') and (7') must not become weaker as these principles are reapplied. This condition seems no less plausible than the iteration principles themselves.

Thus Lewis's entailment claims can be corrected by making explicit the principles (6') and (7'), and by requiring that the combined reasons cited in the premises be sufficiently strong. Of course, even this account might be doubted. Not all epistemologies would accept (6') and (7'). It might be thought that one could have first-order reasons but lack the information or subtlety to have higher-order reasons. Moreover, there is the bothersome Hume, who had inductive reasons to believe the sorts of things everyone else believes, but who doubted that he had those reasons. On Lewis's account of 'have reason to believe' (accepting (6') and (7')), we would have to conclude that Hume was "extremely irrational". So doing would be a mite severe. Whether this difficulty is best accommodated by minor refinements in the interpretation of 'have reason to believe' or by important changes in the iteration principles, I leave as an open question.

22 *Reply to Benejam: 'Thought Experiments and Semantic Competence'*

Antoni Gomila Benejam's paper raises an extremely interesting and important issue—how is it possible that the thought experiments prove useful? More specifically, how are we to understand the semantic or conceptual competence that underlies our ability to recognize the application patterns of our concepts on reflection in the twin-earth thought experiments?

A traditional line is that to have the concept one must "implicitly" understand it. Explicit understanding emerges through reflection, which yields a clear and distinct definitional explication of it.

I have criticized this traditional line at almost every point. I believe that it is possible to have concepts without understanding them at all, even implicitly. Animals and young children have perceptual beliefs, hence concepts, but lack any understanding of the concepts. They have a minimal competence in the use of the concepts. Their concepts mark propositional, including inferential, abilities. Understanding requires a reflective or explicational ability, which includes an ability to objectify concepts, an ability which I think animals and young children lack.

Similarly, the cases of dependence on others that Putnam first brought to prominence and the cases of erroneous explicational understanding—such as the arthritis case—indicate that it is possible to have explicational abilities but not be able even on reflection to produce a correct explication.[1] So individuals have concepts without implicit understanding in these cases as well.

Someone who lacks some aspects of correct understanding cannot attain correct explicit understanding through mere reflection. Animals without any understanding, and people with a partially incorrect understanding, could not engage in the relevant thought experiments.

[1] Compare Hilary Putnam, 'Is Semantics Possible?', originally published 1970, reprinted in Putnam's *Philosophical Papers*, volume II: *Mind, Language and Reality* (Cambridge: Cambridge University Press, 1975); and my 'Individualism and the Mental', *Midwest Studies in Philosophy* 4 (1979), 73–121.

Moreover, where reflection does succeed in yielding better understanding, the success does not in general produce a definitional explication. This is because definitions play only a minor and occasional role in the understanding of concepts.

Even where a simple explication is found (such as chairs are artifacts made or meant for one person to sit upon), there is no internal guarantee that it is complete or correct. It rarely if ever provides a reductive replacement for the concept.

Finally, even experts may not have embedded in their conceptual abilities sufficient material to provide a correct explication, whether definitional or not. This is because of the fundamental point made by anti-individualism that a concept depends in various ways for its individuation on the way objective reality actually is. The similarity conditions that underlie our actual practices may not be knowable except through acquiring additional information.[2] In this respect, anti-individualism provides a genuinely different view of the potential of reflection to provide correct accounts of concepts.

Gomila notes that in my discussion of the linguistic aspect of these matters I distinguish between translational meaning and explicational meaning. He is correct in holding that what I say about these notions requires supplementation if we are to obtain a good account of semantic competence, and of the capacities of reflection through thought experiments to yield deeper understanding of our concepts and meanings.

He is incorrect in holding that I reduce semantic competence to translational meaning or to thoughts like *"arthritis" applies to arthritis*. Gomila quotes a passage from 'Wherein is Language Social?' that is incompatible with such a reductive view. I think that there is nothing in what I write that indicates any such reduction.

I have always accepted that semantic competence, like conceptual competence, is a complex, multi-faceted matter. It frequently involves perceptual, recognitional, applicational abilities. It always involves inferential abilities—some deductive, some inductive. It sometimes involves explicational abilities and meta-semantical abilities. One cannot have a concept, or express a meaning, unless that ability is embedded in a nexus of further representational abilities. A concept and a word's translational meaning simply marks the basic representational ability. It certainly does not exhaust the range of abilities that have to accompany that ability in order for it to be what it is.

This view is also incompatible with the position, which Gomila holds that my writings suggest, that semantic/conceptual competence "amounts to" conformity in actual use with social practice. Conformity at most suggests competence. Moreover, I do not reduce this range of psychological or linguistic abilities to the "externalist" conditions that must be present for concept possession, as Gomila suggests I do. The representational abilities involve competencies in

[2] On this, see my reply to Steven Davis, in M. J. Frapoli and E. Romero (eds.), *Meaning, Basic Self-Knowledge, and Mind: Essays on Tyler Burge* (Stanford: CSLI Publications, 2003).

the individual that are partly individuated in terms of relations to the environment, but not reducible or replaceable by those relations. A theory of ideas or concepts is definitely not incompatible with my anti-individualism.

There are two reasons why my points in this area have been primarily negative. One is that the massive weight of the implicit-understanding tradition, though criticized by others, is very substantial. Getting free of the picture that I presented at the outset of this reply is no mean philosophical task. The second reason is that the positive story is inevitably much more complex than the traditional account indicates. It varies from one type of concept to another. So at present, I think that we are not in a position to provide a very rich and detailed account of the nexus of abilities that go into having a concept or expressing a meaning. The point of the thought experiments is, however, to begin to suggest large contours in certain maps of representational competence.

So what can be said in short compass about what makes the thought experiments possible? Concepts are not made what they are by understanding them. They are grounded in patterns of first-order usage. The usage has counterfactual implications. So it potentially extends to cases of application which never occur. Implicit in this potential is an ability to recognize the cases if they arise. With sufficient sophistication, commonly through language use, an individual may acquire the abilities to imagine counterfactual cases and to think up explications that are meant to cover all cases or simply to generalize about types of cases to which the concept would be applicable. These latter second-order abilities are epistemically posterior to the first-order abilities, even though they may come to accompany the first-order conceptual abilities implicitly.

Much of the ability to test concepts on counterfactual cases depends on empirical knowledge. Knowing that water is a natural kind or that people can use the concept arthritis while misunderstanding it requires either knowledge of others or memory of one's own transition from misunderstanding to understanding. It may be that in some cases particular inferential connections are necessary to having some concepts. Sometimes with sufficient reflection one can come to understand apriori these constraints on minimal conceptual mastery. So understanding as opposed to having the concept of cat probably requires that one believe that if something is a cat it is, if anything, a physical being that occupies space. (I think a child can have a concept of cat before it has any general concept of space, and before it understands anything about the concept.) It may not require that one believe that a cat is, if anything, a mammal or perhaps even an animal.[3]

In mapping the competence associated with thought experiments, it is important, I think, to allow for different levels or types of competence. There is the minimal competence necessary to having the concept. There is the minimal

[3] Compare Hilary Putnam, 'It Ain't Necessarily So', *The Journal of Philosophy* 59 (1962); reprinted in Putnam's *Philosophical Papers*, volume I: *Mathematics, Matter and Method* (Cambridge: Cambridge University Press, 1975).

competence necessary to attempting to understand the concept (or term). Such competence might be compatible with having a thoroughly incomplete or mistaken understanding—which might emerge in, among other ways, poor or uncertain explications. There is the competence sufficient to be socially acceptable. There is the competence of the experts in a given community. This might involve being able to provide short explications that no one in the community has improved on. It might also involve being able to persuade others about the application of the concept (or term) in difficult cases. Finally, there is the competence involved in having an ideal or complete understanding of the concept. As noted, this might go beyond what the experts have. It might depend on coming up with new knowledge about the patterns of usage and the environmental relations that fix the nature of the concept.

Post Daltonian atomic theory is an example of a move from an expert's competence (Dalton's) to a competence that encompasses a deeper knowledge of the environmental relations that fix the identity of states marked by the concept. Post Weierstrassian explication of the fundamental ideas of the calculus is an example of a move from an expert's competence (Newton's or Leibniz's) to a competence that encompasses a deeper knowledge of the mathematical facts and usage that Newton and Leibniz were relying upon and that fixed *their* concepts as well as those of Weierstrass.[4]

I believe that anti-individualism helps explain why thought experiments are productive. Thought experiments center on cases. It is a range of cases, together with a pattern of usage that includes both a nexus of representational abilities (perceptions, inferences, and so on) and a pattern of relations to the environment or subject matter, that fix the representational content of terms or concepts. These content-fixers contrast with definition, or explication, or implicitly understood meta-rule. Definitions, explications, and meta-rules come later—when they come at all. They come to be understood as fitting the patterns of usage and the cases. We do not always carry such matters around in our minds, just by virtue of having the competence to employ a concept or term. That is why finding the general principles that summarize conceptual applications is so much harder than recognizing whether a concept or term applies to a particular imagined case. It is also why the principles commonly turn out to be more complex and less obvious than they would have been if they were implicitly embedded in the concepts in such a way that we were being guided by them. Sometimes very general principles do guide. Usually such principles capture a usage and set of thinker–environmental relations that are antecedently in place.

[4] This sort of point is developed in my 'Intellectual Norms and Foundations of Mind', *The Journal of Philosophy* 83:12 (1986), 697–720; 'Frege on Sense and Linguistic Meaning', in David Bell and Neil Cooper (eds.), *The Analytic Tradition: Meaning, Thought, and Knowledge* (Oxford: Blackwell, 1991), 30–60; and in 'Concepts, Conceptions, Reflective Understanding: Reply to Peacocke', in Martin Hahn and Bjørn Ramberg (eds.), *Reflections and Replies: Essays on the Philosophy of Tyler Burge* (Cambridge, Mass.: MIT Press, 2003), 383–396.

As Descartes emphasized, reflection is very difficult to do well. It is hard to arrive at clear and distinct principles that are associated with our concepts. It requires good judgment. Still, we sometimes manage to do it. A deeper account of the psychological and epistemic bases for our success through reflection on thought experiments is very much to be desired.

23 *Concepts, Conceptions, Reflective Understanding: Reply to Peacocke*

Chris Peacocke's interesting essay is broadly persuasive and broadly congenial to things I have emphasized about incomplete understanding and rationality.[1] It takes these matters in a direction that I have not carried them. It is also congenial to things that I have written about the way the nature of a person's mental states can depend on his relations to his environment. Peacocke's essay investigates the explanation of reflective understanding. He asks how we are to explain an individual's improvement on his understanding of terms or concepts like *chair* or *limit*. He connects these issues in original ways to questions about how we are to explain the dawning of understanding of simple logical or mathematical truths that rests on the basis of an understanding of their component concepts. This is a rich, provocative essay. I will not be able to do justice to it. What I want to do is endorse the spirit of the main proposals about explanation, support the letter of some of them, and raise some questions about points that I have doubts about or that I think may need further development.

I

What seems to me right about the spirit of the proposals is the idea that one can explain many cases of reflective acceptance of "conceptual truths", and the application of incompletely understood concepts, in mentalistic terms. The proposals are also attractive in their appeal to unconscious, "implicit" mental structures—though I will return to the question of what "implicit" should mean here. The idea that these explanations are not only mentalistic explanations but are explanations that bear on understanding our rationality also seems fundamentally on the right track. Finally, the idea that not all implicit conceptions that explain the use of concepts are correct conceptions seems to me fundamentally right and a consequence of the principles that underlie anti-individualism.

[1] Substantially this essay was given as a reply to an earlier version of Christopher Peacocke's essay at the APA in Pittsburgh in April 1997. Page numbers for Peacocke's essay are to his 'Implicit Conceptions, Understanding, and Rationality', in Martin Hahn and Bjørn Ramberg (eds.), *Reflections and Replies. Essays on the Philosophy of Tyler Burge* (Cambridge, Mass.: MIT Press, 2003), 117–152.

What is an implicit conception? I take it that implicit conceptions are unconscious psychological conceptual structures that explain our ability to apply concepts to cases, or to realize that principles involving concepts are true. They explain such applications and principles even though we cannot easily explain to ourselves how we recognize the examples as instances of a concept, or how we explain the dawning of realization that such principles are true. I take it that implicitness is supposed to be compatible with, and in fact demand, psychological reality. And implicitness of a conception is compatible with its being either explicity or implicitly *represented* (p. 137). I am not fully clear about what explicit and implicit representations are. I suppose that the issue turns on whether there are neurally realized syntactically structured *tokens* whose syntax corresponds to the form of conceptual structures that implicit conceptions have. If something like this is what is meant by explicit representation of an implicit conception, it seems to me correct that whether implicit conceptions are explicitly (though unconsciously) represented can be left open.

But I am still not clear what the implicitness of an implicit conception is supposed to consist in. Peacocke seems to indicate that implicit conceptions are conceptions at the subpersonal level (p. 136). I take the subpersonal level to be a level that is not only not conscious, but is not accessible to introspective or reflective consciousness and must be gotten at only theoretically. This is true of the basic grammatical structures underlying our linguistic competence and the information-processing structures underlying our perceptual experience. But elsewhere Peacocke takes implicit conceptions to be difficult but not impossible to make explicit through reflection. This makes it look as if implicit conceptions are real personal-level conceptions, just ones that are unconscious and relatively difficult to articulate in consciously available judgments. My guess is that both sorts of "implicitness" might be relevant to different aspects of explanations of the psychological facts. Perhaps there are even more than two types of unconscious structure here.

A further question I have about the psychological aspect of Peacocke's proposal centers not on the appeal to implicitness but on the presumption that the mentalistic elements in the relevant psychological explanations are always *conceptions* associated with the relevant concepts. It seems to me likely that in many cases, the relevant explicit explicative judgments—and improvements through reflection on previous explicative judgments—will be derived through unconscious mentalistic processes that do not use an unconscious explicative *conception* as material in their transformations. Rather, for example, they may use perceptually stored material, which has not been unified even at any "implicit" level under some explicative, conceptualized principle. The storage may be in the form of purely perceptual judgments about cases. Such perceptually stored material might be used together with certain ("implicit") inductive principles or principles governing relevant similarities to form explicative judgments at the explicit level.

Take the chair case as an example. I discussed this case at length in my 'Intellectual Norms and Foundations of Mind' (1986). On Peacocke's account, those who can arrive at an explicitly articulated definition of 'chair' carry around an implicit "definition" of 'chair' that is often presumably fully correct. I am inclined to think that often we lack such a definition not only at the explicit level of readily accessible conscious judgment, but at any implicit level (both the truly subpersonal level and the unconscious, personal, hard-to-access levels). What enables us to arrive at correct definition is partly the memory of many instances that we have judged to be chairs, or at any rate, dispositions derived from such judgments. So we have stored hard-to-access memories of perceptual judgments (or dispositions to new judgments based on these prior judgments) of things such as ski-lift chairs, deck chairs, living room chairs, and so on.

One might ask how we made these initial judgments (unifying the different perceived examples under the concept *chair*), if we did not have a guiding definition. The answer is that usually we are just told that a ski-lift chair is a chair (at some time in the dim past), or that a deck chair is a chair, without being guided by some antecedent conception of what makes them all chairs. We may never have used even implicitly a conception to include the ski-lift chair under the concept—though we *do* include such chairs under the concept. It does not seem necessary that once we have judged a ski-lift chair to be a chair, we already find (implicitly) a unifying explicative conception that explains what is essentially chairlike in both living room chairs and ski-lift chairs. We may carry only the notion that both are to be sat on, and are to be differentiated from stools, benches, love seats, and sofas—but no conception that yields necessary and sufficient conditions. We may, it seems to me, simply carry unconsciously the perceptual memory of the look of a ski-lift chair together with the unconscious memory that we categorized it as a chair, or simply the present disposition to categorize it as a chair. When we try to form an explicit reflective explicative conception of what chairs are, we may simply use offline—that is, unconscious—"implicit", *inductive* principles to arrive at our explicit explicative conception of our concept *chair*. We may remember the ski-lift chair and realize that it is a counterexample to an explicit explicative conception that held that chairs must have legs. I think it unlikely that such an explicative conception must always be already formed at some unconscious level. So in such cases, it is not conceptions that are "implicit". Rather, it is inductive principles together with a range of examples that are unified under the concept—though not under any conceptualization, or explication, of the concept.

We need not, of course, always work with memories of instances that have actually been categorized under a concept, or with dispositions associated with past categorizations. We may be driven by general unconscious similarity principles from actually categorized cases to include other merely hypothetical cases as well, without having—even at some implicit level—a unifying conceptualization of the cases that is specific to the concept being explicated or conceptualized. An explicative conceptualization may first emerge explicitly, as a product of unconscious processes, at either subpersonal or hard-to-access personal levels, which make use of intentional material that is both more specific and more general than the explicative conceptualization specific to the relevant concept. One could project from one jade sample to the next. One might remain open to the idea of a unifying account of the similarity, and lack any general defining conception. Such a conception might become available only through empirical research, by geologists or philosophers. Here the limits may be set partly by what people have actually applied the jade concept to. Possible similarly looking and feeling minerals, other than jadeite and nephrite, may not count as jade just because they are not in the appropriate actual sample classes.

I find Peacocke's account of the standard model of arithmetic very attractive. But a point similar to the one just made may apply, in more complex form, even to the Leibniz/Newton limit case. It seems to me a stretch—and at any rate, not obviously correct—to think that Leibniz and Newton shared as a stable part of their unconscious repertoire an implicit but fully formed version of the Weierstrass explication. What seems to me more likely is that they had a mastery of the basic calculus techniques, aided perhaps by some geometrical sense of approaching a limit on a line, which yielded correct answers in specific applications. Again, they may have had a not fully conceptualized sense of mathematical similarity, which may or may not be conceptualized into a principle, even implicitly, that explains their ability to project to the cases.

Whether the implicit conception is fully formed seems to me open to investigation, even assuming that a definite concept, that of *limit*, is sharply grasped. The concept is sharply grasped insofar as one applies the concept to exactly the right cases. But what guides the application of the concept might not be purely an implicit conception or rule, but a combination of rules of thumb, paradigm cases, and a sense of mathematical similarities. That is, incomplete conceptualization of a definite concept that is being thought with may be present at both explicit and implicit levels. The individual's ability to get the examples right may be explained by a combination of mental abilities that do not fall at just the level of a correct conceptualization, implicit or explicit. It seems to me doubtful that implicit *conceptions* explain all the phenomena that Peacocke is concerned to explain.

The individual also may not, at the explicit level, make all the right judgments about examples, yet may still grasp a definite concept that includes those examples. It may be that general principles of mathematical practice and rationality can be seen, retrospectively, to warrant inclusion of certain cases under a given concept, even though the individual expert may be disposed to misjudge those cases in individual instances.

I think that this case may be illustrated in the early history of the concept of *set*, when limitative prejudices blocked natural generalizations for at least some experts. I think that the early disputes over the axiom of choice can be seen, at least in some instances, in this light. It is certain that some of those disputes derived from mathematicians having different concepts, while using the same term 'set' to express them. Some had the modern iterative concept of *set*. Some had a concept closer to the modern concept of *class*. But some of the disputes seem to have stemmed from objections to the axiom of choice that were driven not by a noniterative concept of *set*, but by a sense of a need to limit the proliferation of sets by a closer epistemic control on their postulation than the axiom of choice provided.[2] I think that mathematicians with such philosophical views as these had the iterative concept of set, but made mistaken judgments about what counted as a set (rejecting consequences of the axiom of choice), because of philosophical prejudices that interfered with what has come to be seen as sound mathematical practice.

At any rate, it seems to me that it would be a mistake to think that the implicit mental structures that explain explicit judgments *must* themselves always be complete. Sometimes we depend on others. Sometimes we depend on a combination of examples, an unconceptualized *sense* of similarity, and principles at the wrong levels to count as conceptions associated with the concept at issue. Sometimes the limits of a concept are determined partly by similarity principles but partly by whatever happens in actual fact to account for the types of samples that are actually counted as examples. This anti-individualist element in our mastery of concepts combines with the points about the role of non-conceptualized psychological elements in our use of concepts to indicate that the actual correct application of concepts we have is guided by more than implicit conceptions that we associate with the concepts.

II

This psychological point seems to me to bear significantly on our understanding of the epistemology of reflection. Most traditional accounts of apriori reflection have, like Peacocke's, assumed that a conception of the rule associated with application of the concept is—at some implicit and unconscious, or subliminally conscious level—fully formed in any individual that has the concept. Reflection was seen as just a matter of bringing to consciousness and fully articulating a conception or rule that is already present in the mind. Peacocke notes, as traditional rationalists tended not to, that the implicit conception will sometimes be at a subpersonal level. So it will not be accessible to reflection or person-level inference. This seems to me correct. But I think that the distance from traditional conceptions of implicit mastery of concepts goes further.

Kant sometimes writes as if he *identifies* concepts with functions of unity, or rules, for holding cases and subordinate concepts together.[3] He saw the rules as

[2] For a detailed account of the controversies over the axiom of choice see Gregory H. Moore, *Zermelo's Axiom of Choice: Its Origins, Development, and Influence* (New York: Springer Verlag, 1982).

[3] Kant, *Critique of Reason* (B92–93, A103). Ultimately, Kant's view is much closer to mine than these passages suggest.

produced and, at some level, as grasped by the understanding. But the twentieth century has seen an emphasis on the role of instances or examples in individuating concepts. Wittgenstein, Kripke, and Putnam, in their different ways, have indicated that concepts are not entirely fixed by background rules, principles, or descriptions that the individual has grasped. I have tried to develop this idea in my own ways. Implicit in this emphasis on the role of instances or examples in individuating concepts is, I think, a recognition that some of the ways that we have for projecting from examples are stored non-conceptually. The perceptual system and non-conceptualized senses of similarity may guide our projection from central instances to which a concept applies, to further instances. Only with reflective conceptualization of rules that codify these lower-level abilities do we arrive at conceptions that are adequate to explain our application of certain concepts.

Suppose that concepts are not always backed by implicit conceptions—conceptualizations—that explain our application. So reflection on the nature of our concepts is not always a matter of bringing to consciousness a conceptualized rule that guides their application. It is part of the formation of such a rule. But such formation cannot be seen as formation of the concept. For the concept is already fully formed, thought with, and even correctly applied. To think with the concept and even to have a sharp grasp of it, in the sense discussed earlier, it is not necessary that one have an associated descriptive rule for its application.

This point is relevant to understanding philosophical thought experiments. For example, in the Twin Earth cases it is commonly assumed that we all have, unconsciously in mind, descriptions, rules, or principles that guide our use of such concepts as *arthritis, chair, sofa, edge, water*, and so on. It is conceded that these are hard to formulate. Still, it is often assumed that they are always somehow implicitly complete and present in the individual's mind. I believe that this assumption is mistaken. Some philosophers who have sought to refute the thought experiments have added to this view the further mistake of identifying concepts with the supposed underlying descriptions, rules, or principles. They presume that their formulations are themselves environmentally independent. (This presumption is itself unargued and in many cases unconvincing.) They then maintain that these descriptions, rules, or principles guide the individual's use of a term or concept, regardless of the environment.

This line commonly makes further mistakes. But the one that interests me here is the assumption that when an individual has a concept, there is always in the individual an associated, fully formed implicit conception that explains the application of the concept and applies to the same instances that it applies to. Reflection on the nature and application of our concepts seems to me a more complex enterprise than making conscious certain conceptualizations, rules, principles, descriptions, or definitions that are already implicitly in the mind and associated with the concept.

The epistemology of reflection is, I think, correspondingly more complex than traditional philosophy has represented it. It is natural and traditional to see reflection on the nature of concepts as warranted apriori. Let us suppose that 'apriori' means 'independent in justificational force from sense perception or sense-perceptual belief'. Suppose that we are reflecting on the nature of our concept *chair*. Suppose that we recognize that the concept applies—and long has applied—to ski-lift chairs, without legs; but we have no conceptualized principle, even implicitly, for projecting from our standard cases of chairs to these special cases. Thus the connection between the standard cases of chairs and the ski-lift chairs is simply that we have stored perceptual similarities between the cases and have accepted long ago someone's calling a ski-lift chair (or ski-lift chairs) a chair. Thus we are supposing that when we recognize that a ski-lift chair is a chair we do not derive this recognition from a principle or conception that includes specification of the properties of ski-lift chairs that make them chairs. Rather, we make use of a memory of the case and a confirmatory sense of perceptual similarity and generalized conception of functional similarity between the cases. Thus, we note that the ski-lift chairs have a flat seat, accommodate one or two persons, and function to be sat on. But our acceptance of the case is driven not by a generalization but by our memory that *that sort* of object—or perhaps even a particular remembered object—counts as a chair.

Is such a memory warranted apriori? That depends on the nature of the warrant for present tense claims like *that sort of object, which is used in ski-lifts, counts as a type of chair* and *that object counts as a chair*, where the claims are taken to have a role in specifying or teaching the nature of the concept, not merely *ordinary* statements of fact.[4] These questions are complex and multifaceted, and I will not try to answer them here. I want simply to raise them. I do not think that these questions arise for all cases of reflection on the nature of our understanding of concepts. It seems to me that sometimes Peacocke's account of implicit conceptions is correct and fully adequate. I just want to point to what I regard as further complexities that warrant our attention.

III

Let me turn now to issues about explaining rationality, with particular reference to Peacocke's historical points. His invocation of Leibniz and the rationalist tradition seems to me entirely appropriate. I agree with most of what he says in this section. I have here three reservations, two of them of perhaps only minor significance.

As Peacocke notes, Leibniz holds that axioms are evident as soon as their terms are understood. He objects that Leibniz here overstates the ease of the

[4] Of course, I agree with Quine against conventionalists that it is *both* a claim of fact and a claim about meaning or concepts (W. V. Quine, *The Ways of Paradox and Other Essays* (New York: Random House, 1966)). Obviously these questions are close to issues that Kripke raises about the contingent apriori (Saul Kripke 'Naming and Necessity', in Donald Davidson and Gilbert Harman (eds.), *Semantics of Natural Language* (Dordrecht: Reidel, 1972, 253–355)).

discovery of axioms. This may be so. But it is important to remember that Leibniz meant by 'axiom' not just any proposition that might be taken as a starting point for an axiomatic theory. He had the old Euclidean conception of axioms as truths that are basic in a justificational order and that are sufficiently simple that, assuming they are fully understood, there is no need to argue for them or derive them from anything else. I think such a conception has more to be said for it than most modern philosophers presume. Moreover, I think that Leibniz set a very high standard for understanding of the terms. I think that he meant complete, explicit understanding, not merely the sort of understanding sufficient to use the terms and reason with them. So it is not clear to me that there is any mistake in Leibniz's view that given full understanding, one finds basic truths evident. I think the view is virtually definitional of the traditional conception of an axiom. It seems to me that Leibniz is right about a narrow class of truths that might be counted axioms in the old sense. Simple truths of logic and arithmetic seem to me to be so basic that if one understands them, one realizes that they are true. No argument for them could provide them with a justification that adds force that is not already present in understanding them.

The second reservation concerns Peacocke's account of what it is to be clear but not distinct. He notes that according to the traditional view, an idea or concept is clear for a person if the person can use it to recognize instances of the concept. He quotes Leibniz's remark that having a distinct idea lies in the ability to enumerate separately the essential characteristics that distinguish the things the idea applies to from other things. He explicates distinctness in his framework. He claims that a thinker with a distinct idea is one who has succeeded in achieving an explicit formulation of the implicit conception he had when he had only a clear but indistinct idea.

This claim seems only approximately true. For as Peacocke himself remarks, implicit conceptions can be incorrect explications of an individual's idea or concept. In such cases, making them explicit would not be making them distinct. Further, suppose that I am right that even implicit conceptions that are correct as far as they go may not have *conceptualized* the full correct explication of a concept. So merely making such incomplete implicit conceptions explicit will not suffice to make a concept distinct in the old fashioned sense—that is, give it a full, correct explication. Incompleteness of explicational understanding, or indistinctness, can hold at the implicit as well as the explicit level.

A third reservation, the one that interests me most, has to do with Peacocke's discussion of Frege and the rationality of accepting logical axioms. He points out that Frege gave arguments for his axioms from semantical-looking background assumptions for the truth of most of his axioms. Yet he regarded the axioms as self-evident—that is, recognizable as true independently of justifying them through derivation from other truths. In my view, although Frege did not philosophize about this apparent oddity, he knew exactly what he was doing. What is more important, there is a philosophically tenable resolution of the apparent conflict.[5]

First, it should be noticed that the arguments Frege gives that have his axioms as conclusions are fully explicit. So the apparent conflict arises independently of any distinction between implicit and explicit levels. Second, it is important to distinguish between justification of a sentence's expression of an axiom and justification of the axiom itself. Frege believed—and I agree—that the fundamental truths of logic are not strings of symbols, even though strings of symbols express such truths. Frege is interested primarily in the truths, but he is simultaneously setting out and justifying his logical symbolism by showing its adequacy to express the underlying truths. The arguments in question, in *Basic Laws*, bear on both the symbolism and on the logical truths, but in different ways. Close analysis can separate out these points. But Frege is fairly loose in his book about slipping back and forth from semantical discussion about symbols to substantive exposition of his truths. Peacocke also writes sometimes of derivations of sentences (p. 146) and other times of derivations of the logical truths themselves (p. 146). Of course, both are at issue. But the bearing of the semantical arguments is different in the two cases.

It seems to me that the semantical arguments do provide, in a straightforward way, a justification for axiom-expressions and for formal symbolic expressions of the rules of inference. They show rigorously that the symbols are adequate to express what we recognize as axiomatic logical truths and valid inference rules.

But the bearing of the arguments is different on the logical truths and inference rules themselves. It seems to me incontestable that Frege would not have regarded any arguments from language as being capable of justifying language-independent logical truths and rules of inference. In fact, however, most of Frege's semantical-looking arguments for the axioms make no essential reference to language at all. Still, they are arguments from truth-conditions (associated with thought contents) to the language-independent axioms. And the very fact that they are arguments with the axioms as conclusions is already puzzling, since the axioms are supposed to be self-evident and not in need of proof.

What is going on, and what can be shown from comparing several of Frege's texts, is that Frege regards the arguments not as justifying the conclusion but as articulating the content of the conclusion. The arguments are not intended to provide justification for belief in the conclusion by deriving it from premises belief in which is antecedently justified. For, as he says, the conclusions are not in need of proof or justification. Frege means by 'proof' a deductive argument that provides justification from self-evident basic truths as premises.[6] The articulation

[5] For a detailed discussion of these matters, emphasizing the historical point of view, see my 'Frege on Knowing the Foundation', *Mind* 107 (1998), 305–347.

[6] The phrase derives from Leibniz, *New Essays on Human Understanding* (1705, 1765), ed. and trans. P. Remnant and J. Bennett (New York: Cambridge University Press, 1981), e.g. IV, ix, 2; 434; see also Gottlob Frege, *Foundations of Arithmetic*, trans. J. L. Austin (Evanston, IL: Northwestern University Press, 1968), section 3. This notion of proof is discussed at some length in my 'Frege on

of content that the arguments provide is simply an articulation of understanding of the conclusion. So it remains possible for him to hold that the content of the conclusion carries all the evidence needed to recognize the conclusion as true: That is, the conclusions of the arguments, the axioms, are self-evident. Understanding the axioms justifies one in believing them; but full, explicit understanding itself requires an ability to articulate the truth-conditions of the contents that are understood. I want to elaborate this point a bit, since I think it correct. Three background points are important.

One is that any understanding, even understanding of simple logical truths, requires mastery of complex inferential connections. This is one of Frege's greatest contributions to philosophy, and something not present with anything like the same clarity in Leibniz. The point requires that understanding of both terms and propositions is not independent of acceptance of principles and inferential connections. So understanding a logical truth is associated with arguments using the terms or concepts embedded in the truth.

Second, as Frege also famously maintained, what is understood places conditions on truth. So understanding the sense of a sentence, or understanding (grasping) a thought, requires understanding its truth-conditions. Putting the two points together, understanding a logical truth in a fully articulate way requires an ability to articulate through argument its sense or truth-conditions.

Third, it is doubtful that arguments to at least some very simple logical truths—and at least some simple rules of inference—from their truth-conditions are arguments that provide any additional justificational force to that already involved in really understanding those logical truths or rules of inferences. The premises and rules of inference used in the arguments from the lines of a truth table are not any more strongly justified, or more evident, in at least many simple cases, than the logical axioms that they are used to derive. Whatever role Frege's arguments from truth-conditions play, they do not provide any extra rational support or warrant for their conclusions beyond what is involved in understanding the conclusions. The conclusions are, in these cases, at least as rationally evident as the premises. I believe that in Frege's sense of 'justification' and 'proof', the relevant arguments are not justifications or proofs of their conclusion.

Some things in Peacocke's exposition are congenial with this point. He writes of the arguments as *explaining* our rationality in accepting the premises. Such explanation might be distinguished from justification that adds justificational force to our warrant in accepting the conclusion. He also writes of explaining the evidentness of the axiom. Again, such explanations might be seen as articulations of our understanding, not as justifications from more basic premises. Moreover, with certain qualifications I will not try to state here, I join Peacocke and follow Frege in holding that what is understood is to be explicated in terms of truth-conditions.

Knowing the Foundation' and my 'Frege on Apriority', in C. Peacocke and P. Boghossian (eds.), *New Essays on the A Priori* (Oxford: Oxford University Press, 2000), 11–42.

But I do not believe that Frege's arguments for the simplest logical axioms, where these axioms are understood to be thought contents or propositions, were meant as justifications of them. The arguments do not add any justificational force not already involved in complete understanding of the content of the propositions in question. Frege was surely fully aware of this fact.

I think that the view that Frege seems to have held is correct: Understanding the axioms requires an ability to give the sorts of arguments from truth-conditions that he gives in articulating the intentional content of the axioms. But it is the understanding of the axioms themselves, not a justification of them from antecedently understood principles governing truth-conditions, that is fully sufficient for being warranted in believing them. In this sense the axioms are evident *in themselves* and not in need of justification or proof from other truths.

Frege's great contribution is to indicate that because of the dependence of understanding on mastering inferential connections, "in themselves" is a more subtle and complex notion than most traditional philosophers realized. A thought has a definite content, but that content is logically connected to other contents. And thinking with the content necessitates being able to make some of the logical connections. *Understanding* the content (whether minimally or in some deeper way) requires understanding some of the inferential connections. But once understanding is achieved, once one *has* and understands the relevant contents, justification does not require: deriving it from other contents from which it inherits justification. The position seems to me to derive support from the fact that the semantical arguments seem intuitively to add no warrant to full understanding of the conclusion. In fact, they seem to be just argumentative articulations of such understanding.

It is important to bear in mind that there are different possible arguments with the axioms as conclusions that articulate understanding of the axioms. These differences reflect the fact that no one argument is necessary for understanding. Let us consider

$$A \rightarrow (B \rightarrow A)$$

There is, for example, the type of articulation Peacocke outlines. This type appeals to a meta-perspective and explicitly uses a concept of truth that is a predicate of thoughts. Frege gives two other types of arguments, both in the object language: The first one starts with a step we would formalize as

$$\sim (A \rightarrow (B \rightarrow A)) \rightarrow (A \;\&\; (B \;\&\; \sim A))$$

By commutativity and associativity of conjunction, conjunct-elimination, non-contradiction, *modus tollens*, and double-negative removal:

$$A \rightarrow (B \rightarrow A)$$

The second argument (which Frege gives in *Begriffsschrift*) is that if A, then A regardless of whether B, for any B. All of these arguments make use of an understanding of the truth-conditions of the conditional. None seems to do more than articulate what is involved in full understanding of the axiom. And none seems to rest on principles that are clearly more fundamental, or more obviously true than the axiom, or are self-evident in a way that the axiom itself is not.

I am happy to concede that in the case of each principle or rule there is an explanation of the rationality of accepting it. But I am not inclined to think that the most basic principles receive any genuine or needed epistemic warrant—or positive justificational force for believing them—through the arguments. The fundamental warrant for believing them lies in understanding their content. They are self-evident. It is just that any such understanding has to be accompanied by an ability to explain the rationality of accepting the proposition through discursive argument.

I remain attracted to a conception of rational justification of the simplest principles of elementary logic and arithmetic that is broadly similar to the conception shared by Frege and Leibniz. According to this conception there are certain basic truths and rules of inference. Understanding these suffices to warrant belief in them. No argument for them can yield warrant for belief that adds force to the warrant already yielded by understanding them. Such warrant is maximal. The truths "do not need or admit of" any further justification. (See note 6.) In this sense they are basic and self-evident. Arguments of the sort we have been discussing elicit the fact that there are more basic truths and rules of inference (all equally basic) than are needed to develop logic, at least first- and second-order logic. So the foundation is overdetermined.

This feature of overdetermination makes contrasting this sort of foundationalism with a sophisticated coherentism a subtle and perhaps uninteresting terminological matter. I think, however, that warrant is best seen as lying in the understanding of the relevant truths. There is no need to rest the warrant on "coherence" with other truths. But the understanding unquestionably involves abilities to make inferential connections with other truths. And there is, of course, a necessary coherence among the logical and arithmetical truths.

Like Frege, I combine this foundationalist justificational structure with a pluralistic, coherentist conception of understanding. Understanding requires an ability to make inferences to and from the understood intentional contents. Any of various inferential patterns of connections among thoughts can suffice to yield understanding of fundamental logical truths or rules of inference. For basic truths the connections hold both between thoughts and rules of inference that are equally fundamental, and between self-evident ones and some less basic ones.

More needs to be said about the distinction between the role of understanding in rational acceptance of the principles and the role of argument in articulating the understanding. The naturalistic and holistic tendencies that we have inherited from Quine tend to ignore or blur such a distinction. But anyone who refuses simply to reduce understanding or grasp of a thought-content to some *particular* pattern of inferential abilities—while still holding that understanding requires some such pattern of inferential abilities—is in a position to draw it.

More also needs to be said about the traditional notion of basic, self-evident truths. Many have doubted that there remains any use for the idea that some thoughts in logic and mathematics are, from an epistemological or psychological point of view, maximally basic. Even those who do not embrace the empiricist view that logic and arithmetic depend for their justification on their role in empirical science commonly emphasize that there are so many "axiomatizations" of formal theories that finding basic ones is a pointless exercise.[7]

This negative attitude often derives from mixing up the modern conception of *axiom* with the traditional one. The modern conception is centered on what is *taken* as basic in a particular presentation of a theory. Certainly the variety of possible "axioms" in this sense is endless. And certainly some things that are "axioms" in the modern sense are in no way self-evident or epistemically self-sufficient. Some of the axioms of high-level set theory, for example, are certainly not self-evident. They are not even derivable from self-evident truths. Not all of mathematics is derivative from self-evident truths. It does not follow that there are no epistemically fundamental truths in logic or mathematics. In fact, it seems quite obvious that there are truths that are for us epistemically more basic than others. And it seems to me arguable and likely that there are truths that are maximally basic, in the sense that no argument for them could add to the warrant inherent in understanding them.

Whether there are any basic truths that are basic for all finite rational beings is a further question. I am sympathetic to the idea that some truths and rules of inference are necessarily basic for every finite rational being that has the relevant concepts. But I leave this an open question. It is a question that would require extensive and subtle development. Whatever the answer to this question, it seems overwhelmingly likely that, as regards arithmetic and logic, there are broad similarities among human beings in what count for them as simpler and epistemically more fundamental truths, and in what count as more complex and epistemically derivative. It seems to me likely that the line between what is justificationally basic and what is justificationally derivative may be blurred in some cases, for some individuals. And it seems likely that where the line is drawn will vary with the individual. Some individuals may include more truths as basic than others, depending on the depth of their understanding.

The old-fashioned picture of a rational order of truths is out of favor. There are certainly many obstacles to bringing such a picture back into focus. Even a rationalist picture that is much more modest and more qualified than the traditional versions faces numerous obstacles. I think, however, that such a picture is worth developing.

[7] For a contrary view about all this, see Christopher Cherniak, *Minimal Rationality* (Cambridge, Mass.: MIT Press, 1986). None of what follows is meant to do justice to Cherniak's position. I am not, however, persuaded by his arguments. For an interesting discussion of related issues, see Simon Evnine, 'Believing Conjunctions', *Synthese* 118, 201–227.

24 *Reflection*

Classical rationalists appealed to a capacity for reflection. As a first approximation, reflection is a type of rational cognition with four significant features. It goes beyond what is immediately obvious. It is higher-order, in the sense that it involves thought about psychological states or representational contents, although its conclusions need not be about psychological states or representational contents. It aims at constitutive understanding. And it develops such understanding by drawing conclusions, without acquiring new premises, empirical or otherwise, beyond what is already understood or known. Usually reflection aims at improving pre-reflective understanding or knowledge. Much reflection makes use of empirical background knowledge. Where the force of its warrant is independent of sense experience, reflection is apriori.

Some apriori knowledge needs no reflection. Knowledge that 2 + 2 = 4, that cruelty is wrong, and that everything is self-identical seem both too obvious and too first-order to be a product of reflection. Apriori knowledge deriving from first-order reasoning from first-order knowledge also does not derive from reflection. Thus much reasoning in mathematics does not.

Traditionally, reflection was supposed to begin with an attempt to clarify ideas, concepts, or relations among cognitive capacities. It was supposed to arrive at truths by considering thoughts, or principles guiding thoughts—even if the ultimate subject matter was not the thoughts themselves. Thus reflection has played at least an auxiliary role in any both empirical and mathematical sciences, as well as in philosophy.

Obviousness is perhaps a matter of degree. Involvement of a higher-order component is perhaps a matter of emphasis. Clarifying concepts or meanings and development of first-order theory about a subject matter are interdependent. I will not try to sharpen these matters here.

A lot of apriori knowledge, however, seems to utilize and require reflection. Philosophy offers examples of reflection's yielding purported apriori knowledge or apriori warranted belief. Philosophy is not alone. Meta-logic offers examples of reflection's yielding apriori knowledge.[1] Neither of these disciplines invokes

[1] Kant assimilated most or all reflection to reasoning in either general logic or what he called 'transcendental logic'. *Critique of Pure Reason* A261–263/B316–319.

special cognitive powers. They use refinements of forms of understanding and reasoning that are available to common sense and to first-order theorizing.

Although the notion of reflection is only roughly delimited and is incompletely understood by all of us, I take reflection to be a source of knowledge and warranted belief. Some of the knowledge and warrant are, I think, apriori.

I

The classical rationalist conception of reflection holds three main theses. The first is that *in reflection an individual always brings to articulated consciousness steps or conclusions that are implicitly present, subliminally or unconsciously, in the individual's mind before reflection.*

The relevant sense of 'implicit' is *unconscious but capable of being accessed to consciousness through reflection.*[2] The idea is that the material that is accessed through reflection not only marks a mental state that was present before reflection. The mental state so marked must itself be accessible to consciousness through reflection. Reflection brings to full consciousness thought contents and attitudes that were already harbored, though only unconsciously or dimly.

All major classical rationalists thought that all materials for understanding the principles obtained through reflection are present in the mind before and during every stage of reflection.[3] This view was often motivated by the idea that in thought and reasoning, we "must know what we are doing": We must implicitly understand the principles that govern our thoughts. It is these principles that we get at when we succeed in making them explicit through reflection.

The second thesis is that *reflection can yield apriori knowledge of objective subject matters, beyond thoughts that the reflector is engaging in.*

'Apriori' means independent of sense experience for its warranting force. 'Objectivity' carries a variety of senses, depending on the author. A minimum sense is independent of the idiosyncrasies of the individual.

The classical rationalists thought that reflection yields knowledge of subject matters that are not explicitly mental. They thought that even with regard to mind, reflection yields knowledge or cognition of principles that are necessary and universal. So they thought that some of the apriori knowledge applies beyond the

[2] Leibniz, *New Essays on Human Understanding*, Preface, 51, provides a closely related characterization: '. . . reflection is nothing but attention to what is within us . . . '.

[3] Descartes, *Meditations*, V; Leibniz, *New Essays on Human Understanding*, I, i, 73ff.; Kant, *Critique of Pure Reason* A261–263/B316–319. Descartes thought that reflection consists in clear and distinct cognition of natures and their relations. Leibniz thought that it consists in analysis of the structure of ideas or concepts. Kant thought that Leibniz was right about reflection in logic, but that the most fruitful reflection derived from bringing to consciousness a subliminal self-awareness of the procedures governing use of ideas or concepts. Kant went so far as to hold that reflection itself is present, though implicit, in the natural pre-theoretic operations of all thought. See *Critique of Pure Reason*, A103–108; *Anthropology* section 7, AA vii, 141.

psychology of the particular individual doing the reflecting. The thesis entails that reflection can yield knowledge that is distinct from the self-knowledge involved in awareness of what thoughts one is thinking.

Again, different writers explain these epistemic commitments in different ways. Descartes maintains simply that what one reflects upon is relations among ideas, and that these relations include self-evident principles. He holds that the principles emerge when one makes the ideas clear and distinct. But he provides no satisfying account of how one goes about making ideas clear and distinct. For this deficiency he was criticized by all his successors.

Leibniz develops Descartes' views in a particular way: all truths are truths of analytic containment of one idea or concept in another. So for Leibniz reflection takes the form of analyzing ideas or concepts to find definitional relations among them. The definitional relations are taken to reflect an order of epistemic priority.[4]

Kant takes Leibniz's view of reflection to be roughly correct for general logic, but incorrect beyond general logic. He maintains that general logic yields no substantive cognition of a subject matter—but merely presents rules and principles for rational thinking. So for reflective knowledge or cognition that is substantial, he proposes a significantly different picture of what reflection is up to. He holds that reflection can yield cognition of a subject matter only by giving up analysis of the content of thoughts or concepts and by centering on the function and use of mental capacities. He holds that reflection yields cognition of the "form" of such capacities, which he takes to include principles and rules governing their correct use.[5]

All these authors emphasize the normative relevance of principles obtained through reflection. Such principles yield norms for thinking as one "ought"—and not just because the principles are true. The principles are supposed to provide a framework for all of thought. They provide norms for the right direction of the mind in pre-theoretic and unreflective theoretical domains.

Descartes and Leibniz rest huge metaphysical systems on reflection. As is well known, Kant criticized these systems and metaphysical inquiry in general. He gives reflection the role of yielding apriori principles governing the form of cognition. But he agrees with his predecessors in regarding reflection as a source of apriori cognition of necessities with universal application.

The third thesis is that *successful reflection requires skillful reasoning and is difficult: it is not a matter of one-off introspection or intuition*. Descartes thought

[4] For further discussion of these matters, see my 'Frege on Apriority', in C. Peacocke and P. Boghossian (eds.), *New Essays on the A Priori* (Oxford: Oxford University Press, 2000); and 'Frege on Knowing the Foundation', *Mind* 107 (1998), 305–347. Both are reprinted in my *Truth, Thought, Reason: Essays on Frege* (Oxford: Clarendon Press, 2005).

[5] Two further doctrines that are very particular to Kant are (a) that cognition of form falls short of cognition of actual entities that instantiate the form; and (b) that form is made by the mind, not discovered as a mind-independent subject matter. The former view is part of his rejection of traditional metaphysics in favor of a "transcendental" meta-theory about possible cognition. The latter view is fundamental to his idealism.

that most people never come to understand any of their own ideas. Perhaps he thought that most people never reflect. But equally he thought that successful reflection requires great skill.[6]

Classical rationalists hold that even though reflection often yields objectively self-evident principles (their warrant lies in themselves), reflective knowledge is never obvious to just anyone who tries to reflect. One begins with incomplete understanding of what one reflects on, at the conscious, articulate level of thinking. Reflection is held to require following a method. Often the method is dialectical.[7] Reflective method involves following rational principles in a systematic, discursive, often dialectical manner.[8]

II

I accept the second and third theses and reject the first.

The third thesis seems to me exactly right. The general warning that reflection is not to be conflated with simple introspection or with spontaneous judgment is well taken. Reflection follows discursive methods. Some people are better at it than others.

I also accept the second thesis. I think that reflection has been a source of apriori knowledge that is not confined to self-knowledge. Of course, I differ with classical rationalists about the scope of reflection for yielding apriori knowledge. Nearly all philosophers since Kant do. Leibniz and Descartes are wildly over-optimistic regarding how much reflection can tell us about the nature of the physical world, and even of mind. I believe that both philosophers seriously underestimate the role of empirical experiment in natural and psychological science. Kant, who I think understood natural science much better, still overestimates the role of reflection in determining principles for natural science; and he overestimates the scope of reflection in certain other ways. On the other hand, I believe that Kant underestimated the scope for reflection in both logic and philosophy of mind.

I think that Descartes and Leibniz are right to hold that some not immediately obvious principles can be understood to be self-evident through reflection. Leibniz's appeal to analysis of concepts also has a genuine but extremely limited

[6] Descartes, *Rules for the Direction of the Mind*, Rule IX, XII. See also Leibniz, *New Essays*, I, i, sections 23–27, 85–88. Kant, *Critique of Pure Reason*, B1–2. The *hubris* of classical and even Kantian rationalism was slightly mitigated by an emphasis on the difficulty of following rationalist methods, and the need for following the direction of the master teacher.

[7] The Platonic dialogs provided a historical precedent for elaborating the dialectical methodology of reflection.

[8] Descartes and Leibniz favored the Platonic metaphor of vision for describing the understanding arrived at through successful reflection. Although Kant occasionally uses the visual metaphor (*Critique of Pure Reason* A108), he is more like Frege and twentieth-century authors in emphasizing the essentially discursive character of understanding. In *practice*, even Plato, Descartes, and Leibniz reflect much more discursively than their vision-dominated glosses suggest.

role to play in reflection. Kant's centering reflection on the function and use of mental capacities seems to me a fundamental step forward.

Some of the methods classical rationalists used to achieve apriori knowledge through reflection now seem naive. But I believe that they were right to strive for apriori knowledge, and sometimes successful in finding it. Moreover, some special normative status for principles obtained from reflection also seems an inevitable consequence of the second thesis.

My central disagreement with the classical rationalist conception of reflection centers on the first thesis. I will summarize, as background, points I have made in earlier work.

The first thesis maintains that reflecting individuals always bring to consciousness, through discursive methods, what is already implicitly present in their own minds. Anti-individualism has combined with advances in psychology to undermine this thesis.

Anti-individualism is the view that the nature of many representational mental states is constitutively dependent on relations between the individual who has those states and aspects of the environment, or other objective subject matter. The relations can be complex—running through other mental states or through other individuals, including the individual's species-ancestors. The key point is that many mental states are constitutively dependent on there being non-representational, explanatorily relevant relations between the individual and the environment or other subject matter. I have argued for this view elsewhere.[9] The upshot of this view is that the representational states one is in are less a matter of cognitive control and internal mastery, even "implicit" cognitive control and mastery, than classical views assumed.[10]

Two main advances in psychology also bear on the first thesis. One is the fact that capacities of certain animals and young children are best explained in terms of simple propositional attitudes. Since animals and young children seemingly lack the conceptual competence to make their own attitudes the topic of their thinking, they seem to have first-order thought while lacking a capacity for the higher-order thought involved in reflection. Nearly all classical rationalists maintain, on I think no good grounds, that animals lack propositional attitudes. Individuals can think without "knowing what they are doing". They can even lack a capacity to develop a propositional understanding of their thinking. An individual need not be able to represent principles or even inference rules that govern the individual's thinking. Principles and conceptual relations may explain mental activities without guiding them—without the individual's following them

[9] [Added 2011:] See my *Foundations of Mind: Philosophical Essays, Volume II* (Oxford: Oxford University Press, 2007).

[10] This is not to say that the rationalists rejected anti-individualism. I think that Descartes, for example, accepted it. See my 'Descartes on Anti-Individualism' in *Foundations of Mind.* It is more that they did not think through the implications of anti-individualism.

by implicitly knowing them. Not all cognitive activity that is correctly explained as following rules involves a capacity to represent the rules, even unconsciously.

The other advance in psychology is recognition that many principles governing mental activity, even in adult human beings, are modular, or otherwise inaccessible to being reflectively brought to consciousness. Even principles governing non-modular, accessible cognitive activity are often inaccessible in this way.

These facts refute Leibniz's and Kant's extreme claim that all thought constitutively involves reflection. But they do not in themselves refute the first thesis. One could hold that *in those domains, and in those types of reasoners, where successful reflection is possible*, the mental state marked by the relevant principle is in the mind. But such a position is not plausible in light of both psychology and the history of science. Psychology has brought out that accessible higher-order, person-level cognitive control plays a very small role in much of our propositional activity. Principles that best explicate a mathematical or natural-scientific concept are often discoverable only by developing new knowledge, knowledge that it is not psychologically plausible to impute to the unconscious of reasoners before the new knowledge is discovered. Sometimes explication of principles requires development of new concepts. Reflection on one's past usage can yield principles that were not present in the mind before the reflection began. Reflection seems sometimes to produce understanding of a principle that did not mark any antecedent mental state.

Let us develop these considerations at somewhat closer range. In the twentieth century a number of philosophers have emphasized the role of *examples* in attaining illumination through reflection. This emphasis is not new. Some classical rationalists, particularly Socrates and Descartes, used examples prominently in reflection. The aim is to use examples to arrive at principles. The classical idea is that in making judgements about examples, we are guided by principles. The examples help make the principles more explicit. I believe that this idea is sometimes correct. But its scope must be qualified in light of the considerations just advanced.

There is reason to believe that principles that explain judgments about examples are not always stored in the mind of the judge. The person may go on perceptual similarity, where the general principles governing similarity are not yet conceptualized. The principles can explain the interaction of perception and perceptual memory, but be nowhere represented in the individual's psychology. For example, in judging something to be a chair, an individual might operate under principles governing similarity to a set of perceptual paradigms—but might lack conceptualization of the principles. Or, for other reasons, an individual might never have formed a propositional attitude toward the principles, however implicitly.[11]

[11] See my 'Intellectual Norms and Foundations of Mind', *The Journal of Philosophy* 83 (1986), 697–720; reprinted in *Foundations of Mind*; 'Concepts, Conceptions, Reflective Understanding', in

This kind of point applies in non-empirical domains. A mathematician might operate with a sense that two examples are similar enough to fall under a single concept, but lack the concepts necessary to make systematic inferential use of the principles governing the sense of similarity. The concepts in such principles may not yet be formed by the individual. The principles may, for other reasons, not be internal to the individual's psychology. Sub-conceptual senses of similarity may be the psychological basis for the judgments about examples. Conceptualization of the principles that cover the cases may come later.

Interlocution also plays a role in explaining judgments about examples. An individual can make judgments through memory of being told that the example falls under some concept. One might judge a seat in a ski-lift to be a chair because one remembers its being counted a chair—without having formed any general principle that unites the examples that would be counted chairs. Again, what explains and vindicates the individual's application of a concept to cases may not be a conceptualized principle internal to the individual's psychology.[12]

In some cases, the relevant principles governing *correct* application transcend what is available even to a community. In the case of natural kinds, there seems to be some conditional commitment in individuals' psychology to the effect that *if* a concept is a natural kind concept (something that cannot be known apriori), it applies to all instances of the principles that in fact govern the natural kind. These principles may not yet be discovered. The psychological ability to track examples lies in an ability to track superficial properties of the examples that by-and-large line up with the natural kinds in question. I will mention analogous cases in mathematics later.

More broadly, in a wide range of judgments about examples, there is a commitment in thinkers to the objectivity of the subject matter, whether or not the relevant entities instantiate kinds that interest science or mathematics. The correct explication of principles governing the application of concepts can depend on the outcome of investigations that the individual has not engaged in. Even experts may not have internalized principles that explain correct application.[13]

This latter point is supported by consideration not only of thought experiments but also of intellectual history. The correct account of the application conditions for concepts may emerge long after those concepts are in use. This sort of situation is most vividly illustrated in the history of scientific or mathematical concepts. But it is potentially applicable over a wider range.[14] Sometimes, one

M. Hahn and B. Ramberg (eds.), *Reflections and Replies: Essays on the Philosophy of Tyler Burge* (Cambridge, Mass.: MIT Press, 2003); reprinted in this volume.

[12] There is a tendency in classical rationalism to underestimate the role of interlocution and reliance on others in both common sense and scientific domains.

[13] 'Intellectual Norms and Foundations of Mind'.

[14] See 'Intellectual Norms and Foundations of Mind' and 'Frege on Sense and Linguistic Meaning', in D. Bell and N. Cooper (eds.), *The Analytic Tradition: Meaning, Thought, and Knowledge* (Oxford: Blackwell, 1990); reprinted in *Truth, Thought, Reason*.

can make progress in understanding a concept and the phenomonea that it applies to by reflecting on applications that had previously been held together either by a non-conceptualized sense of similarity, or by acceptance of the word of others.

I and others have emphasized that individuals can harbor mistaken explicatory beliefs.[15] Even settled views about principles that govern the application of concepts, purportedly providing necessary and/or sufficient conditions for falling under a concept, are frequently mistaken. In itself, the classical view of reflection accepts such cases. By the third thesis, it emphasizes the difficulty of finding correct principles. It must, however, maintain that the individual's incorrect explications are incompatible with correct principles that are, according to the first thesis, "implicitly" stored in the mind.

Both psychology and anti-individualism have cast incorrect explications in a different light. The correct principles need not be stored in the individual's psychology at all. Explanation of the psychology of incorrect explicatory judgments about examples often should not appeal to implicit internal contradictions. Judgments about examples are often correctly explained without assuming that principles governing the psychology or the correctness of the judgments are the content of some prior psychological state. Yet reflection can sometimes unify what was not previously held together by a principle unconsciously represented.

These various grounds for thinking that explicatory principles that explain correct judgments about examples need not be internalized support a different view of concepts and conceptual psychological ability than that which guided classical views of reflection.

Leibniz maintained that reflection consists in analyzing concepts so as to elicit principles contained within them. Developments in philosophy and psychology have long conspired to undermine the view of Leibniz that most concepts are definable in terms of other concepts, and that having concepts is being in psychological states whose content consists in the definitional relations, which, in turn, constitute principles for the correct application of the defined concepts.

Kant maintained that most concepts are constituted out of a complex conceptual structure contained within them. But as noted earlier, he put little weight on analysis of concepts in his account of the role of reflection in obtaining philosophically interesting results.[16] Like Leibniz, Kant tended to identify concepts with the "function" or rule for its correct application, though he thought that this conceptual structure does not in general suffice to provide sufficient conditions for correct application of concepts.[17] Sensible intuition is often needed over and above concepts. I do not accept Kant's claim that all substantive theoretical cognition requires sensible intuition (perception or singular representations of aspects of the structure of perception) for its warrant. For example, I think that

[15] See Hilary Putnam, 'Is Semantics Possible?' in his *Philosophical Papers*, volume II: *Mind, Language and Reality* (Cambridge: Cambridge University Press, 1975).

[16] See *Critique of Pure Reason* A65–66/B90.

[17] *Critique of Pure Reason* A68–69/B92–94.

cognition in mathematics does not rest on *sensible* intuition.[18] But I think that Kant's invocation of intuition is a major step toward recognizing the centrality and irreducibility of examples in the determination of concepts, hence in reflection.

Kant differed from Leibniz in that he thought that the way to come to understand relevant rules through reflection is not conceptual analysis, but consideration of the inferential and applicational (judgment) capacities marked by the concept, and its relations to sensible intuition. He thought that the relevant principles are fully conceptualized—even though he supposed that the most important ones are meta-rules that connect concepts and principles to sensible intuition. He retained Leibniz's view that having a conceptual ability consists in, or is necessarily associated with, being in a psychological state that has the content of the rule or principle that guides and explains the concept's correct application.

What is the alternative view of conceptual psychological ability? The view is that one can have a conceptual ability to think in a certain way about objective matters, even though the principles explaining that ability are not fully internalized. The principles can involve concepts not available to the individual with the relevant conceptual ability. They may govern not only the thinker's thinking but his perceptual abilities or his sense of mathematical similarity. They may include reference to communicative relations and reliance on the abilities of others. They may involve reference to patterns in the environmental or abstract subject matter that the individual thinks about. The individual's incomplete understanding of the concept—his inability to formulate the principles governing his ability to apply it correctly—can go beyond an inability to make explicit the content of subliminal, but still accessible, psychological states. Thought and the having of thought content are not in general to be explained purely internally. Conceptual ability is not individualistically self-contained.

These considerations show reflection to be more difficult than classical rationalists thought it was. Reflection cannot always expect to find true principles governing pre-theoretic thinking, or unreflective thinking in scientific theory, already present as the content of a psychological state. To obtain answers it sometimes must await or develop materials that go beyond what is available in the individual's current psychology. One way to do this is to use empirical knowledge. Another is to engage in first-order apriori theorizing, perhaps thereby learning new things or even producing new concepts. Even though I think that sometimes an individual can in reflection develop new concepts and principles, it seems clear that, to be successful, reflection sometimes depends on incorporation of discoveries that give it the tools to provide retrospective explications or principles governing a given way of thinking. To obtain answers, it cannot always simply bring to consciousness what is already implicitly known or represented.

[18] See 'Frege on Apriority'.

The same considerations indicate that reflection has a more limited role in acquiring knowledge than the classical rationalists supposed. Since reflection is not guaranteed a subject matter complete in itself, reflection must be given a secondary role. Much knowledge, including apriori knowledge, is obtained without heavy reliance on higher-order consideration of thoughts or psychological states. First-order theorizing is dominant in our acquisition of knowledge.

For all that, I believe that reflection is sometimes a source of knowledge. So far, most of the points that I have made about reflection have been limitative. I would like to make some progress in understanding reflection as a genuine source of knowledge by considering cases. I begin with a brief review of the success of reflection in meta-logic.

III

Reflection produced spectacular, sustained success in the development of meta-logic.[19] I want to consider some salient aspects of the reflection, extended over a long history, that led to the elaboration of certain intuitive concepts as the basis for modern model theory.

The intuitive notions are logical consequence and logical validity. These notions are a product of reflection on good deductive inference. The intuitive notion of *logical consequence* is that of *preservation of truth grounded in or correctly explicable in terms of logical form and logical structure. Logical validity* is simply *truth grounded in or correctly explicable in terms of logical form and logical structure*. Here I distinguish form and structure. Form is a property of sentences and propositional thought contents. Structure is the aspect of the *subject matter* of sentences or thought contents that helps explain systematic preservation of truth.

Let me summarize the development of reflection that issued in these notions and their productive use in leading to modern meta-logic. The oldest discussions of deductive inference center on *necessary connections* between the truth of premises and the truth of conclusions, and on unexplicated notions of demonstratively derived *knowledge*. The intuitive notion of logical consequence emerges in the middle ages. This emergence required isolating truth preservation from knowledge preservation, and separating out a conception of form from the modal notion of necessity as the distinguishing mark of good deductive inference. The point here is not that there was any rejection of modal or epistemic notions. All the relevant writers assumed that in good deductive inference, there is a necessary connection between premises and conclusion. All assumed that

[19] In 'Logic and Analyticity', *Grazer Philosophische Studien* 66 (2003), 199–249, I have traced the emergence of the intuitive concepts of logical consequence and logical validity that became the basis for mainstream modern meta-theory in logic. I will be drawing on points in that article, where the points are developed in greater detail.

argument serves knowledge. Reflection isolated logical consequence and logical validity as phenomena in their own right.

This development included both concept formation and theory building. Yet it derived from reflection on instances of good deductive inference.[20] Reflection on good deductive inference yielded insight into the function of inference, the form of inferences, the type of concept that truth is, and the potential of those forms and their semantical relations to explain good deductive inference.

Reflection on *good* deductive inference is reflection on inference that fulfills a function, aim, or goal. What is the function of deductive inference? There is no single answer, since there are different sorts of function. There are surely social, biological, psychological functions of inference. These are not best understood through reflection. There are, however, other functions that can be understood reflectively. Deductive inference serves belief and judgment. Belief and judgment have a representational function of presenting an individual with true thoughts. Deductive inference functions representationally to preserve truth in argument. The foregoing train of reasoning is part of what underlay development of the intuitive notion of logical consequence.

Three more insights were needed. One was reflection on specific forms of sentences, propositions, or thought contents, in the light of how patterns of good deductive inference are associated with patterns among the forms. A second was the idea that since truth is some sort of fitting with a subject matter—since truth is a semantical notion—deductive inference is associated with patterns not only among the forms but among relations between the forms and structural aspects of a subject matter. The third is that good deductive inference can be explained in terms of patterns of relations between logically relevant forms and structures of subject matters. The flowering of model theory derived from Frege's making these four insights systematic, and Skolem, Gödel, and Tarski's mathematicizing them.

These insights are unquestionably successes for reflection. They may provide clues to finding aspects of reflection that yield success in other areas—in particular, philosophy.

All four insights require the prior recognition of good deductive inference. 'Deductive' and 'inference' need not have been understood precisely. It was enough to recognize a range of cases that serve as paradigms and that hang together. What is striking is the self-evidence of the first-order deductive inferences. This self-evidence supports the meta-characterization of them as good, as fulfilling a representational function.

[20] What is the significance of isolating the key intuitive notions from their metaphysical and epistemic cousins? Necessity was a more metaphysical notion than was needed in logical theory. The epistemic cousin of logical consequence is conceptually posterior and more complex. The simpler, more basic notions are truth and preservation of truth.

The first insight is *into a function of a representational activity*. Reflection on function is closely associated with reflection on standards or norms for fulfilling the function. Recognizing preservation of truth as a relevant good, and relevant function, of inference is closely associated with recognizing what it takes to fulfill that function in deductive inference. This is not a matter of degree. Either one reasons according rules of correct deductive inference or one does not. Still, it is one thing to know what the function is and another to recognize good ways to fulfill it. The relevant function is a certain type of preservation of truth—preservation of truth in virtue of form—a form that potentially reveals structure in the subject matters of inference.

The second insight is *into the form of a representational activity*. Insight into form is also insight *into relations among capacities*. For representational contents, including their forms, help type-individuate representational capacities. On reflection we recognize that the capacity to think that all humans are mortal shares a capacity with the thought that all dogs are mammals.

The third insight is into the *type of concept* that truth is, and the *type of property* that truth is. This insight involves recognizing necessary relations among truth, reference, and truth-of. This is an insight into the domain of application of a concept. Those who came to employ the intuitive concept of logical consequence recognized that truth in general is not just a matter of form. True thoughts characterize a subject matter, and thus bear a relation to it. So the notion of truth is to be understood in conjunction with notions of relation between parts of sentences or thought contents and a subject matter. So deductive preservation of truth is to be understood not only in terms of form (or relations among abilities) but in terms of the reference or the truth-of relations that parts of sentences or thought contents (or the underlying abilities) bear to a subject matter.

The fourth insight is into the *possibility of better understanding one sort of phenomenon in terms of another*. Good deductive inference (the relevant sort of truth preservation) is to be understood in terms of truth preservation grounded in logical form and logical structure. The reflective insight is that the intuitive notion of good deductive inference is to be systematically understood in terms of another intuitive notion, logical consequence, whose explanatory potential is richer. Eventually, the latter notion received systematization through use of the mathematically more precise notions truth of and truth in a model.

All of these insights derived at least partly from reflection. All are apriori. Not all were beyond question. One could rationally question, until the early twentieth century, whether good deductive inference was illuminatingly understood in terms of form and structure—independently of modal notions. I think it now safe to take these insights as successes of apriori reflection. I hope to juxtapose these cases with cases of successful reflection in philosophy.

IV

I want to discuss two examples of successful philosophical reflection.

The first underlies the principle that error presupposes a background of veridicality.[21]

The representational function of belief and perception is to represent veridically. A belief is representationally successful when it is true. A perception is representationally successful when it is correct or accurate. When an individual has a certain belief or perception, the conditions for representational success are set. The belief or perception is partly individuated in terms of its representational content: What it is to be a belief or perception of a particular kind is partly to have the content it has. A representational content is a condition for veridical, successful, representation. For belief, the content is a condition that when met is true. For perception, the content is a condition that when met is correct or accurate. So belief and perception are partly individuated in terms of conditions for their representational success. These are conditions under which their representational functions are fulfilled.

The key idea behind the principle that error presupposes a background of veridicality is that cases in which certain representational successes obtain are in some way more basic than cases under which errors, or representational failures, obtain. Part of the idea is that failure can be understood only as failure at doing something, and the "doing something" constitutes conditions of success. So the notion of failure is given its meaning by conditions for success. This is a relatively easy conceptual point.

But the key idea is stronger. The principle suggests constitutive explanatory connection between *cases* of success and *cases* of failure. The possibility of failure (error) is to be explained in terms of actual conditions that make representational success, cases of veridicality, possible.

I believe that a deeper account of why the principle is true brings in reflection on anti-individualism. But here I mention two initial considerations that support the principle.

One is that the mental states' having representational content is incomprehensible apart from capacities to use the content—associate it with sensory input,

[21] The principle—and its cousin, 'irrationality presupposes a background of rationality'—have antecedents in various formulations of a principle of charity. In *Word and Object* this principle was formulated: '. . . assertions startingly false on the face of them are likely to turn on hidden differences of language'. Quine added, 'The common sense behind the maxim is that one's interlocutor's silliness, beyond a certain point, is less likely than bad translation—or, in the domestic case, linguistic divergence.' *Word and Object* (Cambridge, Mass., MIT Press, 1960), 59. Davidson formulated a similar view: 'Making sense of the utterances and behaviour of others, even their most aberrant behaviour, requires us to find a great deal of reason and truth in them.' 'Belief and the Basis of Meaning' in *Inquiries into Truth and Interpretation*, 2nd edition (Oxford: Oxford University Press, 2001), 153; and '. . . errors take their content from a background of veridical thought and honest assertion', 'The Irreducibility of the Concept of Self', in *Subjective, Intersubjective, Objective* (Oxford: Oxford University Press, 2001), 89.

apply it, use it in action, make inferences with it, and so on. Since the representational content makes reference to a subject matter, the capacities to use it must associate the content to the subject matter that it is about. Erroneous applications would not *per se* relate the content to the subject matter that it is about.[22] So in explaining the having of content, one must advert somewhere to capacities to get things right. Not every such capacity need be realized in every individual. But mental capacities can be attributed only against a background of some actual realizations. The states are part of a system for grounding explanation of actual mental activity. So in the constitutive explanation of errors, there must lie somewhere in the background an appeal to veridical realizations of mental capacities.

A secondary consideration is the long intellectual experience of expecting functions to be either explained by or essentially associated with patterns of actual acts or events. Although I think that not all functions are like this, I think that *specific* mental states' function to represent veridically, involving with *specific* veridicality conditions, must be explanatorily associated with actual acts or events. Such functions are not well understood as brutely inherent in states, acts, or events, completely abstracted from patterns in which they are grounded. The point here is similar to the point that the function of a heart cannot be individuatively understood as an isolated brute fact. One must relate it explanatorily to actual patterns of events that successfully realize needs of the body. Thus having a function of representing a particular subject matter is grounded not only in a network of further relations to the subject matter, but in some pattern of acts or events that successfully realize the function. Only in such a context are failures possible.

Making the error-presupposes-veridicality principle more specific is difficult. How much error and what sort of error are possible? What sort of explanation or grounding—and at what level of specificity—is to be expected? These difficulties are coded in the vagueness of the words 'presupposes' and 'background'—in 'error presupposes a background of veridicality'.

I believe that philosophy is still in the process of reflecting on more specific formulations. Here reflection proceeds dialectically. Consider one of Davidson's formulations: Most of an individual's beliefs have to be true.[23] It is natural to think of possible counterexamples, perhaps even actual ones. We seem to be able coherently to imagine an individual whose every empirical judgment is mistaken—perhaps the result of being put in an extremely abnormal situation. Of course, such counterexamples themselves might involve a mistake. There is scope here for good or bad philosophical judgment. I will, however, take this counterexample as sound. The case forces one to think of the principle in a wider

[22] Veridical aspects of erroneous applications could relate content to subject matter, of course. One could perceive a grey rabbit as a brown rabbit.

[23] Donald Davidson, 'A Coherence Theory of Truth and Knowledge', in E. LePore (ed.), *Truth and Interpretation: Perspectives on the Philosophy of Donald Davidson* (Oxford: Basil Blackwell, 1986), 308, 314.

context. The individual's errors are possible only against a wider background of successes. The individual's perceptual system could not represent the physical shapes that it represents (while being fooled) if the perceptual system had not evolved or been designed in an environment in which the same types of perceptual states in other individuals responded to actual physical shapes. Take another counterexample, there are areas of thought—say, astrology—where error is dominant. What is to prevent an individual's harboring more beliefs about astrology than about anything else? Reflection suggests that certain types of representational success, perhaps perception or simple logical inference, are central to understanding other successes—and the failures.[24]

Progress in sharpening principles such as 'error presupposes a background of veridicality' is slow. Often progress does not issue in scientific theory. But sometimes we do make progress that deepens understanding and yields knowledge, even where system and rigor never emerge.

Let me turn to a second example of successful philosophical reflection—the thought experiments that underlie anti-individualism. These thought experiments proceed in three stages.

The thought experiments begin with a judgment that someone could have thoughts about a given kind or property as such, even though that person is not omniscient about its nature.[25] For example, one can think that arthritis is a painful disease, or that aluminum is a light metal. Or someone could correctly deny that sofas are really religious icons. One could do these things even though one might lack the background knowledge to distinguish those kinds from imposters. One might not realize that arthritis can occur only in joints, or that aluminum has the atomic weight that distinguishes it from other metals that are perceptually and practically similar. One might lack the sophistication to counter a sceptic who incorrectly claims that sofas are not used for sitting, but are religious icons for a sect that fools the population into thinking that sofas are used for sitting.

Then, in the second stage, one imagines a situation like the first one except that a kind or property that is indiscernible (for the individual) from the original kind or property occurs systematically in its place. One also imagines that there is no social network that the individual relies upon that uniquely specifies the kind or property or that interacts with it. For example, one imagines an environment in which no one has isolated arthritis as a disease. Or one imagines an environment in which there is no aluminum, or no sofas. No one in the imagined environment can uniquely specify these entities. In their places are, respectively, another disease that feels about the same, another metal that looks and behaves about the same, and

[24] I believe *de re* representational states are fundamental.

[25] The thought experiments are laid out in detail in 'Individualism and the Mental', *Midwest Studies in Philosophy* 4 (1979), 73–121; 'Other Bodies', in A. Woodfield (ed.), *Thought and Object* (Oxford: Clarendon Press, 1982); and 'Intellectual Norms and Foundations of Mind'; all are reprinted in *Foundations of Mind.* See also the discussion of the thought experiments in that volume's Introduction and in 'Postscript: "Individualism and the Mental"'.

religious icons like those that the sophisticated sceptic mistakenly claims to be present in the original environment.

In the third stage, one judges that in the second environment, the individual could not have thoughts about arthritis, aluminum, sofas, as such. For we have imagined that the individual's own background knowledge does not discriminate arthritis, aluminum, and sofas from their respective substitutes. No one else has isolated arthritis in the imagined environment. No aluminum or sofas occur in the imagined environment for the individual to perceive. There is no one on whom the individual relies who has perceived these entities or who can specify them. One judges that under these circumstances an individual could not have thoughts such as that arthritis is a painful disease, aluminum is a light metal, or sofas are not religious icons.

The conclusion is that what thoughts an individual can have—indeed the nature of the individual's thoughts—depends partly on relations that the individual bears to the relevant environments. For we can imagine the individual's make-up invariant between the actual and counterfactual situations in all other ways pertinent to his psychology. What explains the possibility of thinking the thoughts in the first environment and the impossibility of thinking them in the second is a network of relations that the individual bears to his physical or social surroundings.

We started with recognition that certain cases are possible. One support for this recognition is memory of actual cases that are appropriately similar. The step carries the empirically warranted assumption that arthritis, aluminum, and so on, exist. One natural inference in the first step is from actuality to possibility. Here the first step has an empirical element.[26]

The second step is usually presented as also aimed at establishing a possibility. I think that sometimes possibility is at issue, but I think that it is never the key point of these thought experiments. I think that the key point in the second step is

[26] Is there further insight into possibility that is independent of an empirical assumption that there are relevantly similar actual cases? I do not know. But I will not pursue this question here. The relevant actuality need not be a particular case involving arthritis or aluminum. Although I assume that *arthritis* and *aluminum* are real kinds, the force of the argument does not depend on this assumption's being right. The concepts arthritis and aluminum are epistemic stand-ins for a wide range of successfully referential concepts. The argument works on the assumption that there are relevantly similar concepts that are representationally successful. The reflective power of the thought experiments does not depend on the empirical issue of exactly which concepts apply to real properties or kinds, although I do assume in the thought experiments that arthritis and aluminum are real. That power depends only on the idea that there are certain types of concepts and that a range of them are successful in representing properties or kinds. Thus the reflective judgment is that successful representational exercise of relevantly similar representational abilities cannot be as completely free of environmental connection—as would have to be the case if the same abilities were exercised in the second environmental scenario. One judges that different (equally successful) abilities are being exercised in that scenario. In the background, I think, is the principle that error presupposes a base of veridicality. I think that ultimately this principle is apriori. It may be that in addition to relatively obvious empirical warrants that we have for the first step, there are guiding apriori warrants that apply to whole classes of cases.

actually quite subtle. What is fundamentally at issue, in the second stage, is constitutive explanation, not modality. I shall return to this point.

The impossibility judgment in the third step derives from a fallible understanding of the types of conditions under which the relevant propositional abilities can arise. Specific propositional abilities are necessarily embedded in a network of other capacities and causal relations, evoked by examples in which they are stipulated to be absent.

The thought experiment concludes with a judgment about explaining differences between thoughts in the two environments. The second and third steps evoke a realization that physical differences in the individuals' bodies are not in general explanatory of the differences in thoughts. Putnam stipulates that the bodies are the same even though no water occurs in the counterfactual situation. This is, of course, impossible. But the main point of the step holds. Possibility is not what is fundamentally at issue. The point is to remind one that the nature of an individual's thoughts cannot be explained in terms of constituents of his cells or in terms of minor gravitational effects on his body. Even though, to imagine an individual in a waterless situation, one must imagine no water in the individual's cytoplasm and different gravitational effects on his body, these differences are not themselves relevant to explaining differences in the individual's psychology. Perceptual, communicative, cognitive differences—and macro-causal relations involved in these activities—are relevant resources for explanation. The insight is that whatever differences there may be in the thinker's bodies, these differences will not *in general* be the kind of differences that are relevant to explaining constitutive differences in thoughts. Reflection indicates that phenomenal and behavioral differences (understood in terms of physical movement) are also not key sources of explanation. The differences in thoughts constitutively depend on differences in macro interactions with aspects of relevant environments.

V

I want now to reflect on the kinds of insights involved in the philosophical thought experiments. In the meta-logic case, I cited four types of insights that illuminate inference. I will compare them with insights in the two philosophical cases.

All three cases involve reflective judgment about *the function of representational activity*. In the meta-logic case, the judgment is that the function of good deductive inference is to preserve truth. In the error-presupposes-veridicality case and anti-individualism cases, the judgment is that representational success in perception and belief is veridical specification of a subject matter. Cases in which the individual thinks of aluminum, where there is aluminum in the environment to be thought about, are cases of representational success. Such success fulfills representational function. The functional character of the matters to be explained implicitly guides the eventual judgments about the type of individuative explanations that are necessary.

In the meta-logic and anti-individualism cases, insight into representational function is coupled with insight into generic *ways of fulfilling that function*. We know how to engage in good deductive inference. We know that specifying arthritis or aluminum depends on capacities to theorize about it, perceive it, or communicate with others who do these things.

The meta-logic and anti-individualism cases rely on insights into *the form of a representational activity and into relations among representational abilities*. Thus in the meta-logic case, there is insight into logical form, on which good deductive inference depends. In the anti-individualism case, one isolates from a thought arthritis is a painful disease the part that specifically concerns arthritis. Similarly, whereas the meta-logic case develops specific relations among inferential abilities underlying specific logical forms, the anti-individualism case depends on reflection on relations among cognitive capacities—perception, theory, interlocution.

The error-presupposes-veridicality case differs from the other two in these respects only because the principle being arrived at is so generic that insight into representational form, and into relations among representational abilities, is unnecessary.

All three cases essentially involve elementary insight into *necessary implications regarding the relevant subject matter*. The meta-logical case depends on recognizing relations between the concept of truth and relational semantical concepts like true of (or between the phenomenon of truth and the phenomenon of truth of). The philosophical cases depend on recognizing relations between belief and truth (and in the anti-individualism case, between concepts like arthritis and entities like arthritis).

Finally, all three cases involve insight into *explanatory priority*. The meta-logic case relies on recognizing that preservation of truth is to be explained in terms of semantical relations between logical form and subject-matter structure. The philosophical cases utilize, in different ways, recognition that having propositional attitudes is to be explained in terms of conditions in which such attitudes are successful. So, in the anti-individualism case, one recognizes that having certain thoughts is to be explained in terms of the successful cases in the first environment, and ultimately in terms of the exercise of cognitive capacities and macro-causal relations, underlying such exercise, to entities in that environment. We recognize on reflection that successful exercise of relevantly similar representational abilities cannot be free of environmental connection—as would have to be the case if the same abilities were exercised in the second-stage, counterfactual environment.

All these types of insight are important, and putting them together to yield new principles requires ingenuity and judgment. But the fourth type, insight into explanatory priority, seems to me to be the most striking one. The nature of this insight is somewhat different as between the meta-logic case, on one hand, and the two philosophical cases, on the other.

In the meta-logic case, the idea is that insofar as logical consequence constitutively involves preservation of truth, and insofar as truth necessarily implies and is best constitutively explained in terms of truth-of and relations to a subject matter, logical consequence is to be constitutively explained partly in terms of relational *semantical* notions like true-of. The explanation is ultimately in terms of systematic assignments of entities from the subject matter to formal components of thought contents or sentences.

In the two philosophical cases, these points about truth are assumed as background, since mental states are type-identified partly in terms of their veridicality conditions. But the subject matter is not semantics but mental states. The explanatory relation is between subject matter and mental states. In both cases, the relation involves the primacy of veridical states. In the anti-individualism case, the explanatory claim invokes non-representational relations as well—causal relations in the empirical cases, constitutive relations in the logical cases.

The thought experiments exploit an insight into the nature of representational ability. Representational contents of attitudes help type-identify the attitudes. An ability to represent successfully, veridically, must be associated with conditions that enable that ability to be realized. An ability whose nature is to represent a specific property or kind must be specific to that property or kind—not equally appropriate to other properties or kinds. So the ability must be associated with *conditions* that are specific to that property or kind. The twin earth scenarios exploit this point. They show that there must be actual veridical relations, supplemented by either causal or constitutive relations, between mental states and subject matter if veridical (and hence non-veridical) mental states are to be explained and individuated. I will conclude by reflecting on what unifies these four types of reflective insight.

VI

A principle backing the classical view of reflection is that in having reason, a reasoner must be implicitly aware of norms for reasoning, if the reasoner is to know what he is doing; and to be reasonable, a reasoner must know what he is doing. The principle is supplemented by a metaphor: that reason responsibly sets its own norms and hence is able to reflect on what they are.

The metaphor covers an oversight. Reason is larger than reasoners. *Ideally*, or eventually, reasoners may set norms and reflect on what the norms are. But a given reasoner may not be able to go so far. I have cited two ways in which this shortfall may occur. First, animals and young children may lack the concepts of reason or of propositional attitude necessary to have the relevant insights. Second, even mature reasoners may not have conceptualized or otherwise unified some norms for reasoning, even implicitly. Inaccessibility occurs in aspects of both apriori and empirical reasoning. So the traditional generalizations about why reflection is warranted are flawed. Reason is present phylogentically, ontogenetically, and historically before its full shape can be articulated by reasoners. Still, it seems to me that there are elements of truth in the classical view.

Let us reflect on critical reasoners. A *critical reasoner* is one who can suspend and correct propositional attitudes, conceived *as* attitudes and *as* being corrected, on the basis of reasons, acknowledged *as* reasons. A critical reasoner thus has concepts of truth, falsity, propositional attitude, and reason. By focusing on critical reasoners, I lay aside phylogenetic and ontogenetic difficulties: animals and young children are not capable of reflection. There remains the historical limitation. Resources necessary for successful reflection may require new knowledge or even the development of new concepts. Sometimes an individual develops new knowledge or concepts through reflection. Otherwise, fully successful reflection must await new concepts or knowledge developed by others—sometimes through reflection, often through other means.

We are critical reasoners. We are not in general guaranteed access to norms governing our rational activity. Still, in having the concepts of reason and propositional attitudes, a critical reasoner cannot be fully blind to the nature of his rational activity. Although a critical reasoner need not have a fully formed (implicit or explicit) conception of principles governing any of his critical reasoning, he or she must have a rudimentary grip on norms governing critical reasoning, a grip that involves at least partial conceptualization and some substantial conceptual know-how with the concept reason. If propositional attitudes, reasons, and the applications of reasons are to be conceptualized as such, there must be a capacity to recognize correct applications of reasons in *cases*. This gives a basis for inferring to patterns that constitute norms and principles governing those cases.

More is constitutively involved in being a critical reasoner. To have the concepts of reason, propositional attitude, and truth, many of the largest features of rational activity must be conceptually available to reflection. To get started in thinking about these matters, one must get certain things right. In the cases of these concepts, minimum competence requires a capacity to recognize on reflection certain basic necessary implications associated with these notions, or the phenomena that they apply to. The activity involved in critical reasoning requires having some conception of the point or function of the activity, the states and events that are constituent elements in this activity, and the main norms on and means of carrying out the activity. More concretely, critical reason requires having conceptions of the aims of belief and inference; the form and character of some inference rules and some propositional attitudes; the relations between belief, warrant, and truth; and certain relations among propositional forms and capacities.

All the insights in the meta-logical case and the philosophical cases that I discussed fit in this picture. All the cases are warranted by understanding.[27] I believe that some of them have a further source of warrant. Some are warranted inasmuch as they are necessary elements in the paradigmatically justified activity—critical reasoning. The recognition of the conceptual relations, the relations among psychological states and functions, and the insights into explanatory priority are warranted partly by being constituent elements in the whole critical reasoning process. They are justified inasmuch as they are necessary conditions of the possibility of critical rationality.[28]

A critical reasoner must have some capacity to understand the most fundamental functions and ends of critical reasoning, and must have at least a partial grip on the most general norms governing fulfillment of these functions and ends. To understand these norms, there must be some insight into the conditions under which they can be fulfilled. Insight into shoulds must be accompanied by some insight into cans.

Classical rationalism overestimated how specific and how far-reaching reflective apriori insight can be. I think that so far we have not learned just how much of the structure of critical reasoning is conceptually available to reflection. We have to find the limits of apriori reflection retrospectively—by looking at cases of successful reflection—rather than by trying to determine the scope and limits of such reflection in advance. Much apriori reflection is the result of induction, or genuine discovery, and of the development of new concepts. We can learn what reflection is successful, and what *sorts* of reflection are reliable, mainly by considering cases.[29]

A great deal of metaphysics has, I think, shown itself to be at best a merely speculative adjunct to the empirical or mathematical sciences. Armchair reasoning on the finity or infinity of the world, the nature of matter, problems about impact and continuous bodies, the nature of space and time, parts and wholes, the problems of abstract entities, has been and remain parts of metaphysics. But these enterprises can no longer be reasonably seen as in themselves sources of knowledge, except insofar as they unify what is known from the sciences, including mathematics.[30] They are not areas in which common sense or autonomous apriori intuitive reasoning by itself leads very far.

[27] The particular sub-types of understanding in the different cases, and corresponding particular sub-types of warrant, invite further investigation. At a sufficiently high level of abstraction, all seem to me to be apriori. But one cannot generate conclusions about what specific kinds of things exist from the thought experiments, together with knowing that one has concepts for specific kinds (kinds like *aluminum* and *arthritis*).

[28] This is not to say that there will not be additional sources of warrant for component parts. Since critical reasoning is the supreme authority for reason, its constituent operations obtain authority partly by virtue of being part of it in a way that the Supreme Court's decisions would have authority partly by virtue of its legal supremacy. If only it were an intrinsically rational enterprise!

[29] Obviously, there are many other types of apriori reflection besides those described here.

[30] This is broadly Kant's negative point in *Critique of Pure Reason*. I think that Kant's specific reasons for his view are not correct. But I believe that he was intuitively judging the consequences of

I believe that the meta-cognitive enterprise of reflection is different. Reflection can serve as an adjunct to any enterprise. But I believe it also offers substantive insights that are not parasitic on or merely supplemental to the natural or mathematical sciences. The insights are limited and fallible. Their nature remains to be better understood. But it seems to me that apriori reflection can yield limited autonomous insight, and even knowledge, in certain parts of semantics, philosophy of mind, and epistemology. This is because we are reasoning about reasoning itself. Apriori understanding in these areas is constitutive of understanding fundamental aspects of critical reason, and of us as critical reasoners.

Newtonian physics and mathematics for the methods of traditional metaphysics in areas where the subjects directly overlap. He was also right to believe that there was scope for more reflective methods in epistemology and in certain aspects of the theory of mind. I believe that here too his grounds for this belief were largely mistaken, but his instincts about areas in which apriori reflective knowledge is possible, or at least most likely to be fruitful, were broadly correct.

25 *Living Wages of* Sinn*

Frege's concept sense (*Sinn*) is better understood now than it was fifty years ago. Its value and distinctiveness as a tool for theorizing about language and thought remain, however, underrated in many quarters. Here, against some recent doubts, I argue again for distinguishing sense from any modern notion of linguistic meaning. I renew some criticisms of Frege's *conception* of sense and his employment of the concept sense. I then advocate the power of his concept for understanding thought, and some aspects of language.

I

I assume that there is a concept sense, and that there are senses. I do not assume that everything that Frege thought about senses is true. I impute his mistakes to the conception that he associates with his concept. I take senses to ground certain explanations that he gave the concept a role in. He gave the concept sense four explanatory roles.[1]

First, Frege introduces sense to account for scientific thought expressed in language. Every passage that centers on the concept emphasizes this role. Frege's interest in thought expressed in language is motivated by interest in scientific knowledge. His use of sense does not aim to account for rules of linguistic usage, or literal meanings of words, at least when his conception of scientific thought does not square with ordinary linguistic rules or usage.[2]

Helping to account for scientific thought and cognition is the primary explanatory role of Frege's concept. Senses are representational thought contents. They

* A shorter version of this essay was given as the first annual Saul Kripke Lecture in May 2011 at City University of New York. The paper benefited from audience comments on that occasion.

[1] I made these points in Tyler Burge, 'Belief *De Re*', *The Journal of Philosophy* 74:6 (1977), 338–362; reprinted in my *Foundations of Mind* (New York: Oxford University Press, 2007). I elaborate them in Introduction to my *Truth, Thought, Reason: Essays on Frege* (New York: Oxford University Press, 2005), 31–39.

[2] I made these points in 'Sinning Against Frege', *The Philosophical Review* 88:3 (1979), 398–432; reprinted in *Truth, Thought, Reason*. See also Introduction to *Truth, Thought, Reason*, pp. 35–40. In 'Sinning Against Frege', I wrote that 'linguistic meaning' is multi-purposed and adaptive to theory. One could use *a* notion of linguistic meaning that is the same as Frege's notion of sense. But current notions of linguistic meaning are not, I think, notions of sense.

are ways of thinking as of entities in a subject matter. They are ways that such entities are presented in thought—cognitive routes to them. Senses of declarative sentences are propositional thought contents. Sub-propositional thought contents associated with structural, inferentially relevant parts of sentences are also senses. So senses are contents of thought that are thought with or through language.

Second, senses figure in semantics of scientific thought and language. Denotations are entities in a subject matter that linguistic expressions refer to or otherwise represent. Senses of sentences set truth conditions: if the subject matter meets the condition, the sense is true. Senses of relevant sentential parts *determine* denotations in this sense: if there is a denotation of a linguistic expression of the sense, there is exactly one denotation. For every sense there is at most one denotation. Or in the propositional case, if the sentence or its utterance has a truth value, the sense of the sentence, or sentence utterance, has exactly one truth value.

Third, senses can be denotations when they are subject matter. When I discuss certain thought contents, I refer to senses. I shall do this by underlining the expression of the sense. Thus by '2 + 3 = 5' I refer to the representational content of a belief that 2 + 3 = 5. And '2 + 3 = 5' denotes the representational content 2 + 3 = 5. I call non-propositional thought contents that are not identical with and do not contain occurrence-based applications '*concepts*'.[3] '3' and '=' name the concept 3 and the identity concept, respectively.

Fourth, senses are what are understood by users of scientific language. In understanding language, one associates the language with thought or components of thought. This construal of understanding is to be aligned with the approach to language as a vehicle for thinking, inferring, and knowing, and to the requirement that senses determine denotations. By understanding senses, one normally understands something about denotations. They are ways denotations are presented in thought and ways by which denotations are determined. The language denotes a subject matter because it is associated with thought that represents the subject matter.[4]

Frege's concept sense differs from any ordinary modern concept of linguistic meaning. The concepts are complementary, overlapping tools for theorizing. The distinctiveness of Frege's concept sense is signaled most clearly in his discussion of indexicals.[5] Consider this text:

[3] For discussion of applications—occurrence-based representations—and their fundamental importance to perception and thought, see Burge, 'Belief *De Re*'; 'Postscript to "Belief *De Re*"', in *Foundations of Mind*; and 'Five Theses on *De Re* States and Attitudes', in Joseph Almog and Paolo Leonardi (eds.), *The Philosophy of David Kaplan* (New York: Oxford University Press, 2009), 246–316.

My use of 'concept' differs from Frege's. Frege uses the cognate term '*Begriff*' for functions denoted by predicates. I use 'concept' for certain ways of thinking that may be senses.

[4] Senses determine denotations only because other thinker-subject-matter relations are in place.

[5] Modern notions of linguistic meaning became prominent when attention focused on rules embedded in natural language use. This focus involved a double shift—away from scientific

(A) If someone wants to say today what he expressed yesterday using the word 'today', he will replace this word with 'yesterday'. Although the thought is the same, the verbal expression must be different to compensate for the change of sense that would otherwise be brought about by the different time of utterance (*Sprechens*). The case is the same with words like 'here' and 'there'. In all such cases, the mere wording, as it can be fixed in writing, is not the complete expression of the thought–but one further needs for its correct apprehension also the knowledge (*Kenntnis*) of certain circumstances accompanying the utterance, which are used as means of thought expression (*Gedankenausdrucks*). Fingerpointings, gestures and glances can belong here too. The same wording (*Wortlaut*) containing the word 'I' will express different thoughts in the mouths of different people, of which some may be true and others false.[6]

On any ordinary notion of linguistic meaning–communal or idiolectic–the meanings of sentences (and sentence utterances) containing 'today' and 'yesterday' are different. This point would have been plain to Frege. Frege states, however, that the senses, hence the thoughts expressed, 'will be' the same. So sameness of sense is compatible with difference of linguistic meaning.

Saul Kripke objects to distinguishing Fregean sense from modern notions of linguistic meaning.[7] He builds his interpretation on statements that occur immediately before (A):

(B) If a time indication (*Zeitangabe*) should be made with the present tense, one must know when the sentence was uttered in order to grasp the thought correctly. Therefore the time of utterance is then part of the expression of the thought.[8]

Kripke correctly reads (B) as expressing the view that apart from the time of the utterance, the words 'Today is Thursday' do not express a complete thought. He

language, and away from thought expressed in language, to natural language and to communication. Early Russell uses the term 'meaning'. But he is primarily interested in knowledge and thought. The positivists' notion of meaning also did not center on understanding rules of natural language usage. Frege's notion of sense and modern notions of ordinary linguistic meaning do overlap. Many (cognitive) linguistic meanings are senses.

[6] Gottlob Frege, 'Der Gedanke' (1918–1919) in *Kleine Schriften*, ed. Ignacio Angelelli (Hildesheim, Germany: Georg Olms, 1967), 348. Translations are my responsibility. See also Frege, 'Thought', in *The Frege Reader*, ed. Michael Beaney (Cambridge, Mass.: Blackwell, 2009), 332. I cite pagination in Frege's original in brackets. Passage (A) is from [64].

[7] Saul Kripke, 'Frege's Theory of Sense and Reference: Some Exegetical Notes', *Theoria* 74:3 (2008), 181–218; cf. 201. Kripke thinks that if sense were *not* meaning, it would be odd for Frege to omit introducing a technical term for meaning. Since Frege was not theorizing about ordinary language, or ordinary meaning, this omission is not odd at all. (But see note 44.) Kripke thinks that, on my view, Frege's notion of sense is close to David Kaplan's notion of content, and that Frege, oddly, omits any technical notion analogous to Kaplan's notion of character. Again, I think that the omission is not odd. In fact, I think that Frege's notion of sense is not much like *either* of Kaplan's notions. Cf. David Kaplan, 'Demonstratives', in Almog, John Perry, and Howard Wettstein (eds.), *Themes from Kaplan* (New York: Oxford University Press, 1989), 481–563.

There is much in Kripke's article that I agree with. Although he does not note the point, his view of Frege's hierarchy of senses is very similar to mine. Cf. my 'Postscript: "Frege and the Hierarchy"' in *Foundations of Mind*. As stated below, I also agree with the core of his important criticism of Frege on the sense of names.

[8] Frege, 'Der Gedanke', in *Kleine Schriften*, 348 [64]; 'Thought', in *The Frege Reader*, p. 332.

takes the time itself as well as the verbiage 'Today is Thursday' to be part of the expression of the thought, and even part of the sentence. He regards the time as 'an unrecognized piece of language'.[9] Thus he takes the full sentence to consist of both the verbiage 'Today is Thursday' and the time *t* at which the verbiage is uttered. *t* is taken to refer to itself, and the sense of *t* is 'that of autonomous designation'. Kripke further takes the verbiage 'Today is Thursday' to denote a function from times to truth values, and he takes 'Today' to denote a function mapping each time onto the day containing it.[10]

On this view, the senses of 'Today' and 'Today is Thursday' are what nearly everyone would take to be their linguistic meanings. The only oddity is taking utterance times as self-denoting linguistic expressions. I will not center on this oddity. I discuss it only in the Appendix.

Kripke's interpretation is incorrect. His assumption that Frege thinks that 'today' denotes a function, determined by the sense (meaning) of 'today'—a function from a time to the day containing it—is incompatible with (A), which immediately follows (B).[11] For if, as Frege claims, the utterances, on successive days, of 'Today is Thursday' and 'Yesterday was Thursday' have the same sense, the senses of the utterances cannot be or include their linguistic meanings. For the meanings of the two sentences, and sentence utterances, are plainly different.[12]

Kripke notes that his interpretation is incompatible with (A). He blames Frege for 'confusion'. He writes, 'If Frege really means that we have expressed literally the same thought again, it is very hard for me to see how to reconcile this assertion with his other doctrines'.[13]

Kripke holds that the view stated in (A) is incompatible with two Fregean doctrines. One is that 'today' expresses a sense that determines a function from the time of utterance to the day containing it, and that 'yesterday' expresses a sense that determines an analogous function. Frege nowhere in his writings expresses this "doctrine". He states a view incompatible with it in the very

[9] Kripke, 'Frege's Theory of Sense and Reference', 201–202.

[10] Kripke, 'Frege's Theory of Sense and Reference', 204, 212.

[11] It is also directly incompatible with my interpretation—the very interpretation that Kripke argues against. Thus the assumption begs the question.

[12] Kripke writes that although on the relevant occasions 'Today' and 'Yesterday' denote the same day, 'they plainly pick it out in different ways, paradigmatic cases of difference in sense', 'Frege's Theory of Sense and Reference'. Kripke goes on to claim that on his interpretation, Frege believes that a present-tense thought (or today thought) expressed at time *t* '... cannot be recaptured at any later time. This is simply because the earlier time is never with us again, and immediate acquaintance with it is supposed to be essential to the expression of the thought', *ibid.*, 205. But passage (A) indicates that Frege believed that the same today thought *can* be expressed, and thought, again on the next day—by thinking 'Yesterday was Thursday'. Kripke's belief that recapture is impossible depends not only on his misinterpretation of Frege's account of thought expression, but also on undefended views, which I think questionable, about the relation between modes of presentation in memory and modes of presentation through present acquaintance. Memory can have an anaphoric function in preserving earlier modes of presentation. I will not pursue these issues here.

[13] 'Frege's Theory of Sense and Reference', 204.

paragraph on which Kripke bases his interpretation. So it cannot be invoked in interpreting that paragraph.

The other doctrine that Kripke takes to be incompatible with (A) is that 'the sense of a sentence (the thought) is composed of the senses of its parts'. Kripke reasons that the senses of the parts of 'Today is Thursday' and 'Yesterday is Thursday' differ, because the senses of 'Today' and 'Yesterday' differ; so the senses of the sentences (and sentence utterances) differ. Kripke asks, 'How could anyone argue that the two sentences in question have the same sense?'[14]

As follows: Frege holds the sense-composition principle. But Kripke's application of the principle is not Frege's. As is well known, Frege took his principle to be applicable only once relevant sentence parts are isolated under logical analysis. Logical analysis aims at revealing thought structure. Not just any natural-language word can be assumed to be a sentential part under logical analysis, or to have a sense. Frege repeatedly criticizes natural languages as misleading regarding the structure of thought. He does so in this very passage.

Kripke's argument that 'today' and 'yesterday' sentence utterances cannot have the same sense, because their senses are compounded out of the senses of their parts, *assumes* that 'today' and 'yesterday' are parts that in themselves, *across contexts*, express senses. He states: '"Today" denotes a function which when applied to t_0 (used autonymously) gives the day containing t_0 . . . '.[15] He takes this rule to give the sense of 'today', on Frege's view. Thus Kripke's application of the second doctrine depends on assuming the alleged first doctrine.

Frege does not say that these indexicals *in themselves* express senses or denote anything. He writes of people expressing senses (thoughts) by using sentences that contain indexicals. In (B) he writes of a 'time indication made with present tense. . . . ' But he does not say that the present tense in itself denotes a function, or anything else. Frege never writes that any indexical on its own, cross-contextually, expresses a sense or denotes a function. Yet indexicals have linguistic meaning, analogous to the one Kripke gives for 'today' through the denotation rule.

Frege held the views stated in (A) for over two decades—making it further implausible that they are confused or in conflict with his principle of sense composition. In the cognate passage in 'Logic' 1897 that became 'Thought' 1918–1919, Frege writes that relevant *words* (he cites 'here', 'now', 'I') 'only acquire their full sense always through the circumstances in which they are used'.[16] He targets *words*, not sentence types that contain the words. He states that these words, apart from uses in contexts, do not have *their* full senses. So

[14] 'Frege's Theory of Sense and Reference', 205.

[15] 'Frege's Theory of Sense and Reference', 204.

[16] Frege, 'Logik' (1897) in *Nachgelassene Schriften*, ed. Hans Hermes, Friedrich Kambartel, and Friedrich Kaulbach (Hamburg: Felix Meiner Verlag, 1983, 2nd edition), 146; in English: Frege, 'Logic' (1897), in *Posthumous Writings*, ed. Hermes, Kambartel, and Kaulbach (Chicago: University of Chicago Press, 1979), 135; Frege, 'Logic' (1897), in *The Frege Reader*, 235. I discuss other relevant passages in the Appendix. Note that indexicals do not need contexts to have their full

they cannot denote functions on their own, or across contexts of use, as Kripke's reading requires. Temporal indexicals are not, for Frege, used to contextually determine a time *in a way that corresponds to their linguistic meanings*. Yet they clearly have linguistic meanings across contexts.

So the incompleteness of indexical sentences is *not* that they need a singularly indicated time as argument for a function that, when completed, would yield a value (a time) as denotation, if the sentence utterance is to express a complete thought with a truth value. The contextually used singular term that denotes the time (for example, the present day) is not the result of filling in a functional expression with a singular term that denotes a time of utterance. For Frege, indexical expressions, taken with their cross-contextual linguistic meanings, do not contribute a sense to the complete thought. He thinks that by using 'today' in one context and 'yesterday' in another, an individual can express the same sense and think the same thought.

Frege tends toward treating indexicals as used demonstratives. He takes uses to be backed by a denotation-determining sense, which he takes to be an eternal mode of presentation associated with the "demonstrative" in a context. The mode of presentation, the component of thought content, is linguistically constrained only by the singularity of the "demonstrative" and, with temporal indexicals, by their applying to a time. Such used "demonstratives" take on senses through perceptual belief, or contextual association with definite descriptions, or in other context-dependent ways.[17]

I think that the foregoing points decisively undermine the only serious objection that Kripke raises to taking passage (A) literally. (A) is incompatible with Kripke's interpretation. Frege neglects the linguistic meanings of indexicals because he thinks that they are not relevant to the thought expressed. This position serves his view of senses as elements in thought *ideal for science*. I believe that he thought that indexicals and demonstratives do not occur in a scientifically ideal language.[18] That is why the meanings of indexicals are not, for him, senses.

Frege's emphasis on the timelessness of truth implies that truth is not relative to a time—that *truth*, not truth at a time (or place, or in a model), is the fundamental category of logic. Frege's ontological Platonism about thoughts, and senses, extends this point. Thoughts are in a third realm—dependent on neither the mental

linguistic meanings. What it is to be an indexical is to have the same meaning across contexts, a meaning that allows shifts in referent.

[17] Thus not only the rules that Kripke states, but the rules Kaplan states for determining content and character, do not determine the senses of indexicals for Frege. See note 7 above. I believe that Frege's approach to the senses of indexicals makes his view largely irrelevant to modern projects that aim to give linguistic rules governing indexicals.

[18] See Frege, 'Logik in der Mathematik' (1914), in *Nachgelassene Schriften*, 230; *Posthumous Writings*, 213.

nor the physical.[19] It is plausible, I think, that Frege understood his ontological Platonism about thought content, and his conception of scientific thought, in an even stronger way—a way that excluded indexicals from scientific language.

In 'Logic' 1897, Frege makes these remarks: (1) 'Thought is impersonal'; (2) 'It is of the essence of a thought to be non-temporal and non-spatial'; (3) '. . . the thought is either true, in which case it is always, or better, timelessly, true, or it is false and in that case it is false without qualification. This holds of any particular historical fact: *if it is true, it is true independently of the time at which it is judged to be true*.'[20] In 'Thought' (1918–1919), he writes,

> A property of a thought will be called inessential if it consists in or follows from the fact that this thought is grasped by a thinker.[21]

These passages allow various interpretations. I will develop one way of reading them.

It is plausible that by the rules for the *meaning* of 'today' or present tense, it follows that only *uses* of sentences containing those devices can be true. For example, only uses (either in thought or in public discourse) of 'Today is Thursday' or 'Spain is currently world champion in soccer' can be true. Strictly speaking, only an instantiated use of the sentence is true. And uses require that the meaning is *thought* or *understood* by some thinker.[22] The *meanings*, as units in truth conditions, seem to require *uses* if their truth conditions are to be fulfilled. On this view, although any such truths would not change truth value, they would

[19] See my 'Frege on Knowing the Third Realm', *Mind* NS, 101:404 (1992), 633–649; reprinted in *Truth, Thought, Reason*.

[20] Frege, 'Logik' (1897), 146, 146, and 147, respectively; 'Logic', in *Posthumous Writings*, 134, 135, 135; 'Logic', in *The Frege Reader*, 234, 235, 236. My italics.

[21] Frege, 'Der Gedanke', in *Kleine Schriften*, 361 [76]; 'Thought', *The Frege Reader*, 344.

[22] I am aware that many formal semantical theories for indexicals do not—at least on their face—accept such an implication. They postulate that relevant tensed or indexical *sentence types* are true relative to times (and other parameters), regardless of whether anyone uses the sentences at those times (or other parameters). I think that such postulations are useful idealizations. But I think that truth relative to times (*et alia*) is not the basic semantical notion. I think that truth and falsity are the basic semantical notions. Truth, full stop, is fundamental. Frege himself clearly maintained this view.

Truth is fundamental, I think, because truth conditional semantics is ultimately a psychological theory, and the theory should illuminate truth conditions for belief and judgment, which are among the fundamental psychological states. Belief states and judgment events are true or false, not true-at or false-at. Semantics is a branch of psychology. Frege's motive for holding truth to be fundamental is analogous. He took semantics to be a theory of thought/judgment, but with a difference. He took thoughts and judgments to be idealized event types that meet certain norms in an idealized science, not actual psychological states. Still, he held that in this idealized sense, semantics is ultimately a theory of truth conditions for judgments. And judgments are true or false, full stop—not true-at. So I believe that his reasons for taking truth, not truth-at, to be the fundamental semantical concept are parallel to mine.

I think—and I think Frege thought—that only used tokens of indexical sentences are true or false. Particular times are semantically associated with tensed sentence types through uses of those sentence types at particular times. Thus the natural semantics of indexicals seems to entail that (necessarily) if any structure that contains an indexical element is true, the indexical is used—tokened. Use entails the existence of judgment, or at least belief. This is not to say, of course, that each use is an exercise of judgment or belief. It is to say that there is a necessary, constitutive relation between there being uses and there being judgments or beliefs.

not be essentially non-temporal. Further, they would *not* be independent of times at which they are judged true. The meanings of true sentence-occurrences containing 'I' plausibly require uses by an utterer, if the occurrences are to be true. Analogously, 'I''s meaning rule might be taken not to be "impersonal".

Frege may have reasoned in these ways. In his ontological Platonism, he advocated what I believe to be an unnecessarily strong construal of the non-relativity of truth to time and to thinkers. He may have thought that including indexical meanings in the contents of scientific thought would compromise the transtemporality and objectivity of scientific truth.

Contrary to this Frege-like view, the requirement that truths be true timelessly can be understood *not* to require expulsion of the ordinary meanings of indexicals from the contents of the truths.[23] It is not clear exactly what Frege meant by the four quotations. He does not conceive the meanings of indexicals as part of the senses of sentences or sentence-uses that contain them. I have outlined a motivation that he may have had—a motivation that I do not myself accept.

How to apply modern notions of linguistic meaning to *proper names* is not obvious. So contrasting sense and linguistic meaning in the case of proper names is less straightforward than it is in the case of indexicals. Frege treats proper names and indexicals similarly. He takes the senses of names to shift with context, even for a given name and idiolect. He thinks of such senses in cognitive terms, not in terms of rules governing linguistic usage.[24]

Frege illustrates senses of proper names with senses of definite descriptions. I doubt that Frege held a general theory on the matter. Most definite descriptions in his examples contain names. If he held a general theory, one would expect him to have noted the incompleteness of his examples in illustrating it. Also, in *Begriffsschrift* Frege cites a name guided by intuition to a position in space.[25] He contrasts the intuitive way of determining the position with a descriptive way. I doubt that Frege would have thought that definite descriptions express ways of thinking built on intuitions. The passage suggests that on his later theory of sense,

[23] Taking truth not to vary with time does not entail taking all truths to be timeless entities. For some of my reasons for rejecting Frege's view that true thoughts are in general timeless, mind-independent entities, see *Truth, Thought, Reason*, 50–54.

[24] Kripke seems to think that I deny that Frege holds that communal natural languages are, strictly speaking, amalgams of idiolects. See Kripke, 'Frege's Theory of Sense and Reference', 210. I do not. See Burge, *Truth, Thought, Reason*, Introduction, 38–39; 'Sinning Against Frege', 232–234; 'Wherein Is Language Social?' in Alexander George (ed.), *Reflections on Chomsky* (Oxford: Blackwell, 1989), reprinted in *Foundations of Mind*; and 'Gottlob Frege: Some Forms of Influence', forthcoming in Michael Beaney (ed.), *Oxford Handbook of the History of Analytic Philosophy*.

[25] 'Intuition' in a Kantian sense. Frege, *Begriffsschrift, eine der arithmetischen nachgebildete Formelsprache des reinen Denkens* (Halle, Germany: L. Nebert, 1879), section 8. English translation: *Begriffsschrift, A Formula Language, Modeled upon that of Arithmetic, for Pure Thought* in Jan van Heijenoort (ed.), *From Frege to Gödel: A Source Book in Mathematical Logic, 1879–1931* (Cambridge, Mass.: Harvard University Press, 1981). There is also the famous passage in 'Der Gedanke', in *Kleine Schriften*, 350 [66]; 'Thought', in *The Frege Reader*, 333, in which Frege ascribes to some uses of 'I' a sense private to the speaker that is almost surely not descriptive, or at least not purely descriptive.

Frege may not have agreed that every name has a descriptive sense, much less a purely descriptive sense.

In any case, Frege thinks that senses of many proper names in contexts of use are those of definite descriptions, though perhaps not pure definite descriptions. I think that he took ordinary proper names, as well as indexicals, to be unsuited for an ideal language of scientific thought. He regarded contextual variability of sense as a defect in an expression, for scientific purposes.

I have been writing about Frege's conception of sense. I have not implied that I accept it. I do not.[26] The way to object to Frege's views is to object to his theory of thought, not to claim that he failed to understand linguistic meaning or rules of use in natural language—communal or idiolectic. It is misguided to claim, as some have, that names lack a sense, but then introduce a mode of presentation, a way of thinking associated with them. Modes of presentation associated in certain ways with expressions—even contextually associated with them–*are* senses.

Ordinary linguistic meaning is not confined to, and sometimes hardly includes, the way a user thinks in using language. The meaning of demonstrative constructions commonly does not include the way of thinking that guides a demonstrative, beyond indicating that the way of thinking is in some way context-bound. (Think of perceptually guided demonstratives such as 'that' in 'that is a sphere'. The way of thinking that guides 'that', perhaps a perceptual way, is vastly richer than the meaning expressed by 'that'.) Despite the criticisms, advanced below, of Frege's applications of his concept sense, I think (a) that senses that are thought in uses of language are often different from the linguistic meanings expressed, and (b) that there are senses (especially perceptually based senses) that are different from any meanings that are actually expressed in the thinker's language. Senses are ways of thinking that are at least contextually associated with language in certain explanatorily circumscribed ways that I will discuss later. Frege's concept has a valuable cognitive orientation, not centered on rules of linguistic usage.

As noted, I do not accept Frege's ontological Platonism about senses. Representational content—certainly empirical representational content—depends for its nature on events in space and time. It depends for its nature on mental operations (the psychological counterparts of uses) and, more particularly, on relations of occurrence-based referential applications to particulars. It is unclear whether Frege intended to give a general theory of thought expressed in language, or just a theory of thought ideally suited to science. I have been inclined to attribute

[26] Burge, 'Sinning Against Frege', aims to show that certain objections to Frege's account of the senses of names and other context-dependent devices contain anachronistic elements. Some of these objections are undermined by the anachronism. I was specific about which objections were undermined. I did not and do not think that all objections contain anachronism, or even that all those that do, are mistaken in their conclusions. Kripke and Donnellan showed on epistemic and psychological grounds that reference (or denotation) by names and other context-dependent devices is not semantically determined by description or cognitive resources available to thinkers. Their main points against Frege are, I think, decisive. For elaboration, see Burge, *Truth, Thought, Reason*, Introduction, 35–43ff.; *Origins of Objectivity* (New York: Oxford University Press, 2010), 145–146; 'Frege: Some Forms of Influence', section 6.

the former view. Perhaps, though, he did not exclude indexical meanings and occurrence-bound referential applications from all thought content—only from the content of scientific thought.

Either way, Frege's views on sense cannot be right. For occurrence-based contents (applications) cannot be eliminated from empirical scientific thought, because they cannot be eliminated from perceptual belief. Occurrence-based representational contents depend for their natures on occurrence in time and on being used in occurrent thinking. (See note 3.)

There are further difficulties with Frege's view, conceived as a general theory of thought. For any definite description *the d* and almost any proper name *n*, a thought expressed by a sentence of the form $\underline{n = \textit{the } d}$ is not the thought expressed by a sentence of the form $\underline{n = n}$.[27]

In *Naming and Necessity*, Kripke established this point. He showed that one can use proper names without having a definite description that even purports to be sufficient to denote what the name names. And he showed various epistemic ways in which speakers' uses of a name and any associated definite description can come apart.[28]

The point does not depend on name and description coming apart epistemically. A thought $\underline{n = n}$ is almost never the same as a thought $\underline{n = (\text{the entity that} = n)}$, even assuming that the rigidity of the name exactly matches the rigidity of the definite description. Thinking the pure identity thought with any ordinary proper names engages different cognitive capacities than does thinking any definite-description thought.[29] Thinking of someone as Plato differs from thinking of that person via *any* definite description. The name-type figures essentially in the thought content, and no generalized definite description operator figures in thought with a proper name.[30] Thought contents type-identify cognitive

[27] I am referencing Frege's account in 'Über Sinn und Bedeutung' (1892) in *Kleine Schriften*, p. 144n [27n]; in *The Frege Reader*, p. 153. '=' is read 'is identical with'. Instances of the forms are: 'Aristotle is identical with the teacher of Alexander' and 'Aristotle is identical with Aristotle'. I write 'almost any proper name' to allow for stipulations that names abbreviate definite descriptions. I am inclined to say that these are not genuine cases of proper names. I assume that to be a definite description, an expression's logical form must be governed by a wide-scope uniqueness operator, *not* by a demonstrative or other context-dependent element.

[28] Kripke, *Naming and Necessity* (Cambridge, Mass.: Harvard University Press, 1980). Similar points were made by Keith Donnellan, 'Proper Names and Identifying Descriptions', *Synthese* 21:3/4 (1970), 335–358.

[29] I mean by '*pure identity thought*' any thought of the form of an identity, with the same representational content in each argument place of the identity concept. By '*ordinary proper name*', I intend to exclude "names" stipulated to abbreviate a definite description, where the definite description is perhaps rigidified in certain ways.

I believe that the point that *n* is never epistemically equivalent to $\underline{\text{the entity that} = n}$ holds even if one construes the description—as I think one can—in such a way that it cannot come apart from the name *modally*. See my 'Postscript to "Sinning Against Frege"', in *Truth, Thought, Reason*. My basic point is psychological: uses of names do not involve use of the identity predicate; uses of definite descriptions do involve such use.

[30] The same point applies to the quantificational apparatus that Russell thought played the role of the definite description operator (see note 27). Most individuals learn proper names before mastering such a generalized logical apparatus.

capacities. Since the capacities differ, the thought contents differ. Frege's reflections on identity can be used to show that, in a general theory of thought, his examples of proper names' having the senses of definite descriptions cannot be right.

The same point goes for indexicals. Meanings of indexicals figure in thought. Thinking of a time as today differs from thinking of it via a definite description, or via perception of events linked to the day. The occurrent thoughts today is the one and only *F* and today is the time of [perceptually presented] event *E*, are not the thought today is today (assuming all are true and thought on the same day, and assuming the rigidities match). Testing a sense against a pure identity thought is a powerful tool. What makes the test robust and effective is its connecting sense and thought content with the typing of psychological states and competencies.

A similar test applies to indexicals like tense, without forming identity thoughts.

(a) the leaf is [present tense] green; so the leaf is [present tense] green

does not have the same inferential form as

(b) the leaf is [present tense] green; so the leaf is green at the time I think this very thought.

Assume that, in (a) and (b), citing the times in the conclusions uses anaphora. So the argument cannot be invalid on account of a shift in times. Then (a) and (b) are necessarily and apriori truth preserving. Thinking (a) is not thinking (b). (b) types resources to denote oneself and to think about thought. (a) does not. The inferences differ—and are marked by different logical forms. As with proper names, the sense of an indexical is almost never (see note 27) that of a definite description.

Let me summarize what I have said so far about relations between sense and modern notions of linguistic meaning. The sense of a *demonstrative* on an occasion of use often differs very significantly from the linguistic meaning of the demonstrative on that occasion. Let us bracket the context-bound applications of demonstratives. Applications are part of the representational content of a thought and part of the truth condition of an utterance. For present purposes, I will count such applications neither part of the linguistic meaning nor part of the sense. An ability-general mode of presentation in thought that is appropriately associated with a demonstrative *is* a sense. The ability-general mode of presentation is, very frequently, vastly richer than the linguistic meaning of the demonstrative-governed phrase on the occasion of use.

Occasions on which a demonstrative is backed by perceptual modes of presentation make the point very vividly. Linguistic meaning grounds accounts of communal or idiolectic rules for understanding a linguistic item. Sense is invoked, with certain restrictions to be discussed later, in an account of the ability-general aspect of what a thinker thinks. In the case of use of demonstratives partly backed by perceptual modes of presentation, the two notions usually diverge significantly. For example, the linguistic meaning of 'that cat' (applied in a context in which the utterer is looking at and thinking about a particular cat) centers on rules for understanding the expression. The linguistic meaning is not nearly as rich as the mode of presentation, the sense, that the thinker uses to determine the case visually on a particular occasion. That sense includes a particular perceptual conception of the cat as it appears from a certain angle and distance in a certain type of light. The sense is vastly richer than the linguistic meaning, even the linguistic meaning in the utterer's idiolect. Since such uses of demonstratives are so common, this difference is substantial and theoretically important.[31]

Frege's account of *indexicals* is unacceptable, unless one both focuses on scientific thought and grants that meanings of indexicals are to be banned from scientific thought. In a general account of thought, it must be acknowledged that the linguistic meaning of indexicals is usually a part of the ability-general aspect of what a thinker thinks in using an indexical. (Again, I bracket the occurrent application.) I leave open whether the ability-general aspects of what is thought on particular occasions of use are sometimes richer than linguistic meanings of the indexicals.

Frege's account of *proper names* is mistaken where it assigns them descriptive senses. I think that proper names have an attenuated idiolectic and communal linguistic meaning—what individuals master by mastering rules for using the name. It is clear that uses of names express for individuals cognitive, ability-general modes of presentations—senses. It is clear that these senses are not to be associated with ordinary descriptions. The ability-general aspect of the name is very thin and is associated with the "property" of having the name—being one of the Aristotles or Bobs. (I believe that uses of names, like uses of demonstratives and indexicals, involve context-bound applications. Again, I bracket applications.)[32] I leave open whether the senses of proper names coincide with their linguistic meanings, pending better understanding of the linguistic meaning of names. It seems to me possible that they commonly do coincide.

Distinguishing sense from linguistic meaning helps orient one's approach to *indirect discourse* and *reports of propositional attitudes*. Linguistic rules for such reports allow immense contextual variation with the reporter's purposes. If, however, one focuses on uses in which the reporter's purposes are scientific, or otherwise specifically concern what an individual's psychological states are, one can discern in such uses something approaching reports that designate an individual's senses. On Frege's view, of course, the oblique senses of expressions in (ground-level) oblique contexts determine their ordinary senses; oblique senses are used in attributing the ordinary senses to the individual reported on. In other words, one can use such constructions to designate senses or representational thought contents. In most cases of ordinary reports, however, the meanings of the

[31] I discuss use of the terms 'context-bound', 'ability general', and 'application' in 'Five Theses on *De Re* States and Attitudes', here I think that the intuitive meaning of the terms will suffice. (See note 3.) I make the point that applications are not easily assimilable to Fregean senses in 'Belief *De Re*'. See also *Truth, Thought, Reason*, pp. 50–54.

[32] See my 'Reference and Proper Names', *The Journal of Philosophy* 70:14 (1973), 425–439.

reports fall far short of determining the senses in the thought of the individual who is reported on. Again, Frege was focused on scientific uses of language. He took his account to be a first step in explicating a language for a science of thought. It was.

In the next sections, I develop a further respect in which common uses of the notion of linguistic meaning vary, at least in emphasis, from appropriate uses of the notion of sense. That variation has to do with certain cases of incomplete understanding. To get to that point, I must discuss in more detail Frege's conception of the individuation of senses.

II

Fregean reflection on identity thoughts ought not be applied mechanically. As Frege emphasized, differences in words do not automatically indicate differences in thought content. One cannot read an individual's thought contents off sincere assertions in response to pieces of language. But often, one can use Frege's form of reflection to evoke fine-grainedness in modes of presentation—which mark fine-grainedness in competencies in individual psychologies. What we know about cognitive competence helps in determining the senses of words.

Expositions of Frege's reflections on identity thoughts in 'On Sense and Denotation' often conflate points that he makes about his examples with the main thrust of his reflections, missing their power. It is often said that his point is to distinguish empirical or synthetic thoughts from apriori or analytic ones. Frege does makes these distinctions. His most famous examples (The Morning Star is identical with the Evening Star; The Morning Star is identical with the Morning Star) illustrate them. But these distinctions are not his main point. In writings from the same period, and elsewhere, Frege cites pairs ($2^2 = 4$, $2 + 2 = 4$) both members of which are, by his lights, apriori and derivable from truths of logic, analytic in his sense. He uses the pairs to illustrate differences in sense.[33]

Another common misreading holds that the examples in 'On Sense and Denotation' are meant to contrast statements having cognitive value (*Erkenntniswert*) with

[33] Frege, *Die Grundgesetze der Arithmetik* (1893) (Hildesheim, Germany: Georg Olms, 1962), [7]; partly translated in *The Basic Laws of Arithmetic: Exposition of the System*, trans. and ed. Montgomery Furth (Berkeley: University of California Press, 1964), p. 35. A similar point ($2^4 = 4^2$ and $4 \text{ x } 4 = 4^2$) is made in Frege, 'Funktion und Begriff' (1891) in *Kleine Schriften*, p. 132 [13]; translated as 'Function and Concept' in *The Frege Reader*, p. 138. There are other passages in which Frege contrasts the senses of arithmetic truths, both of which are for him apriori and, at least until late in his career, both of which are for him analytic: ($3^2 = 3^2$ and $2^3 + 1 = 3^2$) Frege to Russell 12/28/1902, *Wissenshaftliche Briefwechsel*, ed. Gottfried Gabriel et al. (Hamburg: Felix Meiner Verlag, 1976), p. 235; *Philosophical and Mathematical Correspondence* (Oxford: Blackwell, 1980), p. 152; ($7 = 3 + 4$ and $7 = 7$) Frege to Russell 11/13/1904, *Wissenschaftliche Briefwechsel*, pp. 245–248; *Philosophical and Mathematical Correspondence*, pp. 163–165; ($5 + 2 = 4 + 3$ and $5 + 2 = 5 + 2$) Frege to Peano (undated), *Wissenschaftliche Briefwechsel*, p. 197; *Philosophical and Mathematical Correspondence*, p. 128; ($5 = 2 + 3$ and $5 = 5$) 'Logik in der Mathematik', in *Nachgelassene Schriften*, p. 242; *Posthumous Writings*, p. 224.

statements lacking it. Frege never writes that statements of the form $a = a$ lack cognitive value.[34] He writes only that they have different cognitive value from statements of the form $a = b$.

An even more common error is to offer a criterion for difference in sense along this line:

A sentence *S* has a different sense from a sentence *S'* if and only if it is possible to understand the senses (thought contents) of *S* and *S'* while having different cognitive attitudes (such as judgment) toward those senses as thought contents.[35]

[34] Frege calls $a = a$ statements 'boring', Frege to Peano undated, *Wissenschaftlicher Briefwechsel*, p. 195; *Philosophical and Mathematical Correspondence*, p. 126. He says of certain statements of the form $a = b$ that they *extend* our knowledge, whereas statements of the form $a = a$ do not, since they are self-evident: 'Logik in der Mathematik', in *Nachgelassene Schriften*, p. 242; *Posthumous Writings*, p. 224. But he never writes that the latter statements do not constitute knowledge, or lack cognitive value. Cf. also Frege to Jourdain undated, Frege to Russell 12/28/1902; Frege to Russell 11/13/1904, *Wissenschaftlicher Briefwechsel*, pp. 128, 234–235, 246–247; *Philosophical and Mathematical Correspondence*, pp. 80, 152, 164–165, respectively. I think that there is no good evidence to believe that he held either of these views.

[35] In view of incomplete understanding, which I discuss shortly, I take 'understand' here to require that the understanding is not defective or incomplete. Close analogs of the principle cited in the text are stated by Michael Dummett and Gareth Evans, except that they write of attitudes toward sentences' being true rather than attitudes toward thought contents. (Frege uses both locutions.) See Michael Dummett, *The Interpretation of Frege's Philosophy* (Cambridge, Mass.: Harvard University Press, 1981), p. 323; Gareth Evans, *Varieties of Reference* (New York: Oxford University Press, 1982), pp. 19–20. Dummett cites the first part of (C), cited below in the text. But he fails to note that the latter part shows that Frege did not accept the criterion for sense difference that Dummett proposes. In interpreting Frege, Evans rightly emphasizes the psychological relevance of sense. But he moves, incautiously, from passages in which Frege states that possible differences in attitude toward contents are sufficient for *differences* in sense to an analog of the criterion stated in the text, an analog that also claims that necessary sameness in attitudes toward thought contents is sufficient-for-*sameness* of sense. Examples of passages in which Frege makes analogs of the weaker, sufficient-for-difference claim are 'On Sense and Denotation', [25–26, 32]; Frege to Russell 12/28/1902, *Wissenshaftlicher Briefwechsel*, p. 236; *Philosophical and Mathematical Correspondence*, p. 153; 'Kurze Übersicht meiner logischen Lehren', in *Nachgelassene Schriften*, p. 212; *Posthumous Writings*, p. 197.

Dummett and Evans rightly require that the sentences be understood. As noted, I take this to mean *fully understood*. Consider this inadequate sufficient-for-sense-difference variant:

(I) If it is possible for an individual with *ordinary* linguistic understanding of *S* and *S'* to take one as true and the other not to be true, then *S* and *S'* express different senses and have different thoughts.

An important theme in Frege's philosophy is to insist that even expert language users can use a sentence with ordinary understanding and lack complete understanding of its sense. Then different attitudes by ordinarily competent language users toward *sentences* that express the same senses would be possible. This point is implied by my 'Frege on Sense and Linguistic Meaning', in David Andrew Bell and Neil Cooper (eds.), *The Analytic Tradition: Meaning, Thought, and Knowledge* (Oxford: Blackwell, 1990), reprinted in *Truth, Thought, Reason.*

In the literature there is some misreading me both as holding that Frege accepted (I) and as holding that Frege *never* thought that an analysis preserves sense—which is contrary to a passage in Frege, 'Logic and Mathematics' (1914), in *Nachgelassene Schriftin*, p. 228, *Posthumous Writings*, p. 211, which I certainly did not overlook since I discussed it in 'Frege on Sense and Linguistic Meaning', in *Truth, Thought, Reason*, pp. 252–253. Neither claim about Frege can be found in my writing.

As I wrote in *ibid.*, I think that many "analyses", including most successful "analyses" in the history of sciences other than semantics—even those culminating in definitions—are explications in which *explicans* and *explicandum* express different senses, different ways of thinking. (I think that Frege

Frege did not accept the 'only if' direction. He thought that if two thoughts are self-evident, understanding them compels sameness of all cognitive attitudes.[36] Perhaps $\underline{2^2 = 4}$ and $\underline{2 + 2 = 4}$ cannot attract different cognitive attitudes, if one understands both. Yet they constitute different thoughts. Frege may not have believed that these arithmetical truths are self-evident. But in *Basic Laws* he clearly holds that there are plural basic logical laws, canonically expressed by different sentences with distinct senses. He clearly holds that all basic logical laws, and even some of their immediate consequences, are self-evident.[37] So he holds that sentences with different senses can compel the same cognitive attitudes. Contrast with pure identity thoughts, or pure identity inferences, is a more powerful differentiating tool than possible contrast in judgment.

The possible-difference-in-cognitive-attitude test is inadequate because it focuses on separability of judgment of whole thoughts, not on thought type. Difference in thought type can hinge on thought-component differences, which mark differences in competencies. Competence with a name is different from competence with a definite description. Competence with squaring is different from competence with addition. These differences remain even if in individual cases they do not yield possible differences in judgment.

The passages that express Frege's sharpest grip on his own concept <u>sense</u> are the one in 'On Sense and Denotation' just discussed and the following passage:

> (C) Now two sentences *A* and *B* can stand in such a relation to one another that anyone who recognizes the content of *A* as true must also recognize that of *B* straightway [*ohne weiteres*] as true, and, also conversely, that anyone who accepts the content of *B* as true must immediately [*unmittelbar*] accept that of *A* (*equipollence*), whereby it is presupposed that the grasp of content of *A* and *B* does not raise any difficulty. . . . I assume that there is

probably underestimated the importance of this point.) Explications and definitions are usually illuminating not because they produce sameness of thought content, or sameness of sense, but because they illuminate applications of the thought content of the *explicandum*. Some might paraphrase these remarks as entailing that there are few interesting "true" conceptual analyses, where 'true analysis' is misleadingly used in the paraphrase to apply to an explication that "unpacks" a complex concept and expresses it more perspicuously.

[36] Frege, 'Gedankengefüge' (1919), in *Kleine Schriften*, p. 393 [50]; *Collected Papers on Mathematics, Logic, and Philosophy*, ed. Brian McGuinness (New York: Wiley-Blackwell, 1984), p. 405. Note that this point shows that another common reading—that the point of Frege's examples is to distinguish informative from uninformative statements—is mistaken. Many statements that express genuine knowledge, but do not '*extend* our knowledge'—and are apriori *and* self-evident—are uninformative in the ordinary sense, and uninformative in the technical sense that accepting them does not affect our subjective probabilities. But many such statements differ in sense.

[37] Frege takes senses of pure identity statements to be self-evident (note 34). These are immediate consequences of the basic logical law of identity. For more on Frege's notion of self-evidence see Burge, 'Frege on Knowing the Foundation', *Mind*, NS, 108:426 (1998), 305–347; reprinted in *Truth, Thought, Reason*. Frege's belief in a plurality of basic logical laws is constant through his career. Cf. *Begriffsschrift*, section 13. (For a caveat, see note 41.) Further evidence that Frege took the senses of sentences expressing different axioms or basic logical laws to be different derives from combining his sense-composition doctrine with remarks about the senses of expressions for logical constants, for example, 'Gedankengefüge' (1919), in *Kleine Schriften*, p. 381 [39]; *Collected Papers on Mathematics, Logic, and Philosophy*, p. 393.

nothing in the content of either of the two equipollent sentences that would have to be at once immediately [*sofort unmittelbar*] recognized as true by anyone who had grasped it correctly.[38]

Frege requires that the content of neither sentence contain a self-evident element (an element that commands immediate acceptance). He does so to block collapse of the senses of all expressions of self-evident thoughts, or thoughts that depend for their truth on self-evident components, into a single thought content.[39] Note the crucial role of immediacy, which is additional to universal co-acceptance. The immediacy condition requires that an individual can recognize the sense of *B* to be true *in* recognizing the truth of *A*, and vice versa, *without inference or reflection*.[40]

This principle is a close analog of the condition that Frege sets out in 'On Sense and Denotation'. If one cannot recognize the truth of a thought *T'* expressed by sentence *B from* the truth of a thought *T* of the form $\underline{a = a}$, expressed by sentence *A*, without inference and without reflection, then thought *T'* has form $\underline{a = b}$, and the senses of *B* and *A* differ—assuming that one understands both thoughts rightly. These are, I think, Frege's best thoughts on sense identity.[41]

[38] Frege, 'Kurze Übersicht meiner logischen Lehren' (1906) in *Nachgelassene Schriften*, p. 213; *Posthumous Writings*, p. 197. Eva Picardi, 'A Note on Dummett and Frege on Sense-Identity', *European Journal of Philosophy* 1:1 (1993), 69–81 (see p. 75), recognizes both points about this passage made in the text. She also correctly notes that immediacy has to do with the form and type of a transition, not its temporal speed. The idea is that there is no inference other than one of the form $\underline{p;\ so\ p}$—which Frege probably would not have counted as an inference.

[39] This point comes clear by comparing (C) with the cognate passage in Frege to Husserl 9/12/1906, *Wissenschaftliche Briefwechsel*, pp. 105–106; *Philosophical and Mathematical Correspondence*, p. 70, which explicitly discusses self-evidence.

[40] Thus *any* inferential transition (other than $\underline{p;\ so\ p}$) suffices to differentiate senses. I believe that this test is the inferential corollary of Frege's account in 'On Sense and Denotation'.

[41] It is well known that not all of Frege's remarks accord with these ideas. In Frege to Husserl 9/12/1906, Frege states a criterion on which all logical truths would have the same sense. Cf. *Wissenschaftliche Briefwechsel*, pp. 105–106; *Philosophical and Mathematical Correspondence*, p. 70. There are a few other remarks, mostly in unpublished writings, that do not accord with (C) and are congenial with the Husserl letter. But the criterion in that letter is incompatible with 'Function and Concept' (1891), [13]—the difference in sense between '$2^4 = 4^2$' and '$4 \times 4 = 4^2$'; *Die Grundgesetze der Arithmetik* (1893), [7]—the difference in sense between '$2^2 = 4$' and '$2 + 2 = 4$'. As noted, note 33, there are many other such passages.

Frege may have unconsciously used two conceptions of sense—a psychologically realistic one and a very idealized one. Carlo Penco so argues in 'Frege, Sense and Limited Rationality', *The Review of Modern Logic* 9:1–2 (November 2001–November 2003), 53–65. I think it more likely that Frege made a mistake. Within a month or so after the Husserl letter, he engages the same issue, using similar language even, but proposes (C)'s restrictive account of sense identity, seemingly correcting the criterion in the Husserl letter. Similar views of the matter are stated by Heijenoort, 'Frege on Sense Identity', *Journal of Philosophical Logic* 6:1 (1977), 103–108; Dummett, *The Interpretation of Frege's Philosophy*, pp. 324–325—'it is hard when a philosopher is held to account for remarks made in purely private correspondence' [it should be noted, however, that not *all* the deviations from (C) are in private correspondence]; and Picardi, 'A Note on Dummett and Frege on Sense-Identity', who plausibly regards (C) as *correcting* the criterion in the Husserl letter. I agree with Picardi that Frege was perhaps sometimes too ready to count an obvious inference as no inference at all. In any case, I regard the approach in 'On Sense and Denotation', 'Function and Concept', *Basic Laws*, and (C), as much the more valuable.

Again, applying such principles requires theoretical acumen. Application is fallible. I believe that whether inference or reflection is required to get from thought *T* to thought *T'* hinges on the psychological capacities that are type-individuated by the senses. We are often entitled to presume that our judgments on these matters have some weight. But ultimately, they are matters for philosophically informed psychology and psycholinguistics.

Frege cites at least four ways in which determining sense is open to complication.

First, although Frege's assimilation of indexicals and proper names to demonstratives and his taking their senses sometimes to be those of definite descriptions are mistaken, his idea that used demonstratives express senses beyond linguistic meaning (say, in perceptual beliefs) is correct. Determining such senses has no recipe. Asking a speaker how he or she conceives the referent is a start. But finding thought content that both type-individuates the relevantly used psychological capacity and determines the demonstrative's referent is a theoretical matter.

Second, Frege's method of formalization requires reflection on patterns of inferences.[42] Formalization seeks to determine the logical structure and inferential capacities of thinkings underlying formalized sentences. Formalization helps individuate senses. Frege showed that one cannot read the formal structure of a thought off ordinary grammar. He claimed that active and passive structures have the same logical structure and sense. Similarly, for interrogative and declarative structures and for other pairs of grammatically different expressions ('Jupiter has four moons'; 'the number of moons of Jupiter is four').[43] Formalization can yield understanding of sense identity or difference that is unobvious to language users.

Third, Frege's distinction of sense from coloring complicates sense determination. Sense sets truth conditions, yields a cognitive route to denotation, figures essentially in recognition of truth, and contributes to logical form. Call these '*cognitive aspects*' of language and psychology. Colorings are psychological elements associated with the expression, or the thinking, of a sense that do not fill these roles. Frege holds that 'and' and 'but' differ in coloring, but not sense.[44]

[42] For extended discussion of Frege's method, see Burge Introduction, *Truth, Thought, Reason*, pp. 7, 11–26.

[43] See Frege, *Grundlagen der Arithmetik* (1884) translated as *The Foundations of Arithmetic*, trans. J. L. Austin (Evanston, IL.: Northwestern University Press, 1968), section 57; cf. also 'Über Begriff und Gegenstand' (1892) in *Kleine Schriften*, [199–200]; 'On Concept and Object', in *The Frege Reader*; 'Der Gedanke', [62]. I am not committed to agreeing with him on these cases.

[44] Frege, 'Gedankengefüge', [43]. Frege sometimes counts coloring part of a sentence's 'content' (*Inhalt*), but not part of its sense. Cf. 'Kurze Übersicht meiner logische Lehren', in *Nachgelassene Schriften*, pp. 213–114; *Posthumous Writings*, pp. 196–197; *The Frege Reader*, p. 300. Here Frege does distinguish an ordinary notion of linguistic meaning (which he terms '*Inhalt*', 'content') and sense. Of course, many modern notions of linguistic meaning follow him—distinguishing a notion of cognitive or literal meaning from a broader notion of linguistic meaning. On this broader notion, 'but' and 'and' do not mean the same. Distinctions between sense and ordinary linguistic meaning remain. These are most prominent in accounting for used demonstratives/indexicals—a hugely important area of understanding the use of language—and, at least in emphasis, in accounting for incomplete understanding.

'But' entails some contrast between the connected clauses. 'And' does not. Despite these psychological differences, their contributions to what is judged true about the subject matter, as opposed to what is expected, and their contributions to formal inference are, he thinks, the same. The sense/coloring distinction is real. Applying it is a theoretical matter.[45]

Fourth, incomplete understanding, which is closely connected to the second and third points, complicates determination of sense.[46] Frege takes *fully* understanding expressions to be a condition on individuating their senses. Of course, a foreigner's inept usage may not indicate much about his or her thoughts. Frege's requirement goes deeper. He held that the sense of an expression might not be fully understood even by expert users.[47] I will dwell on this fourth point.

Frege's discussions of incomplete understanding focus on aspects of language use that are largely ignored in discussions of ordinary notions of linguistic meaning. Frege holds that even the most expert users may not fully understand the sense, communal or idiolectic, of an expression.[48] The view that the most competent and expert users may lack full understanding is at least somewhat unusual.

This view opens a possible gap between how a thinker explains or identifies a sense and what sense the individual expresses. For example, by not fully understanding a simple-seeming expression, one may fail to understand complexity in the expression's sense.[49] In applying a sense identity test that relies on pure identity thoughts or on lack of inference, one must presume that the test is not undermined by incomplete understanding of the thoughts. One is often entitled to the presumption. But one must remain open to correction by theory.

[45] As Grice's development of the idea in his distinction between literal meaning and conversational or conventional implicatures made clear. Paul Grice, *Studies in the Way of Words* (Cambridge, Mass.: Harvard University Press, 1989). The theoretical nature of disputes over whether indicative 'if-then' is the material conditional plus implicatures (a prominent sort of coloring) illustrates the complexity of applying the distinction.

[46] I discuss Frege's notion of incomplete understanding in 'Sinning Against Frege'; 'Frege on Sense and Linguistic Understanding'; and 'Frege on Extensions of Concepts, from 1884 to 1903', *The Philosophical Review* 93:1 (1984), 3–34, reprinted in *Truth, Thought, Reason.* I also discuss incomplete understanding more substantively in *Foundations of Mind*; see the Introduction and subject index. Note that formalization and distinguishing coloring from sense are ways of improving incomplete understanding. Frege's view is more radical than might appear from ordinary cases like these, however. He thought that no one had fully understood the concept number until he discovered his logicist analysis.

[47] Frege's view was not new. It is an application of traditional rationalist views of thought. See Burge, 'Frege on Sense and Linguistic Understanding'.

[48] Such a view runs contrary to most thinking about ordinary linguistic meaning. It is sometimes allowed in the literature that the idiolectic meaning of an individual's expression may depend on meanings of expressions used by others on whom an individual relies. But it is rarely allowed, or at least rarely maintained, that it is possible that no one in a community fully understands the *meanings* of their expressions. Whether Frege's point can be accommodated by modern notions of linguistic meaning is a matter that I shall not try to decide.

[49] Frege considers such a situation in 'Logik in der Mathematik', *Nachgelassene Schriften*, p. 226; *Posthumous Writings*, p. 209.

This issue has tempted some to propose a normal form that specifies *thinking* as a three-place relation—thinker *I* thinks thought content *c* via way *w* of thinking *c*. A *way of thinking* the thought content is added as a parameter over and above the thought content, which is itself a way of thinking its subject matter. On this non-Fregean proposal, modes of presentations of modes of presentations are specified in every specification of thought. The new layer of mode of presentation is taken to aim at sense, the thought content, which is taken to be an object presented in thought.[50]

I believe that this move is misguided. It fails to appreciate the point and power of the idea that senses are thought contents, associated in certain ways with language. And it misconstrues the nature of incomplete understanding. Thought is a real-world psychological phenomenon. Along with the mode of the thought (belief, supposition), thought contents, including their components, type-identify psychological states and events in ways that ground very basic levels of explanation. Senses are aspects of certain psychological kinds. Normally, they are not objects aimed at in thought. Thought content suffices to play the role of individuating cognitively relevant psychological states. The proposed move is epicyclic.

The error of taking thought contents as aimed-at objects of thought is one of the oldest and most persistent in philosophy.[51] Avoiding the error is facilitated by firmly taking thought contents to be an aspect of kinds of psychological states. Avoiding the error is also facilitated by reflecting on the psychology of incomplete understanding.

I will try to explain in more detail why introducing a second layer of mode of presentation is misguided. It is unobjectionable to speak colloquially of someone who incompletely understands a sense as thinking it in a "different way" from someone who fully understands it. But the role of this "way" must be sharply distinguished from the role of senses. This "way" is not a *relatum* in the normal form of specifications of thoughts.

One psychological element that has been thought to play a "second layer" role is the linguistic expression of a sense. Incomplete understanding of expressions can yield mistakes about their sense-relations to other expressions. Incomplete understanding of structural complexity underlying a simple-seeming expression can even lead to conflating what is in fact a pure identity thought or argument

[50] Ideas that motivate this move occur in Felicia Ackerman, 'Analysis, Language and Concepts: The Second Paradox of Analysis', *Philosophical Perspectives* 4 (1990), 535–543. I have seen and heard canonical introduction of modes of presentation (whether linguistic or not) proposed by various others.

[51] Frege occasionally courts the error himself. Cf. 'Der Gedanke', in *Kleine Schriften*, p. 359 [75]; 'Thought', in *The Frege Reader*, p. 342—where he writes of elements in individuals' consciousness "aiming" at thought contents. Moreover, his own metaphor of incomplete understanding as seeing a thought content, or a sense, through a mist invites the error. Cf. *Grundlagen der Arithmetik*, Preface, [vii–viii]; 'Logik in der Mathematik', in *Nachgelassene Schriften*, p. 228; 'Logic in Mathematics', in *Posthumous Writings*, p. 211. Frege's theoretical work—his distinguishing sense from denotation, understanding from reference—provides some of the deepest resources for recognizing and rejecting the error.

with one of another form. Perhaps an example is a mistaken belief that 'Cain hit Abel; so there was a hitting at some time before now, and it was by Cain, and it was of Abel' is not an inference of the form p; so p.[52] There can be structural complexity hidden in an apparently simple expression ('hit'). The surface simplicity of 'hit' can be psychologically significant, even though the word is not part of the sense, and may obscure it. In such cases, the triviality of the argument form and the nature of the underlying cognitive competence are masked by a mistaken meta-linguistic belief. Such errors can be corrected by formalization—the second way in which determination of sense is open to complication. It has been claimed that accounting for the incomplete understanding can be aided by specifying the thinking through the normal form: *I* thinks thought content *c* via e_1—and *I* thinks *c* via e_2.

The proposed normal-form specifications do not correspond to a difference between defective and nondefective understanding. Here the proposed normal form is idle in explaining the incomplete understanding. In both incomplete- and full-understanding cases, the individual thinks the same sense through the same two expressions. The defective understanding can easily be represented in Frege's normal form. The defective understanding consists in a mistaken meta-linguistic belief: *I* thinks: e_1 does not express *c*, and e_2 does.

Another psychological element that has been thought to play a "second layer" role is Fregean coloring.[53] As with expressions, colorings do not play the role of determining sense in the way that senses determine denotations. Colorings do not represent senses or thought contents. They are collateral psychological elements that have some other psychological association—systematic or contextual—with senses. Frege often assimilated colorings to images or feelings, not thought contents. Thus a feeling of disgust might be systematically associated with 'cur' but not 'dog', in an idiolect that otherwise uses the two words identically. Or an image of a rose might first accompany uses of 'flower'. Then the association might lapse. But as Frege occasionally notes, colorings can consist in associated *thoughts* either systematically or contextually associated with the senses of linguistic expressions. Thus thoughts of contrast are colorings systematically associated with 'but', and not 'and'. Thoughts of heroism might be contextually connected with 'steed', but not 'horse'. Such colorings could lead an individual to judge that the senses of the expressions are different, when they are, let us suppose, the same.

[52] I assume, purely for the sake of argument, that the Davidsonian analysis of action sentences, elaborated by Terence Parsons and James Higginbotham, is correct. Donald Davidson, 'The Logical Form of Action Sentences' (1967) in *Essays on Actions and Events* (New York: Oxford University Press, 2001, 2nd edn.); Terence Parsons, *Events in the Semantics of English* (Cambridge, Mass.: MIT, 1990); James Higginbotham, 'On Events in Linguistic Semantics', in James Higginbotham, Fabio Pianesis, and Achille C. Varzi (eds.), *Speaking of Events* (New York: Oxford University Press, 2000), 49–79.

[53] Frege usually writes of colorings as ideas—tokens in individuals' minds. But one can, as I will, think of colorings as psychological types—hence as shareable, as most senses are.

Colorings can be colloquially termed 'ways of thinking'. But they are collateral psychological elements. They are not modes of presentation that determine senses as denotations or *representata* in anything like the way senses are modes of presentation of their subject matters. Colorings are not even parts of literal, full understanding of senses. They bear essentially adventitious relations to them.[54] So they can be specified separately.

Frege never took sense to type-identify every psychologically significant state associated with language, or even every significant aspect of thinking associated with language. In fact, he emphasizes the psychological significance of words in ordinary language—coloring—that does not line up with sense. He stresses that the 'contents of consciousness' are not an immediate guide either to the structure of complex senses or to individuation of non-complex senses.[55]

One cannot infallibly isolate senses or thought contents just by introspecting what is present to consciousness.[56] Senses and thought contents type-individuate psychological competencies, which in turn can be delineated only by reflecting on patterns of use and by applying theoretical concepts of sense and thought content. These concepts center on *cognitive aspects* of psychological kinds—those that are relevant to setting truth conditions, judging a thought true, finding routes to and determining denotations, and making formal inferences.[57]

[54] I believe that other candidates for playing the role of "second layer" modes of presentation are subject to similar points.

[55] Frege, 'Über Sinn und Bedeutung', in *Kleine Schriften*, pp. 145–146 [29]; 'On *Sinn* and *Bedeutung*', in *The Frege Reader*, p. 154: '[An idea] is often drenched [*getränkt*] with feelings; the clarity of [an idea's] individual parts varies and fluctuates. The same idea is not connected with the same sense, even in the same person'; 'Über Begriff und Gegenstand', in *Kleine Schriften*, p. 170 [196]; 'On Concept and Object', in *The Frege Reader*, p. 184; 'Logik' (1897), *Nachgelassene Schriften*, pp. 150–155; 'Logic' (1897), in *Posthumous Writings*, pp. 138–143; *The Frege Reader*, pp. 239–244; 'Einleitung in die Logik' (1906), *Nachgelassene Schriften*, p. 209; 'Introduction to Logic', *Posthumous Writings*, pp. 192–193; *The Frege Reader*, p. 295; 'Kurze Übersicht meiner logische Lehren' (1906) in *Nachgelassene Schriften*, p. 213; 'A Brief Survey of My Logical Doctrines', in *Posthumous Writings*, p. 197; *The Frege Reader*, p. 300; Frege to Hussserl 10/30–11/1/1906, *Wissenschaftlichen Briefwechsel*, p. 102; *Philosophical and Mathematical Correspondence*, p. 67; 'Der Gedanke', in *Kleine Schriften*, pp. 347–348 [63–64]; *The Frege Reader*, pp. 330–331. In all these passages except the first, Frege links coloring with the specific linguistic expression used, and with the problem of formalization.

[56] This remark, and reflection on what senses and thought contents type-individuate, show that requirements that senses and thought contents be "transparent" to the mind through introspection need careful qualification. It has been said that anyone who grasps a sense must be able to differentiate it accurately from any other sense. Assuming that one brackets representational contents of occurrence-based demonstrative applications (as not being senses), such a requirement is plausible for ideal, full understanding. It is unacceptable for minimal understanding, the sort sufficient to think with a sense. One can have the self-knowledge and the control over one's reasoning that can be reasonably expected, while still allowing for vulnerabilities to error regarding senses and thought contents as objects of individuation, and in special cases of environmental or contextual shift. See my 'Individualism and Self-knowledge', *The Journal of Philosophy* 85:11 (1988), 649–663; 'Memory and Self-Knowledge', in Peter Ludlow and Norah Martin (eds.), *Externalism and Self-Knowledge* (Stanford: CSLI, 1998); both reprinted in *Foundations of Mind*. For a different but largely congenial view, see also Mikkel Gerken, 'Conceptual Equivocation and Warrant by Reasoning', *Australasian Journal of Philosophy* 89:3 (September 2011), 381–400.

[57] It has been suggested, contrary to Frege, that different proper names naming the same thing always differ merely in coloring, not in sense. I think that this view is psychologically very

Frege made a bold theoretical postulation. He postulated that thought, conceived in his regimented way, is a psychological kind that grounds powerful explanations. He thought that one could find explanatorily important psychological kinds by reference to *cognitive* aspects of psychological kinds. He thought that normative issues regarding truth conditions, judgment, knowledge, and formally valid inference could be used to help isolate significant psychological kinds. His anti-psychologism can be fruitfully regarded as a protest against understanding not only logic but psychology in terms that disregard what we know about truth conditions and competence for making valid inferences–regardless of coloring.

Frege's postulation has been empirically vindicated. Powerful explanations in semantics, psycho-linguistics, and cognitive psychology are grounded in psychological kinds that are isolated by his conception of sense and thought content.[58] Modern notions of meaning have made use of some of the key guidelines that he used in applying his own concept sense.

I think that "ways of thinking" thought contents involved in incomplete understanding can always be assimilated to collateral, separately specifiable psychological states—conceptual or non-conceptual, cognitive or not—that accompany the relevant thoughts. Incomplete understanding can always be accounted for as failure to accept thoughts—either thoughts about relations between expressions and senses, or first-order thoughts that bear on constitutive explications. Frege's normal form for specifying thoughts need not be altered to account for incomplete understanding.

III

I turn, from here on, to constructive remarks about incomplete understanding. They further support the point of the preceding paragraph. There are many kinds of incomplete understanding, just as there are many kinds of understanding. There are two generic types that share some instances. One—which I call '*linguistic incomplete understanding*'—consists in misapprehension of the relation between an expression and its sense, or inability to explicate an expression's sense. This type may or may not accompany incomplete understanding of the *sense*—that is, of the thought content—itself. The other type is incomplete

implausible. With Frege, I have been ignoring it. Different names correspond to different procedural and epistemic routes to a referent. They have different determination conditions (systematically different causal relations to a referent). They enter into different cognitively or scientifically relevant inferences. Some of their psychological differences are clearly cognitive.

[58] As nearly arbitrarily chosen examples, see Steven Pinker, *The Language Instinct: The New Science of Language and Mind* (London: Penguin, 1994); Lance J. Rips, *The Psychology of Proof: Deductive Reasoning in Human Thinking* (Cambridge, Mass.: MIT, 1994); Gennaro Chierchia and Sally McConnell-Ginet, *Meaning and Grammar: An Introduction to Semantics*, 2nd edn. (Cambridge, Mass.: MIT, 2000).

understanding of the sense or thought content itself. It does not consist in misapprehension of the relation between an expression and the expression's sense. Call this type '*conceptual incomplete understanding*'.

Linguistic incomplete understanding divides into *two species*. In one, the incompleteness can be ameliorated by making conscious what is already fully formed and available in the individual's psychology. In the other, incomplete understanding cannot be improved in that way.

In cases of this *second species* of linguistic incomplete understanding, there is something deficient in the individual's understanding, at all psychological levels, of the meaning of his own linguistic expression. The deficiency lies in not having a veridical, or otherwise fully adequate, general explication of the cases that the meaning in fact applies to. The defect may not be correctable by reflection alone. Since this species often occurs when conceptual incomplete understanding occurs, I do not discuss it separately.

The *first species* of linguistic incomplete understanding divides into several partly overlapping sub-cases. One is failure to distinguish an expression's sense from its coloring. An individual may mistakenly insist that 'and' and 'but' have different senses in the individual's idiolect. The sense type-individuates a different psychological kind from the coloring. The individual needs theory and reflection to draw relevant distinctions. Some differences in coloring center in associated feelings. Others may reside in differing entailments that mark linguistic meanings but not senses. These latter colorings—close to Gricean conventional implicatures—are contents of collateral thoughts, but not cognitive aspects of the use of the relevant word, in the sense indicated above.

Another sub-case is not recognizing logical forms that reflection on usage would reveal. An individual may mistakenly deny that two sentences have the same sense because structural elements in one are not evident in its surface grammar. Not recognizing the putative structure of 'hit' sentences, or of interrogative and declarative forms, may illustrate this sub-case.

A further sub-case involves failure to recognize sense synonymies in certain definitions. An individual may forget a stipulative abbreviation and mistakenly deny sense synonymy purely because of the expressions' surface structural differences, or differences in coloring between *definiendum* and *definiens* that accrued since the abbreviation was established. On some classical philosophical views, abbreviative definitions model definitions in general. I shall argue, shortly, that this model has been overused.

A fourth sub-case involves failure to accept a definition or other constitutive explication, where the *definiens* concept is unconsciously fully formed and guides the *definiendum* concept in the unconscious psychology, and where expressions for *definiens* and *definiendum* differ in sense. An individual may evince this type of incomplete understanding through erroneous stabs at a definition, but may eventually bring to consciousness the *definiens* concept via reflection.

In all these cases, linguistic incomplete understanding consists in an individual's conscious judgments' failing to accord with actual competence. In a sense, these are cases in which performance (conscious judgment that involves incomplete understanding) does not match competence (the understanding that resides in unselfconscious competence, which itself is not incomplete). The nature of linguistic competencies is often not evident to immediate consciousness. Incomplete understanding in these cases is naturally expressed in meta-linguistic form: the individual mistakenly believes that two expressions differ in sense; but the underlying psychology shows a single sense, just differences in coloring.

Conceptual incomplete understanding consists either (a) in the lack, anywhere in the individual's psychology, of a fully formed, adequate conceptual explication, where having such an explication is a relevant kind of understanding, or (b) in the lack of an understanding of the concept's boundaries, vague or otherwise.[59] The individual has the minimal competence-understanding to think with a concept (and sense). But the individual either lacks an adequate explicative conception or lacks an understanding of the scope and limits of the concept's domain of application. Here incompleteness does not consist merely in failing to match sense with expression. Incomplete understanding is of the concept, or sense, itself. There is a deficiency in conceptual understanding. It is not simply a performance problem in connecting expressions to an adequate underlying understanding.

In the sciences—outside semantics—and perhaps in ordinary life, this type of incomplete understanding is the most common and interesting. In nearly all interesting explications, definitions, and "analyses" that remedy incomplete understanding in those sciences that are not mainly concerned with language, the incomplete understanding is not of expressions that express pure identity thoughts, or pure identity inferences. The incomplete understanding does not consist in mistaken meta-linguistic belief. Nor can the incomplete understanding be remedied just by making conscious a fully formed thought content that explicates a sense or thought content. Conceptual incomplete understanding consists in failure to accept, even unconsciously, certain informative sound inferences, or certain informative, true identity thoughts. Such inferences or identity thoughts are those central to explicating thought contents or senses.

I discuss four examples of incomplete understanding, reflecting on their relations to sense identity tests and the psychology of thought. Each exemplifies both *conceptual* incomplete understanding and second-species linguistic incomplete understanding.

[59] The notion of a *fully formed* conception must be understood carefully. In cases where an individual has the component concepts of a conception, or explication, or definition, there is always a sense in which the individual has the conception—because the individual has the competence to put together component concepts in any logically or grammatically admissible way. In this sense, the individual trivially "has the conception" as part of his psychological competence. But in my sense, the individual lacks a fully formed conception. For a conception to be fully formed, it must have an actual psychological function, as a unit, for the individual, or the individual's unconscious. To be "implicit," the conception must meet this condition. And it must be connected in the individual's psychology (perhaps unconsciously) with the concept or expression that the conception provides an understanding of, so that the conception guides that concept's, or that expression's, applications.

First, suppose that an individual cannot—without trial and error—think out a definitional explication of his or her chair concept (or term 'chair').[60] The individual *could* have a fully formed, fully adequate, unconscious *definiens* conception of chairs. Bringing it to consciousness would then be purely a performance problem. Then we have first-species, fourth-case linguistic incomplete understanding. It is more likely that, even if the individual has the component concepts in an adequate, complex explicational *definiens* concept,[61] they have not been composed into that concept—even unconsciously. Often in finding definitions of familiar words, individuals put together a general explicational concept that was not originally fully formed. The sense and meaning of the *definiendum* expression and the nature of the concept are not precisely the same as those of the *definiens* expression or the *definiens* concept. Antecedent uses of the *definiendum* concept are guided by very generic concepts, perceptual templates, and senses of similarity that were not unified into a complex explicational concept. In such cases, the competence typed in the *definiens* concept is different from the competence typed by the *definiendum* concept—for example, the concept chair. The chair concept is used before the complex explicational *definiens* concept is used. Applications of 'chair', and of the concept chair that is its sense, need not rely on a general *definiens* concept. It is often fairly clear that the chair concept and the complex *definiens* concept type-individuate different psychological states and competencies. Reflection may form a concept that explains and unifies applications of the chair concept. The explicating (*definiens*) concept provides a constitutive understanding of the chair concept. Incomplete understanding is remedied by coming to believe the definitional explication. The definitional explication is not a pure identity thought, and the explicating *definiens* concept is not identical with the chair concept.

Second, an individual might have incomplete understanding of arthritis, and of 'arthritis', that is not remediable by the individual's reflection. The individual believes arthritis is arthritis but does not believe—and may even doubt—arthritis is rheumatoid inflammation of the joints.[62] There may be nothing in the individual's inferential patterns that calls for taking 'arthritis' to express inflammation of the joints, and nothing in his or her knowledge base that allows thinking of anything as rheumatoid. Thought contents help type-individuate psychological competencies. Attributing the more complex *definiens* sense to 'arthritis' to account for the individual's thinking may have no psychological basis. The two

[60] I discuss this case from a different angle in 'Intellectual Norms and Foundations of Mind', *The Journal of Philosophy* 83:12 (December 1986), 697–720; reprinted in *Foundations of Mind.*

[61] One can have a chair concept without having certain concepts, like furniture, function, intention, normal, that are probably in a good *definiens* conception. This point in itself would suffice to show that such a chair concept (the 'chair' sense) is not identical with the corresponding complex *definiens* concept/sense.

[62] I discuss this case in 'Individualism and the Mental', *Midwest Studies in Philosophy* 4:1 (1979), 73–121; reprinted with Postscript in *Foundations of Mind.*

thoughts contents are different.[63] The individual's understanding of his concept (and sense) is incomplete because rheumatoid inflammation of the joints is central to explicating arthritis and unifying and justifying the individual's applications—including those that rely on others. The individual lacks an adequate conceptual explication. As with the 'chair' example, the definition does not involve a sense identity. In contrast to the chair example, the individual cannot correct the incomplete understanding purely by reflection.

A third example illustrates incomplete understanding that is unremediable by mere reflection, even by a community. Suppose that a community believes this explication of the concept tomato: the tomato is the most common red vegetable that normally has such and such an appearance. The explication betrays incomplete understanding. Until the community learns that tomatoes are fruits, the error of thinking that tomatoes are vegetables (and not fruits) cannot be remedied. After the community corrects its error, the community continues to think of tomatoes as tomatoes, and continues to believe tomatoes are tomatoes. But it gives up its former explication. Nothing supports the view that thinking of tomatoes as fruits guided thoughts containing the concept tomato during the period of incomplete understanding. So taking the new explication (the most common red fruit that normally has such and such an appearance) to be sense-identical with the sense of 'tomato' and to be a correct articulation of the internal structure of the concept tomato is incompatible with our knowledge of the individuals' psychologies.[64]

In these three cases, accounting for incomplete understanding—and its remedy—does not require introducing modes of presentation of senses in a new normal form for specifying thoughts. The minimal understanding needed to think with the concept or sense is accompanied by conceptual incomplete understanding. The incomplete understanding consists in lack of a good, object-level (not meta-linguistic) explication anywhere in the psychology.

Frege discusses incomplete understanding mainly in connection with technical theorizing, especially in mathematics. Here the matter is more complex, but a similar picture emerges. A fourth example concerns Newton and Leibniz's incomplete

[63] Whether the individual has the same arthritis concept that more knowledgeable members of the community do is not important for the case—or for my use of the case to argue for anti-individualism. See Burge, 'Wherein Is Language Social?' In fact, I think that individuals can and often do share incompletely understood concepts with more knowledgeable communal members. Contents are often preserved in reliance on others. I will not develop the point here.

[64] Hilary Putnam, 'Is Semantics Possible?' (1970) and 'Explanation and Reference' (1973), both reprinted in his *Philosophical Papers*, vol. II: *Mind, Language and Reality* (Cambridge: Cambridge University Press, 1975), made a deep contribution to philosophical understanding of explications of common nouns in everyday life and in scientific theory. He claims in 'Is Semantics Possible?' p. 148, that the meaning of such nouns change if their stereotypical explications change. This may be so for one notion of linguistic meaning. I think that it is not so for sense. And for at least one notion of linguistic meaning (what I have called 'translational meaning'), I think that it is not so. The community's translational meaning of 'tomato' and concept tomato would not change after such a change in dictionary explication. For discussion of translational meaning, see Burge, 'Wherein Is Language Social?'; and 'Concepts, Definitions, and Meaning', *Metaphilosophy* 24:4 (1993), 309–325; reprinted in *Foundations of Mind*.

understanding of concepts in the calculus.[65] Each man favored a different concept of limit, and of the derivative. Newton favored concepts that center on approaching a limit value. Call Newton's *approaching-the-limit limit concept* 'L_N'. Call Newton's *approaching-the-limit derivative* concept 'D_N'. Leibniz favored concepts L_L and D_L that invoke infinitesimals. I believe that each mathematician had a further ur-concept of limit and an ur-concept of the derivative that are less specific than L_N, D_N, L_L, and D_L. Their respective approaching-the-limit concepts and infinitesimal concepts are elaborations of such ur-concepts. The elaborations involve more specific conceptions, and they presage specific (different) definitional explications. Such ur-concepts agree with all elaborations on the central applications, but they are unspecific enough to allow different elaborations. Indeed, the ur-concepts commonly have less sharp extensions than their elaborations. L_N and D_N eventually received clarifying explication in Weierstrass's definition. L_L and D_L later received clarifying explication in Robinson's definition.

I focus on Newton and Weierstrass. Newton's incomplete understanding of D_N showed in failed, even incoherent, explications of his concept.[66] For example, he could not explain why use of D_N does not lead to division by 0 in certain equations. Weierstrass's definition explicates Newton's concepts L_N and D_N, and justifies Newton's applications of them. Let us suppose, for the moment, that D_N and Weierstrass's *definiens* concept D_{W-d} are co-extensive. D_{W-d} provides a generalized explication and justification of the derivative and of approaching limits.

Is the first-order form of Weierstrass's definitional explication a pure identity thought? Is D_N identical with D_{W-d}? I think not. D_N is an element in Newton's thought. It helps mark a kind of psychological state of Newton's. Newton had the component concepts in D_{W-d}.[67] D_{W-d} yields a *mathematical* explanation and

[65] I have discussed this case in 'Concepts, Conceptions, Reflective Understanding: Reply to Peacocke', in Martin Hahn and Bjørn T. Ramberg (eds.), *Reflections and Replies: Essays on the Philosophy of Tyler Burge* (Cambridge, Mass.: MIT, 2003), 383–396, at pp. 385–386; and 'Postscript to "Individualism and the Mental"', in *Foundations of Mind*, pp. 170–172. Newton had infinitesimal-oriented limit and derivative concepts as well as the approaching-a-limit-oriented limit and derivative concepts. There is evidence that Newton may have had other orientations as well. He came to favor his approaching-a-limit concepts. I will follow custom in focusing on his favored versions, and in associating the infinitesimal versions with Leibniz.

[66] Newton had an explication of derivative in terms of approaching a limit that is, at least in modern understandings, correct: the derivative of a function $f(x)$ is the limit as i goes to 0 of $(f(x+i)-f(x)) \div i$. This explication shows something about the nature of Newton's concepts of both derivative and limit. But it was, and was known by Newton to be, an unsatisfying explication. It was unsatisfying because the notion of approaching a limit was notoriously disputed and not well understood—in need of mathematical explication—and because it failed to provide the clarity that would either justify practice or provide a basis for proofs. Satisfying, acceptable mathematical explications must use terms that are at least provisionally sources of illumination, and must be useable in proof. Explications of the concept limit in Newton's day did not meet this standard. Our current use of this explication is backed by a deeper understanding of a limit concept—an understanding provided by Weierstrass's explication.

[67] As noted in Burge, 'Postscript to "Individualism and the Mental"', Leibniz had a concept of infinitesimal that was later explicated by Robinson. But Leibniz lacked some component concepts in the explication. One might think of this case as a fifth example—one in which the community and greatest experts have a concept, but cannot explicate it adequately through mobilizing concepts that

justification of Newton's applications of D_N. But these two points are not enough for D_{W-d} to *be* D_N. It is not enough that Newton *could have* done the reflective mathematical work that Weierstrass did, without learning new concepts. It is not enough that D_{W-d} *mathematically* unifies and justifies Newton's usage of D_N. For D_{W-d} to *be* D_N, D_{W-d} must type the same *psychological* competencies and ground the same psychological explanations that D_N does. D_{W-d} must be fully formed as a unified, used conception in Newton's unconscious psychology, *and* it must be the content of all Newton's thinking expressed with 'derivative'. In scientific explications like Weierstrass's, the sense of the *explicans* almost never does the psychological work that the sense of the *explicandum* did before the explication's discovery.

That Newton had a consciously unarticulated but unconsciously ("implicitly") complete competence with D_{W-d} is not plausible. Distinguishing D_N and D_{W-d} is demanded by recognizing the roles of sense and conceptual content in type-individuating psychological states.

The psychology is often obscured by a classical view of apriori reflection.[68] This view holds that in reflection (the sort that Weierstrass did and that Newton could have done), the individual discovers structure in his concepts, previously hidden below the level of full consciousness.[69] It is not plausible, however, that

they already have. This case is like the Newton case in that the explication was not present in the user's psychology, at *any* level. It differs from the Newton case in that even some *component* concepts in the relevant explication are not present in the user's psychology. Leibniz could not have composed the Robinson explication by using concepts that he already had, either consciously or unconsciously.

[68] This classical conception derives from a particular reading of Plato's *Meno*. It appears in early modern rationalism and in Kant. As applied to definitions and explications in empirical science, the conception is even less plausible than in the mathematical case.

[69] A modern philosopher who relies on this conception of reflection is Christopher Peacocke, 'Implicit Conceptions, Understanding and Rationality', in *Reflections and Replies: Essays on the Philosophy of Tyler Burge*, pp. 117–152; also in Enrique Villaneuva (ed.), *Philosophical Issues 9: Concepts* (Atascadero, CA: Ridgeview, 1998). I have criticized this view for misrepresenting the psychology underlying concept use in the calculus case. See my 'Concepts, Conceptions, Reflective Understanding: Reply to Peacocke', in *Reflections and Replies* (see also note 64). Peacocke expounds substantially the same view, without directly addressing the criticism, in *Truly Understood* (New York: Oxford University Press, 2008), chapter 4. Most of the exposition there is committed to holding that thinkers have tacit *knowledge* of a definition or other generalized explication that is intensionally equivalent with terms or concepts they use. (See, for example, page 133.) Peacocke qualifies this view at one place. There he allows that 'the content of an implicit conception could equally be grounded in the operation of a processor which does not involve, at the subpersonal level, explicit representation of the content of the implicit conception' (page 142). I believe that it is a mistake to take such a case as a case of implicit conception, or a case involving a kind of representation 'operative in' (142) a given thinker. Moreover, if there is no representation of the content at the sub-personal level, there is no tacit knowledge. The case in which a representational content is in the psychology of the thinker (implicitly, tacitly, operatively) and the case in which the content is merely in the laws or principles that govern the operation of the thinker's contentful psychological states are fundamentally different, and cannot be reasonably assimilated to different types of implicit conception, much less tacit knowledge. In the latter case, the content is not a conception by the thinker, or in the thinker, in any sense. It is a conception used in explaining and justifying laws or transitions that occur among the thinker's actual psychological states—whether these are implicit, tacit, sub-personal, unconscious, or not. For discussion of the distinction between implicit representation and contents in the rules or laws governing the operations in a psychological system, see Burge, *Origins of Objectivity*, pp. 403–407, 488–490.

all Newton's uses of the derivative symbol that are on the track toward Weierstrass consisted in unconscious uses of the complex *definiens* concept that Weierstrass assembled. Newton's reasoning can be psychologically accounted for by lower-level rules of thumb, such as apply the same algebraic rules in differentiation as one would with the rational numbers, the derivative of x^2 is $2x$, and the derivative of x^2 is the limit as $i \to 0$ of $((x+i)^2 - x^2) \div i$. I know of no evidence that Newton unconsciously applied Weierstrass's complex concept with stacked quantifiers wherever he used his concept of the derivative, or even that his uses were psychological shorthand for that complex concept. The contrast with the 'Cain hit Abel' case is palpable. The concept that Weierstrass discovered as a mathematical unification and explanation of previous applications of D_N was very probably never even an *unconscious* unit in Newton's psychology.

Mathematical justification and unification are not to be confused with reflective psychological discovery. Apriori reflection is not, in most theoretically interesting cases, assimilable to psychological discoveries of structures already assembled in the unconscious. Mathematical explicative definitions tend to be synthetic apriori, not pure identity thoughts revealed through bringing to consciousness antecedent structures within the explicated concepts.

Is *Weierstrass's* derivative concept D_W identical with D_{W-d}? I will assume, for the sake of keeping exposition simple, that at least before his discovery, Weierstrass's derivative concept was Newton's. Let us call his derivative concept after the discovery 'D_W'. I will discuss the question that I posed in terms of hypothetical possibilities—mainly to bring out considerations that bear on answering it. After the discovery, Weierstrass may have acquired a new concept of the derivative, one that is on the same track as Newton's, but that is strictly speaking different. Let us assume that his explication D_{W-d} and his post-discovery derivative concept D_W are at least co-extensive. Weierstrass's acquisition of a concept D_W different from Newton's D_N might have derived from his utilizing D_{W-d} to settle special issues (differentiate functions) that Newton's concept might not have been fitted for. Given that D_W is coextensive with D_{W-d}, it could not then be coextensive with D_N. So D_N and D_W would be different concepts.

Hypothetical cases in which D_{W-d} settles issues that Newton's concept D_N might not have settled fall into at least two categories. A case (for example, whether a given function is differentiable) might be definitely beyond the domain of application of Newton's concept. (Certain functions might definitely not be in that concept's extension, whereas Weierstrass's definition gives them values.) Then I will say that Weierstrass's concept D_W is a *broadening* of Newton's concept D_N. It differentiates all functions that Newton's concept clearly applies to and adds further ones. Alternatively, a case settled by D_{W-d} might be indeterminate as to whether it is in the domain of application of Newton's concept. Then I will say that Weierstrass's concept D_W is a *sharpening* of Newton's D_N. Any such broadening or sharpening of D_N is the result of D_W's being coextensive with D_{W-d}. If Weierstrass's own post-discovery concept of the derivative D_W is co-extensive with D_{W-d} and Newton's concept D_N is not coextensive with

D_{W-d}, then D_W is not D_N. D_N may be broadened or sharpened (or both) as well as explicated and justified by D_{W-d}. D_W is explicated and justified by D_{W-d}, but not broadened or sharpened by it.

I take it to be an empirical, historical question whether D_W is identical with D_N. If D_W is a broadening or a sharpening of D_N, they are not identical. There may be other reasons why they are not identical, even assuming that D_{W-d} explicates both D_N and D_W. I am not committed on whether D_N and D_W are identical. I know of no decisive reasons on either side. I take it that showing them to be different requires showing a difference in the constitutive minimal competence displayed in applications or inferences in which they were used, other than those applications or inferences that use explication D_{W-d} itself.[70] I think it quite possible that D_N and D_W are identical. My main point is that they need not be identical, even though D_{W-d} explicates both. It explicates both because it provides clarifying understanding and a mathematically acceptable justification for their core uses. It could explicate a concept either while also providing a broadening or sharpening of it or while providing a mathematically equivalent *definiens*.

Let us suppose, for the sake of exposition, that D_N and D_W are not identical. That leaves it open whether D_W is identical with D_{W-d}. Of course, in his mathematical discovery, Weierstrass, unlike Newton, did put together his explicative concept D_{W-d}, and used it as apriori coextensive with his concept D_W. On this basis, one might think that D_W is identical with D_{W-d}.

Even after successful definitional (equivalent) explications, however, the form of the first-order explication is rarely that of a pure identity thought. I believe that D_W is not identical with D_{W-d}. There are two complimentary reasons.

One depends on the epistemic status of definitional explications. Such explications are first order, non-meta-linguistic beliefs. That is, they do not have a form like 'D_W' expresses the same concept as 'D_{W-d}'. They are also not epistemically trivial. Weierstrass's discovery was not primarily a discovery about language. And the explicational equivalence was in principle open to scrutiny and criticism, even after it was discovered, and even after it was taken to be definitional. Pure identity thoughts are not open to such scrutiny or criticism. Counterexamples to the explicational definition can take the form: function *f* is differentiable and is within the natural domain in which the *explicandum* concept is applicable, but is

[70] I say 'other than those applications or inferences that use explication D_{W-d} itself' because I think that explications are responsible for unifying and justifying *antecedent* usage, in applications and inferences (with allowances for broadening and sharpening). Of course, having another person's concept, thinking with it, does not require making all the other person's applications or inferences. There is some minimum usage, comprised of applications and inferences, that constitutes minimum competence with the concept. This minimum might be any of various subsets from a family of uses.

Once one achieves a correct definition, and once one has acknowledged an explication as a definition, as Weierstrass did, there is a commitment to make all usage accord with the definition, at least other things equal. Such a commitment is associated with uses of D_W, but is not associated with any of Newton's uses. I take it that, in formal work, such a commitment is regarded as a normative aspect of the use of the concept. But I doubt that such a commitment is psychologically constitutive to having the concept.

given no derivative by the *explicans* concept. Weierstrass's explication is a remarkable mathematical claim, not a piece of trivial logic or a piece of linguistic analysis. The explication is not a pure identity thought.

A second reason why D_W is not the same concept as D_{W-d} is that, psychologically, the two are differently grounded. Whether or not D_W is identical with D_N, D_W is associated with the tradition of rule-of-thumb problem solving that predated the explication. These uses realize a competence with D_W that is psychologically independent of the *definiens* concept.[71] (See note 69.) That independence grounds epistemic scrutinizing of the explication, just discussed.[72] The explication unifies and justifies an antecedent usage.

This picture is rendered more complicated by the issue of concept broadening and sharpening, noted some paragraphs back. In mathematics, there is a natural progression of finding different, often broader, domains of application for an ur-concept and its successors. Moreover, I think that many natural concepts, including many mathematical concepts, have fuzzy boundaries. That is, there are cases that the concept is neither true of nor false of. Or for concepts applying to functions, there are arguments for which the function provides no value. Concepts commonly apply within a given domain. When new hard cases arise, a concept can undergo elaboration that consists in development of a new concept that is either a broadening of a predecessor or is given a more precise extension (or both). In mathematics, such elaborations usually involve broadening domains of application—wider extensions.

Early concepts of number that applied only to the natural numbers were elaborated to include 0, negative numbers, reals, imaginary numbers, and so on. At each stage, there is a number concept, but number concepts are increasingly

[71] Although Frege's discussion of explicative definitions and stipulative definitions are mostly deep and insightful, his citing the definition of 'integral' as a case of a simple word's having a complex sense is, I think, mistaken. He seems to take the sense of 'integral' to be that of its complex definition. Cf. 'Logik in der Mathematik', in *Nachgelassene Schriften*, p. 226; *Posthumous Writings*, p. 209; *The Frege Reader*, p. 315. Except as a stipulative definition, the non-meta-linguistic form of the standard definition of 'integral' is not the form of a pure identity thought. (It must be remembered that if there is a genuinely stipulative or abbreviative definition in an individual's idiolect, the *definiens* must express a sense that type-individuates the *same* cognitively relevant psychological state that the sense of the *definiendum* does, since the senses are the same: they are the same thought contents. One can use the abbreviating expression as a kind of anaphoric shorthand in thought. But supposing a sense identity, the idea that the individual thinker never thought as a unit the complex sense that the abbreviating expression allegedly abbreviates would be incoherent.) As indicated in the text, I do not say that pure identity thoughts are never the product of informative definitions. Informativeness can involve revelation of unconscious structure in use of and competence with a word. Then the incomplete understanding is primarily linguistic. As a matter of fact, most informative and significant definitions are not of that sort.

[72] These points elaborate the important work by Quine on definition in general and by Putnam on scientific definitions. See W.V. Quine, 'Two Dogmas of Empiricism', in *From a Logical Point of View* (Cambridge, Mass.: Harvard University Press, 1953), section 3; Putnam, 'The Analytic and the Synthetic' (1962), 'Is Semantics Possible?' (1970), and 'Explanation and Reference' (1973), all reprinted in *Philosophical Papers, Volume 2*. Putnam's work reinforces Quine's point that there are many sorts of definitions and "analyses", and that theoretically successful ones in the sciences (including mathematics) are subject to revision—in ways that pure identity thoughts are not.

generalized, and perhaps also made more precise. All the elaboration concepts are on a track with an ur number concept. Each elaboration concept is developed to satisfy a need to apply a number concept in a domain that is suggested by previous numerical operations, but to which earlier number concepts were not clearly applicable.

The development of differentiation concepts manifests this phenomenon. In fact, ur-concepts for derivative and limit underwent several tracks of elaboration, of which those that initially led to Weierstrass and Robinson are just the most famous. Weierstrass's explication fails to differentiate functions that subsequent, broader explications of derivative concepts, in the Weierstrassian tradition, do differentiate.[73] The subsequent *definiens* concepts are not coextensive with D_{W-d}. They provide derivatives for functions that Weierstrass's definition does not. The derivative concepts that they define and are coextensive with are broadenings or sharpenings of D_W. Since the broader or sharper *definiens* concepts are not co-extensive with D_W, they are not identical with it. D_W has Weierstrass's explication D_{W-d}. The more general explications explicate D_W in a different way. They justify it, provide a broadened elaboration of it, unify cases it applies to with further cases, provide insight into its limits, and show how to transcend those limits. The more general explications yield an understanding of D_W that is complimentary to, but fuller than, the understanding provided by D_{W-d}.[74] Unifying, broadening, sharpening, and generalizing are standard functions of scientific explications, even those that provide definitions. A broader or sharper concept $D_{W'}$ is explicated, though not broadened or sharpened, through *its* coextensive *definiens* concept.

Although Weierstrass's definition is not a fully general explication of the approach to differentiation first conceived in D_N, it is a sound definition because it illuminatingly applies to a natural mathematical domain. There are wider (or different) domains that other *definiens* concepts, and other (perhaps broadened or sharpened) derivative concepts apply to.

Of course, in a formal mathematical system a *term* for a derivative can be *stipulated* to abbreviate the *definiens* expression. Then the object-level analog of the definition is a pure identity thought. Such stipulations are usually not explications of antecedent derivative concepts.

So Weierstrass's *definiens* concept D_{W-d} has several uses. It can explicate Newton's concept D_N. Such explication may or may not be a broadening or a sharpening. If it is, the *definiens* is not coextensive with the *explicandum*; and then the *definiens* defines and also explicates a sharpening derivative concept D_W that is (necessarily) coextensive with D_{W-d}, but not with D_N. If it is not, then D_N

[73] I have benefited especially here, but also elsewhere, from reading Sheldon Smith's 'The Derivative and Incomplete Understanding' in draft, and from discussing these matters with him. His marvelous knowledge of post-Weierstrassian developments and of alternatives to Robinson's explication of Leibniz's infinitesimal concept was illuminating for me.

[74] All these points leave open whether $D_W = D_N$, or whether D_W is instead a sharpening of D_N.

and D_W may be identical; and the *definiens* is coextensive with both and provides an explication of both, where the senses of the *definiens* expression and the *definiendum* expression differ. In either case, one can also use the *definiens* D_{W-d} in a stipulative abbreviative definition of 'derivative'. Then the definition correlates with a pure identity thought and does not explicate any concept, though it simplifies a formal mathematical system. In mathematical thinking, such abbreviative definitions are contextual, passing affairs—rarely, if ever, long-term revelations of the natures of explicated concepts.[75]

The explicated concept(s) are *loci* for different cognitive associations. The explicated concept is expressed by a common noun backed by various conceptions. The *explicans* or *definiens* conception functions to unify and justify the other conceptions, and the applications of the *definiendum* concept. Whether it does so successfully is a substantive mathematical question. Theoretical concepts expressed by common nouns like 'derivative' are like concepts expressed by proper names in that, commonly, no descriptive contents (like D_{W-d}) are sense-identical with them. In accord with Kripke's and Putnam's work, the psychology and cognitive function of common-noun-type concepts are almost never those of concepts that explicate them, define them, or give constitutive accounts of their applications. The common-noun-type concepts are open to broadening and sharpening.[76] Some broadenings or sharpenings yield fuller understanding of unbroadened, unsharpened concepts. A track-like continuity often connects elaborations with antecedent concepts. Such continuity and the fact that all the concepts cover clear, basic cases of differentiation justify counting certain non-coextensive concepts all concepts of the derivative.

Incomplete understanding takes many forms. It can be poor recognition of structure, remediable through formalization. It can be failure to distinguish different types of meaning or content. It can be meta-linguistic or object-level. It can be lack of a correct unifying, justifying explication. It can reside in use of a concept associated with procedures that do not suffice to settle relevant cases, so that the concept needs explication that not only unifies and justifies existing usage but broadens or sharpens such usage, so as to cover unforeseen cases. These latter two types of incomplete understanding are not meta-linguistic. They are types that involve failure to hold explicating beliefs that illuminate constitutive aspects of a concept. Correct explication beliefs are neither meta-linguistic beliefs nor pure identity beliefs. As with all conceptual understanding, they involve understanding connections and joints in the world.

[75] Quine, 'Two Dogmas of Empiricism', in *From a Logical Point of View*, section 3; 'Carnap and Logical Truth', in *The Ways of Paradox* (New York: Random House, 1966), section v.

[76] Jim Pryor pointed out that some proper-name concepts may be open to broadening or sharpening. The issue arises, for example, in determining how to think about changing city limits that are not merely a matter of growth or contraction, say, for Berlin. The fundamental distinction between ordinary common nouns (like 'derivative' and 'sofa') and proper names is that the primary representational function of names is to single out a bearer, whereas the primary representational function of ordinary common nouns is to attribute attributes (including relations and kinds).

Newton and his community incompletely understood his ur-concept of the derivative and his approaching-the-limit concept of the derivative, D_N. Weierstrass's explication definition yielded fuller understanding. Newton was perhaps able to produce rules of thumb that fully account for what guided his conceptual and linguistic usage: they constitute the relevant linguistic rules in his idiolect and the conceptions in his psychology that guided his applications of his concept. They might codify his linguistic and conceptual competence in that respect. What was incomplete about his understanding was not (merely) a codification of the linguistic rules governing his usage, or a codification that captured the conceptions that guided his concepts. Incompleteness of understanding lay in his lack of a *correct* general explication that covered and justified the cases that the meaning and concept in fact applied to.

In both the derivative and tomato cases, neither the individual's nor the community's guiding rules of thumb—even the true ones—amount to full understanding of the meaning or concept. Here we see two aspects of conceptual and linguistic understanding. The first aspect concerns codifying rules that guide linguistic usage, and constitute the general conceptions that guide conceptual applications. In the tomato and derivative cases, there was no incompleteness in this aspect of understanding. In those cases, incompleteness lies in the second aspect—the one that concerns not merely codifying usage and the guiding conceptions, but providing a *correct* unifying, potentially guiding, justifying, general explication that *gets right* the nature of the instances to which the concept and the meaning apply.

In the derivative and tomato cases, the senses (the concepts) do not differ from the meanings. But what it is to fully understand the senses (or meanings) is different, depending on whether one focuses on (a) codifying usage or competence—providing explications that are in principle available to the users that best match the general, guiding, rule-like aspects of their competence, or (b) producing a correct, ideally general, potentially guiding, unifying, justifying explication of the senses (or meanings)—one that may invoke knowledge not available to the users, but that is potentially useful and that correctly captures the senses' extensions. Linguistics and modern theories of linguistic meaning are oriented to (a). Scientific explications—and, as the tomato case illustrates, even ordinary nonscientific explications that are influenced by new knowledge—that are concerned with the subject matter of the concepts (senses, meanings) are oriented to (b). Frege's concern with sense, as a component in knowledge and cognition in thought, encouraged (b). It provides a valuable conception of understanding that supplements the tendency in modern theories of meaning to concentrate on codifying and matching actual linguistic explicational or rule-based competence.

Frege's concept sense, as applied in the philosophies of language, logic, and psychology, continues to repay use and reflection, despite a century of censure and misunderstanding. These wages have earned *Sinning* a long and productive life.

APPENDIX

I have criticized Kripke's interpretation of passage (A). I want to make three further points. I begin by re-citing passage (B) together with material just preceding it:

(B') . . . the mere wording, which can be made permanent by writing or the gramophone, does not suffice for the expression of the thought. The *present tense* is used in two ways: first, in order to make a time-indication. . . . If a time indication should be made with the present tense, one must know when the sentence was uttered in order to grasp the thought correctly. Therefore the time of utterance is then part of the expression of the thought.[77]

First, Kripke holds that, in an utterance of 'Today is Thursday', the linguistic *utterance* consists of the just quoted words together with the time *t* when they are uttered: <*W*, *t*>. *W* is supposed to denote a function from times to truth values. Time *t* is supposed to denote itself.

Later in 'Thought', regarding present tense, Frege writes:

(D) To be sure the same verbal expression (*Wortlaut*) can take on another sense on account of the changeability of language with time; but the change then concerns (*betrifft*) [only] the linguistic (*das Sprachliche*).[78]

As quoted earlier, in the 1897 draft 'Logic', an ancestor to 'Thought', Frege writes,

(E) Words like 'here' and 'now' only acquire their full sense always only through the circumstances in which they are used.[79]

(D) can be interpreted in various ways. But it can be naturally read as indicating that linguistic elements change senses with time. (E) states very clearly that particular indexical words take on full senses only through contextual circumstances.

Kripke's proposal is not in accord with this latter point. On his proposal, no expressions, in the relevant cases, ever change senses, as a matter of their ordinary uses. Take the ordered pair <*W*, *t*>, where *W* comprises the 'mere wording' as it can be written down. Suppose that the time is an additional 'unrecognized linguistic expression'. If times are part of the linguistic expression, denoting themselves, they do not change senses. With a change of time, a new expression is used with a new sense. *W* also does not change sense on Kripke's proposal. It always expresses a sense that determines a function from times to truth values. And on that proposal, the component of *W* that consists in present tense, or some other indexical, always denotes the same function in the same

[77] (B') is directly followed by (A). Both passages come from Frege, 'Der Gedanke', in *Kleine Schriften*, pp. 348–349 [64]; 'Thought', in *The Frege Reader*, p. 332.

[78] Frege, 'Der Gedanke', in *Kleine Schriften*, p. 361 [76]; 'Thought', in *The Frege Reader*, p. 344.

[79] Frege, 'Logik' (1897), in *Nachgelassene Schriften*, p. 146; 'Logic' (1897), in *Posthumous Writings*, p. 135; 'Logic' (1897), in *The Frege Reader*, p. 235.

way—perhaps denoting a function from times to times. It does not change sense either.

Certainly in (E), and arguably in (D), Frege implies not only that the mere sentential wording *W* (exemplified by the *type* 'Today is Thursday') can be used to express different complete thoughts. He also implies that indexical words change their senses in different contexts. He thinks that *on their own apart from context*, indexicals never have a full sense.

Kripke's proposal is incompatible with (E) as well as (A).[80] For Frege, expressions regularly change *sense*. Indexicals do not change linguistic *meanings* with context. Indexicality is not ambiguity in linguistic meaning.

Second, in (B') Frege writes of present tense as making a time indication—presumably the time of utterance. We have already seen that present tense does not indicate the time of utterance by denoting a function that maps any given time onto the time of utterance. Such a view, combined with passage (A), would be incompatible with Frege's sense-composition doctrine. Nevertheless, present tense is used to indicate the time of utterance, seemingly in something like the way 'this time' or 'that time' would do so, where the difference between 'this' and 'that' (like that between 'today' and 'yesterday') helps the hearer get on to the relevant sense, but does contribute to the sense. Given that Frege takes present tense to be used as a demonstrative to designate a time, it would be odd for him to take the time as an expression denoting itself. There would be two designators (denoters) of the time: the occurrent use of present tense and the time itself. This point casts some doubt on Kripke's proposal that times denote themselves.

Third, in isolation, the last two sentences of (B')—that is, the sentences in (B)—are naturally read as supporting Kripke's view that times are literally parts of expressions of thoughts, and denote themselves. But much of the surrounding text suggests a different reading.

Frege writes, 'one must know when the sentence was uttered in order to grasp the thought correctly'. Then he states, apparently as the result of an inference, that the time of utterance is part of the thought expression. The exact relation between these remarks is not obvious. He could be thinking of an acquaintance relation to the time, where the time is a symbol denoting itself, as Kripke proposes. Then the first remark would be understood in terms of direct acquaintance with the time—'know the time'. But alternatively, 'the time of the utterance' in the second remark could be shorthand repetition of, anaphorically going back to, 'when the utterance occurred'. Such a reading might only require any circumstance that could be used to specify what time it is. Direct acquaintance with the time would not be required.

I think it plausible, as Kripke says for his proposal, that specification of the time would reveal what time it is. 'Knowing when the sentence was uttered' (Frege's phrase) seems to entail something stronger than having just any way of

[80] There is a similar text that involves 'I' that I will discuss below.

uniquely denoting the time. But on the present view, the requirement is weaker than on Kripke's: no direct acquaintance with the time is required. All that is required is some specification that reveals when the sentence was uttered, assuming that the thinker uses the sense of the specification in thinking the thought. The time of the utterance would be 'part of the thought expression' only in the loose sense that it is specified (hence denoted) by the thinker's use of contextual circumstances to express a mode of presentation that appropriately determines the time. The time could be known through any number of senses or thought components, depending on context.

I see this reading as more in the spirit of the paragraph in which (B') and (A) are contained. Frege seems interested not just in the speaker's grasping the thought, but third-person auditors' grasping it. (The word 'correct' in (A'), just below, suggests this point.) Frege seems more eclectic about what sort of expression can designate the time—or place, in the case of 'here' or 'there'—than invoking the time as a self-designator would suggest. Recall that in (A) after having discussed 'here' and 'there', he summarizes at the end of the paragraph:

(A') In all such cases, the mere wording, as it can be fixed in writing, is not the complete expression of the thought—but one further needs for its correct apprehension also the knowledge (*Kenntnis*) of certain circumstances accompanying the utterance, which are used as means of thought expression (*Gedankenausdrucks*). Fingerpointings, gestures, and glances can belong here too. The same wording (*Wortlaut*) containing the word 'I' will express different thoughts in the mouths of different people, of which some may be true and others false.[81]

In giving examples of thought expressions that supplement indexicals, Frege does not invoke the time or place itself as the means of thought expression. He seems to have in mind knowledge of various other circumstances that might be used to determine time or place. With 'in all such cases', Frege seems to allow any sort of circumstances (including the gestures) that can understood in determining (and specifying) time or place, as expressions of the thought.

As noted earlier, tokens of indexical words tend to be part of the circumstances that determine time or place. In 'On Sense and Denotation' Frege assumes that present tense can be used to designate the present time.[82] In the 1897 passage (D), he takes 'now' to be part of the linguistic expression that singularly denotes a time. I see no reason to doubt that 'today', 'yesterday', 'here', and 'now' in (A) are parts of the language used in specifying time or place. These points and

[81] Frege, 'Der Gedanke', in *Kleine Schriften*, p. 349 [64]; 'Thought', in *The Frege Reader*, p. 332.

[82] Frege, 'Über Sinn und Bedeutung', in *Kleine Schriften*, pp. 155–156 [42–43]; 'On *Sinn* and *Bedeutung*', in *The Frege Reader*, p. 165. Frege takes the present tense to be used as a singular term to designate a time. He does not say in this passage that (unused) present tense denotes a function, as one might expect him to if he believed that it does. He does not say enough to make it *clear* what his view of the semantics is. But what he writes is certainly consistent with what he writes in 1897 and 1918–1919.

examples suggest that Frege understood time determination not to be effected purely by direct acquaintance with a time, considered as a piece of language.[83]

Later in 'Thought' after (B)–(A), Frege writes again that the time of utterance 'belongs to' the thought expression.[84] He infers that the time determination (*Zeitbestimmung*) is thus given. In the same passage, he remarks about a sentence utterance containing 'I' that 'the identity of the speaker is essential to the sense'. I think that his writing that the time is part of the thought expression is to be taken analogously to his remark that the identity of the speaker is essential to the sense of a sentence utterance. The point is not that the speaker is literally part of the sense, or that the time is literally part of the expression of the thought. The point is that the sense must determine the speaker, and the thought expression must denote and specify the time.

Similarly, in the 1897 'Logic', in passage (C) where he comments that words like 'here' and 'now' only acquire their full sense through the circumstances in which they are used, he writes that the time and place 'must be supplied'. The time seems to be 'part of the thought expression' only in the sense that it is supplied, by being specified by the sense of some thought expression. Frege indicates in several places that this "supplying" can take various forms—through present tense, 'now', gestures, indications of events that are associated with the time.[85]

Although (B)–(A), the later passage in 'Thought' (see note 84), and the passage in 'Logic' 1897 (note 85) are compatible with Kripke's reading, I think that Frege probably means that the time is 'part of the thought expression' only in the loose sense that the expression expresses a sense that specifies the time. In these passages, and in the one that discusses present tense in 'On Sense and Denotation' (note 82), Frege writes that time determination can be effected in various ways (for example, by specifying events that occur at the time—'the time of Schleswig Holstein's separation'). So I doubt that Frege thought of times as parts of language. I doubt that he invoked direct acquaintance with times as symbols for themselves. But this issue is not central to the basic point. The basic point is that Frege's notion of sense is not that of linguistic meaning.

[83] See 'Über Sinn und Bedeutung', in *Kleine Schriften*, note 10, [43].

[84] Frege, 'Der Gedanke', in *Kleine Schriften*, [76]; 'Thought', in *The Frege Reader*, p. 344.

[85] Frege, 'Logik' (1897), p. 146; *Posthumous Writings*, p. 135; 'Logic', *The Frege Reader*, p. 235. The remark in 'Der Gedanke', [76] that the identity of the speaker is essential to the sense also occurs in this 1897 passage.

Bibliography

Ackerman, Felicia, 'Analysis, Language, and Concepts: The Second Paradox of Analysis', *Philosophical Perspectives* 4 (1990), 535–543.

Adams, Robert Merrihew, 'Involuntary Sins', *The Philosophical Review* 94:1 (1985), 3–31.

Adelman, Leonard M., 'Molecular Computation of Solutions to Combinatorial Problems', *Science* 266 (1994), 1021–1024.

Appel, K., and Haken, W., 'Every Four Color Map is Colorable, Part I: Discharging', *Illinois Journal of Mathematics* 21 (1977), 429–490.

——'Every Four Color Map is Colorable, Part II: Reducibility', *Illinois Journal of Mathematics* 21 (1977), 491–567.

Armstrong, D. M., *A Materialist Theory of the Mind* (London: Routledge & Kegan Paul, 1968).

Audi, Robert, *The Architecture of Reason: The Structure and Substance of Rationality* (New York and Oxford: Oxford University Press, 2001).

——'Testimony, Credulity, and Veracity', in Jennifer Lackey and Ernest Sosa (eds.), *The Epistemology of Testimony* (Oxford: Oxford University Press, 2006), 25–49.

Baddeley, Alan D., Conway, Martin A., and Aggleton, John P. (eds.), *Episodic Memory: New Directions in Research* (New York and Oxford: Oxford University Press, 2002).

Bailey, David H., and Borwein, Jonathan M., 'Future Prospects for Computer-Assisted Mathematics', *Notes of the Canadian Mathematical Society* 37 (2005), 2–6.

Bermúdez, José L., Marcel, Anthony, and Eilan, Naomi (eds.), *The Body and the Self* (Cambridge, Mass.: MIT Press, 1995).

Bernecker, Sven, 'Externalism and the Attitudinal Component of Self-Knowledge', *Noûs* 30:2 (1996), 262–275.

Bilgrami, Akeel, *Self-Knowledge and Resentment* (Cambridge, Mass.: Harvard University Press, 2006).

Biro, John, 'Testimony and Apriori Knowledge', *Philosophical Issues* 6 (1995), 301–310.

Block, Ned, 'On a Confusion about a Function of Consciousness', *Behavioral and Brain Sciences* 18:2 (1995), 227–247.

Blum, Manuel; De Santis, Alfredo; Micali, Silvio; and Persiano, Giuseppe, 'Non-Interactive Zero-Knowledge', *SIAM Journal of Computing* 20 (1991), 1084–1118.

Boghossian, Paul, 'Content and Self-Knowledge', *Philosophical Topics* 17 (1989), 5–26.

——'Externalism and Inference', *Philosophical Issues* 2 (1992), 11–28.

——'Reply to Schiffer', *Philosophical Issues* 2 (1992), 39–42.
Borwein, J. M., and Bailey, D. H., *Mathematics by Experiment: Plausible Reasoning in the 21st Century* (Wellesley, Mass.: A. K. Peters Ltd, 2004; 2nd edn., 2008).
Brueckner, Anthony L., 'Brains in a Vat', *The Journal of Philosophy* 83:3 (1986), 148–167.
——'Content Externalism, Entitlement, and Reasons', in Sanford C. Goldberg (ed.), *Internalism and Externalism in Semantics and Epistemology* (New York and Oxford: Oxford University Press, 2007), 160–176.
Buraka, Yoram; Roknia, Uri; Meistera, Markus; and Sompolinskya, Haim, 'Bayesian Model of Dynamic Image Stabilization in the Visual System', *Proceedings of the National Academy of Sciences of the United States of America* 107 (2010), 19525–19530.
Burge, J., Fowlkes, C. C., and Banks, M. S., 'Natural Scene Statistics Predict How the Figure-ground Cue of Convexity Affects Human Depth Perception', *Journal of Neuroscience* 30 (2010), 7269–7280.
Burge, Tyler, 'Reference and Proper Names', *The Journal of Philosophy* 70:14 (1973), 425–439.
——'Belief *De Re*', *The Journal of Philosophy* 74:6 (1977), 338–362.
——'Individualism and the Mental', *Midwest Studies in Philosophy* 4 (1979), 73–121.
——'Sinning Against Frege', *The Philosophical Review* 88:3 (1979), 398–432.
——'Other Bodies', in Andrew Woodfield (ed.), *Thought and Object: Essays on Intentionality* (Oxford: Clarendon Press, 1982), 97–120.
——'Russell's Problem and Intentional Identity', in James B. Tomberlin (ed.), *Agent, Language, and the Structure of the World* (Indianapolis: Hackett, 1983), 79–110.
——'Frege on Extensions of Concepts, from 1884 to 1903', *The Philosophical Review* 93:1 (1984), 3–34.
——'Cartesian Error and the Objectivity of Perception', in P. Pettit and J. McDowell (eds.), *Subject, Thought, and Context* (New York and Oxford: Oxford University Press, 1986), 117–136.
——'Individualism and Psychology', *The Philosophical Review* 95:1 (1986), 3–45.
——'Intellectual Norms and Foundations of Mind', *The Journal of Philosophy* 83:12 (1986), 697–720.
——'Individualism and Self-Knowledge', *The Journal of Philosophy* 85 (1988), 649–663.
——'Perceptual Individualism and Authoritative Self-Knowledge', in R. Grimm and D. Merrill (eds.), *Contents of Thought* (Tucson: Arizona University Press, 1988), 86–98.
——'Wherein is Language Social?' in Alexander George (ed.), *Reflections on Chomsky* (Oxford: Blackwell, 1989), 175–192.
——'Frege on Sense and Linguistic Meaning', in David Bell and Neil Cooper (eds.), *The Analytic Tradition: Meaning, Thought, and Knowledge* (Oxford: Blackwell, 1991), 30–60.
——'Frege on Knowing the Third Realm', *Mind* NS, 101:104 (1992), 633–649.
——'Philosophy of Language and Mind: 1950–1990', *The Philosophical Review* 101 (1992), 3–51.
——'Concepts, Definitions, and Meaning', *Metaphilosophy* 24:4 (1993), 309–325.

Burge, Tyler, 'Content Preservation', *The Philosophical Review* 102 (1993), 457–488.
——'Our Entitlement to Self-Knowledge', *Proceedings of the Aristotelian Society* 96 (1996), 91–116.
——'Interlocution, Perception, and Memory', *Philosophical Studies* 86 (1997), 21–47.
——'Two Kinds of Consciousness', in Ned Block, Owen J. Flanagan, and Güven Güzeldere (eds.), *The Nature of Consciousness: Philosophical Debates* (Cambridge, Mass.: MIT Press, 1997), 427–434.
——'Computer Proof, Apriori Knowledge, and Other Minds', *Philosophical Perspectives* 12 (1998), 1–37.
——'Frege on Knowing the Foundation', *Mind* 107 (1998), 305–347.
——'Memory and Self-Knowledge', in Peter Ludlow and Norah Martin (eds.), *Externalism and Self-Knowledge* (Stanford: CSLI Publications, 1998), 351–370.
——'Reason and the First Person', in Crispin Wright, Barry C. Smith, and Cynthia Macdonald (eds.), *Knowing Our Own Minds: Essays on Self-Knowledge* (New York and Oxford: Oxford University Press, 1998), 243–270.
——'A Century of Deflation and a Moment about Self-Knowledge', Presidential Address at Pacific APA, April, *Proceedings of the APA* 73 (1999), 25–46.
——'Comprehension and Interpretation', in L. Hahn (ed.), *The Philosophy of Donald Davidson* (Chicago: Open Court Publishers, 1999), 229–250.
——'Frege on Apriority', in C. Peacocke and P. Boghossian (eds.), *New Essays on the A Priori* (Oxford: Oxford University Press, 2000), 11–42.
——'Concepts, Conceptions, Reflective Understanding: Reply to Peacocke', in Martin Hahn and Bjørn Ramberg (eds.), *Reflections and Replies: Essays on the Philosophy of Tyler Burge* (Cambridge, Mass.: MIT Press, 2003), 383–396.
——'Logic and Analyticity', *Grazer Philosophische Studien* 66 (2003), 199–249.
——'Memory and Persons', *The Philosophical Review* 112:3 (2003), 289–337.
——'Mental Agency and Authoritative Self-Knowledge: Reply to Kobes', in Martin Hahn and Bjørn Ramberg (eds.), *Reflections and Replies: Essays on the Philosophy of Tyler Burge* (Cambridge, Mass.: MIT Press, 2003), 417–434.
——'Perception', *International Journal of Psychoanalysis* 84 (2003), 157–167.
——'Perceptual Entitlement', *Philosophy and Phenomenological Research* 67 (2003), 503–548.
——'Reflections on Scepticism: Reply to Stroud', in Martin Hahn and Bjørn Ramberg (eds.), *Reflections and Replies: Essays on the Philosophy of Tyler Burge* (Cambridge, Mass.: MIT Press, 2003), 335–346.
——'Reply to Benejam: "Thought Experiments and Semantic Competence"', in Maria J. Frapolli and Esther Romero (eds.) *Meaning, Basic Self-Knowledge, and Mind: Essays on Tyler Burge* (Stanford, Ca.: CSLI Publications, 2003).
——'Disjunctivism and Perceptual Psychology', *Philosophical Topics* 33:1 (2005), 1–78.
——*Truth, Thought, Reason: Essays on Frege* (Oxford: Clarendon Press, 2005).
——'Postscript to "Belief *De Re*"' (2006), *Foundations of Mind.*
——'Reflections on Two Kinds of Consciousness' (2006), *Foundations of Mind.*
——*Foundations of Mind: Philosophical Essays, Volume II* (New York and Oxford: Oxford University Press, 2007).
——'Five Theses on *De Re* States and Attitudes', in Joseph Almog and Paolo Leonardi (eds.), *The Philosophy of David Kaplan* (New York and Oxford: Oxford University Press, 2009), 246–316.

——'Primitive Agency', *Philosophy and Phenomenological Research* 74:2 (2009), 251–278.

——'Modest Dualism', in Robert C. Koons and George Bealer (eds.), *The Waning of Materialism* (New York: Oxford University Press, 2010), 233–250.

——*Origins of Objectivity* (Oxford: Oxford University Press, 2010).

——'The Dewey Lectures 2007: Self and Self-Understanding', Lecture I: 'Some Origins of Self'; Lecture II: 'Self and Constitutive Norms'; Lecture III: 'Self-Understanding', *The Journal of Philosophy* 108 (2011), pp. 287–383.

——'Disjunctivism Again', *Philosophical Explorations* 14:1 (2011), 43–80.

——'Origins of Perception', *Disputatio* 4:29 (2011), 1–38.

——'Steps Toward Origins of Propositional Thought', *Disputatio* 4:29 (2011), 39–67.

——'Living Wages of Sinn', *The Journal of Philosophy* 109 (2012), 40–84.

——'Gottlob Frege: Some Forms of Influence', in Michael Beaney (ed.), *The Oxford Handbook of the History of Analytic Philosophy* (New York and Oxford: Oxford University Press, forthcoming).

Butler, J., 'Of Personal Identity'. First Dissertation to *The Analogy of Religion*.

Campbell, John, *Past, Space, and Self* (Cambridge, Mass.: MIT Press, 1995).

Carpenter, Melinda; Nagell, Katherine; and Tomasello, Michael, 'Social Cognition, Joint Attention, and Communicative Competence from 9 to 15 Months of Age', *Monographs of the Society for Research in Child Development*, serial no. 255, 63:4 (1998).

Cassam, Quassim (ed.), *Self-Knowledge* (Oxford: Oxford University Press, 1994).

Castaneda, H. N., 'Indicators and Quasi-Indicators', *American Philosophical Quarterly* 4 (1967), 85–100.

——'On the Logic of Attributions of Self-Knowledge to Others', *The Journal of Philosophy* 65 (1968), 439–456.

Casullo, A. (ed.), *Apriori Knowledge* (Aldershot: Ashgate Publishers, 1999).

Cherniak, Christopher, *Minimal Rationality* (Cambridge, Mass.: MIT Press, 1986).

Cheyney, D., and Seyfarth, R., *How Monkeys See the World* (Chicago: University of Chicago Press, 1990).

Chierchia, Gennaro, and McConnell-Ginet, Sally, *Meaning and Grammar: An Introduction to Semantics*, 2nd edn. (Cambridge, Mass.: MIT Press, 2000).

Chisholm, Roderick M., *Theory of Knowledge*, 2nd edn. (Englewood Cliffs, NJ: Prentice Hall, 1977).

Christensen, David, and Kornblith, Hilary, 'Testimony and the Limits of the *A Priori*', *Philosophical Studies* 86 (1997), 1–20.

Clayton, Nicola S., and Dickinson, Anthony, 'Episode-like Memory during Cache Recovery by Scrub Jays', *Nature* 395 (1998), 272–274.

Coady, A. J., 'Testimony and Observation', *American Philosophical Quarterly* 10 (1973), 149–155.

Conee, Earl, and Feldman, Richard, 'The Generality Problem for Reliabilism', *Philosophical Studies* 89 (1998), 1–29.

Crawford, A. D., Medendorp, W. P., and Marotta, J. J., 'Spatial Transformations for Eye–Hand Coordination', *Journal of Neurophysiology* 92 (2004), 10–19.

Csibra, Gergely, and Gergely, György, 'Social Learning and Social Cognition: The Case for Pedagogy', in Yuko Munakata and Mark H. Johnson (eds.), *Processes of Change in Brain and Cognitive Development: Attention and Performance XXI* (New York and Oxford: Oxford University Press, 2006), 249–274.

Dalla, Sasha R. X.; Giraldeaub, Luc-Alain; Olssonc, Ola; McNamarad, John M.; and Stephense, David W., 'Information and its Use by Animals in Evolutionary Ecology', *Trends in Ecology and Evolution* 20 (2005), 187–193.

Darwin, Charles, *Descent of Man* (1871; Amherst, NY: Prometheus, 1998).

Davidson, Donald, *Essays on Actions and Events* (Oxford: Oxford University Press, 1980; 2nd edn., 2001).

——'Rational Animals', *Dialectica* 36 (1982), 317–327.

——'First Person Authority', *Dialectica* 38 (1984), 101–112.

——*Inquiries into Truth and Interpretation* (Oxford: Oxford University Press, 1984; 2nd edn., 2001).

——'The Social Aspect of Language', in B. F. McGuinness and G. Oliveri (eds.), *The Philosophy of Michael Dummett* (Dordrecht: Kluwer Academic Publishers, 1984), 1–16.

——'A Coherence Theory of Truth and Knowledge', in Ernest LePore (ed.), *Truth and Interpretation: Perspectives on the Philosophy of Donald Davidson* (Oxford: Basil Blackwell, 1986), 307–319.

——'A Nice Derangement of Epitaphs', in Ernest LePore (ed.), *Truth and Interpretation: Perspectives on the Philosophy of Donald Davidson* (Oxford: Basil Blackwell, 1986), 433–446.

——'Knowing One's Own Mind', *Proceedings and Addresses of the American Philosophical Association* 60:3 (1987), 441–458.

——'The Structure and Content of Truth', *The Journal of Philosophy* 87 (1990), 279–328.

——'Epistemology Externalized', *Dialectica* 45 (1991), 191–202.

——'Three Varieties of Knowledge', in A. Phillips Griffiths (ed.), *A. J. Ayer Memorial Essays* (Cambridge: Cambridge University Press, 1991), 153–166.

——*Subjective, Intersubjective, Objective* (New York and Oxford: Oxford University Press, 2001).

Davies, Brian, 'Whither Mathematics?' *Notices of the American Mathematical Society* 52 (2005), 1350–1356.

Dawson, John W., 'The Reception of Gödel's Incompleteness Theorem', *Philosophy of Science Association 1984* 2 (1985), 253–271.

Descartes, René, *The Philosophical Works of Descartes*, trans. E. S. Haldane and G. R. T. Ross, 2 vols. (New York: Dover, 1955).

——*The Philosophical Writings of Descartes*, trans. John Cottingham, Robert Stoothoff, and Dugald Murdoch, 3 vols. (Cambridge: Cambridge University Press, 1985).

Detlefson, Michael, and Luker, Mark, 'The Four-Color Theorem and Mathematical Proof', *The Journal of Philosophy* 77 (1980), 803–820.

Devlin, Keith, 'Test Tube Computing with DNA', *Math Horizons* 2 (1995), 14–21.

Donnellan, Keith, 'Proper Names and Identifying Descriptions', *Synthese* 21:3/4 (1970), 335–358.

Dorigo, M., and Stützle, T., *Ant Colony Optimization* (Cambridge, Mass.: MIT Press, 2004).

Dretske, Fred, 'Epistemic Operators', *The Journal of Philosophy* 67 (1970), 1007–1023.

——*Knowledge and the Flow of Information* (Cambridge, Mass.: MIT Press, 1981).

——'The Pragmatic Dimension of Knowledge', *Philosophical Studies* 40 (1981), 363–378.

Dummett, Michael, *The Interpretation of Frege's Philosophy* (Cambridge, Mass.: Harvard University Press, 1981).

Dummett, Michael, 'Comments on Davidson and Hacking', in Ernest LePore (ed.), *Truth and Interpretation: Perspectives on the Philosophy of Donald Davidson* (Oxford: Basil Blackwell, 1986), 459–476.

Eisenberg, Nancy; Fabes, Richard A.; and Spinrad, Tracy L., 'Prosocial Development', in Nancy Eisenberg, William Damon, and Richard M. Lerner (eds.), *Handbook of Child Psychology, Volume 3: Social, Emotional, and Personality Development* (New York: Wiley, 1998), 646–718.

Evans, Gareth, *The Varieties of Reference*, ed. J. McDowell (Oxford: Oxford University Press, 1982).

Evnine, Simon, 'Believing Conjunctions', *Synthese* 118, 201–227.

Fallis, Don, 'The Epistemic Status of Probabilistic Proof', *The Journal of Philosophy* 94 (1997), 165–186.

Fodor, J. A., *The Language of Thought* (Cambridge, Mass.; Harvard University Press, 1979).

——Bever, T. G., and Garrett, M. F., *The Psychology of Language* (New York: McGraw-Hill, 1974).

Forster, Kenneth I., 'Lexical Processing', in D. Osherson and H. Lasnik (eds.), *Language: An Invitation to Cognitive Science*, vol. 1 (Cambridge, Mass.: MIT Press, 1990), 95–131.

Frapoli, M. J., and Romero, E. (eds.), *Meaning, Basic Self-Knowledge, and Mind: Essays on Tyler Burge* (Stanford: CSLI Publications, 2003).

Frege, Gottlob, *Begriffsschrift, einer der arithmetischen nachgebildete Formelsprache des reinen Denkens* (Halle, Germany: L. Nebert, 1879). Translated as *Begriffsschrift, A Formula Language, Modeled upon that of Arithmetic, for Pure Thought*, in Jan van Hejenoort (ed.), *From Frege to Gödel: A Source Book in Mathematical Logic, 1879–1931* (Cambridge, Mass.: Harvard University Press, 1981).

——*Grundlagen der Arithmetik* (1884), translated as *Foundations of Arithmetic*, trans. J. L. Austin (Evanston, IL: Northwestern University Press, 1968).

——'On Sense and Reference', in *Translations from the Philosophical Writings of Gottlob Frege*, ed. P. Geach and M. Black (Oxford: Blackwell, 1966), 56–79.

——*Die Grundgesetze der Arithmetik* (1893) (Hildesheim, Germany: George Olms, 1962).

——*The Basic Laws of Arithmetic: Exposition of the System*, ed. and trans. Montgomery Furth (Berkeley: University of California Press, 1964).

——*Kleine Schriften*, ed. Ignacio Angelelli (Hildesheim, Germany: Georg Olms, 1967).

——*Wissenschaftlicher Briefwechsel*, ed. Gottfried Gabriel et al. (Hamburg: Felix Meiner Verlag, 1976).

——*Posthumous Writings*, ed. ed. Hans Hermes, Friedrich Kambartel, and Friedrich Kaulbach (Chicago: University of Chicago Press, 1979).

——*Philosophical and Mathematical Correspondence* (Oxford: Blackwell, 1980).

——*Nachgelassene Schriften*, ed. Hans Hermes, Friedrich Kambartel, and Friedrich Kaulbach, 2nd edn. (Hamburg: Felix Meiner Verlag, 1983).

——*Collected Papers on Mathematics, Logic, and Philosophy*, ed. Brian McGuinness (New York: Wiley-Blackwell, 1984).

——*The Frege Reader*, ed. Michael Beaney (Cambridge, Mass.: Blackwell, 2009).

Freud, Sigmund, 'Screen Memories', in *The Standard Edition of the Complete Psychological Works of Sigmund Freud*, vol. 3, ed. James Strachey (London: Hogarth, 1953–1974).

Fricker, Elizabeth, 'Trusting Others in the Sciences: *A Priori* or Empirical Warrant?' *Studies in the History and Philosophy of Science* 33 (2002), 373–383.

——'Martians and Meetings: Against Burge's Neo-Kantian Apriorism about Testimony', *Philosophica* 78 (2006), 69–84.

Frith, Christopher D., *The Cognitive Neuropsychology of Schizophrenia* (Hove, UK: Lawrence Erlbaum, 1992).

Gallese, Vittorio, et al., 'Action Recognition in the Premotor Cortex', *Brain* 119:2 (1996), 593–609.

Gallup, Jr., Gordon G., 'Chimpanzees: Self-Recognition', *Science* 167:3914 (1970), 341–343.

——'Self-Recognition in Primates: A Comparative Approach to the Bidirectional Properties of Consciousness', *American Psychologist* 32:5 (1977), 329–338.

——'Self-Recognition: Research Strategies and Experimental Design', in T. Parker, Robert R. Mitchell, and Maria L. Boccia (eds.), *Self-Awareness in Animals and Humans: Developmental Perspectives* (New York: Cambridge University Press, 1994), 35–50.

Geisler, W. S., 'Visual Perception and the Statistical Properties of Natural Scenes', *Annual Review of Psychology* 59 (2008), 10.1–10.26.

——'Contributions of Ideal Observer Theory to Vision Research', *Vision Research* 51 (2011), 771–781.

Geisler, Wilson S.; Najemnik, Jiri; and Ing, Almon D., 'Optimal Stimulus Encoders for Natural Tasks', *Journal of Vision* 9 (2009), 1–16.

Gerken, Mikkel, 'Conceptual Equivocation and Warrant by Reasoning', *Australasian Journal of Philosophy* 89:3 (2011), 381–400.

Gilbert, Daniel T., 'How Mental Systems Believe', *American Psychologist* 46 (1991), 107–119.

——'The Assent of Man: Mental Representation and the Control of Belief', in D. M. Wegner and J. Pennebaker (eds.), *Handbook of Mental Control* (Englewood Cliffs, NJ: Prentice Hall, 1993), 57–87.

——Krull, Douglas S., and Malone, Patrick S., 'Unbelieving the Unbelievable: Some Problems in the Rejection of False Information', *Journal of Personality and Social Psychology* 59:4 (1990), 601–613.

Goldreich, Oded; Micali, Silvio; and Wigderson, Avi, 'Proofs that Yield Nothing But Their Validity or All Languages in NP Have Zero-Knowledge Proof Systems', *Journal of the Association for Computing Machinery* 38 (1991), 691–729.

Goldwasser, Shaffi; Micali, Silvio; and Rackoff, Charles, 'The Knowledge Complexity of Interactive Proof Systems', *SIAM Journal of Computing* 18 (1989), 186–208.

Gopnik, Alison, and Meltzoff, Andrew N., 'Minds, Bodies, and Persons: Young Children's Understanding of the Self and Others as Reflected in Imitation and Theory of Mind Research', in T. Parker, Robert R. Mitchell, and Maria L. Boccia (eds.), *Self-Awareness in Animals and Humans: Developmental Perspectives* (New York: Cambridge University Press, 1994), 166–186.

Graham, Peter, 'Can Testimony Generate Knowledge?' *Philosophica* 78 (2006), 105–127.

——'Testimonial Justification: Inferential or Non-Inferential?' *The Philosophical Quarterly* 56 (2006), 84–95.

Grice, H. P., 'Personal Identity', *Mind* 50 (1941), 330–351.

Grice, Paul, *Studies in the Way of Words* (Cambridge, Mass.: Harvard University Press, 1989).

Hahn, Martin, and Ramberg, Bjørn (eds.), *Reflections and Replies: Essays on the Philosophy of Tyler Burge* (Cambridge, Mass.: MIT Press, 2003).

Hales, Thomas C., 'A Proof of the Kepler Conjecture', *Discrete & Computational Geometry: An International Journal of Mathematics and Computer Science* 36 (2006), 5–20.

——, and Ferguson, Samuel P., 'A Formulation of the Kepler Conjecture', *Discrete & Computational Geometry: An International Journal of Mathematics and Computer Science* 36 (2006), 21–69.

Hampshire, Stuart, 'Feeling and Expression', in *Freedom of Mind and Other Essays* (Princeton: Princeton University Press, 1971), 143–159.

Hardwig, John, 'Epistemic Dependence', *The Journal of Philosophy* 82 (1985), 335–349.

Hare, B., et al., 'Chimpanzees Know What Conspecifics Do and Do Not See', *Animal Behaviour* 59:4 (2000), 771–786.

Harman, Gilbert, *Thought* (Princeton: Princeton University Press, 1973).

Heijenoort, J. van, 'Frege on Sense Identity', *Journal of Philosophical Logic* 6:1 (1977), 103–108.

Herman, Barbara, 'A Mismatch of Methods', in Derek Parfit, *On What Matters*, vol. 2, ed. Samuel Scheffler (New York and Oxford: Oxford University Press, 2011), 83–115.

Higginbotham, James, 'On Events in Linguistic Semantics', in James Higginbotham, Fabio Pianesis, and Achille C. Varzi (eds.), *Speaking of Events* (New York and Oxford: Oxford University Press, 2000), 49–79.

Hill, Christopher S., 'On Getting to Know Others', *Philosophical Topics* 13 (1985), 257–266.

Hume, David, *A Treatise of Human Nature*.

——*Dialogues Concerning Natural Religion*.

James, William, *The Principles of Psychology*, 2 vols. (Cambridge, Mass.: Harvard University Press, 1981).

Jeannerod, Marc, *The Cognitive Neuroscience of Action* (Cambridge, Mass.: Blackwell, 1997).

——'To Act or Not to Act: Perspectives on the Representation of Actions', *Quarterly Journal of Experimental Psychology* 52A:1 (1999), 1–29.

Kalish, Charles W., and Sheverick, Sean M., 'Children's Reasoning about Norms and Traits as Motives for Behavior', *Cognitive Development* 19:3 (2004), 401–416.

Kant, Immanuel, 'Attempt to Introduce the Concept of Negative Magnitudes into Philosophy (1763), in *Kants Werke*, A.K. II (Berlin: Walter de Gruyter, 1968), 167–204.

——*Practical Philosophy*, ed. Mary J. Gregor (New York: Cambridge University Press, 1996).

——*Lectures on Metaphysics*, ed. Karl Ameriks and Steve Naragon (New York: Cambridge University Press, 1997).

——*Critique of Pure Reason*, ed. and trans. Paul Guyer and Allen W. Wood (Cambridge: Cambridge University Press, 1998).

Kaplan, David, 'Demonstratives', in Joseph Almog, John Perry, and Howard Wettstein (eds.), *Themes from Kaplan* (New York and Oxford: Oxford University Press, 1989), 481–563.

Kemmerling, Andreas, 'Eine reflexive Deutung des Cogito', in Konrad Kramer et al. (eds.), *Theorie der Subjektivität* (Frankfurt: Suhrkamp, 1987), 141–166.

——*Ideen des Ichs: Studien zu Descartes' Philosophie* (Frankfurt: Suhrkamp, 1996).

Kitcher, Patricia, 'Two Normative Rules for Self-Consciousness', in Herbert S. Terrace and Janet Metcalfe (eds.), *The Missing Link in Cognition: Origins of Self-Reflective Consciousness* (New York and Oxford: Oxford University Press, 2005), 174–187.

Kobes, Bernard, 'Mental Content and Hot Self-Knowledge', in M. Hahn and B. Ramberg (eds.), *Reflections and Replies: Essays on the Philosophy of Tyler Burge* (Cambridge, Mass.: MIT Press, 2003), 201–228.

——'Burge's Dualism', *The Waning of Materialism*, R. C. Koons and G. Bealer (eds.), (Oxford: Oxford University Press, 2010).

Korsgaard, Christine, 'Personal Identity and the Unity of Agency: A Kantian Response to Parfit', *Philosophy and Public Affairs* 18 (1989), 101–132.

Koza, J. R., *Genetic Programming: On the Programming of Computers by Means of Natural Selection* (Cambridge, Mass.: MIT Press, 1992).

Kripke, Saul, 'Naming and Necessity', in Donald Davidson and Gilbert Harman (eds.), *Semantics of Natural Language* (Dordrecht: Reidel, 1972), 253–355.

——*Naming and Necessity* (Cambridge, Mass.: Harvard University Press, 1980).

——'Frege's Theory of Sense and Reference: Some Exegetical Notes', *Theoria* 74:3 (2008), 181–218.

Lackey, Jennifer, 'Testimonial Knowledge and Transmission', *The Philosophical Quarterly* 49 (1999), 471–490.

Lakatos, Imre, *Mathematics, Science and Epistemology: Philosophical Papers, Volume 2*, ed. John Worrall and Gregory Currie (Cambridge: Cambridge University Press, 1978).

Lam, C. W. H., 'The Search for a Finite Projective Plane of Order 10', *American Mathematical Monthly* 98 (1991), 305–318.

Leekam, Sue, 'Why Do Children with Autism Have a Joint Attention Impairment?' in N. Eilan et al. (eds.), *Joint Attention: Communication and Other Minds* (New York and Oxford: Oxford University Press, 2005), 205–259.

Leibniz, G. W., *New Essays on Human Understanding*, ed. and trans. P. Remnant and J. Bennett (New York: Cambridge University Press, 1981).

Lewis, David, *Convention* (Cambridge, Mass.: Harvard University Press, 1969).

——'Survival and Identity', in Amélie Rorty (ed.), *The Identities of Persons* (Berkeley: University of California Press, 1976), 17–40.

Liberman, Alvin M., *Speech: A Special Code* (Cambridge, Mass.: MIT Press, 1996).

Lichtenberg, G. C., *Schriften und Briefe* (Munich: Carl Hanser Verlag, 1971).

Locke, John, *An Essay Concerning Human Understanding*, ed. Peter H. Nidditch (Oxford: Clarendon Press, 1975).

McCormack, Teresa, and Hoerl, Christoph, 'Memory and Temporal Perspective: The Role of Temporal Frameworks in Memory Development', *Developmental Review* 19:1 (1999), 154–182.

McDowell, John, *Mind and World* (Cambridge, Mass.: Harvard University Press, 1994).

——'Reductionism and the First-Person', in Jonathan Dancy (ed.), *Reading Parfit* (Oxford Blackwell, 1997), 230–250.

Malcolm, Norman, 'Knowledge of Other Minds', *The Journal of Philosophy* 55 (1958), 969–978.

Martin, Donald M., 'Mathematical Evidence', in H. G. Dales and G. Oliveri (eds.), *Truth in Mathematics* (Oxford: Clarendon Press, 1998), 215–232.

Matsuhashi, Masao, and Hallett, Mark, 'The Timing of the Conscious Intention to Move', *European Journal of Neuroscience* 28:11 (2008), 2344–2351.

Meltzoff, Andrew N., and Moore, M. Keith, 'Imitation of Facial and Manual Gestures by Human Neonates', *Science* 198:4312 (1977), 75–78.

Menzel, Charles, 'Progress in the Study of Chimpanzee Recall and Episodic Memory', in Herbert S. Terrace and Janet Metcalfe (eds.), *The Missing Link in Cognition: Origins of Self-Reflective Consciousness* (New York and Oxford: Oxford University Press, 2005), 188–224.

Mill, J. S., *An Examination of Sir William Hamilton's Philosophy and of the Principal Questions Discussed in His Writings*, 6th edn. (London: Longman, Green, Longman, Roberts, and Green, 1889).

Mitchell, Melanie, An Introduction to Genetic Algorithms (Cambridge, Mass.: MIT Press, 1996).

Moll, Henrike, and Tomasello, Michael, 'Level 1 Perspective-Taking at 24 Months of Age', *British Journal of Development Psychology* 24:3 (2006), 603–613.

Moore, Gregory H., *Zermelo's Axiom of Choice: Its Origins, Development, and Influence* (New York: Springer Verlag, 1982).

Moran, Richard, 'Making Up Your Mind: Self-Interpretation and Self-Constitution', *Ratio* 1:2 (1988), 135–151.

——*Authority and Estrangement: An Essay on Self-Knowledge* (Princeton: Princeton University Press, 2001).

Moser, Paul K. (ed.), *A Priori Knowledge* (Oxford: Oxford University Press, 1987).

Mullins, Simon, and Spence, Sean A., 'Re-examining Thought Insertion: Semi-Structured Literature Review and Conceptual Analysis', *The British Journal of Psychiatry* 182 (2003), 293–298.

Nagel, Thomas, 'What Is It Like to Be a Bat?' *The Philosophical Review* 83:4 (1974), 435–450.

——'Moral Luck' (1976), in *Mortal Questions* (New York: Cambridge University Press, 1979), 24–38.

Naor-Raz, G., Tarr, M. J., and Kersten, D., 'Is Color an Intrinsic Property of Object Representation?' Perception 32 (2003), 667–680.

Nelson, Katherine, 'Emerging Levels of Consciousness in Early Human Development', in Herbert S. Terrace and Janet Metcalfe (eds.), *The Missing Link in Cognition: Origins of Self-Reflective Consciousness* (New York and Oxford: Oxford University Press, 2005), 116–141.

——and Gruendel, Janice, 'Children's Scripts', in Katherine Nelson (ed.), *Event Knowledge: Structure and Function in Development* (Hillsdale, NJ: Lawrence Erlbaum, 1986), 21–46.

Nigro, Georgia, and Neisser, Ulric, 'Point of View in Personal Memories', *Cognitive Psychology* 15:5 (1983), 467–482.

Nonacs, Peter, and Soriano, Joanne L., 'Patch Sampling Behaviour and Future Foraging Expectations in Argentine Ants, *Linepithema humile*', *Animal Behavior* 55 (1998), 519–527.

O'Brien, Lucy, *Self-Knowing Agents* (New York and Oxford: Oxford University Press, 2007).

Olkkonen, M., Hansen, T., and Gegenfurter, K. R., 'Color Appearance of Familiar Objects: Effects of Object Shape, Texture, and Illumination Changes', *Journal of Vision* 8 (2008), 1–16.

O'Shaughnessy, Brian, 'Proprioception and the Body Image', in José L. Bermúdez, Anthony Marcel, and Naomi Eilan (eds.), *The Body and the Self* (Cambridge, Mass.: MIT Press, 1995), 175–204.

Paillard, J., 'Body Scheme and Body Image: A Double Dissociation in Deafferented Patients', in G. N. Gatchev, S. Mori, and J. Massion (eds.), *Motor Control: Today and Tomorrow* (Sofia, Bulgaria: Academic Publishing House, 1999), 197–214.

Parfit, Derek, *Reasons and Persons* (Oxford: Oxford University Press, 1984).

——*On What Matters*, 2 vols., ed. Samuel Scheffler (New York and Oxford: Oxford University Press, 2011).

Parker, T., Mitchell, Robert R., and Boccia, Maria L. (eds.), *Self-Awareness in Animals and Humans: Developmental Perspectives* (New York: Cambridge University Press, 1994).

Parsons, Terence, *Events in the Semantics of English* (Cambridge, Mass.: MIT Press, 1990).

Peacocke, Christopher, 'Our Entitlement to Self-Knowledge: Entitlement, Self-Knowledge, and Conceptual Redeployment', *Proceedings of the Aristotelian Society* 96 (1996), 117–158.

——'Conscious Attitudes, Attention, and Self-Knowledge', in Crispin Wright, Barry C. Smith, and Cynthia Macdonald (eds.), *Knowing Our Own Minds: Essays on Self-Knowledge* (New York and Oxford: Oxford University Press, 1998), 63–98.

——'Implicit Conceptions, Understanding, and Rationality', in Martin Hahn and Bjørn Ramberg (eds.), *Reflections and Replies: Essays on the Philosophy of Tyler Burge* (Cambridge, Mass.: MIT Press, 2003), 117–152.

——'Mental Action and Self-Awareness (I)', in Jonathan D. Cohen and Brian P. McLaughlin (eds.), *Contemporary Debates in the Philosophy of Mind* (Malden, Mass.: Blackwell, 2007), 358–376.

——*Truly Understood* (New York and Oxford: Oxford University Press, 2008).

——'Mental Action and Self-Awareness (II): Epistemology', in Lucy O'Brien and Matthew Soteriou (eds.), *Mental Actions* (New York and Oxford: Oxford University Press, 2009), 192–214.

Penco, Carlo, 'Frege, Sense and Limited Rationality', *The Review of Modern Logic* 9:1–2 (2001–2003), 53–65.

Perry, John, 'The Importance of Being Identical', in Amélie Rorty (ed.), *The Identities of Persons* (Berkeley: University of California Press, 1976), 67–90.

——'The Problem of the Essential Indexical', *Noûs* 13 (1979), 3–21.

——*The Problem of the Essential Indexical and Other Essays* (Oxford: Oxford University Press, 1993).

Picardi, Eva, 'A Note on Dummett and Frege on Sense-Identity', *European Journal of Philosophy* 1:1 (1993), 69–80.

Pinker, Steven, *The Language Instinct: The New Science of Language and Mind* (London: Penguin, 1994).

Polya, George, *Induction and Analogy in Mathematics* (Princeton: Princeton University Press, 1954).

Pomerance, Carl, 'The Search for Prime Numbers', *Scientific American* (December 1982), 136–147.

Price, H. H., 'Our Evidence for the Existence of Other Minds', *Philosophy* 13 (1955), 425–456.

Putnam, Hilary, *Philosophical Papers*, vol. I: *Mathematics, Matter and Method* (Cambridge: Cambridge University Press, 1975).

——*Philosophical Papers*, vol. II: *Mind, Language and Reality* (Cambridge: Cambridge University Press, 1975).

——*Reason, Truth, and History* (New York: Cambridge University Press, 1981).

Pylyshyn, Zenon, 'Is Vision Continuous with Cognition? The Case for Cognitive Impenetrability of Visual Perception', *Behavioral and Brain Sciences* 22 (1999), 341–365.

Quine, W. V., 'Two Dogmas of Empiricism', in *From a Logical Point of View* (Cambridge, Mass.: Harvard University Press, 1953).

——*Word and Object* (Cambridge, Mass.: MIT Press, 1960).

——*The Ways of Paradox and Other Essays* (New York: Random House, 1966).

——*Ontological Relativity and Other Essays* (New York: Columbia University Press, 1969).

Rabin, M. O., 'Probabilistic Algorithms', in J. F. Traub (ed.), *Algorithms and Complexity: New Directions and Recent Results* (New York: Academic Press, 1976), 21–39.

——'Probabilistic Algorithm for Testing Primality', *Journal of Number Theory* 12 (1980), 128–138.

Reddy, Vasudevi, 'Before the "Third Element": Understanding Attention to Self', in N. Eilan et al. (eds.), *Joint Attention: Communication and Other Minds* (New York and Oxford: Oxford University Press, 2005), 85–109.

Reid, Thomas, *An Inquiry into the Human Mind* (Chicago: University of Chicago Press, 1970).

Reis, Diana, and Marino, Lori, 'Mirror Self-Recognition in the Bottlenose Dolphin: A Case of Cognitive Convergence', *Proceedings of the National Academy of Sciences* 98:10 (2001), 5937–5942.

Resnick, Michael, 'Computation and Mathematical Empiricism', *Philosophical Topics* 17 (1989), 129–144.

Reynolds, Steven L., 'Knowing How to Believe with Justification', *Philosophical Studies* 64 (1991), 273–292.

Rips, Lance J., *The Psychology of Proof: Deductive Reasoning in Human Thinking* (Cambridge, Mass.: MIT Press, 1994).

Robinson, John A., and Swanson, Karen L., 'Field and Observer Modes of Remembering', *Memory* 1:3 (1993), 169–184.

Rosenthal, David M. (ed.), *The Nature of Mind* (Oxford: Oxford University Press, 1991).

Ross, James F., 'Testimonial Evidence', in Keith Lehrer (ed.), *Analysis and Metaphysics* (Dordrecht: D. Reidel, 1975), 35–55.

Russell, Bertrand, *Human Knowledge: Its Scope and Limits* (London: George Allen & Unwin, 1948).

Rutland, Adam, et al., 'Social Norms and Self-Presentation: Children's Implicit and Explicit Intergroup Attitudes', *Child Development* 76:2 (2005), 451–466.

Santos, Laurie R., Flombaum, Jonathan I., and Phillips, Webb, 'The Evolution of Human Mindreading: How Nonhuman Primates Can Inform Social Cognitive Neuroscience', in S. M. Platek, J. P. Keenan, and T. K. Shackelford (eds.), *Evolutionary Cognitive Neuroscience* (Cambridge, Mass.: MIT Press, 2007), 433–456.

Scanlon, Thomas M., *Moral Dimensions: Permissibility, Meaning, Blame* (Cambridge, Mass.: Harvard University Press, 2008).

Schachter, Daniel, *Searching for Memory: The Brain, the Mind, and the Past* (New York: Basic Books, 1996).

Schiffer, Stephen, 'Boghossian on Externalism and Inference', *Philosophical Issues* 2 (1992), 29–38.

Schmidt, M., and Lipson, H., 'Distilling Free-Form Natural Laws from Experimental Data', *Science* 324 (2009), 81–85.

Schmitt, Frederick F., 'Justification, Sociality, and Autonomy', *Synthese* 73 (1987), 43–85.

Schwartz, Bennett L., 'Do Nonhuman Primates Have Episodic Memory?' in Herbert S. Terrace and Janet Metcalfe (eds.), *The Missing Link in Cognition: Origins of Self-Reflective Consciousness* (New York and Oxford: Oxford University Press, 2005), 225–241.

Sellars, Wilfrid, *Science, Perception, and Reality* (London: Routledge & Kegan Paul, 1963).

Shaffer, David R., and Kipp, Katherine, *Development Psychology: Childhood and Adolescence* (Belmont, CA: Wadsworth, 2009).

Shoemaker, Sidney, 'Self-Reference and Self-Awareness', *The Journal of Philosophy* 65:19 (1968), 555–567.

——'Persons and Their Pasts', *American Philosophical Quarterly* (1970), 269–285.

——'On Knowing One's Own Mind', *Philosophical Perspectives* 2 (1988), 183–209.

——*The First-Person Perspective and Other Essays* (New York: Cambridge University Press, 1996).

——'Self and Substance', *Philosophical Perspectives* 11 (1997), 283–304.

——*Identity, Cause, and Mind* (New York and Oxford: Oxford University Press, 2004).

——'Careers and Quareers: A Reply to Burge', *The Philosophical Review* 118 (2009), 87–102.

Soteriou, Matthew, 'Mental Action and the Epistemology of Mind', *Noûs* 39:1 (2005), 83–105.

Spinoza, Baruch, *The Collected Works of Spinoza*, vol. 1, ed. and trans. Edward Curley (Princeton: Princeton University Press, 1985).

Strawson, P. F., *Individuals* (1959; London: Routledge, 2002).

——*The Bounds of Sense* (1966; London: Routledge, 2006).

Stroud, Barry, 'Transcendental Arguments', *The Journal of Philosophy* 95 (1968), 241–256.

Teller, Paul, 'Computer Proof', *The Journal of Philosophy* 77 (1980), 797–803.

Tomasello, Michael, 'Do Apes Ape?' in Cecilia M. Hayes and Bennett G. Galef (eds.), *Social Learning in Animals: The Roots of Culture* (New York: Academic Press, 1996), 319–346.

Tulving, E., 'Episodic and Semantic Memory', in E. Tulving and W. Donaldson (eds.), *Organization of Memory* (New York: Academic Press, 1972), 382–402.

——*Elements of Episodic Memory* (Oxford: Oxford University Press, 1983).

——'Multiple Memory Systems and Consciousness', *Human Neurobiology* 6 (1987), 67–80.

——'Remembering and Knowing the Past', *American Scientist* 77:4 (1989), 361–367.

——'Episodic Memory and Autonoesis: Uniquely Human?' in Herbert S. Terrace and Janet Metcalfe (eds.), *The Missing Link in Cognition: Origins of Self-Reflective Consciousness* (New York and Oxford: Oxford University Press, 2005), 3–56.

Tymoczko, Thomas, 'The Four-Color Problem and its Philosophical Significance', *The Journal of Philosophy* 76 (1979), 57–83.

Valone, Thomas J., 'Are Animals Capable of Bayesian Updating? An Empirical Review', *Oikos* 112 (2006), 252–259.

Villanueva, E. (ed.), *Content* (Atascadero: Ridgeview, 1995).

Wegner, Daniel M., and Pennebaker, James W. (eds.), *Handbook of Mental Control* (Englewood Cliffs, NJ: Prentice Hall, 1993).

Welbourne, Michael, *The Community of Knowledge* (Aberdeen: Aberdeen University Press, 1986).

Williams, Bernard, *Descartes: The Project of Pure Enquiry* (Harmondsworth: Penguin, 1978).

Wittgenstein, Ludwig, *Philosophical Investigations*, trans. G. E. M. Anscombe (New York: Macmillan, 1968).

Wolf, Susan, 'Self-Interest and Interest in Selves', *Ethics* 96 (1986), 704–720.

Woodfield, Andrew (ed.), *Thought and Object: Essays on Intentionality* (Oxford: Clarendon Press, 1982).

Author Index

Subject Index

I conceive the index as a fine-grained research tool. There are many more page citations than in most indexes. I have tried to cite essentially all passages in which an entry topic plays a role. Where an entry occurs only in a footnote of a page that is cited, the citation numeral is followed by an 'n'—for example, 287n. *Italicized* numerical citations indicate elementary or basic explications, or fundamental substantive discussions, at least for the more important entry topics. Sometimes italicized citations cite a part of a range of pages in which the entry occurs. In such cases, the italicized citation is situated before the range citation—for example, *489–495*, 489–506. Sometimes an italicized numeral occurs at the beginning or end of a range—for example, *546*–547—signifying that just that page contains an elementary or basic passage. Readers who wish to use the index as a glossary or as a way of quickly looking up basic points can focus on the italicized citations. Readers who want a below-surface understanding of themes in the volume would do well to spend substantial time with this index, exploring topics and relations among topics traced herein.

The manufacturer's authorised representative in the EU for product safety is Oxford University Press España S.A. of el Parque Empresarial San Fernando de Henares, Avenida de Castilla, 2 – 28830 Madrid (www.oup.es/en or product.safety@oup.com). OUP España S.A. also acts as importer into Spain of products made by the manufacturer.

www.ingramcontent.com/pod-product-compliance
Lightning Source LLC
Chambersburg PA
CBHW060048140126
38218CB00003B/49
* 9 7 8 0 1 9 9 6 7 2 0 2 8 *